VII Scientific Conference with International Participation "Information-Measuring Equipment and Technologies" (IME&T 2016)

AA003596

MATEC Web of Conferences Volume 79 (2016)

Tomsk, Russia
25 - 28 May 2016

Editors:

A.V. Yurchenko
V.I. Syryamkin

ISBN: 978-1-5108-3112-4

Printed from e-media with permission by:

Curran Associates, Inc.
57 Morehouse Lane
Red Hook, NY 12571

Some format issues inherent in the e-media version may also appear in this print version.

This work is licensed under a Creative Commons Attribution license:
http://creativecommons.org/licenses/by/2.0/
You are free to:
Share – copy and redistribute the material in any medium or format.
Adapt – remix, transform, and build upon the material for any purpose, even commercial.
The licensor cannot revoke these freedoms as long as you follow the license terms.
Under the following terms:
You must give appropriate credit, provide a link to the license, and indicate if changes were made. You may do so in any reasonable manner, but not in any way that suggests the licensor endorses you or your use. The copyright is retained by the corresponding authors.

Printed by Curran Associates, Inc. (2016)

For additional information, please contact EDP Sciences – Web of Conferences at the address below.

EDP Sciences – Web of Conferences
17, Avenue du Hoggar
Parc d'Activité de Courtabœuf
BP 112
F-91944 Les Ulis Cedex A
France

Phone: +33 (0) 1 69 18 75 75
Fax: +33 (0) 1 69 28 84 91

contact-edps@webofconferences.org

Additional copies of this publication are available from:

Curran Associates, Inc.
57 Morehouse Lane
Red Hook, NY 12571 USA
Phone: 845-758-0400
Fax: 845-758-2633
Email: curran@proceedings.com
Web: www.proceedings.com

TABLE OF CONTENTS

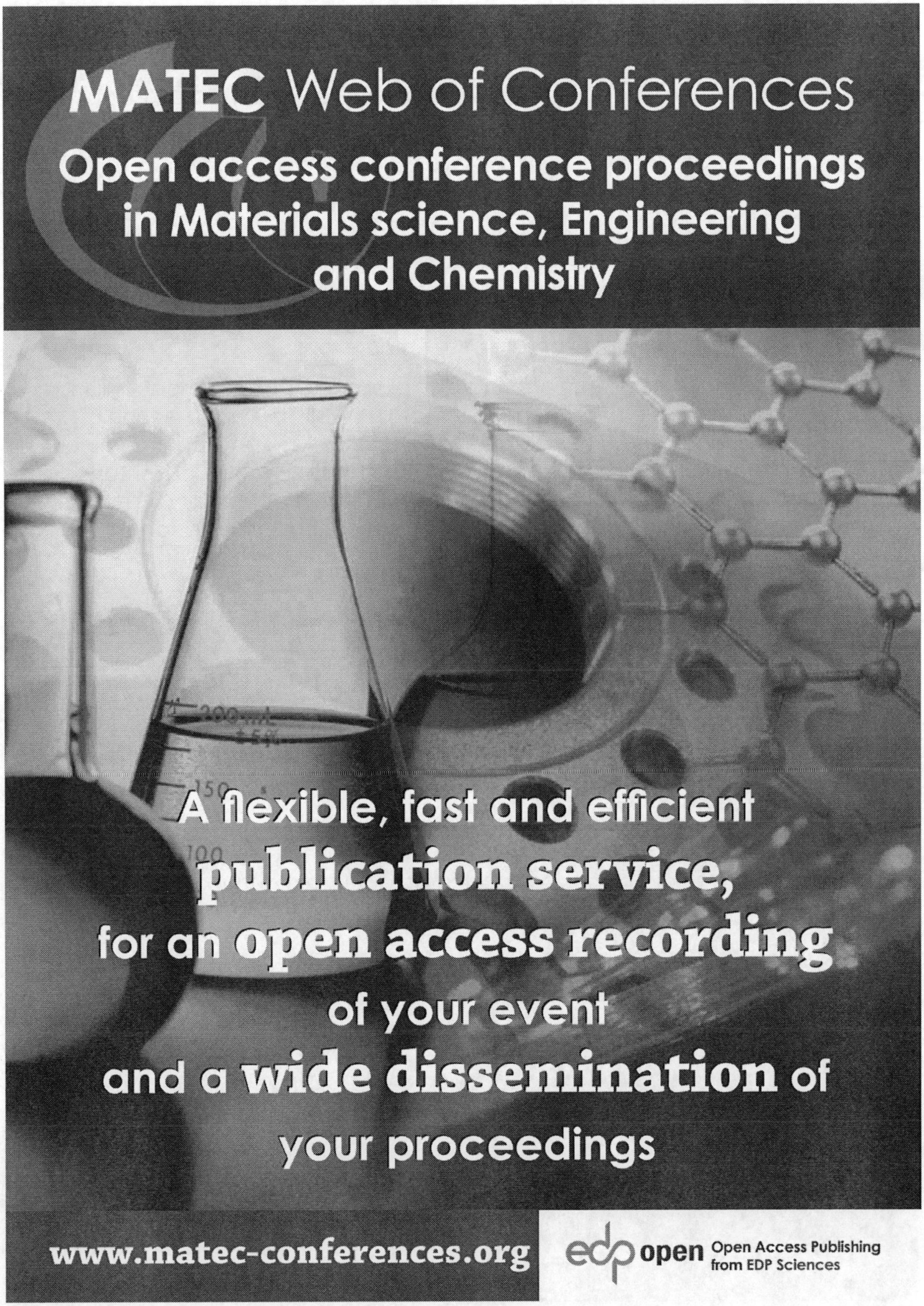
MATEC Web of Conferences
Open access conference proceedings in Materials science, Engineering and Chemistry
A flexible, fast and efficient publication service, for an open access recording of your event and a wide dissemination of your proceedings
www.matec-conferences.org
edp open Open Access Publishing from EDP Sciences

Aims & scope

MATEC Web of Conferences is an open-access journal devoted to the publishing of conference proceedings dealing with Materials science, Engineering and Chemistry.

The journal aims to make conference proceedings visible, citable and widely usable by the scientific community. It offers outstanding services for the publication and dissemination of contents in order to ensure long-term archiving and maximum exposure of scientific events.

Coverage includes all engineering disciplines: civil, naval, mechanical, chemical, and electrical engineering as well as nanotechnology and metrology. The journal also deals with the physical-chemical characterization of materials, their implementation, their resistance in their environment, etc. Others subdisciplines of chemistry, such as analytical chemistry, petrochemistry, organic chemistry, etc., and even pharmacology, are also welcome.

Why publish with MATEC Web of Conferences?

- **Open access:** conference papers available worldwide.
- **Fast publication:** less than 3 months.
- **Total flexibility:** no page restrictions, free unlimited use of colour in the online version, any kind of scientific document welcomed (traditional articles, posters, abstracts and slideshows).
- **High visibility:** all documents indexed in main databases and easily found online.
- **Fair author rights:** CC-BY Creative Commons license.
- **Range of additional publishing options:** printed proceedings, CD-ROMs, copy-editing, etc.

Contact: Isaline Augusto
matec@webofconferences.org

ISSN (electronic edition): 2261-236X

www.matec-conferences.org

edp open Open Access Publishing from EDP Sciences

PRESENTATION OF CONFERENCE:

The conference is held annually since 2010. Conference participants are engineers, professors, students and young scientists.
The conference is devoted to topical issues in the field of development and application of modern information technologies in science and industry. **Topics within the scope of the conference include:**

- Problems of development and application of modern sensors for obtaining reliable data about the object of measurement;
- Construction of measuring instruments and systems based on smart sensors;
- The use of modern measuring systems and information technologies in the production and research activities;
- Experience in application and prospects of development of modern measuring systems;
- Development and application of smart information systems in medicine;
- Smart reconfigurable systems of control, navigation and image processing for autonomous mobile robots;
- Algorithms and mathematical models of high-speed processing and image analysis systems;
- Development and application of radio-wave imaging systems;
- Problems of staff training for modern enterprises which use high-tech systems for data collecting and processing.

The conference was carried out on sections:
1. The use of modern measuring and information devices and technologies.
2. Innovative technologies in radio physics, microwave and semiconductor electronics.
3. Information technologies in medicine.
4. Information and Intelligent systems of mechatronics and robotics.
5. Organizational and methodological issues of staff training in the field of information-measuring devices and technologies. (Use of information technologies in the educational process).

ORGANIZING COMMITTEE

Conference chairman Alexey V. Yurchenko – D.Eng.Sc., professor, Senior Researcher of the international laboratory Vision Systems;
Conference co-chairman Vladimir I. Syryamkin – D.Eng.Sc., professor, Head of the Department of Quality Control of the Faculty of Innovative Technologies, National Research Tomsk State University; President of the Association of Innovative Enterprises and Organizations of Tomsk and Tomsk region; academician of International Academy of Sciences of Higher Education and Russian Academy of Natural Sciences;
Eduard Siemens- professor, Anhalt University of Applied Sciences (Germany);
Anatoliy N. Soldatov – Dr. Sci. in Physics and Mathematics, professor, Head of the Regional Center of the Russian Association of Laser, Honored Inventor of the Russian Federation, dean of the FIT TSU;
Vladimir P. Yakubov – Dr. Sci. in Physics and Mathematics, professor, Head of the Department of Radiophysics, National Research Tomsk State University;
Stanislav V. Shidlovsky – D.Eng.Sc., professor of the Department of Quality Control of the Faculty of Innovative Technologies, National Research Tomsk State University; Head of the Laboratory of high-performance reconfigurable systems;
Ali D. Mehtiyev – Ph.D. in Eng.Sc., Head of the Department of Technologies and Communication Systems of Karaganda State Technical University;
Artem V. Kozlov – academic secretary: Ph.D. in Eng.Sc., research associate of Institute of Atmospheric Optics of Siberian Branch of the Russian Academy of Science (IAO SB RAS);
Sergey E. Shipilov – Ph.D. in Physico-mathematical Sciences, associate professor of the Department of Radiophysics National Research Tomsk State University;
Alexander Yu. Verkhovskiy – Head of the Laboratory of Biochemistry of Transport Systems, Faculty of Innovative Technologies, National Research Tomsk State University;

Maxim V. Syryamkin – director of the interacademic educational and scientific production center Technology Management, National Research Tomsk State University;
Alexander E. Goldstein – D.Eng.Sc., professor of the Department of Physical Methods of Non-Destructive Testing, National Research Tomsk Polytechnic University;
Olga V. Galtseva – Ph.D. in Eng.Sc., associate professor of the Department of Non-Destructive Testing, Science Office the Institute of Non-Destructive Testing, National Research Tomsk Polytechnic University
Irina A. Larioshina – Ph.D. in Eng.Sc., Science Office the Institute of Non-Destructive Testing, National Research Tomsk Polytechnic University;
Artem A. Ananin – Head of the scientific and educational projects of National instruments;
Denis N. Kuznetsov – FIT TSU, Conference Secretary;
Alexey A. Poguda – deputy Dean of the FIT TSU, coordinator of the conference site.

PROGRAM COMMITTEE

Chairman Vladimir I. Syryamkin – D.Eng.Sc., Professor, Head of the Department of Quality Control of the Faculty of Innovative Technologies, National Research Tomsk State University; President of the Association of Innovative Enterprises and Organizations of Tomsk and Tomsk region; academician of International Academy of Sciences of Higher Education and Russian Academy of Natural Sciences;
Nikola Sobutonov – Professor, Academician of the Bulgarian Academy of Sciences;
Anatoliy N. Soldatov – Dr. Sci. in Physics and Mathematics, professor, Head of the Regional Center of the Russian Association of Laser, Honored Inventor of the Russian Federation, dean of the FIT TSU;
Stanislav V. Shidlovsky – D.Eng.Sc., professor of the Department of Quality Control of the Faculty of Innovative Technologies, National Research Tomsk State University; Head of the Laboratory of high-performance reconfigurable systems;
Valery L. Vengrinovich – D.Eng.Sc., professor, Minsk, Belarus;
Sergey B. Suntsov – JSC Information Satellite Systems named after Academician MF Reshetnev. (Krasnoyarsk);
Vladimir P. Yakubov – Dr. Sci. in Physics and Mathematics, professor, Head of the Department of Radiophysics, National Research Tomsk State University;
Ali D. Mehtiyev – Ph.D. in Eng.Sc., Head of the Department of Technologies and Communication Systems of Karaganda State Technical University;
Alexey G. Yakunin – Professor, Head of the Department of Computer Science, Computer Engineering and Information Security, Polzunov Altai State Technical University;
Boris V. Poller – D.Eng.Sc., professor, Head of the Laboratory of Laser Information Systems, Institute of Laser Physics, Siberian Branch of the Russian Academy of Science;
Eduard Siemens – Professor, Dr, Anhalt University of Applied Sciences;
Jörg Bagdahn – Professor, Dr, Anhalt University of Applied Sciences.

SPONSORS:

- National Instruments Corporation
- The Russian Foundation for Fundamental Research

Organizers of the conference:

- National Research Tomsk State University, Tomsk, Russia;
- National Research Tomsk Polytechnic University, Tomsk ,Russia;
- JSC "Scientific-Research Institute of Semiconductor Devices", Tomsk, Russia;
- Polzunov Altai State Technical University, Barnaul, Russia;
- JSC "Information Satellite Systems to them. Reshetnev ", Krasnoyarsk, Russia;
- Karaganda State Technical University, Karaganda, Kazakhstan;
- Anhalt University of Applied Sciences (Germany);

- National Instruments Corporation (USA).

Site of conference: http://tic.tsu.ru/www/pages/iit.php

edp sciences

MATEC Web of Conferences

Statement of Peer review

In submitting conference proceedings to MATEC Web of Conferences, I certify to the Publisher that I adhere to the **Policy on Publishing Integrity** of the journal in order to safeguard good scientific practice in publishing.

1. All articles have been subjected to peer review administered by the proceedings editors.
2. Reviews have been conducted by expert referees, who have been requested to provide unbiased and constructive comments aimed, whenever possible, at improving the work.
3. Proceedings editors have taken all reasonable steps to ensure the quality of the materials they publish and their decision to accept or reject a paper for publication has been based only on the merits of the work and the relevance to the journal.

Title, date and place of the conference

VII Scientific Conference with international participation
«Information-measuring equipment and technologies»

25 – 28 May 2016

National Research Tomsk State University,
634050, 36 Lenin Avenue, Tomsk, Russia.

Proceedings editor(s):

Alexey V. Yurchenko – Dr.Eng.Sc., professor, Senior Researcher of the international laboratory Vision Systems;

Vladimir I. Syryamkin – Dr.Eng.Sc., professor, Head of the Department of Quality Control of the Faculty of Innovative Technologies, National Research Tomsk State University; President of the Association of Innovative Enterprises and Organizations of Tomsk and Tomsk region; academician of International Academy of Sciences of Higher Education and Russian Academy of Natural Sciences;

Date and editor's signature

VISUALIZATION OF BROADBAND SOUND SOURCES

Dmitry Sukhanov[a], *Nadezhda* Erzakova

Radiophysics Faculty, Tomsk State University, 634050, Tomsk, Russia

Abstract. In this paper the method of imaging of wideband audio sources based on the 2D microphone array measurements of the sound field at the same time in all the microphones is proposed. Designed microphone array consists of 160 microphones allowing to digitize signals with a frequency of 7200 Hz. Measured signals are processed using the special algorithm that makes it possible to obtain a flat image of wideband sound sources. It is shown experimentally that the visualization is not dependent on the waveform, but determined by the bandwidth. Developed system allows to visualize sources with a resolution of up to 10 cm.

1 Introduction

Detection and visualization of sound sources is of great practical importance [1, 2] for remote vibration diagnostics, sound quality studies and creation of remote microphones. In addition, it is possible to monitor and determine a coordinates of speaking people [3], which is important for security. The task of arbitrary audio sources imaging is solved on the basis of simultaneous measurement of the sound field at different points in space. It is desirable to place the sensors with spatial gaps less than half of a wavelength. The aperture should be of the order of distance to the test sound source. There are solutions [3], to track speaking people and to distinguish between them by voice. For the calibration of ultrasonic transducers used multiposition scanners. The solution of the inverse problem of wave propagation considering evanescent waves allows to determine the near-zone emitter field [4]. However, in this case, time synchronization is required between the transmitter and receiver. In case of a third-party audio sources detection, a priori about them is no information, and it is impossible to synchronize the transmitter and receiver, and thus provide estimation of the phase at the point of measurement. To preserve the spatial coherence of the measured field necessary to measure simultaneously the signals at various points. Unfortunately, it is impossible to synchronize to arbitrary time source, and if there may be several sources, and each emits its signal. When measured on a flat aperture to rely only on the presence of in-plane resolution, while to obtain range resolution fails. Nevertheless, even the flat images are of great practical importance.

[a] Corresponding author: sdy@mail.tsu.ru

© The Authors, published by EDP Sciences. This is an open access article distributed under the terms of the Creative Commons Attribution License 4.0 (http://creativecommons.org/licenses/by/4.0/).

In this paper, we propose to carry out the focus of the measured field in the object plane at each frequency separately and then sum up the intensities of the received monochromatic images. This approach allows us to visualize any sound sources regardless of the waveform.

2 Visualization of sound sources

The scheme of the broadband measurements on the basis of a two-dimensional array of microphones shown in Figure 1. A two-dimensional array of microphones located in front of the sound source at a distance h. We assume that a third-party source emits a signal with a wide frequency range. Radiated waves fall into the microphone array and simultaneously digitized. It is assumed that the range to the emitter is less than or about the aperture size of the system. Thus, the source is in the near area of array, and therefore applying of wave field focusing it is possible sources visualization with a wavelength resolution.

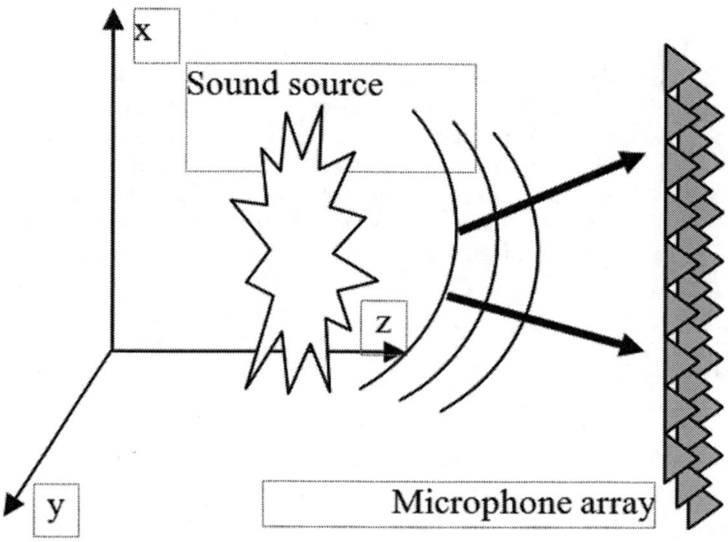

Figure 1. Scheme of ultrasonic measurements.

To solve the inverse problem of accurate sources images recovery we propose a method of broadband spatial matched filtering. First apply the Fourier transform to the signals measured from all microphones:

$$\tilde{P}(x,y,z=0,\omega)=\int_{-\infty}^{\infty}P(x,y,z=0,t)\exp(i\omega t)dt\,, \tag{1}$$

where $P(x,y,z=0,t)$ – signal, measured in point $(x,y,0)$, ω – frequrncy.

Since the measurements are simultaneous, the spatial coherence of signals stored on separate frequencies. Next, we calculate the plane wave spectrum of the wave field at each frequency in the measurement plane:

$$\tilde{\tilde{P}}(k_x,k_y,z=0,\omega)=\int_{-\infty}^{\infty}\int_{-\infty}^{\infty}\tilde{P}(x,y,z=0,\omega)e^{-ik_xx-ik_yy}dxdy\,, \tag{2}$$

where k_x,k_y – spatial frequencies.

By the plane-wave spectrum, we can easily restore the field in the source plane [5–7]. We are restoring the spatial spectrum of the source in the plane at a distance h by a factor $\exp(-ik_z h)$:

$$\widetilde{P}(x, y, z=-h, \omega) = \int_{-\infty}^{\infty}\int_{-\infty}^{\infty} \widetilde{\widetilde{P}}(k_x, k_y, z=0) \exp(-ik_z h)\exp(ik_x x + ik_y y) dk_x dk_y , \quad (3)$$

where $k_z = \sqrt{(\omega/c)^2 - k_x^2 - k_y^2}$. Modifier $\exp(-ik_z h)$ allows us to restore even field and evanescent waves in the near-source zone. Because there is no synchronization of the sound source and the receiver, we can not provide in-phase addition of the signals from the source at all frequencies. It is therefore proposed to sum up the squares of modules of ultrasound images at each frequency. Square module, but not module itself, proposed to use as it will increase the level of the dominant signal relative to the background, which helps visualize concentrated sources. Thus, wideband processing is performed as follows:

$$P(x_s, y_s) = \sum_n \left|\widetilde{P}(x_s, y_s, \omega_n)\right|^2 , \quad (4)$$

The proposed method has been verified by numerical simulations. The numerical model considered two point sources at a distance of 60 cm from the measurement plane. Figure 2 shows the simulated measured field at a frequency of 3500 Hz. It is considered that radiation produced in the frequency band from 500 Hz to 3500. A shape of the signal is random. Two emitters are not synchronized and emit different signals.

Note, however, that the field is not measured at all points in space, but only where the microphones are. Consider the case of the 96 array elements arranged in a rectangular grid 16x6 cells with gaps of 6 cm. Location of the microphones is shown in Figure 3.

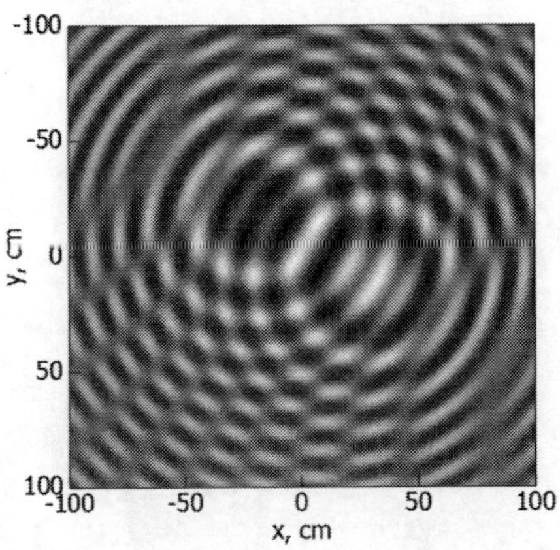

Figure 2. Field of sources on 3500 Hz.

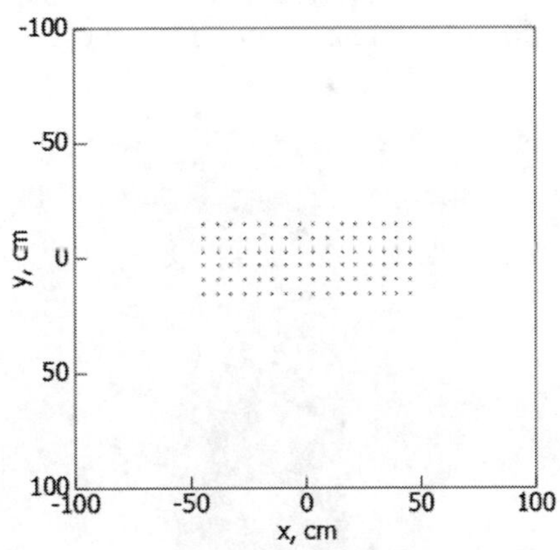

Figure 3. Placement of microphones in receiver array.

After processing by the broadband space - matched filter the intensity image of two sources has been restored (Figure 4). Image has artifacts caused by array sparsity. Image sources localized stronger along the x axis than along the y axis because of the aperture size of x axis is 90 cm and on the axis y is 30 cm. After treatment by the formula (4) the image of two broadband sources was restored (Figure 5).

 DOI: 10.1051/matecconf/20167901001

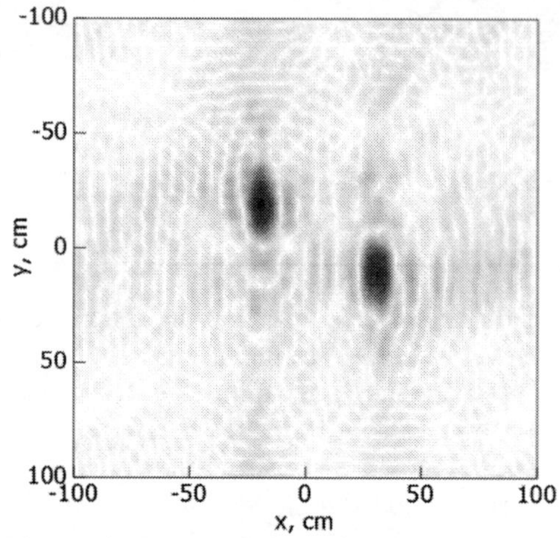

Figure 4. Reconstruction of sources image at a distance 60 cm on frequency 3500 Hz.

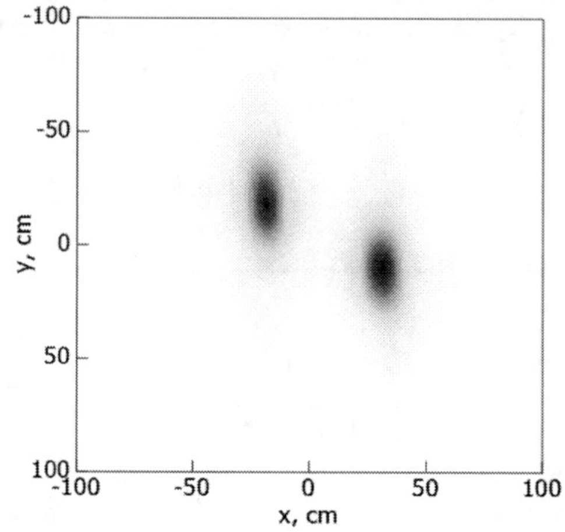

Figure 5. Reconstruction of sources image at a distance 60 cm in an frequency range 500–3500 Hz.

In this image there are no artifacts due to the fact that the sum intensity increases the signal - noise at $\sqrt{N}$ times. numerical simulation results confirm the applicability of the method of broadband space - matched filtering.

3 Experimental research

For experimental studies two-dimensional array of microphones has been developed consisting of 6 linear arrays (Figure 6).

Figure 6. Photo of experimental setup.

Each line contains a set of 16 electret microphones connected to a separate unit with a 16-channel multiplexer and an amplifier. Linear arrays are connected via a multiplexer to the ADC block individual channels. Each electret microphone bias voltage is supplied via a separate resistor. Out of a separate microphone connected via a capacitor to the multiplexer. The multiplexer is connected to the microcontroller STM32F407 and controlled by four binary outputs addressing and one output on / off. Multiplexer output is connected to a microcontroller, which allows the digitized signal from all the microphones in the linear array. This setting allows you to simultaneously digitize signals of a 96 microphones with a frequency of 7200 Hz. Microcontroller transmits digitized signals to a personal computer for further processing. A two-dimensional array of the microphone is positioned vertically to the sound source at a distance of 60 cm. The sound source is a speaker that emits a signal at a particular frequency range and controlled by the computer. As a result of the experiment was measured noise signal (frequency band from 100 to 4000 Hz) being emitted from the speakers and fall into one of the microphones is shown in Figure 7.

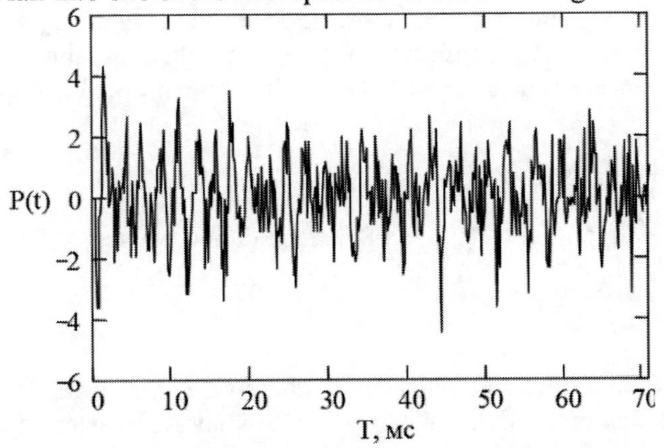

Figure 7. Noise signal in a microphone array from the two oscillators.

It is known that the spectrum of the noise signal is approximately uniform at all frequencies, which means that the signal is broadband. Formula (1–4), whereby the image was restored two sound sources (see Figure 8) have been applied to the measured data.

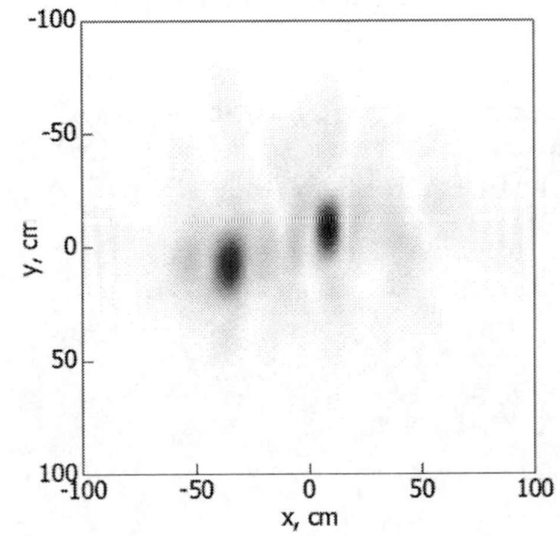

Figure 8. Wideband noise signal Image from the sound sources.

It can be seen that the experimental result is in good agreement with numerical simulations. Point emitters clearly visualized. That is, there is a possibility of detecting the quite arbitrary signal source. It does not require a priori information about the signal form but only enough signal to be broadband. The experimental setup provides a resolution of about 10 cm.

 DOI: 10.1051/matecconf/20167901001

4 Conclusion

The proposed method of recovery of homogeneous and heterogeneous fields near a sound source zone unsynchronized with the receiver, allows you to create a tool for studying the dynamics of the field of near-field ultra-wideband sources. To apply the method requires simultaneous measurements of audio signals of two-dimensional array of microphones. This method has been verified by numerical simulations and experiments. The experimental setup consists of 96 microphones, allowing to digitize signals at a frequency of 7200 Hz. The experimental study confirmed the applicability of the developed method, the experimental data are consistent with the results of the numerical model.

Acknowledgment

The research is supported by Russian Science Foundation project No 16-19-10272.

References

[1] P. Gregush, *Ultrasonic Imaging: Seeing by Sound* (Focal Press, London, 1980)
[2] V. P. Yakubov, S. E. Shipilov; D. Ya. Sukhanov, Russian Physics Journal **55**, 878 (2013) doi: 10.1007/s11182-013-9895-9
[3] A. Quinlan, M. Kawamoto, Y. Matsusaka, *et. al.*, Eurasip Journal on Audio, Speech, and Music Processing **2009,** 673202 (2009) doi : 10.1155/2009/673202
[4] E.G. Williams, J.D. Maynard, Phys. Rev. Lett. **45**, 554 (1980) doi: 10.1103/PhysRevLett.45.554
[5] P.R. Stepanishen, K.S. Benjamin, Journal of the Acoustical Society of America **71**, 803 (1982) doi : 10.1121/1.387606
[6] Yong Thung Cho M.J. Roan, J.S. Bolton, Proc. IMechE Part C: J. Mechanical Engineering Science **223**,819 (2009) doi: 10.1243/09544062JMES1209
[7] D.Ya. Sukhanov, N.N. Erzakova, CriMiCo 2014, 950 (2014) doi: 10.1109/CRMICO.2014.6959709

COLOR SEGMENTATION OF HOMOGENEOUS AREAS ON COLPOSCOPICAL IMAGES

Yana Kosteley[1], *Dmitry* Zhdanov[1,2*], *Artem* Osipov[1]

[1]Diagnostika + LLC, 634055, Tomsk, Russia
[2] National Research Tomsk State University, 634055, Tomsk, Russia

Abstract. The article provides an analysis of image processing and color segmentation applied to the problem of selection of homogeneous regions in the parameters of the color model. Methods of image processing such as Gaussian filter, median filter, histogram equalization and mathematical morphology are considered. The segmentation algorithm with the parameters of color components is presented, followed by isolation of the resulting connected component of a binary segmentation mask. Analysis of methods performed on images colposcopic research.

1 Introduction

In Russia cervical cancer ranks third among cancers of a reproductive system [1]. Colposcopy is one of the methods for early diagnosis of cervical precancer [2]. The images obtained from this procedure are suitable for their analysis aimed at identification of pathological processes on the surface of the cervix. The aim of this study is the selection and implementation of methods for preprocessing image of cervical surface in order to eliminate their artifacts (e.g., glare light in cervical mucosa), and identify homogeneous areas showing changes in the structure of the surface investigated.

2 Research methods

The design team has realized a searching and processing algorithm for flaring areas (image areas with maximum light intensity). At the initial stage, the color model transits from the RGB to HSV (1). After that the flaring areas (where pixel values have up to 15% saturation and not lower than 77% brightness) are searched according to the set parameters of the color model. As a result, a *binary mask*, comprising a glare area, was *constructed* (Figure 1). The image contains many *torn* areas and areas with a low specific weight. To eliminate the problem, pre-processing blurring of an image using *Gaussian filter* can be applied.

* Corresponding author: dim@diagnostic.tom.ru

© The Authors, published by EDP Sciences. This is an open access article distributed under the terms of the Creative Commons Attribution License 4.0 (http://creativecommons.org/licenses/by/4.0/).

$$H=\begin{cases} 0, MAX=MIN \\ 60\cdot\dfrac{G-B}{MAX-MIN}, MAX=R\cap G\geq B \\ 60\cdot\dfrac{G-B}{MAX-MIN}+360, MAX=R\cap G<B \\ 60\cdot\dfrac{B-R}{MAX-MIN}+120, MAX=G \\ 60\cdot\dfrac{R-G}{MAX-MIN}+240, MAX=B \end{cases},$$
$$S=\begin{cases} 0, MAX=0 \\ 1-\dfrac{MIN}{MAX} \end{cases},$$
$$V=MAX \quad (1)$$

where R,G,B – red, green, blue components of a colour model RGB, $R,G,B\in[0,1]$; H,S,V – components of a colour hue, saturation and brightness of the HSV color model, $H\in[0,360]$, $S,V\in[0,1]$; *MAX,MIN* – *RGB*. maximum and minimum value of RGB color components.

Gaussian filter– is one of the most commonly used filter for convolution matrix [3]. The elements of its nucleus are formed by a Gaussian function (2).

Compression is described by function (3). Blurring depends on the size of the compression kernel and value σ (radius). For the correct formation of the kernel *three sigma rule* must be used.

$$f(x,y,\sigma)=\frac{1}{2\pi\sigma^2}e^{\frac{-(x^2+y^2)}{2\sigma^2}}, \quad (2)$$

where *x,y* – coordinates of compression matrix; σ – radius of *Gaussian filter*

$$P^*_{i,j}=\frac{1}{N}\sum_{x=-k/2}^{\frac{k}{2}}\sum_{y=-k/2}^{\frac{k}{2}}P_{i-x,j-y}\cdot M_{x+\frac{k}{2},y+\frac{k}{2}} \quad (3)$$

where $P^*_{i,j}$ – resulting value of the image pixel *i,j*; $P_{i,j}$ – reference value of the image pixel *i,j*; *N* – weight coefficient of compression matrix; *K* – size of the compression kernel; *M* – compression matrix.

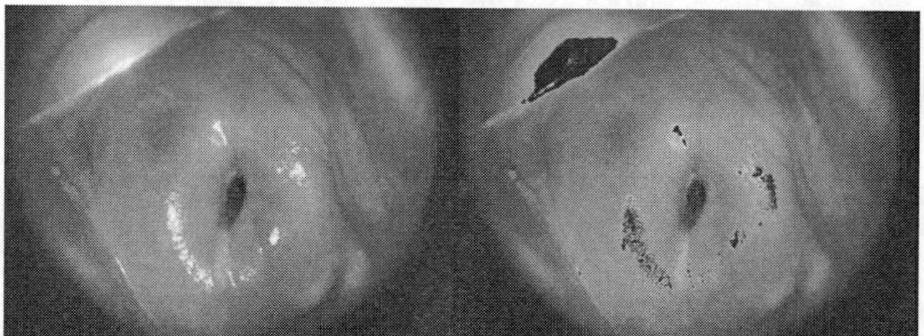

Figure 1. The results of color segmentation with parameters $S\in[0;0.12], V\in[0.77;1]$ for highlighting glare areas.

The described method can be used not only for highlighting glare areas but also for identification of areas containing potential pathology. For example, the areas affected by erosion will be red. After treatment of the epithelium with acetic acid (or iodine), it will not be colored in orange; it will become white. The areas suitable for analysis must undergo further processing of the binary mask obtained using color segmentation. The set of mathematical morphology operations will remove unnecessary small elements and holes, make the outlines smoother, identify shapes of the objects. The paper uses the following operations of mathematical morphology using structural elements:

1. *Expansion.* If the central element and the pixel of the input image are shown in color, the output image pixel and the area around it are colored. It is possible to enlarge an object on a binary mask.
2. *Erosion.* If all elements of a structural element match (also shaded) pixel and its neighborhood, the output image pixel is colored (the current pixel is changing). It is possible to make objects on a binary mask smaller or exclude them.
3. *Disclosure.* Consistent application of *erosion* and *expansion.* eliminates objects which are smaller than the structural element without making the objects smaller.
4. *Closure.* Consistent application of *erosion* and *expansion* makes the objects smoother and combines connected components in the surrounding area.

For visualization of the image segmentation the following sequence of operations is used:

1. Application of erosion with a structural element (Figure 2).
2. Subtraction of the result obtained from the original binary mask.

Figure 2. Structural element used for identification of boundaries of segmented areas.

For recognizing of connected components a modified two-pass algorithm for a serial scanning of eight-bin elements is used. In comparison with a standard method of serial scanning in this approach, each row of the matrix is scanned in two passes – forward and backward ones. The mask of the forward and backward passes is shown in Figure 3, where *e* – the currently analyzed pixel, and *a, b, c, d* – the neighboring pixels.

b	c	d
a	e	

a)

	e	d
a	b	c

b)

Figure 3. Scan-masks: a) sketching of the scan-mask of the forward pass, b) sketching of the scan-mask of the backward pass.

In the direct iteration (forward pass) a local table of equivalences is filled and pre-sketching of the areas is done (4), and in the reverse iteration (backward pass) equivalent values are replaced in accordance with the table of equivalences (5).

$$L[e]=\begin{cases} 0; если - I[e]=0 \\ l, T[l]=l, l=l+1; если - \forall i \in (a,b,c,d) \mid I(i)=0 \\ \min(T[i]), T[L[i]]=L[e]; если - \forall i \in (a,b,c,d) \mid I(i)=1 \end{cases} \quad (4)$$

where *L[i]* –the value in the array of connected components in the pixel *i*; *I[i]* –the matrix of the original binary mask; *l* – the current maximum number of connected components; *T* –the table of equivalence, represented by the union-find structure; *min* –obtaining of the minimum value in the equivalence table.

$$L[e]=\min(T[i]), T[L[i]]=L[e]; если - \forall i \in (a,b,c,d) \mid I(i)=1 \quad (5)$$

This modification reduces the number of scan passes, which greatly increases the processing speed. The result of the operation is a mask with numbered connected areas, allowing to analyze them using pattern recognition algorithms. For improvement of image analysis equalization or linearization of the histogram can be used. It can normalize brightness levels. The median filter can remove some noise. The images of the cervical mucosa usually have brightness differences. Histogram equalization of these images allows to make areas more contrasting, different from main color of the image [4]. This step allows to define the boundaries of the areas more clearly, which greatly improves the quality of segmentation. During screening of the histogram according to the parameters of the RGB model, the components tend to reduce the intensity of their color. Therefore, in order not to change *color temperature*, it is possible to carry out a histogram equalization according to the intensity parameter of HSV color model. The example of histogram equalization is shown in Figure 3.

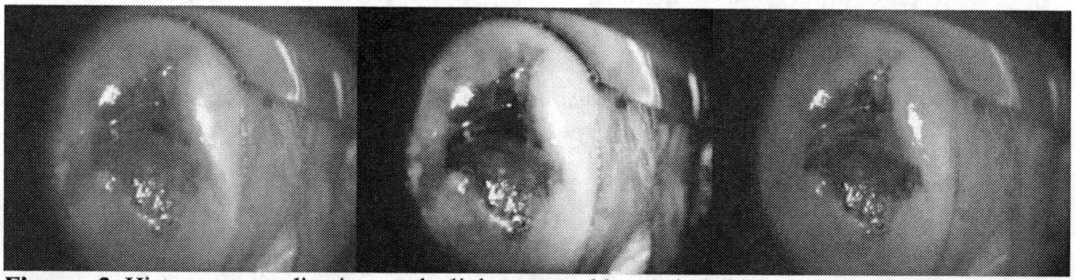

Firgure 3. Histogram equalization on the light-exposed image in RGB color space, and according to S parameter in HSV color space.

The median filter is based on the idea of averaging of pixel values byreplacing them with the median pixel values from the surrounding area [5]. The power of *averaging* will depend on the size of the neighborhood. The example of removal of artifacts is shown in Figure 4.

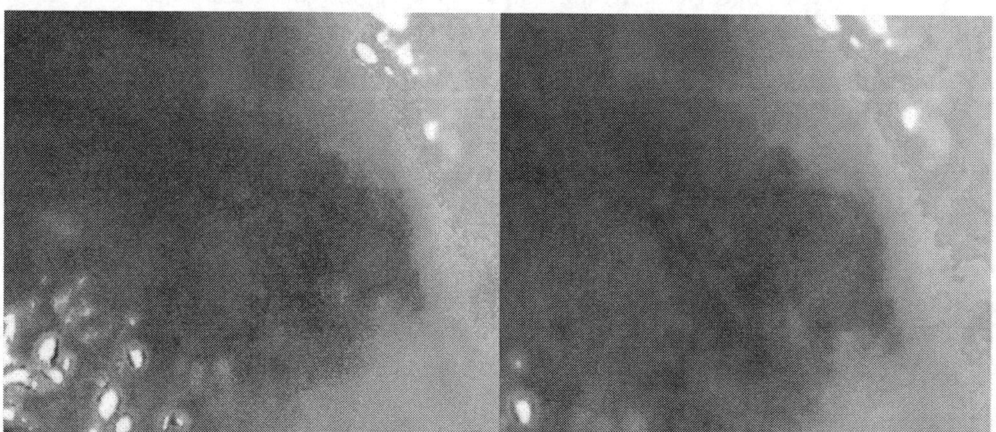

Figure 4. Median filter for removal of artifacts.

It is proposed to use a randomized algorithm for matching the structural editing of images *PatchMatch* (described in the study) for visual better perception of images by doctors [6]. The algorithm applied to the results (Figure 1), is shown in Figure 5. It should be noted that the algorithm can not predict accurate areas, which tend to be overexposed.

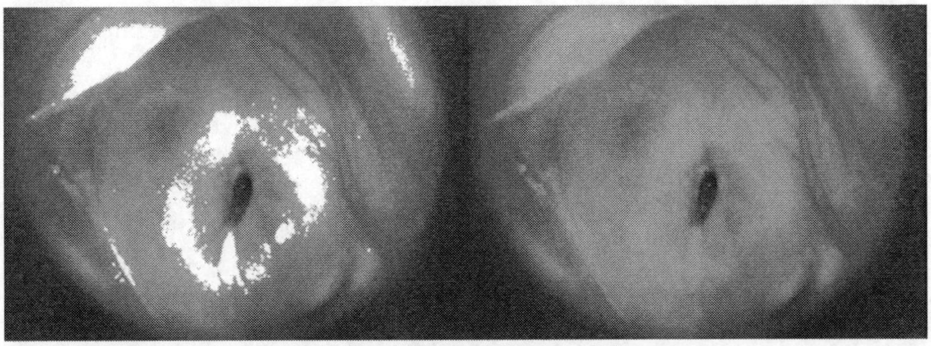

Figure 5. Application of Patch Match algorithm for removal of glare areas.

Figure 6 shows the result of application of the algorithms described in the paper. Application of histogram equalization made it possible to increase the contrast in the erosion areas and convert the hue value to a closer model value. Gauss filtering with a radius 2 and a median filter with a diameter 3 smoothed brightness levels at the boundaries of homogeneous areas. Red areas are marked with a color segmentation using $H \in \left[0;47^o\right]$. Bridging of a binary mask with an element (diameter 9) removed the elements with a low specific gravity and connected potentially simply-connected areas. After that simply-connected areas are searched using a modified two-pass algorithm of sequential scanning. Identification of boundaries of mathematical morphology on the binary mask allowed to mark the areas by applying a binary mask on the original image.

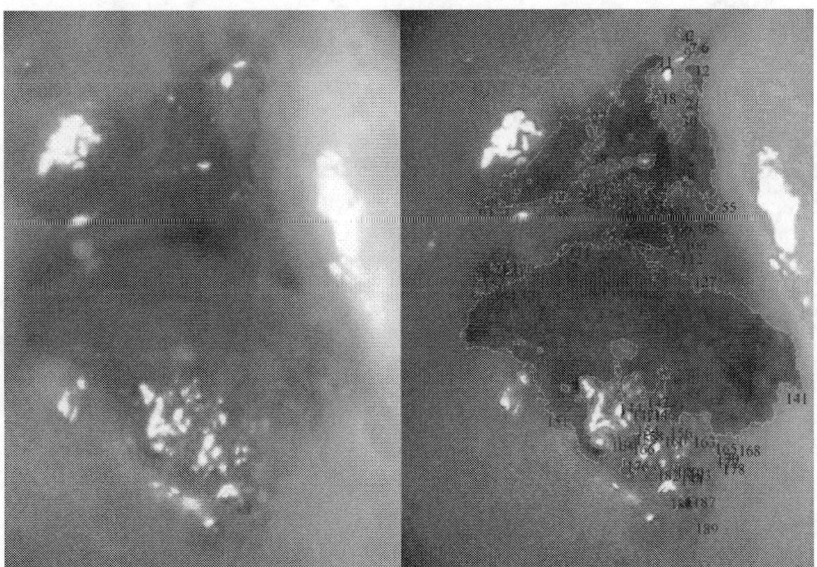

Figure 6. Gaussian filtered image with radius 2, colour segmentation $H \in \left[0;47^o\right]$, erosion and expansion.

3 Conclusion

The described methods have been implemented in computer applications C #. The results of the study, including the analysis of the applicability and implementation of the methods described will be applied in specialized software for the analysis of colposcopic images. In particular, they will be implemented in analytical algorithms and used for improving the quality of colposcopic images and their segmentation. Such algorithms will allow to carry out an automated diagnosis of the surface of the cervix on the basis of its image.

Acknowledgement

The paper was written as part of the research project No. 8.2.31.2015, carried out with the support of the Program "Research Foundation of Tomsk State University named after D.I. Mendeleev" in 2015 – 2016.

References

[1] S.V. Svirskaja, A.T. Egorova, Sibirskoe medicinskoe obozrenie (3), 3 (2011)
[2] S.I. Rogovskaja, N.M. Podzolkova, G.N. Minkina, Ginekologija **13** , 20 (2011) [3] V. I. Syryamkin, D. S.Zhdanov, Telecommunication and Radio Engineering **71**(15),1399 (2012) doi: 10.1615/TelecomRadEng.v71.i15.70
[4] H.D. Cheng, X.J. Shi, Digital Signal Processing: A Review Journal **14**, 158 (2004) doi: 10.1016/j.dsp.2003.07.002
[5] D. C. Lai, V. Verfaille, J. Potenza, Optical Engineering **24**, 1030 (1985)
[6] C. Barnes, E. Shechtman, A. Finkelstein, D. B. Goldman, ACM Transactions on Graphics **28**, 24 (2009) doi: 10.1145/1531326.1531330

MATEC Web of Conferences **79**, 01003 (2016) DOI: 10.1051/matecconf/20167901003
IME&T 2016

COMPUTER CONTROL OF THE SPECTRAL COMPOSITION OF THE POWERFUL LASER SYSTEM IRRADIATION WITH A WIDE RANGE OF LASER TRANSITIONS ON METAL VAPORS

Anatoly Soldatov[1a], *Bogdan* Doroshenko[1], *Ivan* Kostadinov[2], *Yury* Polunin, *Nikola* Sabotinov[3], *Alexey* Shumeiko[1]

[1]Tomsk State University, 634050, Tomsk, Russia
[2]Pulslight Ltd., Sofia, Bulgaria
[3]Bulgarian Academy of Sciences, Sofia, Bulgaria

Abstract. The results of the experimental study cycle of the multiwave metal vapor laser system on the basis of the original configuration of the multimedia laser emitter. The spectral parameters of the setup have been controlled using a personal computer (PC). This allows carrying out their independent optimization according to excitation conditions, and, therefore, promptly allocating the output set of oscillating wavelengths and their relative distribution in power, which makes the system attractive for scientific and technological application.

1 Introduction

Lasers and laser systems are successfully used to solve a lot of applied problems of the atmospheric spectroscopy, in particular remote sensing of atmospheric parameters such as temperature, humidity, concentration, and particle size of various types of aerosol, including man-made products [1, 2]. Among the latter, metal vapor lasers (MVL) are well represented [3, 4]. Currently, the most effective among MVLs are copper vapor lasers [4, 7] and strontium vapor lasers [5, 6], operating with pulse repetition rates from units of hertz up to hundreds of kilohertz. The work [1] describes an oscillator operating on vapors of CuBr, SrIi, and SRII, when two active elements (AE) are placed into a common resonator. The paper presents the design of a laser system in which the system "oscillator-amplifier" is used to increase the average power and the energy in the lasing pulse. Active environments on CuBr and Sr vapors are used in AE. The proposed device allows operating in a multiwave laser system controlled via a personal computer (PC), as the adjustment of the delay time of lasing pulses between the elements of the master oscillator and the amplifier, consisting of AE with different environments, is lightened with the use of autonomous high-voltage power sources. The control and synchronization unit (CSU) is directly connected to the PC.

[a] Corresponding author: general@tic.tsu.ru

© The Authors, published by EDP Sciences. This is an open access article distributed under the terms of the Creative Commons Attribution License 4.0 (http://creativecommons.org/licenses/by/4.0/).

2 Experimental setup

Figure 1 represents a schematic diagram of the experimental setup. The basis of the setup were active laser elements on CuBr and Sr vapors that were united into the classic scheme "oscillator – amplifier". The master oscillator was composed of gas discharge tubes (*1*) with the active zone $L = 800$ mm, $D = 26$ mm, and a gas discharge tube (*2*) with the active zone $L = 550$ mm, $D = 30$ mm. They were located in the common resonator consisting of spherical mirrors: convex (*3*) and concave (*4*). The collimator was comprised of two concave spherical mirrors (*5,6*) coated with aluminum and placed into the common focus of spatial filter mirrors (*7*) with a hole diameter of 0.3 mm, which has allowed to allocate from the oscillator irradiation a spatial component with a spreading close to the diffraction limit. GDT of the amplifier (*7*) with the active zone L=1000 mm and a diameter $D = 35$ mm, and (*8*) L=550 mm and a diameter $D = 30$ mm. have large (as compared to GDT of the generator) diameters of the active channel. Excitation pulses were fed to GDT electrodes from high-voltage power sources (*10 –13*) executed on the basis of converters on powerful transistors with a pulse charge of operating capacitances and hydrogen thyratrons TGI-1 1000/25 as switches, followed by a two-stage compression carried out by ferromagnetic compression cells. The control and synchronization unit (*29*) consisted of the master oscillator (*29a*), the delay module (*29b*), and the synchronization module (*29c*), used for adjusting the relative delay time of the pumping pulse between each active element of the laser system and allowing to change the lasing pulse repetition rate over a wide range. This mode provides stable energy in the lasing pulse, since the pumping parameters and the thermal mode of the active media remained unchanged. The synchronization module had a connector for an external source of computer control (*30*). The laser performance measurement system allowed controlling spectral and power characteristics. Measurement of current-voltage values, such as the amplitude and the duration of high-voltage pumping pulse (voltage on GDT and the current through the tube), with the use of voltage dividers (*14,16,18,20*) and low-resistance shunts (*15,17,19,21*), was carried out to optimize excitation conditions and coordinate active media. Excitation pulses were recorded using the "Tektronix" oscilloscope. The emission spectra were measured with the use of a spectral filter unit (*23*) a monochromator MSDD-1000 (*24*), according to changes in parameters of active media coordination. A power meter "OPHIR" was used to control the output power of the laser. Using a PC (*30*) it is possible to specify commands to the control and synchronization unit (CSU), with a feedback from the controller (*29*), thereby to allocate a region of the spectrum necessary for a particular application from the entire irradiation range. Since the excitation of each AE is carried out with the use of its own high-voltage power source, there is a possibility to carry out an independent optimization of pumping conditions for any selected active medium. It should be noted that a very high gain in active media allows obtaining generation on a large transition of near and infrared media without anti-reflecting coatings on windows.

 DOI: 10.1051/matecconf/20167901003

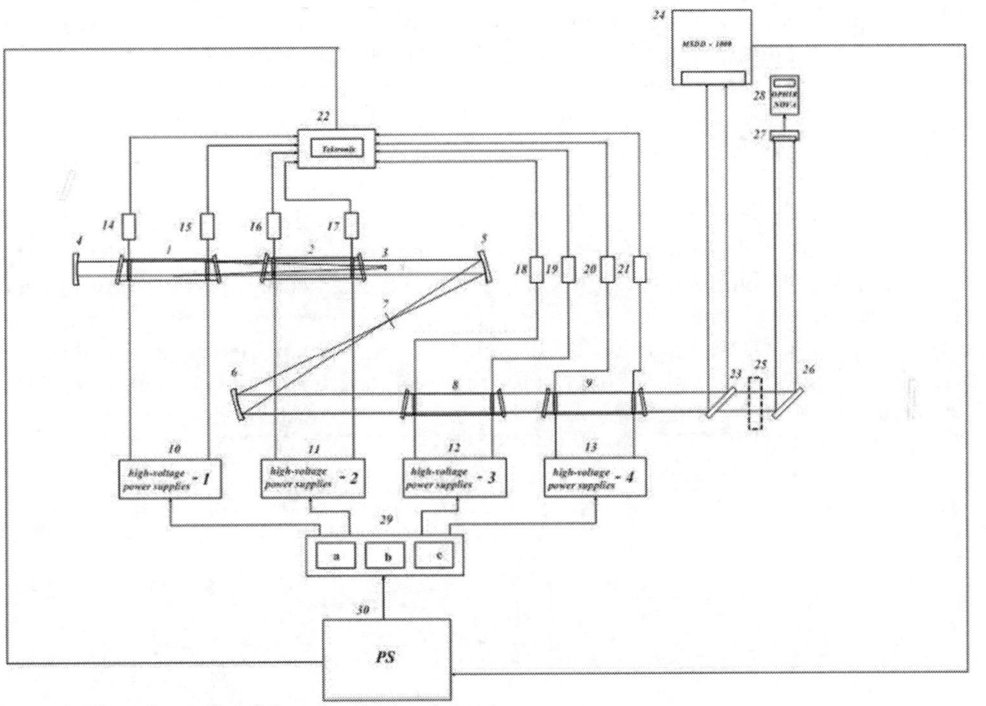

Figure 1. The schematic of the experimental setup:
1 – GDT-1, CuBr; 2 – GDT-2, Sr; 3 – convex mirror; 4 – curved mirror; 5,6 – concave spherical mirrors; 7 – spatial filter; 8 – GDT-3, CuBr; 9 – GDT-4, Sr; 10,11,12,13 – high-voltage power supplies, 14,16,18,20 – voltage dividers; 15, 17,19,21 – low-resistance shunts; 22 – multi-channel oscilloscope "Tektronix"; 23, 26, 30 – rotary mirror; 24 – monochromator MSDD-1000; 25 – unit of spectral filters, 27 – indicator "OPHIR-NOVA"; 28 – gauge "OPHIR-NOVA"; 29 – control and synchronization unit (CSU); 30 – PC.

3 Results and their discussion

The studies on the generation at wavelengths of vapors of the atom CuI (0.51, 0.58), ion SrII (1.03, 1.09) and atom SrI (2.60, 2.69, 2.92, 3.01, 3.06 and 6.45 μm) have been carried out using the setup. Table 1 shows the average power of laser transitions, the average lasing power in individual laser transitions, depending on the delay time of excitation pulses of active elements on CuBr vapors (τ_1) and Sr vapors (τ_2).

Table 1. Average powers of laser transitions (Sr+CuBr).

No	P_{ave}, W (0,51 μm)	P_{ave}, W (0,578 μm)	P_{ave}, W (1.03;1.09 μm)	P_{ave}, W (2.6–3.06 μm)	P_{ave}, W (6.45 μm)	P_{ave}, W (total)	τ_1, ns	τ_2, ns
	CuBr		Sr					
1	14	8	6	9	12	49	20	15
2	0	0	6	9	12	27	-100	15
3	14	8	0	0	0	22	20	-100
4	0	0	0.5	1	8	9,5	-100	30
5	3	1	3	5	2.5	14.5	10	0
6	2	0	0.3	1	8	11.3	0	40
7*	0	0	0.1	12	1	13.1	0	15

The first line of the table presents data in the optimal synchronization mode for two media; this corresponds to one of the states of the computer program. Each of the following lines represents the altered state. The table lists the intermediate values of most explicit changes in the spectral composition of the irradiation. The interelement system filtration mode is shown in the line 7* when the predominant spectral emission range of 2.6 and 3.1 μm.

It is known that the active medium on Cu and Sr vapor are optically transparent interaction. In our case, within the working channel of active element on CuBr vapors present a number of compounds (CuBr, Br_2, etc.), Having a certain structure of molecular absorption bands, which could lead to an increase in the intracavity losses for lasing lines at crossings strontium. Obtained in the third cycle of measurements experimental data, these fears are largely removed. Figure 2 shows the relative positions of the current pulse through the active element (CuBr) and the pulse of the laser of a copper bromide.

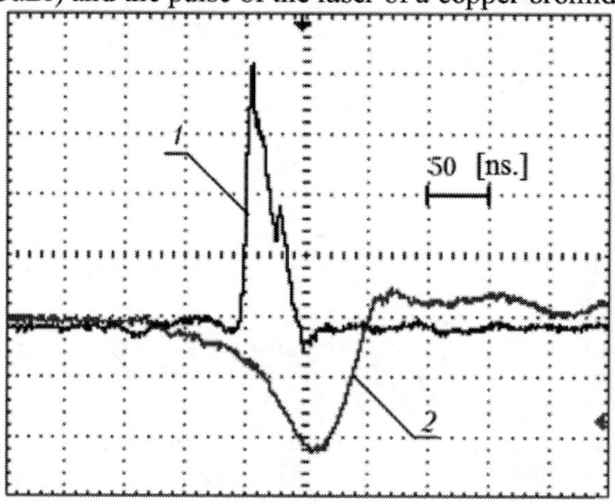

Figure 2. Waveforms generating pulses (1) and a current pulse through AECuBr vapors to (2).

Figure 2 shows the relative positions of the current pulse through the active element (CuBr) and the pulse of the laser of a copper bromide. It can be seen that the latter is located on the front edge of the current pulse, and the duration of the base generation is ~ 40 ns, which is typical for copper vapor lasers and its compounds in pulse repetition rate within 10…20 kHz. Waveforms pulse generation on SrI transitions are shown in Figure 3. Pulse generation duration on lines of strontium ion $\lambda \sim 1$ μm is 50 ns and is close to the pulse width generating CuBr-laser. The pulse duration on the most powerful line of strontium generation ($\lambda = 6.45$ μm) is about 130 ns, the duration of the total generation in the m-m transitions m-m strontium is ~ 70 ns. Generation parameters on the transitions of the atom and strontium ion in multielement laser differs little from generation to one-element Sr-laser.

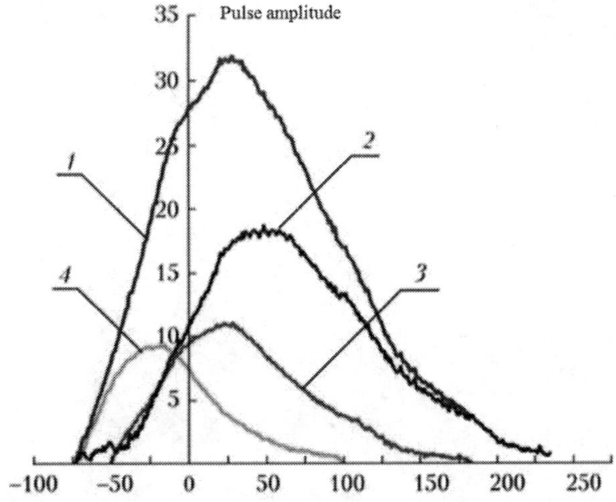

Figure 3. Waveforms of the laser pulses to strontium pairs: 1 – total generating momentum; 2 – impulse generation $\lambda = 6.45$ μm; 3 – impulse generation ~ 3 μm; 4 – generating momentum ~ 1 μm.

4 Conclusion

A computer-controlled system has been presented. The concept allowing controlling the construction of multi-wave laser systems has been proposed. The concept lies in the fact that several gas discharge AE (two or more) are placed into a single (common) resonator. In the proposed device the adjustment of time coordination between elements of the master oscillator and the amplifier consisting of AE with different media and the allocation of the spectral composition of the radiation is carried out using a PC. The scheme of a laser system wherein the oscillator and the amplifier are composed of four active elements operating on CuBr and Sr vapors has been developed. The laser system is capable to operate in a multi-wave mode with a high average oscillating power. The system can operate in multi-wavelength mode with a high average power lasing at 10 spectral lines in the visible and near-infrared range. Such a configuration of the laser system and the system support allow solving applied problems in the area of the atmosphere monitoring: temperature, humidity, concentration, and particle sizes of various kinds of aerosol, including man-made products.

References

[1] E.V. Browell, T.D. Wilkerson, and T.J. McCrath, Appl. Opt. **22**,522 (1983)

[2] A.N. Soldatov, N.V. Sabotinov, E.L. Latush, G.D. Chebotarev, N.K. Vuchkov, N.A. Yudin, *Strontium and calcium vapour lasers. Vol. 1* (Prof. Marin Drinov Academic Publishing House, Sofia, 2013)

[3] P.A. Bokhan, P.P. Gugin, D.E. Zakrevskii, M.A Lavrukhin, M.A. Kazaryan, N.A. Lyabin, Quantum Electronics **43**, 715 (2013) doi: 10.1070/QE2013v043n08ABEH015050

[4] C.E. Little, *Metal Vapour Lasers. Physics Engineering and Applications* (Wiley, United States, 1999)

[5] A.N. Soldatov, S.Ju. Mirza, Ju.P. Polunin, A.S. Shumeyko, I.C. Costadinov, Journal of Applied Spectroscopy **81**, 1025 (2015) doi: 10.1007/s10812-015-0045-8

[6] F. Saleheen, C.-H. Won, 2015 IEEE SENSORS – Proceedings, 7370424 (2015) doi: 10.1109/ICSENS.2015.7370424
[7] G.S Evtushenko, Proceedings of SPIE - The International Society for Optical Engineering **9810**, 98101F (2015) doi: 10.1117/12.2225586

THE DEVELOPMENT OF TECHNOLOGY FOR TOPICAL APPLICATION OF XENON AS PART OF AN EXTERNAL AGENT AND ITS EFFECT ON RAT SKIN WOUND HEALING

Evgeny Petrov[a], *Vladimir* Kozik, *Alexander* Verkhovskiy

National Research Tomsk State University, 634050, Tomsk, Russia

Abstract. We have developed an agent for external use to deliver xenon into the skin. Using the model of skin wound we have studied the effect of xenon on the process of wound healing in rats. It has been shown that the application of xenon-containing cream on wound surface significantly accelerates the healing process as compared to control animals where cream without xenon has been applied.

1 Introduction

Biological effects of xenon mainly studied in the field of anesthesiology and therapy [1, 2]. In previous studies with xenon inhalation have shown that xenon has neuroprotective [3–6], analgetic [7–9], antihypoxic [10] cardioprotective activity [11], exhibits anti-stress and adaptation effect [12], works as cryoprotecting agent in frozen cell cultures [13]. Based on the fact of xenon high solubility in fats and to a lesser extent in water [14, 15], we have developed a technology for producing liquid media with different compositions xenon [16]. Given that the application of liquid media with xenon skin observed its transdermal diffusion [17], it is reasonable to assume that xenon can affect the dynamics of healing, since it is in the area of the wound surface the metabolism is more intense than in the intact adjacent areas. To create a high local concentration of xenon on the skin surface, it is necessary to use agents (creams, ointments, etc.) with high lipophilic content. It is also necessary to take into account the convenience of using such an external agent. The most appropriate means such, in our opinion, would be an emulsion with a high content of lipophilic components. On the basis of such emulsions with a high content of lipophilic substances external agents saturated with xenon can be developed in various forms. In this paper we investigate the effect of topical application xenon on the process of healing of skin lesions (excision wound).

[a] Corresponding author:alverkhovniy@mail.ru

© The Authors, published by EDP Sciences. This is an open access article distributed under the terms of the Creative Commons Attribution License 4.0 (http://creativecommons.org/licenses/by/4.0/).

2 Materials and methods

Experiment on animals were conducted in accordance with WMA Declaration of Helsinki and within a frame of Russian legislation for the use of animals in biomedical research. Work has been done on 30 mixed white male rats weighing 150–180g. All animals were divided into 3 groups of 10. In all animals under light ether anesthesia scalp was removed and in the central region of the upper third of the back full-thickness skin excisions were made. Wounds had a circular shape with a diameter of 6 mm (Figure 1). The first group (control) after excision of skin is not subjected to any stress. In the second group of animals after skin excision on the wound surface 0.2 – 0.3 ml of cream was applied without xenon every morning and evening. In the third (experimental) group after excision of skin daily (morning and evening) was applied xenon-containing cream. Patches of wounded skin were taken for a histological study at 7th and 15th day after wounding.

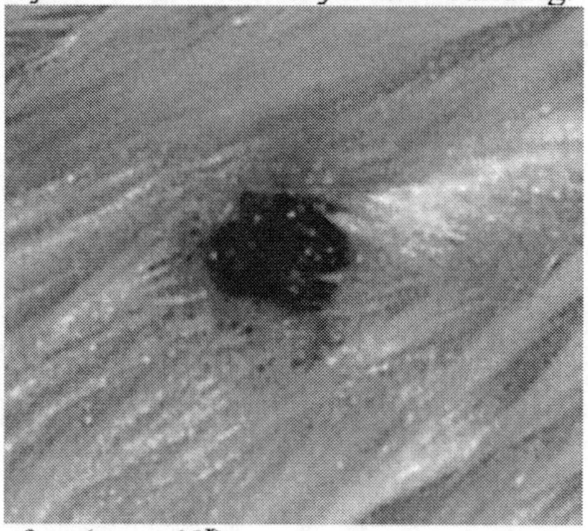

Figure 1. Prepared wound surface (magn.10^x).

The cream for external application was prepared according to the formulation, which includes: olive oil, water, xenon, thickener / emulsifier DC RM 2051 (Sodium polyacrylate, dimethicone, cyclopentasiloxane, trideceth-6, PEG / PPG-18/18 Dimethicone, Dow Corning Corporation, USA), Euxyl 712 preservative (sodium benzoate, potassium sorbate, Shulke & Mayer GmbH, Germany) at room temperature. Using DC RM 2051 allows adding a large amount of oil (40%) in the composition of the cream. Before mixing and homogenizing the components xenon was dissolved in olive oil at ratio 200 cm^3 / 100 ml. Immediately after preparation, the resulting emulsion was poured into a 10 ml sealed containers.

3 Results and discussion

Histology of wounded and intact skin on day 1st is shown in Figure 2. Healing of the wound area occurred in all rats by primary intention with scab formation of characteristic dark brown color. The animals of the first group macroscopically observed a slight decrease in size of the wound for the first 7 days (Figure 3). Morphologically it marked the beginning of the epithelium creeping on wound surface. On day 15 the animals of this group macroscopically observed wound healing with scab rejection (Figure 4). On histological preparations the restoration of the epithelium was observed, however, elements of the dermis of the skin and appendages are underdeveloped.

DOI: 10.1051/matecconf/20167901004

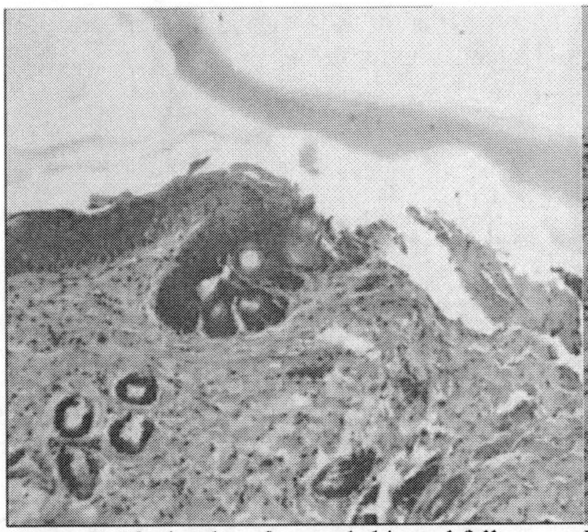

Figure 2. The border of normal skin and full-thickness wounds. (Hematoxylin-eosin staining, magn.120^x).

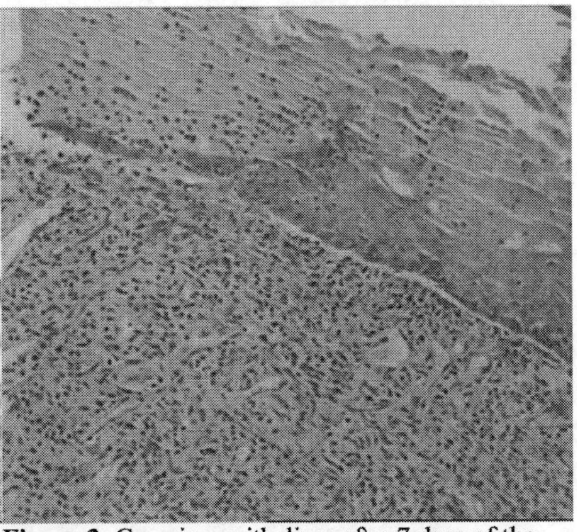

Figure 3. Creeping epithelium after 7 days of the experiment. (Hematoxylin-eosin staining, magn.120^x).

In the second group of rats, the regeneration picture (macro and microscopic) was not significantly different and correspond to the control group. Scab virtually disappeared for 9th –10th day. In the third group on day 7 the wound healing was more pronounced. Wound size was reduced to about 2/3, scab almost disappeared. There was a rapid restoration of hair in the area of the wound. On day 15 the epidermis structure was close to normal, also observed active regeneration of dermis and appendages of skin due to proliferation of dermal cells, vascular neoplasm, and restoration of the structure of sebaceous glands (Figure 5).

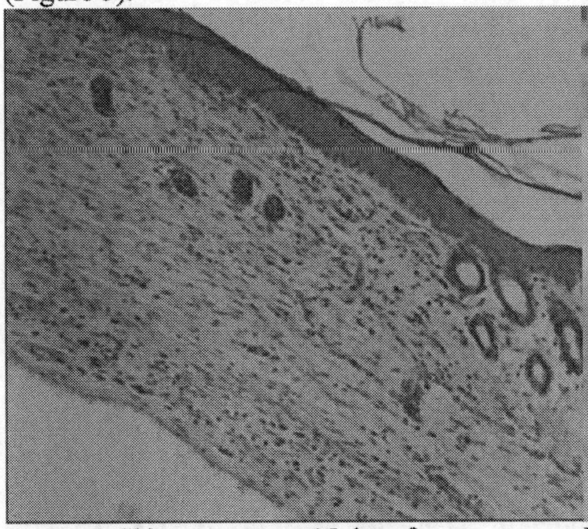

Figure 4. Skin recovery on 15 day of experiment. (Hematoxylin-eosin staining, magn.120^x).

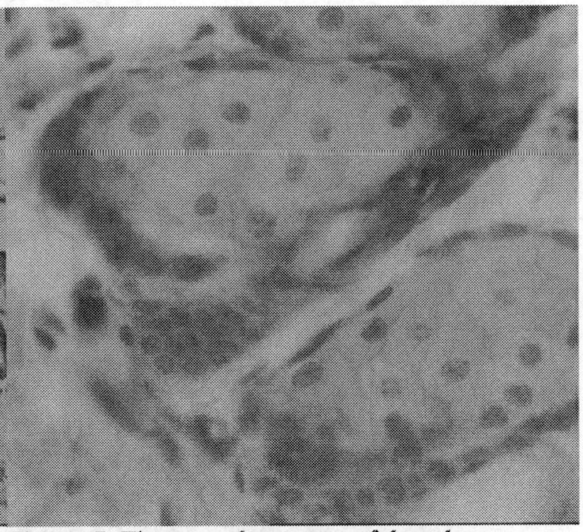

Figure 5. The normal structure of the sebaceous glands on day 15 of the experiment. (Hematoxylin-eosin staining, magn. 400^x).

In this experimental work we have demonstrated the efficacy of xenon present in a cream applied to the wound surface, to enhance the skin regeneration process. Furthermore, it has been observed more rapid recovery of hair. The cream formula used in this study does not include active components used in the treatment of external skin lesions, thus skin wound healing caused directly by xenon dissolved in applied cream. Proposed agents for external use are based on sole mechanism of action - lipophilic vesicles carrying xenon in high concentrations of xenon are readily absorbed onto the surface of epidermis as compared with other components of an external agent (Figure 6).

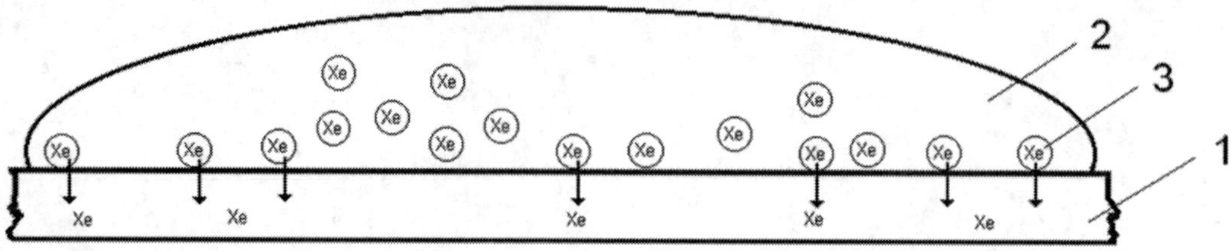

Figure 6. The mechanism of action of xenon-containing external agent.

On epidermis (1) is applied a cream (emulsion), (2), comprising a lipophilic vesicles (3) with a high content of xenon. Absorbed in the epidermis vesicles provide conditions for an effective diffusion of xenon in deeper layers of skin.

4 Conclusions

The xenon-containing emulsion we have developed may serve as a prototype for other external agents in various forms (cream, gel, ointment, etc.). Furthermore, the xenon may be added to already known formulations of means for external use, containing active components, including the form of a patch. Thus, we believe that obtained data can be used in the development of external means used for healing of skin lesions of various origins.

Acknowledgment

The paper was written as part of the research project No. 8.2.31.2015, carried out with the support of the Program "Research Foundation of Tomsk State University named after D.I. Mendeleev" in 2015 – 2016.

References

[1] S.C. Cullen, E.G. Gross, Science **113**, 580 (1951) doi:10.1126/science.113.2942.580

[2] B. Lachmann, S. Armbruster, W. Schairer, M. Landstra, A. Trouwborst, G.J. Van Daal, *et al.*, Lancet **335**, 1413 (1990)

[3] E. Esencan, S. Yuksel, Y.B. Tosun, A. Robinot, I. Solaroglu, J.H. Zhang, Med. Gas Res. **3**,4 (2013) doi:10.1186/2045-9912-3-4

[4] J. Lavaur, M. Lemaire, J. Pype, D. Le Nogue, E.C. Hirsch, P.P. Michel, Cell Death Dis. **7**, e2182 (2016) doi:10.1038/cddis.2016.86

[5] H.M. Homi, N. Yokoo, D. Ma, D.S. Warner, N.P. Franks, M. Maze, *et al.*, Anesthesiology **99**, 876 (2003) doi: 10.1097/00000542-200310000-00020

[6] H.N. David, B. Haelewyn, C. Rouillon, M. Lecoq, L. Chazalviel, G. Apiou, *et al.*, FASEB J. **22**,1275 (2008) doi:10.1096/fj.07-9420com

[7] M. Giacalone, A. Abramo, F. Giunta, F. Forfori, Clin. J. Pain. **29**(7), 639 (2013) doi:10.1097/AJP.0b013e31826b12f5

[8] T. Fukuda, C. Nishimoto, S. Hisano, M. Miyabe, H. Toyooka, Anesth. Analg. **95**, 1300 (2002)

[9] L. Yang, C. a Tulk, D.D. Klug, I.L. Moudrakovski, C.I. Ratcliffe, J. a Ripmeester, *et al.,* Proceedings of the National Academy of Sciences of the United States of America **106**, 6060 (2009) doi:10.1073/pnas.0809342106

[10] V.N. Ananev, Receptor mechanisms of neutral gas in oxygen consumption, Fundam. Res. (11),11 (2013)

[11] W. Liu, Y. Liu, H. Chen, K. Liu, H. Tao, X. Sun, Med. Gas Res. **3**, 3 (2013) doi:10.1186/2045-9912-3-3

[12] V. Potievskaya, N. Stryapko, T. Sazontova, I. Vdovina, Y. Archipenko, I. Molchanov, Eur. J. Anaesthesiol. **31,** 159 (2014)

[13] K. Seino, K. Sato, M. Ujihira, IEEE Eng. Med. Biol. **22**, 1989 (2008) http://embc.embs.org/files/2013/3025_FI.pdf

[14] E. Overton, *Studien über die Narkose, zurgleich eing Bertrag zur allgemeinen Pharmakologie* (Verlag Gustav Fischer, Jena, 1901)

[15] H. Meyer, Arch. Für Exp. Pathol. Und Pharmakologie **42**, 109 (1899) doi:10.1007/BF01834479

[16] A. Verkhovsky, *A device for gas saturation of liquid media* Patent № 156912, 2015

[17] A.Y. Verkhovskiy, D.N. Atochin, S.N. Udintsev, S.I. Tverdokhlebov, Y. Anfinogenova, V.Y. Serebrov, Dermatology Asp. **3**, 2 (2015) doi:10.7243/2053-5309-3-2

COMPUTER SPATIALLY ORIENTED RECONSTRUCTION OF A 3D HEART SHAPE BASED ON ITS TOMOGRAPHIC IMAGING

Natalya Natalinova[a], *Diana* Avdeeva, *Veniamin* Kazakov, *Vladimir* Baranov, *Olga* Galtseva, *Denis* Ivashkov

Tomsk Polytechnic University, 634050, Tomsk, Russia

Abstract. Diagnostics of cardiovascular system conditions and diseases is considered the most important task of electrocardiology. The aim of the study is to define geometric parameters of the patient heart and the synthesis of a realistic three-dimensional heart image based on a series of two-dimensional images obtained by computer tomography. The preparation and further processing of medical image data is an important initial step for further study of the heart electrodynamic activity. The problem is solved with the help of computer tomography method by preparing a series of images of the patient heart. The technique of volumetric rendering is applied to represent certain anatomical structures in a three-dimensional (3D) graphical form. It is concluded that the character of the obtained three-dimensional model of the patient heart is determined by the quality of the input data, the resolution of the tomograph, the tomographic slice thickness, the accuracy of the object boundaries determining the segmentation process and the peculiarities of medical image processing by the software applied.

1 Introduction

The cardiovascular system is regarded as a complex multi-level mechanism in its functioning. Thus, a problem of ensuring an accurate diagnosis might often be solved only by experienced medical specialists. This study was conducted as part of the development of non-invasive methods of diagnosis causes of sudden cardiac death. One reason is presumed to be the presence of a potential delayed depolarization of the heart ventricles. The research objective is to develop a computer model of the heart, which in combination with the highly sensitive ECG measurement methods can be used to diagnose potential delayed depolarization of the ventricles of the heart. A creation of numerical model of a heart will allow solving the inverse problem of reconstructing the characteristics of cardiac tissue on the basis of the ECG signal with the high-resolution data in the hearts and computed tomography of the heart geometry. The virtual model heart application is a visual representation of the cardiovascular system and enables the doctor-diagnostician to compare the electrocardiogram (ECG) changes with the damaged myocardial tissue on the realistic three-dimensional heart images, in particular, at the level of microvolts. Therefore

[a] Corresponding author : natalinova@tpu.ru

© The Authors, published by EDP Sciences. This is an open access article distributed under the terms of the Creative Commons Attribution License 4.0 (http://creativecommons.org/licenses/by/4.0/).

there is a vital need to obtain an apparent three-dimensional visualization of the patient heart for further research of the electrodynamic heart activity by means of the finite element method of CAD software [1, 2].

The paper aims to define geometric parameters of the patient heart and the synthesis of a realistic three-dimensional heart image based on a series of two-dimensional images obtained by computed tomography. Thus, obtaining and further processing of medical image data is viewed as an important step for further research of electrodynamic heart activity.

Figure 1 shows the algorithm for obtaining a 3D heart model.

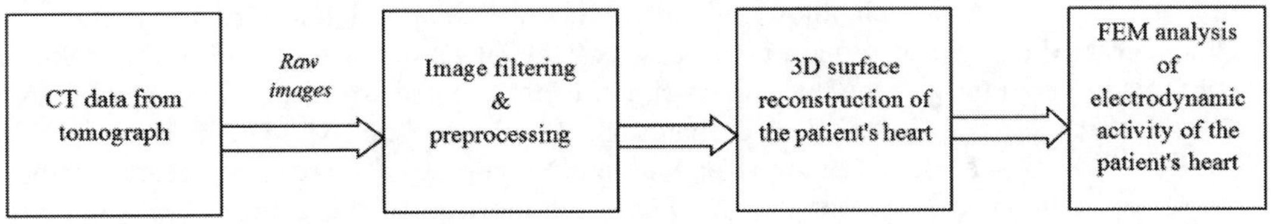

Figure 1. Processing of medical imaging data based on computed tomography slices.

2 Method of 3D reconstruction of tomographic images

The initial information obtained by computer tomography is a sequence of two-dimensional (2D) digital images. Each pixel in a 2D computed tomography (CT) image is a function of X-ray density of the object of study at the corresponding point q (x, y, z), represented in gray scale. These images are stored and transmitted as Digital Imaging and Communications in Medicine (DICOM) file format series. Tomographic slices have an equal thickness and the same number of pixels per slice. The tomographic slices are further ordered; there is created the volume image represented by a regular grid of voxels in the diagnostic software (tomographic reconstruction) which is characterized by different absorption intensity and reflects the structure of the research object. Figure 2 shows the series of 2D images of the patient's heart obtained from CT. The visual representation of individual anatomical structures in virtual three-dimensional objects is obtained by a 3D volume rendering technique. For this purpose the operator determines the threshold X-ray density (e.g., those corresponding to the density of the bone tissue, or cardiac tissue).

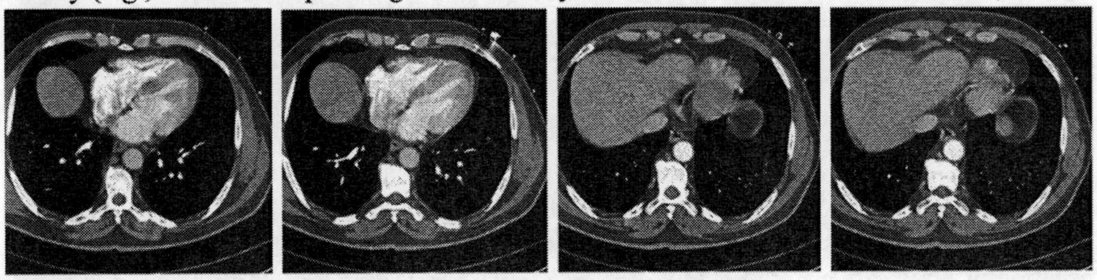

Figure 2. Series of 2D images of the patient's heart obtained from CT.

The three-dimensional model is then built according to a given range of X-ray density in the software. There exist numerous image processing packages, either commercial (Amira, Visage Imaging; MeVisLab, Mevis; Imaris, BitPlane) or open source (VOXX, 3D Slicer, BioImage XD, etc.). These packages offer excellent solutions for the specific problems they were designed to solve [1].

To understand the process of three-dimensional image reconstruction, consider the algorithm which is used in the software 3D Slicer. By changing the threshold density X-ray techniques and graphic image processing, it is possible to create a complex, multicomponent model consisting of individual elements, which correspond to the bones, muscles, lungs, blood vessels, etc. The method of Marching cubes is used to transform a set of voxels into the polygonal model as the standard mathematical algorithm. Some 3D models are emphasized by different colors and transparency changes for highest visibility and easy image perception of the research object.

3 Results

Figure 3 shows the logical structure of the software algorithm implementing the numerical model of the heart, which allows to solve the inverse problem of reconstructing the characteristics of cardiac tissue on the basis of the ECG signal data and high-resolution computed tomography of the heart geometry. Software consists of the following modules: 1 – the module of geometric data preparation, 2 – the module of heart bidomain model configuration, 3 – the model of numerical modelling on base of the obtained heart bidomain model configuration, 4 – the control module which determines to continue to seek the error minimum, or to stop, 5 – the module of error calculation between heart model ECG and realistic ECG with the high-resolution [3–5].

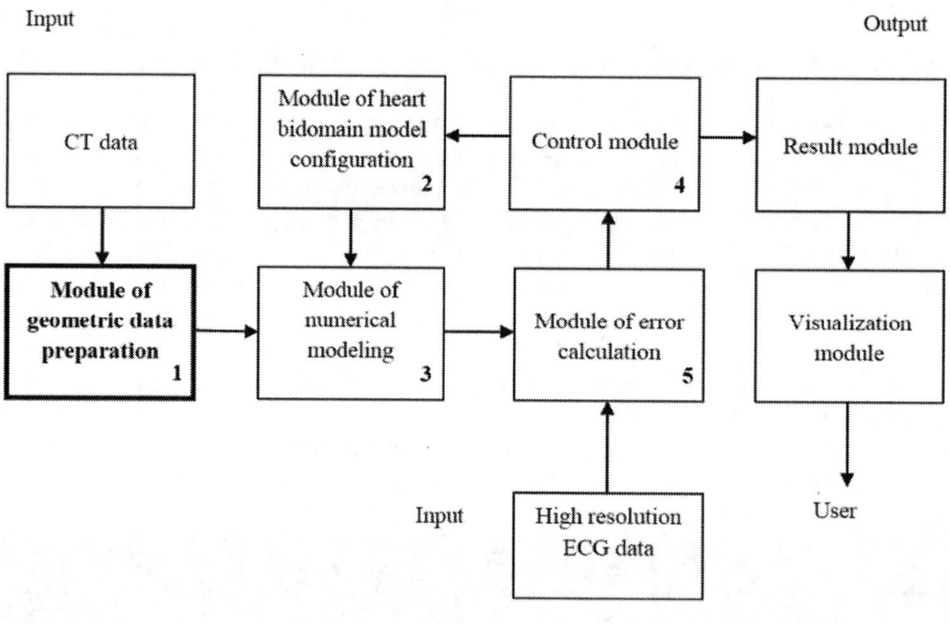

Figure 3. The logical structure of the algorithm.

The input data for the program are the data of the patient CT tomograph and an ECG signal obtained via a hardware software system for measuring heart micropotential. The creation of three-dimensional model of the patient's heart (preparation of the geometry data) is a preparatory stage for the diagnosis of heart disease at an early stage .

The first step of the process of creating a three-dimensional model of the patient heart covered the uploading of the tomographic images in DICOM format. Then the range of gray shades corresponding to the density of the heart was selected, and, finally, a geometric model was reproduced based on these data, Figure 4 shows the resulting image of the individual realistic model of the patient heart.

DOI: 10.1051/matecconf/20167901005

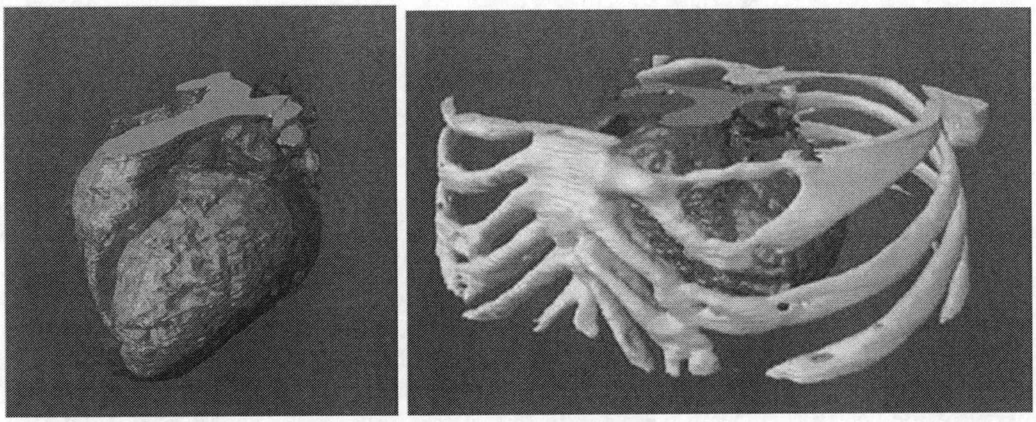

Figure 4. An individual realistic model of the patient heart.

The resulting 3D virtual model of all the elements of the patient torso areas were removed not to effect further calculations of electrodynamic heart activity. The resulting model was saved in the STL format, which allowed importing it into the CAD complex «Chaste» [2, 11] for further analysis by the finite element method.

The quality of the resulting model is determined by the quality of the input data resolution images, the tomographic slice thickness, the precision of the object boundary definition during the segmentation process, some peculiarities of the model handling features in the medical imaging software [6–12].

The great importance in the process of creating virtual 3D-models is a preliminary assessment of the tomographic image quality, the identification of existing artifacts and their further elimination in the process of 3D rendering. The quality of the resulting model is determined by several features, such as: the quality of input data resolution of a tomograph, the tomographic slice thickness within the definition of the research object boundaries in the segmentation process, model handling features in software for medical imaging. Although a computer tomography is a precise method of medical research, its implementation might be affected by various artifacts, which in their turn can significantly influence the quality of the resulting 3D-model. Among the artifacts there are the following ones:

- linear artifacts that occur near objects which absorb most of the x-rays;
- the effect of the image blurring in the sharp edge areas and abrupt transition zone from high to low x-ray density. To reduce this negative effect it is possible to increase a number of sensors and decrease the tomographic slice thickness.
- occurrence of one or more "rings" in the image. This effect is usually associated with mechanical breakdown of the detector;
- "a noise artifact" that shows grainy images. This effect is usually associated with the low ratio of signal and noise or insufficient capacity of an X-ray tube;
- artifacts associated with the movement of the research object, appear blurred or with linear distortions. These artifacts are minimized in modern tomographs by means of reducing the time spent on processing a single slice.
- "steps" in the images. This effect is usually associated with large thickness of the tomographic slice, etc. [11].

Obtained geometrical model is a part of the software designed for the diagnosis of low-amplitude biopotentials of the heart, such as ventricular late potentials (VLP), arising due to delayed depolarization of the ventricles of the heart. The heart electrodynamic model allows solving the problem of detection and localization of necrotic tissue sections of heart, leading to the appearance of low-amplitude high frequency components in the ECG signal by varying the size distribution and the characteristics of the conductivity tensor of the transmembrane action potential.

This computer heart model is combined with highly sensitive methods of ECG measurement and it is part of the hardware-software complex diagnostic potential delayed depolarization of the ventricles of the heart.

4 Conclusion

Image processing system with modeling of 3D patient's heart makes possible to get comprehensive information on the condition of the cardiovascular system, its shape, a location and geometric dimensions. A computer model of a 3D heart shape based on its tomographic imaging duplicates the human heart as accurate as possible. The computer model of the heart provides an opportunity for further study of the electrodynamic activity of the heart using finite element method based on the anatomical characteristics of a particular patient's heart. Thus, creating a heart model is an important step in the process of non-invasive method of heart disease diagnostics. Due to this, scientists will be able to determine abnormalities in the heart, even at early stages and early diagnosis of ordinary citizens to reduce the number of severe and fatal outcomes of chronic heart disease.

Acknowledgment

The research was financially supported by the Federal Targeted Programme "Research and Development in Priority Fields of S&T Complex of Russia in 2014-2020", the Agreement No. 14.578.21.0032 dated 05.06.2014 "Development of the experimental sample of a hardware-software complex for noninvasive recording of heart micropotentials in a wide frequency band without filtering and averaging in real time to early detect the symptoms of sudden cardiac death", the unique identifier of the contract: RFMEF157814X0032.

References

[1] S. Sovilj, R. Magjarevic, N. Lovell, S. Dokos, CinC **40**, 377 (2013)
[2] J. Pitt-Francis, P. Pathmanathan, M.O. Bernabeu, R. Bordas, J. Cooper, A.G. Fletcher, G.R. Mirams, P. Murray, J.M. Osborne, A. Walter, S.J. Chapman, A. Garny, I.M.M. van Leeuwen, P.K. Maini, B. Rodriguez, S.L. Waters, J.P. Whiteley, H.M. Byrne, D.J. Gavaghan, Comput. Phys. Commun. **180**, 2452 (2009) doi: 10.1016/j.cpc.2009.07.019
[3] V. Y. Kazakov, D. K. Avdeeva, M. G. Grigoriev, N. M. Natalinova, I. V. Maksimov, M. V. Balahonova, BLM **7**, 1 (2015)
[4] D. K. Avdeeva, V. Y. Kazakov, N. M. Natalinova, M. L. Ivanov, M. A. Yuzhakova, N. V. Turushev, Eur. J. Phys. and Health Educ. **6**, BM-015-14 (2014)
[5] D. K. Avdeeva, V. Y. Kazakov, N. M. Natalinova, I. V. Maksimov, M. V. Balahonova, BLM **6**, BM-025 (2014)
[6] P. Pathmanathan, R.A. Gray, Int J Numer Method Biomed Eng. **30**, 525 (2014) doi: 10.1002/cnm.2615
[7] S.A. Niederer, A.K. Shetty, G. Plank, J. Bostock, R. Razavi, N.P. Smith, C.A. Rinaldi, PACE **35**, 204 (2012) doi: 10.1111/j.1540-8159.2011.03243.x
[8] D. Noble, A. Garny, P.J. Noble, J Physiol. **590**, 2613 (2012) doi: 10.1113/jphysiol.2011.224238
[9] V. N. Borikov, P. F. Baranov, A. D. Bezshlyakh, IEEE SIBCON-2009, 5044870 (2009)

[10] X.H. Wang, X.Y. Li, X.J. Zhang, IFMBE Proc. **31 IFMBE,** 1503 (2010) doi: 10.1007/978-3-642-14515-5_383
[11] G.R. Mirams, C.J. Arthurs, M.O. Bernabeu, R. Bordas, J. Cooper, A. Corrias, Y. Davit, S.-J. Dunn, A.G. Fletcher, D.G. Harvey, M.E. Marsh, J.M. Osborne, P. Pathmanathan, J. Pitt-Francis, J. Southern, N. Zemzemi, D.J. Gavaghan, PLoS Comput. Biol. **9** (3), e1002970 (2013) DOI: 10.1371/journal.pcbi.1002970
[12] E.J. Vigmond, M. Hughes, G. Plank, L.J. Leon, J Electrocardiol. **36**, 69 (2003) doi: 10.1016/j.jelectrocard.2003.09.017

VIRTUAL ANALYSIS OF TECHNICAL EQUIPMENT OPERATIONAL STATES VARIABILITY

Tatiana Aleksandrova[1,*], *Evgeny* Gromakov[1,2], *Roman* Chugunov[1] and *Vladimir* Syramkin[1,2]

[1]Tomsk Polytechnic University, 634050, Tomsk, Russia
[2]Tomsk State University, 634050, Tomsk, Russia

Abstract. The efficiency of technological equipment workout is determined by urgency of its maintenance. Perspective method of urgent maintenance is based on control of equipment state variability. To analyse the condition of the equipment in the real-time proposed the virtual control of equipment state variability based on Neyman-Pearson criterion. The results of data processing of oil pump unit showed the ability to detect faults at an early stage of their occurrence.

1 Introduction

Oil pumping stations (OPS) are complex set of oil pumps in main oil pipeline. The necessary condition for operation of trunk pipelines is ensuring efficient and reliable operation of the pump stations. The problem of providing efficient, reliable and safe operation of main oil pipelines becomes quite relevant in connection with the fall in oil prices and the depreciation of the main technological equipment, in particular, the main pump units, as the most energy-intensive OPS equipment.

Conventionally we can group all the faults of the pump unit on three types: fault associated with impaired mounting rigidity of the pump unit and its components; defects of electromagnetic origin; failure of mechanical and hydrodynamic origin.

Continuous operational control of the process equipment operational state is usually carried out with monitoring of a number of indirect parameters that are characterizing technical condition. Usually these parameters are vibration, temperature, pressure at the inlet and outlet. However, in some cases the results of online monitoring analysis do not have the necessary level of completeness and timeliness. Practical experience with the data shows that their validity is often insufficient in cases of operational control.

The main reason is the difficulty of identifying system variability on small sample volumes of controlled signals of hydrodynamic origin. But it is critical to process this data because it contain the information about a false error in deciding to stop the pumping unit. For example, the coincidence of the natural oscillation frequencies of working vanes and pump components with frequencies of oscillations induced by cavitation, especially at low

* Corresponding author: tata09@mail.ru

© The Authors, published by EDP Sciences. This is an open access article distributed under the terms of the Creative Commons Attribution License 4.0 (http://creativecommons.org/licenses/by/4.0/).

feed rates, the appearance of intense oscillations of the blades and the rotor, piping and frame that can be easily removed without stopping the pumping unit, changing its mode of operation.

Therefore, there is a scientific and technical problem of improving the completeness, timeliness and accuracy of maintenance services information support by creation of real-time monitoring the variability parameters.

2 Virtual analysis

The aim of this work is to develop a virtual analyzer (VA) of the operational states' variability of the process equipment based on operational statistical control.

During the operation centrifugal pumps installed in the main oil pipelines, are constantly tense in deformed state. Also the wheel and blades pumps are subjected by combined effect from the transported oil.

Virtual OPS states analyzers can be used to support operator's decision-making as a supplement to the control system that form rapid warning and forecast maintenance requirements in real-time operation of the process. Virtual condition analyzers of technological equipment belong to the display systems class.

Equipment indication systems carry out only the definition of object technical condition without warning instructions for the type of defect. Indication of OPS situation gives to the operator either directly in the graphic control or through a stand-alone interface. Algorithms for subsequent identification of the defect based on the additional information: expertise or regulatory documents. In many cases, performance standards, state and industry standards, examples of which are [1] used as such documents.

OPS scheme typically includes four pump units, one or more may be in operation (Figure 1).

In accordance with recommendations of the regulatory documents that normalize OPS operation, virtual analyzers should display simultaneously at least two trends on any selected measured parameters with the following recommended time intervals (on operator handling time):

- 40 days with 2 hours steps;
- 1 year with 1 day steps;
- 9 years with 7 days steps.

Because all pumps are connected each other and to the collector of receiving and outtake pipes, relations is statistically inhomogeneous perturbations arising from the performance of each pump is transferred to other units and line pipe, making it difficult to identify inconsistencies pump unit from the group pumps still working.

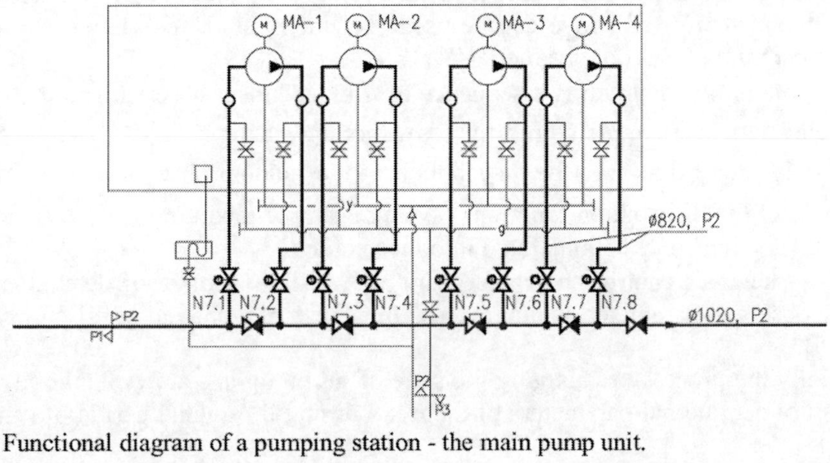

Figure 1. Functional diagram of a pumping station - the main pump unit.

Visual monitoring of operational diagnostic paramaters helds by OPS operator every two hours with control display [2] for the principal and retaining support of pump units. Data is written to the log (statement) of operational diagnostic controls by OPS personnel on duty. Evaluation of the retaining pumps main vibration is carried out on the control and vibration equipment signal. Register magnitude of vibration is made at least once per shift for each monitored point at steady state

The main technical parameters of the main oil pump units are:

P_{in} is an inlet pressure (inlet side) of the pump, MPa;

P_{out} is the discharge pressure (discharge pipe) of the pump, MPa;

T_k is the temperature of the pump housing, °C;

T_{el} is a motor bearing temperature, °C;

T_p is a pump bearing temperature, °C;

V_{el} is a vibration motor (vibration velocity), mm/s.

To process the data of hydrodynamic origin with help of statistical analysis we use the method based on the Neyman-Pearson criterion. According to this criterion we select detection rule that provides the minimum value of the signal crossing the probability of unacceptable change of technological equipment for a given probability of false alarm.

Normative documents established the following classification by value of sudden failure:

1. low risk missing;
2. average risk missing;
3. high risk missing.

Low-risk missing systems must have the value of the risk of sudden failure less than 5%.

Medium-risk missing systems must have the value of the risk of sudden failure in the range 5 – 30%.

High-risk missing systems must have the value of the risk of sudden failure more than 30%.

As shown in [3] special statistical processing of regular measurements in the form of Shewhart cards solves the task of identifying specific (in this case, potentially dangerous) reasons for variations in equipment condition with error probability (the probability of a false rejection) $\alpha = 0.0027 = 0.27\%$. This is far below the standards of low risk passes sudden failure. The statistical algorithms can be supplemented with Shewhart algorithms vibration control equipment, providing additional confidence in the assessment, both at the operational and medium-term horizons of technical tinning and repair planning

Control card for state of the pump unit monitoring is a graph of regular measurement samples of a parameter which characterizes the state of the pumping unit ***X***. As ***X*** one of the oil pump parameters listed above can be used. Telemetry statistical data collected can be combined in a sample chart on the map *g(X)*.

The sample in this diagram is a sequence of n independent observations $x_1, x_2, ..., x_n$ of X index related to a certain period of time (day, month, year).

Suppose $f(x,\theta)$ describes the distribution of the random observations variable $\boldsymbol{x}$ at a certain value of the distribution functions (average sample medium medium samples, span, etc.). These functions may be integrated into the vector θ.

Using a Shewhart control chart posed and solved the problem of detecting unnatural variability observed parameter, which may be the cause of a change of oil pump operating mode [4, 5].

Especially the process of changing the state of oil pump is a slow change in variability in the form of operational parameters of the trend during the annual period of time and fast as the appearance of a series of unusually grouped outlets controlled parameter on the map in real time characterized by a second period of the survey measurement sensors. We assume that the statistical properties of the time series of measurements, and the reasons for generating changes in vector functions distribution θ remain unchanged or change slowly in the selected intervals. Monitoring of the oil pump status reduced to the control of samples measurements and detection of the controlled parameter changes the observed properties of the vector θ with values from θ_0 to θ_1.

Changing the oil pump operating mode may occur at an unknown time t_0. This event will be called discord operation process. Useful problems solved automated condition monitoring on a system is to detect a disorder in the early stages of its development, operational control of operating parameters and predict operational oil pump state.

Reviewed range of practical problems associated with the operational oil pump states, characterized by a random sequence $\{x_1, x_2, ..., x_n\} = \{x_1^n\}$, which at the time t_0 changing their properties, clearly defined parameters vector θ, $\dim\{\theta\} = r$. Since t_0 the vector θ is set to change point θ_1 $(\theta = \theta_1)$. The moment of disorder t_0 is detected to establish the criteria for evaluating the sequence $\{x_1^n\}$, $N \to \infty$, with the appearance of the next point x_n.

Each time a new sample to test the hypothesis $\theta = \theta_0$, pointing to the absence of the unnatural variability [6].

Because of oil pump hydrodynamic parameters controlling in depending of ambient temperature and oil viscosity their values are changed in considerable limits, the observed parameter is expedient to use the input / output relationship of each pressure pump unit for sampling time averaged. These current values enter the virtual analyzer, which is undergoing continuous analysis of oil pump, which construct control charts for the current mode of his work for a controlled period of time. Data are grouped in the sample and compared with control limits, which are calculated according to the formula (1) [7].

$$UCL, LCL = \overline{\overline{X}} \pm A_2(n) \cdot \overline{R}, \tag{1}$$

where *UCL, LCL* are upper and lower control limits,

$\overline{\overline{X}}$ is an average of medium,

$A_2(n)$ is a coefficient depending on the sample size *n*, equals 0.266 when $n = 12$, because 12 measurements per day,

$\overline{R}$ is an average Range.

For a retrospective analysis of the technological equipment state, the data of one separately working oil pump in period from February to March 2015 is used. The resulting $\overline{X} - R$ chart is shown in Figure 2.

Then from calculations values of current parameters measured in the initial 72 hours after installation or repair of the pump were excluded, as well as measurements of switches of the measuring lines on oil account nodes and start/stop of controlled pump or neighboring oil pump units of OPS.

By introducing these amendments (the data were excluded for 20, 25 February and 5, 14 and 16 of March), the control $\overline{X} - R$ chart of variability of medium and dispersion of sample data for a single oil pump shown in Figure 3.

It was decided to set the upper and lower control limits equal to 3.09σ. This corresponds to the fact that at the exit for the border at least one point of the sample data on

 DOI: 10.1051/matecconf/20167901006

the oil pump state probability of an erroneous conclusion about the absence of irregularities in his work does not exceed 0.001.

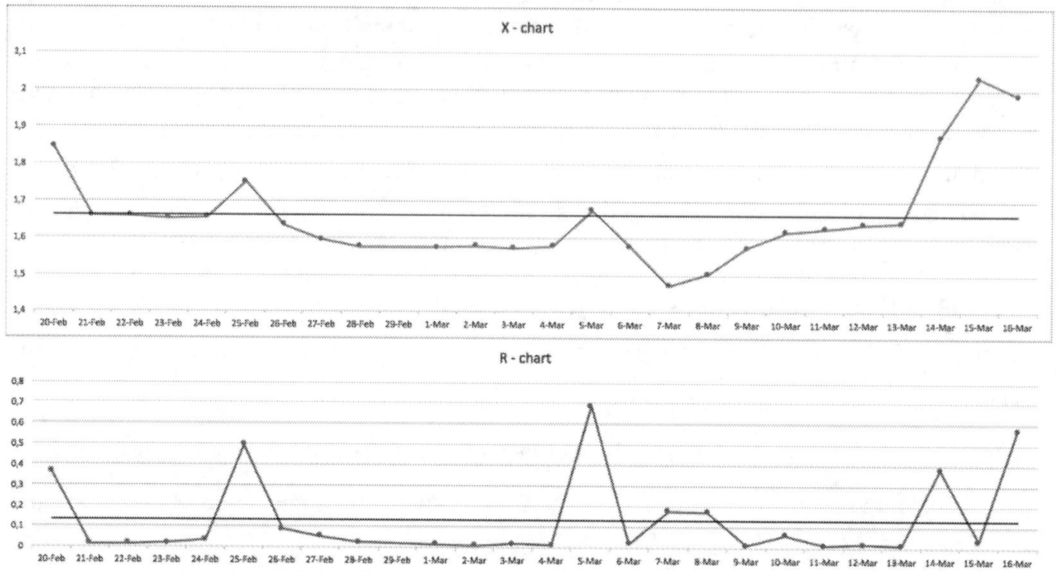

Figure 2. $\overline{X}-R$ Shewhart control chart.

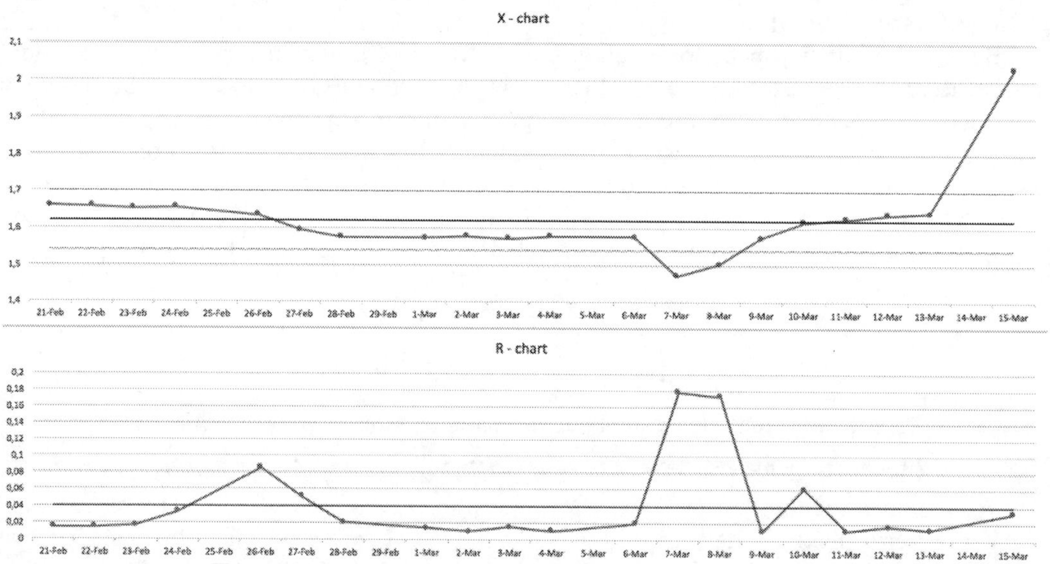

Figure 3. Adjusted $\overline{X}-R$ Shewhart control chart.

Analyzing the arrangement of the control points of the chart in relation to the control limits are consistently on the charts scale and the average values in the samples, we can conclude that on March 6, there were signs of abnormal operation of the pump unit.

It should be noted that the vibration control data in this period has not yet foreshadowed discrepancies in the work. However, March 15 pumping unit has been shut down for repair service.

Thus, using the state on the control chart, we can watch the development of the inconsistencies in the pumps at an early stage, to begin to identify the reasons for using online monitoring tools and start preparing for a stop unit.

3 Conclusion

Checking the virtual analyzer performance was carried out on the data, which were obtained in the actual operating equipment (trunk unit MN7000x210).

It was found that Shewhart chart is quite informative for the assessment of the overall technical condition of OPS pumps.

They allow to identify the early stages of oil pump disruption.

The proposed virtual evaluation of equipment on operational state in online mode allows to identify the controlled samples variability in the quantitative control of the pump unit state.

The proposed method of virtual analysis of the OPS pump state variability allows to realize the maintenance system on equipment actual condition.

Acknowledgements

The paper was written as part of the research project No. 8.2.31.2015, carried out with the support of the Program "Research Foundation of Tomsk State University named after D.I. Mendeleev" in 2015 – 2016, grant RFBR No. 16-29-04388/16.

References

[1] RD-35.240.00-KTN-207-08 "Automation and remote control of the main oil pipelines. Fundamentals" of OJSC "AK Transneft"
[2] OR-07.00-29.12.90-KTN-002-2-00 "Maintenance schedule and diagnostic monitoring of pump units" of OJSC "AK Transneft"
[3] STN ISO 7870-2 Control charts – Part 2: Shewhart control charts TC/SC: ISO/TC 69/SC 4 (2013)
[4] T. V. Aleksandrova, E. I. Gromakov, IOP Conf. Ser.: Mater. Sci. Eng. **93**, 012053 (2015) doi : 10.1088/1757-899X/93/1/012053
[5] E.I. Gromakov, T.V. Aleksandrova and B.P. Ivanenko, IOP Conf. Ser.: Mater. Sci. Eng., **124**, 012015 (2016) doi : 10.1088/1757-899X/124/1/012015
[6] A Wald, *Sequential analysis* (Courier Corporation, Mineola, 2004)
[7] D Wheeler, D Chambers, *Understanding Statistical Process Control* (SPC Press, Knoxville, 1992)

COMPLEX SECURITY SYSTEM FOR PREMISES UNDER CONDITIONS OF LARGE VOLUME OF PASSENGER TRAFFIC

Vladimir Yakubov, *Sergey* Shipilov[1], *Grigoriy* Parvatov, *Vladimir* Antipov, *Timur* Muksunov

National Research Tomsk State University, the Department of Radiophysics, 634050 Tomsk, Russia

Abstract. Subsystems of the design of a complex security system for premises under conditions of large volume of passenger traffic are considered. These subsystems provide video- and thermal imaging control, radio wave tomography, and gas analysis. Simultaneous application of all examined variants will essentially increase the probability of timely prevention of dangerous situations with the probability of false alarm as low as possible. It is important that finally, this will provide protection of population and will facilitate the work of intelligence services.

1 Introduction

The urgency of the development of a complex security system in transport with intensive passenger traffic is obvious nowadays. Among the most important places that require special attention are entrances at railway stations, airports, and subway stations. A paramount problem here is real-time detection and neutralization of various items forbidden for transportation including fire and cold steel arms, explosives, and toxicants. Constantly increasing threat of terrorist attacks is a reality of modern life. Both luggage and passengers are inspected. Whereas x-ray apparatuses are suitable for luggage inspection, the problem is much more complicated for passenger inspection. Even low doses of x-ray radiation can be accumulated in multiple trips and can cause various diseases. The same is true for protection of intelligent service personnel from radiation exposure. However, the security problem is acute not only in transport, but also in protection of premises and territories of manufacture and storage of dangerous substances and their components, protection of banks and public institutions, places of mass events, as well as schools, high schools, etc. Nevertheless, transport remains the most vulnerable link in the chain of all problems under conditions of large volume of passenger traffic.

What are the basic requirements to security systems? First of all, the security system should operate remotely, but be selective. This means that danger should be detected as soon as possible, before it will be manifested itself. Furthermore, the system operation must be hidden – it should neither provoke potential terrorists nor create excessive inconveniences for ordinary passengers. The next requirement is that the system should

[1]Corresponding author: s.shipilov@gmail.com

© The Authors, published by EDP Sciences. This is an open access article distributed under the terms of the Creative Commons Attribution License 4.0 (http://creativecommons.org/licenses/by/4.0/).

 DOI: 10.1051/matecconf/20167901007

operate quickly, that is, detection should be carried out in real time and create no jams. Parallel inspection can be used here. Finally, there is the requirement of highly reliable danger detection, with reasonably low level of false alarms. This requirement leads inevitably to a multi-parametric system, that is, to a complex system when different physical methods of sensing are used simultaneously or sequentially [1–6].

2 Video and thermal vision subsystems

Video control is the first integral part of any complex security system. The complexity of this subsystem (Figure 1) is a huge flow of quickly changing information that must be analyzed using supercomputer technologies and special programs of face control by reference points. Anyway, a database of undesirable or potentially dangerous persons must be preliminary created here [4]. In this subsystem, an analysis of the dynamics of movements of suspected person limbs – hands and feet – and head can be used. This direction is now being intensively developed and promises great results. It is important to note that detection can be performed just outside the entrance to the premises being protected. Sufficient range of detection is 50 m.

Figure 1. An example of operation of a system of video observation and identification of metro visitors.

The thermal vision subsystem allows potentially dangerous or forbidden items hidden under clothes to be identified (Figure 2). Here the contrast between the human body temperature and an extraneous item hidden under clothes is used. The maximal range of detection is 30 m. The item is identified by its shape. Figure 2 shows an example of visualization of a cellular telephone under clothes using a thermal vision subsystem.

3 Doppler subsystem

The Doppler subsystem additionally equipped with a focusing system is capable of detecting quickly moving people against the background of a quietly moving crowd. Narrow-band radio radiation, for example, at a frequency of 24 GHz is used. It is suffice to perform spectral analysis of reflected signals. This subsystem is easily combined with the video and thermal vision subsystems. It is important that the Doppler subsystem can be used to record the relief of the radio image of a hidden item. Figure 3 shows the results of restoration of the image of a test object shaped as two metal plates with planes mutually shifted by approximately 4 mm and deviated from the vertical. The geometrical sizes of plates were 35 × 15 cm. This test object does not lie in one plane and imitates an object with a relief. It can be seen that the deviation of plates and their mutual shift in the image depth are unambiguously reconstructed.

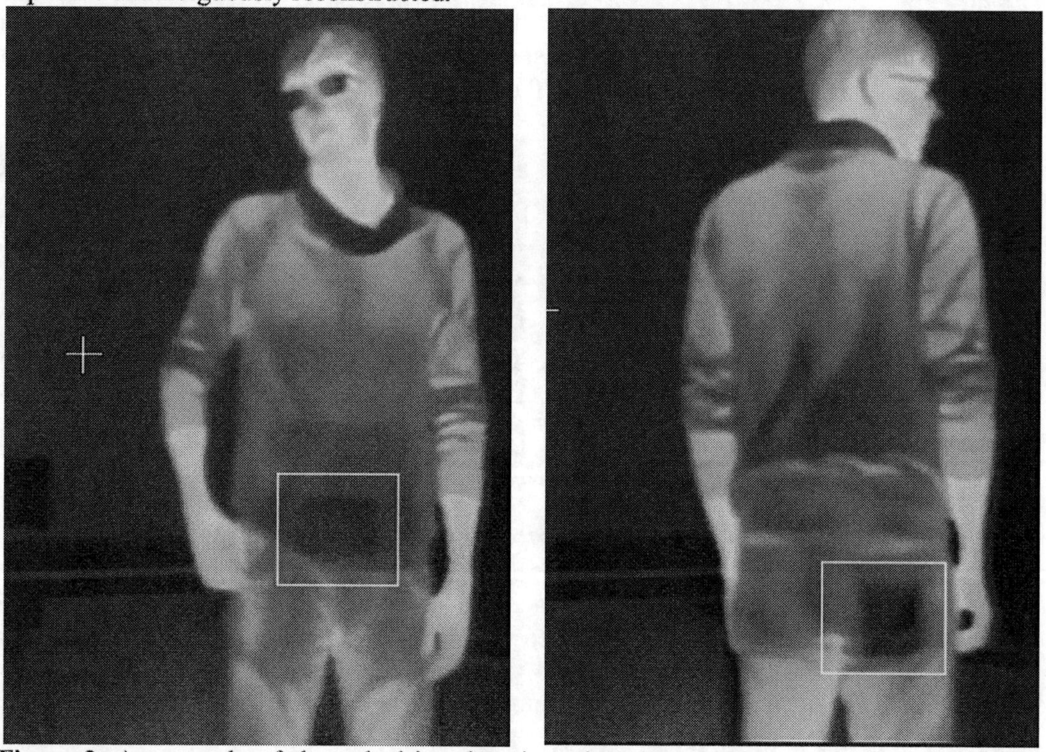

Figure 2. An example of thermal vision detection of an item hidden under clothes using the developed software.

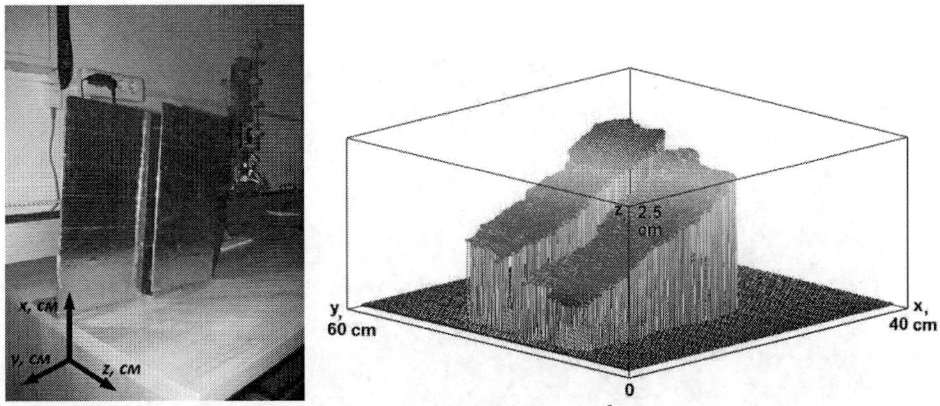

a b

Figure 3. Restoration of the relief of a test object: photograph (a) and reconstructed 3D relief (b).

The phase method of relief reconstruction was used. The concept is based on the idea of creation of a phased antenna array (PAA) of radio range with electronic control. According to this method, radiation is focused in a preset region of space, and subsequent scanning by the focus point over the entire examined area is performed. Radiation focusing can be implemented on the physical level – using lenses and mirrors – and on the computer level – using the technology of wide aperture synthesis. A combination of these levels is the most perspective direction from the viewpoint of fast response, adaptability to manufacture, and commercial attractiveness. Figure 4 shows as a flat reflective aperture can focus radiation. It is suffice to create the corresponding distribution of the reflected signal phase, for example, using its binary values $\pm\pi$.

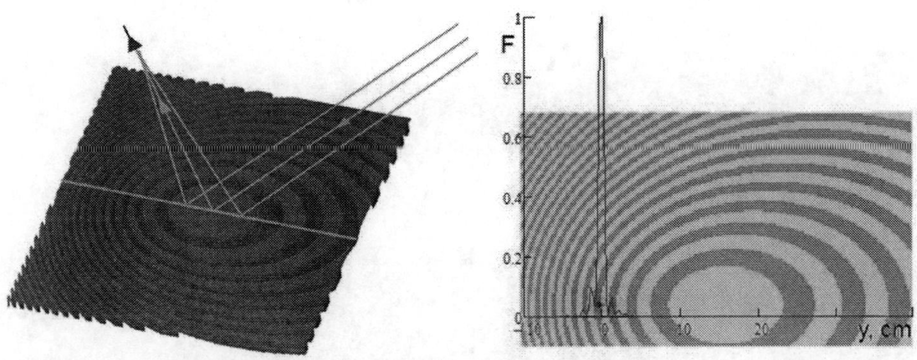

Figure 4. Radiation focusing with a flat reflector.

Fast switching of the collective reflector phase distribution will allow one to control effectively the direction of reflected radiation and hence to obtain the desired reflective properties. The formation of required distributions for any base surface allows methods of radio tomography to be used to develop the technology of the so-called "Clever Skin."

The suggested technology based on reflective PAA is not exotic. This principle has already been declared in a tomography Smith Detection's eqo people screener (USA – Jürgen Sachs –Handbook of Ultra-Wideband Short-Range Sensing: Theory, Sensors, Applications – Wiley-VCH –2012). Its external view is shown in Figure 5. In modern radiolocation, reflecting and transmitting waveguide-type antenna arrays have already been used.

DOI: 10.1051/matecconf/20167901007

a b c

Figure 5. Tomography Smith Detection's eqo people screener (USA) with reflective PAA (a), panel from reflective elements (b), and controllable reflecting waveguide-type unit developed at TSU (c).

4 Ultra-wideband subsystem

The next subsystem that deserves special attention is also the radio-wave ultra-wideband (UWB) subsystem being in its essence a radio-wave tomograph (Figure 6). This subsystem uses ultrashort sensing impulses, a clocked switchable linear UWB transceiving antenna array, and technology of aperture synthesis for signals processing. This radio-wave tomograph allows one to visualize movement of people behind barriers or through walls. The attained video-image change rate was 10 frames/s. This makes it possible to estimate the emotional state of a person from its breathing frequency.

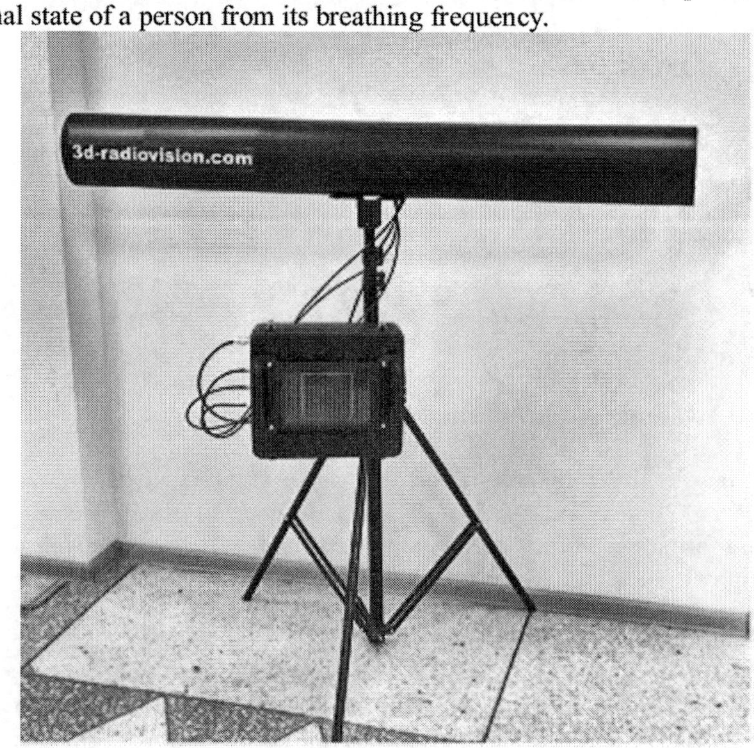

Figure 6. Linear radio-wave tomograph "Radiodozor" intended to control movement of people through barriers.

DOI: 10.1051/matecconf/20167901007

Figure 7 shows an example of possible application of this subsystem in conducting an anti-terrorist operation.

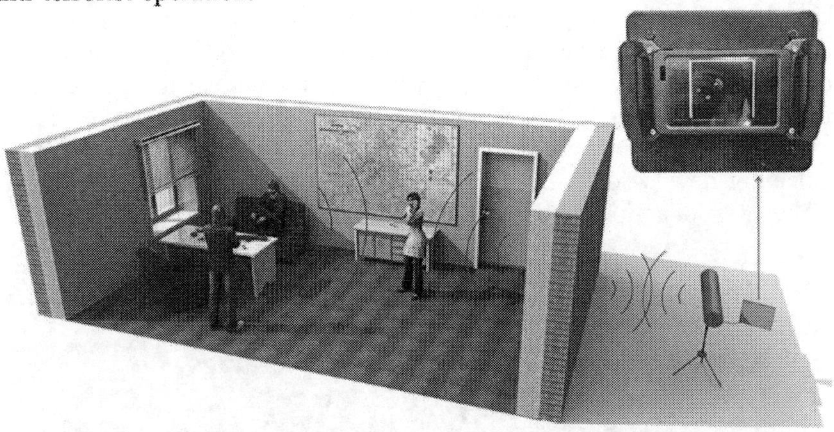

Figure 7. An example of possible application of "Radiodozor" to conduct an anti-terrorist operation.

The next device of the tomographic type is a portable one intended for real-time and safe inspection of items forgotten in transport. Figure 8 shows an example of radio-wave tomography of 4 coins hidden in a box located behind a wall. It can be seen in which layer these coins are located. The device represents an UWB radio-wave tomograph with a planar clocked antenna array. The time of measuring and recording the complete 3D tomogram is 8 s.

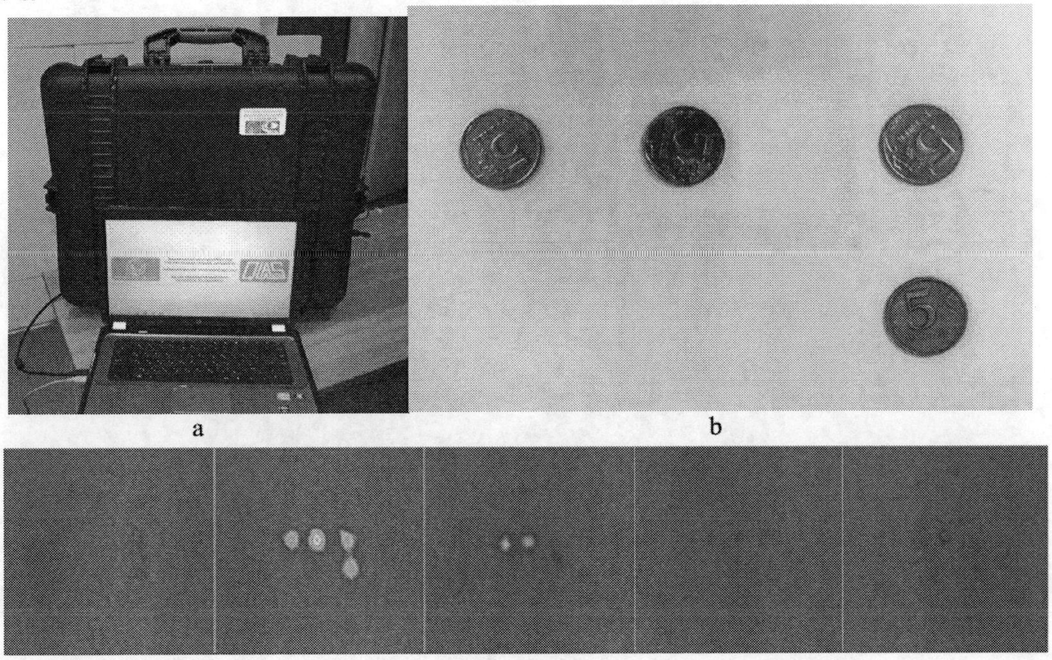

a b

c

Figure 8. Portable UWB radio-wave tomograph (a), hidden test object in the form of coins (b), and its radio-wave tomogram (c).

5 Nonlinear radar subsystem

The nonlinear radar subsystem (Figure 9) is sufficiently promising for security provision. It allows hidden radio electronic devices (cellular telephones, detonators, transmitters, miniature video cameras, taps etc.), detonators, and killing metal elements to be detected on a human body or in a hand luggage. To increase the information content, the radar is capable to conduct additionally zonal inspection of a person. This makes it possible to localize radio engineering devices on a human body and to estimate the degree of their danger (Figure 9). The range of action of the radar is 10–25 m.

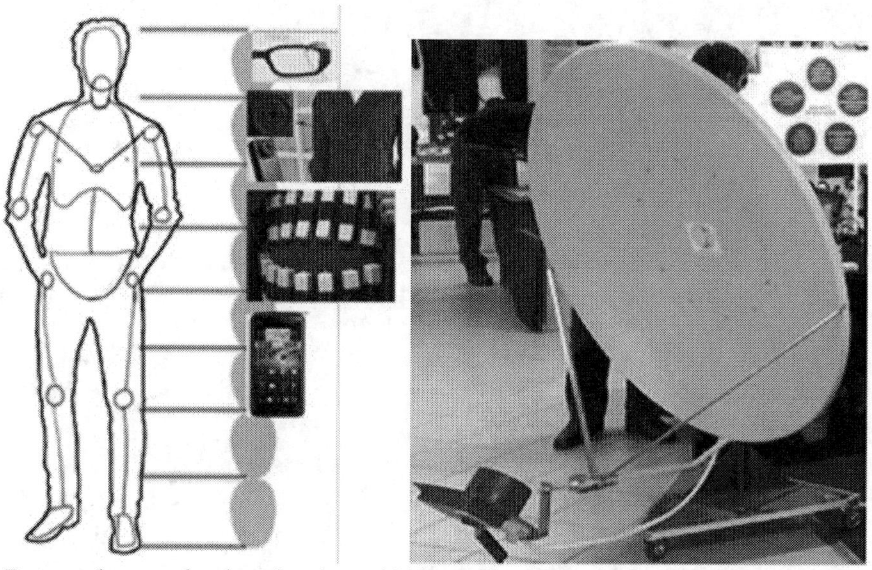

Figure 9. Remote detector developed to detect hidden radio electronic devices.

The next security subsystem is a multi-sensor system of highly sensitive semiconductor gas sensors (Figure 10) that allows ultra-low concentrations of explosives and toxicants to be detected in air and a source (carrier) of dangerous substances to be identified (Figure 11) [7]. Components of the system can be mounted in the ceiling and can be invisible to visitors (Figure 12).

Figure 10. Multisensor developed for detection of explosives and toxicants.

DOI: 10.1051/matecconf/20167901007

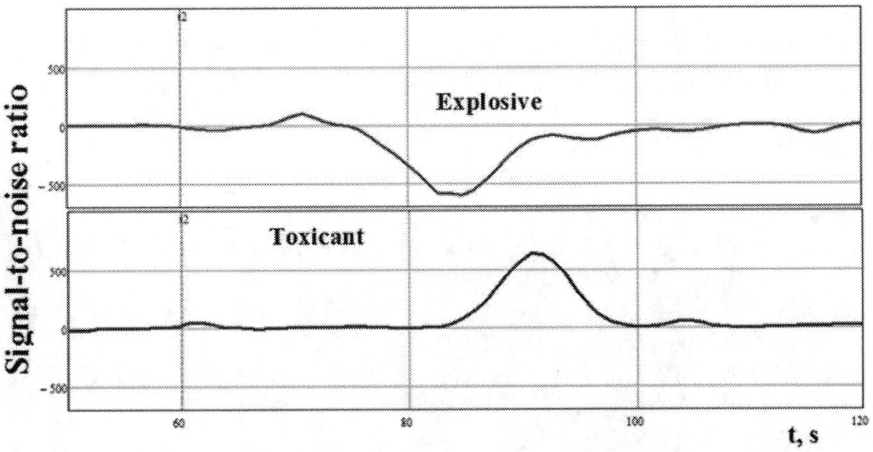

Figure 11. Multisensor response to small concentrations of explosives and toxicants.

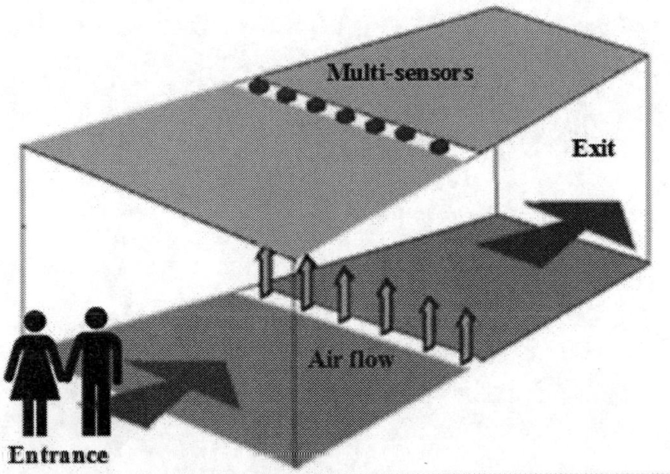

Figure 12. Example of arrangement in the ceiling of the entrance of a premise of gas multi-sensors intended for detection traces of explosives and toxicants on visitors.

6 Conclusion

The subsystems considered above provide the basis for the complex security system for premises under conditions of a large volume of passenger traffic. Combined application of all subsystems will essentially increase the probability of timely prevention of dangerous situations with the probability of false alarm as low as is wished. Finally, this ensures the protection of people and facilitates the work of intelligent services.

Acknowledgment

This research is conducted as part of the Program of Scientific Foundation n. a. D.I. Mendeleev in Tomsk State University (Project No. 8.2.48.2015).

References

[1] V.P. Yakubov, S.E. Shipilov, R.N. Satarov, A.V. Yurchenko, Technical Physics. The Russ. J. of Applied Phys. **60**, 279 (2015) doi: 10.1134/S1063784215020267

[2] S. Kharkovsky, R. Zoughi, IEEE Instrum. Meas. Mag. **10**, 26 (2007) doi:10.1109/MIM.2007.364985

[3] Y. Kan, Y. Zhu, Q. Fu, MATEC Web of Conferences **56**, 02001 (2016) doi10.1051/matecconf/20165602001

[4] L. Lianlin, W. Zhang, F. Li, IEEE Trans. Geosci. Remote Sens. E **48**, 415 (2010) doi: 10.1109/TGRS.2009.2024690

[5] Tan Guo, Xiao Heng Tan, Chao Chen Xie, MATEC Web of Conferences **44**, 01039 (2016) doi: 10.1051/matecconf/20164401039

[6] Gregory J. Mazzaro, F. Martone Anthony, M. McNamara, IEEE Transactions on Aerospace and Electronic Systems **50**, 477 (2014) doi:10.1109/TAES.2013.120798

[7] T.R. Muksunov, N.K. Maksimova, E.J. Sevast'anov, S.E. Shipilov, V.P. Yakubov, Russ. Phys. J. **57**, 1287 (2015) doi: 10.1007/s11182-015-0376-1

DIFFERENTIAL SENSOR FOR PH MONITORING OF ENVIRONMENTAL OBJECTS

Sergey Romanenko, *Timofey* Radenkov, *Egor* Newsky and *Artur* Kagirov[a]

Ecology and Life Safety Department, Non-destructive control institute, Tomsk Polytechnic University, 634050, Tomsk, Russia

Abstract. Differential pH sensor is proposed. Reference electrode and measuring electrode are the same type. Reference electrode is immersed in standard buffer solution with known pH value. The differential pH sensor has longer service life as compared with the traditionally used sensors with silver chloride reference electrode. Ultrasonic cleaning system is proposed to clean the primary measuring transducer from pollution that form as result of silting during long-term operation with the sensor.

1 Introduction

Currently, one of the most common methods of environmental monitoring is an online autonomous monitoring. In Russia, the most common monitoring system is sampling and subsequent analysis in the laboratory, and, in rare cases, the mobile monitoring. In turn, the autonomous remote monitoring is a promising step in the development of environmental objects control methods [1–4].

Development of the automated analytical complex for monitoring of key indicators of water quality (after cleaning) and control of the techno logical cleaning process is challenging because it needs selectivity and sensitivity of the determination of the components, their diversity, as well as ensuring a reliable long-term operation of sensors and the preservation of their sensitivity and calibration at an acceptable cost.

The main impediment to the creation of devices for autonomous monitoring, is the lack of transducers providing reliable long-term operation, while maintaining selectivity and sensitivity [5].

pH control is an integral part of the enviromental objects monitoring, as in many cases, the pH value is a informative indicator of the state of the environment. Measuring of pH value is realized in different laboratories, both stationary and mobile, including a those working in the field conditions [6–9]. Potentiometric analysis methods are effective for the control of the environment. These methods are characterized by high sensitivity and selectivity, high speed of response to changes in the composition of the object being analyzed, ease of automation and possibility of remote control. Finally, they do not require expensive analytical equipment and can be used in laboratory, production and field conditions [10, 11].

[a] Corresponding author: kagirov@tpu.ru

© The Authors, published by EDP Sciences. This is an open access article distributed under the terms of the Creative Commons Attribution License 4.0 (http://creativecommons.org/licenses/by/4.0/).

The aim of the study is to develop a differential sensor for pH monitoring of environmental objects. The main objectives are:

— procuring of pH at-site analysis and simplification of sensor calibration procedure;
— ensuring long life (few months) of the sensor without the need for its servicing.

A wide variety of sensors for pH-meters is currently produced [12, 13]. Most of them take into consideration the dependence of the potential on the temperature, but their use involves the provision of laboratory conditions to run correctly.

Thus, it nesessary a search of pH sensor for continuous monitoring of environmental objects having indications stability in long time and strength sufficient for use in production and field conditions.

Also, there is the problem of finding the reference electrode having a stable potential. Silver chloride electrode is most widely used today [14]. It is known that the main factor reducing the operating life of pH sensor is the instability of the reference electrode potential, in particular silver chloride electrode. Its disadvantages are:

— a relatively high rate of electrolyte expiration from electrode;
— a favorable medium for microorganisms growing;
— a dilution of internal solution, which leads to change of the electrode potential caused by change of the internal concentration of the solution.

Thus, for increasing the lifetime of the pH sensor has been proposed the differential measurement scheme, which avoids the shortcomings of the silver chloride reference electrode.

2 Main partition

Differential pH sensor (Figure 1) with the ion-selective electrode comprises a body 1 with a measuring glass electrode 2 and the reference electrode 4. The measuring glass electrode 2 and the reference electrode 4 are connected with the secondary transducer 5. The reference electrode 4 is glass electrode 3, immersed in reference solution 6 having a buffer capacity. The electrodes are mounted in the body by sealing elements 7, 8. In order to determine the pH value, body of sensor with electrodes is placed in the analyzed medium 9.

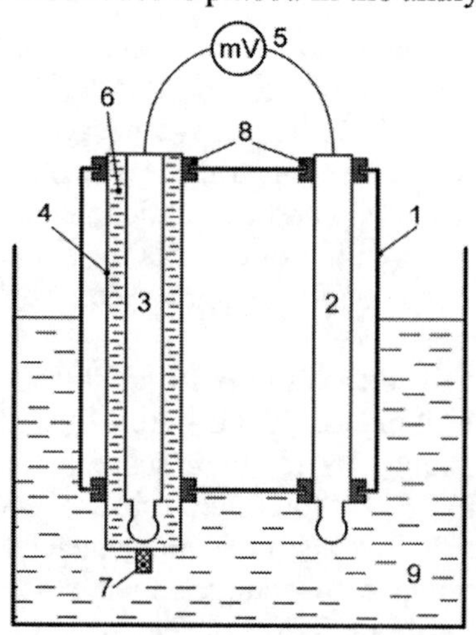

Figure1. Functional scheme of differential pH sensor with ion-selective glass electrodes.

The sensor works like this: when it is immersed in the analyzed medium on the measuring glass electrode appears the potential, which is proportional to the pH value. Since the glass electrode 3 is placed in a solution 6 having a stable pH value, its potential is constant. The porous membrane eliminates mixing analyzed medium and a reference solution having a buffer capacity.

The potential of a glass electrode depends linearly on the pH:

$$E = E^{\circ} - b\,\mathrm{pH}, \tag{1}$$

where: E°–the standard potential of the glass electrode; $b = RT\ln 10 / F$–factor depending on the temperature; pH–pH value.

Then the potential of the measuring electrode is equal:

$$E_1 = E_1^{\circ} - b\,\mathrm{pH} \tag{2}$$

Reference electrode potential is equal:

$$E_2 = E_2^{\circ} - b\,\mathrm{pH}_{\mathrm{ref}}, \tag{3}$$

where $\mathrm{pH_{ref}}$— potential of the reference solution.

Since the indicator electrode and the reference electrode are the same type, the E_1° and E_2° is also the same. Thus, the potential difference between the indicator electrode and the reference electrode, measured by secondary transducer is determined by the following relation:

$$E = b(\mathrm{pH}_{\mathrm{ref}} - \mathrm{pH}). \tag{4}$$

Since the solution having a buffer capacity, has a known pH, there is no need to calibrate the electrodes on a series of buffers [15]. As a solution having a buffer capacity, we used phosphate buffer (pH = 6.86), in which 25-fold dilution causes a change in its pH not more than 0.05 pH-units.

The *b* coefficient is the only parameter used in the recalculation potential *E* to pH. The steepness of the calibration curve b depend of temperature, the influence of which can be take into account in secondary transducer.

Calibration of the differential pH sensor with ion-selective electrodes can be made on one point. For this purpose, the device is immersed in the analyzed medium with a known pH value and measured value of the potential difference *E* between the measuring glass electrode and auxiliary reference electrode. Then calculate coefficient *b*:

$$b = \frac{E}{\mathrm{pH}_{\mathrm{ref}} - \mathrm{pH}}. \tag{5}$$

To improve the accuracy of b when the device calibration is performed is necessary that the difference in pH between the sample solution and $\mathrm{pH_{ref}}$ was maximum. For example, when the internal solution $\mathrm{pH_{ref}}$ = 6.86 buffers with 1.65 or 12.43 pH can be used for calibration.

When used as a first transducer glass electrode measuring scheme must satisfy a number of conditions. On the water you need to use operational amplifiers with input impedance more than 10 GOhm and low leakage currents. As the input amplifier was selected AD8606. Simplified measuring scheme of differential pH sensor is shown in Figure 2.

The measuring electrode is connected to the noninverting input of the operational amplifier AD8606. The reference electrode is connected to the output of the operational amplifier, the potential of which relative to the internal ground potential as half the supply voltage. After RC chain with a cutoff frequency of 20 Hz filtered signal is input to an analog-to-digital converter AD1115. In I2C bus microcontroller receives a digital signal corresponding to the potential difference between the indicator and reference electrode.

DOI: 10.1051/matecconf/20167901008

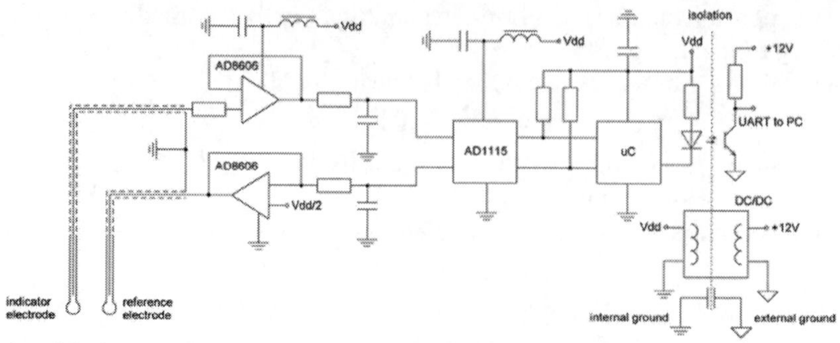

Figure 2. Simplified measuring scheme of differential pH sensor.

To account for the temperature dependence of the electromotive force sensor has a temperature measurement channel on a digital thermometer DS1820. At measured potential difference and temperature microcontroller calculates pH from equation

$$pH = pH_{ref} - \frac{298E}{b'T}, \tag{6}$$

where b' — experimentally received slope of calibration curve at 298 K.

The result is transmitted by UART to a computer via photocoupler, which provides galvanic isolation between internal and external ground. Isolating power supply converts the input voltage of 12 V to output 7 V, then this voltage is supplied to the linear stabilizer 7805 to generate 5V voltage for analog and digital circuits V_{dd}. To reduce electromagnetic noises internal and external ground are connected by capacitor.

Calibration characteristics of the differential pH sensor with the new electrodes and after six months of operation are shown in Figure 3.

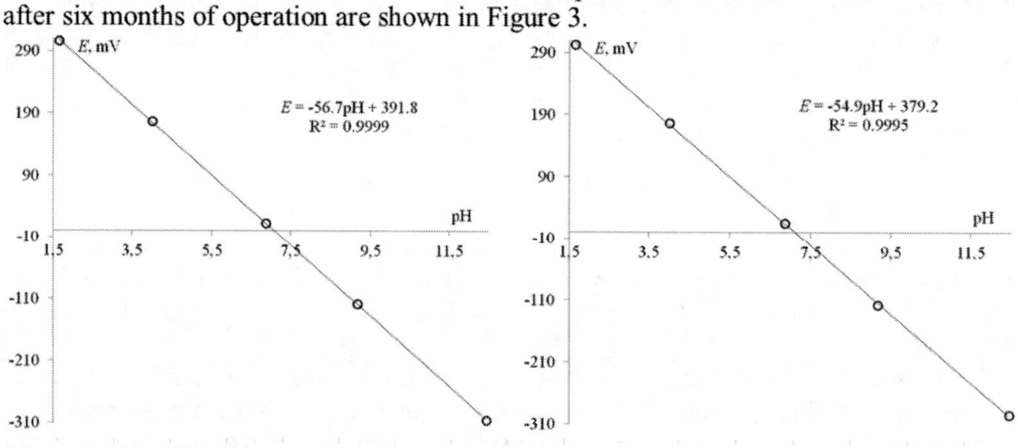

Figure 3. The calibration characteristics of the differential pH sensor with new ion-selective electrodes (left) and after six months of operation (right).

It should be noted important property of calibration characteristics: only its steepness (b) is changed in continuous operation of the differential sensor, while the isopotential point coordinates remain constant E = 0, pH = 6.86.

During prolonged continuous operation in natural water sensor can accrete silt or algae, leading to measurement error. To ensure the purity of the transducers ultrasonic cleaning system was used [16, 17], shown in photos differential sensor (Figure 4).

Figure 4. Photo of the differential pH sensor with ultrasonic cleaning system (bottom left).

Ultrasonic cleaning system is a pulsed boost converter supplying the piezoelectric element 40 mm in diameter. The microcontroller switches on cleaning system once every five minutes for 20 seconds, this mode allows to maintain a clean transducer even in the most adverse conditions. The power consumed by the sensor during cleaning, is not more than 30 W, and in the measurement mode without purification less than 1W. The diameter of sensor 100 mm, length 350 mm, weight 1.5 kg.

3 Conclusion

Advantages of the proposed design of the sensor, realizing the differential measuring scheme:

— simplifying the calibration procedure (one-point, instead of the traditional two-point);

— increasing the duration of continuous operation through the use of a glass electrode in phosphate buffer pH 6.86 as the reference electrode;

— ultrasonic cleaning system that supports transducers clean and does not allow the ion-selective electrodes to grow algae or sludge.

During the six months of continuous operation of the sensor the steepness of calibration function has decreased from 56.7 to 54.9 mV/pH, which corresponds to the measurement error in the range 0 – 12 pH, no more than 0.1 pH. An important feature of the differential pH measurement scheme is a long-term stability of the isopotential point (pH = 6.86).

Acknowledgment

The reported study was supported by RFBR, research project No. 16-48-700230.

References

[1] P. Jaikang, K. Grudpan, T. Kanyanee, Talanta **132**, 884 (2015) doi: 10.1016/j.talanta.2014.10.046

[2] A. Kagirov, S. Romanenko, Testing. Diagnostics (Sp), 157 (2011)

[3] S.H.A. Hassan, S.W. Van Ginkel, M.A.M. Hussein, R. Abskharon, S.-E. Oh, Environment International **92–93**, 106 (2016) doi: 10.1016/j.envint.2016.03.003
[4] A. Kagirov, D. Kalashnikova, Testing. Diagnostics (13), 73 (2014)
[5] B.S. Echols , R.J. Currie, D.S. Cherry, J.R. Voshell, Environmental Monitoring and Assessment **185**, 1341 (2013) doi: 10.1007/s10661-012-2636-7
[6] C. Manjarrés, D. Garizado, M. Obregon, N. Socarras, M. Calle, C. Jimenez-Jorquera, Journal of Applied Research and Technology **14**, 1 (2016) doi: 10.1016/j.jart.2016.01.003
[7] T. Radenkov, S. Romanenko, A. Kagirov, Testing. Diagnostics (Sp), 146 (2011)
[8] A. Kulasekaran, G. Andal, R. Lakshimipathy, J. John Alexander, International Journal of ChemTech Research **8**, 16 (2015)
[9] F. Scholz, T. Steinhardt, H. Kahlert, J. Behnert, Electroanalysis **16**, 2058 (2004) doi: 10.1002/elan.200403059
[10] J. Plambeck, *Electroanalytical chemistry: basic principles and applications* (Wiley, New York, 1982)
[11] M. Sander, T.B. Hofstetter, C.A. Gorski, Environmental Science and Technology **49**, 5862 (2015) doi: 10.1021/acs.est.5b00006
[12] I.M. Perez De Vargas Sansalvador, C.D. Fay, J. Cleary, A.M. Nightingale, M.C. Mowlem, , D. Diamond, Sensors and Actuators, B: Chemical **225**, 369 (2016) doi: 10.1016/j.snb.2015.11.057
[13] M. Schirrmann, R. Gebbers, E. Kramer, J. Seidel, Sensors **11**, 573 (2011) doi: 10.3390/s110100573
[14] A. Mroz, M. Borchardt, C. Diekmann, K. Cammann, M. Knoll, C. Dumschat, Analyst **123**, 1373 (1998) doi: 10.1039/a708992i
[15] P. Spitzer, B. Werner, Analytical and Bioanalytical Chemistry **374**, 787 (2002) doi: 10.1007/s00216-002-1453-1
[16] Yi Hu, Shu Wang, Ming Zhu, MATEC Web of Conferences **59**, 01006 (2016) doi: 10.1051/matecconf/20165901006
[17] Y.-S. Li, L.-C. Shi, X.-F. Gao, J.-G. Huang, Desalination **390**, 62 (2016) doi: 10.1016/j.desal.2016.04.008

DOI: 10.1051/matecconf/20167901009

CAPACITANCE CONTROL ON THE WIRE PRODUCTION LINE

Alexander Goldshtein, *Galina* Vavilova*, *Sergey* Mazikov

Tomsk Polytechnic University, Institute of Non-Destructive Testing, 634050 Tomsk, Russia

Abstract. The paper presents technical implementation of the electrical-capacitive method to perform in-process measurement of the capacitance per unit length of a single-core electric wire. The design of the electro-capacitive measuring transducer is proposed. The block diagram of the device CAP-10 developed to implement the proposed method is presented. The appearance of the device CAP-10 is showed, and its operating principle is described. It is shown that the change in water conductivity has a significant impact of the measurement result of the wire capacitance per unit length. The techniques to offset from the impact of water conductivity variation on measurement results are proposed. The technique of the device CAP-10 initial adjustment is proposed. It provides the desired function of the output signal transformation. The technique of the "operating" adjustment to correct measurement results through systematic measurement error elimination is offered.

1 Introduction

The capacitance per unit length is one of the main normalized values. Its stability over the entire length determines the cable products quality. It is known that the wire capacitance per unit length is largely dependent on the geometric dimensions of the wire and insulation electrical properties. Therefore, the variation of the capacitance per unit length indicates change in the wire geometric dimensions or insulation electrical properties. The need to ensure the stability of the electrical and geometrical parameters of the entire length of the cable products requires continuous in-process control of these parameters [1–5]. A continuous capacitance control can solve this problem.

The wire capacitance depends on the characteristics of in-process insulation applying; therefore, the capacitance control is to be carried out during extrusion [1, 2, 6–8, 10]. The only possible embodiment of the measuring transducer in this case is the use of the tubular electrode [11] immersed in cooling water together with the test wire. This technique is widely used by the leading companies in the cable industry: Sikora (Germany), Zumbach (Switzerland) and others [12–14].

* Corresponding author: wgw@tpu.ru

© The Authors, published by EDP Sciences. This is an open access article distributed under the terms of the Creative Commons Attribution License 4.0 (http://creativecommons.org/licenses/by/4.0/).

2 The design of the measuring transducer

Figure 1 shows the electro-capacitive measuring transducer (ECMT) used to implement the proposed method of the capacitive control. Its design is described in detail in [9, 15–17]. The main elements of ECMT are tubular measuring electrode 1, a pair of tubular guard electrodes 2 and 3 placed in metal housing 4 and isolated from it by dielectric layer 5. The guard electrodes are used to ensure high field uniformity in the test zone. The electrodes are connected to the alternating voltage generator 6. The wire core and the ECMT housing are grounded. The ECMT is immersed in the cooling water bath installed in the extrusion line. The test wire 7 is permanently moved inside the ECMT.

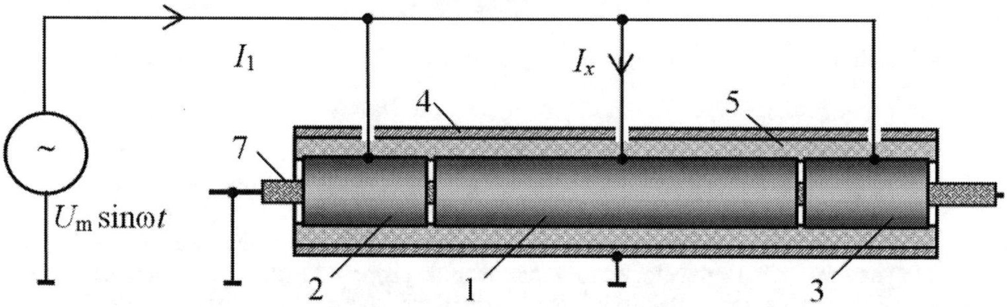

Figure 1. The ECMT design.

To minimize the dependence of the wire capacitance test result on its geometric parameters, electrical insulation properties and the change in water conductivity, it is necessary to provide an optimal ECMT design. The optimal design values for the ECMT can be selected through computer simulation using the COMSOL Multiphysics. The basic ECMT design values are the inner diameter of the tubular electrodes, the length of the measuring and guard electrodes, the distance (gap) between the measuring and guard electrodes, the inner diameter of the ECMT cylindrical housing [16].

An optimal ECMT design provides high longitudinal (axial) uniformity of the electric field in the test zone. The electric field is uniform if the values of the radial component E_r, V/m, and the electric field intensity vector $\overline{E}$ are similar, and the longitudinal component E_x, V/m, (directed along the cable axis x) is equal to zero at any fixed value of the radius r in the measuring electrode (Figure 2).

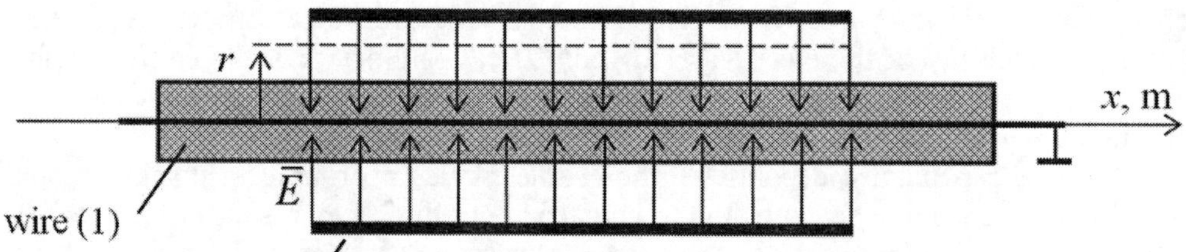

Figure 2. The field inside the transducer.

The criterion for the optimal design is the equality of the coefficient β to one. The coefficient β is the ratio of the capacitance per unit length in the central part of the measuring electrode C_1, pF/m, to the capacitance per unit length along the total length of the measuring electrode C_2, pF/m, (in the central part of the measuring electrode, the electric field is known to be uniform):

$$\beta = \frac{C_2}{C_1} \quad (1)$$

The results of computer simulation show that high uniformity of the electric field inside the test zone is provided by the guard electrodes introduced into the ECMT design. In this case, the electric field spreads to the distant edges of the guard electrodes.

Figure 3 shows the distribution of the electric potentials for interaction of the ECMT electric field with the test wire plotted in equipotential lines for the case of guard electrodes of R radius (equal to the inner radius of the measuring electrode), of $1R$ length and 1 mm distance (gap) between the measuring electrode and the guard electrode [17].

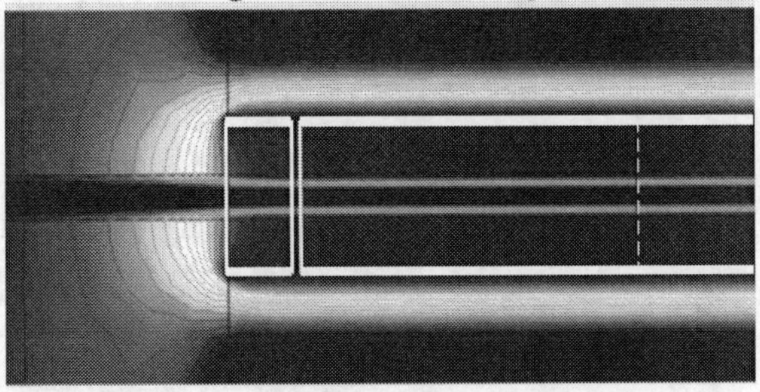

Figure 3. The electric field created by the ECMT with guard electrodes.

The basic design parameters of the optimal ECMT chosen through computer simulation [16] are as follows:

- the length of the measuring electrode is (200…300) mm;
- the length of the guard electrodes is (0.5…1)R;
- the inner radius of the electrodes is $R_{el} \geq 2r$;
- the inner diameter of the housing is$R_h \geq 2R_{el}$;
- the gap between the measuring electrode and the guard electrode is (1…3) mm,

where r is the radius of the outer wire insulation;

R is the inner radius of the measuring electrode.

3 The device CAP-10 for measuring the wire capacitance per unit length

The developed ECMT is the main element of the device CAP-10 which is intended for in-process control of the capacitance per unit length for single-core electric wires.

The main technical characteristics of the device CAP-10 are as follows:

- the test object is a single-core insulated wire;
- the range of the measured capacitance per unit length is (50…500) pF/m;
- the outer diameter of the wire insulation varies from 0.5 mm to 12 mm;
- the maximum permissible measurement error for the capacitance per unit length is 2.5% of its nominal value within the range of the change in water conductivity equivalent to the variation in the NaCl concentration range (0 ... 4) g/l.

The appearance of the device CAP-10 is presented in Figure 4 and its block diagram is shown in Figure 5.

The value of test wire capacitance in ECMT 1 is transformed into the current value I_x. This signal is applied to input analogue conversion unit 2, which includes current transformer (CT) 3, current-to-voltage converter (CVC) 4 and amplitude-phase detector (APD) 5. CT is used for galvanic isolation of the electrode measuring circuit and the signal analogue conversion circuit. The CT output current I_x is applied to CVC, where it is converted into voltage, and then to the APD input, in which the real and imaginary parts of the signal measurement information are differentiated [1, 9, 15].

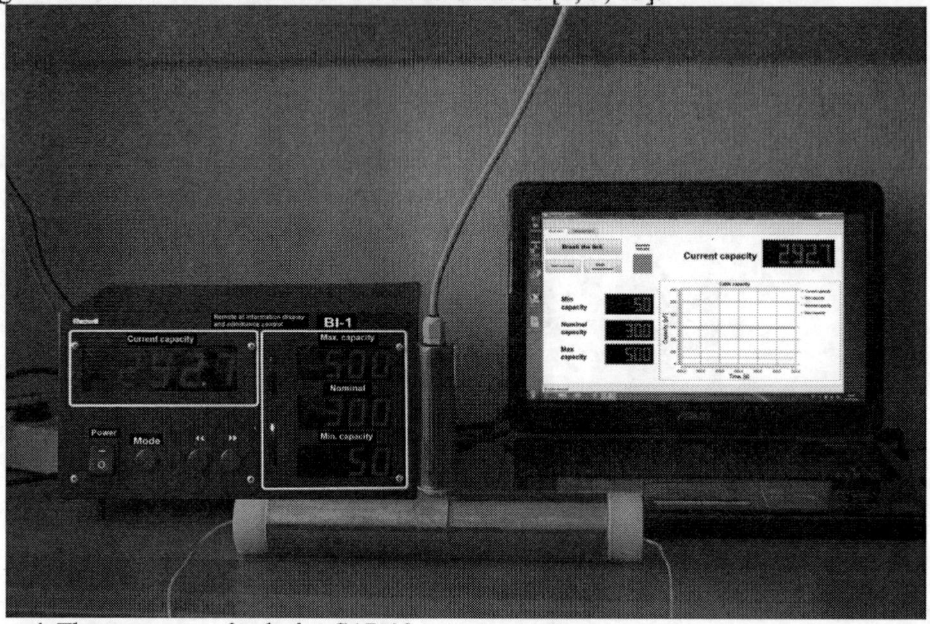

Figure 4. The appearance the device CAP-10.

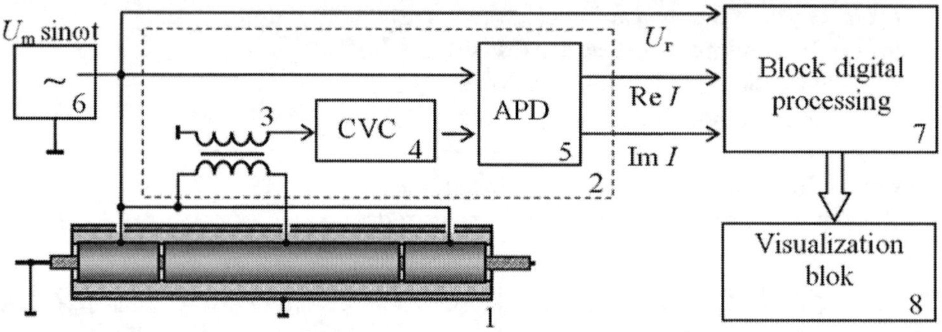

Figure 5. The block diagram of CAP-10.

The APD reference signal is the voltage of generator 6. The APD output voltages (Re U и Im U) proportional to the amplitude values of the complex parts of the current in the measuring electrode circuit are applied to digital processing unit 7 based on the microcontroller. Digital processing unit 7 performs filtering, measurement data averaging within 1 second and digital processing.

Visualization block unit 9 provides display of the measurement results in the form convenient for an operator. Information on the measured parameters is displayed on the panel BI-1 presented in Figure 6, *a*.

A personal computer is used to record an array of data on capacitance per unit length within the set time interval and to display the measured data. The virtual front panel of the device CAP-10 is showed in Figure 6, *b*.

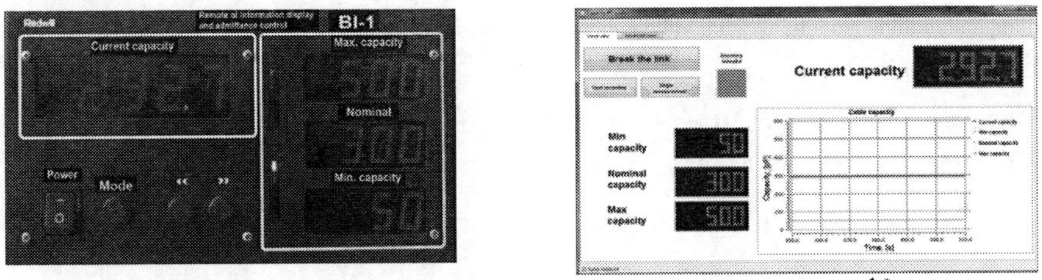

a) b)

Figure 6. Visualization of measured data by the device CAP-10.

4 Detection of local defects using CAP-10

The primary function of the wire capacitance control by CAP-10 is continuous monitoring of local defects (a cracks, a voids in a insulation etc.) in wire insulation.

Local defects in wire insulation cause sharp abrupt change in the wire capacitance as a result of sharp change in geometric dimensions and electrical insulation. The device CAP-10 records the change in the capacitance. When it detects a defect, the parameters are displayed (Figure 7). The presented capability of the device CAP-10 extends the scope of its application and provides additional capabilities of spark testing in non-destructive testing in the cable industry [1, 2].

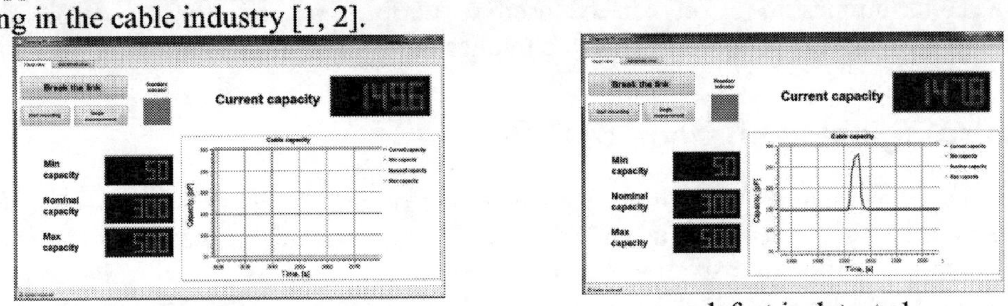

no defect a defect is detected

Figure 7. Test results for the sample free of defects and that with a detected defect.

5 The techniques to offset from the impact of change in water conductivity

The algorithm for the ECMT output signal transformation in the capacitance is based on the linear function of the measured current I_x, mA, transformation into capacitance per unit length C_w, pF/m:

$$C_w = C_0 + k \cdot I_x \qquad (2)$$

where C_0, pF/m, is a constant component, k, pF/(m·A), is a proportionality factor. These parameters depend on the ECMT design parameters.

The studies [1, 9, 15, 17] indicate that the change in water conductivity due to change in water temperature and concentration of impurities have a significant impact on the result of the wire capacitance test. Consequently, the coefficients C0, pF/m, and k, pF/(m·A), are the functions of the current water conductivity values. The wire capacitance test without consideration of the current water conductivity causes a measurement error that can reach 20 %.

The decreased impact of the change in water conductivity is implemented through the offsetting techniques based on indirect measurements of the current water conductivity. To perform the offset, the phase of the current of the measuring electrode equal to

$t = \mathrm{Re}\dot{I}_x / \mathrm{Im}\dot{I}_x$ is determined [9, 15] and the generator current is measured (Figure 1) [18].

The values C_0, pF/m, and k, pF/(m·A), in functions (2) are replaced with their respective functions $C_0(t)$ and $k(t)$ or $C_0(I)$ and $k(I)$:

$$C_w = C_0(t) + k(t) \cdot I_x . \tag{3}$$

or

$$C_w = C_0(I_1) + k(I_1) \cdot I_x \tag{4}$$

The functions $C_0(t)$, $k(t)$, $C_0(I_1)$ and $k(I_1)$ are described by the second order polynomials. The coefficients of these polynomials are determined by the design ECMT parameters.

The device CAP-10 implements both techniques to offset from the impact of change in water conductivity. The main of the techniques is based on determination of the measuring electrode current phase.

The techniques based on determination of the current generator I_1 should be used only when the phase of the measured current change not only due to change in water conductivity, but also because of the change in the ratio of the wire electric parameters C_x, pF/m, and R_x, kOm/m.

These techniques to offset can be used to perform an individual control of wire parameters. Switching between the two techniques is performed automatically.

6 Metrological assurance of CAP-10

To ensure compliance with the characterictics provided in the CAP-10 technical documentation, the device is adjusted. For this purpose, specially prepared segments of the tested wire with different insulation materials and known values of the capacitance per unit length within the limit of CAP-10 measurement are used [2, 4]. The actual values of the wire capacitance per unit length were determined in compliance with GOST 27893-88 "Communication Cables. Methods of Tests" [19].

At the manufacturing stage, the device CAP-10 is initially adjusted, i.e. the coefficients of the function of the ECMT output signal transformation into capacitance per unit length (2) are found for each implementation of the offsetting techniques. The Initial adjustment is carried out in normal weather conditions, according to GOST 27893-88 [19].

If the device CAP-10 installed in the wire production line (for the first time or after its repair) exhibits excessive capacitance measurement error (2.5%), the obtained result can be adjusted. The "operating" adjustment is automated. This adjustment is required to eliminate the additive and a multiplicative components of the systematic error. During "operating" adjustment, the device CAP-10 is placed in the cooling bath installed in the wire production line, in which the manufactured wire is placed after its extrusion.

The task of the "operating" adjustment is to determine the capacitance ΔC, pF/m. This value is calculated as follows:

$$\Delta C = \Delta C_1 + \frac{\Delta C_2 - \Delta C_1}{C_{x2} - C_{x1}} (C_x - C_{x1}). \tag{5}$$

The corrected value of the capacitance per length for each wire sample is calculated as follows:

$$C_x^* = C_x - \Delta C . \tag{6}$$

The quality of the "operating" adjustment is estimated using a test wire segment with known valid capacitance.

7 Conclusions

The developed ECMT design allows selection of optimum design parameters to provide a minimum methodical measurement error caused by the influence of the wire geometric dimensions, electrical properties of insulation and change in water conductivity.

The techniques to offset from the impact of change in water conductivity and their algorithms are proposed and implemented based on determination of the measuring electrode current phase and the generator current. These techniques reduce the measurement error of the wire capacitance per unit length in the range from 50 to 500 pF/m in conditions of significant change in the water conductivity.

The feasibility of the device CAP-10 to detect local defects (a cracks, a voids in a insulation and others) in the wire insulation is shown. It expands the scope of the spark method of cable product control.

The technique of the device CAP-10 initial adjustment to provide the desired function of the output signal transformation is offered. The technique of the device CAP-10 "operating" adjustment to correct measurement results through systematic measurement error elimination is proposed.

References

[1] N.S. Starikova, V.V. Redko, G.V. Vavilova, J. Phys.: Conf. Ser. **671**, 012056 (2016) doi: 10.1088/1742-6596/671/1/012056
[2] E.M. Fedorov, I.D. Bortnikov, Tech. Phys. **60**, 1689 (2015) doi: 10.1134/S1063784215110110
[3] A.P. Surzhikov, V.V. Peshev, A.M. Pritulov, S.A. Gyngazov, Russ Phys J. **42**, 490 (1999)
[4] O.V. Galtseva, S.V. Bordunov, S.N. Torgaev, IOP Conf. Ser.: Mater. Sci. Eng. **110**, 012094 (2016) doi: 10.1088/1757-899X/110/1/012094
[5] D.D. Burleigh, M.V. Kuimova, Ponte **72** , 2 (2016)
[6] R. Kodermyatov, M. Ivanov, M. Yuzhakov, V. Kuznetsov, M. Yuzhakova, E. Timofeeva, MATEC Web of Conferences **48**, 05004 (2016) doi: 10.1051/confmatec/20164805004
[7] A. I. Nizhegorodov, A.N. Gavrilin, B. B. Moyzes, J. Phys.: Conf. Ser. **671**, 012037 (2016) DOI: 10.1088/1742-6596/671/1/012037
[8] A.A. Bespalko, L.V. Yavorovich, P.I. Fedotov, Russ. J. Nondestr. Test. **47**, 680 (2011) doi: 10.1134/S1061830911100068
[9] A.E. Goldshtein, G.V. Vavilova, Int. J. Syst. Appl., Eng. & Dev, **10**, 217 (2016)
[10] A.P. Surzhikov, E.N. Lysenko, V.A. Vlasov, A.V. Malyshev, E.V. Nikolaev, RUSS PHYS J. **56** (6), 681 (2013) doi: 10.1007/s11182-013-0085-6
[11] P. Fleming, L.R. Coleman, *A system for monitoring fluctuations in the thickness of a cable insulating sheath* GB Patent 2358928 (2001)
[12] CAPACITANCE 2000: Capacitance measuring devices for cable production lines. URL: http://sikora.net/wp-content/uploads/2016/03/CAPACITANCE_2000_GB.pdf (2016)
[13] CAPAC®/FFT: Capacitance measurement systems for cable production lines. URL: http://www.zumbach.com/pdf/Literature/CAPAC/CAPAC_Familie_CAPA.002.0002.E.pdf (2016)
[14] W. Blohm, Proceedings Intl. IWMA Conf. Economical Processing of Rod to Wire & Cable, 67 (1999)
[15] A.E. Goldshtein, G.V. Vavilova, S.V. Mazikov, J. Phys.: Conf. Ser. **671**, 012062 (2016) doi: 10.1088/1742-6596/671/1/012062

[16] P. Fleming, L.R. Coleman, *Capacitance monitoring systems*, US Patent 20030128038 (2003)
[17] A.E. Goldshtein, G.V. Vavilova, V.Yu. Belyankov, Russ. J. Nondestr. Test. **51**, 86 (2015) doi: 10.1134/S1061830915020047
[18] A.E. Goldshtein, G.V. Vavilova, V.V. Redko, *Method of measurement capacitance of a solid strand electrical wire* RU Patent 2578658 (2016)
[19] GOST 27893-88 *Telephone cables, Test methods* (Standartinform, Moscow, 2010)

MATEC Web of Conferences 79, 01010 (2016) DOI: 10.1051/matecconf/20167901010
IME&T 2016

THE DEVELOPMENT OF MONITORING AND CONTROL SYSTEM OF THE LOW PV/T SOLAR SYSTEM

Alena Okhorzina [1,2,*], *Alexander* Bikbulatov [1], *Alexey* Yurchenko [1,2], *Norbert* Bernhard [3], and *Oksana* Aldoshina [1]

[1]National Research Tomsk Polytechnic University, 634050, Tomsk, Russia
[2]National Research Tomsk State University, 634050, Tomsk, Russia
[3]Anhalt University of Applied Sciences, 06366, Köthen, Germany

Abstract. The article presents an autonomous PV/T solar installation. Installing converts solar energy into electricity and heat. The description of its components and elements that enhance its effectiveness shows. The description of the control program and control of the installation is given. The control system provides for tracking the sun and cooling the photovoltaic module.

1 Introduction

One of the decisive factors, preventing solar energy from being widely used, are the costs of solar energy systems. Cost reduction of photoelectric systems, generating electrical energy, is possible by reducing the costs of photoelectric modules or increasing the efficiency of such modules and the solar stations, they are built in [1].

One method of increasing the energy production is to use solar concentrators. Systems with built-in concentrators use mirrors and lenses to focus the incoming sunlight. The construction of photoelectric concentrators comes with several manufacturing problems: it has to reflect light to a given direction, be adequate to mass production, be able to keep high concentrations, be insensitive to manufacturing inaccuracies and installation, and be able to ensure consistent illumination of the photovoltaic modules surface. Concentrator based solar systems turn out to be highly efficient, when steadily cooled in form of heat dissipation.

Heretofore there was no necessity for a widespread use of alternative energy sources in Russia, since Russia occupies a huge amount of traditional energy sources, such as oil, coal, gas, water and atomic energy, and it has a fitting infrastructure. As a result of the financial crisis, though, the delivery of combustible materials to remote and elusive places suffers.

Therefore, the authors see great prospects for alternative energy in providing energy resources of small settlements and private houses in remote settlements and holiday villages (for example, a village in the forest, forest wardens houses, national parks and reserves).

* Corresponding author: ohra.avit@gmail.com

© The Authors, published by EDP Sciences. This is an open access article distributed under the terms of the Creative Commons Attribution License 4.0 (http://creativecommons.org/licenses/by/4.0/).

2 Installation description

The installation is a hybrid system (Figure 1), which converts solar energy into electricity and heat. The reflected sunlight incident on the photovoltaic module is converted into electrical energy. Heat production is due to the cooling of the PV.

To increase the energy production is added in the low reflective concentrator. The concentrator is a reflecting surface that consists of three segments. The concentrator provides increased density of the incoming solar energy flow in 2…3 times, thereby increasing the efficiency and the production of electricity. Also because of the increase in incoming solar flux on the surface of the PV module is heated; which leads to loss of efficiency and, when heated to a temperature above 130 ° C, delamination or deterioration of elements of ohmic contacts photocells [2, 3].

It is therefore necessary to add cooling system reliability. It is possible to use different radiators or liquid cooling. Heat transferred by the cooling liquid, may be used for further purposes (ex for water heating). Thus, a combined system is obtained, which can generate both electricity and heat.

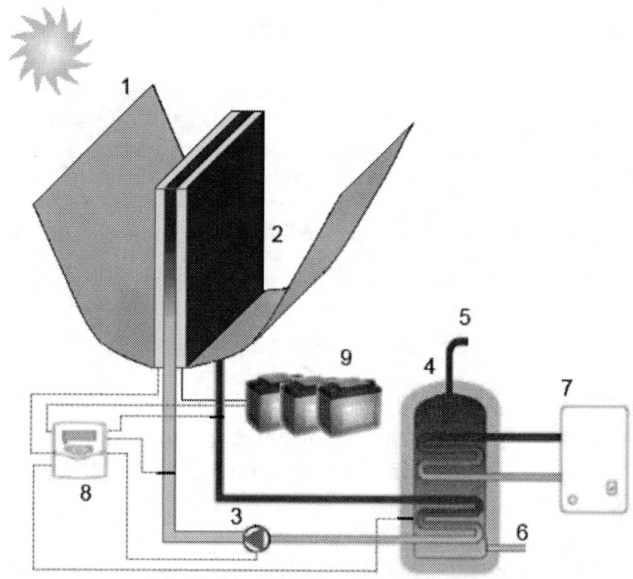

Figure 1. PV concentrator system with a heat sink: 1 – Concentrator, 2 – PV module, 3 – Pump, 4 – Twin coal, 5 – Output, 6 – Input, 7 – Boiler, 8 – Control system, 9 – Accumulator.

The application concentration and tracking the sun can increase the efficiency of the solar energy system, resulting in a more uniform production of electricity from sunrise to sunset.

3 Monitoring and control system

To ensure the end-consumer a simply and steadily working solar station, it is necessary to have a monitoring and control system. The given system ensures flawless functioning of the solar station and requires intervention only in the case of an error.

Overall control of the system will be based on a single-board computer Beaglebon. Beaglebon comprises a processor Sitara AM3359AZCZ100 series with clock speed of 1 GHz, also processing core Cortex-A8 and SGX530 graphics accelerator. The board uses 512MB DDR3L RAM, Flash memory eMMC volume of 2 GB [4]. Data features are more than enough for a wide range of monitoring and control tasks.

At the start of the system several actions are carried out, such as the initialization of the single-board computer Beaglebone and the start of the monitoring program.
The functional principles of the program algorithm are shown in Figure 2.

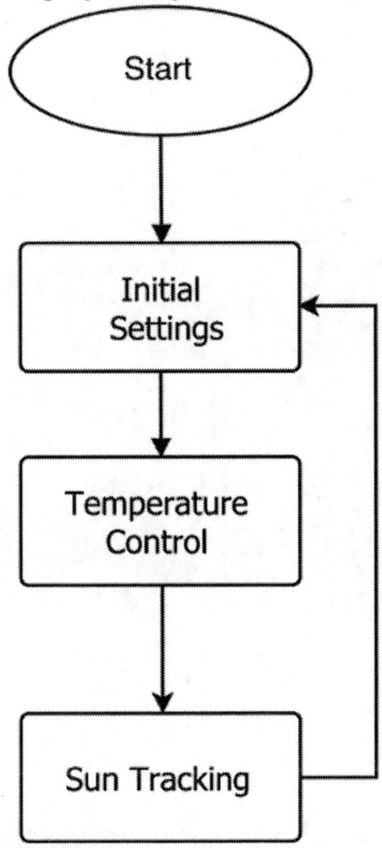

Figure 2. The program algorithm.

The given program monitors and controls the temperature of the PV by using liquid cooling. The computer controls the pump, which facilitates different speeds of circulation of the liquid. The higher the temperature of the PV, the higher is the speed of circulation of the liquid. Another job of the monitoring and control system is to orientate the whole station towards the sun to increase the received energy flux coming from the sun. This leads to a higher and more homogeneous energy input during the day. The controlling algorithm and the sensor of the orientation system are used for the orientation of the authors' earlier tracking system [5]

In the following the algorithm of the temperature-controlling subprogram will be explained in detail. The algorithm is shown in Figure 3.

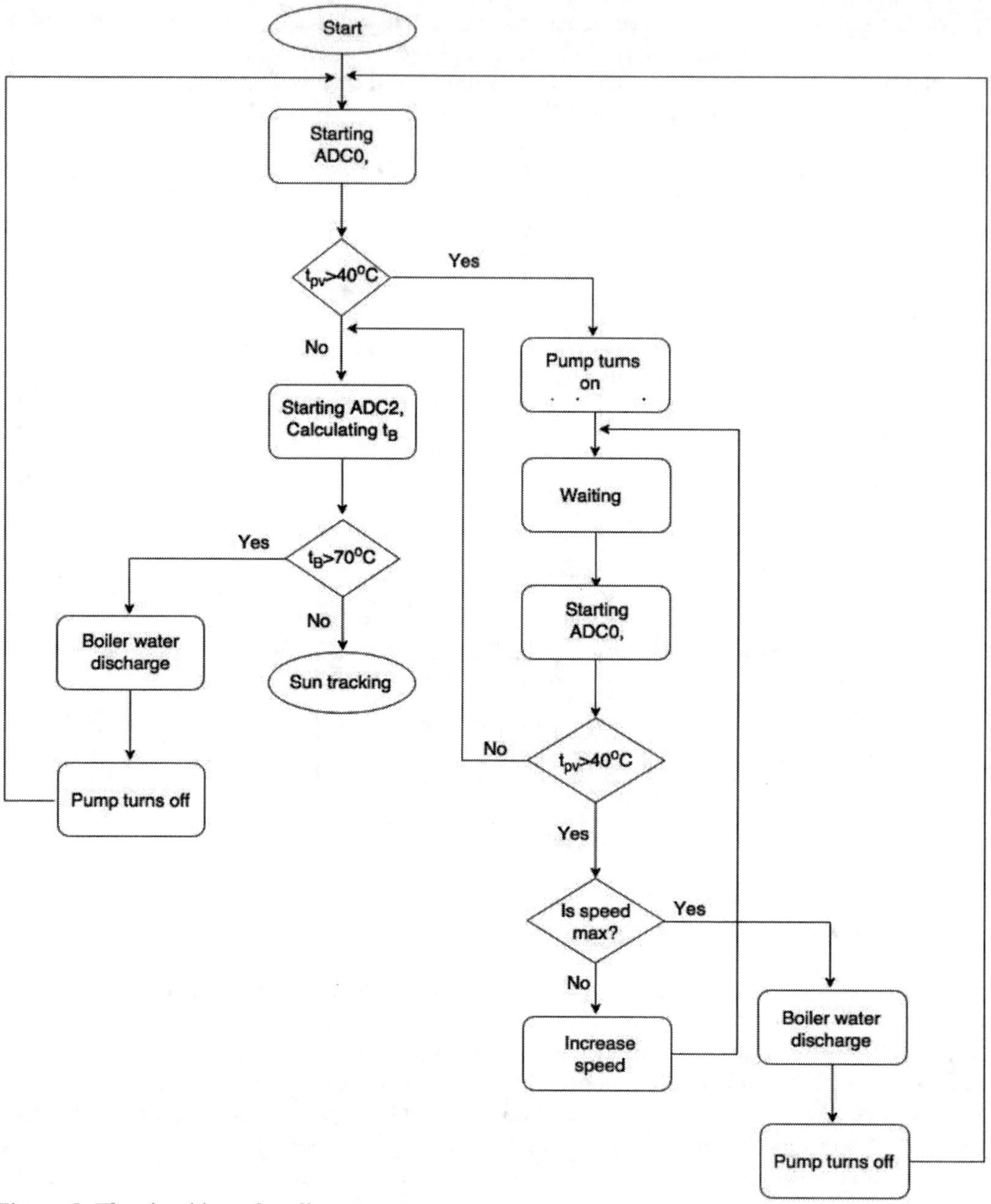

Figure 3. The algorithm of cooling management.

At the beginning the temperature values from the ADC are read from the sensor. Based on these values the temperature of the PV is calculated. Depending on the temperature of the PV the operating mode of the pump is selected. If the temperature of the PV is less than 40°C, the PV will not be cooled, because it is not necessary. With increasing temperature the speed of circulation of the liquid also increases, which automatically leads to a more intensive heat conduction and, thus, cooling of the system. Once the system temperature reached the threshold of 70°C and the speed of circulation its maximum value, water is let out of the boiler. This process is necessary to exchanging hot water cold one, which in turn increases the intensity of the temperature exchange.

4 The first experiment

As can be seen from the field experiment (Figure 4), the heating of the PV surface is a strong (about 85 ° C). When the cooling is turned on there is a significant cooling (about 30 ° C).

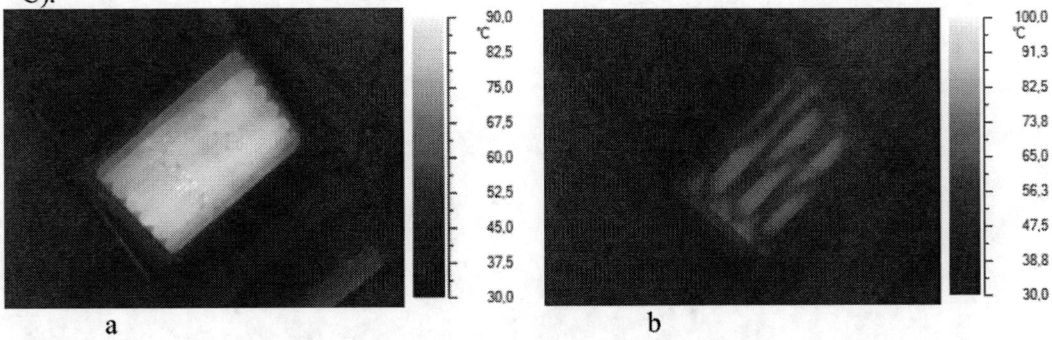

Figure 4. The results of the field test: a–without cooling; b–with cooling.

To increase the cooling uniformity in the design can be changed or to add absorber, as well to augment the number of turns of the coil.

According to the results of the experiment can be said that the cooling system is functioning well.

5 Summary

The main target group, which this station will be interesting for, are residents of remote and elusive places, tourist accommodations, which are located far away from current paths, foresters, workers of natural reserves, as well as vacationers and gardeners.

It is presented a description of a new hybrid solar system, which uses concentrated sunlight and orientates itself towards the sun. These additional features increase the gain of electrical energy and make it possible to obtain thermal energy. The stored thermal energy may be used in everyday life.

The earlier explained algorithm controls the hybrid solar system. It implements the cooling of the PV and the orientation of the station towards the sun. Particular attention is to pay to the part of the algorithm, which controls the cooling of the PV.

The functioning of an experimental station was proved by an experiment.

Acknowledgment

The following research has been financially supported by Ministry of Education and Science of the Russian Federation, Agreement № 14.577.21.0018 (RFMEFI57714X0018).

References

[1] C.-Y. Lee, P.-C. Chou, C.-M. Chiang, C.-F. Lin, Sensors **9**, 3875 (2009) doi:10.3390/s90503875

[2] Y.-K. Wua and S.-M. Chang, MATEC Web of Conferences **55**, 06001 (2016) doi: 10.1051/matecconf/20165506001

[3] H.A. Zondag, Renew Sustain Energy Rev. **12**, 891 (2008). doi:10.1016/j.rser.2005.12.012

[4] BeagleBone Black, description URL: http://beagleboard.org/black

[5] A. Yurchenko, A. Kozloff, A. Okhorzina, M. Kitaeva, A. Purgin, Polzunovsky vestnik (2), 79 (2014)

PRINCIPLES OF A PROTOTYPE MODEL OF A DEVICE FOR AUTOMATIC CONTROL OF CARDIO-PULMONARY RESUSCITATION PROCEDURES

Artem Bureev, *Ekaterina* Dikman[a], *Ivan* Zemlyakov, *Artem* Osipov, *Anton* Seleznev, *Liudmila* Khohlova

National Research Tomsk State University, 634050, Tomsk, Russia

Abstract. The article is to present a prototype of a device intended for automatic control of the cardiopulmonary resuscitation outside the hospital. For the assessment of a patient's condition the device uses the vital parameters: respiratory noises and vibrations of the walls of the carotid arteries. The device is made in the form of a «surgical collar» and does not requires extra actions to fix the sensors to a patient's body. Resuscitation is performed with sound and light signals of different types for cardiac compressions and artificial respiration.

1 Introduction

Practically applied devices assess the correctness of ongoing resuscitation by means of indirect indicators, measuring the pressure force on the chest during compressions, or vertical offset. The feedback of a state of patient's vital functions during the procedure is only available on the bedside monitor. However, these conditions are only accessible in a medical institution. Ambulance and emergency personnel, mine-rescue workers and other professionals working in the field, do not have such assistance.

Thus, there is a significant demand in the development of easy-to-use, compact, reliable and low-cost self-powered devices for use outside hospitals for control of a resuscitated patient's vital functions and technical support during the cardiopulmonary resuscitation (CPR) procedure.

2 Research methods

Previously the design team proposed a layout of hardware and software for monitoring of CPR procedure [1].The device developed controlled the officially recommended evaluation items [2]: dynamic pattern of the diameter of the apple of the eye and air movement in the trachea. A video camera with the USB-output, fixed to the eye socket of the resuscitated patient, and a microphone installed in the larynx were used to take the readings.

[a] Corresponding author: eka.kisleyova@gmail.com

© The Authors, published by EDP Sciences. This is an open access article distributed under the terms of the Creative Commons Attribution License 4.0 (http://creativecommons.org/licenses/by/4.0/).

Subsequently, it was decided to abandon the use of video cameras, since fixing of sensors on the human body was excessively time consuming [3].

Currently, the developed device for CPR control is based on the analysis of acoustic signals recorded at the bifurcation of carotid arteries (blood flow noise and pressure pulse of vessel walls) and larynx (respiratory breath sounds). CPR procedures focus on the specifications developed by Russian experts and recommended by the Ministry of Health of Russia[2, 4, 5], as well as the recommendations of the European Resuscitation Council [6] for CPR and emergency aid in case of cardiovascular diseases.

The design of the device developed is similar to an orthopedic surgical collar with an inflexible splint for a cervical spine. This technical solution allows to perform a *Safar's Method* (opening the resuscitated patient's airway), and to begin intensive care [3] (Figure 1). Fast easy to put on device meets the principle of strict time limit, known in medicine of emergency conditions as *early access* and *early start*. It should be noted that the survival rate reduces by 10% each minute of the CPR delay [2].

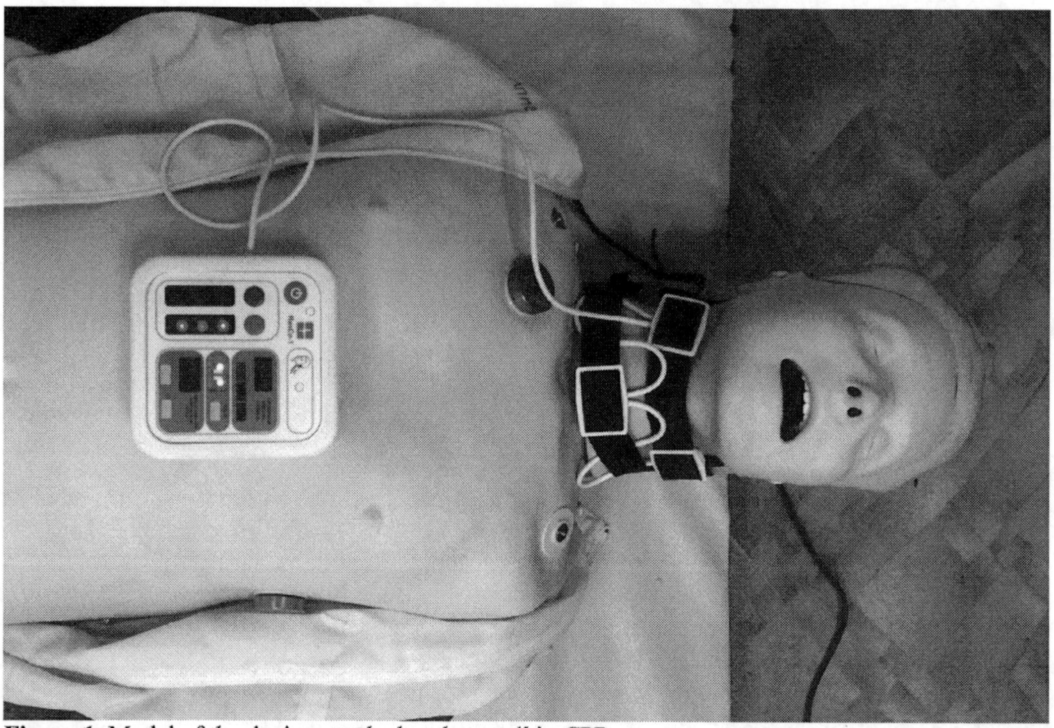

Figure 1. Model of the device attached to the manikin CPR.

In the place where the device is adjacent to the front of the neck there are three acoustic sensors: two on the left and right projections of the carotid artery, and one - in the larynx area (Figure 2). Each sensor is connected to the input of the converter *charge-voltage* with a shielded-conductor. After filtration a cable signal enters a multi-channel 12-charging analog-to-digital converter. The digitized signal is processed by a high-speed 32-charging microcontroller ARM Cortex M4, where algorithms of selection and recognition of respiratory and blood flow noise are implemented. Operation of the device is cyclical (Figure 3).

Figure 2. The piezoelectric sensor.

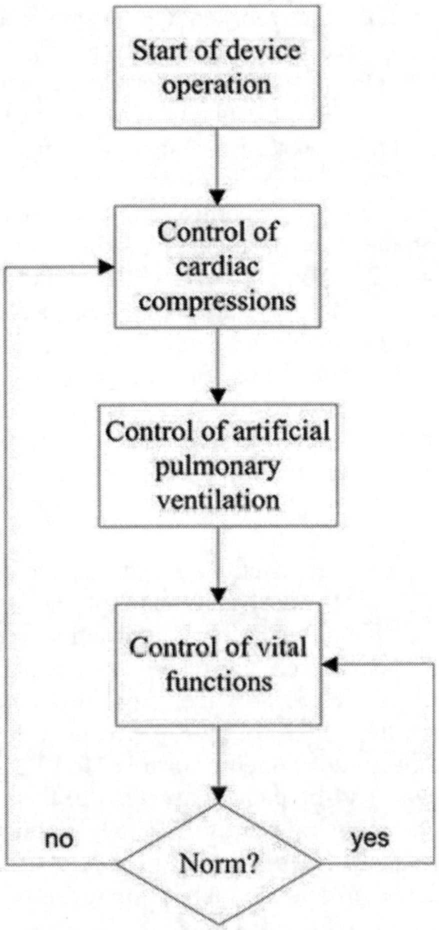

Figure 3. Operation algorithm of the device for CPR monitoring.

The implementation of the algorithm is as follows. After placing the device on the neck of the resuscitated patient and switching it begins to control sensors status (presence of signals) and starts the timers of CPR control subsystems. The CPR control mode consists of three phases [2, 5].

Phase 1: Control of cardiac compressions. The goal is to ensure the blood supply to the brain. Control of the CPR procedure is performed by transmitting 30 short signals with a frequency of 1 kHz and an interval of 600 ms to a built-in speaker. This corresponds to the recommended intensity of chest compression of 100...120 times a minute [2, 5]. The end of the cycle of cardiac compressions is indicated by a signal of double duration. To set the rate of artificial pulmonary ventilation process two signals with frequency of 500 Hz are given to to the built-in speaker.

Phase 2: control of artificial pulmonary ventilation. The goal is oxygen saturation of blood. According to the recommendations [2, 5], at this stage the rescuer has to do two acts of artificial respiration mouth-to-mouth with a duration from 1 second with an interval of 3 seconds. The built-in speaker receives two signals with a frequency of 500 Hz.

Phase 3: control of vital functions. Signals from the acoustic sensors are recorded and analyzed. Upon occurrence of independent heart rates in the carotid arteries periodic signals with low frequency (15…100 Hz), corresponding to pulse fluctuations of the vascular wall. Upon occurrence of breathing in the area of the larynx of the resuscitated patient periodic breathing noises of medium frequency range (300...800 Hz), corresponding to the respiratory cycles *inhale – exhale,* are recorded. Cycle time monitoring of vital functions is 5 seconds (recommended no more than 10 seconds [2, 5]). In case the device did not register any spontaneous breathing and heart contractile activity (Norm?), phase 3 is followed by a continuous signal with a frequency of 1 kHz, warning about the beginning of a new CPR cycle and demanding to start cardiac compressions.

In case, if the device recorded signs of spontaneous breathing and heartbeats, the device automatically switches from CPR control mode to Control of vital functions mode. It is followed by monitoring the patient's spontaneous breathing and heartbeats. According to current recommendations [2, 5], the criteria of the CPR ending are awakening of a patient, opening of his/her eyes, and normal breathing, observed visually and aurally. If signals of the functions stop coming from sensors for 15 seconds, the device prevents the occurrence of a critical situation with a three-time longer signal automatically returns to the CPR control mode. For the convenience of the resuscitation team audio signals generated by the device are accompanied by super-bright red and green LEDs.

In order to display breathing and heart beats it was decided not to use a color LCD-display due to: 1) complexity of data perception in bright light; 2) data redundancy; 3) mechanical fragility of the display.

In the absence of signals of vital functions on the device sensors both indicator lights are red. The appearance of typical noise of blood flow and respiration observed during cardiac compressions and artificial pulmonary ventilation procedures leads to periodic yellow light signals, synchronized with the recorded signals. Independent breathing and heartbeats are indicated by periodic green light signals, also synchronized with the recorded signals. For the reporting of the CPR procedure performed device immediately begins to protocol signals received the time of their transmission. To store the date and time of current signals the device uses real-time clock as a part of the microcontroller. It continues to function even when the device is powered off. The report of proceedings is saved to the nonvolatile backup memory and, if necessary, can be sent to the doctor's computer work station via USB, decrypted, and presented in a printing form.

3 Conclusion

During the study performed a layout of the device for quick assessment the vital functions of the affected and to some extent control the course of cardiopulmonary resuscitation procedures outside a medical institution has been developed.

Correct CPR procedures performed by rescuers without medical education, but under the control of professional ones increase the chances of survival of a victim by approximately fifty percent [4, 7]. The developed device is intended for replacing (to some extent) a specialist's control of the CPR procedure.

In the future, a voice synthesizer will be included in the device for autonomous CPR control.

Acknowledgement

The study was performed within the framework of the Agreement № № 14.578.21.0078 (unique identifier for the project – RFMEFI57814X0078) entitled «The device for monitoring CPR procedures of a human being», signed between National Research Tomsk State University and the Ministry of Education and Science of the Russian Federation, and the program of improving competitiveness of the TSU (project No. 8.2.31.2015).

References

[1] V. I. Syryamkin, A. Sh. Bureev, A. V. Osipov, S. A. Klestov, Medicinskaja tehnika, (3), 27 (2012)

[2] L. J. Doyle, R.A. Saltsman, L. J. Doyle, R.A. Saltsman, *Cardiopulmonary resuscitation : procedures and* challenges (Nova Science Publishers, Hauppauge,2012)

[3] V. I. Syryamkin, A. Sh. Bureev, I.Yu. Zemlyakov, A method for controlling CPR and device for its implementation: the patent. Registration number 2012129757 (2012)

[4] P. Safar, J.O. Elam, *Advances in Cardiopulmonary Resuscitation* (Springer-Verlag, New York; 1977)

[5] J. M. Field, P.J. Kudenchuk, R. O'Connor, T. VandenHoek, *The Textbook of Emergency Cardiovascular Care and CPR* (Philadelphia, Lippincott Williams & Wilkins, 2009)

[6] C. D. Deakin, J. P. Nolan, J. Soar, K. Sunde, R. W. Koster, G. B. Smith, G. D. Perkins, Resuscitation **82**, 140 (2011) doi: 10.1016/j.resuscitation.2010.11.009

[7] H. Halperin, Critical Care Clinics **28** ,167 (2012) doi: 10.1016/j.ccc.2012.02.002

WIDEBAND ANTENNA FOR HPM MEASUREMENTS

Ivan Kurkan[1,2,*], *Nicolay* Bykov[1], *Nataliya* Kurkan[2] , *Alexandr* Ryumkin[2]

[1] Institute of High Current Electronics, 634055, Tomsk, Russia
[2] National Research Tomsk Polytechnic University, 634050, Tomsk, Russia

Abstract. The measurements of microwave pulses of gigawatt power level have a lot of constraints. A receiving antenna is a starting and core point of the measurement system. Waveguide based and dipole antennas have a limited wide bands, while the use of commercially available wideband antennas is restricted by their maximum peak power acceptances. The design of the wide band antenna with the small effective area was proposed. The characteristics of prototype were obtained in numerical simulations with ANSYS HFSS software and by calibration tests in the frequency band of 1–13 GHz. It has the effective area about the 1 mm2 in X-band and square-law dependence on the wavelength in a wide band. The cross polarization rate is more than 60 dB at the centre position and not less than 30 dB within the range of $\pm5^0$ in azimuth and elevation angle. The wide beam radiation pattern forces a user to discriminate reflected signals. This antenna could greatly simplify the measurement system, replacing a set of narrow band antennas that connected to several recording channels.

1 Introduction

The progress in pulsed power technologies resulted in development of high power microwave (HPM) sources for the last decades. These HPM generators from decimetres to millimetres frequency bands reach the output peak power of $10^8 \ldots 10^9$ W with pulse duration of 1…100 ns [1, 2]. The challenges of such pulses measurement are naturally caused by their high power level, short pulse duration and relatively wide band comparing to continuously operating sources. There are other negative factors featured by feeding sources: the high level of EM noise and bremsstrahlung radiation caused by MeV electron beam. So, a facility requires the presence of radiation shielding and screen room for operators and measurement instruments. It usually makes some constraints on the room available for placement of receiving antenna and requires distant transmission lines to remote screen room. The most of HPM sources operates in the single shot mode or at rather low repetition rates. The measurement of output microwave radiation includes power amplitude, pulse shape, pulse energy, radiation spectrum and pattern. The latter one is directly connected to the operating mode of microwave generator and the internal configuration of electromagnetic fields.

* Corresponding author: ikk@lfe.hcei.tsc.ru

© The Authors, published by EDP Sciences. This is an open access article distributed under the terms of the Creative Commons Attribution License 4.0 (http://creativecommons.org/licenses/by/4.0/).

A receiving antenna is a starting and core point of microwave measurement system. It should be capable to operate at high power density level (~ 100 kW/cm^2). So, it should be of the low effective area to prevent a breakdown in the transmitting lines, that limits one to use commercially available wide band antennas. It is preferred to be linearly polarized to detect radiation polarization. Horns and waveguide based antennas are widely used as receiving ones at X-band and higher frequencies (Figure 1, *a*) [2]. The coaxial based dipole antennas are preferable at L, S and C-bands as a sizes of waveguides rise with a wavelength (Figure 1, *b*). Both types of these receiving antennas has the limited operating band. It is restricted by the cut-off frequency and appearing of other than fundamental mode for waveguide. The construction of dipole requires the λ/4- cuts for symmetric operation.

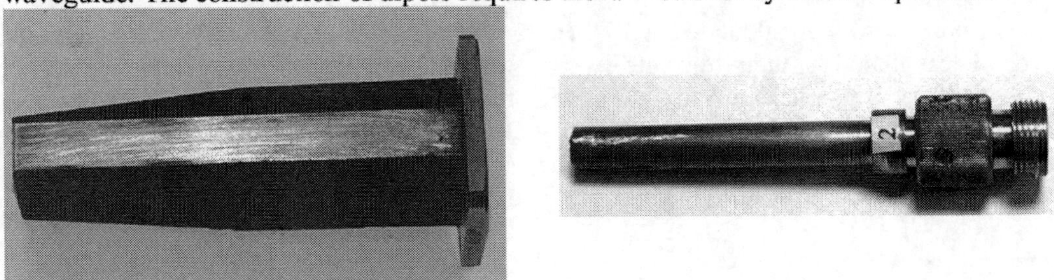

a) *b)*

Figure 1. The receiving antenna based on 23x10 mm waveguide – *a*; the short symmetric dipole with N-type connector – *b*.

During a testing and tuning a HPM generator, experimenters commonly deal with some specific modes of generation those have few concurrent frequencies or valuable harmonics in the output spectrum. These frequencies and particularly harmonics might be out of working band of receiving antenna. So, the usage of a wide band antenna is quite reasonable in this case.

We propose the antenna with the effective area of square millimetres. The main idea was to greatly reduce the receiving element size L comparing to the wavelength λ, so the effective area would have smooth dependence on the frequency, rising as its square root. It stays correct until $L < \lambda/4$. We took a coaxial cable as the base and shortcut the central electrode to the shielding with the minimal gap as it is presented in Figure 2. There is the semi-rigid cable FlexiForm 402 by Habia here, but other kinds of 18 GHz semi-rigid or rigid (with the full copper jacket) were also tested and approved.

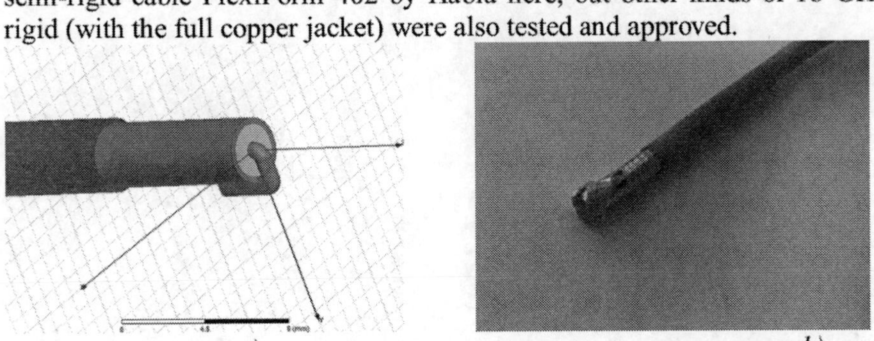

a) *b)*

Figure 2. The model of receiving element of antenna – *a*; the receiving element based on semi-rigid cable FlexiForm 402 by Habia – *b*.

We use the pyramidal microwave absorber to cover a cable and connector to prevent surface currents which result in standing waves between the receiving end and the connector. Adsorbing plates on the waveguide based antenna do the same [2].

In the paper, the results of numerical simulation and measured characteristics of real antenna we proposed are presented.

2 Numerical simulation results

Numerical simulations of this antenna were performed with ANSYS HFSS [3] software in 3D. The antenna was placed in coordinate system as shown in the Figure 2, *a*, so the working polarization was the Y one. The task was to prove the selective polarization properties of the antenna and to check its sensitivity to the positioning error within small angles about zero.

The core setup was calculated at 10 GHz by the iterative solver to satisfy S- parameters error of less than 1%, then "discrete sweep" with the same mesh was made to calculate fields at lower frequencies up to 1 GHz with the step of 0.5 GHz. The far field radiation patterns were plotted and analysed at calculated frequency points. Figure 3 shows radiation patterns of the antenna at several frequencies as well as pattern at 4 GHz for antenna without the absorbing pyramid to demonstrate its role.

a) *b)* *c)* *d)*

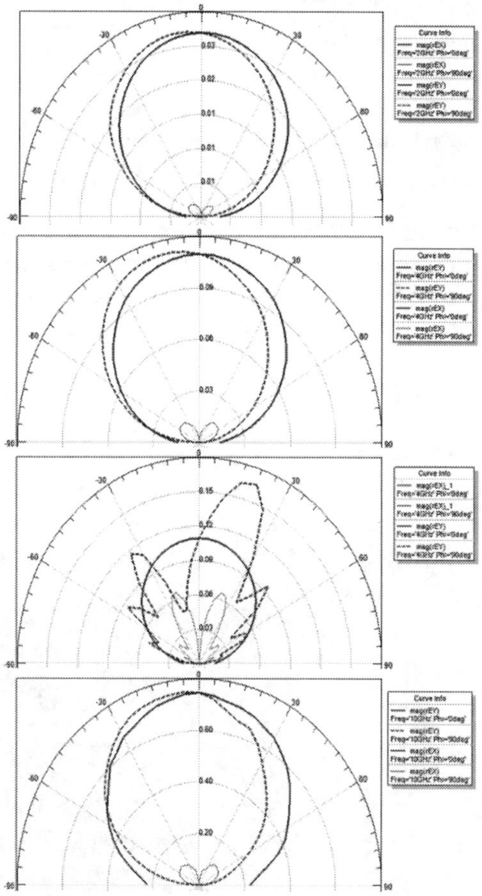

Figure 3. Calculated radiation patterns of the antenna. *a* – 2 GHz with absorber; *b* – 4 GHz with absorber; *c* – 4 GHz w/o absorber; *d* – 10 GHz with absorber. Legend: bold solid(red) line – rEy component in azimuth plane, bold dashed(brown) line – rEy component in elevation plane, thin solid(magenta) line – rEx component in azimuth plane.

As we are interested in characteristics within small angles about the main direction, the use of Cartesian system for the field description is feasible. The pattern for working Ey field polarization is symmetric in the azimuth plane and slightly raised in the elevation plane. The sensitivity within small angles about the working position is stable in the band. The cross polarization rate is more than 60dB at the centre position and not worse than 30 dB within the conus $\pm 5^0$. As this antenna has the wide beam radiation pattern, a user should supress reflected signals, using microwave absorbers at the test site.

Figure 3, *c* represents the radiation pattern for antenna without absorbing cover at frequency of 4 GHz. Patterns for other frequencies differ by the number of side lobes. Although the working polarization pattern in azimuthal plane is the same as for absorber covered antenna, patterns in elevation plane and for cross polarized field in azimuth are changed. This antenna became very sensitive to the error in inclination angle positioning and its cross polarization selectivity is much worse, comparing to the absorber covered antenna.

3 Experiment

Several antennas based on different type of 18 GHz approved cables were designed, produced and calibrated. All antennas but receiving ends were covered by carbon foam pyramids that were used in microwave anechoic chamber. The main characteristics used by HPM experimenters is the effective area of antenna, which allows one to calculate a power density. The calibration procedure was similar to the one described in the paper [2]. The network analyser Agilent 8719ET was used in the measurement. The difference was in replacing narrow band horn antennas by two ETS-Lindgren's Model 3117 double-ridged waveguide horn antennas with the similar characteristics within the band 1-18 GHz [4]. All standard measures recommended by Agilent to decrease a measurement error were involved: slow frequency scan, 32 passes averaging.

The antennas produced with different types of cables in fact are very close in dimensions, which are predefined by cables frequency band. So, their effective areas are about the same. The effective area dependence on frequency for based on semi-rigid cable FlexiForm 402 by Habia represented at the Figure 4.

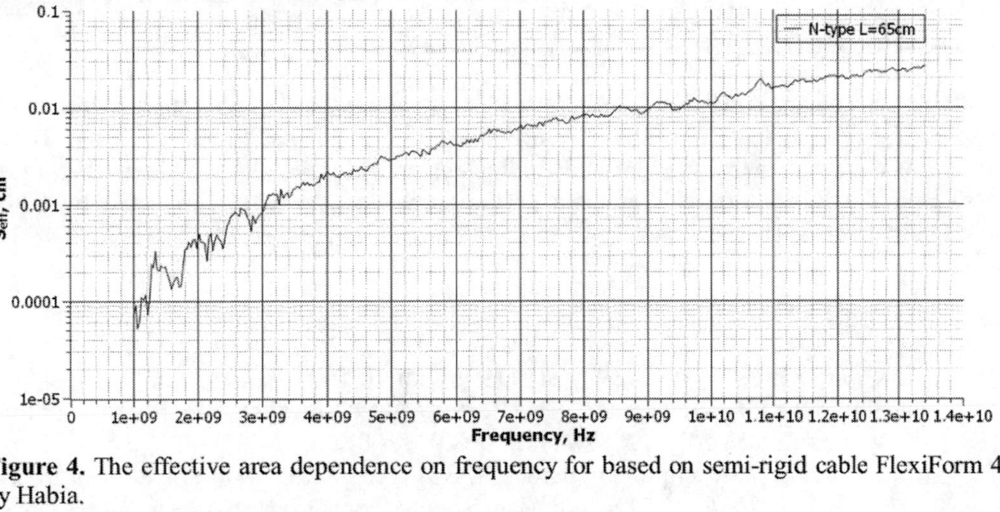

Figure 4. The effective area dependence on frequency for based on semi-rigid cable FlexiForm 402 by Habia.

It confirms the square-law for the effective area dependence on the wavelength. The antenna effective area in X-band is about 1 mm^2.

The antenna of this type was used in the experiments with the vircator based on plasma-filled diode, allowing us to monitor its emission in wide band [5] as well as in the experiments with the oversized Cherenkov type generator in X-band. The external view of the antenna at the test site is presented in Figure 5.

Figure 5. The antenna at the test site.

4 Conclusion

The proposed antenna properties were confirmed in numerical simulations and calibration tests. It has the effective area about the 1 mm^2 in X-band and square-low dependence on the wavelength in a wide band. The cross polarization rate is more than 60dB at the centre position and not less than 30 dB within the range of $\pm 5^0$ in azimuth and elevation angle. It has rather wide main lobe of the directivity pattern, so the special measures should be taken to discriminate the reflected signals. It became the advantage, while working in microwave anechoic chamber, as some positioning errors do not result in to a valuable measurement error. This antenna could simplify the measurement of the wide band HPM radiation characteristics, using the single antenna instead of a set of narrow band ones. The effective area rise with the frequency could be partially compensated by a long transmission line attenuation, which also rises with the frequency also, increasing a dynamic resolution of the measurement system as a whole.

Acknowledgements

We thank our colleague Alexey I. Klimov for assistance with the calibration methodology and plentiful discussions of results. We are also grateful to our colleagues from Tomsk State University of Control System and Radioelectronics, especially to Alexey V. Fateev, for sharing the ETS-Lindgren's antennas. This work was partially supported by RFBR under Grant #14-08-01123a.

References

[1] A.I. Klimov, Russ. Phys. J. **39**, 1241 (1996) doi:10.1007/BF02436167
[2] A.I. Klimov, O.B. Kovalchuk, V.V. Rostov, A.N. Sinyakov, IEEE Trans. Plasma Sci. **36**, 661 (2008) doi: 10.1109/TPS.2008.917300
[3] High Frequency Structure Simulator (HFSS) (Ansys Corporation, Pittsburgh, PA, 15317), May 2016. URL: http://www.ansys.com/Products/Electronics/ANSYS-HFSS
[4] EMC antennas. Double-ridged waveguide horn. Model 3117. (ETS-Lindgren Inc., 1301 Arrow Point Drive, Cedar Park, Texas 78613, USA) URL: http://www.ets-lindgren.com/pdf/3117.pdf
[5] A.A. Zherlitsyn, I.K. Kurkan, Russ. Phys. J. (to be published)

MATEC Web of Conferences **79**, 01013 (2016) DOI: 10.1051/matecconf/20167901013
IME&T 2016

UNDERLYING SURFACE REMOTE SENSING BY THE MICROWAVE RADIOMETER WITH HIGH MEASUREMENT RATE

Anton Ubaichin[1*], *Egor* Alexeev[2], *Gregory* Zhuk[2], *Inna* Plotnikova[1], *Evgeniya* Timofeeva[1], *Tilekbek* Abdirasul uulu[2], *Daniil* Danilov[2] and *Abdulaziz* Tashhodgaev[2]

[1]Tomsk Polytechnic University, Institute of Non-Destructive Testing, 634050, Tomsk, Russia
[2] Tomsk State University of Control Systems and Radioelectronics, 634050, Tomsk, Russia

Abstract. The paper describes a new approach to microwave radiometer design. The approach implies simultaneous using both modified zero measurement method and multi-receiver technique. Simultaneous using increases the operating characteristics of airborne microwave radiometers for aircrafts with self-contained power supply. The block diagram of the onboard Earth remote sensing microwave radiometric system is presented. The block diagram and operating timing diagrams of the designed radiometer are shown. An original technique to design a fiducial noise source for transfer characteristics is discussed. The advantages of the designed radiometer in comparison with the state of the art zero-type microwave radiometer are described.

1 Introduction

The radiometric methods are widely used in various branches of national economy. The technique level of the microwave radiometric system is the major factor of measurement accuracy, and it determines the metrological level of measurement. Unlike the mobile system, the problem of the development and design of stationary radiometric system is completely solved by means of classical measurement methods [1].

Application of microwave radiometric systems in aircrafts and other mobile objects with self-contained power is of current relevance. The main feature of the microwave radiometric system is a limited capacity of the power cell. The necessity of temperature control and power supply of the receiver gain control system leads to high energy requirements [2].

The urgency of the airborne microwave radiometric system is beyond question. One of the most important and challenging applications is monitoring of the underlying surface for search for "cold" or "hot" specific emission areas. The specific emission area can be formed as result of the fire or a hidden metallic object [3, 4].

* Corresponding author:ubaychin@tpu.ru

© The Authors, published by EDP Sciences. This is an open access article distributed under the terms of the Creative Commons Attribution License 4.0 (http://creativecommons.org/licenses/by/4.0/).

 DOI: 10.1051/matecconf/20167901013

2 Theory

We have designed a new type of the airborne microwave radiometric system for the *Ka*-range. The microwave radiometric system has low energy consumption, high measurement rate and accuracy, which are inherent to the zero-type microwave radiometer.

The designed radiometric system is based on the modified zero measurement method, multi-receiver techniques and an original technical solution.

The block diagram of the radiometric system is shown in Figure 1.

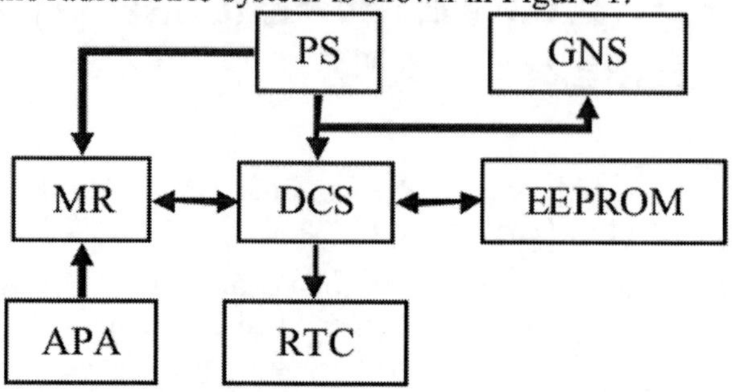

Figure 1. Block diagram of the radiomentric system.

The radiometric system consists of the antenna phased array APA, microwave radiometer MR, digital control system (DCS), electrically erasable programmable read-only memory EEPROM, radio transmitting channel RTC, self-contained power supply PS and global navigation system module GNS.

Figure 2 shows the block diagram of the two-receiver zero-type radiometer with reduced energy consumption.

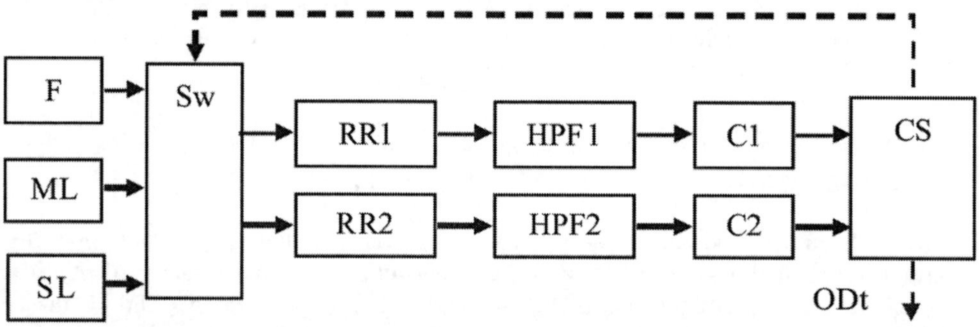

Figure 2. Designed radiometer block diagram.

The radiometer consists of antenna feeder F, the output of which is connected to the input of microwave switch Sw, the corresponding outputs of which are connected to matched load ML and short load SL, herewith, the first Sw output is connected to the first input of control system CS via the first radiometric receiver RR1, high pass filter HPF1 and the first zero-crossing comparator C1 connected in series. The second Sw output is connected to the second CS input via the second radiometric receiver RR2, high pass filter HPF2 and second zero-crossing comparator C2 connected in series. The control system has a single control data bus, which connects the corresponding CS output with the control input data bus of the Sw, and the second CS data bus is the output data bus ODt of the radiometer.

 DOI: 10.1051/matecconf/20167901013

3 Operating principle

The microwave radiometer operates according to the timing diagrams shown in Figure 3 as follows.

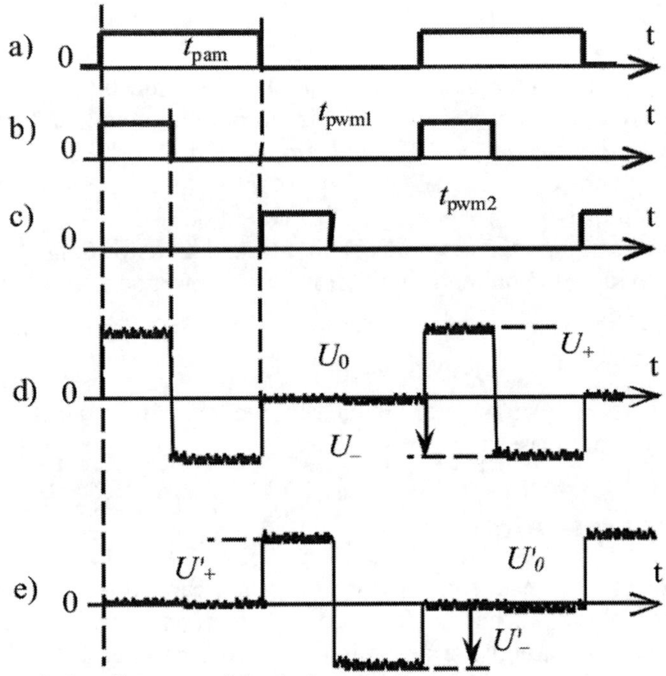

Figure 3. Operating timing diagrams of the designed radiometer.

The radiometer operates synchronously with three logical control signal t_{pam},t_{pwm1} and t_{pwm2} generated by CS. Sequential connection of the antenna phased array to the inputs of the first and second radiometric receivers is carried out by control signal t_{pam} in the microwave switch. While the t_{pam} is high state, the antenna phased array is connected to the input of the second receiver, low state of the t_{pam} corresponds to the antenna phased array connections to the input of the first receiver. The high state of control signals t_{pwm1} and t_{pwm2} corresponds to the matched load connections to the inputs of the first and second receivers respectively. While t_{pwm1} is low and t_{pam} is high the short load is connected to the input of the first receiver. While both t_{pwm2} and t_{pam} is low the short load is connected to the input of the second receiver. In Figure 3, *d* and Figure 3, *e* diagrams corresponds to the voltage waveform at the input of the first and second comparators respectively.

The microwave radiometer operating algorithm is as fallows. Assume that the matched load has noise temperature T_{ML}, the noise of first radiometric receiver is T_{NR1} and the noise temperature of the antenna phased array is T_A. In this case, the amplitude of positive and negative pulses (see Figure 3, *d*) denoted as U_+ and U_-, respectively, and the voltage level corresponding to the antenna array connection to the receiver input is U_0.

$$U_+ \approx k\Delta f_1 G_1(T_{ML}+T_{NR1}) \quad (1)$$

$$U_- \approx k\Delta f_1 G_1(T_{NR1}) \quad (2)$$

$$U_0 \approx k\Delta f_1 G_1(T_{NR1}+ T_A) \quad (3)$$

where$k$ is Boltzmann constant, Δf_1 is the bandwidth of the first radiometric receiver, G_1 is the gain of the first radiometric receiver.

An original technique of negative pulse generation in the designed block diagram is of special interest. It implies reflection of the radiometric receiver noise in the short load. The fulfillment of condition (4) leads to the equality of the voltage-time areas [5]:

$$(U_0 - U_0)(t_{\mathrm{pam}} - t_{\mathrm{pwml}}) = (U_+ - U_0)t_{\mathrm{pwml}} \tag{4}$$

Condition (4) is fulfilled by pulse-width signal duration adjusting, while the comparator performs voltage continuous monitoring at the amplitude-modulated pulse half period corresponding to the antenna array connection to the radiometric receiver input. Determination of the zero voltage level crossings of the detected pulse series with excluded direct component in the high pass filter [3] fulfills condition (4).

Substituting expressions (1), (2) and (3) in (4) and transforming (4) relative to T_A, subtract T_{NR1} and dividing the left and right terms by the common multiplier $k\Delta f_1 G_1$ obtain

$$T_A = T_{ML}(t_{\mathrm{pwml}}/t_{\mathrm{pam}}) \tag{5}$$

Equality (5) shows the invariance of the measurement results to the changes in both gain and noise temperature of the receiver.

4 Experimental research

To check the theory the experimental research series is carried out. For implement it the zero type microwave radiometer is designed in accordance with block diagram (see Figure 2). For microwave radiometer designing are used computer-aided design systems of technological documentation (Solid Works) and electromagnetic analysis (AWR Microwave Office, CST Microwave Studio). The measurement characteristics of the designed microwave modules is carried out by special equipment produced by LTD "Planar" (vector network analyzer Obzor-804/1)and JSC "Micran" (scalar network analyzer R2M-18A). Low frequency modules of the designed radiometer are tested by oscilloscope Tektronix TDS 2024B.The designed radiometer part is shown in Figure 4.

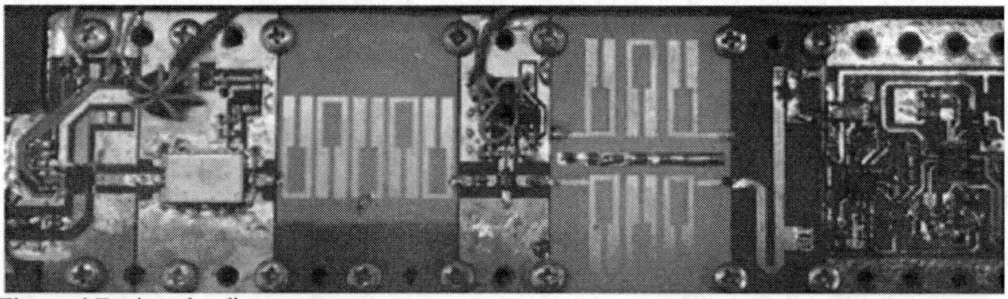

Figure 4. Designed radiometer part.

For the microwave modules designing is used foiled dielectric Flan with dielectric permittivity ε=3.8. This material has low cost, high technological effectiveness and satisfactorily stable of electric characteristics in compare with other dielectric material, for example ceramics. LPKF ProtoMat S100 milling machine is used for microwave printed circuit boards manufacturing.

Figure 5 shows the diagram of the low frequency part of the radiometer in the research of the negative pulse generation principle.

 DOI: 10.1051/matecconf/20167901013

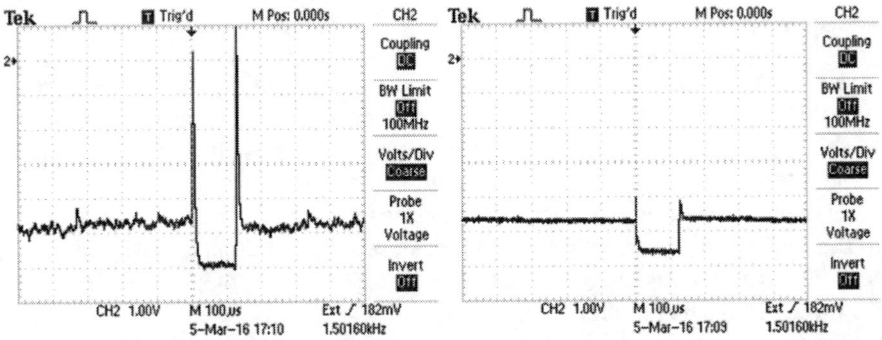

a) b)

Figure 5. Negative pulse generation during the microwave radiometer calibration:
a) without analog averaging, b) with analog averaging.

The technique of the negative pulse generation is a one task of the research. The test technique consist in operating algorithm realization, mentioned above, when antenna equivalent connected to the system (for example, temperature controlled matched load) with noise temperature close to noise temperature of the matched load which is a part of the microwave radiometer. Wherein, the output diagrams of the low frequency part of the radiometer should have only negative pulse, which is proportional to the system noise temperature.

The high state of the signal in Figure 5 corresponds to the fiducial connection to the system input (matched load and antenna equivalent) with noise temperature close to 300 K, low state corresponds to system input connection to the microwave short load during the time interval t_{PWM}. Analog averaging is performed by low pass filters with RC time τ=24 ms.

Figure 6 shows the time diagram of zero balance achieving in the system. Wherein, at the t_{PWM2} time interval, fiducial with noise temperature T_{Nlow} = 150 K is connected to the system input.

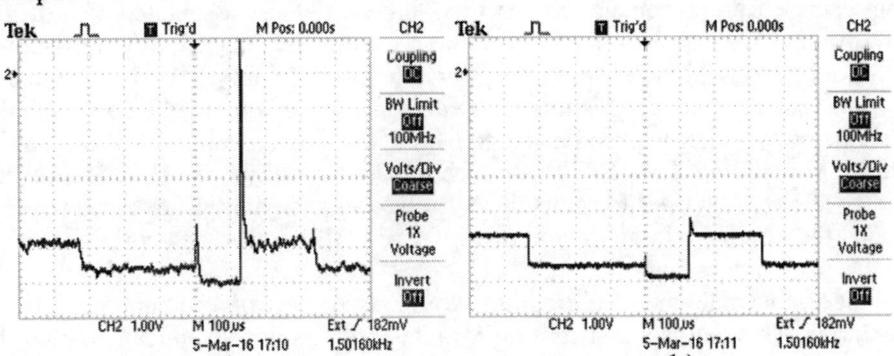

a) b)

Figure 6. Time diagram of zero balance achieving in the system:
a) without analog averaging, b)with analog averaging.

The high state of the signal corresponds to connection of the matched load with noise temperature T_{ML}=300 K to the system input, middle level corresponds to the antenna equivalent signal with noise temperature $T_{N\ low}$=150 K, low state corresponds to signal, that proportional to system noise when microwave short is connected to the input. Analog averaging performed similar with diagram at Figure 5.

5 Discussion

The analysis of the diagrams of the low frequency part of radiometric system (see Figure 5 and Figure 6) and compare with theory (Figure 3) shows the full operability of the designed method. The technique to negative pulse generation make possible to realize the zero type measurement without using additional active noise fiducial sources.

The researches of designed system shown the system dynamic range is 0...300 K, wherein, the upper limit determined by temperature of the matched load which is a part of the system. System noise is about 75 K, which is satisfactorily with considering of applied technology and electronic components.

Interesting results were obtained by analysis of the diagram without averaging. There are specific voltage spikes when noise references are switching. The analysis of the spikes and their characteristics leads to conclusion about the need to changing a synchronization technique of low pass filters and microwave switch which in the system input, more thorough screening of the analog measuring parts or using a digital filtering.

6 Conclusion

A new approach to design the microwave radiometer by the zero measurement method and multi-receiver technique is presented. The designed two-receiver zero-type microwave radiometer uses the receiver noise temperature and matched load as fiducial points for transfer characteristics. This feature allows us to avoid the use of the fiducial noise generators based on the impact avalanche and transit-time diode and Gunn diode. In comparison with the radiometer described in [5], the designed radiometer has a simplified structure as it excludes the directional coupler, noise generator and second microwave switch. The described radiometer does not need the matched load temperature control for measurement result correction.

The generation of the fiducial noise signal by the reflection of the receiver noise is associated with a number of the features described, for example, in [5]. In the considered radiometer, these features can also be observed, including those associated with the receiver noise interference. The method to eliminate these features is protected by "know-how".

In the designed radiometer microwave switch based on the Peregrine Semiconductor PE4257 microchip, it make possible to realize mode of alternately connection of the antenna, microwave short load and matched load to the receiver input. Low noise amplifiers realized by Minicirquits PMA5453+ microchip. The Schottkydiode HSMS-2850 together with matched circuit is used for detection of microwave signals. Low frequency modules based on the Analog Devices microchips. Digital control system is realized by AT91SAM7S256 microchip.

For microwave radiometer designing we are used electronic components of foreign manufacturing. This fact is significant disadvantage of this new technical solution. In spite of this, designing concept implies using only domestic electronic components in the case of serial production of the designed system after successfully passed of field tests series.

Besides of a microwave measuring parts, the designing on board system is supplemented by new sensors to improve in formativeness of remote sensing. These include the infrared cameras. Operating principle, design technique and integration to the measurement system will take place in the near future.

In future we want to carry out field and laboratory tests. Laboratory tests including the researches of stability and fluctuation sensitivity, field tests means the researches of Siberian Federal District areas to estimation of the designing system efficiency during the different types of underlying surface researching.

Acknowledgments

The reported study was partially supported by RFBR, research project No. 16-37-00237 mol_a.

References

[1] A. Camps, J.Tarongi, Rem. Sens. **2**, 191(2010) doi:10.3390/rs2010191
[2] A. Harvey, R. Appleby, Aeron. J. **107**, 87 (2003)
[3] Y. Li , H. Zhao, J. Fan, MATEC Web of Conferences **22**, 04008 (2015) doi: 10.1051/matecconf/20152204008
[4] A. Bespalko, A. Surzhikov, L.Yavorovich, P. Fedotov, Russ. J. Nondestr. Test. **48**, 221 (2012) doi: 10.1134/S1061830912040043
[5] A.V. Filatov, A.V. Ubaichin, D.E. Paraev, Inst. and Exp. Tech. **55**, 59 (2012) doi: 10.1134/S0020441211060066

BOOLEAN DIFFERENTIATION EQUATIONS APPLICABLE IN RECONFIGURABLE COMPUTATIONAL MEDIUM

Stanislav Shidlovskiy[1,2]

[1]National Research Tomsk State University, 634050, Tomsk, Russia
[2]National Research Tomsk Polytechnic University, 634050, Tomsk, Russia

Abstract. High performance computing environment synthesis with parallel architecture reconstructing throughout the process itself is described. Synthesized computational medium involving Boolean differential equation calculations so as to function in real-time image processing. Automaton imaging was illustrated involving the rearrangement of every processing medium element to calculate the partial differentials of n-th order in respect to Boolean function variables. The method of obtaining setting codes for each element was also described. An example in calculating 2nd -order Boolean derivative to two differentials in respect to Boolean functions, depending on three arguments within the reconstructible computational medium of 8x8 processing elements was given.

1 Introduction

Architecting updated robotic systems functioning autonomously in a-priori unknown medium involves directly positioned in- robot computer systems in order to process a considerable body of sematic information. Real-time processing of this information attributable to classical computing architecture results in increasing robot weight and structure design.

One possible solution of above-mentioned problem could be applying the reconfigurable computing medium (RCM) [1–6]. Hardware-in-the-loop RCM design and its functioning is based on grouped evaluator model:

$$S = < C, G, A(P(D)) > ,$$

where, $C = \{c_i\}$ – set of processing elements c_i, $i = 0, 1, \ldots, N - 1$; G – intercomputing linkage matrix structure (grid); A – processing element algorithm set C interrelated with G, in hardware parallel support program P of data processing D.

Grouped evaluator construct $H = < C, G >$ reflects the following architectural concept:

1) parallelism in processing (parallel processing element algorithm set C intercomputing with G);

2) programmability structure;

3) homogeneity of structure H (homogeneity of processing elements $c_i \in C$ and macrostructure G).

© The Authors, published by EDP Sciences. This is an open access article distributed under the terms of the Creative Commons Attribution License 4.0 (http://creativecommons.org/licenses/by/4.0/).

Let's consider the case when functioning processing element algorithm c_i is the basis {AND, OR, NOT}, thus acquiring the hardware invariance property.

Algorithm design in terms of solving one specific problem-oriented task via RCM, bearing on this limitation is associated with significant peformance. This paper describes the synthesis of such RCM for Boolean differentiation implementation and integration.

2 Problem statement

Usually, binary data is represented as {0, 1} and would look like arbitrary finite dimension matrix. By zero extension it will be representable as $2^n \times 2^n$ matrix $\mathbf{R}_{2^n}$, where n – positive integer number. This matrix $\mathbf{R}_{2^n}$ could be interpreted as microsequencing or code of Boolean function system, binary image (animated object frame), dynamic binary system state at the present sampling time, operating technical system component interaction graph. Further, this matrix will be considered as a system of 2^n Boolean functions (BF) $f_j(X)$, $j = \overline{0,\ 2^n - 1}$, n – variables x_1, x_2, ..., x_n, Each BF has its own vector value $\mathbf{X}_j = \left[x_j^{(0)}\ x_j^{(1)}\ \text{K}\ x_j^{(2^n-1)}\right]^{\text{T}}$, where T – transposition symbol; element $x_j^{(t)} \in (0,1)$ – j-value of BF for set of variables $x_1^{t_1}$, $x_2^{t_2}$, ..., $x_n^{t_n}$; t_1, t_2, ..., t_n – binary representation parameter $t = \overline{0, 2^n - 1}$; $x^0 = \bar{x}$, $x^1 = x$. The system of 2^n BF $f_j(X)$, having its own vector value $\mathbf{X}_j$, can be written in the form of the matrix

$$\mathbf{R}_{2^n} = \left[\mathbf{X}_{2^n-1} \vdots \cdots \vdots \mathbf{X}_1 \vdots \mathbf{X}_0\right].$$

In [7] the algorithm of Boolean derivative calculations by private variable is:

$$\frac{\partial^{(k)}\mathbf{X}_j}{\partial x_1 \partial x_2 \ldots \partial x_k} = \mathbf{D}_{2^n}^{(k)} \ldots \mathbf{D}_{2^n}^{(2)} \mathbf{D}_{2^n}^{(1)} \mathbf{X}_j \ (\mathrm{mod}\, 2), \tag{1}$$

$$\mathbf{D}_{2^n}^{(i)} = \mathbf{I}_{2^{i-1}} \otimes \begin{bmatrix} 1 & 1 \\ 1 & 1 \end{bmatrix} \otimes \mathbf{I}_{2^{n-i}}. \tag{2}$$

where **I** – identity matrix.

The analysis of the algorithm shows that RCM processing element should implement one of the following automaton-based display image (Figure 1).

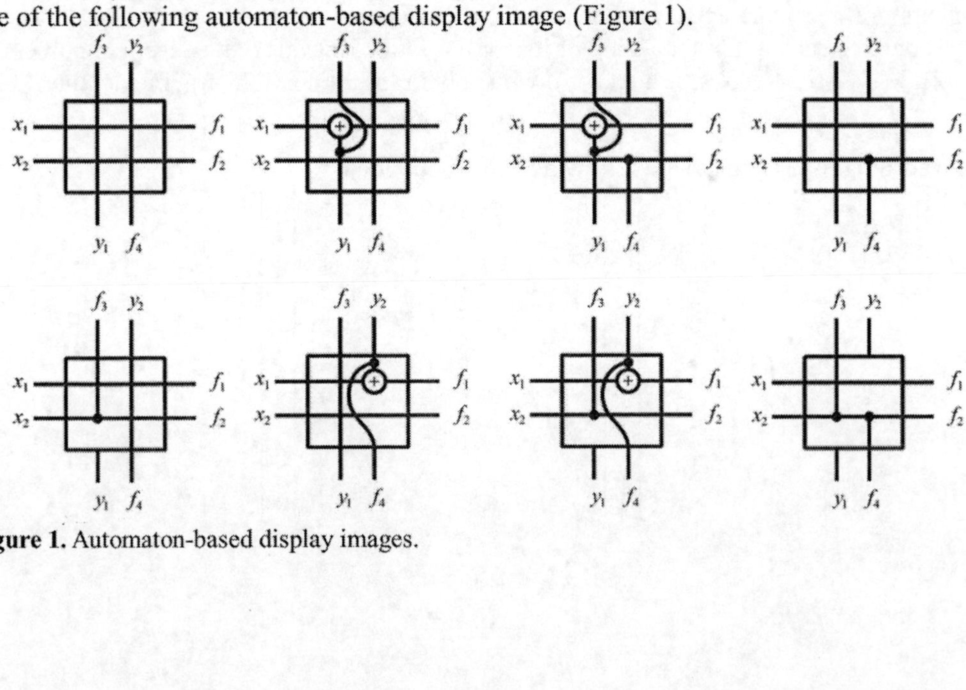

Figure 1. Automaton-based display images.

Applying automaton-structured method [5] RCM functioning processing element algorithm could be synthesized:

$$\begin{cases} f_1 = (x_1\bar{y}_1\bar{z}_2) + (x_1\bar{y}_2 z_2) + (x_1 z_3) + (\bar{x}_1 y_1 \bar{z}_2 \bar{z}_3) + (\bar{x}_1 y_2 z_2 \bar{z}_3), \\ f_2 = x_2, \\ f_3 = (x_2 z_1 z_2 \bar{z}_3) + (x_2 z_1 \bar{z}_2 z_3) + (x_2 \bar{z}_1 z_2 z_3) + (y_1 z_1 z_2 z_3) + (y_1 \bar{z}_1 \bar{z}_2) + (y_1 \bar{z}_1 \bar{z}_3) + (y_2 z_1 \bar{z}_2 \bar{z}_3), \\ f_4 = (x_2 z_1 \bar{z}_2 \bar{z}_3) + (x_2 \bar{z}_1 z_3) + (y_2 z_1 z_2) + (y_2 z_1 z_3) + (y_2 \bar{z}_1 \bar{z}_3). \end{cases}$$

where, x_1, x_2, y_1, y_2 – data inputs; $f_1,\ldots, f_4$ – outputs; $z_1,\ldots, z_3$ – setup inputs of RCM processing element [3, 5].

Realization of algorithm (1) into RCM via evaluators, implementing automaton images (fig.1) are plotted as the matrix structure of computing-based system. Determining the matrix Z configuration is based on the setting up each processing element within one possible automaton image:

$$\mathbf{Z} = \sum_i \mathbf{M}_i \tag{3}$$

$$\mathbf{M}_1 = (\mathbf{A} \circ \mathbf{I}_n)k_1,$$
$$\mathbf{M}_2 = (\mathbf{A} \circ \bar{\mathbf{I}}_n \circ \mathbf{H}_n)k_2,$$
$$\mathbf{M}_3 = (\mathbf{A} \circ (\bar{\mathbf{I}}_n \circ \mathbf{H}_n)^{\mathrm{T}})k_3,$$
$$\mathbf{M}_4 = \bar{\mathbf{A}} k_4,$$

$$\mathbf{A} = \mathbf{D}_{2^n}^{(k)} \ldots \mathbf{D}_{2^n}^{(2)} \mathbf{D}_{2^n}^{(1)} \pmod 2 \tag{4}$$

where, o – Hadamard product; $\mathbf{I}_n$ – n x n identity matrix; $\mathbf{H}_n$ –n x n upper triangle matrix of one; vertical overline (¯) means x-element inversion in binary matrix; ^{T} – transposition matrix symbol; k_i – automaton image code.

3 Example

It is required to calculate Boolean second-order derivative with respect to the BF variables x_1 and x_2, as in the following expression

$$f_i(\mathbf{X}) = x_1 x_2 \vee x_3 . \tag{5}$$

To implement Boolean derivative calculation by private variable four basic evaluator elements are necessary (fig. 1). Each element is denoted by k_i (i = 1, 2,…, 4). Value k_i corresponds to each automaton code of its configuration, introduced as decimal numeration (as follows $k_1 = 2$; $k_2 = 3$, $k_3 = 4$, $k_4 = 5$). We apply in (5) all possible argument values:

$$\mathbf{X}_j = [0\ \ 1\ \ 0\ \ 1\ \ 0\ \ 1\ \ 1\ \ 1]^{\mathrm{T}}.$$

According to (3) and (4) the following can be obtained:

$$\mathbf{A} = \begin{bmatrix} 1 & & 1 & & & & & \\ & 1 & & 1 & & & & \\ 1 & & 1 & & & & & \\ & 1 & & 1 & & & & \\ & & & & 1 & & 1 & \\ & & & & & 1 & & 1 \\ & & & & 1 & & 1 & \\ & & & & & 1 & & 1 \end{bmatrix} \oplus \begin{bmatrix} 1 & & & & 1 & & & \\ & 1 & & & & 1 & & \\ & & 1 & & & & 1 & \\ & & & 1 & & & & 1 \\ 1 & & & & 1 & & & \\ & 1 & & & & 1 & & \\ & & 1 & & & & 1 & \\ & & & 1 & & & & 1 \end{bmatrix} = \begin{bmatrix} 1 & & 1 & & 1 & & 1 & \\ & 1 & & 1 & & 1 & & 1 \\ 1 & & 1 & & 1 & & 1 & \\ & 1 & & 1 & & 1 & & 1 \\ 1 & & 1 & & 1 & & 1 & \\ & 1 & & 1 & & 1 & & 1 \\ 1 & & 1 & & 1 & & 1 & \\ & 1 & & 1 & & 1 & & 1 \end{bmatrix}.$$

$$\mathbf{Z}=\begin{bmatrix}2&5&3&5&3&5&3&5\\5&2&5&3&5&3&5&3\\4&5&2&5&3&5&3&5\\5&4&5&2&5&3&5&3\\4&5&4&5&2&5&3&5\\5&4&5&4&5&2&5&3\\4&5&4&5&4&5&2&5\\5&4&5&4&5&4&5&2\end{bmatrix}.$$

Obtained matrix of Z configuration corresponds to RCM, depicted in Figure 2.

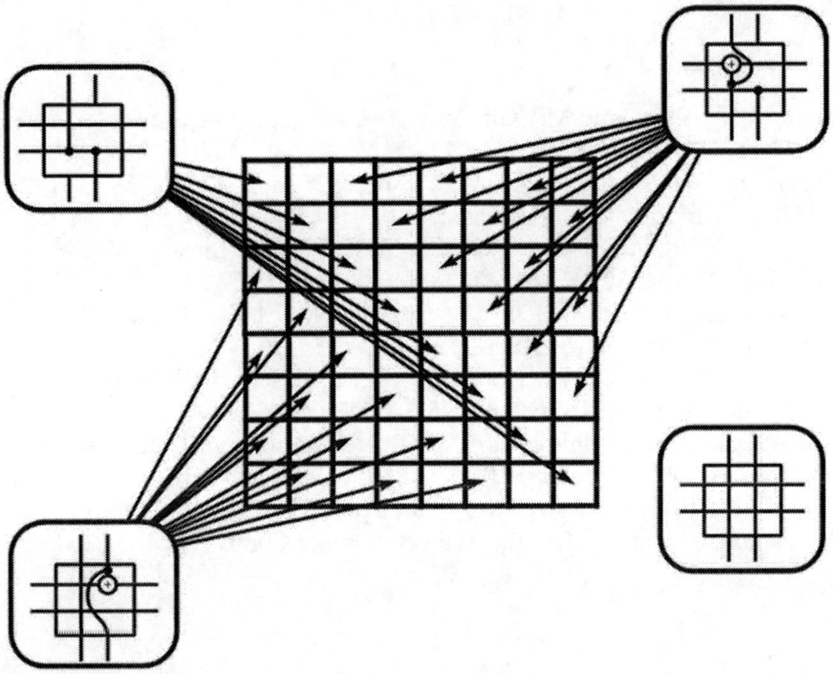

Figure 2. RCM configuration.

Having generated the signals corresponding to (6) the result providing the BF calculation by private variables (5) x_1 and x_2 is recorded on data inputs of obtained RCM and **on medium return:**

$$\frac{\partial^2\mathbf{X}_j}{\partial x_1\partial x_2}=\begin{bmatrix}1&0&1&0&1&0&1&0\end{bmatrix}^{\mathrm{T}}.$$

4 Conclusion

RCM algorithms were synthesized to provide the calculation of BF private variables. These algorithms meet the specified requirements applicable for VLSI process. In this case, such algorithms involve regular and localized computation communication, oriented on pipeline computations. Proposed algorithm configuration (3), involving specific setting code inputs for each RCM processing element, is responsible of calculating Boolean derivatives to different argumants within one and the same RCM.

According to the reference description, the synthesis method of specific Boolean function evaluators is based on the operating diagram of the previously selected Boolean base (AND, OR, NOT) and is housed in synthesized-based medium. It should be noted that the prescribed functional diagram or logical network within selected medium is 'fitted'. Obtained algorithms of operating RCM (reconstructible computational medium) could be the basis of developing special-purpose evaluators producing basic operations of morphologically processing digital images.

Acknowledgment

The following research has been financially supported by Russian Foundation for Basic Research (RFBR) 16-07-01138 A "Intelligent Reconfigurable Control Systems, Navigation and Image Processing for Autonomous Mobile Robots."

References

[1] I. Kulagin, A. Paznikov, M. Kurnosov, Lecture Notes in Computer Science **9251**, 405 (2015) doi: 10.1007/978-3-319-21909-7_39
[2] N. Maa, Sh. Wang, Ali S. Mohsin, X. Cui, Yu. Peng, MATEC Web of Conferences **45**, 05001 (2016) doi: 10.1051/matecconf/20164505001
[3] S.V . Shidlovskii, J. Comp. Sys. Sci. Inter. **45**, 282 (2006) doi: 10.1134/S1064230706020122.
[4] A.I. Kalyaev, I.A. Kalyaev, J. Comp. Sys. Sci. Inter. **55**, 211 (2016) doi: 10.1134/S1064230716010081
[5] I.A. Kalyaev, I.I. Levin, A.I. Dordopulo, L.M. Slasten, IFAC Proceedings Volumes **46**, 210 (2013) doi: 10.3182/20130925-3-CZ-3023.00009
[6] D.V. Shashev, S.V. Shidlovskiy, Optoelectronics, Instrumentation and Data Processing **51**, 227 (2015) doi: 10.3103/S8756699015030036
[7] S. Janushkevich, D. Bokhmann, R. Stankovich, G. Tosic, V. Shmerko, Avtomatika I Telemeknanika (Journal of Automation & Remote Control) **6**, 155 (2004).

MATEC Web of Conferences 79, 01015 (2016) DOI: 10.1051/matecconf/20167901015
IME&T 2016

TESTING SYSTEM FOR ANALOG DEVICES DIRECT DIGITAL SYNTHESIZER

Ilia Loskutov*, *Ivan* Shvetsov-Shilovskiy, *Dmitry* Boychenko

National Research Nuclear University MEPhI (Moscow Engineering Physics Institute), 115409, Moscow, Russia

Abstract. The paper is devoted to the setup for controlling and testing of Direct Digital Synthesizer (DDS) Integrated Circuits (ICs). Control and measurement setup is designed on the basis of National Instruments module equipment PXI-4110, PXI-7841R and LabVIEW development environment. Block diagram of the developed system and software structure are depicted as well as test results for several ICs.

1 Introduction

Different electronic devices often need an IC that generates signals of various frequencies and allows controlling parameters of such a signal [1]. Direct Digital Synthesizer IC can play the role of the generator of this kind. The device produces analog signal by means of generating series of samples in digital form and then converting these to the analog signal using ADC [2, 3]. The main blocks of DDS are phase accumulator, phase-to-amplitude converter (PAC), DAC and low-pass filter. Block diagram of the simplest DDS IC is presented on Figure 1. Adjustment of the IC's parameters (settling the desired frequency, phase, amplitude, form etc.) is made by means of special build-in registers. These registers are available for read and write to via different interface protocols, e.g. SPI.

One of the features of DDC ICs is high frequency setting resolution and high precision of frequency rendering. Some modern digital synthesizers are capable of generating signals with frequency of 1 Hz up to 400 MHz with the step size of 0.00001 Hz.

At the moment DDS are applied in two main areas: signal generation in communication systems and signal analysis in industrial and biomedical applications.

While developing devices of various purposes, often raises an aim to test the functionality of its separate components. Thus, for example, it is necessary to test the functionality of the DDS ICs. This paper concerns the design of the control and measurement setup [4, 5] for Analog Devices DDS ICs.

* Corresponding author: iolos@spels.ru

© The Authors, published by EDP Sciences. This is an open access article distributed under the terms of the Creative Commons Attribution License 4.0 (http://creativecommons.org/licenses/by/4.0/).

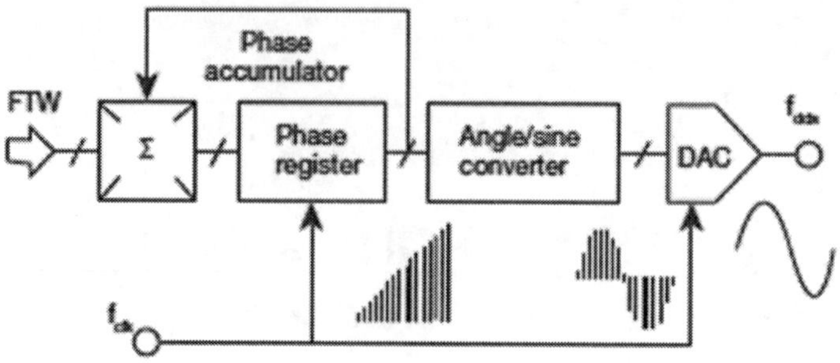

Figure 1. DDS block diagram.

2 Device under test and setup requirements

Devices Under Test (DUTs) were AD9851, AD9910 and AD9912 Analog Devices DDS IC.

AD9851 is a device that is capable to operate in voltage supply range from 2.7 V to 5.25 V; clock frequency can be up to 180 MHz. AD9851 also includes a high-speed 10-bit DAC.

AD9910 and AD9912 are ICs with integrated 14-digit DAC; voltage lies within range of 1.8 V to 3.3 V; output frequency is up to 400 MHz [6].

One of the distinctive features of the abovementioned ICs is a similar SPI communication interface that has made possible to design universal control and measurement setup.

3 Solution

Control and measurement system was designed on the basis of automated setup that included National Instruments PXI module devices and software developed in LabVIEW development environment [7–14].

Hardware included following components:

- PXI-1033: 5-slot chassis (data transfer rate of 110 MB/s, noise less than 38 dB, power 400 W).
- PXI-4110: programmable power supply (three channels with a voltage range from 0 V to 6 V, from 0 V to 20 V and from 0 V to -20 V accordingly).
- PXI-7841R: multifunction input/output board (based on Virtex-5R FPGA, 96 bidirectional digital ports, 8 analog inputs, 8 analog outputs) [7].
- universal communication PCB.
- PCB with the IC.
- Agilent 54642D oscilloscope.

PXI-4110 module allowed controlling supply voltage and measuring supply current in range up to 1 A and with precision of 0.02 mA.

Desired communication interface between test equipment and ICs was implemented by means of the PXI-7841R module. PXI-7841R embedded Programmable Logic Device (PLD) has a high operation rate; base frequency is 40 MHz that can be altered due to build-in Phase-Locked Loop (PLL). The module also includes analog lines (8 analog inputs and 8 analog outputs). This gives the opportunity to measure logical output signal level with precision of 0.3 mV.

Universal adapter PCB has pins for all PXI-7841 digital and analog signal lines and channels of PXI-4110 power supply module.

DUT PCBs were developed for each DDS type according to the reference schematic presented in technical documentation. The photo of the universal communication board and printed circuit board for testing is shown on Figure 2.

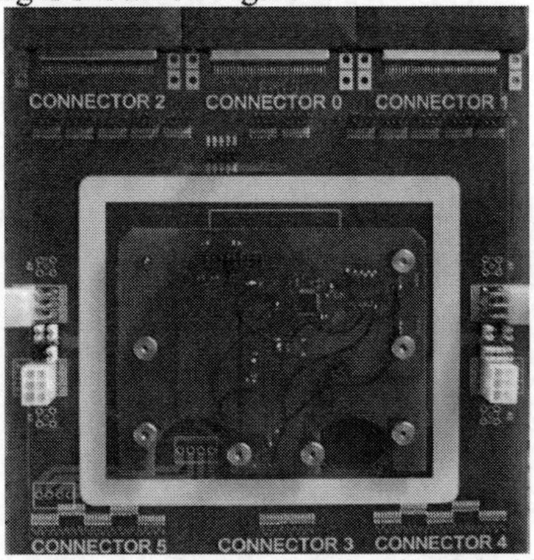

Figure 2. Appearance of the universal communication board and board with the PCB.

Agilent 54642D oscilloscope was used to control output signal's frequency. Block diagram of the measurement setup is presented on Figure 3.

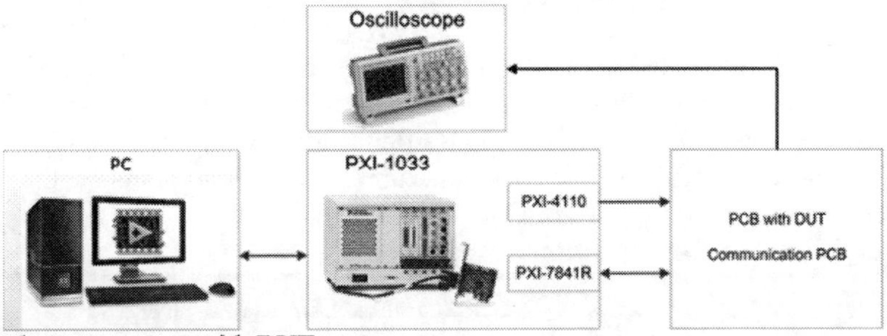

Figure 3. Hardware structure with DUT.

Software part was made in LabVIEW 2013 development environment.

This software allowed controlling PXI-4110 and PXI-7841R modules, as well as create user interface.

The program was divided into two main parts: HOST part and FPGA part. The first one was run on PC with relatively slow execution rate and provided the automatization of testing process; it set supply voltage levels, logged the results and initialized the FPGA module [8].

The second part was programmed into PLD and had a high operation rate. Its task was to set the clock frequency, transmit and receive data [9–18].

To run the DDS IC one often needs a reference frequency; PXI-7841R module allowed to generate such high-frequency periodic signal. Block diagram of the Virtual Instrument (VI) that is responsible for frequency generation is shown on Figure 4 [9].

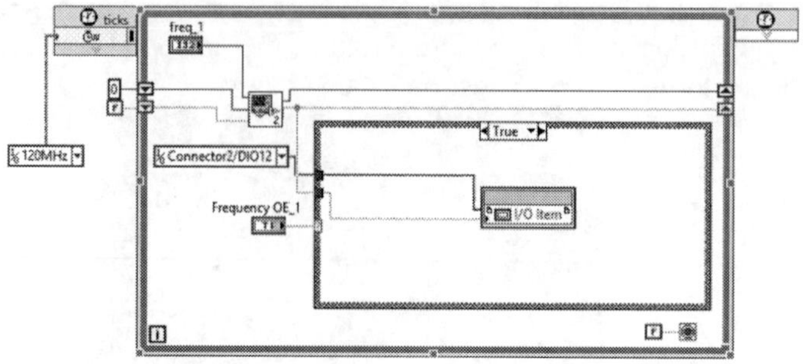

Figure 4. PXI-7841R frequency generator block diagram.

Prior to test procedure user is asked for communication interface parameters, such as register address, byte number, transfer direction, bits order and timing parameters. The example of write timing diagram is presented on Figure 5.

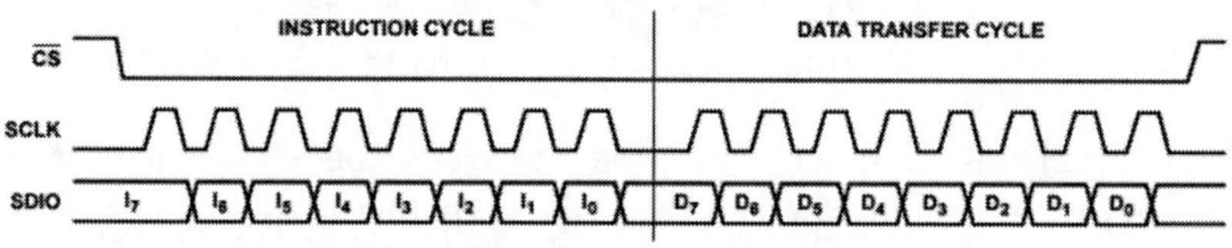

Figure 5. Data write timing diagram for AD9910 IC.

Low level on CS indicates the beginning of data communication. SDIO line is used to transfer 8-bit address first and then 8 bits of data; the state of the line is latched at the rising edge on the SCLK line. The transmission ends after sending a high level on CS.

Block diagram of the LabVIEW virtual instrument that implements sending the data via SPI interface is presented on Figure 6. One should notice the use of cluster data type, which allows fast digital lines reconfiguration.

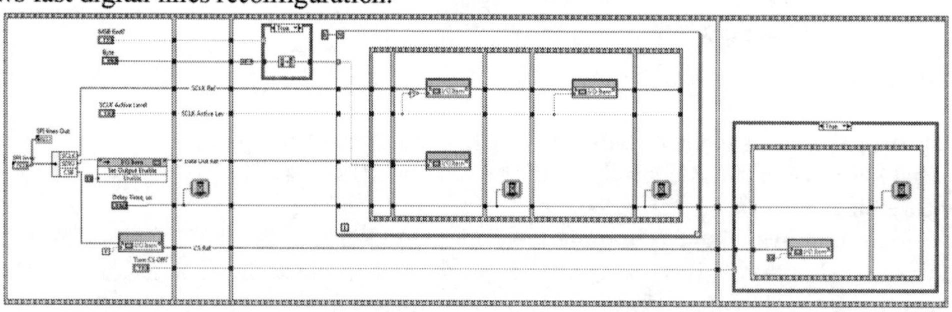

Figure 6. Block diagram of the VI that sends one byte via SPI communication interface.

Test procedure consisted of the following steps:

- IC current supply control by means of ammeter embedded in NI PXI-4110 module.
- Output logic levels control.
- Writing and reading the control registers.
- Taking measurements of the output frequency with certain base frequency division and multiplication coefficients. An example of the output signal is presented on Figure 7.

 DOI: 10.1051/matecconf/20167901015

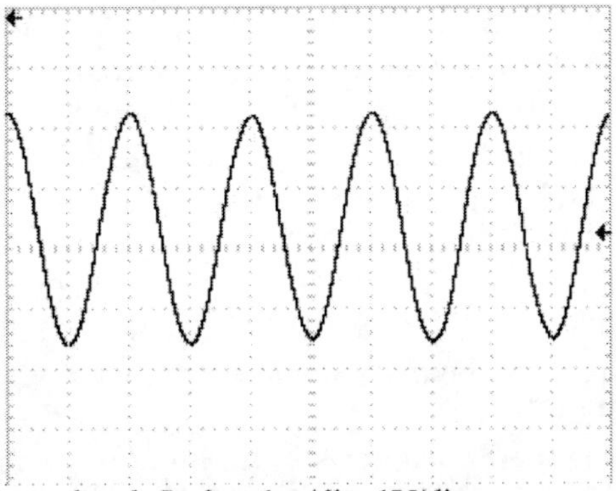

Figure 7. Waveform of the output signal. Scales: 1ns/div; 1V/div.

Convenient user interface (Figure 8) made the system easy to use.

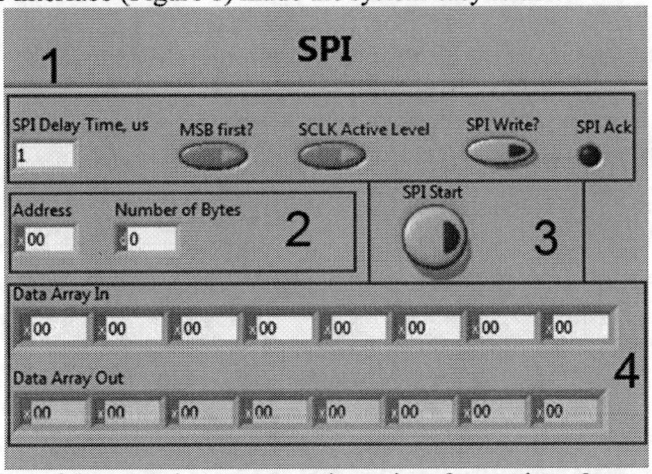

Figure 8. User interface of the control system:1 – primary interface settings; 2 – controls for entering register address and byte number to receive or transmit; 3 – communication start button; 4 – received and transmitted data.

4 Conclusions

AD9851, AD9910, AD9912 ICs were successfully tested by means of foregoing measurement setup. Measurement setup has made possible to fulfill the functional and parametrical testing. It is planned to add several other standard interface libraries such as 1-wire, I2C etc. that will allow controlling and testing other devices.

Acknowledgements

The work was performed as part of the grant of Russian Ministry of Education and MEPhI № 14.578.21.0075.

References

[1] D. Wang, J. Dong and S. Xiao, MATEC Web of Conferences **22,** 02005 (2015) doi: 10.1051/matecconf/20152202005

[2] N. E. Aristova, A. Y. Borisov, A. I. Tararaksin, L. N. Kessarinskiy, A. V. Yanenko, 2015 International Siberian Conference on Control and Communications (SIBCON), 7146984 (2015) doi: 10.1109/SIBCON.2015.7146984
[3] E. Murphy, C. Slattery, Analog Dialogue **38**, 1 (2005)
[4] W. Fan, Y. An, D. Liu and S.Guan, MATEC Web of Conferences **40**, 07002 (2016) doi: 10.1051/matecconf/20164007002
[5] Hang Bai, Yunlai Huang, Di Lu, Ping Lai and Hongzhi Fan, MATEC Web of Conferences **40**, 07006 (2016) doi: 10.1051/matecconf/20164007006
[6] Analog Devices URL: http://www.analog.com/ru/products/clock-and-timing/direct-digital-synthesis.html
[7] PXI Platform - National Instruments URL: http://www.ni.com/pxi/
[8] V. A. Marfin, P.V. Nekrasov, O. A. Kalashnikov, K. A. Kagirina, 2015 International Siberian Conference on Control and Communications (SIBCON) , 7147142 (2015) doi: 10.1109/SIBCON.2015.7147142
[9] I. O. Loskutov, A. B. Karakozov, P.V. Nekrasov, A.Y. Nikiforov, 2015 International Siberian Conference on Control and Communications (SIBCON) , 7147128 (2015) doi: 10.1109/SIBCON.2015.7147128
[10] M. P. Belova, D. V. Pechenkina, A. Y. Borisov, L. N. Kessarinskiy, D.V. Boychenko, 2015 International Siberian Conference on Control and Communications (SIBCON), 7146995 (2015) doi: 10.1109/SIBCON.2015.7146995
[11] I. I. Shvetsov-Shilovskiy, P.V. Nekrasov, A.V. Ulanova, A. A.Smolin, A. V. Sogoyan, 2015 International Siberian Conference on Control and Communications (SIBCON), 7147282 (2015) doi: 10.1109/SIBCON.2015.7147282
[12] A. B. Boruzdina, A. A. Orlov, A. V. Ulanova, N. G. Grigor'ev, A.Y. Nikiforov, 2015 International Siberian Conference on Control and Communications (SIBCON), 7147007 (2015) doi: 10.1109/SIBCON.2015.7147007
[13] G. G. Davydov, A. S. Kolosova, L. N. Kessarinskiy, D.V. Boychenko, 2015 International Siberian Conference on Control and Communications (SIBCON), 7147094 (2015) doi: 10.1109/SIBCON.2015.7147094
[14] N. E. Aristova, A. I. Tararaksin, L. N. Kessarinskiy, A. Y. Borisov, A.Y. Nikiforov, 2015 International Siberian Conference on Control and Communications (SIBCON), 7146985 (2015) doi: 10.1109/SIBCON.2015.7146985
[15] N. E. Aristova, A. Y. Borisov, A. I. Tararaksin, L. N. Kessarinskiy, and A. V. Yanenko, 2015 International Siberian Conference on Control and Communications (SIBCON), 7146984 (2015) doi: 10.1109/SIBCON.2015.7146984
[16] A.V. Demidova, A. Y. Borisov, L. N. Kessarinskiy, and D.V. Boychenko, 2015 International Siberian Conference on Control and Communications (SIBCON), 7147036 (2015) doi: 10.1109/SIBCON.2015.7147036
[17] V.V. Elesin, A. G. Kuznetsov and D. I. Sotskov, 2014 24th International Crimean Conference Microwave and Telecommunication Technology (CriMiCo), 6959664 (2014) doi: 10.1109/CRMICO.2014.6959664
[18] P. Nekrasov, A. Demidov, and O. Kalashnikov, Instruments and Experimental Techniques **52**, 196 (2009) doi: 10.1134/S0020441209020092

NUMERICAL MODELING AND EXPERIMENTAL INVESTIGATION OF THE BREADBOARD MODEL OF A NEAR-FIELD INTERFERENCE MICROSCOPE

Victor Belichenko[a], *Andrey* Zapasnoy, *Aleksandr* Mironchev

National Research Tomsk State University, 634050, Tomsk, Russia

Abstract. A circuit design for technology of near-field interference microwave microscopy using evanescent fields of several active radiators (probes) is suggested and substantiated. The feasibility of providing the spatial overlap of the evanescent fields of two probes representing expanding coaxial conical horns is demonstrated. It is shown that the structure of the electromagnetic field in the region of overlap can effectively be controlled by changing the phase difference of oscillations arriving at inputs of the probes. This technology can be used for diagnostics of liquid media comprising dielectric inhomogeneities, flaw detection, and hygroscopy.

1 Introduction

Recently more and more researchers direct their attention to investigation of possibilities of near-field microwave diagnostics [1–7]. Such diagnostics involves medical (detection of human tumor formations) and industrial applications (flaw detection and hygroscopy), but in both cases restriction is imposed on the depth of field penetration into the examined medium or object. The penetration depth is of the order of the near field zone at the maximum wavelength from the frequency band of a probing signal. This necessitates a rigid connection of the receiver and transmitter of the probing system, since the field scattered by the objects in the near field zone of such system must be fixed. For such unilateral technology of sensing, the choice of the signal frequency band, receiving antenna aperture area, and transceiving system scanning altitude is very important.

In this work, an alternative sensing technology of bilateral near-field sensing using overlapping evanescent (non-propagating) microwave fields generated by spatially distributed systems of radiators is considered. Previously we conducted theoretical research [8] showing the significant role of the interference of energy fluxes arising due to overlapping (superimposition) of near fields of several radiators. Some considerations concerning the method of control over these fluxes for sensing of media, materials, and objects were also presented.

[a]Corresponding author: bvp@mail.tsu.ru

© The Authors, published by EDP Sciences. This is an open access article distributed under the terms of the Creative Commons Attribution License 4.0 (http://creativecommons.org/licenses/by/4.0/).

2 The prototype and the technique of experimental research

By virtue of the foregoing, attention here is focused on the one hand, on detailed numerical modeling aimed at working out the design of the most important microscope units – probes, and on the other hand, on experimental investigations of the possibilities of a new version of the near-field microscope. Computer modeling was performed using the software package CST Microwave Studio. The model probing system comprised two identical conical coaxial horns each 36 cm long facing each other; the diameters of the central and outer conductors in the horn aperture were 3 and 7 cm, respectively, and the gap between the horns was 5 cm. The horns were excited by waveguide ports with identical characteristics, but the phase of the signal of the second port was shifted relative to the signal of the first port. The electric and magnetic field distributions were calculated in the frequency band from 0.2 GHz to 10 GHz.

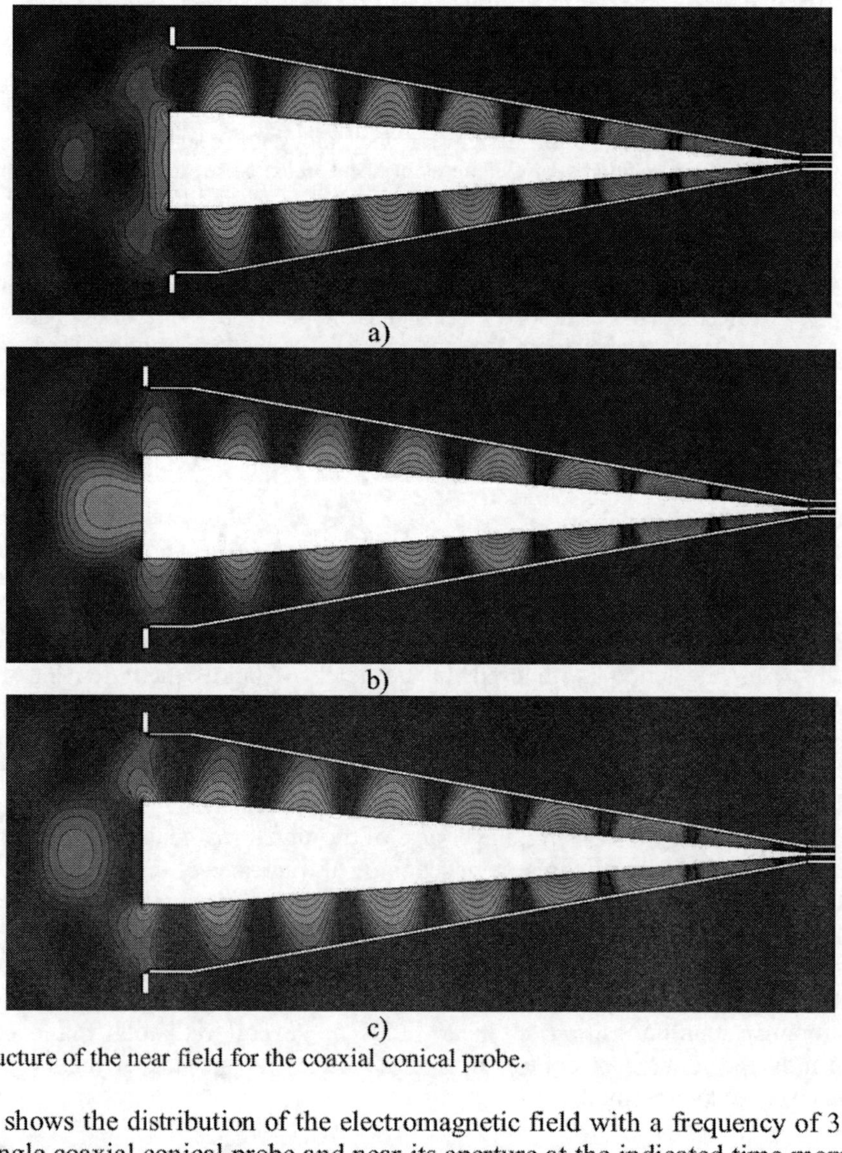

Figure 1. Structure of the near field for the coaxial conical probe.

Figure 1 shows the distribution of the electromagnetic field with a frequency of 3 GHz inside the single coaxial conical probe and near its aperture at the indicated time moments. At the initial moment of time (Figure 1, *a*) the region with high field concentration was localized mainly inside the probe, near its aperture. Figure 1, *b* illustrates the moment of maximum field localization already outside of the probe. It can be seen that the field is localized near the end of the central conductor of the probe, and the sizes of the field localization region are of the order of the conductor diameter. Figure 1, *c* shows the formation of the radiation field and the noticeable field attenuation near the end of the central conductor of the probe. The above-described special features of the field localization are in quite good agreement with the data presented in [9]. In particular, the conclusion that the spatial resolution of the probe should be of the order of the diameter of the inner conductor of the coaxial probe is qualitatively confirmed. We note that the frequency band used by us and the probe design allow the regime of object diagnostics using both evanescent and radiated waves to be realized.

The patterns of field distributions in the probing system modeled by two coaxial conical probes facing one another are shown below (Figure 2). By analogy with the single probe, modeling was performed for a frequency of 3 GHz. Its main purpose was to show the dynamics of forming the overlapping evanescent fields in the near-field zone of the probes and the dependence of their behavior on the initial phase difference ($\Delta\varphi$) of the fields in the probes: a) $\Delta\varphi$=0°, b) $\Delta\varphi$=15°, c) $\Delta\varphi$=30°, d) $\Delta\varphi$=45°, and e) $\Delta\varphi$=90°. It can be seen that by analogy with the case of the single probe, the TEM wave propagates in each probe. In the plane of the probe aperture, owing to a sharp break of its central conductor, the radiation fields start to form. In the immediate proximity to the ends of the central conductors of the probes, the near field is localized. It can be seen that for $0° < \Delta\varphi < 30°$, the field localization differs significantly from that for the single probe: the near field occupies much larger spatial volume. Already for $\Delta\varphi$=45°, the transformation of this volume starts to be manifested clearly, which is accompanied by predominant localization of the near field in the vicinity of the end of the central conductor of the left probe. Moreover, the highest degree of localization is observed at $\Delta\varphi$=90°. The results of modeling presented here agree quite well with conclusions of work [8]. In this work, the theoretical investigations of the structure of the near field for several elementary radiating systems revealed the feasibility of efficient control over the interference parameters of energy fluxes formed in near overlapping fields of the radiators.

This feasibility has also been confirmed in the present work based on rigorous electrodynamic modeling of an actual probing system. It is important to emphasize that the parameters of the interference energy flux formed in the gap between the probe apertures can be controlled; moreover, they can be controlled electrically, namely, by changing the phase difference between the probe fields. This control is realized via "accentuating" one of the interference components of the Pointing vector. This circumstance means that a sufficiently small selected volume of the examined object or medium can be detected via flexible change of the regime of its irradiation by the evanescent fields created by a radiator system (moreover, the motionless system).

DOI: 10.1051/matecconf/20167901016

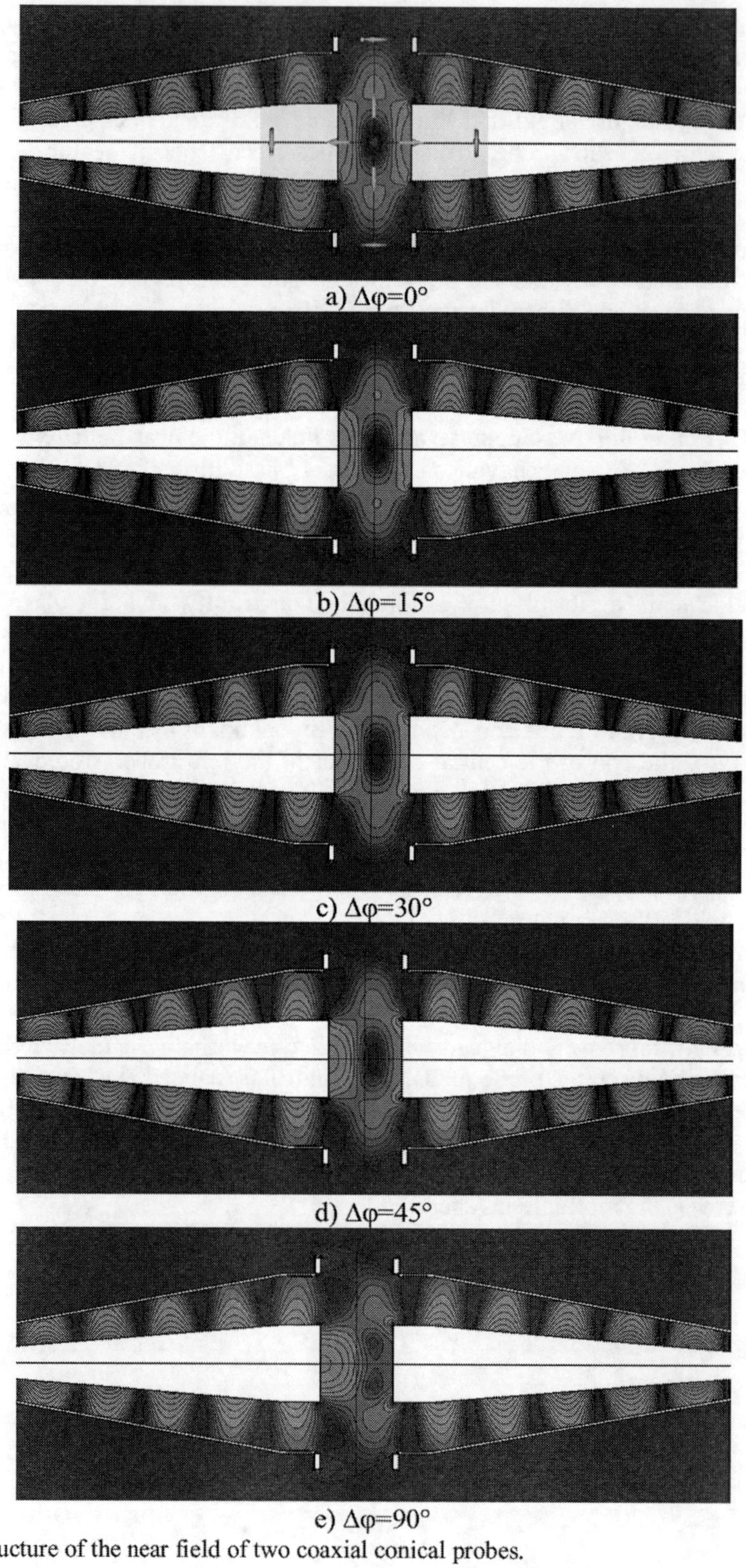

Figure 2. Structure of the near field of two coaxial conical probes.

The results of numerical modeling were used to propose a circuit design for the technology of near-field interference sensing simplified as much as possible. Based on this circuit design, a working breadboard model of the device was developed and tested experimentally to estimate its resolution and sensitivity.

3 Experimental results: analysis and interpretation

The breadboard model of the microwave interference microscope comprised the following main units: a N5230C PNA-L Network Analyzer (Agilent Technologies) with the working frequency band 10 MHz – 40 GHz, two coaxial conical horns (probes), power divider, and additional sections of transmission lines (Figure 3). The network analyzer allowed measurements to be performed with high speed and high accuracy, and the proper calibration provided correction for twelve error components.

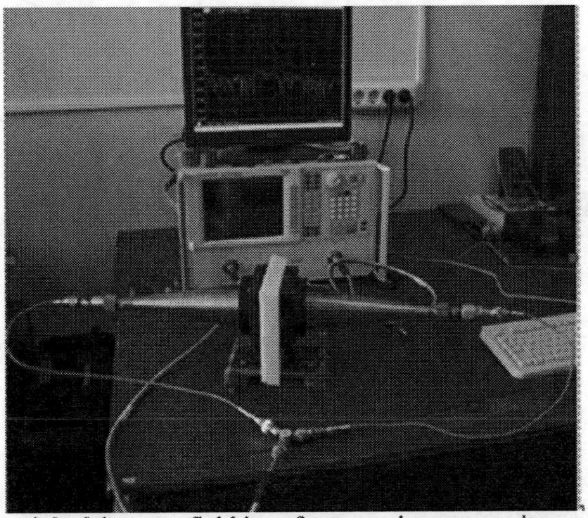

Figure 3. Breadboard model of the near-field interference microwave microscope.

For data registered by the network analyzer, the frequency dependence of the complex reflection coefficient (S11) of one of the probes was chosen. The frequency band during measurements was 500 MHz – 1 GHz. The digitization step was 250 kHz since measurements in the chosen frequency band were performed in 2000 points. For convenience of perception, the initial dependences were normalized by the reflection coefficient of the objects without foreign components or defects. As objects of research, we had chosen a laminate board having defects in the form of through holes of different diameters, diesel fuel with different proportions of kerosene and synthetic engine oil impurities.

The first series of experiments was aimed at elucidation of the sensitivity of the breadboard model of the interference microwave microscope for diagnostics of the material quality in comparison with conventional systems of unilateral sensing. As an example, results of diagnostics of the laminate board having defects in the form of through holes 3, 4, and 10 mm in diameter (curves 1, 2, and 3 in Figure 4, *a*) are shown. To demonstrate the differences between the conventional sensing technology and the bilateral technology of near-field sensing by the overlapping evanescent microwaves, an experiment on diagnostics of defects of the laminate board using evanescent fields of only one probe was carried out (Figure 4, *b*). Attention is drawn to the fact that the presence of a defect (even 10 mm in diameter, see curve 3) is very weakly manifested in the Figure (does not exceed 0.11 dB).

DOI: 10.1051/matecconf/20167901016

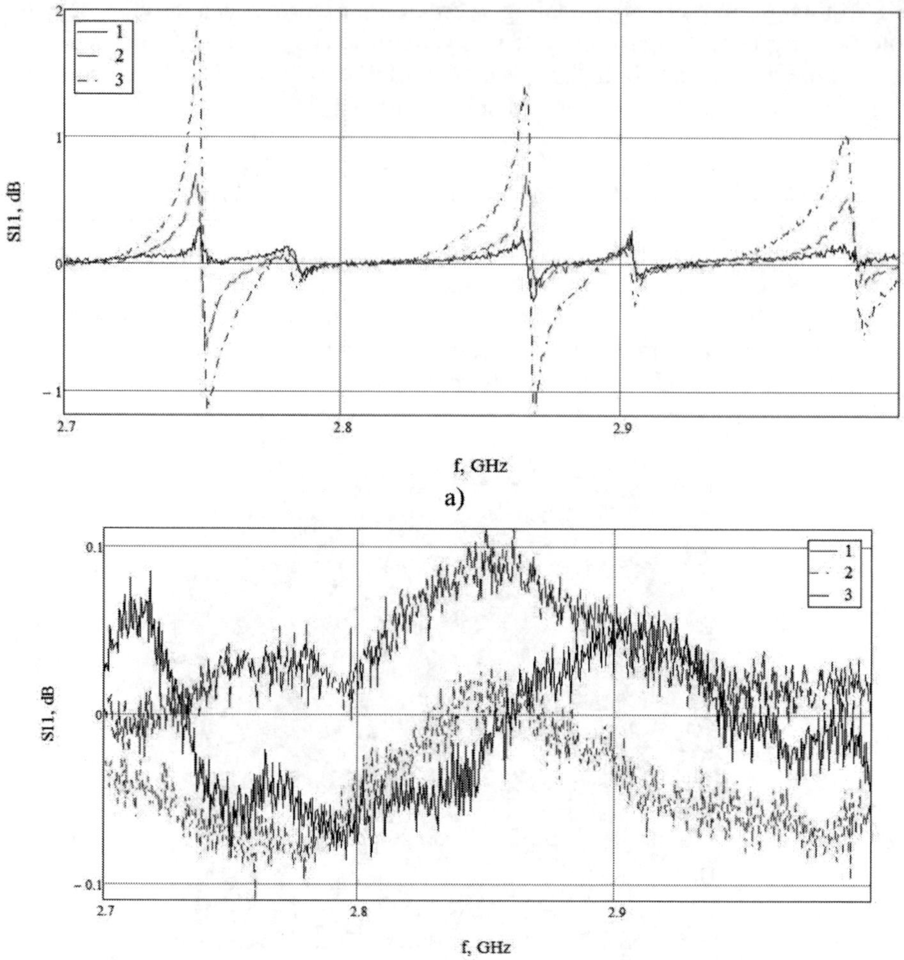

a)

b)

Figure 4. Dependence of the coefficient of reflection from the laminate board 5 mm thick in the frequency band. The hole diameters were 3 (curve 1), 4 (curve 2), and 10 mm (curve 3).

Figure 5 shows the spectrum of the reflection coefficient for mixtures of diesel fuel and foreign impurity in the form of kerosene in proportions 10 and 20 ml of kerosene in 200 ml of diesel fuel (curves 2 and 3). Curve 1 here is for the diesel fuel without impurity. Curve 1 is shown to demonstrate changes arising in the signal after adding the impurity. The presence of a 10 ml of the impurity is manifested sufficiently clearly at a number of frequencies, in particular, at frequencies of 0.77 and 0.89 GHz. The impurity added in double amount is quite reliably fixed not only at the above-indicated frequencies, but also at a lower frequency of 0.65 GHz.

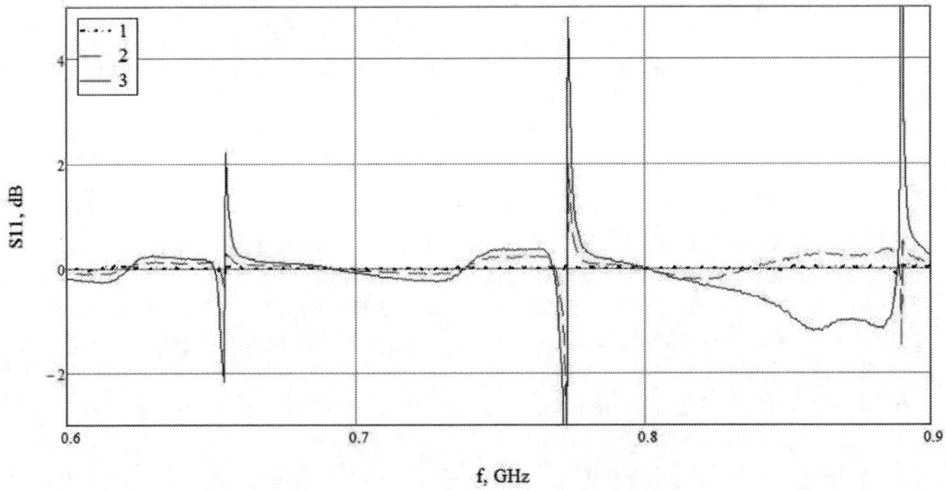

Figure 5. Frequency dependences of the reflection coefficient for mixtures of diesel fuel and kerosene.

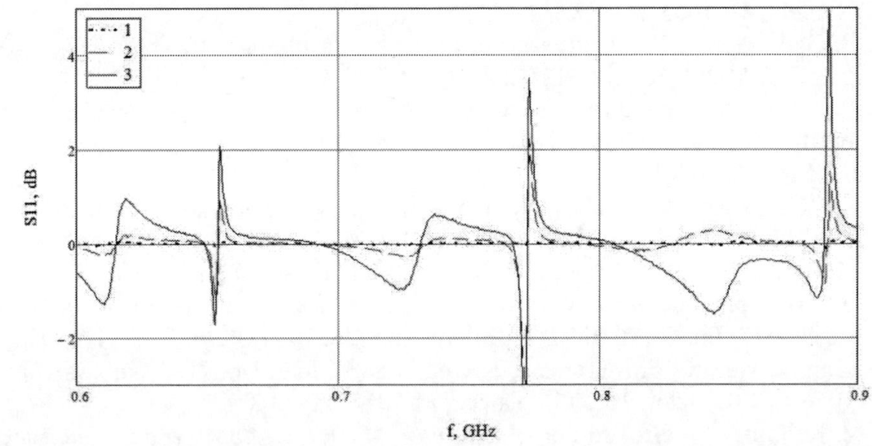

Figure 6. Frequency dependences of the reflection coefficients for mixtures of diesel fuel and oil.

The frequency dependences of the normalized reflection coefficient for mixtures of diesel fuel and synthetic engine oil are shown in Figure 6. Here curves 3 and 6 are for 10 and 20 ml oil and 200 ml diesel fuel, respectively; and curve 1 is for diesel fuel without impurity.

4 Conclusion

The plots shown here demonstrate the sensitivity of the breadboard model of the interference microwave microscope using overlapping counter-propagating evanescent fields of two coaxial conical probes. Numerical modeling performed by us demonstrated that the sensitivity of the microscope with one probe was much worse. This can be explained by lacking of the additional interference energy flux formed in the counter-propagating evanescent fields of two probes [8] and providing deeper penetration into the examined medium as well as by the increased sensitivity in comparison with conventional technologies of measurements [10, 11].

Acknowledgment

This research carried out in 2016 was supported by «Russian Science Foundation» grant (No. 16-19-10272).

References

[1] M. Tabib-Azar, J.L. Katz, S.R. LeClair, IEEE Transactions on Instrumentation and Measurement **48**, 1111 (1999) doi:10.1109/19.816123

[2] D.M. Hagl, D. Popovic, S.C. Hagness, J.H. Booske, M. Okoniewski, IEEE Transactions on Microwave Theory and Techniques **51**, 1194 (2003) doi:10.1109/TMTT.2003.809626

[3] A.N. Reznik, N.V. Yurasova, Technical Physics **51**, 86 (2006) doi: 10.1134/S1063784206010142

[4] C. Gao, B. Hu, I. Takeuchi, K.-S. Chang, X.-D. Xiang, G. Wang, Measurement science and technology **16**(1), 248 (2005) doi: 10-1088/0957-0233/16/1/033

[5] A.O. Oladipo, A. Lucibello, M. Kasper, S. Lavdas, G.M. Sardi, E. Proietti, F. Kienberger, R. Marcelli, N.C. Panoiu, Applied Physics Letters **105**, 133112 (2014) doi: 10.1063/1.4897278

[6] K.P. Gaikovich, P.K. Gaikovich, Inverse Problems **26**, 125013 (2010) doi: 10.1088/0266-5611/26/12/125013

[7] K.P. Gaikovich, Ye.S. Maksimovitch, Journal of radio electronics: electronic scientific journal, **1** (2015) [URL: http://jre.cplire.ru/jre/jan15/2/text.pdf (reference date of 14.06.2016)]

[8] V.P. Belichenko, A.S. Zapasnoy, P.V. Shestakov, Optics and Spectroscopy **118**, 605 (2015) doi: 10.1134/S0030400X15040037

[9] S.M. Anlage, F.C. Wellstood, K. Vlahacos, D.E. Steinhauer, *Near-field scanning microwave microscope having a transmission line with an open end* Patent US No. 5900618, Int. Cl.^{6} G01J 1/20, Filed: Aug. 26, 1997, Publ.: May 4, 1999

[10] L. Cui, N. Marturi, E. Marchand, S. Dembélé, N. Piat, MATEC Web of Conferences **32**, 05003 (2015) doi: 10.1051/matecconf/20153205003

[11] S.V. Kalinin, A. Gruverman, *Scanning Probe Microscopy: Electrical and Electromechanical Phenomena at the Nanoscale* (Volume 1), (New York: Springer, 2007)

PROGRAMMABLE CONTROL OF THE PULSE REPETITION RATE IN THE MULTIWAVE STRONTIUM VAPOR LASER SYSTEM

Anatoly Soldatov[1a], *Anna* Vasilieva[1], *Ivan* Kostadinov[2], *Yana* Loeva, *Yury* Polunin[1], *Nikola* Sabotinov[3], *Alexey* Schumeiko[1]

[1]Tomsk State University, 634050, Tomsk, Russia
[2]Pulslight Ltd., Sofia, Bulgaria
[3]Bulgarian Academy of Sciences, Sofia, Bulgaria

Abstract. The aim of the present work was the development of laser systems for ablation of biological tissues with a programmable control over the lasing pulse repetition rate in a wide range. A two-stage laser system consisting of a master oscillator and a power amplifier based on strontium vapor laser has been developed. The operation of the laser system in a single-pulse mode operation, multipulse mode operation, and with a pulse repetition rate up to 20 kHz has been technically implemented. The possibility of a bone tissue ablation with no visible thermal damage is shown.

1 Introduction

Currently, the use of lasers lies at the heart of many technologies. The unique properties of laser radiation allow achieving high precision of material processing and small sizes of the treated area. This is relevant for the use of lasers in medicine as well. The use of lasers for ablation allows performing tissue removal without damaging surrounding areas. That is, when a laser radiation is absorbed by a biological tissue there is a conversion of the photon energy into the vibrational energy of molecules leading to their destruction, which are the basis for the subsequent photothermal ablation. The process begins immediately after reaching the required radiation power density.

At any interaction of laser radiation with a matter, an important role is played by properties of the substance itself, as well as by properties of laser radiation and the environment. Plenty of works are devoted to the study of parameters of laser radiation on the material [1–4]. However there are no separate studies on the effect of the pulse repetition rate on the ablation process and, in some cases, this parameter is a result of preselected durations and energies of the laser pulse.

It shall be noted that pulse repetition rate determines, to a large extent, the collateral thermal effect, or in other words the thermal damage zone (TDZ) of the target. When only biological tissues are considered as a target, the value of the interpulse interval is of greater

[a] Corresponding author: general@tic.tsu.ru

© The Authors, published by EDP Sciences. This is an open access article distributed under the terms of the Creative Commons Attribution License 4.0 (http://creativecommons.org/licenses/by/4.0/).

importance than for any other material, because during the action of a laser pulse the accumulated heat in tissues must dissipate during the inerpulse interval. Otherwise, the accumulation of heat in the areas adjacent to the radiation zone will lead to tissue coagulation followed by carbonization, i.e. the tissue loses its vitality. Therefore, for a more effective impact on the biological target to produce the ablation it is necessary to vary the characteristics both of the oscillator pulse and the repetition rate of these pulses in a fairly wide range, selecting optimal conditions for each specific material. As studies show [5], strontium vapor laser (SVL) is a promising laser source for medicine or, for example, for processing of polymers, because SVL radiation enters the absorption bands of polymers and biological tissues. The advantages of using strontium vapor laser for ablation of biological tissues: resonance wavelength, nanosecond pulse duration, minimum TDZ.

2 Experimental setup

A laser system consisting of an oscillator and an amplifier based on active elements of a strontium vapor laser (Figure 1), providing the ablation mode of hard biological tissues, has been developed. Specifications are given in Table 1.

Table 1. Specifications of experimental setup.

Parameter	Value
Wavelength, μm	1.03; 1.09; 2.60; 2.69; 2.73; 2.92, 3.01 and 6.45
Spreading	0.5
Pulse duration:	
6.45 um	20
~ 3 um	150
~ 1 um	80
Maximum pulse energy, mJ,	1.7
Beam diameter, mm	35
Average power, W	1…28
Pulse repetition rate, kHz	15…20
Cooling	water

Laser active elements on strontium vapors were used as the master oscillator (1) and the power amplifier (7). The master oscillator was a cylindrical gas discharge tube (GDT) with a diameter and a length of the operating channel 26 and 800 mm, respectively. The telescopic unstable resonator (with M = 35) consisted of two spherical mirrors, a concave (2) and a convex (3) with curvature radii of 3500 mm and 100 mm, respectively. The spatial filter collimator (with M = 1.5) included two concave spherical mirrors (4,5) with an aluminum coating with radii of 3000 mm and 2000 mm, and a spatial filter (6) with a hole diameter of 0.3 mm, which has allowed to allocate from the oscillator radiation a spatial component with a spreading close to the diffraction limit (~ 0.5 mrad).

GDT with a diameter of 35 mm and a length of 1000 mm was used as the amplifier. Windows of both GDTs were made of CaF_2, an optical material with low absorption in the mid IR-range and good performance characteristics. Heating and excitation of active elements was carried out using pulse power supplies (8,9), in which thyratrons TGI1-1000/25 were used as a switchboard. The control and synchronization unit (10) consisted of the master oscillator (10a), the delay module (10b), and allowed to adjust the relative delay time of oscillator and amplifier pumping pulses.

 DOI: 10.1051/matecconf/20167901017

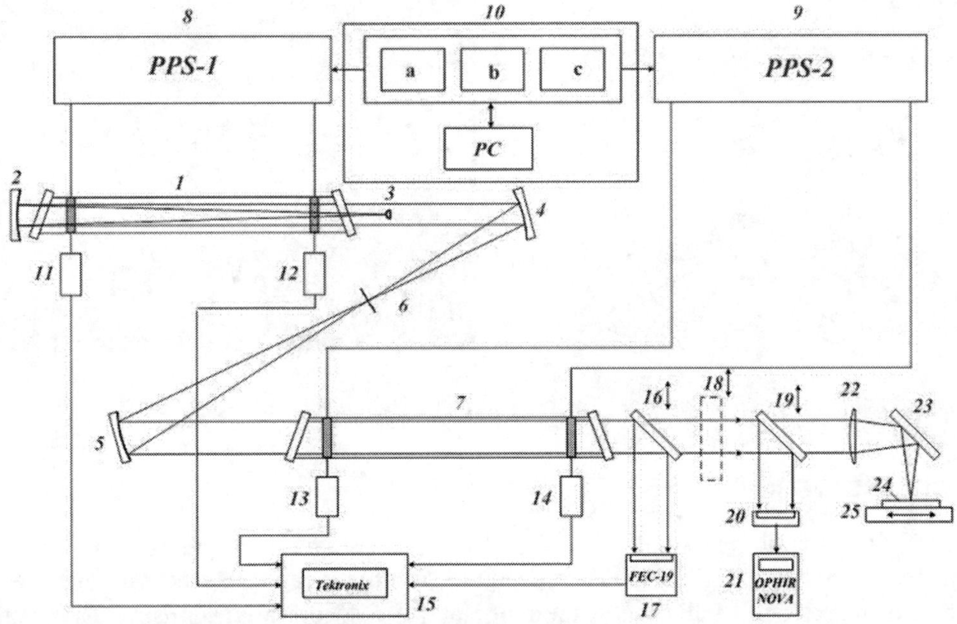

Figure 1. The schematic of the experimental setup:
1 – GDT-1; *2* – rear mirror; *3* – output mirror; *4* – input mirror; *5* – output mirror; *6* – spatial filter; *7* – GDT-2; *8, 9* – high-voltage power supplies; *10* – control and synchronization unit (CSU); *11, 13* – voltage dividers; *12, 14* – low-resistance shunts; *15* – multi-channel oscilloscope; *16, 19, 23* – rotary mirrors; *17* – FEC-19; *18* – unit of spectral filters; *20, 21* – power meter; *22* – spherical lens; *24* – target; *25* –coordinate table.

3 Discussion and results

A feature of the operation of active elements of metal vapor lasers is their operation in the self-heating mode [6, 7]. Heating of the active medium and its excitation takes place during the excitation pulse. Thus, pulse repetition rate cannot be reduced unlimitedly, because it is necessary to maintain a certain concentration of metal vapors. On the other hand, PRR cannot be increased indefinitely, since it leads to overheating of the active medium. Therefore, pulse-periodic discharges with the excitation pulse repetition rate in the range of 15…20 kHz were formed in active elements of the oscillator-amplifier, maintaining the necessary concentration of strontium vapors. Setting on a PC the time shift of oscillator and amplifier launch pulses included the cutoff of the laser radiation for a predetermined period of time, the multipulse (tandem) oscillation mode and the single-pulse mode were set as well. Thus, the possibility to set the lasing pulse repetition rate under the action of radiation on the target while keeping the pulse energy unchanged has been implemented.

Figure 2 shows examples of ablated areas of hard biotissues made with the use of the developed setup.

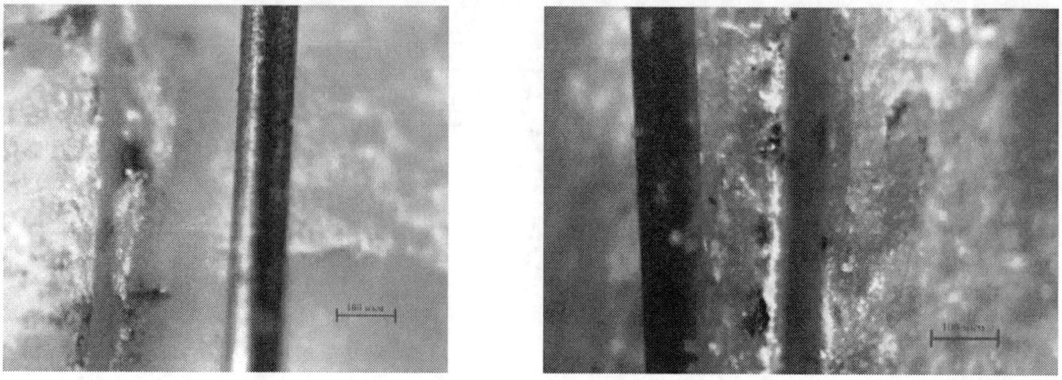

Figure 2. Photo of the ablated area of a pig cortical bone, 16 mm/sec., 1.5 mJ, 1 pass.

4 Conclusion

Currently, the search for small-sized laser sources operating in the mid-IR range for medicine is carried out. From this viewpoint, strontium vapor lasers with nanosecond duration of pulses and oscillating in the range of 1 – 6.45 μm are promising laser sources for laser ablation.

A two-stage laser system consisting of a master oscillator and a power amplifier based on strontium vapor laser has been developed. The control range of the lasing pulse repetition rate of the setup, determined by a PC user, is from units of hertz up to tens of kilohertz. The possibility of a bone tissue ablation without visible thermal damage has been demonstrated using the given system.

Acknowledgment

The paper was written as part of the research project No. 8.2.32.2015, carried out with the support of the Program "Research Foundation of Tomsk State University named after D.I. Mendeleev" in 2015 – 2016.

References

[1] M.R. Johnson, P.J. Codd, W.M. Hill, T. Boettcher, Lasers in Surgery and Medicine **47**, 839 (2015) doi: 10.1002/lsm.22424

[2] X. Zhang, Z. Zhan, H. Liu, H. Zhao, S. Xie , Q. Ye, Journal of Biomedical Optics **17**, 038003 (2012) doi: 10.1117/1.JBO.17.3.038003

[3] T. Bilici, S. Mutlu, H. Kalaycioglu, A. Kurt, A. Sennaroglu, M. Gulsoy, Lasers in Medical Science **26**, 699 (2011) doi: 10.1007/s10103-011-0915-0

[4] C.S. Nathala, A. Daskalova, I. Bliznakova, S. Lueftenegger, A. Zhelyazkova, S. Enikoe, T. Ganz , W. Husinsky, MATEC Web of Conferences **8**, 02005 (2013) doi: 10.1051/matecconf/20130803009

[5] A.N. Soldatov, Proceedings of SPIE – The International Society for Optical Engineering **7751**, 77510C (2010) doi: 10.1117/12.881003

[6] Yu.V. Gulyaev, M.A. Kazaryan, N.A. Lyabin, Yu.M. Mokrushin, O.V. Shakin, A.G. Tamanyan, Lasers in Engineering **15**, 293 (2005)

[7] K.A. Temelkov, S.I. Slaveeva, N.K.Vuchkov, Proceedings of SPIE - The International Society for Optical Engineering **8770**, 87701L (2013) doi:10.1117/12.2012663

MATEC Web of Conferences **79**, 01018 (2016) DOI: 10.1051/matecconf/20167901018
IME&T 2016

DEVELOPMENT OF SOCIAL AND ENTREPRENEURIAL SKILLS OF STUDENTS OF ENGINEERING AND TECHNICAL SPECIALTIES IN THE MODERN UNIVERSITY

Evgeniya Liventsova, *Tat'yana* Rumyantseva[a], *Ekaterina* Syriamkina

National Research Tomsk State University, 634050, Tomsk, Russia

Abstract. The article analyses experience of the National Research Tomsk State University on development of social and entrepreneurial skills of students of engineering and technical specialties. The experience includes description of activity of the Park of social and humanitarian technologies, which is a TSU's structural unit. The article is based on analysis of global trends of university development, analysis of documents and practical experience in developing students' social and entrepreneurial skills and includes project- research experiment. The authors conclude that in current conditions training of students of engineering and technical specialties requires modern dealing with knowledge, focus on the synergetic development of the personality, integration of information with innovative methods of education and various forms of social and cultural activities.

1 Importance of students' social and entrepreneurial competencies for innovative sphere

Decree of the Government of the Russian Federation of March 16, 2013 № 211 «On measures of state support of the leading Russian universities in order to enhance their competitiveness among the world's leading research and education centers» and other strategic documents of the Russian Federation in the sphere of education and science, as well as competitiveness enhancement programs of the leading Russian universities provide for creation of favorable environment and conditions for the development of priority international areas of basic and applied research. The Strategy of development of information technology industry in the Russian Federation for 2014 – 2020 years and up to 2025 (the order of the Government of the Russian Federation of November 1, 2013 № 2036-r) emphasizes that institutions of professional and higher education should focus on development of business and entrepreneurial skills among students of engineering specialties. The Atlas of new professions – 2020 (the project, issued by Moscow School of Management «Skolkovo» and the Agency for Strategic Initiatives) stresses that «the modern world is changing so quickly that we can no longer afford to study theoretical disciplines for five years, and then take some time to master the profession for account of

[a] Corresponding author: rtb98@mail.ru

© The Authors, published by EDP Sciences. This is an open access article distributed under the terms of the Creative Commons Attribution License 4.0 (http://creativecommons.org/licenses/by/4.0/).

an employer. Therefore, education, especially for university students, becomes more practical-oriented. This means that the emphasis is shifting from the theory to real students' projects, including their start-ups».

National Research Tomsk State University (TSU) is among the leading universities of Russia, whose competitiveness enhancement program received the highest assessment of the Council on improving the competitiveness of the leading universities of the Russian Federation. According to the plan of measures on realization of competitiveness enhancement program for 2013–2020, TSU has identified six key strategic areas of development. One of them is strategic direction for implementation of models and practices of successful international research universities.

The main activities under this direction include the following:

— creation and development of TSU centers of excellence on the basis of networking and interdisciplinary studies to provide international leadership in the field of research of quality of human life and society;
— development of network of engineering centers and collective use centers;
— organization of events on the federal and regional levels in the field of innovation, technology and social entrepreneurship;
— formation of entrepreneurial spirit and intensification of innovative activity among young people.

Development of youth entrepreneurship is also one of the priorities of the state program «Development of Entrepreneurship in Tomsk region». The program provides for allocation of significant budget funds to support and develop youth entrepreneurship and includes 14 measures aimed at involving young people in business. Activities include development of skills and knowledge in the field of entrepreneurship, enhancing of entrepreneurial competences, formation of a positive image of entrepreneurship among young people, testing and motivation to start entrepreneurial activity, creation of new business entities, access to finance, information and consulting support, development and support of youth entrepreneurship associations, and other activities.

Serious attention of public authorities to the development of youth entrepreneurship also reflects the results of studies of recent years, which indicate emerging difficulties in attracting staff for the small business sector. According to Tomsk oblast Administration Decree "On Approval of the State Program "Development of Entrepreneurship in Tomsk Oblast" of December 12, 2014 № 492a, these difficulties are associated with changes in the preferences of the most part of young people. Not only in Tomsk region, but also in Russia as a whole, young people tend to build future career with a view to employment in the public service or public corporation. This is because of apparent stability of this sector of the Russian economy, and, at the same time because of the nature of business risks.

A similar trend is also confirmed by the results obtained during the study «Investment Activity of University Students of Tomsk», conducted by specialists of the departments of sociology and social work of the Faculty of Philosophy of Tomsk State University in 2013 (Figures 1 and 2).

According to the study, the proportion of students who already had their own business was low (2%); only one of ten students (12%) were going to start their own business in the near future. Less than a tenth part of students pointed their unwillingness to develop own business (9%). The vast majority did not plan to develop own business, but considered this opportunity in the future (68%). Other students realized that they would be employees in the public or private institutions. Among those who were planning their own business, the majority (60%) focused on traditional sectors of economy – the sphere of trade, services, etc. The rest were ready to consider business opportunities in innovative sphere (40%) [1].

DOI: 10.1051/matecconf/20167901018

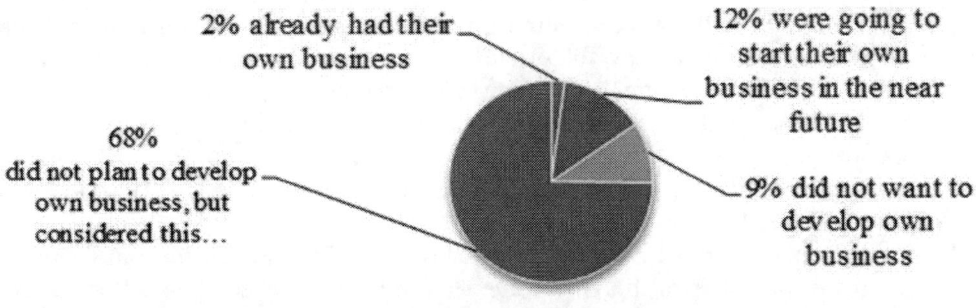

Figure 1. Attitude to own business of students of Tomsk universities.

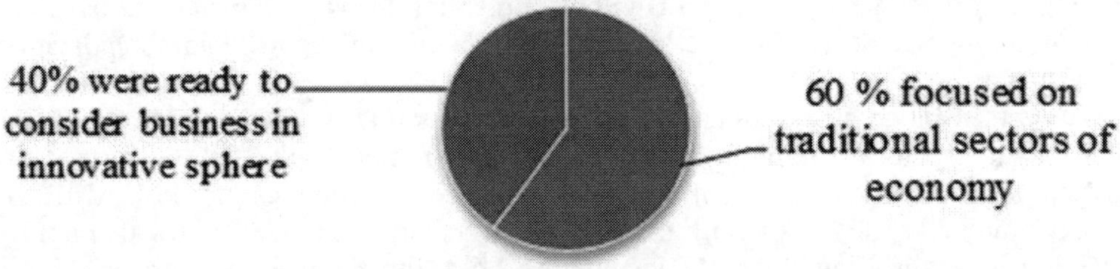

Figure 2. Preference to own business direction of students of Tomsk universities.

Thus, the local community, and especially the universities, face the problem of propaganda of self-employment and a positive image of entrepreneurship among the youth, as well as formation of social, professional skills of students of engineering and technical specialties and provision of information and consulting support.

A system of business support infrastructure is formed in the Russian Federation in order to prepare young people for entrepreneurship. Under the RF Federal Law «On Development of Small and Medium-Size Business in the Russian Federation» of July 24, 2007, hundreds of business incubators, dozens of venture funds, special programs for small and medium-sized enterprises operate all over the country at the federal, regional and municipal levels. The characteristic feature of the Russian infrastructure of support of small and medium-sized enterprises is focus on support of enterprises in scientific, technical and investment spheres, «a shift towards innovative small businesses». Analysis of the structure of the organizations forming the infrastructure of business support in the regions also confirms this fact. For example, in Tomsk region there are more than fifty infrastructure elements, including special economic zone, engineering center, cluster development center, nanotechnology center, center for stimulating demand for innovative products, student and technological business incubators, offices for commercialization development at universities and academic institutions, youth innovative creativity centers, technology transfer centers, Tomsk venture fund, consortium of Tomsk universities and research organizations, a network of venture financing, consulting companies and other organizations. This fact is proved by Tomsk oblast Administration Decree "On Approval of the State Program «Development of Innovative Activity and Science in Tomsk oblast» of October 30, 2014 № 414a.

Some experts believe that the current Russian system of business support does not correspond to sectorial structure of the national economy and social sphere of the country. Experience of some foreign countries on formation of business support policy, Great Britain, in particular, shows that policy aimed mainly at supporting specific industries that could positively affect the development of the region in the future, over time changes in the direction of creating favorable conditions for business development on the whole [2].

Report on the main results of the system of higher and professional education in Tomsk region in 2014 shows the tendency for orientation towards innovative small business in institutions of higher education, including universities. They create student and technological business incubators, commercialization offices, technology transfer centers, small companies form an innovative university belt. The total number of innovative companies of National Research Tomsk State University was 51 in 2014, including 14 companies created in 2014.

Preparing students to innovative and entrepreneurial activity in science and technology is also in the list of priorities of the Government of the Russian Federation. For nearly 10 years the Foundation for assistance to small innovative enterprises in scientific and technological sphere has implemented «UMNIK» program aimed at promoting mass participation of youth in science, technology and innovation. The program provides support to young scientists seeking to realize their potential through innovation, as well as stimulates young scientists to create small innovative enterprises for commercialization of research results.

However, in modern conditions, it becomes obvious that the higher education training system in addition to providing traditional knowledge in a particular field of science needs to prepare specialists with broad outlook, more advanced communication skills, willingness for cooperation and life-long training. Many world leading countries follow this approach to training in engineering. In particular, in the USA the number of basic principles of engineering education includes the need to provide strong foundations of knowledge in mathematics, physics and engineering sciences; the importance of experimental design and laboratory work; the need to develop social skills of engineers; the need to include social, economic and human sciences in the curriculum; the need to prepare students for lifelong education.

TSU professors V. Klochko and E. Galazhinskiy have studied various empirical indicators and personal criteria related to the processes of motivation to innovate activity by exploring persons who exhibited inclination to entrepreneurship and innovation or achieved significant results in this sphere. The researchers tried to determine personal qualities, potentials, resources necessary for a person to be capable to carry out innovative and entrepreneurial function. According to the results of the study, innovative potential includes the following:

1. Personal qualities such as tolerance to uncertainty, ability to take justified risk, responsibility, need for self-realization, motivation to achievements, reflexivity, creativity (the quality of intelligence, intellectual initiative).
2. Competence, including project-design, communicative and information competence.
3. Vitality (special features of value-semantic organization of the life-world, viability, sovereignty, ability to work, mobilization potential, level of self-control, human orientation towards a certain quality of life) [3].

2 Role of the modern university in development of students' social and entrepreneurial competence

Today, institutions of higher education should focus on the global and national trends in order to increase their competitiveness.

1. Modern society increasingly requires active participation of universities in economic and social development. The process of education system reforming must consider this trend. Member countries and partner countries of the organization for economic cooperation and Development (OECD) consider the third role of universities as one of the key indicators of higher education development.

Program of TSU strategic development as a national research university for 2010–2019 highlights socio-humanitarian knowledge and technologies in modernization of economy and social sphere as one of the main direction. Moreover, among provincial universities with the status of national research university, only TSU has announced humanitarian studies as one of the priority research areas.

2. The document prepared based on results of the «University and Community» project conducted by «New Eurasia» Foundation, describes the conceptual model of socio-entrepreneurial university and highlights its main characteristics. The idea is that an university earns not only on traditional educational programs for the traditional target group, but also initiates and implements innovative social and business projects together with representatives of the local community [4]. First, the university should be open to the community. Openness here means both involvement of representatives of non-academic environment in the educational process and inclusion of direct experience of work in community in the educational process. For example, during a training course a student gains practical experience in a business company under the supervision of a tutor. Such programs are common in the US and Europe and are highly popular among teachers and students. They enable to find and propose solutions to various problems of the local community during the educational process.
3. Influence of the university environment to innovative activity is reflected in the theory of the «Triple Helix», introduced in England and the Netherlands by Henry Etzkowitz, professor of Newcastle University, and Loyetom Leydesdorf, professor of Amsterdam University, at the beginning of the XXI century. The triple helix symbolizes alliance between the government, business and universities, which are key elements of the innovation system of any country. This theory shows their interaction at every stage of creation of an innovative product. Government and university interact on the initial stage of knowledge generation; university works with business during the process of technology transfer; government together with business bring the result to the market. The role of the university in the triple helix model is undeniable.

Based on the great experience of training specialists in social and humanitarian sphere and taking into account global trends, Tomsk State University started to pay more attention to development of social skills, including students of engineering and technical specialties. This means business activity in the social sphere, which leads to the development of interpersonal skills, system thinking, formation of an active lifestyle and professional positions, ability for effective team communication, learning of the basic principles of behavior on the labor market.

3 Development of social and entrepreneurial competence of students of engineering and technical specialties

Currently, the basic list of subjects in university curriculum does not include learning the basics of social, project activities, and entrepreneurial culture. Meanwhile, any person with appropriate inclinations can be an entrepreneur. Therefore, this activity is a subject to strong influence of accidents, a variety of different sources of replenishment, is joined with a number of other activities and involves intensive exchange of experience, skills, methods, knowledge and tools. All this contributes to the viability of businesses. In the business environment, there is a place for development of personal creativity, strong-willed, intelligent and energetic properties.

A system of education, especially professional, influence greatly the desire to become an entrepreneur (in any field). Results of a study conducted in Canada showed that people with low level of education tended to have a low income and did not seek to create their own businesses, while highly educated people tended to open their businesses, both for getting sufficient income, and for achievement of self-realization and independence. People with average level of education, by contrast, did not consider their own business as best option, and preferred to be hired employees. The study also found that university education increased people's desire for self-employment by 8.6% on the average, including men – by 12.9%, women – by 6.1%. In general, the results obtained by researchers established a clear relationship between self-employment and a person's level of education: lower education significantly reduced the likelihood of self-employment [5].

Higher education institutions face a number of problems in order to generate a new wave of entrepreneurs. First, it is recognition of persons capable of business activities in order to provide further assistance to them in improving the appropriate inclinations. Because talent or ability to entrepreneurship can be not found, and will just disappear. The solution to this problem requires for thoughtful individual work with students. Second, establishment of creative atmosphere in educational institutions, in order to provide conditions for realization of manufacturing independence, initiative, activity; striving for identification of problems that can be of great practical importance and a subject for students' serious inspiration. Third, a broad demonstration of various forms and types of businesses, lives and practices of successful entrepreneurs. It is important to acquaint students with work and experience of entrepreneurs in various spheres. Finally, fourth, development of education through various forms of students' universal social competences, which allow being competitive in any field.

Humanitarian sciences play a major role in formation of social competence of students of technical specialties. When acquiring knowledge in humanitarian sciences along with technical science, a student gets world outlook and values and becomes a subject of his own life. According to British psychologist J. Raven, involvement in a certain activity is critical for competence development; this means that it is important to have a purpose, for which a person will act, the intensity of individual motivation determines the level of competence manifestation [6]. Thus, the development of any activity must be based on strong ideas about its social and subjective significance, usefulness of the results. The entrepreneurial component becomes an integral part of any professional activity and provides prestige of any profession, as it allows people to adapt it to the conditions of today's market. The desire of young professionals, including engineers, to create and implement their own business projects, is an indicator of their social and economic competence, manifestation of an active life position, readiness to solve the urgent problems facing society, or simply to create a product necessary to the consumer.

Realizing that the concepts of «innovation potential» and «entrepreneurial resource» are largely connected with the individual's creativity, TSU has created a structural unit - Park of social and humanitarian technologies (hereinafter - the Park) aimed to implement the above-mentioned technologies of work with students both of social and humanitarian sciences and engineering and technical sciences. Another aim of the Park's activity is to involve students in social and entrepreneurial activity, as well as to attract professional experience of members of the local community (including business) to the university.

The Park's goal is to organize environment for innovative development of university students in order to enhance social and humanitarian technologies, social entrepreneurship. To achieve this goal, the Park tries to solve such problems as formation and development of conditions for commercialization of innovative projects based on the results of social and humanitarian research; involvement of students and employees of TSU to the process of creation, development and implementation of innovative humanitarian projects; involvement of TSU students and employees in entrepreneurial activities; coordination of interaction between society, government and business organizations in development and implementation of social and humanitarian technologies. During 2012–2015, the Park's staff has implemented the strategy of development of entrepreneurial competencies of students of various specialties. Now, the Park successfully implements technologies of generating innovations in the social sphere.

The Park builds its work with students in three structurally and substantively related areas:

1) «From an idea to a project» (identifying student's creative ideas and raising them to the project activities).

2) «From a project to a technology» (expert evaluation of student's projects and developing them in social technologies).

3) «From technology to commercialization» (launch of technologies to the local community market, development of social entrepreneurship among university students).

All three stages contribute to the development of students' social and entrepreneurial skills. In this logic, the first direction is very important, in which initiation and incubation of students' ideas occur.

Among examples of the Park's initiation of students' project ideas is an annual internet competition of creative ideas «Student's Tomsk», purpose of which is search for student's creative ideas (including technical and IT – oriented ideas) relevant to solution of social problems Tomsk.

A social tutor (usually, the term «social tutor» is used in connection with socialization of graduates of boarding schools. The Park uses this term in connection with tutoring of social initiatives, in the meaning of «socialization» and promotion of students' ideas and projects) starts to work with winners of the competition. The first thing is joint drafting of a road map for the development of the project idea with designation of educational, competence and resource deficiency in the idea promotion. This allows the idea to become a full and long-term project.

At the stage of «From design to technology», experts support sustainable projects and analyze the project in terms of its possible technologizing. Representatives of relevant agencies and local community work as experts and think over the possible scaling and replication of technologies, as well as over their launching in social or innovative business mode.

The Park's direction «From technology to commercialization» is a priority. The Park organizes annual social investment contest aimed at investors' selecting and funding of social innovative technologies and services of student organizations-winners. The contest organizers encourage any form of interaction suitable for entrepreneurs: tutoring of a student association by a business support teacher, individual investment in activities of socially oriented student associations. Thus, long-term mechanism for promotion of university youth's innovative social services on the regional market is working.

Competitive mechanisms contribute to the development of students' entrepreneurial competences and social investment, even in a non-traditional form, enables to hold the students' interest for the further development and implementation of their projects. There are several reasons for this. First, the project team receives professional evaluation from experts. Second, these competitions are rather directed on the developing students' business and communication skills (the ability to negotiate with a sponsor or an expert on joint activity) than involve financial rewards. Third, the success of competitions largely depends on the training component of the student teams. An obligatory condition for students' participation in competitions is taking additional educational programs aimed to develop social and entrepreneurial skills of students of different specialties.

During 2012–2015, in cooperation with experts from local community the Park has realized a new educational program «Social accelerator» aimed to get acquainted TSU students with the phenomenon of social entrepreneurship. Educational subjects presented in the program are unique to all university specialties. This program is created for youth teams who have elaborated the idea of social startup as well as for youth teams who are implementing a social startup. Representatives of social business and socially responsible companies, who are ready to share their successful experience, representatives of government and social-educational institutions work as experts and mentors within the educational program.

The Park's specialists and experts have developed entrepreneurial campus training course «Three Steps to Entrepreneurship» for students who do not yet participate in the Park's activities. The educational program of the course consists of three main steps: «immersion», «learning», «testing». This course is universal. It has been successfully adapted to different areas of students training in TSU.

Thus, today the Park successfully implements active educational technologies for development social and entrepreneurial competencies of students of different specialties. Table 1 shows the summary data.

Table 1. Formation of social and entrepreneurial skills of students of engineering and technical areas inside and outside the educational institution (university).

Educational forms (workshops, training courses, accelerators)	**Competitive mechanisms** (Social, business and research competitions in the university and outside)	**Project technologies** (Initiation and implementation of own social and business projects in volunteer centers, youth public and scientific associations, in the local community)	**Forms of support** (tutoring, expert evaluation, mentoring)
Contribute to successful development of social competence of future engineers, as the nature and structure of students' cognitive activity fundamentally changes during the learning process; it leads to development of personal creative potential	Stimulate the emergence of students entrepreneurial skills, develop such qualities as risk tolerance, self-organization, consistency, communication skills, and other; enables to undertake entrepreneurial testing	Enable to expand students' skills for project development, skills for teamwork and other important skills; enable to support and stimulate active students' interest to generation of new ideas and bringing them to the project activity stage	Allow students to pass through all stages of social and business design more successfully

4 Conclusion

Thus, along with development of students' professional competence in the specific field of activity we draw attention to importance of development of personal (initiative, achievement drive, etc.) and social competence (ability to work in a team, to communicate effectively with the target groups, etc.). Practice-oriented training contributes to development of social and entrepreneurial skills of students of engineering and technical areas of training in a modern university. In turn, currently developing mechanisms such as social investment and strategic partnership of business community with universities provide benefits to each participant. Students get the possibility to carry out professional tests during the basic education; the university get possibility to apply innovative approaches in education; business gets human resources and valuable information in the form of non-standard student ideas and solutions.

In conditions of transition of the global economy to the sixth technological cycle, promotion and commercialization of technologies require technicians and managers «of a new breed»; therefore, there is need to redefine teaching methods. High education is becoming one of the leading areas of activity in the sixth technological cycle, in which the value of a complexly organized person increases significantly. Therefore, personnel training should pay special attention to «creative practice». Motivation and interests of students are the key and most expensive resources for education efficiency. Renewal of education requires modern dealing with knowledge, focus on the synergetic development of the personality, integration of information with innovative methods of education and various forms of social and cultural activities [7]. The totality of the abovementioned forms of educational technology allows motivating students to obtain entrepreneurial skills. In this case, it is important for a student to determine the result of his efforts as possible realization in practice of important to society social and entrepreneurial skills that he has obtained at the university. In addition, this approach (use of aggregate educational technology) can overcome the gap between the theoretical training of students within the framework of the basic educational programs of the university and the practical experience, since any activity loses sense if it does not meet the needs of a modern society.

Acknowledgements

The paper was written as part of the research project No. 8.2.31.2015, carried out with the support of the Program "Research Foundation of Tomsk State University named after D.I. Mendeleev" in 2015 – 2016, grant RFBR No. 16-29-12858.

References

[1] V. Mikhailov, Expert-online Journal (2013) URL: http://expert.ru/siberia/2013/33/investor-na-shest-let/ (Assessed 01.02.2016).
[2] N. Williamsand, T. Vorley, Entrepreneurship and Regional Development **26**, 257 (2014) doi: 10.1080/08985626.2014.894129
[3] V.E. Klochko, E.V Galazhinsky, Bulletin of the Tomsk State University (325), 146 (2009)
[4] J. Boshee, Nonprofit World **19**, 15 (2001)
[5] O. Masakure, Education Economics **23**, 693 (2015) doi: 10.1080/09645292.2014.891003
[6] Dzh. Raven, *Competence in modern society* (Kogito Center, Moscow, 2002)
[7] V.I. Syryamkin, E.G. Syryamkina, Proceedings of the International Conference on Research Paradigms Transformation in Social Sciences 2014 (RPTSS-2014) **166**, 468 (2015) doi: 10.1016/j.sbspro.2014.12.556

MATEC Web of Conferences **79**, 01019 (2016) DOI: 10.1051/matecconf/20167901019
IME&T 2016

OPTIMIZATION OF THE STABILIZATION SYSTEM FOR ELECTROMAGNETIC SUSPENSION IN ACTIVE VIBRATION ISOLATION DEVICES

Inna Plotnikova[1,*], *Maxim* Vedyashkin[1], *Raisa* Mustafina[2], *Igor* Plotnikov[3], *Olga* Tchaikovskaya[4], *Josefa* Bastida[5], *Mikhail* Verpeta[1]

[1]Tomsk Polytechnic University, Institute of Non-Destructive Testing, 634050, Tomsk, Russia
[2]S.Toraighyrov Pavlodar State University, Department Electrical Engineering and Automation, 140000, Pavlodar, Kazakhstan
[3]Tomsk Polytechnic University, Institute of Power Engineering, 634050, Tomsk, Russia
[4]Tomsk State University, Faculty of Physics, 634050, Tomsk, Russia
[5]Universidad de Murcia, Department of Chemical Engineering, 30001, Murcia, Spain

Abstract. The research aims to study the vibration isolation device using the elastic forces of the electromagnetic suspension. The optimum gain values of the signal feedback circuit, which minimize the impact of noise introduced by the elements of the active vibration isolation device, are determined. The possibility of developing the vibration isolation device with the control circuit coefficients found on the boundary of the stability region is considered. The paper provides the amplitude-frequency characteristics of a single-mass electromagnetic suspension.

1 Introduction

The consideration of additional loads (static, vibration, seismic), which can lead to structural instability [1, 2] or even destruction [3-5], is relevant under the elevated static, mechanical and thermal loads. Nowadays, electromagnetic suspension systems are widely used in various fields of technology [6, 7].

It should be noted that both Russian [8, 9] and foreign researchers [10–12] have succeeded in stabilizing the electromagnetic suspension in active vibration protection devices. However, these results cannot be considered as a basis for the development of universal facilities capable of choosing the optimum structural solutions for vibration isolation devices with due regard to the functional area [13, 14]. Further investigation of the optimization of the stabilization system for electromagnetic suspension is needed to obtain rational structural solutions of the object-oriented nature.

The research aims to determine the optimum parameters of the stabilization system under the law of change of the control voltage of the active vibration isolation device

[*]Corresponding author: inna@tpu.ru

© The Authors, published by EDP Sciences. This is an open access article distributed under the terms of the Creative Commons Attribution License 4.0 (http://creativecommons.org/licenses/by/4.0/).

(AVID). With this, elastic forces of the electromagnetic suspension (EMS) are used to minimize the errors introduced by the elements of the vibration isolation device.

2 Methodology

In the general case, random disturbances affect the AVID that means that random signals will occur in sensors, voltage sources and electromagnets of the vibration isolation system [15]. A random process $X(t)$ is represented as $X(t) = m_x(t) + \overset{0}{X}(t)$, where $m_x(t)$ denotes the mathematical expectation (mean, systematic value) of a random process, and $\overset{0}{X}(t)$ is a centered random processes. For example, the voltage converter generates the random signal *u*(*t*), and this voltage can be represented as $u = U_0 + \Delta u_n$, where U_0, reference voltage, is the system response to external disturbance and Δu_n, a random component (fluctuation) of the supply voltage, is the error introduced by the voltage source. Thus, the signals from other elements of the vibration isolation device can be represented in a similar way.

Therefore, we need to optimize the AVID supplied with EMS to reduce the impact of noise introduced by the elements of the vibration isolation device, i.e. to perform optimization in order to minimize the centered component of random signals. Variance is a characteristic of the centered random process

$$D_{ex} = \sigma_x^2 = M\{X(t) - m_x(t)\}^2 = M\{\overset{0}{X}(t)\}^2 . \tag{1}$$

Variance is a measure of the scattering process $X(t)$ around the average value.

Take a minimum error variance introduced by the elements of the vibration isolation system as an optimization criterion. Determine the dependence of the selected criterion D_{ex} on the stabilization system parameters α, β, γ.

Due to the limitations imposed on the system parameters α, β, γ with regard to a specific area of the steady operation of the vibration isolation device, the AVID optimization is performed using the method of nonlinear programming. Conditions for the stable AVID operation are as follows: $\alpha > \frac{Rb}{a}$, $\beta > \frac{bL_0 - a^2}{a}$, $\gamma > \frac{-mR}{a}$, where R is the resistance of the EMS electromagnet winding; L_0 is the EMS inductance corresponding to the equilibrium point of the suspended body; m is the mass of the vibration-isolated object; *a* is a linearization coefficient of the electromagnet traction, which depends on the electromagnet current and relative displacement of the object raised to the power (-2); *b* is a linearization coefficient of the electromagnet traction, depending on the electromagnet current and relative displacement of the object raised to the power (-3) [16–18].

3 Results and discussion

3.1 Choosing the active vibration isolation device parameters

Perform the system optimization to reduce the noise introduced by the voltage converter into the AVID supplied with a single-mass EMS. Assume that the noise is a stationary random process such as noise with zero mean and constant spectral density $S_u = S_0$. In this case, variance of the control voltage fluctuations is determined by the formula:

$$D_{eu} = \int_{-\infty}^{\infty} |W_u(j\omega)|^2 S_u d\omega = S_0 \int_{-\infty}^{\infty} |W_u(j\omega)|^2 d\omega = min\,, \tag{2}$$

where W_u is the closed-loop transfer function that relates the output value $\delta(t)$ to the noise introduced by the voltage regulator (Figure 1).

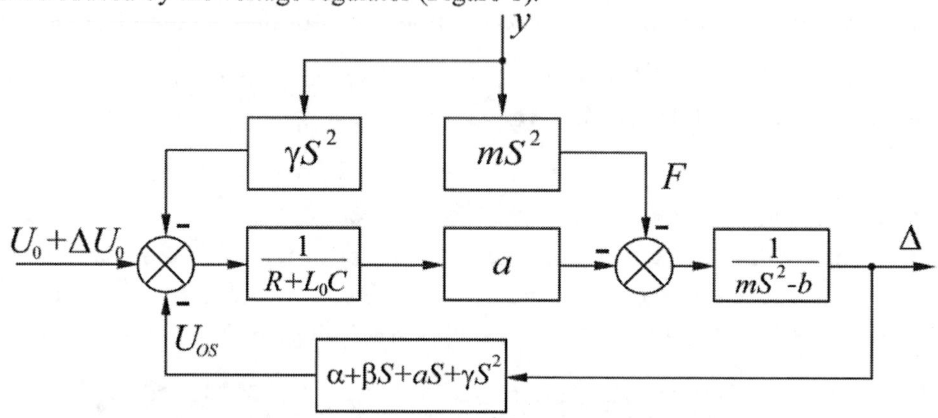

Figure 1. Block-diagram of the AVID supplied with a single-mass EMS.

$$W_u(s) = \frac{-a}{mL_0S^3 + (mR + a\gamma)S^2 + (a^2 + a\beta - bL_0)S + (a\alpha - Rb)} \tag{3}$$

Expression (2) with regard to the transfer function (3) takes the form

$$D_{eu} = \int_{-\infty}^{\infty} |W_u(j\omega)|^2 S_u d\omega = S_0 \int_{-\infty}^{\infty} |W_u(j\omega)|^2 d\omega = min\,. \tag{4}$$

Integrand (4) is a fractional rational function of the frequency ω. Thus, the standard integral with the known solution is used to determine the error variance D_{eu} [19].

The standard integral for this case is written as

$$I_3 = \frac{1}{2\pi j} \int_{-\infty}^{\infty} \frac{B_3(\omega)}{A_3(\omega)A_3(-\omega)} d\omega\,, \tag{5}$$

where $A_3(\omega) = a_0\omega^3 + a_1\omega^2 + a_2\omega + a_3$; $B_3(\omega) = b_0\omega^4 + b_1\omega^2 + b_2$.

The solution of the standard integral I_3 can be obtained from the expression

$$I_3 = \frac{-a_2b_0 + a_0b_1 - \frac{a_0a_1b_2}{a_3}}{2a_0(a_0a_3 - a_1a_2)}, \tag{6}$$

in the considered case $a_0 = -jL_0m$; $a_1 = -(R_m + a\gamma)$; $a_2 = j(a^2 + a\beta - bL_0)$; $a_3 = (a\alpha - Rb)$; $b_0 = 0$; $b_1 = 0$; $b_2 = a^2$.

The desired expression for the error variance with regard to expressions (5) and (6) can be represented as

$$D_{eu} = \frac{\pi a^2 S_{ou}(mR + a\gamma)}{(a\alpha - Rb)\left[(mR + a\gamma)(a^2 + a\beta - bL_0) - (L_0m(a\alpha - Rb))\right]}. \tag{7}$$

Following the procedure of nonlinear programming, write Lagrangian function with regard to the limitations imposed on the system parameters

$$F = D_{eu}(\alpha;\beta;\gamma) + \lambda_1\left(\frac{Rb}{a} - \alpha\right) + \lambda_2\left(\frac{bL_0 - a^2}{a} - \beta\right) + \lambda_3\left(\frac{-mR}{a} - \gamma\right), \tag{8}$$

where $\lambda_i(i = 1,2,3)$ is Lagrangian multiplier.

When using the method of nonlinear programming, the optimum values of the coefficients α^*, β^*, and γ^* are determined from the system of equations:

$$\begin{cases} \alpha^*\left[\dfrac{\partial D_{eu}}{\partial \alpha} - \lambda_1^*\right] = 0 \\ \beta^*\left[\dfrac{\partial D_{eu}}{\partial \beta} - \lambda_2^*\right] = 0; \\ \gamma^*\left[\dfrac{\partial D_{eu}}{\partial \gamma} - \lambda_3^*\right] = 0 \end{cases} \qquad \begin{cases} \lambda_1^*\left[\dfrac{Rb}{a} - \alpha^*\right] = 0 \\ \lambda_2^*\left[\dfrac{bL_0 - a^2}{a} - \beta^*\right] = 0 \\ \lambda_3^*\left[\dfrac{-mR}{a} - \gamma^*\right] = 0 \end{cases} \tag{9}$$

Among the possible solutions of system (9), choose those that meet the conditions of existence of the extreme point of the function considered [1]. Thus, we obtain $\lambda_1^* = \lambda_2^* = \lambda_3^* = 0$, and the optimum gain values are calculated from the relations:

$$\frac{\partial D_{eu}}{\partial \alpha} = 0; \ \frac{\partial D_{eu}}{\partial \beta} = 0; \ \frac{\partial D_{eu}}{\partial \gamma} = 0. \tag{10}$$

Substituting relation (7) obtained for the variance into expression (10), we find the system of equations to determine the optimum values of the gains α, β, and γ, which ensure the fulfillment of condition (2):

$$\begin{cases} -\pi a^2 S_0(mR + a\gamma)\left[a(mR + a\gamma)(a^2 + a\beta - bL_0) - 2L_0 ma(a\alpha - bR)\right] = 0; \\ -\pi a^2 S_0(mR + a\gamma)\left[a(mR + a\gamma)(a\alpha - bR)\right] = 0; \\ S_0\pi a^3(a\alpha - bR)\{\left[(mR + a\gamma)(a^2 + a\beta - bL_0) - L_0 m(a\alpha - bR)\right] - \\ -S_0\pi a^3(mR + a\gamma)(a^2 + a\beta - bL_0)\} = 0 \end{cases} \tag{11}$$

The optimum gain values obtained on the border of the system stability follow from the joint solution of equations (11)

$$\alpha^* = \frac{Rb}{a}; \ \beta^* = \frac{bL_0 - a^2}{a}; \ \gamma^* = -\frac{mR}{a}. \tag{12}$$

Therefore, we take the gain values based on the conditions of stable operation of the vibration isolation device.

Perform optimization of the considered AVID supplied with EMS to reduce the errors introduced by the position, velocity and acceleration sensors.

The optimization criterion is written similarly to (2)

$$D_{ed} = \int_{-\infty}^{\infty} |W_d(j\omega)|^2 S_d d\omega = min, \tag{13}$$

where $W_d(p)$ is the transfer function that relates the output value Δ(t) to the noise introduced by the position sensors

$$W_d(s) = \frac{a\alpha + a(a+\beta)S + a\gamma S^2}{mL_0 S^3 + (mR + a\gamma)S^2 + (a^2 + a\beta - bL_0)S + (a\alpha - Rb)}. \quad (14)$$

Assume that the considered noise is a stationary random process such as white noise with zero mean and constant spectral density $S_d(\omega) = S_{d0} = const$. Then expression (13) with regard to the conditions and formula (14) is written as:

$$D_{ed} = S_{d0} \int_{-\infty}^{\infty} \frac{\left| a\alpha - \omega^2 a\gamma + j\omega a(a+\beta) \right|^2}{\left| -j\omega^3 L_0 m - \omega^2 (mR + a\gamma) + j\omega(a^2 + a\beta - bL_0) + (a\alpha - Rb) \right|^2} d\omega. \quad (15)$$

Using the solution of the standard integral $I_3(\omega)$, we find that

$$D_{ed} = \pi S_{d0} \frac{(a\alpha - Rb)\left[a^2\gamma^2(a^2 + a\beta - bL_0) + L_0 m(a^2(a+\beta)^2 - 2a^2\alpha\gamma\right]}{L_0 m\left[(mR + a\gamma)(a^2 + a\beta - bL_0) - L_0 m(a\alpha - Rb)\right](a\alpha - Rb)}. \quad (16)$$

3.2 Optimum values of amplification coefficient

A comparison of the characteristics of Figures 2 and 3 reveals that the acceleration sensor installation onto the vibration-isolated object and introduction of the additional relation to the acceleration output value reduce the initial frequency value, which marks the beginning of the effective vibration isolation of the object, by several hertz.

With regard to the imposed limitations, the optimum gain values α^*, β^* and γ^*, reducing the effect of the sensor errors on the output signal circuit $\delta(t)$, are determined similarly to that of the error reduction from fluctuations in the supply voltage.

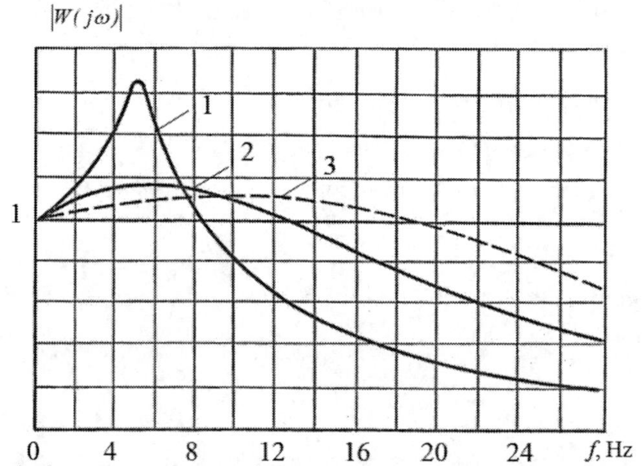

Figure 2. Amplitude-frequency characteristics (transfer coefficients) of the single-mass EMS ($\alpha = const$).

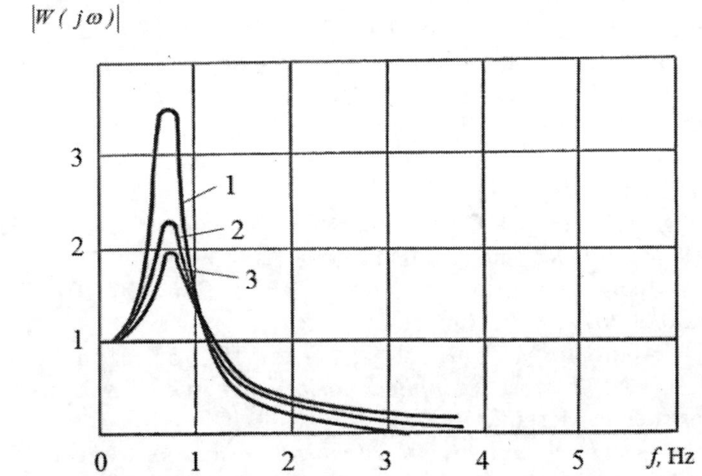

Figure 3. Amplitude-frequency characteristics of the EMS with the additional relation to acceleration of the inertial element (α $- const$).

4 Conclusion

The optimum gain values of the signal feedback circuit, which minimize the impact of noise introduced by the elements of the active vibration isolation device, are obtained on the boundary of stability. The vibration isolation devices, which control circuit coefficients are found on the boundary of the stability region, can be developed during the variable structure system design. Due to the fact that the system development implies a number of difficulties, the values of the feedback loop coefficients are to be selected on the basis of stable operation of the vibration isolation device and be close to the optimum values to design the vibration isolation devices with fixed structure. In this case, a suboptimal device is developed, but not the entirely optimum device of the active vibration isolation.

References

[1] A.P. Surzhikov, O.V. Galtseva, E.A. Vasendina, V.A. Vlasov, E.V. Nikolaev, IOP Conf. Ser.: Mater. Sci. Eng. **110**, 012002 (2016) doi : 10.1088/1757-899X/110/1/012002

[2] E.N. Lysenko, A.P. Surzhikov, S.P. Zhuravkov, V.A. Vlasov, A.V. Pustovalov, N.A. Yavorovsky, J. Therm. Anal. Calorim. **115**(2), 1447 (2014) doi: 10.1007/s10973-013-3456-x

[3] Y.A. Chursin, E.M. Fedorov, Optics&Laser Technology **67**, 86 (2015) doi: 10.1016/j.optlastec.2014.09.017

[4] M.V. Vedyashkin, O.P. Muravlev, Russ. Electrical Eng. **84**(7), 397 (2013) doi: 10.3103/S1068371213070146

[5] O.V. Galtseva, S.V. Bordunov, S.N. Torgaev, IOP Conf. Ser.: Mater. Sci. Eng. **110**, 012094 (2016) doi: 10.1088/1757-899X/110/1/012094

[6] R.M. Mustafina, I.A. Plotnikov, I.V. Plotnikova, O.N. Tchaikovskaya, J. Phys: Conf. Ser. **671,** 012046 (2016) doi: 10.1088/1742-6596/671/1/012046

[7] O.V. Galtseva, G.V. Vavilova, I.V. Plotnikova, SENSIG '15, 91 (2015)

[8] I.J. Bezbah, V.A. Melik-Shahnazarov, D.V. Fiyanchuk, V.I. Strelov, SCIENCE&EDUCATION **9**, 317 (2012)

[9] V. A. Melik-Shakhnazarov, V. I. Strelov, D. V. Sofiyanchuk, I. Zh. Bezbakh, Techn. Phys. Lett. **38** (3), 283 (2012) doi: 10.1134/S1063785012030303
[10] N. Docquier, A. Poncelet, M. Delannoy, P. Fisette, V. Sys. Dyn. **48**, 1439 (2010)
[11] S.V. Bordunov, O.V. Galtseva, J. Phys: Conf. Ser. **671,** 012009 (2016) doi:10.1088/1742-6596/671/1/012009
[12] M. Jamil , A.A. Janjua, I. Rafique, S.I. Butt, Y. Ayaz, S.O. Gilani, Life Sc. J. **10** (SPLISSUE 12), 653 (2013)
[13] K. Kitching, D.J. Cole, D. Cebon, Journal of Dynamic Systems, Measurement and Control, Transactions of the ASME **122**,498 (2000)
[14] J. Lei, Z. Jiang, Y.-L. Li, W.-X. Li, Int. J. Cont. **87**(10), 2081 (2014) doi:10.1080/00207179.2014.901564
[15] I. Mihai, F. Andronic, Mech. **20**, 64 (2014) doi: 10.5755/j01.mech.20.1.6591
[16] K. Frolov, A. Furman, *Applied theory of vibration protection systems* (Mashinostroenie, Moscow, 1980)
[17] R. I. Furundjiev, *Designing of optimal vibration isolation systems* (Visshaya shkola, Minsk, 1971)
[18] X.Wen, C. Zhu, R. Hu, Ch. Song, J. Zhang, MATEC Web of Conferences **35**, 02010 (2015) doi: 10.1051/matecconf/20153502010
[19] M.F. Rosin, V.S. Buligin, *Statistical dynamics and efficiency systems theory* (Mashinostroenie, Moscow, 1981)

FUNCTIONAL AND PARAMETRICAL CONTROL OF TEMPERATURE SENSORS WITH THE NI MYRIO MODULE

Kirill Moskalenko[a], *Kuandyk* Omar, *Pavel* Nekrasov, *Andrey* Karakozov, *Dmitry* Boychenko

National Research Nuclear University MEPhI (Moscow Engineering Physics Institute), 115409, Moscow, Russia

Abstract. Developed automated setup for functional and parametrical control of temperature sensors is presented. The hardware setup is based on the NI myRIO operated under LabVIEW software. Two ways of communication with the temperature sensor, advantages and disadvantages are described. The experimental results functional and parametrical control for three temperature sensors are presented.

1 Introduction

The digital temperature sensor is one of the most used elements in different electrical systems [1–4]. It determines the object or environment temperatures with the required accuracy [5, 6]. The advantages of the sensor with digital interface [7, 8] are noise immunity when receiving data and the possibility to connect several devices to a single bus.

Usually the input control of the digital or analog integrated circuit is performed by automated systems or complexes of the functional and parametric control [9–11]. In this paper we describe the system of the parametrical and functional control for temperature sensor DS1624S+ [12].

2 Device under test

The temperature sensor DS1624S+ is packaged in PDIP package and consisted of two separate functional blocks: 256B EEPROM and digital temperature sensor. The interaction with the integrated circuit occurs through I2C protocol. The read and converted temperature is loaded in a 12-bit word. The temperature range is from -55 to +125 degrees Celsius with the accuracy of 0.0625 degrees. Maximum conversion time is 200 ms.

The developed system has to provide the following features:

- a connection and control of up to 4 temperature sensors;
- an independent power supply from 2.7V up to 5.5V on every integrated circuit;
- measurements of the current consumption with accuracy up to 0.1uA;
- measurements of the output logic levels V_{OL} and V_{OH};

[a] Corresponding author: shaxtyr@yandex.ru

© The Authors, published by EDP Sciences. This is an open access article distributed under the terms of the Creative Commons Attribution License 4.0 (http://creativecommons.org/licenses/by/4.0/).

- writing and reading data via serial interface.

3 Setup requirements

Hardware part of the system consists of National Instruments (NI) myRIO module and SANWA PC5000 multimeter. Application circuit for the temperature sensor is shown on Figure 1. Power supply, data exchange and output logic levels measurement are implemented on the base of NI myRIO input/output lines [13].

Multimeter is used for current measurement in the power supply circuit and is controlled via RS-232 interface from computer. Custom software was developed in LabVIEW development system.

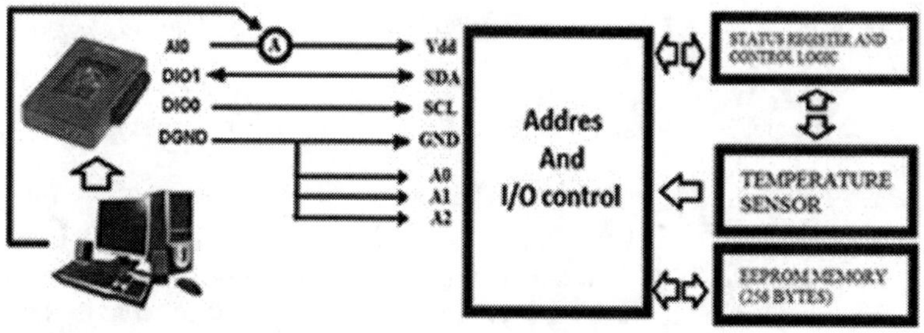

Figure 1. Temperature sensor circuit.

4 Solution

Serial interface for data exchange was be implemented to provide the functional control (Figure 2). Each sensor must be connected to a separate bus to exclude the mutual influence of the remaining sensors during testing.

Two ways of communication with the sensor were implemented:

- built-in hardware I2C interface of the myRIO module;
- NI myRIO FPGA [14, 15] based implementation of the I2C interface.

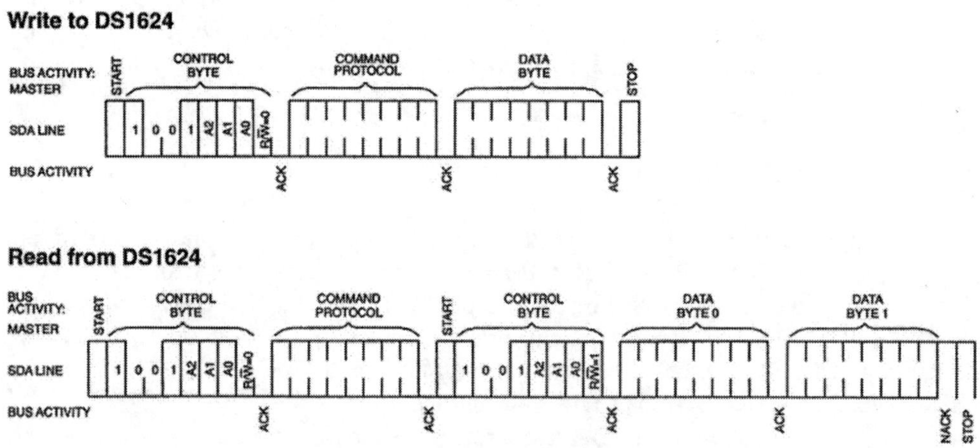

Figure 2. Algorithm of reading and writing to DS1624S+ .

The built-in hardware I2C interface module is an advantage of the myRIO system (Figure 3). This virtual instrument implemented on the target machine (the target machine is realized as such construction: a structure "case" is in a loop "while"; a variant of "case" is chose by constant "enum") has the capability of single and continuous temperature acquisition, reading/writing data to memory, the ability to write the configuration register and read data from it (Figure 4). The front panel that virtual instrument is shown on Figure 5.

However it allows independent simultaneous connection of only 2 sensors because of a number of ports on myRIO for I2C is limited.

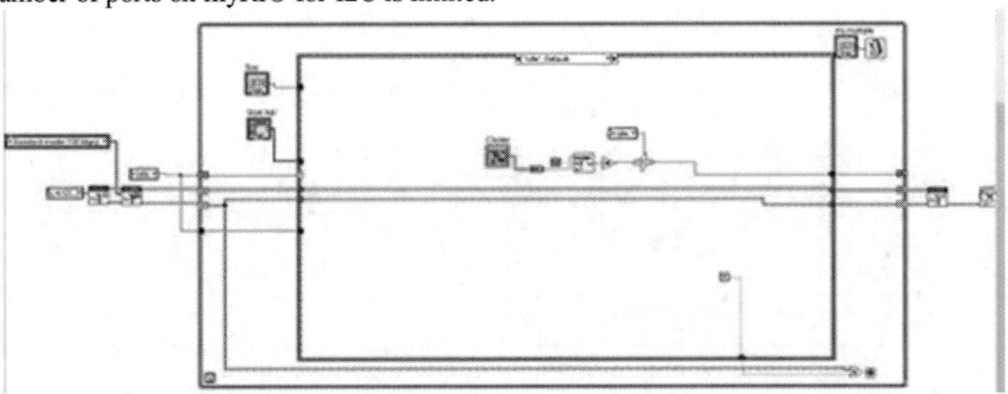

Figure 3. Main state machin.

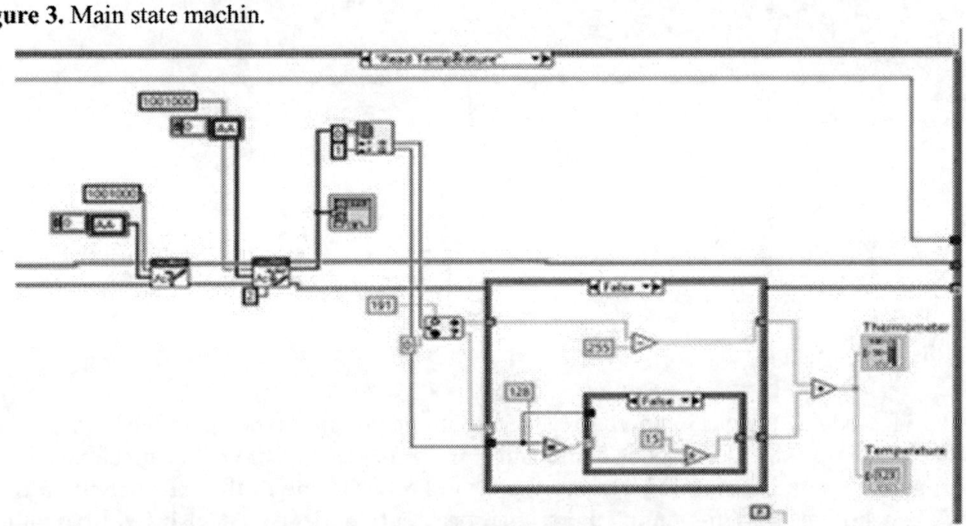

Figure 4. Block diagram of the VI responsible for temperature reading.

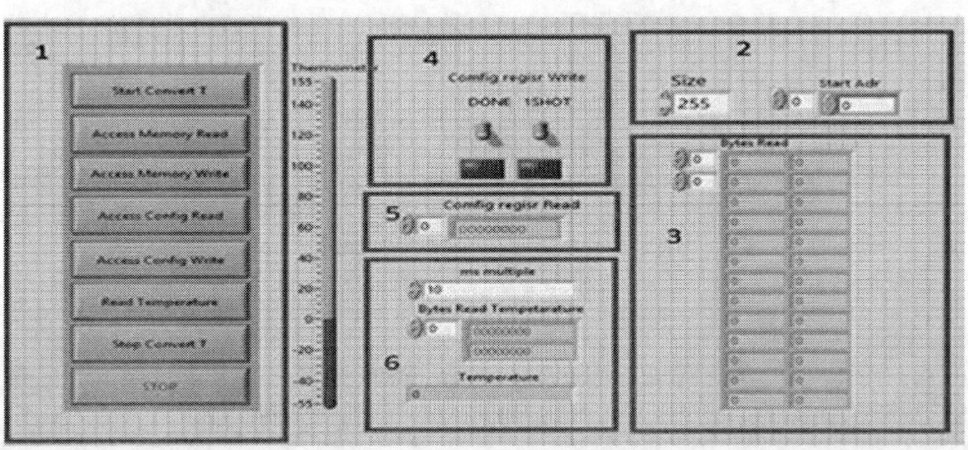

Figure 5. Front panel of the VI: 1–operation select; 2-3–size of the memory to write and to read to; 4–configuration register setup; 5–reading settings for the configuration register; 6–temperature value.

The other solution for problem was based on FPGA. The possibility to implement about 20 independent digital interfaces I2C is advantage of that method.

Computer software consists of 2 paths – HOST.vi and FPGA.vi. FPGA implements data exchange via serial interface on low level according to the sensor documentation. Transfer algorithm of the data frame was implemented by using developed hardware reading (Figure 6) and writing (Figure 7) subsidiary virtual instruments. These subsidiary virtual instruments provide time characteristics according to the I2C protocol.

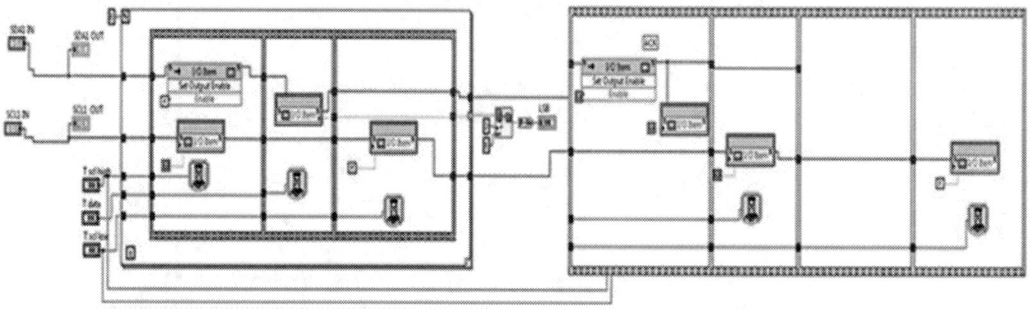

Figure 6. Reading byte VI.

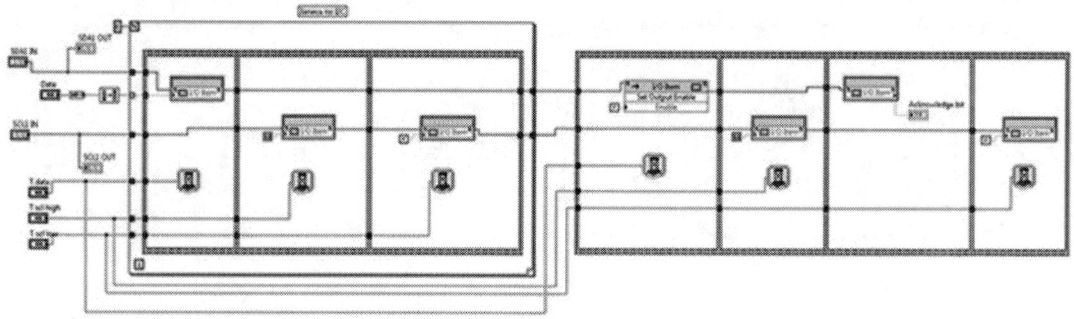

Figure 7. Writing byte VI.

HOST module controls temperature sensor parameters, implements on the high level data protocols (Figure 8). Everyone commands are represented their hexadecimal number according to documentation [5]. During the testing process one of the four operation modes could be chosen: waiting, temperature measurement in degrees Celsius, also reading / writing from / in EEPROM a control command. Information is shown on indicators.

During the regime "a measurement temperature" reset of information is happened and protocol "writing information in DS1624S+" is implemented. First a writing control byte which last bit is "0" is realized using under virtual instrument "writing of byte". Then a command which signals about beginning of conversion temperature is written. Later the protocol "reading information from DS1624S+" is realized. Writing a control byte which last bit is "0", is realized using under virtual instruments "writing byte" and "reading byte", then a command which signals about beginning of reading temperature is realized and a control byte which last bit is "1" according to protocols of exchange information.

The measured temperature was written in two bytes: MSB - integer part, and LSB - factorial part which four least significant bits are null according to documentation [5]. The front panel of HOST.vi is showed on Figure 9.

A power supply was provided by analog power lines of NI myRIO. They can provide up to 3mA. This is enough to ensure the sufficient current consumption of the sensor, which peak value should not exceed 1.2 mA.

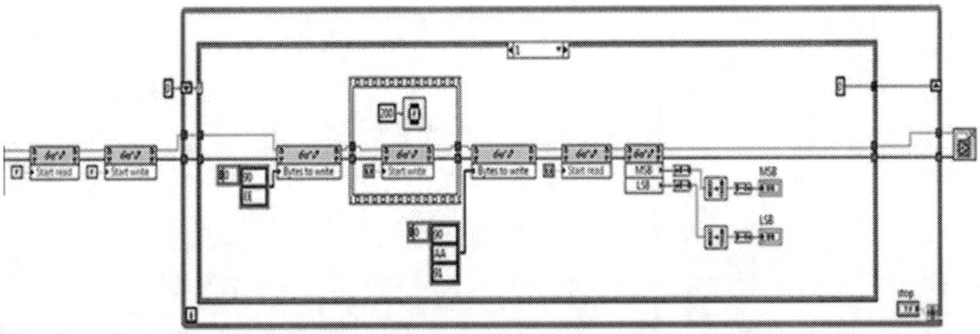

Figure 8. HOST.vi Block-diagram.

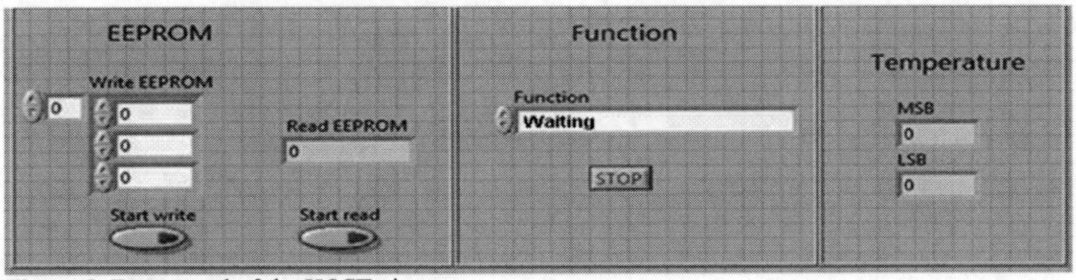

Figure 9. Front panel of the HOST.vi.

5 Experimental results

The experimental research including functional and parametrical control of the operation of the temperature sensors under the destabilizing influence was conducted for the DS1624S+ temperature sensor. Only 3 samples were examined. Two samples measured a stabilized environmental temperature, i.e. 21°C. Third sample measured the temperature of a hot airflow which was 50°C…60°C. The plots of measured temperature on experiment time are shown on Figure 10. End of plot line determines fault of the temperature reading from sensor. Internal flash fault happened about $1.4 \cdot 10^4$ sec. from the beginning of the experiment. Flash test consisted of sequential writing and reading of reference data. On the fault time it revealed loss of possibility to rewrite the information; it remained the same after the write procedure. Measured sensors' dynamic and standby currents vs time curves are shown on Figure 11 and Figure 12 respectively.

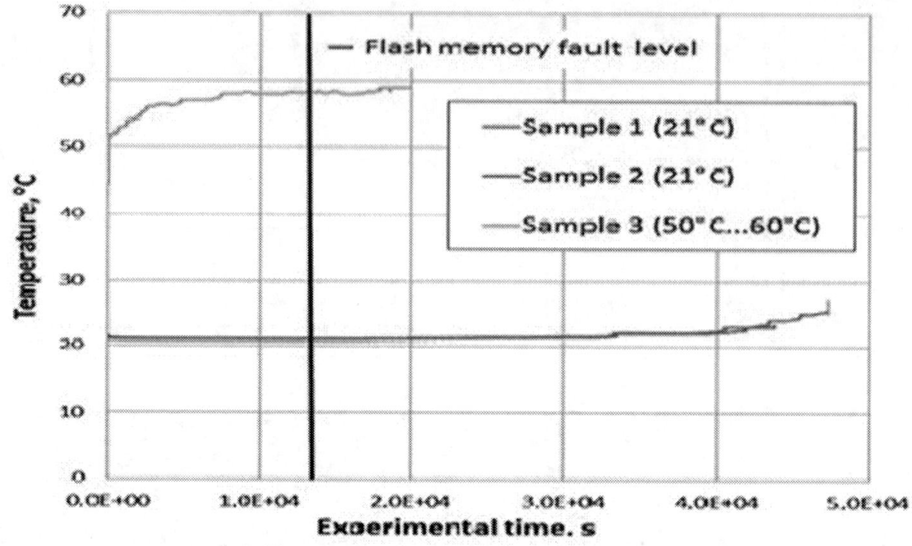

Figure 10. Sensors' temperature vs time.

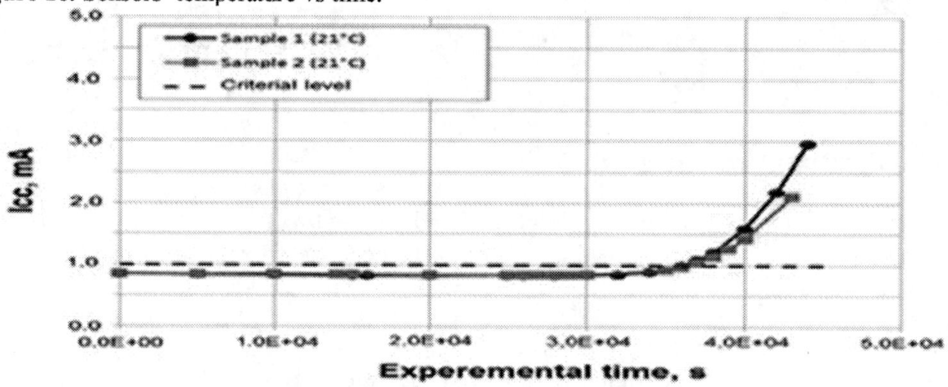

Figure 11. Sensors' dynamic current vs time.

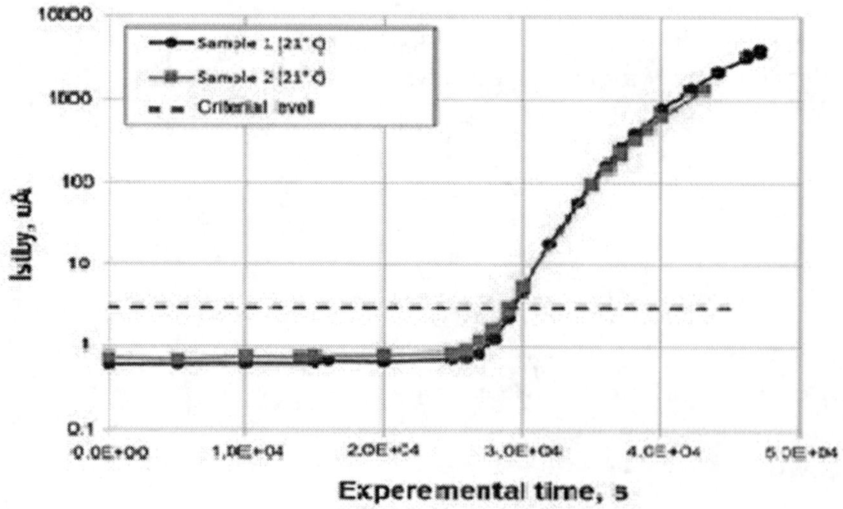

Figure 12. Sensors' standby current vs time.

6 Conclusions

The digital temperature sensor DS1624S+ is one of the most used elements in different electrical systems. Two ways of communication with the temperature sensor DS1624S+ were designed. The developed systems can handle data from two independently connected temperature sensors in case of using built-in hardware I2C interface of the myRIO module and four independently connected temperature sensors in case of using NI myRIO FPGA based implementation of the I2C interface, measure the current consumption and the output voltage of the chips. The number of sensors is limited to the number of analog output channels. The experimental results are showed. Systems have been tested in the preparation of studies of temperature sensors in the AO "ENPO SPELS" and IEPE MEPhI.

References

[1] B. Karakozov, O. V. Korneev, P. V. Nekrasov, M. N. Sokolov, D. A. Zagryadsky, Proceedings of the European Conference on Radiation and its Effects on Components and Systems (RADECS), 6937407 (2013) doi: 10.1109/RADECS.2013.6937407

[2] G. I. Zebrev, D. Y. Pavlov, V. S. Pershenkov, A.Y. Nikiforov, A. V. Sogoyan, D.V. Boychenko, V. N. Ulimov, V. V. Emelyanov, IEEE Transactions on Nuclear Science **53**, 1981 (2006) doi: 10.1109/TNS.2006.877851

[3] G. V. Chukov, V.V. Elesin, G.N. Nazarova, A.Y. Nikiforov, D.V. Boychenko, V. A. Telets, A. G. Kuznetsov, K.M. Amburkin, IEEE Radiation Effects Data Workshop, 7004589 (2015) doi: 10.1109/REDW.2014.7004589

[4] O. A. Kalashnikov, V.V. Elesin, D.V. Gromov, 2014 24th International Crimean Conference Microwave and Telecommunication Technology Conference Proceedings (CriMiCo 2014), 6959666 (2014) doi: 10.1109/CRMICO.2014.6959666

[5] P. K. Skorobogatov, K. A. Epifantsev, O. A. Gerasimchuk, Russian Microelectronics **44**, 40 (2015) doi: 10.1134/S1063739715010059

[6] P. K. Skorobogatov, A.Y. Nikiforov, O. B. Mavritsky, A. N. Egorov, A. V. Kirgizova, European Space Agency, (Special Publication) ESA SP (536), BP4 (2004)

[7] A. V. Grebenkina, D.V. Boychenko, A.Y. Nikiforov, V. A. Telets, V. V. Amelichev, V.A.Kharitonov, Proceedings of the International Conference on Microelectronics (ICM), 6842103 (2014) doi: 10.1109/MIEL.2014.6842103

[8] G. G. Davydov, A. S. Kolosova, L. N. Kessarinskiy, D.V. Boychenko, 2015 International Siberian Conference on Control and Communications (SIBCON 2015), 7147094 (2015) doi: 10.1109/SIBCON.2015.7147094

[9] N. E. Aristova, A. Y. Borisov, A. I. Tararaksin, L. N. Kessarinskiy, A. V. Yanenko, 2015 International Siberian Conference on Control and Communications (SIBCON 2015), 7146984 (2015) doi: 10.1109/SIBCON.2015.7146984

[10] I. O. Loskutov, A. B. Karakozov, P.V. Nekrasov, A.Y. Nikiforov, 2015 International Siberian Conference on Control and Communications (SIBCON 2015), 7147128 (2015) doi: 10.1109/SIBCON.2015.7147128

[11] I.I.Shvetsov-Shilovskiy, P.V. Nekrasov, A. V. Ulanova, A. A.Smolin, A. V. Sogoyan, 2015 International Siberian Conference on Control and Communications (SIBCON 2015), 7147282 (2015) doi: 10.1109/SIBCON.2015.7147282

[12] DS1624 Digital Thermometer and Memory Data Sheet URL: http://www.mouser.com/ds/2/256/DS1624-27266.pdf

[13] NI myRIO-1900 user guide and specifications URL:http://www.ni.com/pdf/manuals/376047a.pdf
[14] V.A.Marfin, P.V. Nekrasov, O. A. Kalashnikov, K. A. Kagirina, 2015 International Siberian Conference on Control and Communications (SIBCON 2015), 7147142 (2015)
[15] D. V. Bobrovskii, O. A. Kalashnikov, P. V. Nekrasov, Russian Microelectronics **41**, 226 (2012) doi: 10.1134/S1063739712040038

A NEURAL NETWORK MEASUREMENT OF THE LEVEL AND PACE OF GLOBAL ECONOMIC AND TECHNOLOGICAL DEVELOPMENT

Sergey Gorbachev[a] and *Nadezhda* Gorbacheva

National research Tomsk State University, 634050, Tomsk, Russia

Abstract. The article considers the problem of measuring the level and rate of technical and economic development of countries in terms of technological change. Advantages of cognitive neural network approach to monitoring and quality analysis for integrating into a single model of economic, scientific-technological, innovative and other quantitative and qualitative components of growth, not amenable to traditional statistical analysis, with calculation of the forecast evaluation time of the reference trajectory of technical and economic development. Presents the results of the calculations. To improve the accuracy of the model trajectories are encouraged to use self-organizing Kohonen maps.

1 Introduction

A correct assessment of the level of technical development of the national economy, its position in the global energy required for the development of rational scientific-technical policy. Due to the unique features of each national economy has its own rational trajectory of future energy resources. Planning, possible only in the form of rough scenarios should be based on global trends of technical and economic development (TED) with the above national peculiarities. The most important of them is the situation in cross-country hierarchy TED. For this assessment is relevant model reference trajectory calculations, which sets the common for all countries of the reference system and scale, and serves as the basis for measuring the technical development of national economies.

Many researchers note that the global technological change generated by countries that are leaders in the life cycle of the relevant technological structures [1]. Although technological developments that make up the content of the life cycle of each technological structures occur in the world market, the economic structure of the countries most closely reflects the structure of the relevant specifications, and the dynamics of their TED - evolution of these specifications. Therefore, as the reference trajectory calculations can be considered the actual trajectory calculations leaders in their respective specifications.

Virtue of the laws of reproduction of social capital life cycle of a technological structure in a market economy is reflected in the specific form of Long Waves of economic conditions. The so-called Long wave, or Kondratieff waves, are the subject of study of

[a] Corresponding author: gsv@mail.tsu.ru

© The Authors, published by EDP Sciences. This is an open access article distributed under the terms of the Creative Commons Attribution License 4.0 (http://creativecommons.org/licenses/by/4.0/).

special areas of economic research - analysis of long-term processes of social reproduction [2]. In modern periods allocate 5 Kondratieff long waves (Figure 1).

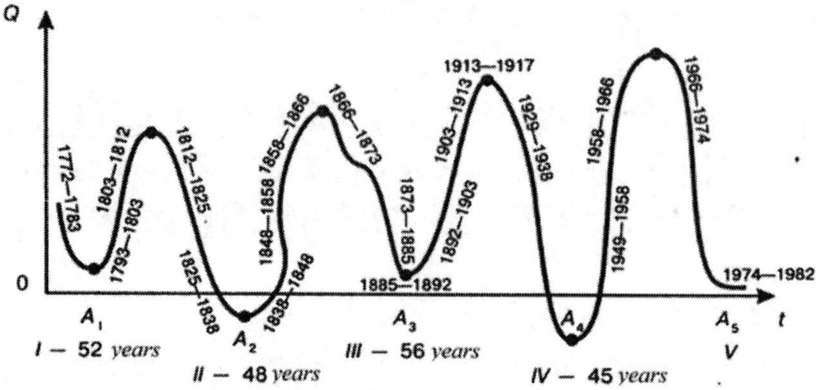

Figure 1. Modern periodization Kondratieff long waves.

In empirical studies of the Long waves was established in principle unidirectionality of what is happening in different countries, technological changes, the similarity of national trajectories calculations, as well as the tendency to synchronization of macroeconomic fluctuations and technological changes. In particular, the same form of the trajectories of energy resources in countries with market-based and policy-driven economy was identified in the structure of energy consumption in metallurgy and mining industry, in the dynamics of the transport infrastructure and other sectors of the economy [1]. The unidirectionality of energy resources in different countries, as well as the formation of a unified rhythm of the global economic system, due to the formation of the global market and rapid expansion of international economic relations since the industrial revolution. Experience macroeconomic research says about the possibility and fruitfulness of using cross-country comparisons for obtaining high-quality and fairly accurate quantitative conclusions, including the forecast of nature [1].

Research related to the modeling of global socio-economic processes in connection with technological development and resource limitations, were launched in the 1970s, the works of J.M. Forrester [3] and D. Meadows [4]. They had a great impact on researchers, and public opinion began to develop ecology, the spread of the idea of sustainable (sustainable - self-sustaining development. Although the research was a huge success, they at the same time were criticized by mathematicians, futurists, economists and sociologists. According to experts, the macro model of the world dynamics, created by the authors, was too mechanistic: it is not documented regional structure of the object, ignored political and social factors, there were differences with theories of economic growth, were not taken into account the adaptive capacity of the global system through social and scientific-technical progress. Following the publication of Forrester and meadows group was followed by the work of American, German, British, Japanese, Polish, Bulgarian and other researchers who tried to take into account the existing criticism. Next generation models were Multiregional description global development and description of the world market.

In Russia, studies were continued at the Computing center (under the direction of academician N.N. Moiseeev) and the Institute of system analysis (under the guidance of academician V.A. Gelovani). In global modeling can be distinguished publication [5] the staff of the Institute of system analysis of Russian Academy of Sciences S. Dubovsky and V. Britkov that have worked closely with D. Meadows and Club of Rome in 70-ies of the last century, made under the direction of academician V.A. Gelovani. In a macroscopic model of the system dynamics Forrester and Meadows were introduced control actions that will significantly expand and explore the range of possible scenarios of world development.

Significant results of scenario studies received in 2009-2010 as a result of execution of the subprogram "Mathematical modeling and systems analysis of the global dynamics" of the program of fundamental research of the Presidium of the Russian Academy of Sciences "Economics and sociology of knowledge" (scientific leaders, academicians, G.V. Osipov and V.A. Sadovnichy, a leading organization - Institute of applied mathematics to them M.V. Keldysh of RAS). The purpose of these studies is the analysis of the emerging trends of world development on the basis of mathematical modeling, the study of scenarios of development of the world system to 2030-2050 years, and arising in this context, alternatives.

In a broader theoretical framework for the issues in the institutionalization of socio-economic and environmental processes are closely linked to the adequacy of the models. Methods of research using global models, they can be divided into two groups - simulation and optimization models. The simulation model describes a closed system of equations, i.e., all functional relationships, the values of parameters and exogenous variables (including control actions) set in advance to the functioning of the model. The study simulated the system using a simulation model is to determine the impact of different assumptions about the functional relationships and the numerical values of the parameters and control actions on the system behavior. The system of equations optimization models is not closed - the part of the exogenous variables (control actions) is not set. The study of an object using an optimization model is to find the values of these variables, ensuring the achievement of pre a particular purpose, as a rule, optimizing a given functionality.

From well-known global models to simulation models include Forrester, Meadows and model of the Mesarovich-Pestel. The first three models represent a system of nonlinear differential equations. With this description in the model included pre-selected constant control mechanism, which is essentially a decision-making model, identified on the historical development of the system. In this case, the research system is to search the various "policies" and the analysis of their impact on system behavior. Under "policies" refers to various combinations of hypotheses about functional relationships, the structure of the steering mechanism, the numerical values of the parameters. This approach helps in the first stage to better understand the dynamics of the investigated processes, to identify possible critical trends. However, the fixation of the steering mechanism does not involve conscious of the changes humans have made to the existing trends of development and does not allow to analyze all the possibilities of the system when a valid control effects.

Thus, the construction of global models and research through various scenarios of world development has formed the research direction called "Global modeling". It can be defined as a set of mathematical, social and economic and other methods on the basis of information technologies applied to the study of global processes. The basic perspective and the novelty of the development of mathematical methods for global modeling, according to the authors, consists in the use and development of cognitive neural network technologies, which are defined as interdisciplinary intellectual solve semi-structured problems, with the aim of developing predictive solutions in the social and economic sphere. They are designed to improve those processes that involve people.

Experts differ in opinion, in which technological environment is now Russia is in the 4th or in the 5th? What potential growth opportunities to use? Where to look for the reserves? How to "jump" in the 6-th technological structure? The best way to overcome the crisis and prolonged depression caused by the change of Kondratieff cycles, as shown by the prominent German economist, Mens [6] is an innovative breakthrough, through the timely development and dissemination of basic technologies for the next 6th Kondratieff cycle. The period from 2010 to 2020 is thus the most favorable time to introduce a new wave of underlying technologies. In our opinion, a significant leap in understanding to build global models lies in finding an adequate way of measuring the level and pace of economic and technological development of countries and define meaningful indicators of the technical-economic growth. Note that in the scientific literature there is no common notion of "energy" or "the pace TED", with virtually no publications, of which we could understand how to measure these growth parameters.

2 Purpose and method

The purpose of this research is to develop the technique of multi-criteria neural network measuring the level and pace of global economic and technological development on complex of quantitative and qualitative indicators. To build a predictive model for the neural network is no requirement of stationarity of the process. Neural networks as universal tool that can accurately detect nonlinear patterns and relationships between the components of the multidimensional random processes.

3 Results and discussion

Anticipating the results of the comparative analysis of energy resources, it is necessary to highlight two factors. On the one hand, objects and territories of the States, regardless of the size and role in the overall economic development, are not self-contained entities and, despite the internal trends and relationships cannot plan its development from its own internal dynamics and proportions. On the other hand, although the countries, united international division of labour, develop according to the general directions of the feasibility of evolution, they vary significantly in terms of absolute level of performance and energy consumption (measured in relative units - per capita or per unit of national income), even if you are on the same level of energy. This is due to the historical, scientific, educational, cultural, psychological, climatic and other characteristics of each country, which are reflected in its economic structure. Therefore, to construct an adequate model for the trajectory calculations in the group of investigated parameters we introduce the qualitative component of growth is the productivity index of primary resources, which is measured as the ratio of GDP to the cost consumed by the economy is the primary raw material resources [7].

Primary resources – the totality of primary organic and inorganic resources, which in droves are used both for consumption and for further processing in the process of material production (food, construction materials, fuel and energy resources). Most developed countries with the lowest energy - consumption by main types of resources, nevertheless consume per capita 1.5-2.5 times more primary raw material resources than the state of the “second echelon” - Russia, Brazil, India and China. Productivity as an indicator of resource efficiency reflects the progress of science and technology. Even in the recession period (1991–1998) in Russia was quite intensive processes to improve the quality of products and services (Figure 2).

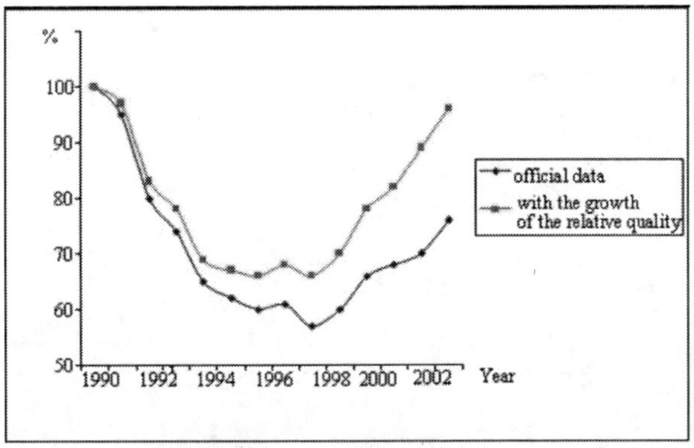

Figure 2. Russia GDP dynamics in 1990–2003, to the 1990 level.

This is reflected in the fact that the productivity of primary resources were increased on average by 2.3% per year. Considering the fact that during this period significantly decreased the dynamics of the qualitative components of growth for the United States (to 0.95% per year, compared with 2.3% per year on average over the previous 30 years), the index improved proportions exchange for the Russian economy amounted to $ 1,016. Since 1998, in the period of recovery of recovery, the dynamics of the qualitative components of growth for Russia grew to nearly 4% per year. However, because in this period also increased the qualitative component of growth for the United States (2.4% per year), the index of the changing proportions of the currency changed slightly to 1,017.

Therefore, the gap in economic development of Russia and the United States could be overcome provided that the rapid growth of productivity (science and technology) in Russia. When modeling the trajectory calculations will take into account that innovation reveal and define the essence of the technological process at the present stage of history. Selected quantitative and qualitative indicators of the feasibility of the development will complement the indicators of innovation potential of the country, divided into 4 groups:

1) indicators characterizing the financial component of the innovation potential (the share of investment relative to GDP, the share of investment in total and so on);

2) indicators characterizing the material component of the innovation sphere (the number of organizations involved in scientific activities, the cost of their fixed assets, the share of scientific organizations);

3) indicators of staffing component of innovative capacity (number of employees involved in innovation, their average wage, average age, their share toy);

4) indicators characterizing the resulting component of innovation capacity as a factor of economic growth and showing his ability to bring the effect (sales, costs and profit associated with innovative activity, the number of created, exported and imported technologies, patents filed and granted patents, their share).

Thus, as the trajectory calculations simulated multivariate time series that represents the combination of several one-dimensional time series, each of which describes the change over time of any of the listed features that characterize the object of the study (Figure 3).

DOI: 10.1051/matecconf/20167901021

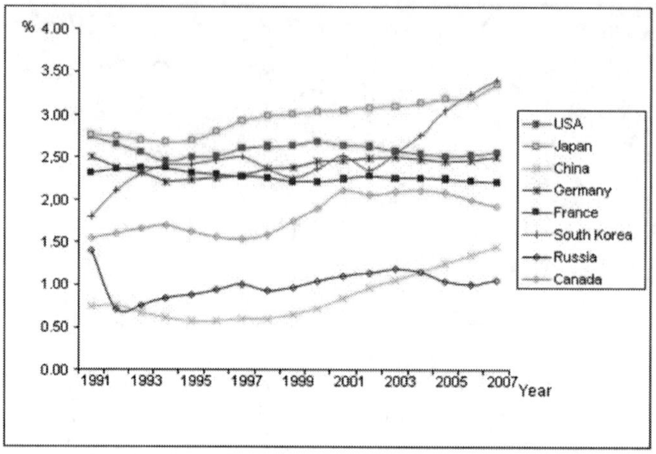

Figure 3. One-dimensional time series (indicator - the share of R & d expenditures in GDP) across countries.

For each year t_i a separate country is represented by a point in a multidimensional (Figure 4 – three-dimensional) space of quantitative and qualitative characteristics.

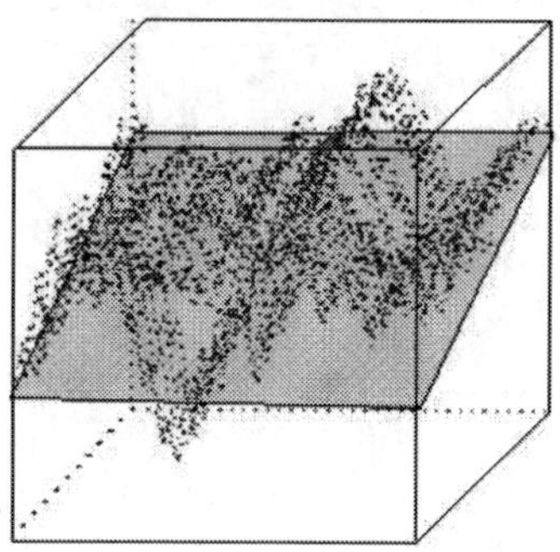

Figure 4. The three-dimensional space of input features.

Linear statistical processing of the data is not able to describe the curve calculations with reasonable accuracy. Used to solve problems of prediction of the statistical apparatus of the correlation analysis, which builds upon the previous values of one-dimensional random process, it is not possible to establish a causal connection settings predictable system as a whole. In addition, the building for each country its generalized curve TED (by linear compression space of input vectors) and their cross-national comparative the analysis is not entirely incorrect due, first, the "fluid" nonequilibrium nature of techno-economic development, with waves fall and rise (Kondratieff waves); and secondly, it is very important input parameters for constructing trajectories, in our opinion, should not be hardcoded - they depend on the national system of accounts in each country, and from historical, cultural, psychological, climatic and other characteristics of the country at the time of observation t, while the importance of the same parameters to build the model will be different for different technological structures.

Thus, the probabilistic approach cannot be considered as reliable and adequate tool for solving semi-structured problems. Limitations and disadvantages of the application of "classical" formal methods in solving semi-structured problems are the result formulated by the founder of the theory of fuzzy sets L. Zadeh, "the principle of incompatibility": "...the closer we come to the solution of real world problems, the more obvious that by increasing the complexity of the system, our ability to make accurate and confident conclusions about its behavior diminishes until a certain threshold beyond which precision and confidence become almost mutually exclusive concepts."

To improve the quality of the modeling, the authors propose to conduct a comprehensive neural network analysis of quantitative and qualitative indicators, which are in the process of learning are ranked weighting factors according to the level of significance for solving the problem. According to the available data samples are built and trained a time series Kohonen neural networks [8] , showing the dynamics of global economic and technological development, which is then used to calculate estimates of the level of energy, performance and energy consumption of individual countries and for forecasting future values of a multidimensional process. Their mapping on the reference and national trajectories TED allows to obtain a reliable assessment of not only speed, but also the level of technical and economic development of each country.

The main difference between self-organizing Kohonen maps from other models is clarity and ease of use (Figure 5).

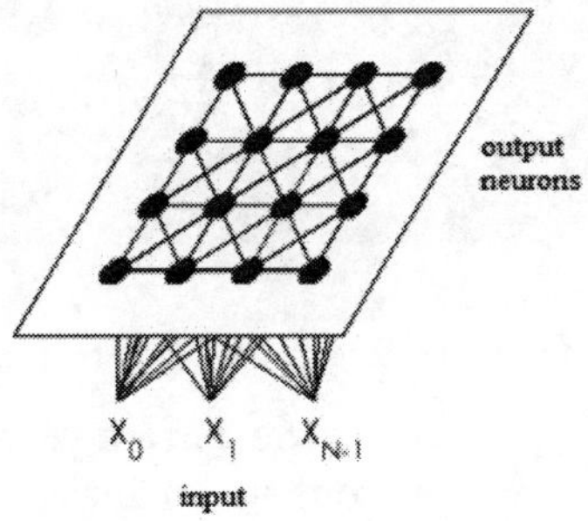

Figure 5. A neural network model of Kohonen.

These networks allow to simplify the multi-dimensional structure, they can be considered one of the methods of projection of the multidimensional space into a space with lower dimensionality. Figure 6 depicts a neural network model of the global techno-economic portrait, reflecting the hierarchy of countries in 2005 (for clarity, here we applied two dimensional spline interpolation).

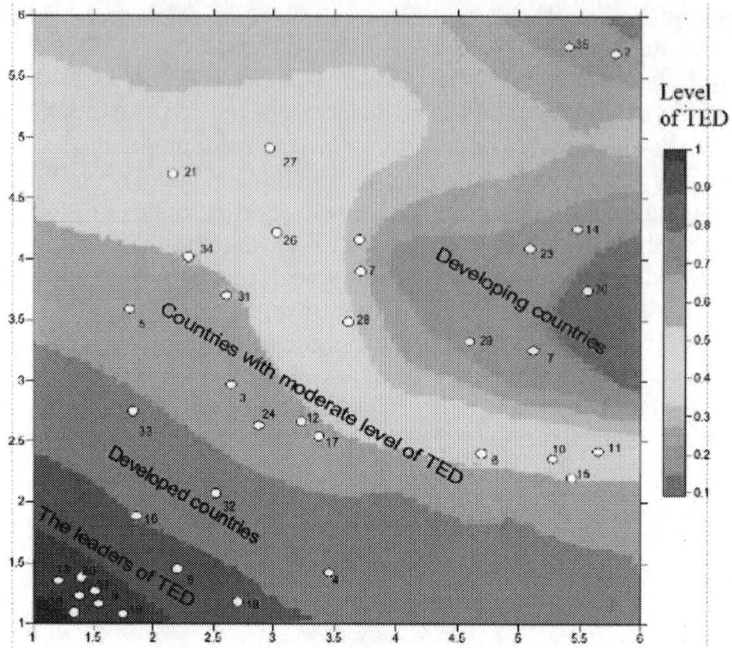

Figure 6. Neural network model of the global technical and economic portrait (2005).

Each country in terms of their energy in the current year is displayed specific cell on the map. Cells with identical coordinates contain countries with similar state energy. The farther to the map coordinates of the countries, the more different from each other, their feasibility portrait. In this case, the trajectory of the energy we can produce dynamic range of Kohonen neural networks (Figure 7), reflecting the global techno-economic portrait in the *i*-th year, each country is characterized by a set of parameters (Y_i, A_i, B_i, C_i), where:

Y_i € [0;1] is the level of energy and fuel resources of the country in the *i*-th year;

A_i is the actual distance (the number of years elapsed from the moment when the reference level parameter TED corresponded to the level of the country under consideration in the *i*-th year of observation);

B_i – promising distance (the number of years it will take for this country, starting from the *i*-th year, to achieve the reference level of technical development in the *i*-th year of observation);

C_i – conditional distance (the number of years required for the output to the reference trajectory).

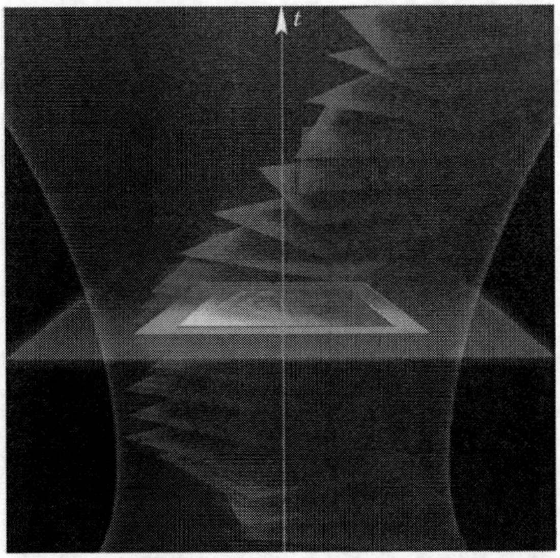

Figure 7. The model trajectory calculations as a time series of trained Kohonen neural networks in the multidimensional feature space.

4 Insights

Neural network clustering data (2005–2010) shows that the analyzed countries are divided into 4 main categories in terms of their energy (Table.1), which takes values from 1 to a reference level of TED to 0 for the minimum level of development. For each country we calculated the growth rate of technical-economic activity over a five-year period. For each five-year period the calculation is based on the output values of the trained neural network Kohonen, reflecting changes in the level of energy in terms of absolute changes in macroeconomic and innovation indicators during this period.

Table 1. The category of countries, selected on the level and pace TED by neural networks modeling (2005–2010).

Group	**Leaders of growth (> 4 %)**	**Countries with an average growth rate (2 – 4 %)**	**Countries with low growth rate (1 – 2 %)**
Leaders of TED	Canada	USA, UK, Italy, France, Germany	Japan
Developed countries	Ireland, Denmark, Sweden, Switzerland	Iceland, Slovenia	Austria, Belgium, Luxembourg, Netherlands
Countries with moderate levels of TED	India, Brazil, China, Russia, Cyprus, Estonia	Hungary, Lithuania, Poland, Portugal, Finland, Slovakia	Norway, Spain, Greece, Malta
Developing countries	Turkey, Bulgaria, Romania	Greece, Latvia	Croatia, Serbia

In terms of global crisis, the canadian economy has demonstrated a high degree of stability, improving its position in the list of world leaders TED to GDP and investment attractiveness. Relatively high growth rates and energy consumption for the developed countries account activation business with effective state regulation and export growth. In addition, the canadian economy is very favorable to the proximity of the U.S. and various trade agreements such as the automotive agreement (1965–2001), canadian-American free trade agreement of 1989 (FTA) and the North American free trade agreement 1994 (NAFTA). GDP per capita (population of Canada is little more than 34.2 million people) in 2010 reached 45.9 thousand dollars. USA with stable consumer spending. In this regard, Canada is on par with the United States (47,1 $ thousand), but surpasses all other countries included in the "seven" leaders TED (Japan, UK, Germany, France and Italy).

In terms of global crisis, the canadian economy has demonstrated a high degree of stability, improving its position in the list of world leaders TED to GDP and investment attractiveness. Relatively high growth rates and energy consumption for the developed countries account activation business with effective state regulation and export growth. In addition, the canadian economy is very favorable to the proximity of the U.S. and various trade agreements such as the automotive agreement (1965 – 2001), canadian-American free trade agreement of 1989 (FTA) and the North American free trade agreement 1994 (NAFTA). GDP per capita (population of Canada is little more than 34.2 million people) in 2010 reached 45.9 thousand dollars. USA with stable consumer spending. In this regard, Canada is on par with the United States (47,1 $ thousand), but surpasses all other countries included in the "seven" leaders TED (Japan, UK, Germany, France and Italy).

Among the developing countries among the growth leaders became Turkey – due to the low level of public debt and the banking system, not by the global financial crisis, Turkish companies will be able to benefit from the current low costs and competition in the manufacturing sector.

In total GDP Russia close behind Brazil. But most importantly, Russia is much closer to the level of the UK, Italy and France, behind them now within just 20 %, showing a higher growth rate TED. GDP per capita lags behind the US in 4 times, from Western Europe 3 times, from Canada and Japan - 3.1 - 3.3 times, from Germany - 3 times, from the UK, Italy and France less than 3 times, but ahead of Mexico 18%, Brazil - 22%, China -1.8 times, India - 3.2 times. Attention is drawn to the strong position of China. Chinese GDP already exceeds 60 % of the US GDP, and Japan's GDP is now only about half of China's GDP.

The calculation results for "actual distance" is presented in the Table 2.

Table 2. The category of countries, selected by the parameter "Actual distance" of neural networks modeling (2005-2010).

The actual length	Country
< 10 years	Switzerland, Finland, Germany, Denmark, Sweden, UK
10 – 20 years	Russia, Czech Republic, Greece, Malta, Portugal, Hungary, Lithuania, Bulgaria, Poland, Slovakia, Italy, Norway, Spain
> 20 years	Romania, Latvia, Turkey, Croatia

Neural network analysis identifies the following indicators of the technical-economic growth, the significance level which exceeds 0.7 units (the GDP, the share in global GDP of major economies, the cost of consumption of primary raw materials (food, construction materials, fuel and energy resources), the volume of industrial production, total expenditure on research and development, the share of R & d expenditures in GDP, the index of quality of life, index of enrolment, people who have received an academic degree and higher education per 1,000 population aged 20 to 29 years, the share of research and development costs for the technology and medium-high level (% of manufacturing costs), expenditure on innovation (% of total turnover), the share of high tech exports in total exports, patents per million population and others).

Thus, the significance of the parameter "Total expenditure on R & d" is 0.83% In 2007 all world expenditure on R & d amounted to approximately 1.1 trillion dollars, including the U.S. share corresponded to approximately 33 % of total, Japan was ranked second among the countries in the world in R & d expenditures, its share is 13 %, China's share is about 9 %. Germany and France took fourth and fifth place with 6 and 4%, respectively, Russia - 1 %.

To achieve the reference level of energy among the main directions of social policy in relation to the income it would be advisable to include the development of complex of measures on improvement of quality of life as a factor of GDP growth, in particular and energy in general. Social policy should focus on achieving a high quality of human capital characteristics, variety and alternatives provided for opportunities for self-improvement, realization of creative potential, the ability of quick adaptation, versatility perception of the world, health and longevity.

5 Conclusion

On the basis of many national and international scientific research developed the technique of multi-criteria neural network measurement of the level and pace of global economic and technological development on complex quantitative and qualitative indicators. On the basis thereof a flexible neural network model of the trajectory of the global tehnical and economic development in the form of dynamic range cluster Kohonen maps with relatively simple but powerful strategy for improving the accuracy of clustering without prejudice interpretability, with evaluation of the level of TED, performance and TED consumption of individual countries and for forecasting future values of a multidimensional process. Their mapping on the reference and national trajectories TED allows to obtain a reliable assessment of not only speed, but also the level of technical and economic development of each country.

On the basis of many national and international scientific research developed the technique of multi-criteria neural network measurement of the level and pace of global economic and technological development on complex quantitative and qualitative indicators. On the basis thereof a flexible neural network model of the trajectory of the global tehnical and economic development in the form of dynamic range cluster Kohonen maps with relatively simple but powerful strategy for improving the accuracy of clustering without prejudice interpretability, with evaluation of the level of TED, performance and TED consumption of individual countries and for forecasting future values of a multidimensional process. Their mapping on the reference and national trajectories TED allows to obtain a reliable assessment of not only speed, but also the level of technical and economic development of each country.

On the basis of many national and international scientific research developed the technique of multi-criteria neural network measurement of the level and pace of global economic and technological development on complex quantitative and qualitative indicators. On the basis thereof a flexible neural network model of the trajectory of the global tehnical and economic development in the form of dynamic range cluster Kohonen maps with relatively simple but powerful strategy for improving the accuracy of clustering without prejudice interpretability, with evaluation of the level of TED, performance and TED consumption of individual countries and for forecasting future values of a multidimensional process. Their mapping on the reference and national trajectories TED allows to obtain a reliable assessment of not only speed, but also the level of technical and economic development of each country.

Acknowledgment

The paper was written as part of the research project No. 8.2.31.2015, carried out with the support of the Program "Research Foundation of Tomsk State University named after D.I. Mendeleev" in 2015 – 2016., grant RFBR No. 16-29-12858.

References

[1] S.K. Glaziev, *Theory of long-term techno-economic development* (Moscow, 1993)
[2] N.D. Kondratiev, *Big cycles of conjuncture and the theory of foresight* (Economy, Moscow, 2002)
[3] J.W. Forrester, *World dynamics* (Wright Allen Press, Cambridge, 1971)
[4] D.L. Meadows, *Dynamics of Growth in a Finite World* (Wright Allen Press, Cambridge, 1974)
[5] V.A. Gelovani, V.B. Britkov and S.V. Dubovsky, *USSR and Russia in the global system (1985-2030). The results of the global modeling* (Editorial URSS, Moscow, 2009)
[6] G. Mensch, *If this Long Wave Steeps-Up and Breaks: What then? Kondratieff Waves* (IOS Press, Amsterdam, 2006)
[7] M.N. Uzykov, Probl. of forecasting **3**, 25 (2004)
[8] I.V. Zaencev, *Neural networks: basic models* (Voronezh state University, Voronezh, 1999)

MATEC Web of Conferences 79, 01022 (2016) DOI: 10.1051/matecconf/20167901022
IME&T 2016

JITTER IN GEOLOCATION DATA PROCESSING

Andrey Klokov[1,*], *Andrey* Zapasnoy[1], *Svetlana* Shipilova[2]

[1]National Research Tomsk State University, 634050, Tomsk, Russia
[2]National Research Tomsk Polytechnic University, 634050, Tomsk, Russia

Abstract. Technology of increasing resolution of an "OKO-2" industrial ground penetrating radar with jitter application is considered on an example of geolocation of tunnels of underground animals. In the present work we suggest to use the alternative approach based on separation of jitter. To separate jitter, the domain of Fourier frequencies and the window functions that separate high-frequency signal components are used. When the jitter was separated, focusing is performed.

1 Introduction

Nowadays systems of subsurface sensing – ground penetrating radars – are widely used. These systems are used for nondestructive testing of the state of water and gas supply pipes, quality of laying asphalt at public roads, inspections of historical burial places, search for artifacts of any type, etc [1–5]. In this regard, a problem arises of quick and high-quality processing of ground penetrating radar data. To obtain high resolution, the approach based on an artificial increase in the antenna aperture – the method of synthetic aperture – is used. Its essence consists in scanning of the examined volume from different positions and then application of focusing operation to the obtained wave projections. The subsurface radio tomography is based on the application of this method.

2 Theoretical principles of geolocation with jitter application

From physical reasons it is clear that the spatial resolution is determined primarily by the upper frequency of the radiation spectrum. However, on the other hand, the higher the frequency, the smaller is the depth of radiation penetration into the examined medium. Therefore, both domestic (Logic Systems Ltd and JSC VNIISMI) and foreign manufacturers (Geoscanners AB, MALA GeoScience AB (Sweden), and OYO Corporation (Japan)) of ground penetrating radars offer to end-users a wide variety of antenna units with different sensing depths and different resolutions. For example, Logic systems Ltd produces antenna units for the "OKO-2" ground penetrating radar with central frequency from 90 MHz to 2 GHz. The AB-90 antenna unit has a sensing depth of 16 m and a resolution of only 0.5 m and the AB-1700 antenna unit has a sensing depth of only 1 m, but its resolution is 10 times higher and makes 0.03 m.

* Corresponding author: 701-kav@mail.tsu.ru

© The Authors, published by EDP Sciences. This is an open access article distributed under the terms of the Creative Commons Attribution License 4.0 (http://creativecommons.org/licenses/by/4.0/).

 DOI: 10.1051/matecconf/20167901022

From the literature (for example, see [6]) it is well known that as a result of diffraction, each point scatterer in a homogeneous medium is displayed by a hyperbole (Figure 1).

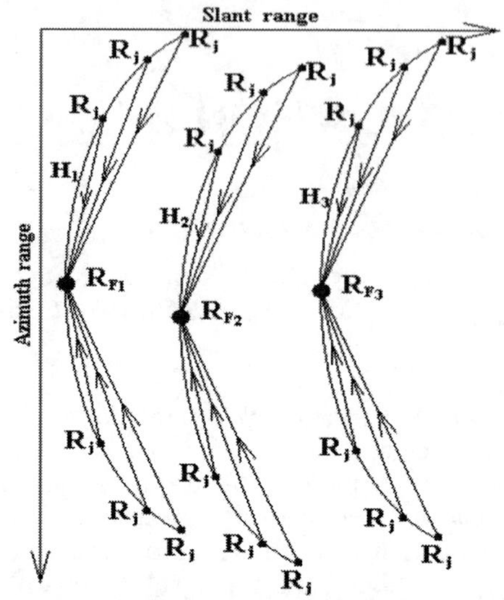

Figure 1. Reflections from point objects in a homogeneous medium.

In this case, the true position of the scatterer corresponds to the hyperbola apex. Focusing displaces reflections to their true positions eliminating the diffraction effect and increasing the spatial resolution. If at the focusing point there is an actual inhomogeneity, the response from a stronger inhomogeneity will be more pronounced. Changing the position of the focusing point, all probed subsurface space can be scanned and an image of the inhomogeneous medium can be obtained. The focusing method can be implemented using different algorithms, for example, diffraction summation along hyperboles, the Stolt migration, and focusing by the phase shift method. Some of these algorithms are implemented in the time domain and others in the domain of spatial frequencies. Two focusing methods are most widely used for data processing, including the diffraction summation method in the time domain [6] and the Stolt migration method in the spectral domain [6, 7].

The foregoing is common to all geolocation methods. The transverse and longitudinal resolutions are determined by the minimum radiation wavelength and the spectrum width. A contradiction arises: the smaller the wavelength (the higher the frequency), the higher is the resolution; however, the corresponding frequencies are quickly attenuated in the medium, and their penetration depth is limited by the skin layer. Low frequencies penetrate relatively deep into the ground, but give low spatial resolution. At present high-power ground penetrating radars whose high-frequency spectral components have sufficient energy are used to resolve this contradiction. In the present work we suggest to use the alternative approach based on separation of jitter.

Detectors of modern microwave ground penetrating radars are stroboscopic oscilloscopes. Their operation principle consists in stroboscopic transformation of the examined signal into its LF analog, that is, a sample from successive signal parts is used to record ultrashort signals. Only one signal reading is taken from each signal with a fixed time step (strobe). As a result of sensing of an object, a medium, or an inhomogeneity, the informative impulse appears to be covered by a noise track (Figure 2) that reduces the signal/noise ratio. To eliminate the effect of the jitter and to increase the signal/noise ratio, signal accumulation (Figure 3) and subsequent averaging (Figure 4) are used. At first sight this noise is harmful and useless; moreover, it is amplified in the regions of maximum values of impulse derivatives on impulse fronts and decays, as can be seen from Figure 5. This Figure shows the standard deviation for ten realizations of an ultra wideband (UWB) impulse. After closer examination it becomes clear that the total signal is the result of jitter modulation by the informative signal. The spectrum of the jitter being separated has higher frequencies than the average informative signal and at the same time, the spectrum is coherent with the signal since it is modulated by the signal. In this case, the seemingly negative factor is that the jitter is not averaged but is separated and subsequently used to increase the spatial resolution. To separate jitter, the domain of Fourier frequencies and the window functions that separate high-frequency signal components are used.

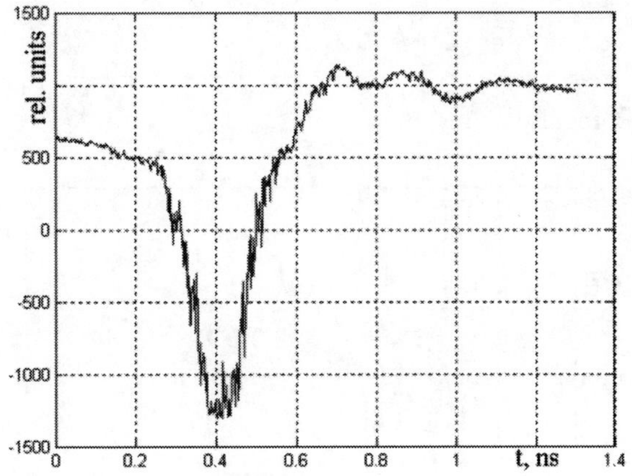

Figure 2. UWB impulse for stroboscopic detection.

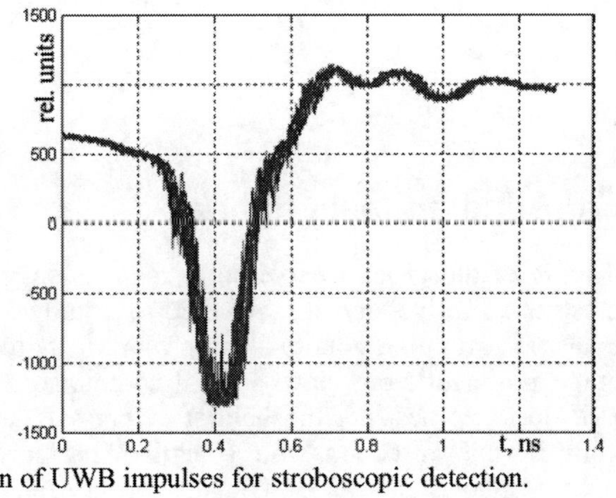

Figure 3. Accumulation of UWB impulses for stroboscopic detection.

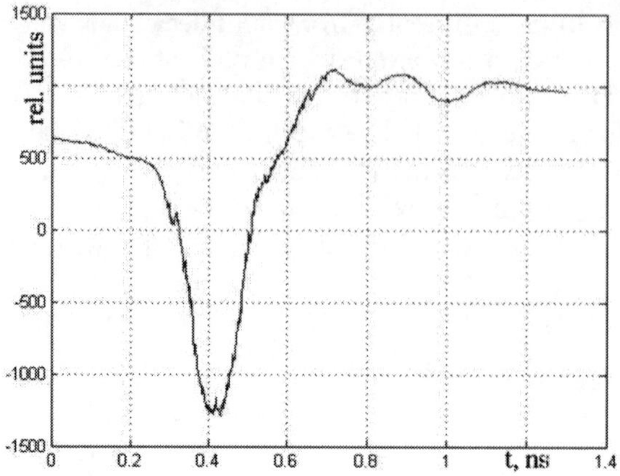

Figure 4. Average UWB impulse.

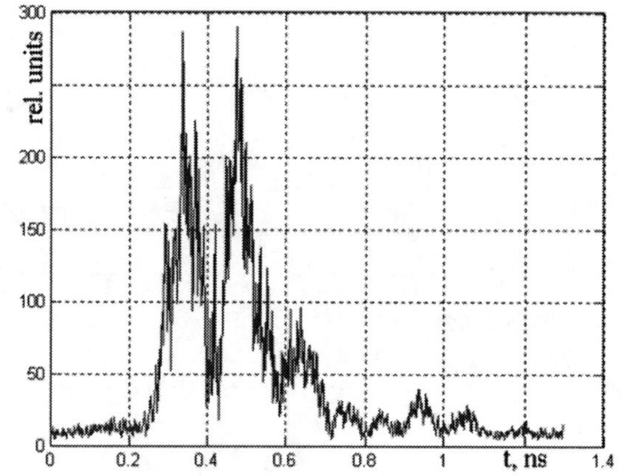

Figure 5. Standard deviation of the UWB impulse.

3 Study of underground animals habitat

Absolutely new application of the ground penetrating radar is suggested by scientists of National Research Tomsk State University. It is devoted to a study of burrows and tunnels of underground mammals, in particular, zokors – mole-rat rodents living underground and digging tunnels with their powerful front claws [8]. In comparison with the conventional methods, such approach does not destruct the habitat of animals and hence the existing natural balance. Acting in this direction, the scientists performed investigations of agricultural fields in Kozhevnikovo District of Tomsk Region near Novopokrovka village with the ground penetrating radar in August 2015. Several plots with clearly pronounced signs of zokor activity – obvious signs of zokor coming on the surface – were investigated. The plots represented squares with the edge 7.5 m long. To ensure high resolution, the radar subsystem "OKO-2" with bipolar probing radio impulses having radiation spectrum with an average frequency of 1700 MHz was chosen. The estimated sensing depth was of the order of one meter, and the depth resolution was about three centimeters [9]. This resolution allowed the zokor tunnels to be imaged.

To obtain the subsurface image, a precise position of the ground penetrating radar on the surface must be recorded. The relative radar position along the sensing path was determined by radar tools – a control wheel combined with a displacement sensor. The transverse motion should be controlled by the user; therefore, each plot was marked before sensing. The scanning pitch in the longitudinal direction was 0.022 m, and in the transverse direction it was 0.2 m. The ground was marked with a nylon cord. When scanning a relatively large area, several uncertainties arise, one of which is referred to the zero leveling, since the ground penetrating radar is not tightly adjacent to the ground, and the second is caused by the need for data smoothing along the direction of transverse radar motion. In this case, to minimize errors in determining the coordinates, a steel tape ruler was stretched along one of the square edges transverse to the direction of radar motion. Since the tape ruler was tightly adjacent to the ground, it was considered as the zero level of the ground surface along the vertical direction and as the origin of radar traces in the longitudinal direction. A total of 37 parallel longitudinal traces were marked on each plot of the ground. Readings of the scattered field were obtained at 350 points on each trace. To increase the signal-to-noise ratio (SNR), measurements were performed 9 times at each point, thereby leading to a threefold increase of the SNR. The row data obtained with the ground penetrating radar "OKO-2"are shown in Figure 6.

Figure 6. Raw geolocation data (the position of the tape ruler is marked by the oval).

It can be seen that there are shifts in the zero level signal from the tape ruler (marked with the oval) along the radar trajectory. This shift occurred due to different initial radar positions on the transverse traces. To interpret the data correctly, the traces should be adjusted against the tape ruler position. To perform the necessary adjustment, an area around the tape ruler was selected and the correlation coefficients were calculated for traces relative each other. The latter were transformed into readings of trace shifts. The result of such adjustment is shown in Figure 7. In this Figure signals from the tape ruler are arranged perpendicular to the traces without significant shifts.

DOI: 10.1051/matecconf/20167901022

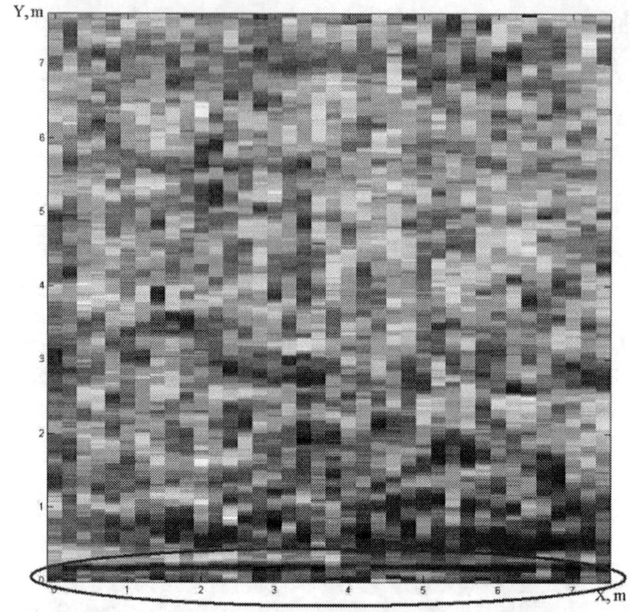

Figure 7. Geolocation profile after data smoothing (the position of the tape ruler is marked by the oval).

For further processing of the data of the ground penetrating radar, it is necessary to consider the exponential attenuation of radiation with depth of penetration into soil [10]. The soil being probed contains a certain amount of moisture which causes the attenuation of radio signals. It can be considered only on the average for the background medium. The correction for this attenuation via renormalization of all measured signals by the average exponential decay allows the contribution of all layers to be leveled. Then the jitter is separated by the method described above and level-by-level focusing is performed. Results of processing with and without jitter are shown in Figures 8 and 9, respectively.

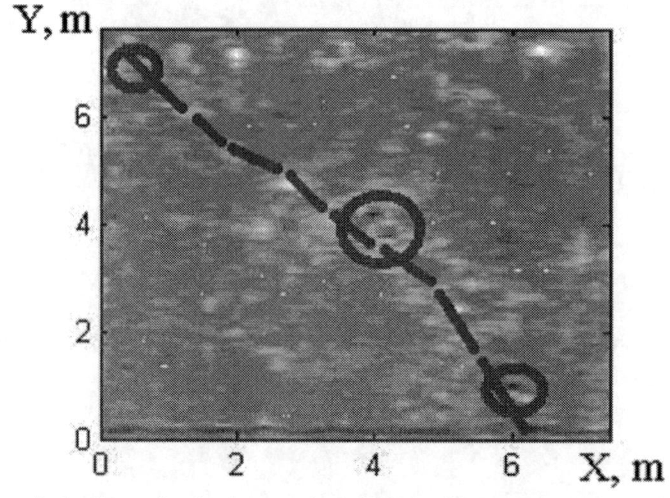

Figure 8. Crosscut of a horizontal zokor's tunnel at a depth of 0.3 m; data processing with jitter.

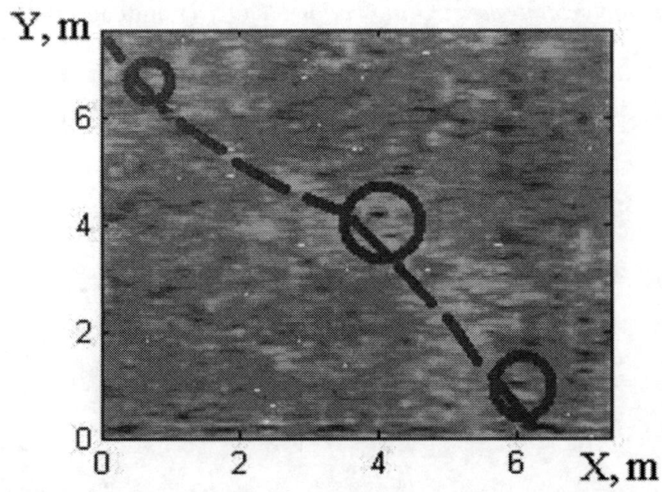

Figure 9. Crosscut of a horizontal zokor's tunnel at a depth of 3 m; data processing without jitter.

4 Conclusion

Analyzing the results of data processing shown in Figures 8 and 9, we can conclude that the jitter application for processing of geolocation data allows fine details of the relief to be detected, artifacts to be smoothed or eliminated at all, and more reliable results to be obtained.

The main purpose of this research is to develop a robust method for detection and recognition of form of subsurface objects and also full three-dimension tomography of subsurface. In the paper presents the alternative approach for geolocation data processing based on separation of jitter. This method can improve resolution of a subsurface image.

Acknowledgment

This work was supported in part by the Russian Foundation for Basic Research (grant (mol_a) No. 16-32-00619 "Study possibilities for applying jitter to increase resolution in the subsurface tomography".

References

[1] W.L. Lai, T. Kind, H. Wiggenhauser, NDT and E International **44**, 111 (2011) doi: 10.1016/j.ndteint.2010.10.002
[2] Ó. Pueyo-Anchuela, A.M. Casas-Sainz, M.A. Soriano, A. Pocoví-Juan, Physics and Chemistry of the Earth **36**(16), 1211 (2011) doi: 10.1016/j.pce.2011.05.010
[3] C. Tinelli, A. Ribolini, G. Bianucci, M. Bini, W. Landini, Comptes Rendus - Palevol **11**, 445 (2012) doi: 10.1016/j.crpv.2012.04.002
[4] L. Song, W. Yang, J. Huang, H. Li, X. Zhang, Journal of Geophysics and Engineering, **10**, 034004 (2013) doi: 10.1088/1742-2132/10/3/034004
[5] R. Sala, R. Tamba, E. Garcia-Garcia, Elements **12**, 19 (2016) doi: 10.2113/gselements.12.1.19
[6] O. Yilmaz, M. Stephen, Doherty *Seismic Data Analysis: Processing, Inversion and Interpretation of Seismic Data* (Society of Exploration Geophysicists, 2000)
[7] R.H. Stolt, Geophysics **43**, 23 (1978)
[8] I.P. Laptev, *Mammals of Western Siberia Taiga* (TSU Publishing, Tomsk, 1958)

[9] *Manual for the ground penetrating radar OKO-2* (Logitcheskiye systemy, Moscow, 2011)

[10] V.P. Yakubov, S.E. Shipilov, D.Ya. Sukhanov, A.V. Klokov, *Radio-wave tomography: achievements and prospects* (NTL,Tomsk, 2014)

MATEC Web of Conferences 79, 01023 (2016) DOI: 10.1051/matecconf/20167901023
IME&T 2016

AUTOMATED COCONUT CHOPPING MACHINE

AhilaRajan Jayasankar[1*], *Fedor* Gubarev[1], *Sadasivan* Theekay Veetil[2], *Samyak* Samar[2], *Ankit* Kumar[2]

[1]National Research Tomsk Polytechnic University, 634050, Tomsk, Russia
[2]Manipal University Jaipur, VPO Dehmikalan, Jaipur, Rajasthan 303007, India

Abstract. The paper explains the purpose, working principle, advantages and applications of the new machine named "Automated Coconut Chopper". The main purpose to design this machine is to automate an activity of a farmer/labor i.e. chopping of green coconut in order to consume water inside it and also designed considering the two main factors safety and cost.

1 Introduction

In India, there are huge numbers of coconut tree cultivation and wide range of its consumption, one of the major consumption is drinking the fluid inside the coconut. Usually a labor is trained to cut the coconut. As the labor /farmers are disappearing gradually, this machine helps to overcome or to substitute this action.

In India, coconut is cultivated mainly in the coastal tracts of Kerala, Tamil Nadu, Karnataka, Andhra Pradesh, Orissa, West Bengal, Pondicherry, and Maharashtra and in the islands of Lakshadweep, Andaman and Nicobar. By the late 1970s it accounted for some 68% of total production in India and at one stage some 899, 198 hectares were reportedly under cultivation. Today Kerala produces roughly 45% of India's coconuts, with some 92% of total production lying in the southern Indian states and Kerala's neighbors [1].

On the increased necessity of time and lack of skilled labor the field of automation and automatic devices is developing in a wide range. To automate a day to day routine activity or to easier an activity, we designed an automated machine "Automated coconut chopper" the combination of various basic mechanical mechanism and only with the scrub materials. The main purpose of this machine is to make "Green coconut" directly available for consuming with reduced intermediate labors.

On the list of various catering machines like SIRMAN vegetable cutter, peeling machine, grinders, roaster, cashew peeling machines from various manufactures we designed a new machine which make the chopping of green coconut easier and also make it economically feasible by all kind of man [2].

* Corresponding author: ahilarajanj@gmail.com

© The Authors, published by EDP Sciences. This is an open access article distributed under the terms of the Creative Commons Attribution License 4.0 (http://creativecommons.org/licenses/by/4.0/).

2 Description of Mechanical system

The general principles of machine design are given in books [3–10]. According to the purposes of projected device, we have suggested and made the following mechanical parts. The overall schematic diagram is presented in Figure 1.

BASE PLATFORM is the base of entire system made of wood with the size of (5×4) inches. All the components like conveyor, cutting stand mechanism, drilling mechanism, all other motors and electrical components are all mounted on it. All the performance or actions of the machine are held under this area or within the base platform.

COCONUT HOLDER is a device where the green coconut is placed and holds for the entire process under taken by the machine. And it is suited on the conveyor system. The overall mechanical view of the coconut holder is shown in the Figure 2, *a*. The base structure is a rigid component of the entire mechanism, wood as a base material is considered due to the load factor. And only this portion of the coconut holder is fastened to the conveyor belt. Holding jaw is the only component is in direct contact with the green coconut. There are two holding jaws in which one is rigid and other is movable. Material considered for this element is wood strongly coated with rubber. Coated rubber helps in strong grip. Rotating rod is fixed at the mid attaching to the base structure and movable jaw. Rod is made of steel and externally threaded for certain region. The movable jaw is assembled to this in order to perform translate action. The working principle or the mechanism is plain screw vice mechanism. Basically all components are rigid except movable jaw and connecting rod. The movable jaw moves to and fro on rotating the connecting rod.

CONVEYOR is used to transfer the coconut to different task and return back to its original position automatically. Flat belt conveyor is used. Power transmission from the motor is done by chain drives. It is considered upon the factor heavy load obtained by the objects and operations performed in the machine. Two chain drives are placed on both sides of the conveyor and driven by a DC motor which is controlled by the controller. Belt conveyor is a machine transporting material in a continuous way by friction drive. It is mainly composed by rack, conveyor belt, belt roll, tensioning device and gearing. Belt conveyor is composed by two endpoint pulleys and a closed conveyor belt. The pulley that drives conveyor belt rotating is called drive pulley or transmission drum; the other one–only used to change conveyor belt movement direction–is called bend pulley. Drive pulley is driven by the motor through reducer, and conveyor belt dragging relies on the friction drag between the drive pulley and the conveyor belt. The drive pulleys are generally installed at the discharge end in order to increase traction and be easy to drag. Material is fed on the feed-side and landed on the rotating conveyor belt, then rely on the conveyor belt friction to be delivered to discharge end.

DOI: 10.1051/matecconf/20167901023

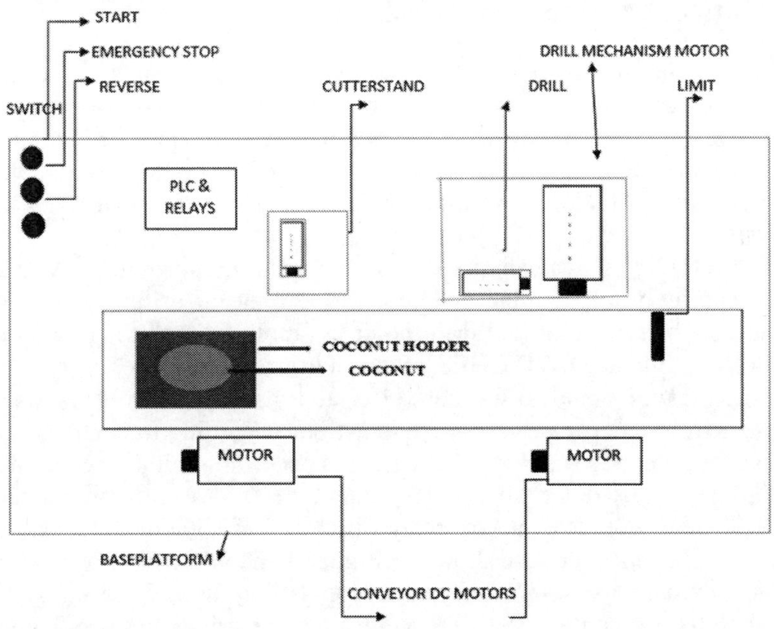

Figure 1. Overall schematic diagram.

CUTTING MECHANISM is used in system for chopping the certain top layer of the coconut. This element consists of rigid stand made of steel fastened to the base platform rigidly holding the cutting motor with the support of holding plate rigidly. The holding plate is adjustable to any height (depends upon the size of coconut).

CHAINDRIVES are common bicycle drives as the project was done by scrub.

DRILLING MECHANISM is used to give automatic feed to the drilling motor. The drill mechanism is shown in Figure 2, *b*. It consists of base structure made of steel fixed to the base platform rigidly to support the entire operation. Holding plate holds the drill motor and it is assembled to the rotating rod in order to perform up and down movement. The connecting rod is also steel and externally threaded to the certain portion where the holding plate is assembled. The rotary motion of this element is done by chain drives which are connected to the DC motor controlled by the conveyor. The working principle of this mechanism is the connecting rod is rotated by motor connected with the chain drive and due to this rotation the holding plates tends to move up and down. This leads as the automatic feed of the drill motor.

a *b* *c*

Figure 2. Parts of the coconut chopping machine: a) coconut holder; b) drill mechanism; c) final assembled machine.

3 Description of Electrical system

The BRAIN of the automation system in this machine is the PLC controller (Siemens S7-200 [11]) which controls all kinds of inputs and output accordingly. Depending on the load of the conveyor system two DC motors are used as the project is done from the scrub materials old wiper motors are used as CONVEYOR MOTORS and also the same kind of motor is used as DRILL MECHANISM MOTOR in which the chain sprocket is assembled on their shafts.

DRILL MOTOR is a small sized portable drilling machine (AC motor) which is modified accordingly to assemble on the hanging plate of drilling mechanism. CUTTER MOTOR is also the same motor of drill motor but replaced with the rotary cutting blades in place of drilling spindle. SWITCHES, push buttons are used for start, stop, and reverse switches which gives signal to the controller and also limit switch is also placed on the conveyor system in order to perform certain actions according to controller. Relays are also used to change the power supply to the actuators according to their requirements. Normally single coil relay is used for all actions except the motor reversing operation and this operation is done by double contactor relay which is more practically feasible to use.

The electrical connections are shown in Figure 3. The PLC controller is switched ON by 12 V DC supply on considering the types of inputs/outputs and the safety precautions. All inputs (switches) are connected to the controller according to specifications. Before all connection to the output relays are used according to the requirement. The main purpose on using single contact relay on cutter and drill motor is to switch the Powėr supply as the output from the controller is in DC and the motors are AC, Relay is required to switch the supply to the motor. Reversing of conveyor motor and drill mechanism motor is required as the machine needs the conveyor move to and fro and also the drill motor up and down.

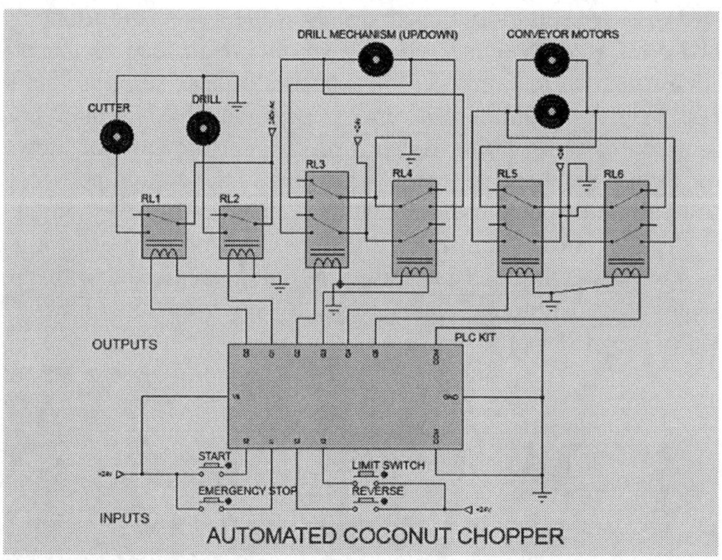

Figure 3. Electrical connections.

This circuit uses a DPDT (Double Pole Double Throw) relay to switch which direction the motor is turning. The motor is connected to both normally closed and normally open (in reverse) sides of the relay. This in effect reverses the wiring whenever the relay is turned on or off. The minimum voltage to drive this circuit is determined by the "pickup" voltage of the relay. This is listed as 9.6 V – but we have found it to function properly on direct connection with the controller.

DOI: 10.1051/matecconf/20167901023

4 Software and operation

The algorithm of the entire machine is shown in the Figure 4. The algorithm and program is designed according to the operations performed by the machine. The sequence of the operation performed by the machine is explained below in details.

1. Coconut is fixed/assembled to the holder (manually).
2. Start button is pressed /Machine is started.
3. The cutter motor starts simultaneously the conveyor motor rotates in forward resulting in forward movement of the conveyor.
4. Conveyor stops exact in front of drilling motor for drilling operation. This action is performed as the limit switch switches ON upon the arrival of coconut holder.
5. Drill motor starts and the drill mechanism motor rotates forward resulting as the automatic feed to the coconut. This operation is held for 10 seconds.
6. Now, drill mechanism motor is reversed in order to bring back the drilling motor to its original position. This operation is timed to 13 seconds.
7. Conveyor motor rotates in reverse resulting in rotating conveyor in reverse direction as the coconut holder is back to its original position and above action is timed to 16 seconds.
8. If the Emergency Stop button is ON all the actions are stopped suddenly and this design is considered on safety precautions.
9. If the Reverse button is ON, the conveyor rotates in reverse resulting in reverse movement of conveyor (used only after emergency stop).

According to the design and manufacturer of the PLC controller, the programming software is used in which the commanding of PLC and various main performances are done. The ladder diagram of the PLC program is shown in Figure 5. Ladder diagrams are specialized schematics commonly used to document industrial control logic systems. Hence used in this machine also.

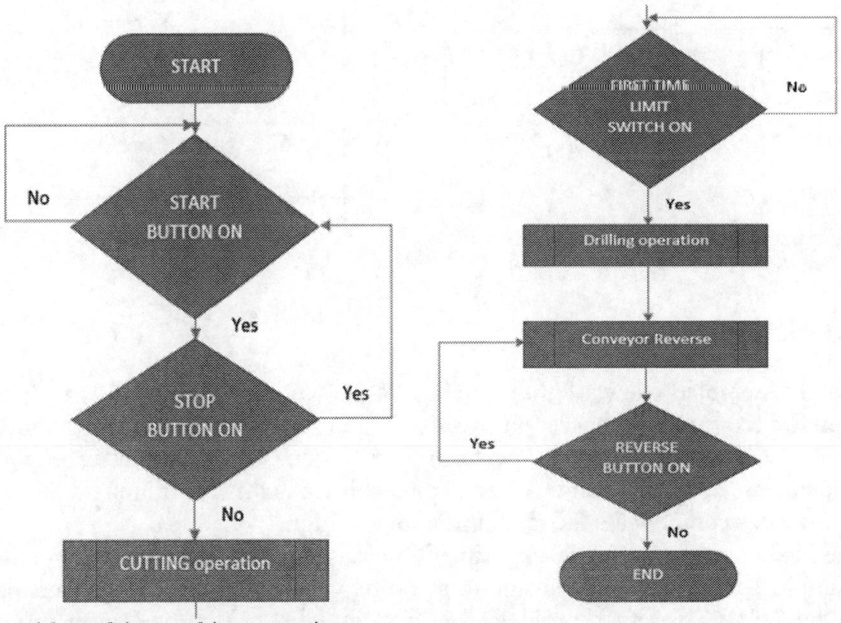

Figure 4. Algorithm of the machine operation.

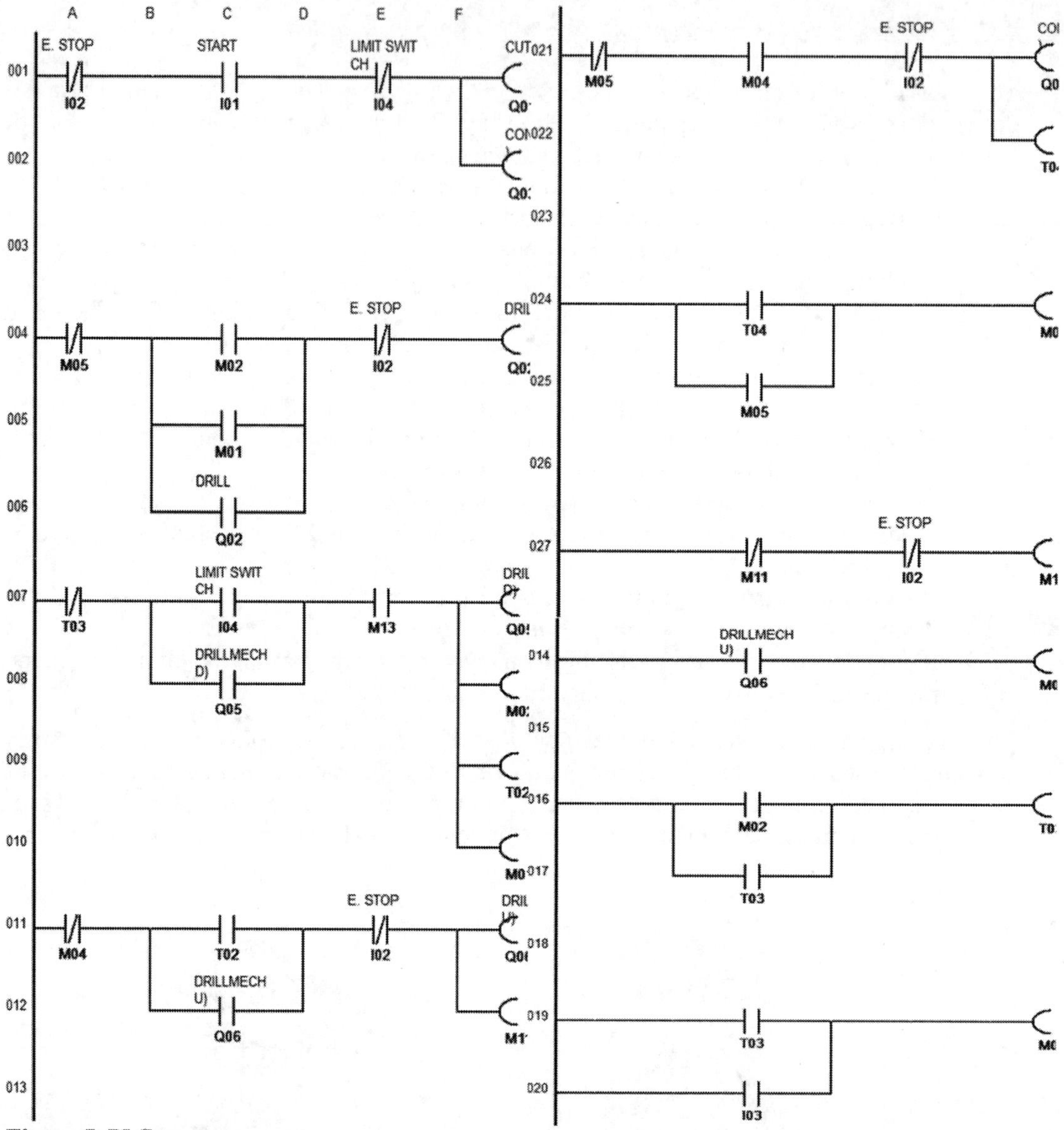

Figure 5. PLC program.

4 Conclusion

The final assembled view of the machine is shown in the Figure 2, *c*. The total time required to perform entire operation by this machine is around 60 seconds which is equal to the time required for labor to perform this action. The time can be reduced if all components are specially designed for this machine. As this machine is done only from the scrub materials/components the machining time is high.

The challenging part while experimenting was the rotation of conveyor due to heavy loads and balance of the belt also and this problem was cleared by having a supportive plate under the belts. The other problem was feeding the drill motor automatically, lots of wobbling was occurred and this problem was over came by adding additional supports to the drill mechanism. And the other problem was conveyor stops while the cutting operation takes place and this leads to damage of coconut also. This problem leads use to design additional motors to the conveyor movements.

This machine costs around 300USD excluding the component PLC. As this method of construction of machine was hit and trail this amount of money is used in many experimental process. On considering from the manufacturer view the machine can be manufactured for around 100USD including the controller where the PLC is replaced by microcontrollers.

Now, there are no intermediate labors/farmers are required to perform this action and also safety priorities are increased with this machine. Usually in India, only on the road sides or outside the buildings the green coconut is available for consumption. But on installing "Automatic money collecting system "with this machine helps to overcome the above problem and this machine can be placed inside the corporate canteens also like automatic coffee machines etc. Due to economic reasons the fixing of coconut is not automated but it can be updated within the expense of 100USD.

References

[1] M.A. Oommen, *Kerala economy since independence* (Oxford & IBH Pub. Co., New Delhi, 1979).
[2] Raw Cashew and Machinery. URL: http://www.cashewmachineryindia.in (2016)
[3] R.K. Singal, Mridual Singal, Rishi Singal, *Basics of Mechanical Engineering* (I.K. International Publishing House Pvt. Limited, New Delhi, 2007)
[4] R.K. Rajput, *A Textbook of Manufacturing Technology (*Firewall Media, New Delhi, 2007)
[5] J.E. Shigley, C.R. Mischke, *Standard Handbook of Machine Design* (McGraw-Hill, New York, 1986)
[6] T. Bartelt, *Industrial Automated Systems: Instrumentation and Motion Control. 1st Edition* (Cram101 Textbook Reviews, 2011)
[7] L.A. Bryan, E.A. Bryan, *Programmable controllers. Theory and implementation* (Industrial Text Company, 1997)
[8] A.J. Tripp, *Basic electricity* (Van Valkenburgh, Nooger & Neville, Inc., 1992)
[9] A.K. Gupta, S.K. Arora, *Industrial Automation and Robotics* (Laxmi Publications, New Delhi, 2007)
[10] D.W. Pessen, *Industrial Automation: Circuit Design and Components* (John Wiley & Sons, New York, 1989)
[11] SIMATIC S7-200. URL: http://w3.siemens.com/mcms/programmable-logic-controller/en/simatic-s7-controller/ s7-200/pages/default.aspx (2016)

DETERMINATION OF THE DEPTH OF CLOSED BLIND CRACKS IN NON-METAL CHECK SAMPLES AND TEST PANELS FOR PENETRANT TESTING

Nikolay Kalinichenko[1], *Aleksey* Kalinichenko[1], *Irina* Lobanova[1*], *Anna* Zaitseva[1], *Egor* Loboda[2]

[1] National Research Tomsk Polytechnic University, 634050, Tomsk, Russia
[2] National Research Tomsk State University, 634050, Tomsk, Russia

Abstract. Check samples and test panels are used to quality assessment and comparing sensitivity of sets of penetrant material. Authors made research in field of manufacturing non-metal test-objects with specified parameters, such as length a width. The paper considers a technique to determine the depth of blind closed cracks in check samples and non-metal test panels for penetrant inspection with a special tool to measure the depth of the undercut depth or a gauge indicator on the outboard support.

1 Introduction

A new version of the International Standard for non-destructive penetrant testing was approved in the late 2006s. The new standard allows the use of different check samples which satisfy specified requirements. However, the most common check samples which meet the International standard Non-destructive testing. Penetrant testing. Part 3: Reference test blocks (EN ISO 3452-3) standard in the European Union are expensive, and in many cases impracticable because they lead to high values of sensitivity levels [1, 2].

Among check samples used abroad, the one to be considered is a set of four plates with nickel-chromium coating with a thickness of 10, 20, 30 and 50 μm (type 1 sample) [3].

The accuracy of the coating thickness uniformity is about 10%. Transverse cracks on each panel are produced by longitudinal stretching of the panel. The plates with a coating thickness of 10, 20 and 30 μm are used to determine the sensitivity level of the fluorescent penetrant systems. The sensitivity of the systems based on color-contrast penetrants is determined by the 30 and 50 μm plates.

The check samples corresponding to EN ISO 3452-3 [3] used to determine the level of sensitivity of flaw detection sets may result in high values. This is due to the U-shaped form of cracks in these samples which have entry to the side of the open cracks. When the penetrant is applied to these samples, the air in the crack cavity is squeezed, and the liquid penetrant easily enters the cavity. Therefore, in some cases, even low sensitivity penetrant

*Corresponding author: konarevai007@tpu.ru

© The Authors, published by EDP Sciences. This is an open access article distributed under the terms of the Creative Commons Attribution License 4.0 (http://creativecommons.org/licenses/by/4.0/).

can provide sufficiently bright contrasting indicator traces referred to a higher sensitivity class.

Penetration of the penetrant in the cavity of the cracks formed in the central part on the check sample surface, i.e. closed cracks, proceeds in a different manner. In this case, the air closed up in the defect cavity prevents its rapid completion by the penetrant, and this is a sufficiently easier way to identify the penetration characteristics during a series of operations [1].

Typically, the passports of check samples for penetrant testing made of metal through nitriding [4–7] with the following exposure to loading indicate that a crack depth is equal to the thickness of the nitrided layer with no numerical values provided. In recent years, the manufacturers of check samples do not report any data on the crack depth since the measurement of the crack depth in the range of tens to hundreds of micrometers is problematic.

The technique that employs non-metal check samples and test panels to eliminate this drawback is considered below.

2 Technology of non-metal test sample manufacturing

The technology of non-metal test sample manufacturing reported in [8] allows manufacturing defects of the designed parameters: length, depth and disclosure, and variation of these parameters.

When implementing an open crack which extends from one side edge to the other side edge of the test sample, this technology allows measurement of the crack depth on both sample edges and calculation of an average depth value.

The variability of the crack depth during manufacturing exhibits the following advantages of the proposed technology. As an example, after manufacturing of the check sample using a 1 mm foil strip, the crack depth can be reduced in further treatment to virtually any desired value (10 – 500) μm, which will determine the defect depth after etching.

The studies that address the problem of manufacturing non-metal check samples allow provide with the technology of manufacturing test panel with both open and closed blind cracks for penetrant testing. However, several cracks of different sensitivity classes may be implemented on the test surface [9].

As mentioned above, the problem to be faced when manufacturing check samples and test panels with closed blind cracks is measuring the crack depths.

Consider one of the following techniques of manufacturing this type of cracks in non-metal test panels with the ability to control the crack depth. A non-metal (based on epoxy adhesive) bar of the size smaller than that of the future test panel is prepared. Then, metal strips of equal length and width are cut. In this case, the strip workpieces were made of the following materials:
- mica plate with a deposited layer of silver;
- silver leaf;
- metal foil.

The problem of implementation of test panels, particularly of sensitivity class I, is the absence of foil of required thickness (less than 1 μm). However, mica plates coated with the silver layer (approximately 1 micron) can be used instead.

The strips are adhered in pairs (of equal thickness) opposite to each other on the sides of the bar in its lower (bottom) part with epoxy adhesive (Figure 1).

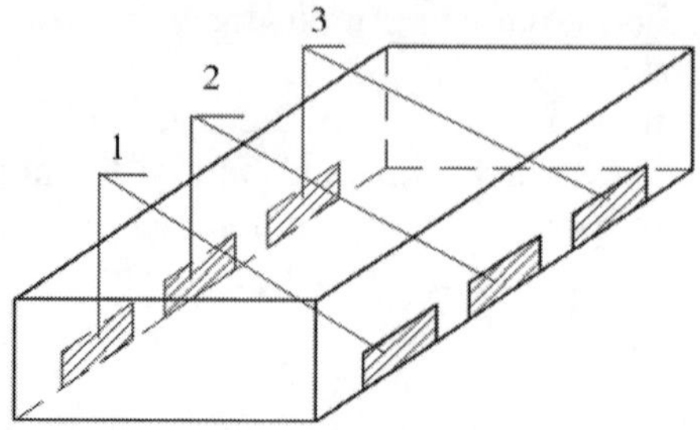

Figure 1. Rectangular non-metal bar with fixed strips 1–3.

Then, the bar is placed in the pre-prepared larger mold that will determine the size of the test panel. The mold is filled with epoxy adhesive up to the desired level. Within the first minutes of its solidification, an ultrasonic bath is used to reduce the amount of bubbles.

After epoxy solidification, the workpiece is removed from the mold, grinded to the required parameters of cracks (especially depth) and polished. In this case, the bottom surface of the panel is its test surface.

The technique to control the crack depth in the investigated test panel is proposed below. Several 2 – 3 mm diameter holes are to be drilled in the workpiece test (bottom) surface, as shown in Figure 2.

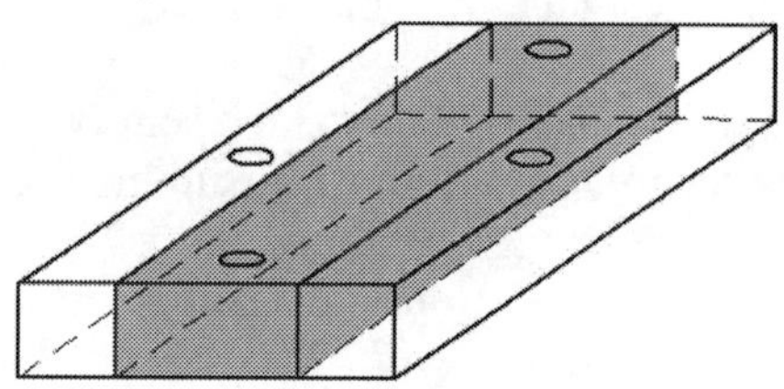

Figure 2. Rectangular non-metal test panel with drilled holes.

The hole depth is equal to the width of the adhered metal strips. During treatment, the hole depth and thus the remaining strip width may be controlled using a special tool to measure the undercut depth [10], i.e. the dial depth gauge with outboard support (Figure 3, *a*, *b*). These tools employ an indicating gage with the measurement limit of 0 – 10 mm and resolution of 10 μm as a reading device. You can also use a high-precision digital vernier caliper designed for welds (Figure 3, *c*).

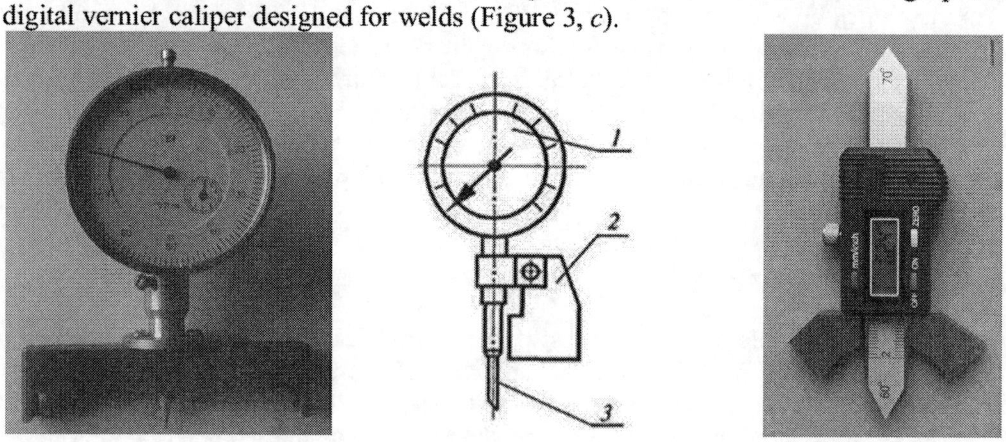

a b c

Figure 3. Dial depth gauge with outboard support (a, b, c); tool to measure the depth of undercuts: indicator "0 – 10" with a rotating scale (1); 2 – support bracket; 3 – measuring pin.

Excessive grinding of the sample material during treatment leads to unevenness of both the strip width and the crack depth. The deviation of the crack depth unevenness (defect) flatness can be significantly reduced by using the technique of non-metal check sample manufacturing considered further on. In addition to compliance with the standards recommended for the metal section manufacturing [11], metal strips should be placed at the workpiece edge maximum distance. Since the strips are located in the center of the workpiece at a considerable distance from the workpiece edge, the width of the remaining metal strips after treatment is almost equal. This makes the crack depths virtually equal after etching of the remaining strips.

When the width of metal strips reduces as a result of the workpiece test surface treatment, the hole depth decreases. The width of the metal strips is controlled by measuring the hole depth using a dial depth gauge or the tool to measure the depth of undercuts. The holes in different areas of the test surface makes possible to control the width of metal strips to make it uniform. In the end of treatment, the required width is achieved in metal strips, which ultimately determine the crack depth in the test panel. The resulting treatment of the bar test surface must be performed with the abrasive material of a minimal grit to provide a minimal test surface when using a test panel for further work.

After that, the groove is cut in the test surface to divide the test panel into two symmetrical parts (Figure 4). With this, traces of holes in the central part of the panel are eliminated. The remaining holes will not interfere with the analysis of the test results, as the holes are wide and shallow and do not add any indicator traces.

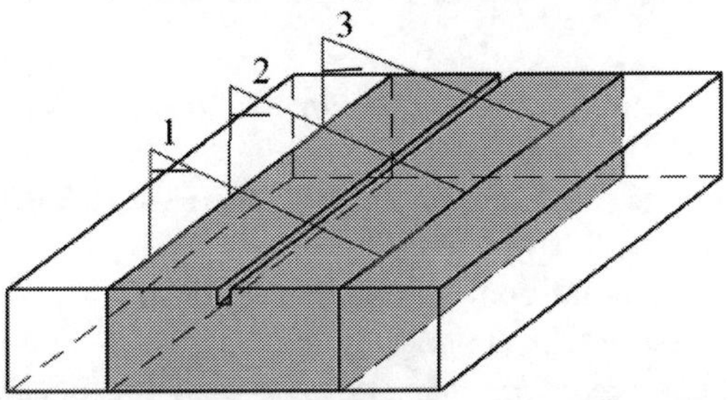

Figure 4. Test panel according to sensitivity classes I–III: 1 – crack of sensitivity class III; 2 – crack of sensitivity class II; 3 – crack of sensitivity class I.

The remains of the metal foil are etched with ferric chloride solution. The obtained cracks are tested for the opening width.

Consider another technique to determine the depth of cracks. Since the dividing groove implementation is required for manufacturing of these test panels, the groove can be used to control the crack depth.

In doing so, the dividing groove of uniform depth, which is not less than the width of metal strips, is cut before the treatment of the bottom surface of the sample. Further treatment of the sample surface enables the control of the groove depth along its entire length, and therefore the width of strips to obtain the required values. In this embodiment, no control holes are required.

Consider another technique for manufacturing the test panel according to different sensitivity classes with the ability to control the depth of blind closed cracks. In our opinion, it is the most manufacturable one. To implement the technique, we need to:
- make the mold of the size equal to that of the future test panel;
- prepare metal strips to implement the cracks with the required parameters;
- adhere the pairs of metal strips to the bottom surface opposite to each other;
- fill the mold with epoxy adhesive;

- cut a groove of the depth more than the strip width (2 – 3 mm) in the center of the bar bottom surface between the rows of the strips after epoxy curing;
- treat the bottom surface unless the required strip width (the depth of future cracks) is achieved;
- etch the remaining metal strips;
- control the parameters of the obtained dimensions of the blind closed cracks.

The parameters of the test panel implemented according to the technique proposed are as follows:

- opening width of cracks: No. 1 – 0.9 μm, No. 2 – 2 μm, No. 3 – 15 μm;
- crack length: 25 mm;
- crack depth: 100 μm.

3 Conclusion

The techniques proposed to control the depth of blind cracks in check samples and non-metal test panels enable a high accuracy in determination of their depth.

Manufacturing of samples with a wide range of crack depths allows for various practical studies on hydrodynamics of the defect backfill and development, and on the nature of the impact of external factors (temperature, humidity, etc.) on the defect parameters.

References

[1] N.P. Migun, N.B. Gnusin, *Heat exposure at the capillary non-destructive testing* (Nauka, Minsk, 2011)
[2] Zh. Zhang, H. Feng, W. Zhao, MATEC Web of Conferences **40**, 04001 (2016) doi: 10.1051/matecconf/20164004001
[3] ISO 3452-3–1998 International standard Non-destructive testing. Penetrant testing. Part 3: Reference test blocks (ITD)
[4] J.M. William, *Test piece for inspection penetrant performance assessment and comparison* US Patent 6729175 B2 (2004)
[5] F. J. Vicki, S. Shimizu, A *Surface penetrant inspection test piece having varying thickness plating* US Patent 4610157 (1986)
[6] W. J. Martin, *Test piece for inspection penetrant performance assessment and comparison* Patent US 6311538 B1 (2001)
[7] P. Dunnwald, *Non-destructive liquid penetrant inspection process integrity verification test panel* Patent WO 2011097146 A1 (2011)
[8] N.P. Kalinichenko, A.N. Kalinichenko, I.S. Lobanova, A.Y. Popova, S.S. Borisov, Meas. Tech. **57**, 484 (2014) doi: 10.1007/s11018-014-0485-1
[9] N.P. Kalinichenko, A.N. Kalinichenko, I.S. Lobanova, A.A. Zaitseva, E.L. Loboda, J. Phys.: Conf. Ser. **671**, 012033 (2016) doi: 10.1088/1742-6596/671/1/012033
[10] RD 03-606-03, Instruction on visual testing (Gosgortechnadzor of Russia)
[11] Yu.P. Egorov, *Collection of guidelines for laboratory works on the course "Materials"* (Tomsk Polytechnic University, Tomsk, 2014)

MATEC Web of Conferences **79**, 01025 (2016) DOI: 10.1051/matecconf/20167901025
IME&T 2016

COMPARISON OF THE INFLUENCE OF "SOLVENT/NON-SOLVENT" TREATMENT FOR THE ATTACHMENT OF SIGNAL MOLECULES ON THE STRUCTURE OF ELECTROSPUN PCL AND PLLA BIODEGRADABLE SCAFFOLDS

Semen Goreninskii[a], *Ksenia* Stankevich, *Evgeny* Bolbasov, *Nadezhda* Danilenko, *Victor* Filimonov, *Sergei* Tverdokhlebov

[1] National Research Tomsk Polytechnic University, 634050, Tomsk, Russia

Abstract. Electrospun biodegradable scaffolds (matrixes) made of poly(ε-caprolactone) (PCL) and poly(L-lactic acid) (PLLA) are three-dimensional fibrous structures that are commonly used in regenerative medicine and drug delivery systems. Modification of such structures allows manipulating with biological and immune response. Previously, our team suggested a number of surface modification strategies for thin films made of PLLA. One of the proposed strategies are based on treatment of the material with "solvent/non-solvent" mixture that allows absorbing biologically active molecules or linkers on the surface of the sample. The aim of this work was to compare the influence of "solvent/non-solvent" treatment on the structure and crystallinity of the elecrospun biodegradable PCL and PLLA scaffolds. For that purpose, original PCL and PLLA scaffolds were treated with mixture of toluene and ethanol in different proportions. Morphology of the obtained samples was studied using scanning electron microscopy. It was shown that "solvent/non-solvent" treatment doesn't lead to changes in scaffolds morphology such as gluing or cutting of the matrix fibers. By means of X-ray diffraction analysis it was shown that treatment of the samples with selected mixtures doesn't change material crystallinity. Thus, it was demonstrated that proposed composition of the "solvent/non-solvent" mixture can be used for the modification of electrospun PCL and PLLA scaffolds.

1 Introduction

Poly(L-lactic) acid (PLLA) and poly(ε-caprolactone) are biodegradable polymers widely applied in modern medicine [1]. Because of their outstanding properties, such as degradability [2], processability [3] and biocompatibility [4] these materials are used in production of the wide range of biomedical devices such as catheters [5], coronary stents [6], and implants [7].

[a] Corresponding author: semgor93@gmail.com

© The Authors, published by EDP Sciences. This is an open access article distributed under the terms of the Creative Commons Attribution License 4.0 (http://creativecommons.org/licenses/by/4.0/).

However, a number of complications arise due to the relative chemical inertness of these materials including a lack of biological activity, irritation and inflammation processes. In order to improve the surface properties of PLLA thin films both physical [8] and chemical [9] methods may be applied.

Solution blow spinning (SBS) [10] and electrospinning (ES) [11] are common techniques allow for the production of free-dimensional fibrous structures called matrixes (scaffolds). Because of their specific architecture, biodegradable scaffolds are being used in tissue engineering as an extracellular matrix for support of the cell growth [12]. Moreover, their biodegradability and controllable degradation rate give an opportunity for the target drug delivery [13].

Previously our team reported a number of PLLA modification strategies. One of them allows attachment of biologically active molecules and reactive groups on the surface of the material and based on treatment of the polymer with "solvent/non-solvent" mixture (Figure 1) [14, 15]. It was suggested that proposed modification method may be applied on the other biodegradable polymers such as PCL.

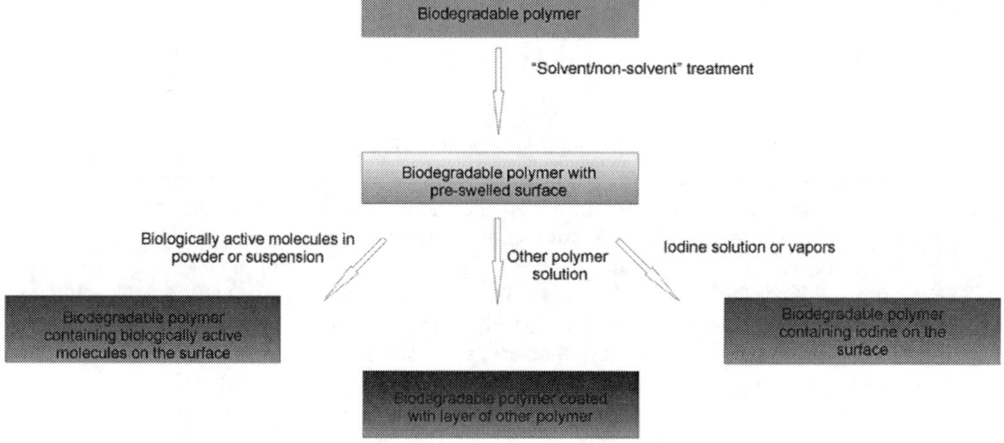

Figure 1. Proposed method for the modification of biodegradable polymers.

The aim of our research was to identify the possibility of application of such method for the modification of electrospun PCL and PLLA scaffolds and to reveal convenient modification conditions. The influence of the treatment on the structure of both materials was compared. In order to investigate the influence of "solvent/non-solvent" treatment on the morphology of the samples scanning electron microscopy (SEM) was used. In addition, X-ray diffraction analysis was applied for investigation of the changes of the material crystallinity after the treatment.

2 Materials and methods

2.1 Preparation of nonwoven materials

All polymer solutions were prepared in a sealed glass reactor at a room temperature. Electrospinning process was performed on NANON-01A (Mecc Co., Japan) equipment. The electrospun nanofibers were placed on a frame cylinder collector (a diameter of 100 mm, a length of 210 mm) with a rotation speed of 50 rpm. The distance between syringe tip and collector was 150 mm for PLLA (PL 38, Purasorb, M_w=670000 g/mol) and 190 mm for PCL (PCL, Sigma-Aldrich, M_w=70000–90000 g/mol). PLLA matrixes were prepared from 3% (w/w) solution in chloroform under the following conditions: voltage of 27 kV, feed rate of 6 mL/h. For PCL matrixes production the following parameters were used: voltage of 20 kV, feed rate of 6 mL/h.

2.2 "Solvent/non-solvent" treatment of the electrospun scaffolds

In order to investigate the influence of the "solvent/non-solvent" treatment on the properties of PLLA and PCL electrospun scaffolds, the original materials were soaked in the mixture of toluene and ethanol in different proportions (1/9, 0.9/9.1, 0.8/9.2, 0.7/9.3, 0.6/9.4, 0.5/9.5, v/v) for 10 minutes. Then, all the samples were placed in vacuum for 24 hours to remove residual solvents.

2.3 Scanning electron microscopy

For analysis of the scaffolds morphology scanning electron microscopy (SEM, ESEM Quanta 400 FEG, FEI, USA) was used. All the samples were pre-coated with gold using Smart Coater (Jeol, Japan) sputtering system. The scanning was carried out at a high vacuum, with accelerating voltage of 20 kV and magnification of 1000 and 21000 times. Diameter of material fibers was measured by using ImageJ 1.38 software (National Institutes of Health, USA).

2.4 X-ray diffraction analysis

The crystalline structure of the samples was investigated by means of X-ray diffraction analysis (DRON-3M, Russia). During the analysis all samples were irradiated with Cu K-radiation with a wavelength of 1.54056 Å under following parameters: accelerating voltage of 35 kV, beam current of 25 mA and scanning rate of 4°/min. Scanning angle range was from 12° to 22° and from 21° to 26° for PLLA and PCL matrixes respectively. The average crystal size of the samples was calculated according to Debye–Scherrer equation:

$$l_c = \frac{\kappa\lambda}{cos\theta\sqrt{\beta^2-\beta_r{}^2}}, \quad (1)$$

where κ=0.9, λ is the wavelength if the incident radiation, β is the width of the reflection at a half height, β_r is the broadening reflex of the apparatus, θ is the angle of diffraction.

2.5 Statistics

Statistical analysis was performed in Prism (GraphPad Software, USA) software. Non-parametric Mann–Whitney test was used.

3 Results and discussion

3.1 Scaffolds morphology

SEM images of the original and treated PCL and PLLA scaffolds are shown in the Figure 2.

DOI: 10.1051/matecconf/20167901025

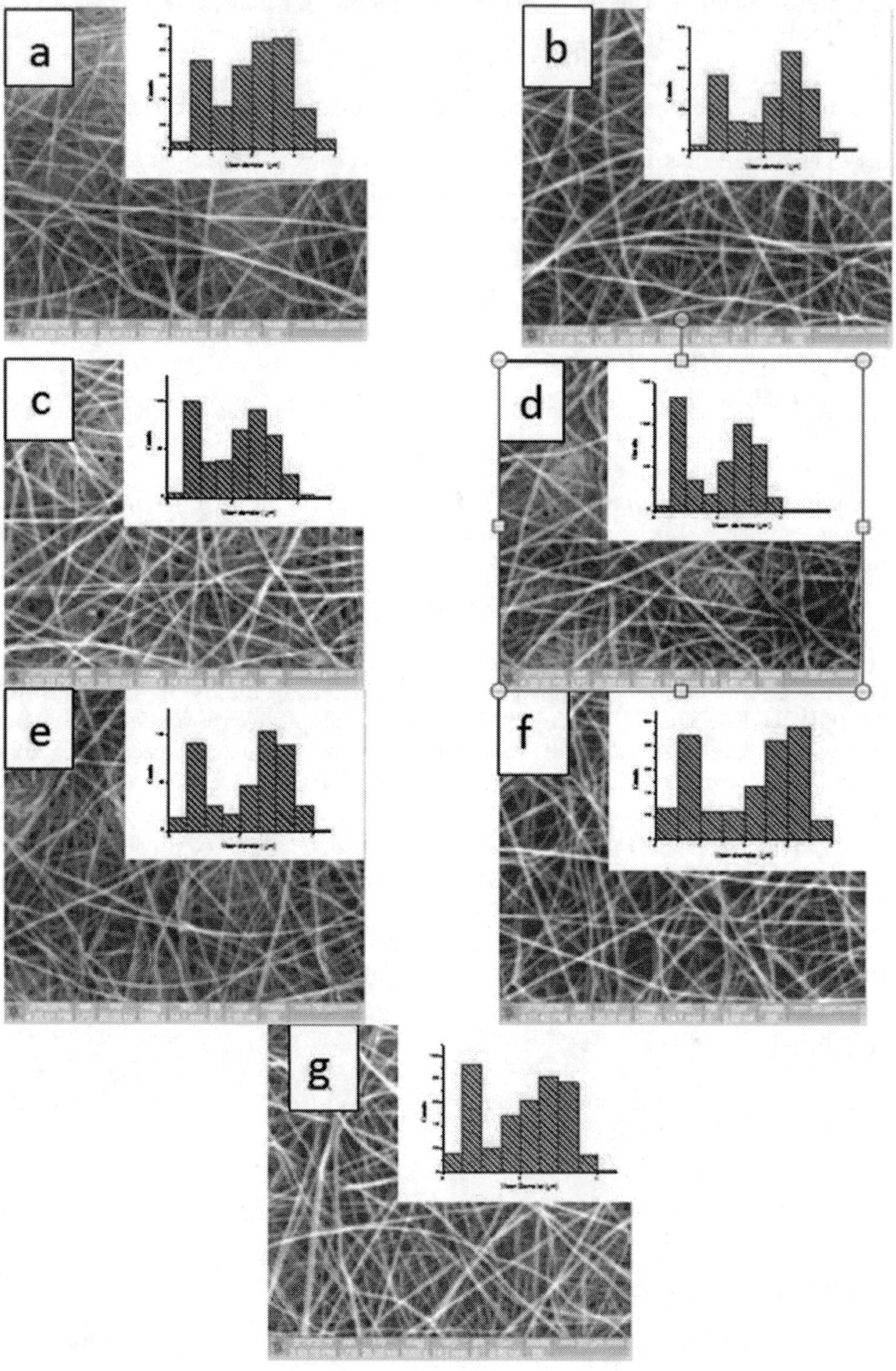

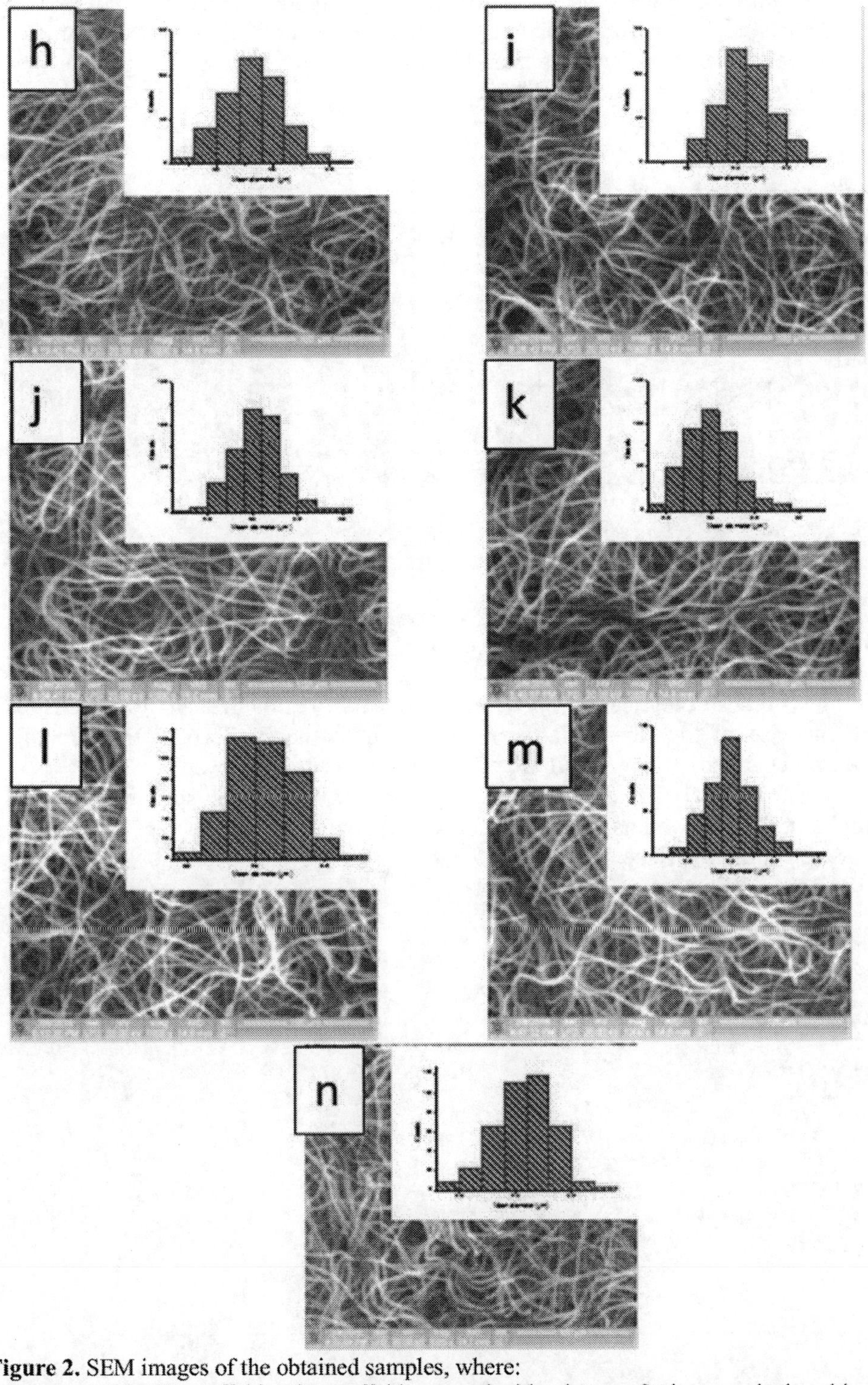

Figure 2. SEM images of the obtained samples, where:
a – non-treated PCL scaffold, PCL scaffolds, treated with mixture of toluene and ethanol in different proportions: b – 0.5/9.5, v/v, c – 0.6/9.4, v/v, d – 0.7/9.3, v/v, e -0.8/9.2, v/v, f – 0.9/9.1, v/v, g – 1/9, v/v; h – non-treated PLLA scaffold, PLLA scaffolds, treated with mixture of toluene and ethanol in different proportions: i – 0.5/9.5, v/v, j – 0.6/9.4, v/v, k – 0.7/9.3, v/v, l – 0.8/9.2, v/v, m – 0.9/9.1, v/v, n – 1/9, v/v.

It was found that produced PCL scaffolds have bimodal distribution of the fiber diameter, whereas PLLA scaffolds demonstrate unimodal distribution. Mean diameter of the fibers was calculated from not less than 400 measurements on each image. Changes of the mean diameter of the fibers are presented in Figure 3.

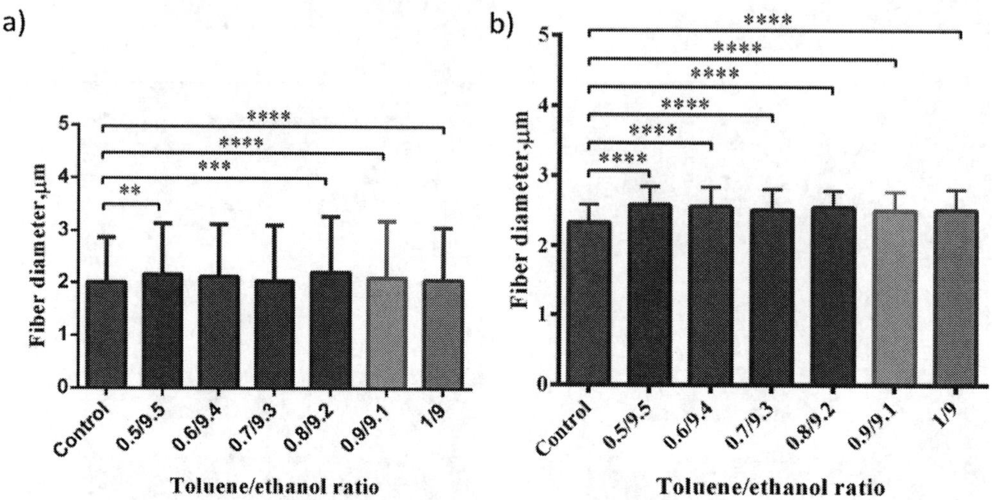

Figure 3. Mean diameter of the scaffolds after the treatment with mixture of toluene and ethanol in different proportions: a) PCL b) PLLA (**** p<0.0001, *** p<0.001, ** p<0.01).

It was found that in most cases "solvent/non-solvent" treatment leads to a statistically significant increase of the fibers diameter. On the other hand, fibrous structure of the material wasn't changed: gluing or cutting of the fibers was not observed.

Thus, toluene/ethanol mixture in proportion of 1/9 (v/v) was selected for further modifications of PCL as well as PLLA scaffolds as it contains more solvent for better pre-swelling of the polymer surface.

3.2 X-ray diffraction analysis

Figure 4 show representative XRD patterns of the obtained samples of PCL and PLLA scaffolds.

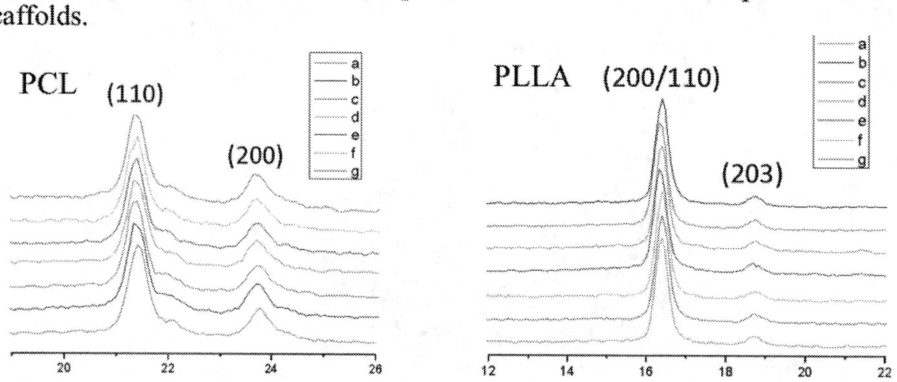

Figure 4. XRD patterns of PCL and PLLA scaffolds, where:
PCL: *a* – non-treated PCL scaffold, PCL scaffolds, treated with mixture of toluene and ethanol in different proportions: *b* – 0.5/9.5, v/v, *c* – 0.6/9.4, v/v, *d* – 0.7/9.3, v/v, *e* -0.8/9.2, v/v, *f* – 0.9/9.1, v/v, *g* – 1/9, v/v; PLLA: *a* – non-treated PLLA scaffold, PLLA scaffolds, treated with mixture of toluene and ethanol in different proportions: *b* – 0.5/9.5, v/v, *c* – 0.6/9.4, v/v, *d* – 0.7/9.3, v/v, *e* – 0.8/9.2, v/v, *f* – 0.9/9.1, v/v, *g* – 1/9, v/v.

XRD patterns of the produced PCL scaffolds are presented by two reflections in 21.3° and 23.7° regions, corresponding to (110) and (200) plates. No additional reflexes were found after the treatment. Average size of PCL and PLLA crystallites, calculated according to the equation (1), is shown in the Table 1. In XRD patterns of PLLA scaffolds reflexes appearing in two areas: 16.4° and 18.8°, corresponding to (200/110) and (203) plates. No additional reflexes were found after the treatment. For PCL and PLLA scaffolds "solvent/non-solvent" treatment had no influence on the phase structure of the polymer. Thus, according to the obtained data, "solvent/non-solvent" treatment didn't lead to the formation of additional phases.

Table 1. Average size of PCL and PLLA crystallites.

Proportions of toluene and ethanol, v/v	Average crystallite size of PCL scaffolds, treated with mixture of toluene and ethanol, nm	Average crystallite size of PLLA scaffolds, treated with mixture of toluene and ethanol, nm
Non-treated	19.35±0.61	21.61±0.38
0.5/9.5	19.80±0.65	21.05±0.32
0.6/9.4	19.35±0.59	21.05±0.31
0.7/9.3	19.51±0.70	21.24±0.55
0.8/9.2	19.48±0.21	21.61±0.33
0.9/9.1	19.35±0.61	20.69±0.53
1/9	18.89±0.28	21.07±0.83

Statistical analysis indicated that treatment with "solvent/non-solvent" mixture with all the selected compositions has no significant effect on the size of PCL crystals. Size of PLLA crystals also didn't change significantly after "solvent/non-solvent" treatment.

As no influence on the crystal structure of the polymer was observed with toluene concentration increase, the mixture of toluene and ethanol (1/9, v/v) was chosen for further experiments.

4 Conclusion

The influence of "solvent/non-solvent" treatment on the structure of electrospun nanofibrous scaffolds made of poly(ε-caprolactone) and poly(L-lactic) acid was compared. It was shown, that treatment with "solvent/non-solvent" mixtures containing toluene and ethanol in different proportions has no crucial effect on the structure and crystallinity of both materials. Thus, this solvent system will be used for the attachment of signal molecules on the biodegradable scaffolds surface for programming and regulating of biological and immune response.

Acknowledgements

This research was funded by Russian Science Foundation (project № 16-13-10239) and performed in Tomsk Polytechnic University.

References

[1] H. Tian, Z. Tang, X. Zhuang, X. Chen, X. Jing, Prog. Polym. Sci. **37**, 237 (2012) doi: 10.1016/j.progpolymsci.2011.06.004

[2] T. Ishigaki, W. Sugano, A. Nakanishi, M. Tateda, M. Ike, M. Fujita, Chemosphere **54**, 255 (2004) doi: 10.1016/S0045-6535(03)00750-1

[3] S. Frackowiak, J. Ludwiczak, K. Leluk, M. Kozlowski, Compos Part B-Eng. **92,** 1 (2016) doi: 10.1016/j.compositesb.2016.02.036

[4] L. Wu, L. Wang, X. Wang, K. Xu, Acta Biomater **6**, 1079 (2010)

[5] I. Gradus-Pizlo, R.L. Wilensky, K.L. March, N. Fineberg, M. Michaels, G.E. Sandusky, D.R. Hathaway, J. Am. Col. Cardiol. **26**, 1549 (1995) doi: 10.1016/0735-1097(95)00345-2

[6] C.M. Bünger, N. Grabow, K. Sternberg, C. Kröger, L. Ketner, K.-P. Schmitz, H.J. Kreutzer, H. Ince, C.A. Nienaber, E. Klar, W. Schareck, J. Surg. Res. **139**, 77 (2007) doi: 10.1016/j.jss.2006.07.035

[7] K.E. Uhrich, D. Abdelhamid, Biosynthetic Polymers for Medical Applications, 63 (2016) doi: 10.1016/B978-1-78242-105-4.00003-1

[8] R. Morent, N. De Geyter, T. Desmet, P. Dubruel, C. Leys, Plasma Process. Polym. **8**,171 (2011) doi: 10.1002/ppap.201000153

[9] R.M. Rasal, A.V. Janorkar, D.E. Hirt, Prog. Polym. Sci. **35**, 338 (2010) doi: 10.1016/j.progpolymsci.2009.12.003

[10] E.N. Bolbasov, Y.G. Anissimov, A.V. Pustovoytov, I.A. Khlusov, A.A. Zaitsev, K.V. Zaitsev, I.N. Lapin, S.I. Tverdokhlebov, Mater. Sci. Eng. C **40**, 32 (2014) doi: 10.1016/j.msec.2014.03.038

[11] A. Moheman, M.S. Alam, A. Mohammad, Adv. Colloid. Interfac. **229**, 1 (2016) doi: 10.1016/j.cis.2015.12.003

[12] S. Beke, R. Barenghi, B. Farkas, I. Romano, L. Kőrösi, S. Scaglione, F. Brandi, Mater. Sci. Eng. C **44**, 38 (2014) doi: 10.1016/j.msec.2014.07.008

[13] G.N. Abdelrasoul, B. Farkas, I. Romano, A. Diaspro, S. Beke, Mater. Sci. Eng. C **56**, 305 (2015) doi: 10.1016/j.msec.2015.06.037

[14] K.S. Stankevich, A. Gudima, V.D. Filimonov, H. Klüter, E.M. Mamontova, S.I. Tverdokhlebov, J. Kzhyshkowska, Mater. Sci. Eng. C **51**, 117 (2015) doi: 10.1016/j.msec.2015.02.047

[15] V. D. Filimonov, K. S. Stankevich, S. I. Goreninsky, N. V. Danilenko, S. I. Tverdokhlebov, 3rd International Conference on Chemical Technology: Book of Abstracts, 49 (2015)

A MOBILE FULL-TIME DAILY SYSTEM FOR FETAL MONITORING

Artem Bureev[1], *Elena* Vaganova[2], *Dmitry* Zhdanov [1,2*], *Ivan* Zemlyakov[1], *Ekaterina* Dikman [1]

[1] Diagnostika + LLC , 634055,Tomsk , Russia
[2] National Research Tomsk State University, 634050, Tomsk, Russia

Abstract. The article describes a mobile hardware and software system designed for daily monitoring of the state of fetal and maternal cardiovascular systems. The assessment is carried out by means of recording and further online analysis of acoustic data, obtained from the abdominal surface of a pregnant woman's body. The components and operating principles of the hardware and software system designed are described. The results of experimental studies aimed at assessing the applicability of a method of acoustic data analysis implemented in the system developed are shown. The results obtained have been compared with the results obtained using cardiotocography.

1 Introduction

Fetal monitoring of the state of the cardiovascular system in the antenatal period is the most common non-invasive method of a fetus wellbeing control. The relevance of the method is confirmed by numerous publications (both professional and popular scientific) in Russian and foreign medical literature. Currently used methods are based on the analysis of trends in fetal heart rate, resulting in the construction of a *rhythmogram* of *between systolic* time *intervals*, known as a cardiotocogram. The procedure for its construction is called cardiotocography. Currently, this technique is based on the ultrasound examination for outpatients, where physical heart contractions of a fetus are assessed with a Doppler sonography. Frequent use of ultrasound in medicine is not considered to be fully secure. Until now there is no definite answer whether the ultrasound acoustic power (ALARA) within 30–40 minutes is safe for fetal development [1, 2]. There is very little information on possible hidden biological effects of ultrasound on the development of a human embryo or fetus. There is evidence that even low power ultrasound can affect the development of brains and visceral organs of small animals. However, direct adaptation of these results to human beings appears invalid [1, 2].

In Russia according to the federal standard of care for women with normal pregnancy (Order of the Ministry of Health of the Russian Federation, Sept. 14, 2006 № 662) the number of recommended sessions of cardiotocography should not exceed three times. Some researchers still believe that the level of ultrasound power during cardiotocography is quite

* Corresponding author: dim@diagnostic.tom.ru

© The Authors, published by EDP Sciences. This is an open access article distributed under the terms of the Creative Commons Attribution License 4.0 (http://creativecommons.org/licenses/by/4.0/).

low and the technique is not contraindicated for pregnant women, even in case of five or more sessions [1]. However the predictive value of cardiotocogram does not surpass two weeks. Thus long-term monitoring in the case of a pathology of fetal development may need to involve additional safer methods.

At the present time, cardiotocograms obtained by means of complex electrodes applied to the anterior abdominal wall of the pregnant woman, are increasingly frequently used as an alternative to ultrasonic methods.

The procedure for obtaining cardiotocograms is usually performed in a hospital, under the condition of a complete physical rest of a pregnant. This is due to the fact that the amplitude of the cardiotocogram teeth, recorded from the abdominal wall of a mother, is about 100 times smaller than the amplitude of its electrocardiogram. Nevertheless there are many cases when portable devices for fetal daily monitoring are necessary.

Monica Healthcare (GB) has developed a wearable device for daily monitoring of fetal heart rate Monica AN24. The device measures the fetal and maternal heart rate, and other parameters by recording fetal and maternal electrocardiograms as well as electromyograms of maternal uterine muscles. The device is a hardware system consisting of a portable battery-powered recorder (about 350 gr.), and the doctor's computer work station on the basis of a personal computer with specialized software. It practically helps doctors to avoid false decisions [3]. However despite the possibility of the continuous monitoring, reliability and safety of the method, there is a certain complexity of fetal electrocardiogram recording and fetal heart rate assessment. Receiving of high quality cardiotocograms using MONICA AN24 is only possible with limited mobility of the patient (even better in the supine position) [3, 4].

Certainly fetal electrocardiogram has undeniable advantages. However, separating the basic QRS-complex (or at least R-wave fetal electrocardiogram) against the background electromiotic activity skeletal and respiratory muscles, as well as the maternal electrocardiogram is challenging and requires significant computing resources. It makes difficult to assess fetal heart rates in real time. In addition, for improvement of the signal / noise ratio patient's preparation for the electrode application requires careful processing of the skin surface of her skin with a special fluid, removing the horny layer of the epidermis [3].

2 Methods

The team of project developers have proposed a portable fetal monitor using passive acoustic detection of fetal heart tones recorded (phonocardiography) [5]. Currently there are no commercially available or being under research and development wearable daily monitors for control of fetal and maternal heart rates using phonocardiography. Various device for home use, manufactured in China and other countries of Southeast Asia and allow to listen to the fetal heart tones. However due to due to external factors they can not be assigned to *medical daily monitors*.

The developed prototype of a mobile daily monitor of fetal and maternal heart rates consists of four units (Figure 1):

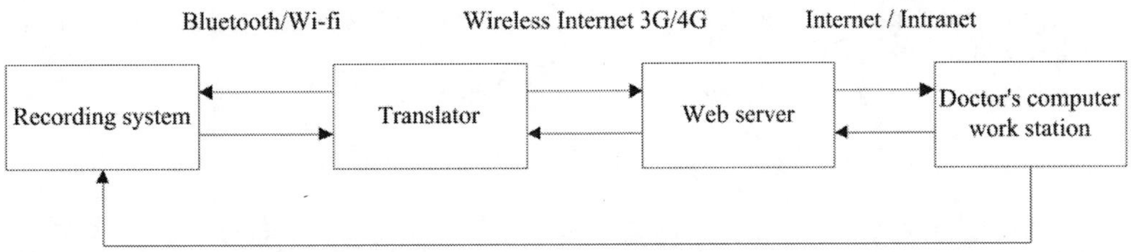

Figure 1. Block scheme of a prototype of a mobile daily monitor.

A wearable recorder is a compact battery-powered device (95×115×30 mm). Fetal heart tones are recorded from the anterior abdominal wall of a pregnant woman by piezoceramic transducers adjusted by a doctor. Their location depends on the fetal attitude (Figure 2).

In the developed mobile version of the system a smart/mobile phone (with Internet access), is a transitional translator. Special software receives and processed signals and sends them to a remote server via wireless Internet. Before sending the data they are encoded and archived in accordance with the Federal Law of the Russian Federation regarding personal data protection (№153, 27.07.2006). Ddepersonalization of data transmissed is performed by a system of personal codes and markers. The experimental results showed that the computing capacity of a Android 4.x smartphone of a low price range (less than $ 50) is enough for the task.

Besides reception, processing and transmission of data, a smartphone allows to get information about the recording system. The processed signals are presented in text and graphic form. The mode of voice and text communication with your doctor is available. In case of emergency signals from the recording system are transmitted to the doctor's computer work station through a translator and server online.

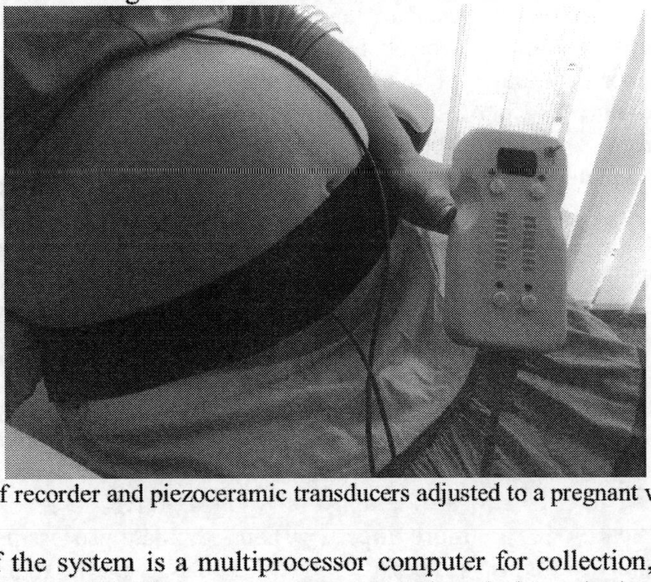

Figure 2. Layout of recorder and piezoceramic transducers adjusted to a pregnant woman's body.

The server of the system is a multiprocessor computer for collection, systematization, gathering, storage, modification, processing, analysis of data obtained from portable recorders. To ensure high-quality connection with translators the server has a fiber-optic communication link with the provider. The server is based on complex programs. They search, identify, process, filter and place the data obtained in a database. All the necessary processing procedures are done automatically. They involves identification and removing artifacts from acoustic signals of heart tones, evaluation of *between systolic* time *intervals*, and building of cardiotocograms.

After evaluating fetal heart rate using cardiotocography, analytical complex of the server begins evaluation of well-being of the fetus. If it is within normal range, a brief conclusion will be sent to a patient's smartphone.

A cardiotocograms, built automatically, is shown in Figure 3; duration 62 minutes.

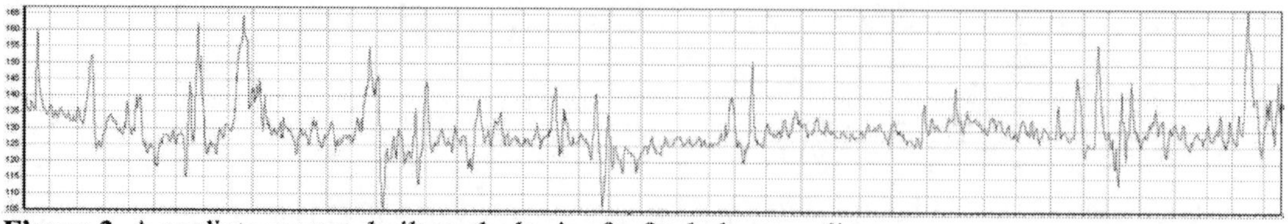

Figure 3. A cardiotocogram, built on the basis of a fetal phonocardiogram.

Comparison of cardiotocograms, built on the basis of a fetal phonocardiogram and built on the basis of an ultrasound investigation, showed imperceptible (no more than 3%) differences in the assessment of *between systolic* time *intervals*.

In case of deviations, anticipated in the settings of the analytical program complex or an abnormal situation, not included in the settings, a relevant entry is registered. indicating a pregnant woman, giving a preliminary conclusion, and coordinating the fragment obtained.

The doctor in charge can track a dynamic pattern of maternal and fetal condition using the doctor's computer work station, view phonocardiogram curves, and analyze calculated assessment parameters.

3 Results and discussions

The team of authors has proposed an operation scheme of the mobile system in which the portable units (recorder and the translator) are battery-powered and supports data exchange via a secure wire-free communication Stationary high efficiency elements of the system, the server and the doctor's computer work station are connected by digital links.

Similar in structure and purpose schemes had been proposed previously [6]. But they were intended for use in a hospitals. Digital data were transmitted over a wireless communication channel or through a specialized translator. In the presented device smartphone (serial production) with pre-installed specialized software is used as an intermediary. The proposed operation scheme is less expensive and easier to manufacture. It is not geographically limited due to availability of a mobile communication. Patent applications for the proposed methodology and technical solutions have been filed [7].

The accuracy of recognition algorithms for maternal and fetal heart rates, in comparison with the reference method of ultrasonic cardiotocography, was acceptable: the average error of evaluation ranged from 3 to 6 beats per minute, or less than 3%. Multiple processing of the same phonocardiogram by a complex of analytical programs gave identical results. That indicates a high stability of the algorithms used.

Compared with fetal electrocardiography of MONICA AN24, recognition algorithms for allocation of heart tones are more effective. Thus, the data processing of a workstation with dual-core Intel Pentium i5 processor with 8GB of RAM, was mainly determined by the duration of the original record. The signal duration of 18 minutes was processed and analyzed fully automatically from 25 to 30 seconds and signal duration of 8 hours was processed from 250 to 300 seconds. Such time expenditures of signal processing and analysis are significantly lower than when using fetal electrocardiogram. For example, for processing and analysis of a *moderately noisy signal* lasting 35 minutes MONICA AN24 needs 60–65 minutes with an error probability of no more than 15% [3].

4 Conclusion

The present study focuses on practical solution to the problem of creating a mobile system for continuous daily monitoring of fetal cardiovascular activity. The developed prototype of a hardware and software system is proceeding to preclinical trials at Tomsk medical institutions. There are all necessary prerequisites for the realization of an individual portable daily monitor. Computing capacities of microcontrollers with low power consumption are enough for solving the above mentioned problems in real-time.

Acknowledgement

The study was performed within the framework of the Agreement № 14.579.21.0019 (unique identifier for the project – RFMEFI57914X0019) on the «Development of the device for daily monitoring of the fetus and the mother during pregnancy by controlling the parameters of the cardiovascular system based on the acoustic data», signed between the R&D company «Diagnostika +» LLC and the Ministry of Education and Science of the Russian Federation, and the program of improving competitiveness of the TSU (project No. 8.2.31.2015).

References

[1] E.N. Zelenko *Analysis of amplitude-frequency parameters of heart rate fluctuations in diagnostic testing of fetal distress during pregnancy and childbirth: Ph.D. thesis.* (Belarusian State Medical University Press, Minsk,2006)

[2] J. Adam, Rev. Obstet. Gynecol **5**(3–4), e132 (2012)

[3] Zubair, Z. Alfirevic, American journal of obstetrics and gynecology **204,** 262 (2011) doi: 10.1016/j.ajog.2010.10.686

[4] S. L. Cibull, G. R. Harris, D. M. Nell, J Ultrasound Med **32**, 1921 (2013) doi : 10.7863/ultra.32.11.1921

[5] Sh. Bureev, D. S. Zhdanov, I. Ju. Zemlyakov, *Method of remote monitoring of physiological parameters of a human body and a mobile system for its implementation* Patent, Registration number 2015133653/08 (051 772) (2015)

[6] A. M. Povodator, V. U. Godlevskij, V. L. Stepanov, F. N. Abuzjarov, Ju. V. Naberezhnev, I. V. Isaev, A. D. Zakutskij, *Device for monitoring of physiological functions*,Patent RU 2200463 C2 (2003)

[7] S. Matyaschuk Home Child: online journal. URL: http://www.domrebenok.ru/blog/bezopasnost-meditsinskogo-ultrazvuka-ofitsialnoe-zaklyuchenie/?sphrase_id=3681610

ASSESSMENT OF LEFT VENTRICULAR MECHANICS IN PATIENTS WITH ISCHEMIC HEART DISEASE

Natalia Gladkikh[a1], *Alexey* Evtushenko[1,2], *Vladimir* Evtushenko[1,2], *Elena* Pavlyukova[1]

[1]Federal State Budgetary Scientific Institution "Research Institute for Cardiology", 634050, Tomsk, Russia
[2]National Research Tomsk Polytechnic University, 634050, Tomsk, Russia

Abstract. The paper presents new ultrasound technology for assessment of left ventricular (LV) strain in the longitudinal, radial, and circumferential directions and the LV area strain at rest and during dobutamine stress-echocardiography in patients with intermediate pretest probability of ischemic heart disease (IHD). Analysis was performed for the first time using real-time three-dimensional echocardiography (4D Strain). It was shown that in patients with single-artery left anterior descending artery (LAD) and left circumflex coronary artery (LCX) stenosis of 50% and more, the decreases in the global LV strain in the longitudinal, radial, and circumferential directions as well as in the LV strain area were observed at peak dobutamine stress in 4D Strain mode. Assessment of global LV strain in 4D Strain mode during dobutamine test had low sensitivity, specificity, and diagnostic accuracy in diagnosing single-vessel coronary artery disease.

1 Assessment of left ventricular mechanics in patients with ischemic heart disease

Myocardial strain is currently considered a parameter characterizing left ventricular (LV) contractility [1]. Assessment of global LV strain in the longitudinal direction has been widely implemented in clinical practice based on speckle tracking imaging technology (2D Strain) [2]. Recently, three-dimensional real time imaging (4D Strain) was developed for assessment of LV strain. This approach consists in speckle tracking in three dimensions through time which requires the absence of motions beyond a plane of scanning. Due to this, 4D Strain can be considered a natural extension of 2D Strain [3]. Advantages of 4D Strain over 2D Strain are obvious: lower time expenditure due to feasibility of single-step strain assessment in all directions: longitudinal, radial, and circumferential [4], as well as supposedly better accuracy due to the absence of the fallout effect [3, 5]. Therefore, 4D Strain study seems promising for early diagnosis of IHD. However, there is currently a lack of works showing the sensitivity, specificity, and diagnostic accuracy of the myocardial strain assessment by stress echocardiography in 4D Strain mode.

[a] Corresponding author: gladkikh100@mail.ru

© The Authors, published by EDP Sciences. This is an open access article distributed under the terms of the Creative Commons Attribution License 4.0 (http://creativecommons.org/licenses/by/4.0/).

The aim of our study was to assess the LV strain in the longitudinal, radial, and circumferential directions as well as the LV strain area in patients with intermediate pretest IHD probability by dobutamine stress echocardiography with real time three-dimensional imaging in 4D Strain mode.

2 Materials and Methods

The study comprised 59 patients (41 men) aged 55.81 ± 6.97 years (Me = 57.00 years) with suspected IHD who had intermediate pretest probability for IHD (46.56 ± 30.76%; Me = 32.40%). Single-vessel coronary artery disease with stenosis of 50% and more was detected in 32 (54.24%) out of 59 patients. Out of these patients, single-vessel coronary artery disease was diagnosed in 11 individuals with LAD stenosis, in 9 patients with RCA stenosis, and in 7 patients with LCX stenosis. Clinical characteristics are presented in Table 1.

Control group comprised 11 healthy volunteers (8 men) aged 45.55 ± 10.11 years (Me = 47.00 years). None of them were smokers and none presented with complaints characteristic for cardiovascular pathology. Healthy volunteers did not have arterial hypertension (AH), obesity, abnormal blood lipids, carbohydrate metabolism disorders, cardiac rhythm and conduction disorders, and heart valve disease. Echocardiography was performed by using Vivid E9 cardiovascular ultrasound system (*GE, Healthcare*) in two-dimensional mode according to standard method from parasternal (LV short axis views at annulus fibrosus of the mitral valve and at the apex) and apical (LV long axis 4- and 2-chamber views) positions with M5S matrix array probe (1.5–4.8 MHz). Echocardiography parameters of patients with suspected IHD and of healthy volunteers are presented in Table 1. Real time three-dimensional images were acquired from apical position by using 4VD probe under the condition of simultaneous visualization of all walls of the left ventricle with 29 frames per second. Scanning was performed in Multi-Slice mode. To obtain cine-loop with optimal characteristics for post-processing, recording was done during six cardiac cycles at an expiratory breath hold to exclude bonding artefacts caused by breathing movements. Upon completion of automatic analysis, a dynamic 17-segment bulls-eye diagram was obtained for the LV strain in the longitudinal (GLS4D), radial (GRS4D), and circumferential (GCS4D) directions as well as for the strain area (GAS_{4D}) during cardiac cycle. Image was automatically frozen at the end of systole; as a result, we obtained the values of the global LV strain and the strains of individual segments including the LV apex.

Stress echocardiography was performed according to standard protocol. Assessments of commonly-accepted parameters, local contractility abnormalities in grayscale, and strain parameters in 4D Strain mode were performed at rest and at peak of dobutamine stress test.

3 Statistical data analysis

Testing the hypothesis of Gaussian distribution by Kolmogorov-Smirnov criteria in the form of Lilliefors and Shapiro-Wilk tests discarded this hypothesis. Therefore, the Mann-Whitney U test and the Wilcoxon test were performed. Correlation relationships between the pairs of qualitative variables were evaluated by using nonparametric Spearman's rank correlation coefficient. In all procedures of statistical analysis, the level of significance was chosen at 0.05. The results are presented as M ± SD (M is arithmetic mean, SD is mean root square deviation) and lower and upper quartiles. Sensitivity, specificity, and diagnostic accuracy were assessed by using standard formulas.

4 Results

No clinically significant complications were observed during stress test. The images adequate for the analysis in 4D Strain mode at peak of dobutamine stress test were obtained from 69 patients. One patient was excluded from the analysis because of sub-optimal visualization of endocardial boundaries. Therefore, the feasibility rate of the method was 98.5%.

The values of the global LV strain in the longitudinal, radial, and circumferential directions as well as the LV strain area in patients with coronary artery disease and in healthy volunteers are presented in Table 2. No significant differences in the values of GLS_{4D}, GCS_{4D}, GRS_{4D}, and GAS_{4D} at rest between patients with single-vessel coronary artery disease and healthy volunteers were detected.

In dobutamine stress test, the global LV strain in the longitudinal, radial, and circumferential directions as well as the LV strain area in 4D Strain mode did not change in healthy individuals (Table 2) whereas, in the presence of LAD stenosis of 50% and more, we observed the decreases in GLS_{4D}, GRS_{4D}, GCS_{4D}, and GAS_{4D} (Table 2). Similar changes were observed in LCX disease (Table 2). Clinical case of dynamics in the LV strain in 4D Strain mode in dobutamine stress test in patient with angiographically-significant coronary atherosclerosis is presented in Figure 1.

In IHD patients with RCA disease, the LV strain in the longitudinal, radial, and circumferential directions as well as the strain area at peak of stress echocardiography did not significantly change (Table 2).

The LV strain in 4D Strain mode during dobutamine stress test showed low sensitivity, specificity, and diagnostic accuracy (Table 3). Similarly low sensitivity, specificity, and diagnostic accuracy were found for LAD, RCA, and LCX disease (Table 3).

Table 1. Clinical and echocardiography data of patients with ischemic heart disease and in healthy volunteers.

Parameters	Healthy volunteers		LAD		RCA		LCX	
	M±SD	Me	M±SD	Me	M±SD	Me	M±SD	Me
Age, years	45.55±10.11	47.00	58.80±5.51	61.00	55.70±6.45	55.50	58.33±4.50	59.00
BMI, kg/m^2	23.32±2.66	23.44	27.65±4.08	28.04	26.74±3.48	27.61	29.72±2.97	28.70
Glucose, mmol/L	5.66±0.37	5.90	6.12±2.05	5.60	6.35±2.64	5.45	6.80±2.57	6.00
Total cholesterol, mmol/L	4.90±0.76	4.50	5.19±2.05	5.33	4.98±1.39	5.15	5.43±2.50	5.67
LDL cholesterol, mmol/L	3.14±0.78	3.41	3.49±2.11	3.53	3.15±1.43	3.16	3.99±2.85	4.24
Gender ratio: m/f	8:3 (72.73%/ 27.27%)		7:4 (63.63%/36.36%)		7:2 (77.78%/22.22%)		6:1 (85.71%/ 14.29%)	
AH, n (%)	-		9 (81.81%)		6 (66.67%)		5 (71.43%)	
Diabetes mellitus, n (%)	-		2 (18.18%)		1 (11.11%)		2 (28.57)	
ACT, n (%)	-		4 (36.36%)		2 (22.22%)		3 (42.85%)	
Syntax score, c.u.	-	-	15.07±9.62	14.00	15.83±11.74	16.00	19.17±9.56	15.00
$EDV_{(Simpson)}$, mL	84.55±26.02	83.00	99.40±24.54	97.00	97.00±26.14	100.50	99.11±22.28	108.00
$ESV_{(Simpson)}$, mL	23.81±6.08	25.00	30.00±10.78	31.00	29.60±8.26	32.50	30.22±8.28	31.00
LV EF, %	71.36±3.53	70.00	69.67±6.84	67.00	68.90±4.2	67.00	69.44±4.33	68.00
ALC index, c.u.	1.00±0.00	1.00	1.02±0.05	1.00	1.04±0.06	1.00	1.04±0.06	1.00
IVS, mm	7.07±0.92	6.60	9.67±1.37	10.00	9.96±1.46	10.35	10.59±1.20	10.70
LVPW, mm	7.82±1.23	7.00	9.27±1.22	9.00	9.45±1.15	9.25	9.99±1.13	10.40
E_{mitr}, cm/s	77.91±11.82	81.00	65.40±17.11	66.00	66.20±14.34	63.50	65.67±16.28	69.00
A_{mitr}, cm/s	55.67±8.89	53.00	69.60±18.60	63.00	63.50±14.88	60.50	79.11±21.40	85.00
E/A_{mitr}, c.u.	1.61±0.24	1.62	0.98±0.32	0.86	1.08±0.31	1.03	0.85±0.21	0.83
E_m cm/s	10.50±3.10	10.50	8.77±2.38	9.00	9.60±2.12	10.00	7.63±1.99	8.00
E_{mitr}/E_m, c.u.	7.54±1.25	7.50	6.89±2.01	6.83	6.54±1.69	6.83	6.98±2.38	5.93
$LA_{(Simpson)}$, mL/m^2	18.59±3.82	17.73	24.55±8.25	25.48	25.23±4.02	25.54	28.25±8.39	27.16

M: arithmetic mean; SD: standard deviation; AH: arterial hypertension; ALC: abnormal local contractility; ACT: abnormal carbohydrate tolerance; SAP: systolic arterial pressure; DAP: diastolic arterial pressure; EDV: end diastolic volume; ESV: end systolic volume; LV EF: left ventricular ejection fraction; LA: left atrium; CA: coronary arteries; LAD: left anterior descending artery; RCA: right coronary artery; LCX: left circumflex coronary artery; E_{mitr}: maximum transmitral flow velocity during early diastole, cm/s; A_{mitr}: maximum transmitral flow velocity during atrial systole, cm/s; E_m: mitral annulus velocity at lateral wall of the left ventricle during early diastole, cm/s.

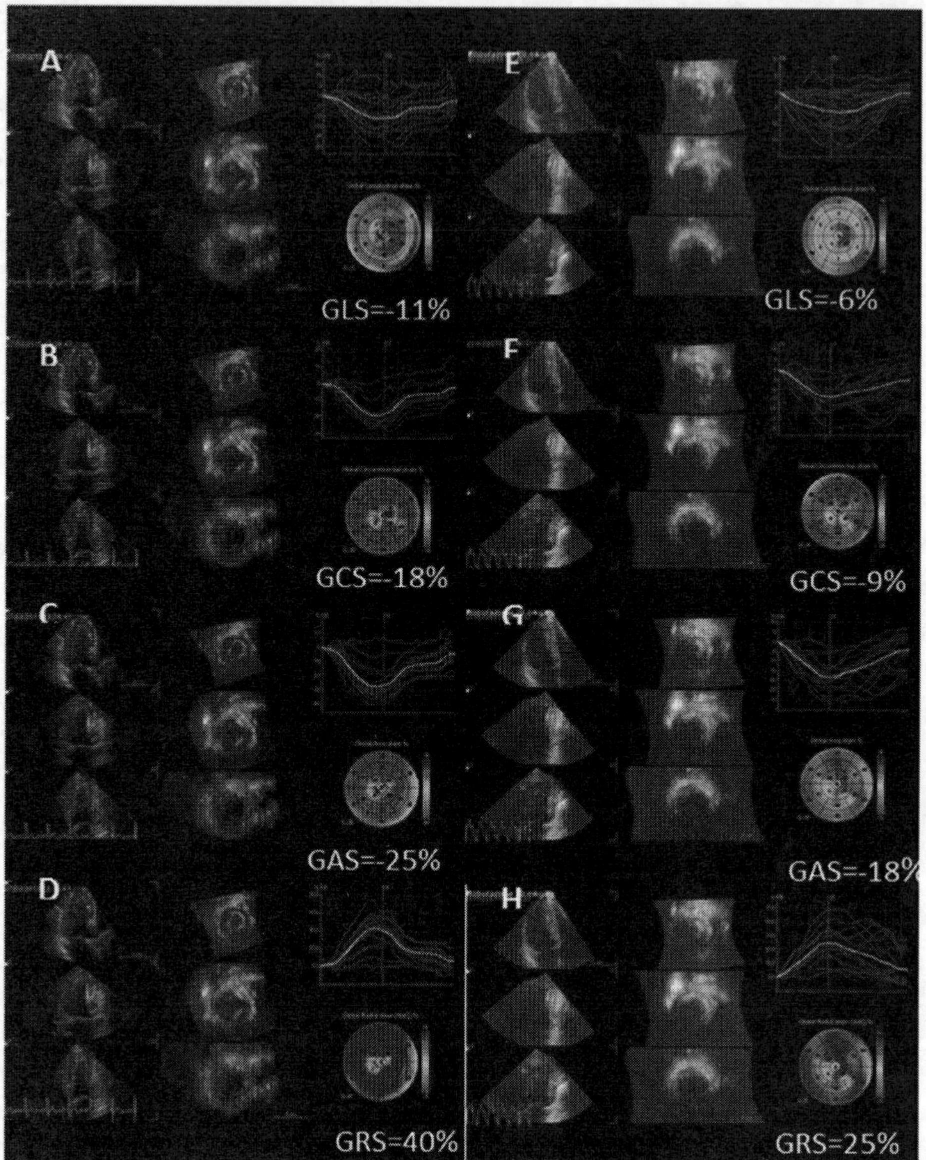

Figure 1. Clinical case. Patient P, 61 y.o., LCX stenosis of 70%. Left ventricular strain at rest (A-D) and after infusion of 20 μg/kg (E–H), heart rate of 134 bpm. Dynamics in the global and segmental LV strains in the longitudinal (A, E), circumferential (B, F), and radial (D, H) directions as well as in the strain area (C, G).

Table 2. Global left ventricular strain in 4D Strain mode at rest and during dobutamine stress echocardiography in patients with significant coronary artery disease and in healthy volunteers.

Group		GLS_{4D}, %	GCS_{4D}, %	GAS_{4D}, %	GRS_{4D}, %
		M±SD	M±SD	M±SD	M±SD
Healthy volunteers	At rest	-13.55±1.04	-14.36±0.81	-25.27±1.62	37.09±3.11
	Dobutamine	-9.82±4.53	-16.55±4.16	-24.27±3.82 &€	36.73±7.58 £¥
LAD	At rest	-11.15±3.96	-16.60±5.13	-24.20±6.55	37.60±13.85

	Dobutamine	-8.00±2.94 #	-13.30±3.74 #	-19.22±5.63 #&	28.30±9.42 #£
RCA	At rest	-11.83±4.02	-19.00±4.65	-27.17±6.65	44.00±14.81
	Dobutamine	-10.20±6.91	-17.40±5.32	-24.00±8.89	39.60±16.07
LCX	At rest	-10.25±3.92	-15.75±4.53	-23.25±6.20	34.88±13.13
	Dobutamine	-7.13±3.79 #	-13.13±3.98 #	-18.13±6.58 # €	26.00±11.83 #¥

#$p<0.05$ compared with the values at rest; £U = 27.000, Z_{adj}=-2.151, p=0.03; §U=11.000, Z_{adj}=-2.376, p=0.017; €U = 19.000, Z_{adj}=2.224, p=0.02; ¥U=19.000, Z_{adj}=-2.198, p=0.02.

Table 3. Sensitivity, specificity, and diagnostic accuracy of the parameters of left ventricular strain in 4D Strain mode in detection of coronary atherosclerosis by dobutamine stress echocardiography.

Parameter	Sensitivity, %		Specificity, %		Diagnostic accuracy, %	
	Strain	ALC	Strain	ALC	Strain	ALC
Stenosis over 50%						
GLS_{4D}	62.50	33.33	21.05	89.79	50.74	72.86
GCS_{4D}	57.44		31.57		50.00	
GAS_{4D}	68.08		27.77		56.92	
GRS_{4D}	66.67		31.57		56.71	
LAD stenosis over 50%						
GLS_{4D}	78.57	26.67	35.84	85.45	44.78	72.85
GCS_{4D}	64.28		40.38		50.00	
GAS_{4D}	69.23		30.76		38.46	
GRS_{4D}	64.28		32.07		35.82	
RCA stenosis over 50%						
GLS_{4D}	75.00	40	33.89	86.67	38.80	80
GCS_{4D}	62.50		39.65		42.42	
GAS_{4D}	75.00		31.57		36.92	
GRS_{4D}	62.50		32.20		37.35	
LCX stenosis over 50%						
GLS_{4D}	77.78	55.55	34.48	88.52	40.29	84.28
GCS_{4D}	66.67		40.35		43.93	
GAS_{4D}	77.78		32.14		40.00	
GRS_{4D}	66.67		32.75		37.71	

4 Discussion

As of today, national and foreign literature lacks data on the use of 4D Strain mode in stress echocardiography. It is worthy to note that only a few publications report data on the strain assessment by stress echocardiography in 2D Strain mode. Data of clinical and experimental studies demonstrate the possibility of assessing the parameters of the myocardial strain in stress echocardiography [6–8]. There is evidence that assessments of longitudinal and radial strains with Tissue Doppler Imaging [6, 9] are sensitive to ischemia.

According to experimental data, the strain in the longitudinal, radial, and circumferential directions increases at peak of stress test in the absence of ischemia [10]. The experiments show that a decrease in the strain in these directions is not observed in stress echocardiography in the presence of hemodynamically insignificant coronary artery stenosis. The presence of hemodynamically significant stenosis leads to a decrease in the strain by 39% in the longitudinal, by 28% in the radial, and by 26% in the circumferential directions [10]. Kodali V. *et al.* demonstrated that assessment of two-dimensional strain in longitudinal direction by dobutamine stress echocardiography shows higher sensitivity compared with expert evaluation of abnormal local contractility at global deformation critical level of -8% [11]. Data concerning dynamics of the LV strain in 4D Strain mode at peak of stress tests are occasional in literature [12]. Based on theoretical advantages of the method, we hypothesized that the strain assessment in 4D Strain mode is more sensitive to changes in inotropic function of the left ventricle compared with the values of 2D Strain. We observed a decrease in the parameters of global LV deformation in 4D Strain mode in patients with angiographically significant coronary artery stenosis. Takano M. *et al.* [13] also observed a decrease in the values of 4D Strain in adenosine stress echocardiography and it was considered a sign of myocardial ischemia. According to these data, sensitivity and specificity of 4D Strain mode were 62.5% and 83.6%, respectively [13]. Authors of this publication believe that the values of the circumferential LV strain are more adequate for a detection of myocardial ischemia. We obtained lower values of sensitivity and specificity and, accordingly, of diagnostic accuracy for the strain assessment in 4D Strain mode. However, Takano M. *et al.* considered coronary artery stenosis significant if it was over 75% [13].

It is necessary to note that myocardial strain is reflective of contractility of the part or of the entire left ventricle i.e. it is reflective of the LV function whereas the presence of coronary artery stenosis according to data of coronary angiography merely demonstrates angiographic significance of stenosis and its potential ability to limit coronary flow to the part of the LV myocardium. A true hemodynamic significance of coronary artery stenosis can be determined based on the assessment of fractional flow reserve [14]. Limitation of our study was the lack of opportunity to determine this parameter. Perhaps, diagnostic accuracy of the LV strain assessment including that in 4D Strain mode is higher in regard to a detection of coronary atherosclerosis as an anatomical substrate.

5 Conclusions

In patients with single-vessel LAD and LCX stenosis of 50% and more, decreases in the global LV strain in the longitudinal, radial, and circumferential directions as well as in the LV strain area were observed at peak of dobutamine stress echocardiography. The global LV strain in 4D Strain mode in dobutamine stress echocardiography had low sensitivity, specificity, and diagnostic accuracy. Unfortunately, we have not performed dobutamine stress echocardiography in patients with a high pretest probability of IHD, because there are not indications for functional tests for these patients, according to the ESC/EACTS Guidelines on myocardial revascularization (2014). We believe that the sensitivity, specificity and diagnostic accuracy will be much higher if this group of patients will be examined.

Acknowledgements

The study was supported by a grant from the Russian Science Foundation (№ 15-15-10016), and the program of improving competitiveness of the TSU (project No. 8.2.31.2015).

References

[1] G. R. Sutherland, L. Hatle, P. Claus, J. D'hooge, B. H. Bijnens, *Doppler myocardial imaging* (BSWK-Hassel, Zagreb, 2006)

[2] R. M. Lang, L. P. Badano, V. Mor-Avi, J. Afilalo, A. Armstrong, L. Ernande, F. A. Flachskampf, E. Foster, S. A. Goldstein, T. Kuznetsova, P. Lancellotti, D. Muraru, M. H. Picard, E. R. Rietzschel, L. Rudski, K. T. Spencer, W. Tsang, J.-U. Voigt, J Am Soc Echocardiogr. **28**, 1 (2015) doi: 10.1016/j.echo.2014.10.003

[3] V. C.-C. Wu, M. Takeuchi, K. Otani, N. Haruki, H. Yoshitani, M. Tamura, H. Abe, F.-C. Lin, Y. Otsuji, J Am Soc Echocardiogr. **26**, 1274 (2013) doi: 10.1016/j.echo.2013.07.006

[4] S. Buccheri, I. Monte, S. Mangiafico, V. Bottari, S. Leggio, C. Tamburino, BioMed Research Intern. **2013**, 297895 (2013) doi: 10.1155/2013/297895

[5] J. A. Urbano-Moral, A. R. Patel, M. S. Maron, J. A. Arias-Godinez , N.G. Pandian, Echocardiography **29**, 997 (2012) doi: 10.1111/j.1540-8175.2012.01773.x

[6] C. B. Ingul, A. Stoylen, S. A. Slordahl, R. Wiseth, M. Burgess, T.H. Marwick, JACC **49**, 1651 (2007) doi: 10.1016/j.jacc.2007.01.059

[7] G. Derumeaux, M. Ovize, J. Loufoua, X. André-Fouet , Y. Minaire, A. Cribier, B. Letac, Circulation **97**, 1970 (1998)

[8] J. Gorcsan, D. P. Strum, W. A. Mandarino, V. K. Gulati, M. R. Pinsky, Circulation **95**, 2423 (1997)

[9] B. H. Amundsen, T. Helle-Valle, T. Edvardsen, H. Torp, J. Crosby, E. Lyseggen, A. Støylen, H. Ihlen, J. A. C. Lima, O. A. Smiseth, S. A. Slørdahl, JACC **47**, 789 (2006) doi: 10.1016/j.jacc.2005.10.040

[10] P. Reant, L. Labroussc, S. Lafittc, P. Bordachar , X. Pillois , L. Tariosse , S. Bonoron-Adele , P. Padois , C. Deville , R. Roudaut , P. Dos Santos, JACC **51**, 149 (2008) doi: 10.1016/j.jacc.2007.07.088

[11] V. Kodali, S. K. Saha, K. Parcham-Azad, R. S. Toole, A. S. Gopal, JACC **57**, E. 767 (2011)

[12] P. Reant, L. Labrousse, S. Lafitte, L. Tariosse, S. Bonoron-Adele, P. Padois, R. Roudaut, P. Dos Santos, and A. DeMaria, J Am Soc Echocardiogr. **23**, 95 (2010) doi: 10.1016/j.echo.2009.10.001

[13] M. Iwai-Takano, T. Watanabe, JACC **61**, E. 1019 (2013)

[14] S. Windecker, P. Kolh, F. Alfonso, J.-P. Collet, J. Cremer, V. Falk, G. Filippatos, C. Hamm, S. J. Head, P. Jüni, A. P. Kappetein, A. Kastrati, J. Knuuti, U. Landmesser, G. Laufer, F.-J. Neumann, D. J. Richter, P. Schauerte, M. S. Uva, G. G. Stefanini, D. P. Taggart, L. Torracca, M. Valgimigli, W. Wijns, and A. Witkowski, Eur Heart J. . **35**, 2541 (2014) doi:10.1093/eurheartj/ehu278

NI BASED SYSTEM FOR SEU TESTING OF MEMORY CHIPS FOR AVIONICS

Anna Boruzdina[a], *Andrey* Yanenko, *Georgy* Tikhomirov, *Vitaly* Tumanov

National Research Nuclear University MEPhI (Moscow Engineering Physics Institute), 115409, Moscow, Russia

Abstract. This paper presents the results of implementation of National Instrument based system for Single Event Upset testing of memory chips into neutron generator experimental facility, which used for SEU tests for avionics purposes. Basic SEU testing algorithm with error correction and constant errors detection is presented. The issues of radiation shielding of NI based system are discussed and solved. The examples of experimental results show the applicability of the presented system for SEU memory testing under neutrons influence.

1 Introduction

The problem of Single Event Upset (SEU) in memory under influence of atmospheric neutrons and also under secondary neutrons of physics facilities is well known [1–3]. This problem is increasing along the feature sizes of critical electronic components (processors, memories) of control systems became smaller and demands a high level of attention [4–8], especially referring to avionics safety.

Solving of the problem of SEU in memory under influence of neutrons is in prediction of the electronic components sensitivity to neutron irradiation [9] and in implementation of error correction methods utilizing the level of sensitivity. To obtain sensitivity parameters one should perform radiation tests using experimental facilities, such as proton accelerators and neutron generators. One of these facilities is the neutron generator in MEPhI (Moscow, Russia) which can produce neutron fluxes of energy about 2 MeV or 14 MeV.

Memory is an essential component of modern control systems, in which memory is usually represented as is and also as a part of processors, programmable logic etc. Considering SEU, the memory seems to be the most sensitive part of the systems, due to its highest integration level. Therefore, we have chosen memory chips as the object of our experiments.

To perform functional control of memory chips during SEU tests the special test algorithms should be used. The algorithms should include an information safety check in each memory cell, and also an operability check of each cell. The latter is important to be sure that the error of the cell is the SEU rather than constant failure (imprinting effect) due to total ionizing dose (TID). The special algorithm and the hardware were developed on the basis of National Instrumen (NI) platform and NI LabVIEW 11.0 software.

[a] Corresponding author: abbor@spels.ru

© The Authors, published by EDP Sciences. This is an open access article distributed under the terms of the Creative Commons Attribution License 4.0 (http://creativecommons.org/licenses/by/4.0/).

Another part of the work is to supply safety operation of NI based system under influence of neutron fluxes of neutron generator. The distance between memory chip and NI based test system is about 2 meters and could not be larger taking into consideration the length of standard cables. At this distance the level of irradiation is rather high to leave it unsolved: the risk of system failure due to neutron flux and total dose irradiation should be decreased to negligible levels. Keeping in mind the high penetration ability of neutrons there are only two ways to decrease the level of irradiation: (1) to increase the distance between the system crate and the neutron source, which is impossible for the length of cables, and (2) to mount an appropriate shield between them, which was one of the tasks of this work.

2 Hardware and software

The special algorithm and the LabVIEW software for memory chip SEU testing along with simultaneous operability check of each memory cell was developed using NI LabVIEW 11.0. The software was developed to be running on the NI PXI platform which was based on PXI-1033 chassis, or so called "crate", with a field-programmable gate array (FPGA) module PXI-7951R supplemented with an adapter module NI 6581 and a PXI-4110 power supply which all together were "the measuring unit".

The exterior view of the VI toolbox managing operational modes of the memory chip under test is shown in Figure 1. The parametric control panel allows to command two power supply channels of memory chip under test, measure supply current and detect single event latchup (SEL) in each channel separately. The functional control panel allows choosing a functional test pattern, performing writing and reading operations start and stop SEU test and indicating the number of both SEU and constant errors. At the bottom of the VI toolbox an operator can input the filenames for log files.

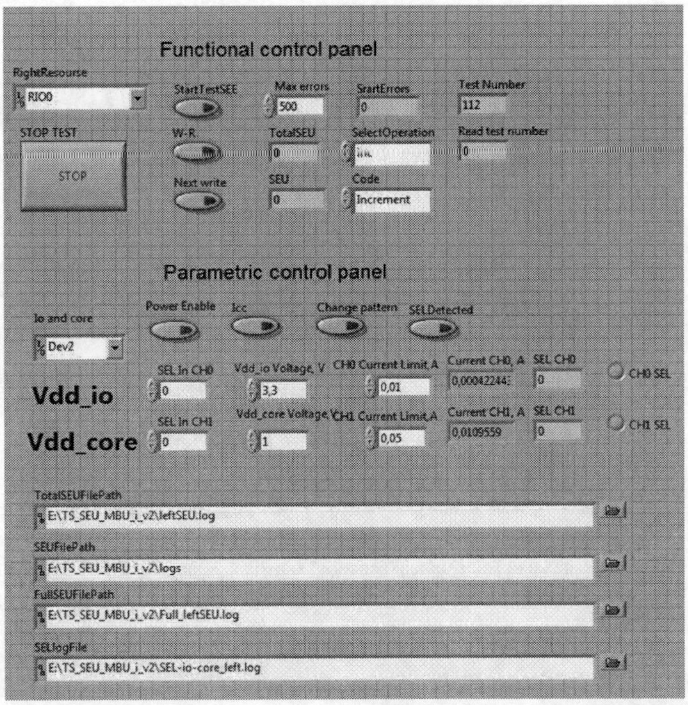

Figure 1. The exterior view of the VI toolbox managing operational modes of the memory chip under test.

The measuring unit was connected to the memory chip under a test with a standard cable providing suitable frequencies for the test. The controlling computer was located 7 meters away from the measuring unit, in a safe area where no irradiation was available.

3 Test algoritm

The block diagram shown in Figure 2 describes the special algorithm for SEU testing developed to control both the SEU under neutron irradiation and constant hard errors in each memory sell of the memory chip under test. The algorithm includes the following steps:

1) Writing the test pattern P2 which is inverse to the initial pattern P1.
2) Reading and comparing the read data with the pattern P2 to evaluate constant errors.
3) Writing the initial test pattern (P1) to the memory chip.
4) Reading and comparing the read data with the pattern P1 to evaluate constant errors.
5) Writing the initial test pattern (P1) to the memory chip.

SEU testing starting from the address 0x0 up to the maximum logical address performing the following operation for each cell in the logical word:

1) reading the logical word and comparing the read data with the corresponding data W1 from pattern P1 for preliminary SEU registration;

2) writing the data W2 which is inverse to W1 to the same logical word;

3) reading the same logical word again and comparing the read with W2 to evaluate constant hard errors (imprinted W1 data) in each bit;

4) writing W1 to the same logical word;

5) reading the same logical word again and comparing the read data with W1 to evaluate constant hard errors (imprinted W2 data) in each bit;

6) calculating true SEU data in each bit using preliminary SEU data excluding imprinted bits evaluated in intermediate steps.

The selection of initial test pattern is essential. Plain patterns such as “all ones” or “all zeroes” are not effective for detecting errors in an address decoder on the memory chip. The better choice is random pattern but this kind of pattern is sophisticated for hardware implementation. The compromise is the so called “incremental” pattern which is irregular enough to reveal address decoder problems and at the same time regular enough to be implemented to hardware using simple counter registers. The view of the fragment of the VI block diagram shown in Figure 3. performs the algorithm described above.
The SEU testing is looped until the appropriate number of SEUs will be counted.

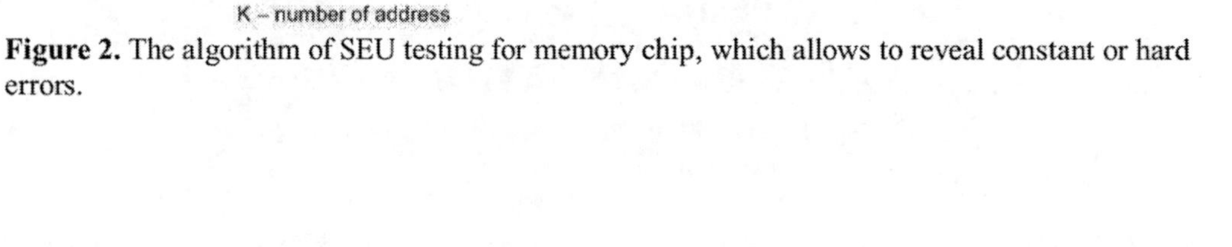

Figure 2. The algorithm of SEU testing for memory chip, which allows to reveal constant or hard errors.

DOI: 10.1051/matecconf/20167901028

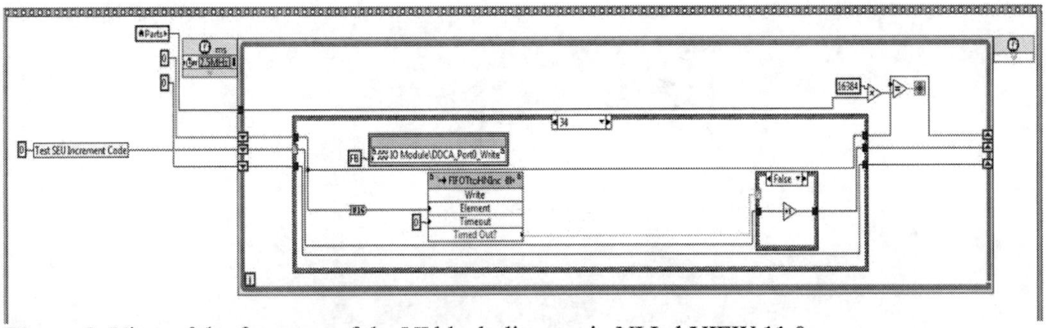

Figure 3. View of the fragment of the VI block diagram in NI LabVIEW 11.0.

4 Developing the shield for NI based test system in a harsh environment of neutron generator

One of the tasks of this work is to supply the NI based test system with appropriate shield to minimize the influence of radiation from the neutron source of the neutron generator.

The neutron generator consists of linear accelerator of deuterons and the tritium (for 14 MeV neutrons) or deuteron (for 2 MeV neutrons) target (Figure 4). The resulting emission of neutrons is almost homogeneous on the sphere surface. The neutron flux is decreasing along the distance from the target approximately inversely proportional to the square of the distance.

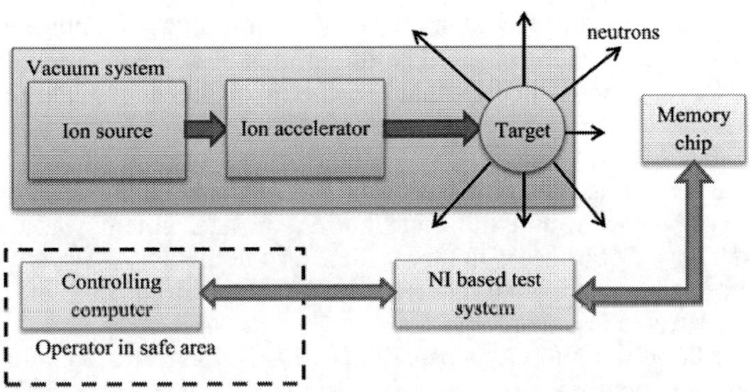

Figure 4. Basic schematic of neutron generator facility and NI based test system installation.

The shielding of NI based test system by moving it away from the target is not effective and limited by the length of the cable between the NI crate and the memory chip under test. Therefore, we had to develop heavy passive shield (Figure 5) from available material, which was steel bricks.

The design of the steel shield allows us to mount the memory chip under test very near to the neutron generator target and, at the same time, to create enough shadow to prevent direct influence of the neutrons emitted from the target on the NI crate. It should be noted that using steel as a shielding material leads to additional emission of secondary low energy neutrons and gamma-photons. Therefore, the neutron fluxes distortion and TID should be estimated at the place of disposal of the memory chip under test as well as at the place of NI crate disposal.

 DOI: 10.1051/matecconf/20167901028

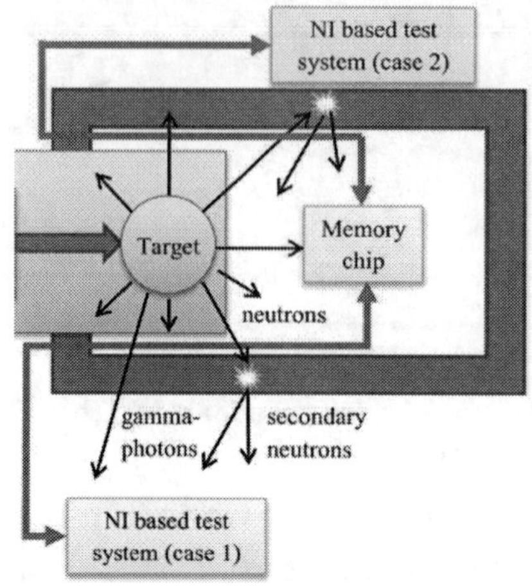

Figure 5. The shilding of the target and the memory chip under test leads to producing secondary neutrons and gamma-photons, but the flux of primary neutrons becomes significanlty less.

The estimation of neutron fluxes and the TID in the place of the NI crate disposal have been performed taking into consideration both primary and secondary irradiation. The calculations were performed using MCNP software utilizing Monte Carlo codes. Calculations were performed step by step starting with a very simple environment model (dotted neutron source approximation) to a complete environment model taking into consideration the materials and the geometry of the tritium target, approximated geometry and material of the memory chip mounted on the printed circuit board and the steel shield.

The results of calculations showed that the fluxes of secondary neutrons and gamma-photons on the memory chip under test are not negligible. The neutron flux increase, in the energy range from 1 MeV to 14 MeV, was about 8% on account of the geometry and the material of the target, while the contribution of the memory chip material and the printed circuit board was about 5%, and the contribution of steel shield was about 0.5%. The corresponding decreases of 14MeV neutrons flux on the memory chip are about 11%, 6% and 1%. Therefore, the aggregated uncertainty of the neutron flux of 14 MeV on the memory chip under test is less than 20%.

The neutron flux on the NI crate was calculated by taking into account that the typical neutron flux on the memory chip needed during SEU test experiment is about 10^{11} neutron/cm^2. Keeping in mind that the distance from the neutron source is essential for the neutron flux we considered two cases (see Figure 5) of the geometry allowed by the length of the cable between the NI crate and the memory chip under test: (1) the NI crate is situated as far as possible from the target of the neutron generator, but still in the "shadow" created by the steel shield, and (2) the NI crate is situated about in the middle of a "shadow" but very close to the walls of the steel shield.

The results of calculations of the neutron fluxes on the NI crate under different conditions are shown in the Figure 6. As it follows from the calculation results the steel shield effectively absorbs initial 14 Mev neutrons irradiation but at the same time it becomes the sourse of a huge amount of low energy neutrons especially with the energy of less then 1 MeV, but these kinds of particles can not produce any upsets in electronics and may be assumed harmless. The place of disposal do not play essential role while using the steel shield, but case (1) is more effective for low energy neutrons, because in this case the NI crate is situated more far from the steel shiels which is the source of secondary neutrons in this case.

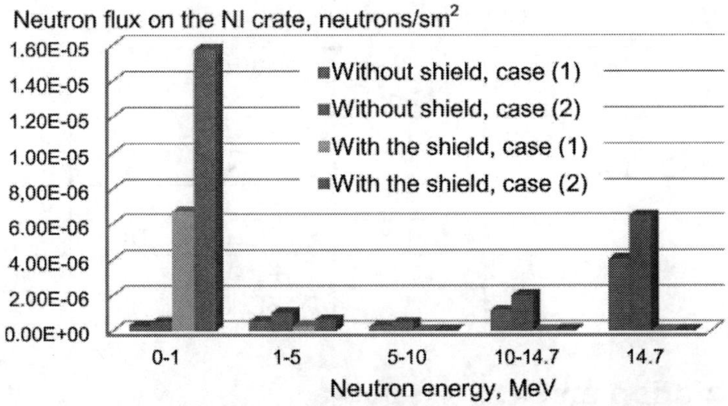

Figure 6. The results of calculations of the neutron fluxes on the NI crate under different conditions.

The TID on the memory chip was also calculated in correspondence to the flux of about 10^{11} neutron/cm^2. According to the calculations the TID level is about 20 rad per one experiment which is quite a negligible value for memory chips. The TID level in the place of NI crate disposal was several orders of magnitude less than on the memory chip being tested, which allows us to use the NI crate for thousand experiments without any fears of TID failure.

Developed external memory controller interface was successfully used for testing a variety of microprocessors which include general purpose microprocessors (IDT79RC64474 with MIPS architecture, AT697F with the SPARC architecture, Intel 80386, etc.) and DSPs(processor's family by Texas Instruments - TMS320C2x, TMS320C3x, TMS320C5x , TMS320C2xx, TMS320C5000, TMS320C6000 and many others)., this unit requires minimal setup that significantly reduces the time of preparation.

Note that this implementation allows us to significantly increase the speed (up to 100 MHz) and doesn't contain uncontrolled time delays. The minimum time interval is equal to the period of the clock signal for the Time Loop cycle. At the same time, this implementation of an external memory interface keeps such advantages as flexibility, standardization and flexibility of the developed external memory interface.

The work was performed as part of the grant RFBR № 14-29-09210.

5 Experimental results

The NI based SEU test system was verified during several experiments with two types of memory chips (Table 1) under the influence of 14 MeV neutrons. Both memory types had supply voltage of 5V.

Table 1. Memory chips used in the experiment.

Type	Manufacturer	Size	Feature size
AS7C4096	Alliance	512Kx8	0.18 um
K6X1008C2D	Samsung	128Kx8	0.13 um

The results of SEU tests (Table 2) show an applicability of NI based SEU test system for the experiments of that sort. The numbers of SEU are in good correlation with measured neutron fluxes. No constant errors N_{CE} in memory cells were observed during the experiment, it agrees with negligible levels of TID.

Table 2. Results of SEU tests

Type of memory chip	Number of sample	Neutron flux	N_{SEU}	N_{CE}
AS7C4096	1	2.20E+09	251	0
	2	5.48E+09	535	0
	3	5.19E+09	496	0
K6X1008C2D	1	8.02E+09	101	0
	2	7.70E+09	102	0
	3	7.54E+09	78	0

6 Implementation and its prospects

The automated NI based system developed in this work allows performance of SEU tests in memory chips during neutron irradiation. The appropriate shielding was developed to minimize total dose and neutron flux influences on NI crate. This NI based system was successfully implemented at JSC "ENPO SPELS" and NRNU MEPHI (Moscow). Future plans are to elaborate the control logic for dynamic types of memory.

Acknowledgment

This paper is a part of the research done within the government assignment #8.826.2014/K given by the ministry of education and science of Russian Federation.

References

[1] J. F. Ziegler, W. A. Lanford, Science **206**(4420), 776 (1979)

[2] E. Normand, IEEE Trans. Nucl. Sci. **43**, 461 (1996) doi: 10.1109/23.490893

[3] E. Normand, International Journal of High Speed Electronics and Systems **14**, 285 (1996) doi: 10.1142/S0129156404002351

[4] A. I. Chumakov, Russian Microelectronics **44**, 27 (2015) doi: 10.1134/S1063739715010047

[5] F. L. Kastensmidt, L. Tambara, D. V. Bobrovskiy, A. A. Pechenkin, A. Y. Nikiforov, IEEE Trans. Nucl. Sci. **61**, 3130 (2014) doi: 10.1109/TNS.2014.2369008

[6] A. V. Sogoyan, V. A. Polunin, Russian Microelectronics **44**, 54 (2015) doi: 10.1134/S1063739715010102

[7] A. O. Akhmetov, A. V. Yanenko, A. I. Bazhan, Proc. 14 th European Conf. on Radiation and its Effects on Components and Systems (RADECS-2013), 6937364 (2013)

[8] A. B. Boruzdina, A. V. Ulanova, A.G. Petrov, V. A. Telets, P. Reviriego and J. A. Maestro, Proceedings of the European Conference on Radiation and its Effects on Components and Systems (RADECS), 6937393 (2013) doi: 10.1109/RADECS.2013.6937393

[9] N. E. Aristova, A. I. Tararaksin, L. N. Kessarinskiy, A. Y. Borisov, A.Y. Nikiforov, in Proc.2015 International Siberian Conference on Control and Communications (SIBCON 2015), 7146985 (2015) doi: 10.1109/SIBCON.2015.7146985

MATEC Web of Conferences 79, 01029 (2016) DOI: 10.1051/matecconf/20167901029
IME&T 2016

ANTI-JAMMING ALGORITHM FOR DETECTION OF QRS AND ST SEGMENTS ON ELECTROCARDIOGRAM

Kirill Overchuk, *Alexander* Uvarov, *Inna* Lezhnina*

Tomsk Polytechnic University, Institute of Non-Destructive Testing, 634050, Tomsk, Russia

Abstract. As it was earlier described, capacitive electrodes could be widely used in electrocardiography, especially, in the state-of-the-art personal devices as they employ major advantages of these sensors. But such a way of electrocardiogram measurement usually generates quite noisy and unstable signals, which respectively must be processed with some anti-jamming algorithm. At the same time algorithm developers must take into account low computing power of portable personal devices, so designed methods must be "light" and "fast". Such an algorithm is developed by authors and is further described in this paper.

1 Introduction

In the publications presented by the authors earlier [1], they published the results of performance tests for the capacitive electrocardiographic electrodes from "Plessey Semiconductors" as applied to portable systems. The study showed that the capacitive electrodes are appropriate for use in those areas where the need for speed and convenience of measurement prevail over the requirements to the signal / noise ratio (SNR) and measurement stability and requires repeated application. Personal electrocardiogram (ECG) devices are one of the best applications as their popularity keeps growing all over the world. In this case, the received signals are often more noisy than those from the standard devices, and portable implementation imposes restrictions on the analog and digital processing methods. Resulting it requires to develop an anti-jamming algorithm for analysis of ECG with low computational complexity. The following is a description of the algorithm developed by the authors and testing results. The algorithm for detection of QRS and ST segments is chosen for implementation as the most important for further processing.

2 Main part

Here and following the standard ECG analysis algorithms are used as a base in the development [2–16]. Algorithms are combined together and some original solutions are

*Corresponding author: inna84-08@mail.ru

© The Authors, published by EDP Sciences. This is an open access article distributed under the terms of the Creative Commons Attribution License 4.0 (http://creativecommons.org/licenses/by/4.0/).

added. The algorithm consists of a block chain, which process input signals, convert and transform it for further calculations.
The input data processing block is a bandpass filter derived from Pan - Tompkins algorithm [2–5]. The filter was simulated if MATLAB software for testing and optimization of its parameters, as it shown on Figure 1.

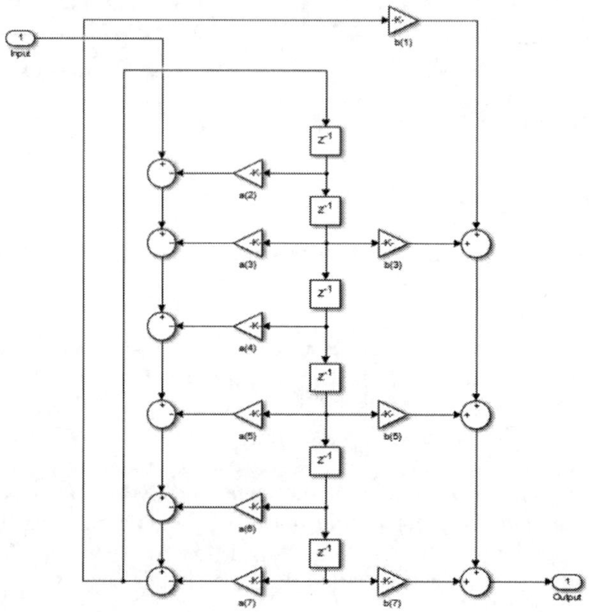

Figure1. Structure of Pan – Tompkins digital filter.

The filter order and bandwidth is matched experimentally. As most efficient from point of highest QRS amplitude and smallest transition time, the 6th order FIR filter with 5 Hz and 15 Hz cutoff frequencies is used for QRS complex selection. Table 1 shows the coefficient values for optimized digital filter.

Table 1. The coefficients of input FIR filter.

a	b
1	0.00021960621122537265
- 5.7147813373165643	0
13.644496911440541	- 0.00065881863367611791
- 17.421690919797662	0
12.546689097011155	0.00065881863367611791
4.8323508451445623	0
0.77763856023807842	- 0.00021960621122537265

The designed filter allocates bandwidth comprising a QRS complex, while efficiently suppressing other frequency ranges and does not introduce additional distortion. The filter operation results without and with input noise is shown on Figure 2.

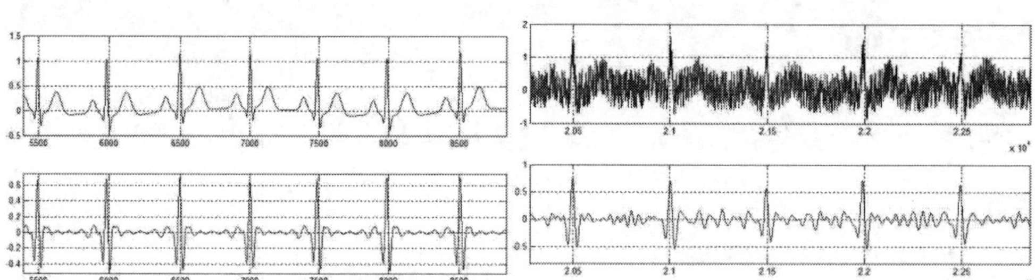

a) b)

Figure 2. The outputs of input (stage 1) filter: a) without input noise; b) in presence of noise.

Input filter operating by the algorithm of Pan - Tompkins efficiently allocates QRS complex in the absence and in the presence of large noise, which in turn provides a high reliability extraction algorithm peaks.

The next stage is a function of non-linear transformation which is intended to allocate QRS complex signal. For this purpose there are two well-known functions.

The first one – is the derivative operator from Pan - Tompkins algorithm with equation (1).

$$y(n) = 1/8 \cdot [2x(n) + x(n-1) - x(n-3) - 2x(n-4)]^2 \quad (1)$$

This transformation efficiently allocates QRS complex, but the output signal is too complex for further analysis without additional processing. Figure 3 shows the results without input noise (Figure 3, *a*) and with its presence (Figure 3, *b*). In this case, it is seen that the presence of noise has no significant effect on the output, which means good noise immunity. The disadvantage of method is that for using of a threshold detector, the received signal must be smoothed. Smooth function in this case must have a large order, which leads to a large signal delay and becomes impractical.

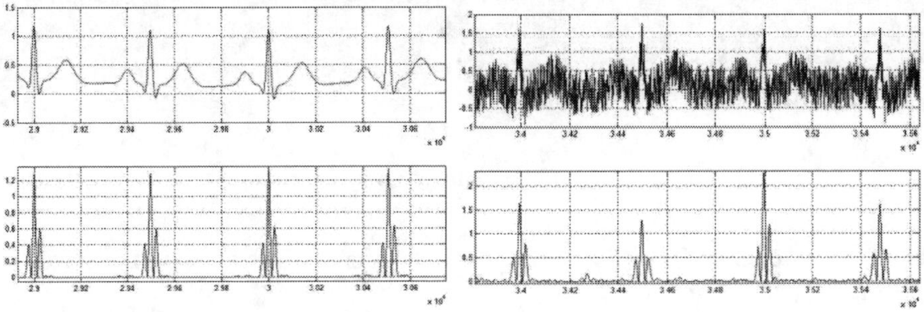

a) b)

Figure 3. The outputs of peak (stage 2) filter with Pan – Tompkins algorithm: a) without noise; b) in presence of noise.

An alternative function that can be applied to detect the QRS complex is the equation of Murthy and Rangaraj published in [2] with the equation (2).

$$y(n) = \sum_{i=1}^{N} |x(n-i+1) - x(n-1)|^2 (N-i+1) \quad (2)$$

Operation results of function (2) are shown on Figure 4.

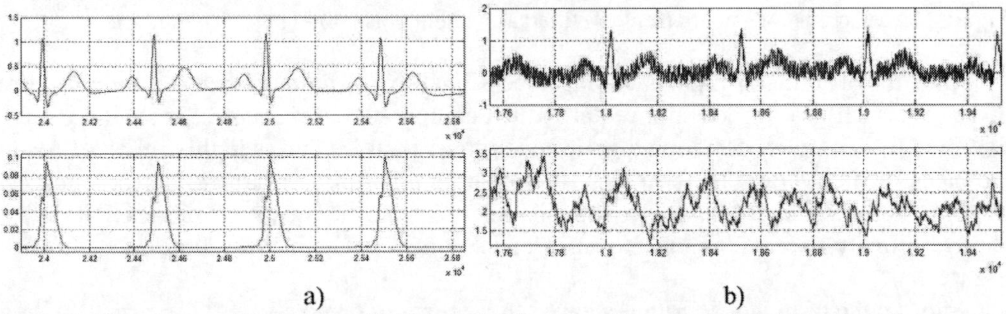

a) b)

Figure 4. The outputs of peak (stage 2) filter with Murthy and Rangaraj algorithm: a) without noise; b) in presence of noise.

The given Figures clearly shows that the equation (2) allows to obtain smooth peak, which is good shaped for further analysis with threshold detector and low-order smooth pre-processing function, if necessary. However, it is worth noting that this method isn't resistant to interference, which can be seen on Figure 4, *b*: the presence of a small noise makes it impossible observe any QRS peaks.

For obtaining of easy for recognition output peaks features combined with good anti-jamming performance, it was decided to combine the functions (1) and (2). The result is formula (3). Combining the two functions has allowed compensating disadvantages of each and getting the optimal solution, providing a stable discharge peaks and noise immunity.

$$y(n) = \sum_{i=1}^{N} | 2x(n) + x(n-1) - x(n-3) - 2x(n-4) |^2 (N - i + 1) \quad (3)$$

The results of function (3) operation are shown on Figure 5.

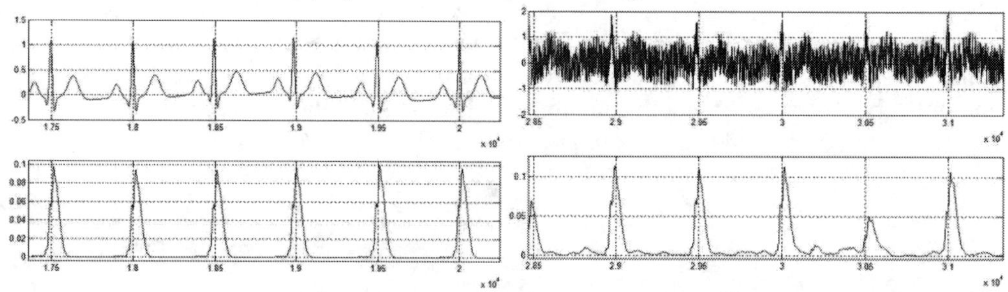

a) b)

Figure 5. The outputs of peak (stage 2) filter with authors' algorithm: a) without noise; b) in presence of noise.

For obtaining smooth, well recognizable peaks on the next stage, an integrating filter is used. Filter output is shown in Figure 6.

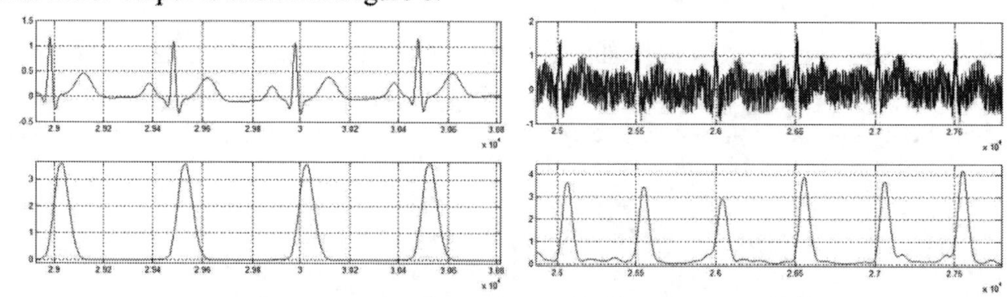

a) b)

Figure 6. The outputs of smooth (stage 3) filter: a) without noise; b) in presence of noise.

After the all transformations the received signal may be analyzed by using a threshold detector, which in turn does not require much computational resources. It should be noted that the QRS complex detection algorithm is robust to noise and is highly reliable. Also on the basis of the given transformation, the separate algorithm is built to detect sinus heart rhythm. It detects episodes of rhythm disorders as the moments of sudden RR interval change. Finally it provides the sequence of RR intervals against the time as it shown on Figure 7.

After all the tests the algorithm showed the following results:

- sensitivity – 95%;
- specificity – 92%.

Moreover, other analysis methods such as ST depression / elevation detectors are based on the given results. ST segment position is calculated relative to the position of the QRS complex. Figure 8 shows the results of such a ST-segment detector, red markers pick the segments positions.

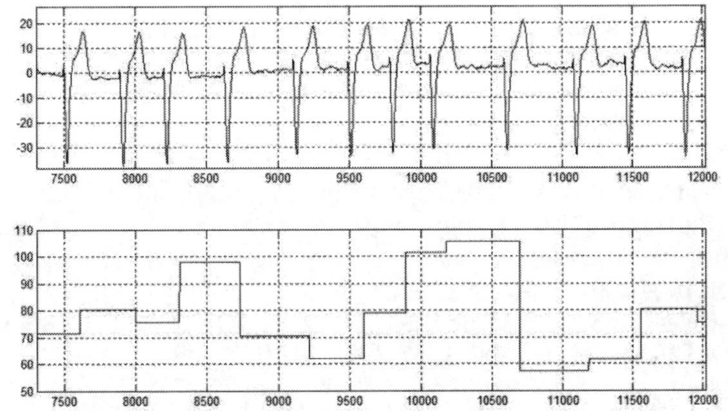

Figure 7. The output of the algorithm for rhythm disorders detection.

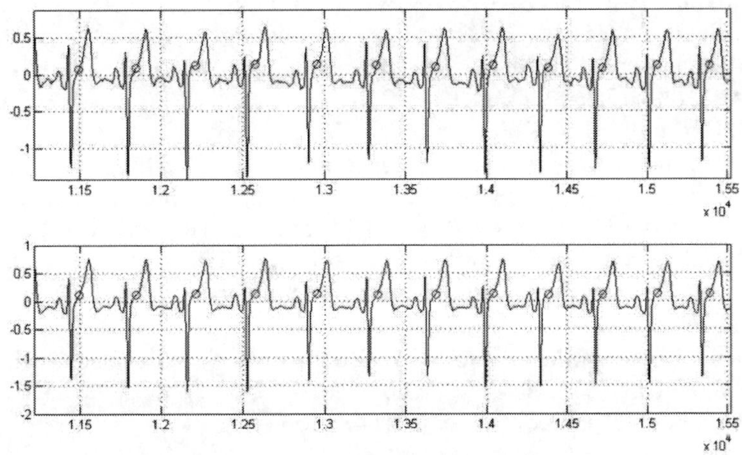

Figure 8. The result of the algorithm for the detection of ST segment.

3 Conclusion

The combination of two existing methods for QRS complex detection (Pan-Tompkins and Murthy-Rangaraj algorithms) provides the new method with much more stable operation in noisy conditions as well as smooth shaped and obtainable output peaks for further analysis. The result is a low-order difference function having the low computational complexity and, which is quite important for user-friendly devices, a minimum delay of the output. The function provides clear peaks that can be easily processed further threshold detector. These results are used to design special personal equipment based on the capacitive sensors.

References

[1] A.A. Uvarov , I.A. Lezhnina, K.V. Overchuk, A.S. Starchak, S.D. Akhmedov, I.A. Larioshina, IOP Conf. Ser. **671**, 012032 (2016) doi: 10.1088/1742-6596/671/1/012032

[2] R.M. Rangayyan, *Biomedical Signal Analysis* (IEEE Press/Wiley, New York, 2015)

[3] G.D. Clifford, URL: ibme.ox.ac.uk/research/biomedical-signal-processing-instrumentation/prof-l-tarassenko/publications/pdf/gdcliffordthesis.pdf. (handling date: 15.06.2016)

[4] W.J. Tompkins, *Biomedical Digital Signal Processing* (Upper Saddle River, NJ: Prentice Hall, 1995)

[5] J. Pan, W.J. Tompkins, IEEE Trans. Biomed. Eng. **BME-32**, 230 (1985) doi: 10.1109/TBME.1985.325532

[6] A. Diery, D. Rowlands, D.A. James, T. Cutmore, URL: aprs.org.au/anziis2003/Papers/paper173.pdf (handling date: 25.03.2016)

[7] A.P. Surzhikov, T.S. Frangulyan, S.A. Ghyngazov, E.N. Lysenko, Journal of thermal analysis and calorimetry **102**, 883 (2010) doi: 10.1007/s10973-010-0912-8

[8] O.V. Galtseva, S.V. Bordunov, S.N. Torgaev, IOP Conf. Ser.: Mater. Sci. Eng. **110**, 012094 (2016) doi: 10.1088/1757-899X/110/1/012094

[9] A.A. Bespalko, A.P. Surzhikov, L.V. Yavorovich, P.I. Fedotov, Russ. J. Nondestr. Test. **48**, 221 (2012) doi: 10.1134/S1061830912040043

[10] L. Kozlova, E. Bolovin, L. Payuk, IOP Conf. Ser.: Mater. Sci. Eng. **132**, 012005 (2016) doi: 10.1088/1757-899X/132/1/012005

[11] A.P. Surzhikov, O.V. Galtseva, E.A. Vasendina, V.A. Vlasov, E.V. Nikolaev, IOP Conf. Ser.: Mater. Sci. Eng. **110**, 012002 (2016) doi: 10.1088/1757-899X/110/1/012002

[12] D.D. Burleigh, M.V. Kuimova, Ponte **72**, 2 (2016)

[13] A. Uvarov, I. Lezhnina, K. Overchuk, A. Starchak, Proceedings of 2014 International Conference on Mechanical Engineering, Automation and Control Systems, MEACS 2014, 6986935 (2014) doi: 10.1109/MEACS.2014.6986935

[14] O.V. Galtseva, G.V. Vavilova, I.V. Plotnikova, SENSIG '15, 91 (2015)

[15] N.M. Natalinova, O. V. Galtseva, E.A. Moldovanova, IEEE 3rd Int. Conf. EECEA, 52 (2016)

[16] R. Kodermyatov, M. Ivanov, M. Yuzhakov, V. Kuznetsov, M. Yuzhakova, E.Timofeeva, MATEC Web of Conferences **48**, 05004 (2016) doi: 10.1051/confmatec/20164805004

METHODOLOGICAL ASPECTS OF THE DEVELOPMENT OF TECHNOLOGICAL ENTREPRENEURSHIP AND IMPLEMENTATION OF FINANCIAL SUPPORT TOOLS IN RUSSIAN UNIVERSITIES

Irina Babkina, *Anna* Bokova, *Maria* Bulygina, *Dmitry* Sukhanov, *Maxim* Syryamkin, *Michail* Chikov*

National Research Tomsk State University, 634050, 36 Lenin Avenue, Tomsk, Russia

Abstract. This article describes the development of methods of identification and support of communication between a developer and a technological entrepreneur. It is to promote university research activities. Special attention is paid to stimulating inventive activity and a university need in evolution of a developer from an inventor to series developmental engineer. The importance of an entrepreneurial path has been empathized. Financial tools of attraction of alternative funding for university innovative projects (e.g. endowment fund) have been analyzed.

1 Introduction

Today the world's leading universities focus on an effective system involving the integration of scientific, technical and production areas in the process of socio-economic development of society. Universities are no longer just a place of concentration of scientific research and fundamental knowledge. They play the role of the most important entities that determine the pace of development, structure and processes of formation of an innovative field. The main difficulties of creating integration relations include poor material and technical base of universities, source of funding of innovative activity [1–3].

2 Some aspects of the development of technological entrepreneurship in Russian universities

The key factor of innovative entrepreneurship - is technology transfer. Reorganization of universities is only possible through the implementation of the technology transfer mechanism, which is necessary to motivate inventor to invent, and an entrepreneur to run a business. Thus, it is necessary to identify principles of the formation of innovative environment.

* Corresponding author: maximus70sir@gmail.ru

© The Authors, published by EDP Sciences. This is an open access article distributed under the terms of the Creative Commons Attribution License 4.0 (http://creativecommons.org/licenses/by/4.0/).

Firstly, the principle of *equal opportunities* - equal access to research, innovative and entrepreneurial activities for students, teachers, researchers and other university personnel. Secondly, the principle of *self-organizing project teams* - creation of conditions for teams necessary to create and implement projects. Thirdly, the principle of an *entrepreneurial path* - the absence of any obstacles and limitations for those wishing to combine teaching and / or research activities with an entrepreneurial one within the walls of one university. Young cosmopolitan scientists will be able to implement their ideas, organize their research team, establish collaboration, earn money, which will keep them at the university. This *mechanism* must be flexible, but stable.

Today Russian universities participating in the project 5–100 (a comprehensive academic excellence initiative that unites top-tier Russian universities) take a pattern by teams of well-known and foreign scientists. Rather than focusing on the faculties, laboratories and departments, the university focuses on the stimulation and motivation of project teams, which must accumulate young cosmopolitan and competent local scientists, and students. Such teams are able organize themselves to carry out projects with the possibility to connect laboratories and departments. They can also be included in international project groups. Such teams mostly focus on commercially attractive projects, which include contract research (applied research) and research and development, market research. Gradually, in order to support and motivate project and business teams and facilitate the exchange of inventive and entrepreneurial knowledge and technologies as well as to create spin-off companies, special tools must be established.

One of the main conditions of the international competitiveness of modern universities is a well-functioning system of financial support of their activities. It includes a steady access to alternative funding sources. This ensures the independence of the university and determines the degree of its adaptability during constant changes.

According to the *Statistics of Science and Education* [4] the main source of university research and development funding is the federal budget (over 50%), resources of the business sector are about 25%, the funds of the universities are less than 10%. Financial support of innovation and research activities of universities is provided in the form of grants and subsidies through competitive tools of federal programs and projects.

The Ministry of Science and Education of the Russian Federation implements the project 5–100, which involves funding aimed at selecting at least five universities conforming to level of the top 100 world universities according to the most authoritative world rankings - Quacquarelli Symonds, Times Higher Education or Academic Ranking of World Universities.

The human factor (human capital) is key to success of R&D. Thus investment in science is, first of all, investment in the human capital of the country [5].

3 Some features of tools of financial support of innovative projects in the Russian universities

Funding of innovative activity in Russian universities depends on the type of R&S and is financially supported from different sources.

Applied research is mainly ordered by enterprises of different sectors of the economy. For universities signing these kind of contracts for R&D is the most significant and preferable. This type of funding is very important for the development of applied science in general. An example of successful collaboration between enterprises and universities is creation of a scientific and educational center *Radio Electronics and Microwave* at the Tomsk State University in cooperation with Research & Production Company *Micran*.

There are several industries that are actively supported by the government: space, military, railway, energy. This kind of scientific research is funded by the Advanced Research Foundation, Russian Fundamental Research Fund, Russian Science Foundation, Russian Humanitarian Scientific Fund. Though Russian venture industry is still underdeveloped and venture capital is poorly integrated in the innovative ecosystem of universities, it becomes more affordable for innovative companies. In 2006 OJSC Russian Venture Company (RVC) was established. With the participation of RVC there were formed eight venture capital funds, in which 51% belong private investors [6].

Many Russian universities are actively developing endowment funds. The existence of such funds does not guarantee quick results. They require a long time, a professional team and fundraising activities effectively integrating education, research and innovation policies of universities. The university must have at least a few endowment funds: educational, research, innovative, and others.

In Russia first endowment funds were established in 2007. Now the program of development of endowment funds in regions is supported by the Vladimir Potanin Charity Fund.

However high potential of such funds as a tool of additional budgeting of universities is not realized. According to open sources, in 2015 the size of the endowment fund of one of the largest universities in the Russian Federation (Moscow State University of Foreign Affairs) amounted more than 1.5 billion rubles. At the same time, the funds of most universities in Russia, have less than 20 million rubles. Such amounts are not enough for effective activity, mainly because universities can use the income received from the endowment capital, not the endowment capital itself. According to experts, in order to be profitable for the university the minimum size of its endowment fund should be at least 100 - 150 million rubles.

European and American researchers are publishing data on the financial stability of endowment funds [7–9], risk prevention [10], and investment policy [11]. In Russian publications devoted to endowment funds special attention is paid to necessary changes in the institutional environment of the Russian economy and education, undeveloped legal framework. Another problem is business is not ready to invest in such funds, because it believes targeted aid to be more efficient.

In order to solve this complex problem the government can provide the following support.

Firstly, the establishment of tax incentives for profit completely or partially to the donors (individuals, legal entities, and specialized management organizations). It should be noted that the law regarding endowment capital allows making donations not only in money assets but also in security papers and real estate. However, endowment funds are not property tax free. Thus donated real estate can become a major article of expenditure instead of the source of income. Thus it is necessary to establish tax incentives for donated real assets (at least for a certain period of time, or until it reaches a certain amount of money, for example, 20 million rubles).

Second, the costs associated with the formation and replenishment of endowment capital of universities must be reduced. The government can assist in reducing of banking commission in case of transferring substantial sums of money, at least for banks with state participation.

Third problem in the functioning of endowment funds in Russia is that the endowment capital of universities are increased by donations of successful businessmen and politicians. But this resource is often not available for universities, located in the regions. Such universities have no access to the head offices and management of donating corporations. They are mostly based in Moscow, and in most cases, cooperate with Moscow universities.

Finally, activities aimed at supporting universities by alumni are not common. In general, philanthropy in universities is not very popular. In Russia charity, is mainly aimed at helping sick children. Thus, it is important to develop a strategy for the development of academic fundraising [12] with the support of university leaders and government officials [13, 14].

In Tomsk State University an endowment fund assisting projects in education, science and culture was established in 2010. The fund provides scholarship programs and necessary help in publishing activities. Also the programs of support of the Siberian Botanical Garden and Creation of Educational Systems in the Field of Culture and Art in TSU are being implemented. A group developing fundraising strategy has been created. It is responsible for special events (including different marketing and communication activities) aimed at raising funds from different categories of donors. Special attention is paid at work with TSU alumni. The main directions of TSU fundraising are collaboration with majors donors (alumni, philanthropists, entrepreneurs) and the development of public fund-raising.

4 Conclusion

Any separate invention acquires its value only because of its possible development in a long process chain of the innovation process: from an idea, to manufacturing. Many *players* are involved in this process. The major player in the process is a technological entrepreneur. Even though in the Russian reality, he is not considered to be really important.

Entrepreneurial path does not exist in the Russian university as a matter of discussion. Jobs in spin-off companies, or in application-oriented projects ordered by an enterprise are not appreciated in the academic community. University teachers do not support and encourage students successful in entrepreneurial activities whose academic achievements are not very high. Entrepreneurial indicators not taken in the laboratory, department, faculty, university KPI. There are no special academic programs for identification of entrepreneurial skills and abilities.

There are no financial tools for attracting alternative funding of innovative projects in universities (e.g. endowment funds). Economic stability and independence of R&D centers are impossible without collaboration with technological entrepreneurs, who can help engineers to commercialize as many inventions as possible in an industrial turnover. They involve relevant businessmen and companies in the task setting process. They also assist in sharing production costs between interested parties, making the technology financially accessible.

Acknowledgment

The paper was written as part of the research project No. 8.2.31.2015, carried out with the support of the Program "Research Foundation of Tomsk State University named after D.I.Mendeleev" in 2015 – 2016, grant RFBR No. 16-29–12858.

References

[1] I. A. MacLeod, Engineering, construction and architectural management **17**, 21 (2010) doi: 10.1108/09699981011011294

[2] M. Almeida, International journal of technology management & sustainable development **7**, 39(2008) doi: 10.1386/ijtm.7.1.39_1

[3] P. Jones, R. Patz, B. Thomas, S. McCarthy, Industry and Higher Education **28**, 39 (2014)
[4] Science and Education Statistics (6), (2015) URL: http://www.csrs.ru/archive/stat_2015_finance/finance_2015.pdf
[5] A. Vyzhigin, SHS Web Conf. **29**, 02045 (2016) doi:10.1051/shsconf/20162902045
[6] The official website of OJSC «Russian Venture Company» URL: http://www.rusventure.ru/ru/
[7] M. Drouvelisa, J. Sonnemans, The endowment effect in games URL: http://papers.tinbergen.nl/15114.pdf
[8] S. N. Gary, Journal of College and University Law, Forthcoming URL:http://papers.ssrn.com/sol3/papers.cfm?abstract_id=2656640
[9] C. Goodhart, C. Osorio, D. Tsomocos , CESifo Working Paper Series No. 2885, (2009)
[10] S.G. Dimmock, Review of economics and statistics **94,** 789 (2012) doi: 10.1162/REST_a_00180
[11] R. C. Merton, *Optimal investment strategies for University endowment funds*, In *Studies of supply and demand in higher education*, edited by C. Clotfelter and M. Rothschild (University of Chicago Press, Chicago, 1993)
[12] S. M. Pritchard, Libraries and the Academy **11**, 595 (2011) doi: 10.1353/pla.2011.0018
[13] B. L. Ciconte, and J.G. Jacob, *Fundraising Basics: A Complete Guide*, second edition (MA: Jones and Bartlett Publishers, Sudbury, 2001)
[14] R.R. Kharisova, S.V. Kiselev, N.M. Chikisheva, V.L. Grankina, L.V. Khabibullina, Journal of sustainable development **8,** 194 (2015) doi: 10.5539/jsd.v8n3p194

HARDWARE-SOFTWARE COMPLEX FOR FUNCTIONAL AND PARAMETRIC TESTS OF ARM MICROCONTROLLERS STM32F1XX

Aleksey Egorov[a], *Pavel* Nekrasov, *Oleg* Kalashnikov

[1]National Research Nuclear University MEPhI, 115409, Moscow, Russia

Abstract. The article presents the hardware-software complex for functional and parametric tests of ARM microcontrollers STM32F1XX. The complex is based on PXI devices by National Instruments and LabVIEW software environment. Data exchange procedure between a microcontroller under test and the complex hardware is describes. Some test results are also presented.

1 Introduction

Detection of functional and parametric failures is the main objective during testing of very large-scale integration (VLSI). Specialized technique is required which is able to provide functional tests of basic elements of a VLSI under test, and to measure its parameters with required accuracy. A microcontroller (MCU) is a VLSI which consists of a microprocessor core and a large set of peripheral units such as memory, universal ports, timers/counters, interrupt controller, interface ports, etc. Testing of these units is quite complex procedure and is implemented by specialized algorithms. A hardware-software complex is required to provide a MCU testing [1]. It is important to receive and to store test results information. This article provides a description of the hardware-software complex which is based on PXI equipment by National Instruments and designed for functional and parametric testing of ARM MCU STM32F1XX.

2 Object of the research

STM32F103C8T6 [2] has been chosen as a test sample; it is a 32-bit Cortex-M3 MCU by STMicroelectronics, which provides the following characteristics:

- Operation frequency up to 72 Mhz;
- 20 Kbyte SRAM;
- 64 Kbyte FLASH;
- 3 timers/counters;
- 3 USART interfaces;
- 2 SPI interfaces;
- 2 I_2C interfaces;

[a] Corresponding author: ayegor@spels.ru

© The Authors, published by EDP Sciences. This is an open access article distributed under the terms of the Creative Commons Attribution License 4.0 (http://creativecommons.org/licenses/by/4.0/).

- 1 USB interface;
- 1 CAN interface;
- 37 GPIO;
- 2 10-channel 12-bit ADC.

The basic requirements for hardware-software complex were formulated based on the MCU parameters:

- MCU main functional blocks for test: RAM, FLASH, GPIO, timers/counters, USART, SPI, interrupt controller;
- supply current measurements in static and dynamic mode with accuracy not less than 0.1 mA;
- testing frequency range from 8 to 72 MHz;
- internal ADC testing – full-scale transfer function measurement and conversion parameters (integral and differential nonlinearity, offset and gain errors) calculation.

3 Equipment and software

To implement the hardware-software complex National Instruments PXI module equipment [3] was used:

- PXI-4110, power supply board [4];
- PXI-7841R, FPGA-based fee collection data reconfigurable board [5];
- PXI-4071, multimeter board [6].

PXI-4110, PXI-7841R and PXI-4071 units are placed into PXI-1073 chassis which provides connection between the complex and a PC. The complex hardware is controlled by LabVIEW software environment with use of LabVIEW FPGA module.

The hardware-software complex also includes a sample chip PCB snap-in and a switching board.

The complex designed can be used not only for MCU testing but also for wide range of electronics: microprocessors, analog ICS [7], optoelectronic devices [8], transistor test structures [9] and so on.

4 Solution

A sample of MCU placed onto the PCB snap-in is connected to the switchboard, which provides connection between MCU pins and NI system channels, and also allows to set any logical state onto a MCU pins (pull-up, pull-down, high logical level, low logical level, connection to another line, Z-state) without NI system involving [10].

Hardware structure with MCU under test is presented in Figure 1.

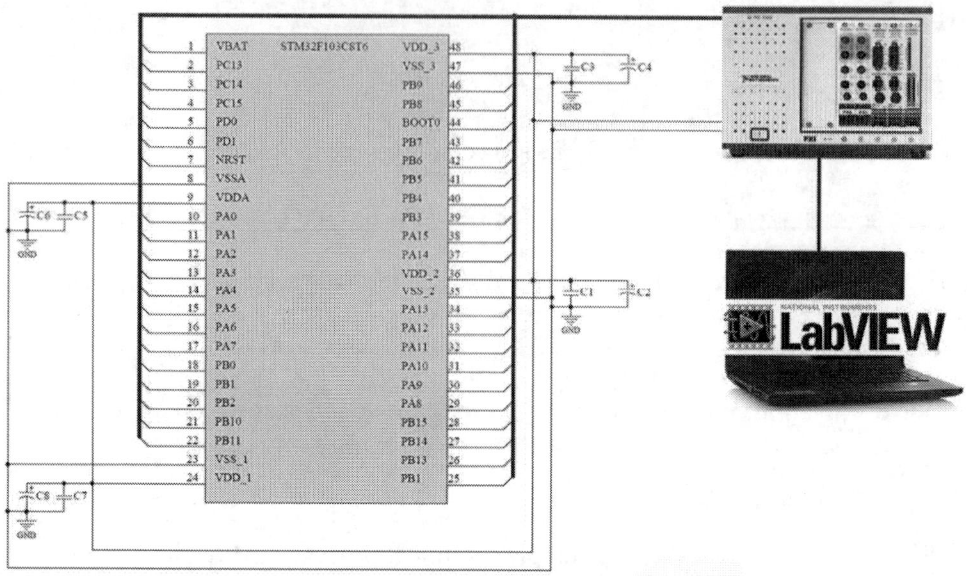

Figure 1. Hardware structure with MCU.

The virtual device has been implemented in the LabVIEW software environment. This software controls MCU under test and provides information from NI complex gathering and processing. The virtual device contains the following software blocks:
- reception and primary processing of information;
- errors counting;
- supply current monitor ;
- restart control;
- hangs control;
- internal ADC testing.

MCU internal USART and SPI interfaces are used to transmit the information about test results from MCU under test to LabVIEW virtual device. These interfaces are implemented by PXI-7841R board based on FPGA. MCU supply current monitor is performed by two PXI-4110 boards under LabVIEW virtual device control. Input and output currents measurement is carried out by PXI-4071 multimeter board. Output voltages of low and high logical levels are measured by integrated analog input channels of PXI-7841R board.

USART and SPI testing were performed using of a control word transfer correctness checking. One USART interface was configured as an output for the information transfer from MCU to PXI board. Two remaining USART interfaces are connected to each other - input of one unit is connected to output of another and vice versa. Two SPI interfaces are also connected to each other. One SPI interface can transfer the information from MCU to PXI board without affecting to the test results. Chess code, inversion chess code, field of zeros and field of ones are used as control words. Difference between received and transmitted information in a same operating cycle is considered as an interface module fault.

Hardware-software complex satisfies requirements of speed and reliability for synchronization with MCU USART and SPI interfaces (no errors when receiving data blocks coming from a test chip to the virtual device). To ensure the reliability of data transmission it is possible to transmit information immediately in parallel with USART block and SPI block. This procedure completely excludes the possibility of transmission failure. Selection of the package is carried out by using of gate signals that indicate usefulness of the parcel and type of the test block. Interface blocks operate in the following speeds: USART – 115200 bit/s; SPI – 4000000 bit/s.

RAM and FLASH memory testing consists of writing a 32-bit control word to all memory cells with subsequent reading [11, 12]. Read word is compared to a reference; if they are not equal failure of a memory cell is recorded. After testing all memory with one word, the test is continued with another word. There are four types of control word: chess code, inversion chess code, field of zeros and field of ones. 10 Kbytes of SRAM and 40 Kbytes of FLASH are tested.

For the interrupt controller testing an external signal supplies from the PXI board to the one of MCU input which operates in the external interrupt input mode. The MCU interrupt controller handles the interrupt and provides a pulse to one of I/O ports which configured as an output. The pulse is detected by the PXI board and is transmitted to LabVIEW virtual device. Absence of a pulse at an output informs about an error of interrupt controller.

Two timers/counters send meander signal to the PXI board and then to the virtual device. Virtual device compares frequency of pulses from the timers/counters with reference frequency. In case of difference between two frequencies an error is registered.

The MCU internal ADC testing procedure is also developed. It is necessary to perform a large number of USART and SPI transmission cycles to obtain the full-scale transfer function on the one hand, and relatively large number of samples required to improve accuracy on the other hand. We decided to implement the procedure of ADC sampling averaging by MCU software code, and to transfer the averaged data. This procedure allows to reduce the amount of USART and SPI transmission cycles, as well as gives opportunity to unload partially the ADC data processing unit which is implemented in the LabVIEW software environment. As a result the full-scale transfer function is measured for 11 seconds against 47 seconds with LabVIEW ADC sampling averaging.

Linearly increasing voltage (from 0 V to 3.3 V) is applied from one of PXI analog output channels to ADC input. ADC samples input voltage and sends the resulting code back to the virtual device through PXI board using USART or SPI. Using the transfer function the conversion parameters of ADC can be calculated. A part of measured transfer function together with the ideal one is shown in Figure 2.

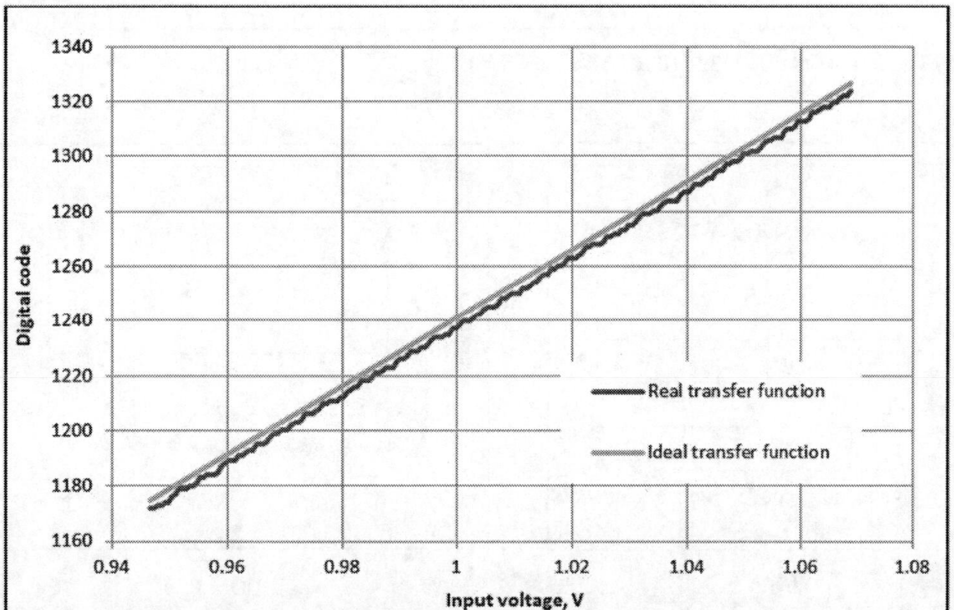

Figure 2. Real and ideal transfer functions of ADC.

The supply current monitor block allows to adjust supply voltage of MCU under test, as well as to monitor the current consumption. In case of exceeding the power consumption the block detects an error and turns off the power supply.

5 Application and development

The hardware-software complex has been successfully implemented in NRNU MEPhI, Moscow. Seven MCU samples were tested in different applications and environment conditions by using this complex [13–15]. Some test results obtained by using the hardware-software complex are presented below.

The time dependence of STM32F103C8T6 internal ADC integral nonlinearity is presented in Figure 3. Figure 4 show time dependencies of dynamic (a) and static (b) supply current of the STM32F103C8T6 respectively.

The hardware-software complex can be adapted for testing of other MCU types by minor changes.

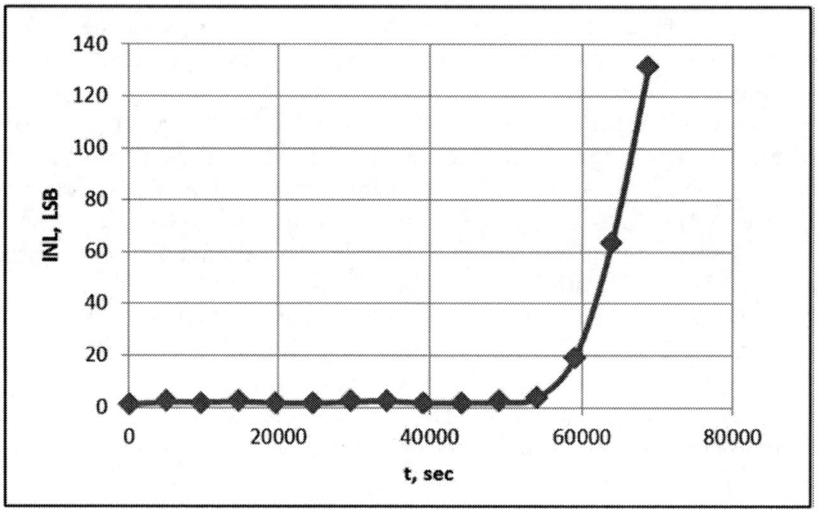

Figure 3. The INL of the STM32F103C8T6 internal ADC.

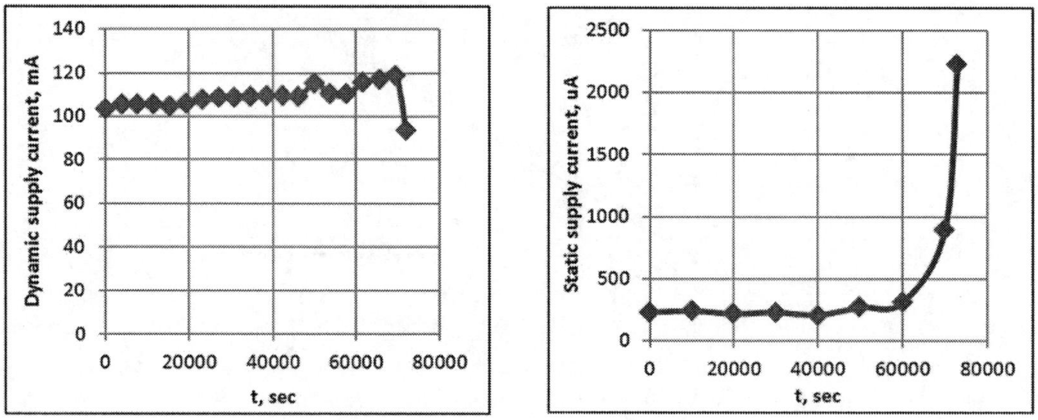

(a)　　(b)

Figure 4. Time dependence of STM32F103C8T6 dynamic(a) and static(b) supply current.

6 Conclusions

The hardware-software complex for ARM microcontroller STM32F1XX functional and parametric test is developed. The complex is implemented based on National Instruments equipment on a PXI-1073 chassis. The software is developed in LabVIEW.

The hardware-software complex has been successfully tested on the STM32F103C8T6 microcontroller. It is planned to adapt this system to test other types of microcontrollers.

References

[1] P.V. Nekrasov, A.A. Demidov, O.A. Kalashnikov, Proceedings of the Instruments and Experimental Techniques **52** , 196 (2009) doi: 10.1134/S0020441209020092

[2] STM32F103x8/STM32F103xB datasheet URL: http://www.st.com/web/en/resource/technical/document/datasheet/CD00161566.pdf

[3] PXI Platform URL: http://www.ni.com/pxi

[4] NI PXI-4110 URL: http://www.prompribors.ru/images/catalog/NIPXI4110.pdf

[5] NI PCIe-7841R URL:http://zone.ni.com/reference/en-XX/help/373197D-01/target2devicehelp/pcie-7841r/

[6] NI PXI-4071: https://www.valuetronics.com/Manuals/NI-PXI-4071.pdf

[7] A.V. Demidova, A. Y. Borisov, L. N. Kessarinskiy, D.V. Boychenko, Proceedings of the International Siberian Conference on Control and Communications (SIBCON), 7147036 (2015) doi: 10.1109/SIBCON.2015.7147036

[8] M.E. Cherniak , A.V. Ulanova, A.Y. Nikiforov, Proceedings of the International Conference on Microelectronics (ICM), 195 (2014) doi: 10.1109/MIEL.2014.6842119

[9] I. I. Shvetsov-Shilovskiy, P.V. Nekrasov, A. V. Ulanova, A. A. Smolin, A. V. Sogoyan, Proceedings of the International Siberian Conference on Control and Communications (SIBCON), 7147282 (2015) doi: 10.1109/SIBCON.2015.7147282

[10] I. O. Loskutov, A. B. Karakozov, P.V. Nekrasov, A.Y. Nikiforov, Proceedings of the International Siberian Conference on Control and Communications (SIBCON), 7147128 (2015) doi: 10.1109/SIBCON.2015.7147128

[11] A. B. Boruzdina, A. A. Orlov, A. V. Ulanova, N. G. Grigor'ev, A.Y. Nikiforov, Proceedings of the International Siberian Conference on Control and Communications (SIBCON), 7147007 (2015) doi: 10.1109/SIBCON.2015.7147007

[12] A. G. Petrov, A. L. Vasil'ev, A.V. Ulanova, A. I. Chumakov, and A.Y. Nikiforov, Central European Journal of Physics **12**, 725 (2014) doi: 10.2478/s11534-014-0503-6

[13] A.O. Akhmetov, D.V.Boychenko, D.V.Bobrovskiy, A.I.Chumakov, O. A. Kalashnikov, A.Y. Nikiforov and P.V.Nekrasov, Proceedings of the International Conference on Microelectronics (ICM), 329 (2014) doi: 10.1109/MIEL.2014.6842156

[14] A.S. Artamonov, A.A. Sangalov, A.Y. Nikiforov, V. A. Telets, and D.V. Boychenko, IEEE Radiation Effects Data Workshop **2015-January**, 7004600 (2014) doi: 10.1109/REDW.2014.7004600

[15] A. Sogoyan, A. Artamonov, A. Nikiforov, D. Boychenko, Facta universitatis Series: Electronics and Energetics **27**, 329 (2014) doi: 10.2298/FUEE1403329S

MATEC Web of Conferences **79**, 01032 (2016) DOI: 10.1051/matecconf/20167901032
IME&T 2016

EFFECT OF TEMPERATURE IN ELECTROCAPACITY IN-PROCESS CONTROL OF CABLE PRODUCTS

Nadezda Starikova*, *Vitaly* Redko, *Ludmila* Redko and *Nikolay* Ermoshin

Tomsk Polytechnic University, Institute of Non-Destructive Testing, 634050, Tomsk, Russia

Abstract. In this paper the analysis of temperature effect on the value of cable capacitance per unit length during its manufacture is provided. The physical model of cable product is developed. This model involves heat and electric components. The single-wire electric cable NV-3 insulated with plastic compound of PVC is chosen as a research object.

1 Introduction

In-process control is one of the most important stages of cable manufacture in factories. During the manufacture it is necessary to control such parameters as eccentricity [1], external diameter [2], length of cable product, dielectric strength of insulation and cable capacitance per unit length [3].

Cable capacitance per unit length is informative parameter for cable insulation quality control. The cable product (wire) can be represented as cylindrical capacitor. The first plate of it is core, the second plate is conductor or conducting medium on the insulation surface and the dielectric is wire insulation. According to the known formulae a capacity of cylindrical capacitor depends on geometrical dimensions and dielectric constant of the capacitor insulation material disposed between the plates. Thus, the core diameter, the diameter of the wire insulation, the length of controlled area of wire and the wire insulation material affect the value of cable capacitance.

2 Research problem

Nowadays the electrocapacity method is a widely used method of in-process control of cable insulation quality, which is based on measuring the value of capacitance per unit length of a manufactured wire [3–7]. The new method of in-process control of cable insulation quality was developed previously [8]. This method consists in measuring the capacitance per unit length of produced wire during spark test. The main point of both methods is an applying voltage between wire insulation surface and core, which is grounded. Then the capacitance per unit length is measured and the difference between measured and nominal values is the indicator of cable insulation quality.

* Corresponding author: starikovanadezda@mail.ru

© The Authors, published by EDP Sciences. This is an open access article distributed under the terms of the Creative Commons Attribution License 4.0 (http://creativecommons.org/licenses/by/4.0/).

The nominal value of the capacitance per unit length is measured under normal conditions. In this case the ambient temperature is about 20°C. But the technological regimes require significant heating a cable product and in-process control is carried out under insulation and wire temperatures which differs from normal. Consequently, the formula for the capacitance of a cylindrical capacitor can be presented as

$$C(T) = 2\pi\varepsilon_0\varepsilon_r(T)\frac{l}{\ln[R_2(T)/R_1(T)]} \quad (1)$$

$$R_1(T) = R_1 + \Delta R_1 = R_1 + \alpha_1 \cdot \Delta T \cdot R_1 \quad (2)$$

$$R_2(T) = R_2 + \Delta R_2 = R_2 + \alpha_2 \cdot \Delta T \cdot R_2 \quad (3)$$

where C is a capacitance, ε_r is a relative permittivity, l is a length of wire, R_1 is a core radius, R_2 is a external insulation radius, α_1 is a coefficients of linear thermal expansion of a core material, α_2 is a coefficients of linear thermal expansion of a insulation material.

Thus, it is necessary to assess the effect of temperature on the value of the capacitance per unit length. The single-wire electric cable NV-3 insulated with plastic compound of PVC is chosen as a research object.

3 Assessment of the temperature operating range for cooling process during the manufacture

The equipment for in-process control of capacitance per unit length can be placed in of extrusion line, after the cooling bath and in the stages of twinning and repackaging. The place of the control equipment depends on the realized method of in-process control.

During the extrusion of insulation with application of a screw barrel the stock material is melted in the melt zone of the extruder. Then the plastic compound is push out through the die on moving core. The heating mode of extruder is chosen based on the type of insulation material. The temperature of the wire insulated with plastic compound of PVC after the die is 160–180°C [9]. After extrusion stage a wire is passing through the cooling bath for solidifying of insulation material. There is not strict requirement for cooling process of wire insulated with a plastic compound of PVC because of small value of the coefficients of linear thermal expansion ($20 \cdot 10^{-5}$ 1/°C [10]) and low turn for void and crack formation in insulation [9, 11]. The cooling process is provided by means of cold tap water which temperature is about 20°C.

After the cooling stage the insulation temperature from inner to upper layer shall not exceed temperature of a softening of the insulation material. This condition is necessary for manufacture of quality cable product. The insufficient cooling leads to formation such defects as surface cut, changing of external insulation diameter and thickness, wire eccentricity.

Because of amorphous properties the softening temperature of plastic compound of PVC is the glass transition temperature or temperature of transition from glassy state into a molten or rubber-like state. The glass transition temperature of plastic compound of PVC is 70–80°C [12]. But for the safety of the cable factory staff a cable product is cooled till 40 – 50°C [9, 10] in practice.

Hence, in the beginning of cooling stage the temperature of the wire insulated with plastic compound of PVC is 160–180°C and after this stage the temperature is 40–50°C.

4 The physical model

To analyze the temperature effect on the value of cable capacitance per unit length the complex physical model was developed by means of the finite element analysis software COMSOL Multiphysics [13]. This mode simulates changing of the temperature field distribution during the insulation material cooling and also considers the electric field which is needed for the capacitance per unit length measuring.

4.1 The thermal model

The heated wire cooling process in the cooling bath is described, in general, by two types heat exchange: heat conductivity and convection. Radiative heat loses are negligible in comparison with the listed thermal exchange types. Heat conductivity provides the heat exchange from the centre of the cooled product to its surface. The water, which contacts with the surface of the cable, induces the cooling of the insulation by the forced external convection.

The initial conditions of the model (in cylindrical coordinates) are:

$$T(r, \varphi, t=0) = 180\ ^{\circ}\mathrm{C} \tag{4}$$

Suppose that the cooling process can be described by the Newton's law of cooling [14]. The boundary conditions for the insulation surface are:

$$-n\cdot(-\lambda\nabla T) = \alpha\cdot(T_w - T) \tag{5}$$

$$\alpha = f(L, V_w, T_w), \tag{6}$$

where λ is the coefficient of thermal conductivity, α is the heat transfer coefficient, L is the length of the cooling bath, T_w is the water temperature of the cooling bath, V_w is the linear velocity of the core insulating. Suppose that the linear velocity of the core insulating is 1m/s.

4.2 The electrical model

To measure the capacitance the alternative voltage is applied to the insulation surface, the core is grounded (Figure 1). Then the capacitance is measured by means of different methods (resonant method, I-V method [3]).

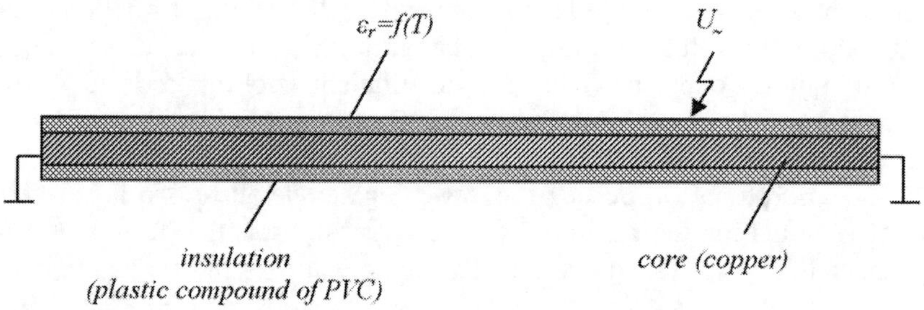

Figure 1. The electrical model of a wire.

The boundary conditions for the electrical model are:

$$V(r = r_c, \phi, t) = 0\ \mathrm{V} \tag{7}$$

$$V(r = r_{ins}, \phi, t) = 10 \cdot \sin(62800 \cdot t) \cdot V, \qquad (8)$$

where r_c is core radius, r_{ins} is insulation radius.

The dependence of the relative permittivity of the insulation material (plastic compound of PVC) from temperature is presented in the next graph (Figure 2) [12].

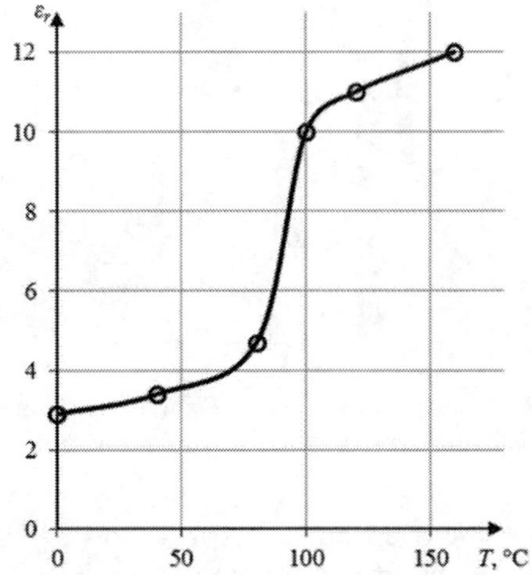

Figure 2. The dependence of the relative permittivity of the plastic compound of PVC from temperature.

It can be noted that the relative permittivity of the plastic compound of PVC is significantly affected by the temperature, which is considered in the model.

5 Analysis of the of physical modelling results

During the work the analysis of the effect of core and insulation thermal expansions on the capacitance per unit length is carried out. This effect is negligible because the coefficients of linear thermal expansions of core and insulation materials are small and leads to capacitance changing which is not exceed 1%.

As a result of simulation the electric and heat fields distributions for insulation cross section are obtained. The Figures 3, 4 show the electric and heat fields distributions at the initial time moment after the wire to be immersed into water of the cooling bath.

According to the heat field distribution at the initial time moment (Figure 3) it can be noted that almost all thermal gradient is in near-surface layers. It is based on change of heat transfer mechanisms at the insulation surface. During following cooling process the thermal gradient is more uniform through the insulation.

The electric field distribution is nearly constant during the cooling process because the insulation thickness is small and the thermal gradient through the insulation is insignificant [15]. An exception is the electric field distribution for the glass transition temperature range of plastic compound of PVC.

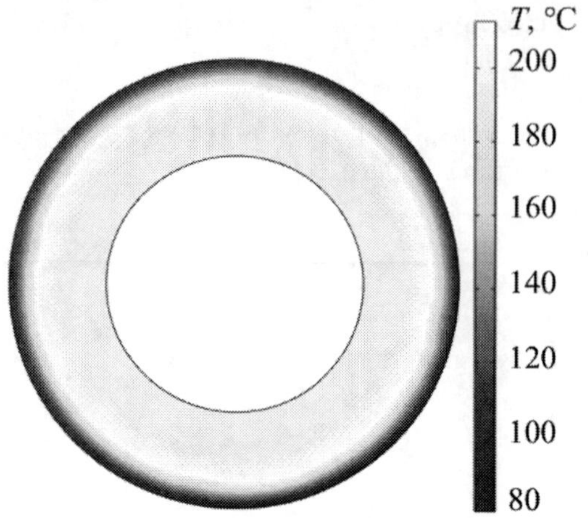

Figure 3. The heat field distribution for insulation cross section at the initial time moment.

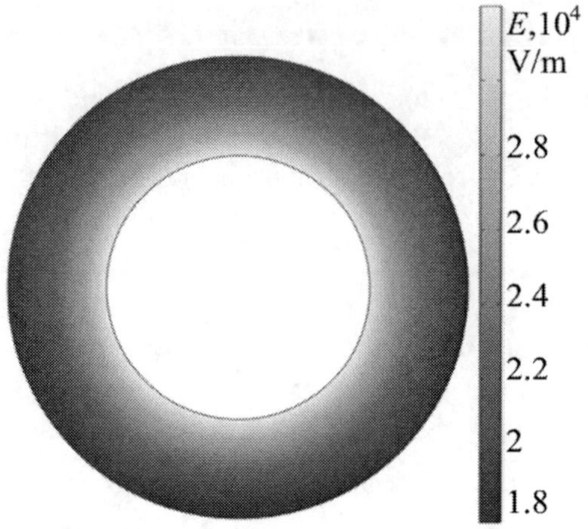

Figure 4. The electric field distribution for insulation cross section at the initial time moment.

In the next figure the obtained dependence $\delta(\tau)$ is shown (Figure 5), where δ is deviation of the capacitance per unit length from the nominal value, τ is cooling time, expressed in relative units in such way that the insulation material have transited into glassy state and the temperature of insulation surface becomes lower 40°C at the relative time τ=1. The value of $\delta(\tau)$ is always positive because the value of the capacitance per unit length increases while heating.

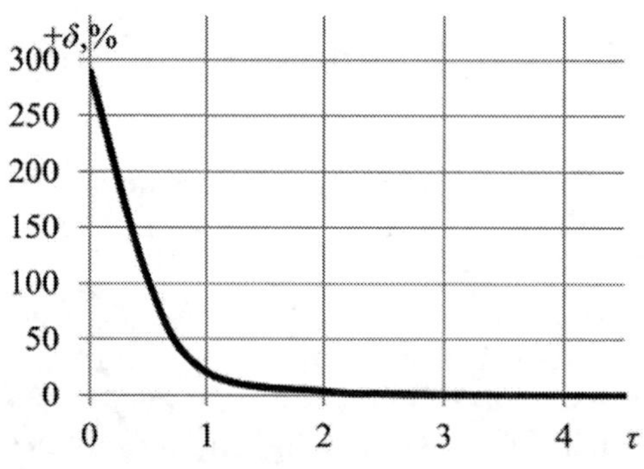

Figure 5. The dependence of the capacitance per unit length deviation from the cooling time.

According to this graph it can be noted that the deviation of the capacitance per unit length is 20% at the relative time τ=1.

6 Conclusion

It is important to evaluate correctly the capacitance per unit length of the cable product for the reliable in-process control. For the correct evaluation of capacitance per unit length the location of the control device must be considered.

On the twinning and repackaging stages, when the temperature of the product is constant and equals to the ambient temperature, thermal effects are negligible for nominal value definition. For the measure of the capacitance per unit length in the cooing bath or after the cooling process it is necessary to consider the cooling regime of the insulation material and introduce the correction coefficients for the nominal value definition. Taking into account the thermal effects in the insulation material can lead to the measurement accuracy enhancement and hence improve the reliability of the cable product in-process quality control.

References

[1] A.E. Goldshtein, E.M. Fedorov, Rus. J. Nondestr. Test. **46,** 424 (2010) doi: 10.1134/S1061830910060069
[2] E.M. Fedorov, I.D. Bortnikov, Tech. Phys. **60**, 1689 (2015) doi: 10.1134/S1063784215110110
[3] A.E. Goldshteyn, G.V. Vavilova, V.Y. Belyankov, Russ. J. Nondestr. Test. **51**, 86 (2015) doi: 10.1134/S1061830915020047
[4] A.M. Pritulov, R.U. Usmanov, O.V. Gal'Tseva, A.A. Kondratyuk, V.V. Bezuglov, V.I. Serbin, Russ Phys Journal **50**, 187 (2007) doi: 10.1007/s11182-007-0026-3
[5] O.V. Vasileva, Y.N. Isaev, A.A. Budko, A.I. Filkov, J. Phys.: Conf. Ser. **552**, 012026 (2014) doi: 10.1088/1742-6596/552/1/012026
[6] A.A. Bespalko, A.P. Surzhikov, L.V. Yavorovich, P.I. Fedotov, Russ. J. Nondestr. Test. **48**, 221 (2012) doi: 10.1134/S1061830912040043
[7] A.A. Bespalko, L.V. Yavorovich, P.I. Fedotov, Russ. J. Nondestr. Test. **47**, 680 (2011) doi: 10.1134/S1061830911100068
[8] N.S. Starikova, V.V. Redko, G.V. Vavilova, J. Phys.: Conf. Ser. **671**, 012056 (2016) doi: 10.1088/1742-6596/671/1/012056
[9] V.M. Leonov, I.B. Peshkov, I.B. Ryazanov, S.D. Holodnyj, *Osnovy kabel'noj tekhniki* (Akademiya, Moscow, 2006)
[10] I.D.Troickij, *Proizvodstvo kabel'nyh izdelij* (Vysshaya shkola, Moscow, 1988)
[11] V.A. Privezencev, *Osnovy kabel'noj tekhniki* (Energiya, Moscow, 1975)
[12] P.P. Nikotin, *Materialy kabel'nogo proizvodstva* (Gosehnergoizdat, Moscow, 1963)
[13] A.V. Abramova, E.V. Ivanova, P.A. Strizhak, MATEC Web of Conferences **23**, 01001 (2015) doi: 10.1051/matecconf/20152301001
[14] T. Sebesi, P. Brehdshou, *Konvektivnyj teploobmen* (Mir, Moscow, 1987)
[15] V.N. Mitroshin, J. of Samara State Technical University : Ser. Phys. and Math. Sc. **38,** (2005)

MATEC Web of Conferences **79**, 01033 (2016) DOI: 10.1051/matecconf/20167901033
IME&T 2016

THERMODYNAMIC MODELLING OF HIGH-TEMPERATURE SYNTHESIS OF THE TITAN AND CHROME CARBIDES ON AN ALLOYED STEEL FOR ELECTRON-BEAM MELTING OF MODIFYING COATINGS

Sergey Kornilov[1], *Nikolay* Rempe[1], *Natalia* Smirnyagina[1,2,*], *Valentina* Khaltanova[2,3], *Anna* Lapina[2]

[1]Tomsk State University of Control Systems and Radioelectronics, 634050, Tomsk, Russia
[2]Institute of Physical Materials Science SB of the RAS, 670047, Ulan-Ude, Russia
[3]Buryat State University, 670000, Ulan-Ude, Russia

Abstract. Adiabatic temperatures of interaction and equilibrium phase compositions in the 300–2673 K temperature range are determined by thermodynamic calculations made for a interaction of titan and chrome oxides with carbon on a surface of the alloyed steel 321 at non vacuum electron-beam melting. Synthesized phases are found to be thermodynamically stable refractory compounds – oxides, carbides, nitrides Cr, Ti, Fe and, stable in contact with the solid-state metallic base.

1 Introduction

Thermodynamic modeling (TM) is a universal method of research, which is widely applied at formation of new metal, ceramic and composite materials and work out new effective, recourse - and energy saving up methods of their synthesis. For example, for calculation of binary and multicomponent diagrams of phase balance, which are a basis for running out of new alloys and make one of the major bases of materials science, method CALPHAD (Calculation of Phase Diagrams) [1–3] is used. Modeling of an equilibrium status in multicomponent multiphase chemically reacting systems is applied to process out and optimization of many processes of synthesis of inorganic substances in materials technology and various areas of chemical technology [4, 5]. Thus, TM spend in various modes, for example, isobar-isothermal, adiabatic, etc. also use various computer methods and specialized programs [2, 4, 5]. TM allows to define equilibrium composition of interaction products (gaseous and condensed) in multicomponent heterogeneous systems; temperature of chemically reacting system in adiabatic conditions (i.e. without taking into account heat of loss in environment) to estimate influence of various factors on an exit of a product and to create a basis for description of the probable mechanism of physical and

* Corresponding author: smirnyagina09@mail.ru

© The Authors, published by EDP Sciences. This is an open access article distributed under the terms of the Creative Commons Attribution License 4.0 (http://creativecommons.org/licenses/by/4.0/).

chemical interaction, phase – and structure formation in difficult systems at synthesis and operation of new materials.

With position to synthesis of new materials, TM it has been successfully used by working out and optimization of reaction processes of multicomponent powder sating environments by an aluminothermy method for the subsequent formation of wear proof carbides coatings on carbonaceous steels by thermochemical treatment [6]. The thermodynamic modeling applied in present study, is the important stage at formation of composite materials and the processes of their reception based on the phenomenon of self-extending high-temperature synthesis (SHS) [7–9].

The purpose of the work was interaction consideration between components of heterogeneous powder compositions on a basis the titan and chrome oxides with carbon on a surface of the alloyed steel at atmospheric pressure, an establishment of equilibrium phase composition and forecasting of thermal stability of the titan and chrome carbides in a wide interval of the temperatures which top value corresponds to a metal-basis melting point. Modelling of possible phase and chemical transformations at SHS process and electron-beam melting of SHS products on an alloyed steel surface at formation of a composite layer of the titan and chrome carbides of the thickness 100 microns on air.

2 Substantiation of a thermodynamic modelling technique

The adiabatic temperature T_{ad} of interaction is the important characteristic in SHS processes. Last years for TM use computer programs with thermodynamic databases (TD) which are based on search of a minimum of Gibbs energy of multiphase multicomponent system [4, 5, 10–13]. Thus, TM it is reduced to search in TD compounds, which can be formed of the elements, which are a part of initial substances, and to formation of their quantitative parity responding balance under set conditions. For SHS processes in multicomponent systems at the Institute of Structural Macrokinetics, Russian Academy of Sciences and problems of materials technology of the Russian Academy of Sciences (Russian acronym ISMAN) develops specialized program ISMAN-TERMO which allows to spend TM only in adiabatic mode [8, 9].

2.1 Universal program ASTRA-4

Developed in Bauman state technical university the universal program ASTRA-4 carries out search of the equilibrium composition responding a condition of a local maximum of entropy of system S in the presence restrictions (a condition of weight continuation of each element and full internal energy of system) for the set initial composition and a thermodynamic parameter [10, 11]. TD of programs ASTRA-4 it is based on reference books on thermodynamics of individual substances [12–19] and contains of the big number of inorganic compounds.

The application technique of program ASTRA-4 for TM synthesis of materials in a adiabatic mode has been developed earlier with reference to classical SHS processes [20] and to synthesis of powder compositions for thermochemical treatment of steels. It contains on carrying out of a calculations series taking into account all possible binary diagram's [21, 22] and introduction in TD new compounds (absent in a standard configuration). The composition of the products received at system cooling after synthesis from temperature T_{ad} to room, can differ from equilibrium composition at T_{ad}. Therefore, further it is spent TM in isobar-isothermal conditions at temperature below points of phase transformations in the multicomponent system. At each stage calculations are carried out some times, i.e. it is iterative, taking into account temperature areas of existence of those or other phases-products.

2.2 Computer program TERRA

In the present work for TM computer program TERRA which represents the new version of program ASTRA-4 is used. Program TERRA is intended for work in Windows system and possesses new options for situation with TD, more expedient interface and the extended possibilities for presentation of calculations [23].

The software of complex Terra for calculation of chemical reactions starts with various conditions of balance: achievements of a minimum of Gibbs energy and an entropy maximum.

In the first case, the optimizing problem presented in general, a kind condition

$$\Delta G^{o}_{reaction} \rightarrow min; \sum_{i,j}^{N n} C_{i,j} X_j \leq b_i, \tag{1}$$

where b_i certain quantity of reacting chemical element B_i; X_i stoichiometric factor of j compound, $C_{i,j}$ – quantity of atoms B_i in compound j, N – quantity of the chemical elements participating in reaction; n – quantity of reaction products.

In the second case, calculation is worked out, preceding from conditions of entropy maximum $S \rightarrow S_{max}$, a constancy of full internal energy, presentation of the continuation law of weight of all chemical elements, acquiescence of a gas phase to the composition equation for a mix of ideal gases.

A problem formulation of thermodynamic modelling demands to appoint two conditions of balance of studied system with environment. These conditions can be or numerical values of thermodynamic characteristics of equilibrium, or functional parities between parameters of this status. In work, a pressure and temperature were varied as parameters of system equilibrium. For the description of the system as material object, the maintenance of chemical elements forming it was necessary the nobility only. Internal and interphase interactions are described by modelling thermodynamic parities for which short circuit properties only individual substances - equilibrium components are used.

Thanks to simplicity of the statement, modeling problem the program complex TERRA allows to use a thermodynamic method for studying of the big number of the diversified high-temperature conditions and processes.

In the present study used interface of a program complex:

TERRA. The limiting number of chemical elements from which the investigated system can consist, is equal 50; the number of the condensed phases considered during one calculation, is limited 200, and the quantity of components of the gas phase formed in balance (number of individual substances), can reach 800. At carrying out of calculations for heterogeneous systems probably use, both models of one-component not mixing up phases, and models of the condensed solutions.

TRIANGLE. Calculations are possible in two modes:

- For isothermal conditions when system equilibrium is set by value of temperature (T, K) and pressure (p, MPa)
- For conditions of adiabatic equilibrium transformation (burning) set in each point of calculation by values of pressure (p, MPa) and enthalpy of reaction of initial substances (I, kJ/kg).

RECTANGLE. Areas with an identical set of the one-component condensed phases as a result of repeated calculations in the set temperature range for all possible parities of initial substances come to light.

Thermal behavior of oxides, carbides and nitrides investigated in the range of temperatures from 273 K to 2673 K at atmospheric pressure (10^5 Pa). Research of stoichiometric composition in which carbides of transitive metals are formed, carried out in the same conditions. Possible interactions with participation Ti, Cr, C and the alloyed steel

(Fe70.88C0.12Cr18Ni10Ti1) in atmosphere of inert gas (He, Ar) and air (N53.91O14.48Ar0.32) were investigated.

In calculations considered following phases: nitrides TiN, CrN, Cr_2N, Fe_4N and Fe_3N_4, oxides Cr_2O_3, TiO_2, Ti_2O_3, Ti_4O_7, Ti_3O_5, TiO, FeO, Fe_3O_4, Fe_2O_3, NiO and carbides TiC, Cr_3C_2, Cr_7C_3, $Cr_{23}C_6$ and Fe_3C.

The special attention is given research of phase equilibrium and to revealing of phase transformations sequence and fields of crystallization of co-existing phases in system TiO_2-Cr_2O_3-C. Phase balances in all concentration area through 1–5 mol. % were investigated. Isothermal cuts in the range of temperatures from 773 to 2573 K and pressure $p=10^5$ Pa have been simulated. For revealing of the tendency of crystallization of this or that phase, in a concentration triangle put isotherms crystalline fields at one pressure or on isothermal section specified areas of phases taking into account various pressure.
Developed materials are intended for reception of composite heat resisting layers of the titan and chrome carbides on a surface of the alloyed steels at non-vacuum electron-beam melting.

3 Results of thermodynamic modelling and their discussion

Into composition of an initial mix entered substances which at heterogeneous interaction in the course of electron-beam melting are capable to form with each other and with a metal basis (Fe-Cr-Ni-Ti) a number of refractory compounds.

3.1 Synthesis of carbide of titan TiC

Initially have been simulated reactions in system Ti-C-O_2. In a temperature interval from 1273 K to 1673 K, it is observed changes of thermodynamic properties (in three stages) on which basis it is possible to estimate size of thermal effect of chemical interaction, a component ΔH equal to -486 kJ/mol.

Interaction the titan oxide$TiO_2$with carbon begins with reduction. In mixes have been appear titan oxides Ti_4O_7 (1298 K) and Ti_2O_3 (1548 K). In the condensed phase, an occurrence of the titan carbide is observed at 1948 K, which concentration does not vary to temperature 2173 K, and further there is a decomposition and transition of titan and carbon vapors in a gas phase (Figure 1). And, occurrence of steams of the titan is fixed from temperature 1473 K, and since 1228 K, the gas phase is ionized (Figure 1).

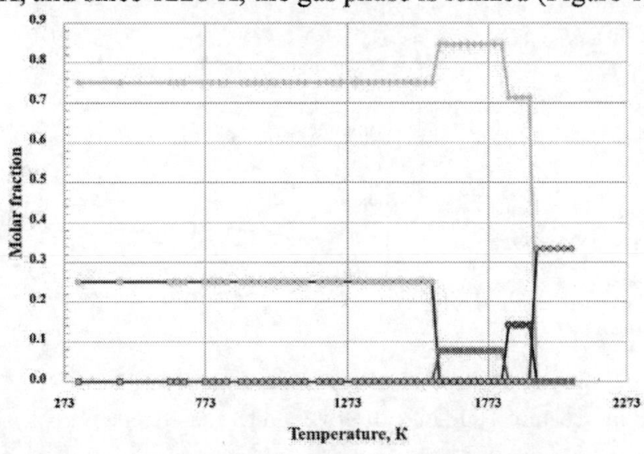

Figure 1. Change of phase composition in the condensed phase in mix TiO_2:2C at atmospheric pressure.

 DOI: 10.1051/matecconf/20167901033

Synthesis of titan carbide TiC on air begins at temperature 1248 K (Figure 2). At this temperature, it is formed oxide Ti_4O_7, which turns in oxide Ti_2O_3 (1548 K).

In s stoichiometric mixture TiO2:2C on air a titan nitride TiN can be formed at temperature 1598 K. The titan nitride is an impurity phase for titan carbide (Figure 2, *a*). Surplus of carbon C is necessary the formation of single-phase carbide. Introduction of surplus in number of 0.1 (2.1 mole) leads to formation of single-phase carbide without the nitride TiN, and it is accompanied by increase in a quantity of carbide. It is necessary to notice, that is thus observed, insignificant quantity of the carbon impurity in process of rise in temperature. Possibly, it should call formation of nonstoichiometric carbide TiC_{1+x}.

Synthesis of titan carbide on alloy steel surface on air also occurs multistage. It is possible to allocate stages of interaction initial oxide TiO_2 with carbon with formationoxide Ti_4O_7 (1298 K), oxide Ti_2O_3 (1548 K). Last oxide reacted with nitrogen with formation of titan nitride TiN (1573K) (Figure 3).

Formation of carbide TiC begins at temperature 1648 K.

Especially it is necessary to note behaviour of an alloying component of a steel, namely, chrome. By thermodynamic calculations chrome in reactionary mixes should be the oxide Cr_2O_3 (upon temperatures 1348 K). Further, oxide are reacted with carbon by formation the carbide Cr_3C_2 which then forms carbide Cr_7C_3 (1973 K).

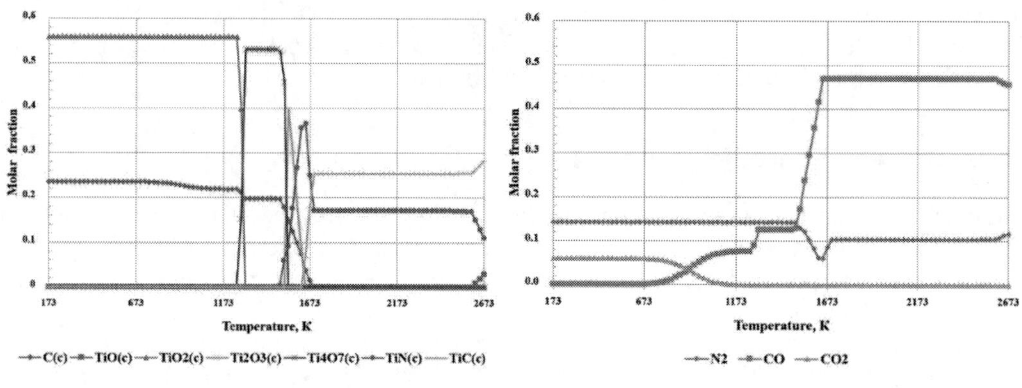

a b

Figure 2. Interaction in mixture TiO_2:2C on air: a– condensed phases, b–gas phase.

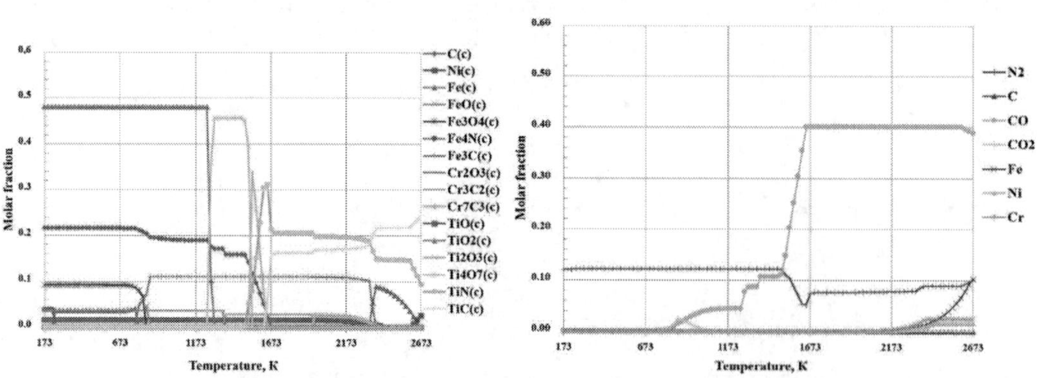

a b

Figure 3. Interaction in structure TiO_2:2C –alloy steel on air: a– condensed phases, b–gas phase.

As a result of chemical interaction of a reactionary mix with a alloy steel surface on air at temperatures above 1673 K in products should be present thetitan nitride TiN, carbides TiC, Fe_3C and Cr_3C_2 which then passes in carbide Cr_7C_3.

Introduction of superfluous quantity of carbon leads to reduction of the maintenance of the titan nitride, and its finishing to 2.2 mole carbon (against 2.1 asking simply on air) calls increase in an exit of carbide titan the from 30.25 to 33.17 mass. % in a mix of carbides Fe_3C and Cr_3C_2. It is necessary to notice, that carbide Cr_3C_2 remains steady and does not pass in carbides of other structures (Cr_7C_3 and C_6Cr_{23}) (Figure 3).

Thus, the presented thermodynamic calculations show on possibility of synthesis of the titancarbide on a surface of the alloyed steel on air from reactionary mix TiO_2:2.2C. Carbide TiC should be in a mix with carbides Fe_3C and Cr_3C_2.

3.2 Composites synthesis on the basis of titan and chrome carbides

As the maintenance of an alloying impurity (chrome) in the alloyed steel is high (18 mass. %) represents to stop on interaction of the titan and chrome oxides with carbon. Threefold system TiO_2-Cr_2O_3-C has been investigated. It is necessary to notice, that system TiO_2-Cr_2O_3 is studied [24].

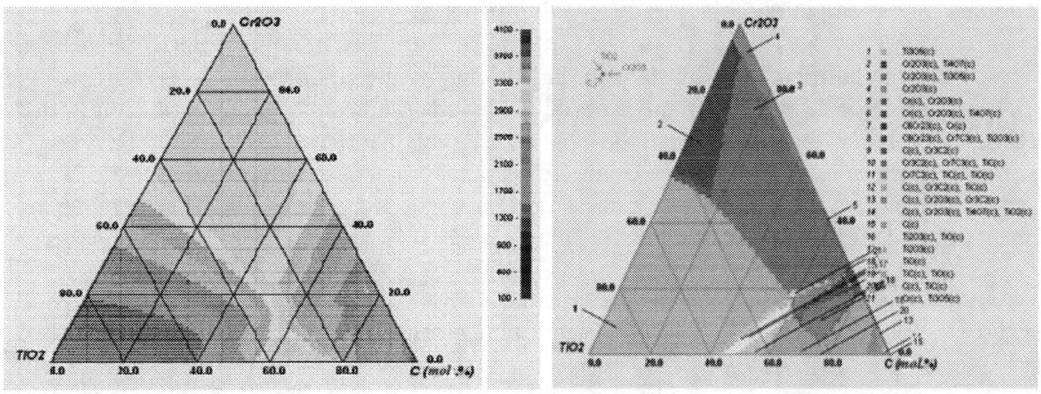

a b

Figure 4. Interaction in system TiO_2-Cr_2O_3-C ($p=10^5$ Pa): a – isotherms; b – equilibrium composition of the condensed phase.

As a result of thermodynamic calculation values of adiabatic temperature Tad and equilibrium composition of products of heterogeneous interaction are defined at this temperature (Figure 4, *a*). The Adiabatic temperature of the titan and chrome oxides interaction with carbon reaches 3700 K. Phase composition re in an equilibrium conditions is combined (Figure 4, *b*).

In Table 1 composition in which formation of carbides of the titan and chrome is possible are resulted.

Table 1. The maintenance of phases (mass. %) in the condensed phase at 1673 K on air.

Composition of a mix for a composite	TiC	Cr_3C_2	TiN	Fe_3C	C	Ti_2O_3
TiO_2:0.3 Cr_2O_3:3.3C	33.95	37.21		14.38	7.54	
0.3TiO_2: Cr_2O_3:3.3C		23.08	7.10	8.92		19.82

Thus, strengthening phases should represent themselves thermodynamic stable refractory compounds of carbides and nitrides of chrome, the titan and iron. Hence, at chemical interaction in the given system the probability of development exothermic interactions in a mode of thermal explosion is rather great.

3.4 Formation of composite coating on alloy steel by non-vacuum electron-beam melting

For experimental test of the thermodynamic calculations by standard methods of x-ray diffraction and scanning electron microscopy researches of phase composition and structure of the composite layers generated non-vacuum electron-beam melting are carried out. The reactionary daub was piled up a previously preformed surface of samples with thickness of 0.5–1 mm. Oxides (Cr_2O_3 and TiO_2), carbon C (birch charcoal) and the organic binding were entered into the daub composition. Researches spent on samples, the size 80×80 mm and height 7 mm executed from the alloyed steel.

The layer thickness made by non-vacuum electron-beam melting was 120–150 microns. Layers are uniform on all thickness. The accurate border is observed has undressed a layer-basis.

According the x-ray analysis (powder diffractometer Phaser 2D Bruker, CuK α-radiation) composite layers have been consists of crystal phases (72.9 mass. % (sample №1) and 74.5 mass. % (sample №2) and also contains 27.1 mass. % and 25.5 mass. % X-ray amorphous phases, accordingly). It is necessary to notice, that at all investigated samples on roentgenograms there are the reflexes belonging to nickel. The nickel maintenance fluctuates from 1.2 to 2.18 mass. %. Besides, it is possible to observe and lines, which intentificated as martensite, but the maintenance does not exceed them 2 mas %. Occurrence of martensite plates can be observed in austenite grains of an initial metal basis. Occurrence martensite can be connected with hardening processes at cooling of samples after removal of influence of an electron beam. Occurrence in a composite layer of chrome carbides Cr_7C_3 and $Cr_{23}C_6$ also testifies about high temperature in samples as a result of influence of an electron beam (Table 2).

Table 2. Phase composition of layers from reactionary daub on the alloyed steel after non-vacuum electron-beam melting (mass. %).

№ sample	TiC 00-006-0693	Fe_3C 00-003-0400	Austenite 00-023-0298	Fe 01-077-7904	Fe_3O_4 01-086-1344	Fe_2O_3 01-088-2359	Cr_3C_2 00-001-1186	Cr_7C_3 01-071-3789	$Cr_{23}C_6$ 00-035-0783
1.TiO_2+0.3Cr_2O_3+3.3C	13.16	6.33	9.40	4.97	6.11	43.95	3.31	3.85	6.80
2.TiO_2+2.2C	36.12	13.19	11.34	3.55	2.38	11.57	-	-	-

Besides, as synthesis of the titan and chrome carbides in reactionary daubs on air an oxidation of a metal basis is possible. According to X-ray analysis in samples presence iron oxides and Fe_3O_4 and Fe_2O_3 is fixed.

Thus, experimental data confirm efficiency of the above-stated method of thermodynamic modelling for receipt of layers by non-vacuum electron-beam melting.

4 Conclusions

1. Computer program Terra (working out Bauman state technical university, Trusov B.G.), is effective for thermodynamic modelling and optimization of structure of reactionary mixes of the titan and chrome oxides and carbon for synthesis and out of vacuum electron-beam melting for heat resisting layers on the basis of carbides of transitive metals on the alloyed steels.

2. Interaction between initial oxides with carbon is accompanied by heat significant amount. Calculated adiabatic temperature can reach 3700 K in reactionary mixes. At formation of strengthening layers on a surface of the alloyed steels have been are self-extending high-temperature synthesis proceeds.

3. According to experimentally confirmed results of the thermodynamic modelling, formed in process non-vacuum electron-beam melting the strengthening phases represent thermodynamic stable refractory compounds of the titan and chrome carbides and nitrides.

Acknowledgment

Work supported by the Ministry of Education and Science of Russian Federation under Agreement number 14.577.21.0018 (RFMEFI57714X0018).

References

[1] B.S. Bokstein M.I. Mendelev, D.J. Srolovitz, *Thermodynamics and Kinetics in Materials Science: A Short Cours* (Oxford University Press, Oxford 2005)
[2] N. Saunders, A.P. Miodownik, N. Saunders, *CALPHAD (Calculation of Phase Diagrams): A Comprehensive Guide* (Elsevier Science Ltd., Cambridge, 1998)
[3] M. Hillert, *Phase Equilibria, Phase Diagrams and Phase Transformations: Their thermodynamic Basis* (Cambridge University Press, Cambridge, 2008)
[4] V. Belov, *Thermodynamic modeling: methods, algorithms, programs* (Nauchny Mir, Moscow, 2002)
[5] P. Mason, M. Mignanelli, Adv. Mat. Proc. **153**, 21 (1998)
[6] L.G. Voroshnin, *Multi-component diffusion carbide coatings on iron-carbon alloys* (BNTU, Minsk, 2007)
[7] N.P. Novikov, I.P. Borovinskaya, A.G. Merzhanov, *Problems of combustion in chemical technology and metallurgy* (OICF, Chernogolovka, 1975)
[8] S.S. Mamyan, A.A. Shiryaev, A.G. Merzhanov, J. Eng. Phys. **65**, 431 (1993)
[9] A. Shiryaev, International Journal of SHS **4**, 351 (1995)
[10] G.B. Sinyarev, *Usage of computers for thermodynamic calculations of metallurgic processes* (Nauka, Moscow, 1982)
[11] N.A. Vatolin, G.K. Moiseyev, B.G. Trusov, *Thermodynamic modeling in high-temperature inorganic systems* (Metallurgy, Moscow, 1994)
[12] I. Barin, O. Knacke, *Thermochemical Properties of Inorganic Substances* (Springer-Verlag, Berlin, 1973)
[13] I. Barin, O. Knacke, O. Kubaschevski, *Thermochemical Properties of Inorganic Substances. Supplement* (Springer-Verlag, Berlin, 1977)
[14] U.D. Veryatin, *Thermodynamic properties of inorganic substances* (Atomizdat, Moscow, 1965)
[15] V.P. Glushko, *Thermal constants of substances* (VINITI, Moscow, 1979)
[16] V.P. Glushko, *Thermodynamic properties of individual compounds* (Nauka, Moscow, 1982)
[17] I. Barin, *Thermochemical Data of Pure Substances* (VCH Publishers, New York, 1995)
[18] M.W. Chase Jr., NIST-JANAF Thermochemical Tables, 4th Ed., J. Phys. Chem. Ref. Data, Monograph 9 (1998)
[19] M. Binnewies, E. Milke, *Thermochemical Data of Elements and Compounds* (Wiley-VCH Verlag GmbH, Weinheim, 2002)
[20] B.B. Khina, A.F. Ilyushchenko, A.V. Belyayev, Powder metallurgy **26**, 35 (2003)
[21] T.B. Massalski, H. Okamoto, P.R. Subramanian, L. Kacprzak, *Binary Alloy Phase Diagrams* (ASM International, Materials Park, 1990)
[22] N.P. Lyakishev, *Binary metal systems diagrams* (Mashinostroenie, Moscow, 1999)
[23] N.I. Ilinykh, and V.P. Vassiliev, MATEC Web of Conferences **3**, 01074 (2013)
[24] S. Riyas, P.N. Mohan Das, Brit. Ceram. Transactions **103**, 23 (2004)

MATEC Web of Conferences **79**, 01034 (2016) DOI: 10.1051/matecconf/20167901034
IME&T 2016

INDUSTRIAL ROBOT AUTOMATION IN SOLVING NON-VACUUM ELECTRON-BEAM WELDING PROBLEMS

Vasiliy Borovik [1,3], *Vladislav* Shatravin[1,3], *Il'jas* Junusov [2], *Dmitri* Shashev [3], *Sergei* Kornilov [1], *Nikolai* Rempe[1], *Stanislav* Shidlovskiy [2,3,a]

[1] Tomsk State University of Control Systems and Radioelectronics, 634050, Tomsk, Russia
[2] National research Tomsk polytechnic university, 634050, Tomsk, Russia
[3] National research Tomsk state university, 634050, Tomsk, Russia

Abstract. The paper describes one modeling automated hardware-software system to develop additive technologically - based products. Algorithm synthesis for industrial robot control was performed. In this case, the operating tool is an electron -beam projector with plasma emitter. The model testing results of thermal processes (in 3D-enqueuing) proceeding within additive substrate material system have been described. The modeling results revealed those required parameters of emission power which would provide the melting of additive material (TiC) excluding boiling.

1 Introduction

As great emphasis is placed on such factors as production rate and cost, the application of additive technologies in developing industrial products furthers the possible implementation of economically- effective methods in the production of items based on consumable patterns. This technology implements the basic principles of developing new-generation materials and embraces an innovative approach in designing and producing updated products comparable to the conventional methods, i.e. casting and mechanical metal-cutting.

This paper describes one method in architecturing automated hardware-software system for designing additive technologically-based [1, 2] items. This system includes electron-beam projector with plasma emitter, industrial robot and hardware-software appliance ensuring mutually-agreed interaction of the whole system itself. Electron-beam projector with differential-pumping system [3], intended for non-vacuum material processing and operating as a tool for the industrial robot [4, 5]. The latter also provides the positioning of the electron-beam projector relative to the workpiece.

[a] Corresponding author: shidlovskiysv@mail.ru

© The Authors, published by EDP Sciences. This is an open access article distributed under the terms of the Creative Commons Attribution License 4.0 (http://creativecommons.org/licenses/by/4.0/).

2 Electron-beam projector with plasma emitter

Low-voltage gas discharge mode – hollow-cathode reflective discharge [6] is used in projectors with plasma emitter which are designed to generate focused high-brightness E-beams. Electrons exposed to electric field are emitted from the gas-discharge plasma [6].

The basic projector structure unit is the metal-ceramic unit – discharge chamber (Figure1, *a*, marked region).

Discharge generates within the discharge cell, which, in its turn, includes internal hollow cathode surfaces (3), cylindrical anode (4) and emitting cathode (1) (Figure1, *a*). Axially-symmetric field including the axial induction of about 0.1T is formed between the hollow and emitting cathodes of the permanent magnet.

Plasma-supporting gas pressure in the discharge chamber is approximately $5 \cdot 10^{-2}$Torr. The same pressure of about 10÷20cm^{3}atm/h is created by dosed gas puffing supplied through the channel into the hollow cathode and pumped into the axial channel of emitting cathode. The glowing discharge is continuous under the voltage of 350÷450 V and current from 0.1 to 0.5A. This discharge current range renders electron beam of one to several hundred milliamphere.

Electron emission from the plasma occurs through the orifice in the emitting cathode of discharge chamber to accelerating field zone between this cathode and grounded high-voltage anode (extractor).

Produced electron beam in the accelerating gap is focused by the magnetc field and transported to the vacuum chamber [7] and / or atmosphere [3].

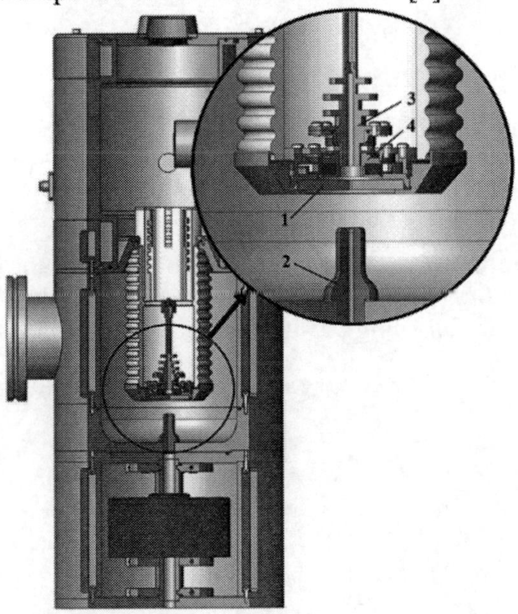

a

b

Figure 1. Electrode diagram (a) and external view (b) of the projector with plasma emitter [8]: 1 – emitting cathode; 2 – extractor; 3 – hollow cathode; 4 – anode.

External projector view, application design designed with beam extractor in atmosphere (Figure1, *b*). This projector produces focused electron beams of 120 keV and current up to 200 mA.

3 Automated industrial robot

3D printing involves designing the actual (physical) product based on its digital model. Such a model could be a standard component designed within the framework of solid modeling and printed to STL. STL format is a 3D model as a list representation of triangular faces describing its surface, as well as its standard specifications. Further applying the programming support "slicer" the model is transformed into G-code, representing the vocabulary for 3D printer and relevant Cartesian coordinates. Therefore, manipulator movement algorithm involves Cartesian coordinates and G-code commands as initial data, i.e. to solve specific instance of inverse kinematic problems.

Let's consider 3DOF robotic manipulator. The simpliest calculation is the geometrical method in solving inverse kinematic problems [9].

The first problem solution step is selecting initial manipulator position, including first manipulator link (segment) perpendicular to the earth; second link (segment) – straight along the axis *OX*. Thus, initial-values of related manipulator angles are:

$$\alpha_0 = 0^\circ,\ \beta_0 = 90^\circ,\ \gamma_0 = 0^\circ,$$

where, α_i – second link angle of rotation to axis *OX* in the plane parallel to *OXY*; β_i – angle between horizontal plane surface of installed manipulator and first link angle after movement in i – point; γ_i – angle between first link angle and second link angle after movement in i-point.

Lateral view of the horizontal plane movement axis is denoted *OU*, where the points lie in the Pythagorean theorem:

$$u = \sqrt{x^2 + y^2}\ ;\ \forall x \in X\ ;\ \forall y \in Y \qquad (1)$$

Vertical axis – axis *OZ* of initial data.

Initial point- starting first manipulator link. The fact that the manipulator is installed on a mobile base will be considered in the addition of correction coefficients to obtained result.

Consequently, the point of manipulator gripper is described by the top-view coordinates (x_i, y_i), of which the side-view angles α_i, and (u_i, z_i) are determined, and, whereof angles β_i and γ_i are calculated. It is obvious that the initial point of the gripper could be determined in terms of the manipulator link length. So, the first link length l_1 = 1350mm, second link l_2 = 1500 mm. In this case, the initial point has the following top-view coordinates (1500; 0) and side-view coordinates (1500; 1350).

Let's consider the top-view (Figure 2).

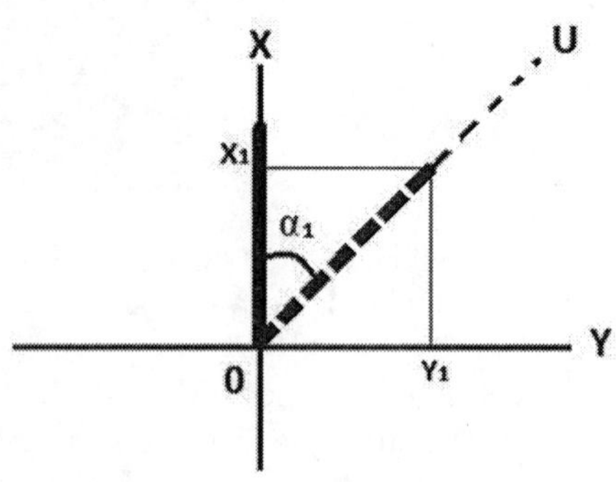

Figure 2. Coordinate system *XY.*

It is evident that according to the Pythagorean theorem moving from the initial point (x_0, y_0) to point (x_1, y_1), the manipulator rotates through an angle α_1 of system defined:

$$\begin{cases} \sin(\alpha_i) = x_i / \sqrt{a^2 + y^2}, \\ \cos(\alpha_i) = y_i / \sqrt{a^2 + y^2}, \end{cases} \quad (\tag{2}$$

or, transforms into:

$$\sin(\alpha_i) - \cos(\alpha_i) = (x_i - y_i) / \sqrt{x^2 + y^2}. \tag{3}$$

The problem-solving result of this system is the sought-for angle α_1.

Let's consider the side-view movement in plane *OUZ* (Figure 3).

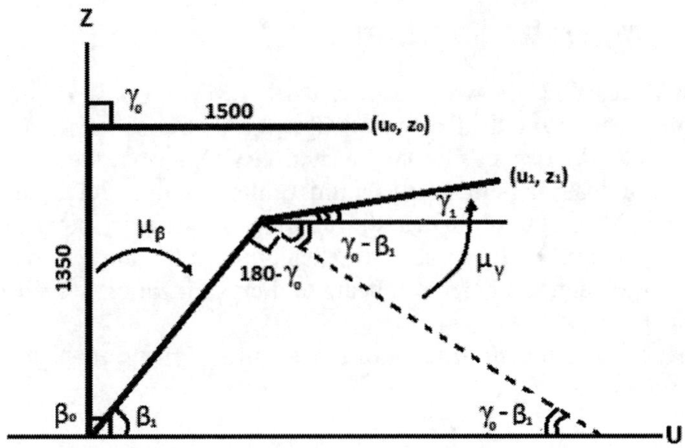

Figure 3. Coordinate system *ZU.*

Let us assume that the manipulator is moved to a point (u_1, z_1). As is seen from the above diagram to simplify the calculations the movement links are divided into two steps:

1) rotation of first link to an angle μ_β relative to the origin of the coordinates;
2) rotation of second link to an angle μ_γ relative to the conjugation point of the first link.

To determine these angles the following algorithm is used: the dependence of each coordinate to the finite point of each link is described through the length of these links l_1 and l_2 and angles β_i and γ_i. The following system of nonlinear equations was calculated:

$$\begin{cases} l_1 \cos(\beta_i) + l_2 \cos(\gamma_i) = u_i, \\ l_1 \sin(\beta_i) + l_2 \sin(\gamma_i) = z_i. \end{cases} \tag{4}$$

The solution of one system involves the angles β_i and γ_i. However, according to selected algorithm, β_i and γ_i are not angles to μ_β and μ_γ; but, based on this fact and using the initial values of adjoint angles β_0 and γ_0 it can be expressed by:

$$\mu_\beta = \beta_0 - \beta_1, \tag{5}$$

$$\mu_\gamma = \gamma_0 - \beta_1 + \gamma_1. \tag{6}$$

The links, moving to the calculated angles μ_β and μ_γ, center the gripper in the specified point. It is obvious that this algorithm could be iterative and one could calculate as many points as possible. However, it is necessary to recalculate the angle γ_i after each iteration, accordingly to the following relation:

$$\gamma_i = \gamma_{i-1} - \mu_\gamma$$

This is due to the fact that the system solution is the internal narrow angles between the links, and, according to specified definition, this angle is counted clockwise from the axis of the previous link.

The main disadvantage of described algorithm is the strong dependency of system solution equation results to the initial approximation task. To obtain correct results it is necessary to make an appropriate assumption of the angle dimension at the initial stage.

Obtained control algorithm is implemented in Matlab and applying extension packs of Simscape Multibody. Accuracy testing of examined algorithm results was conducted with consideration of original effector position gravity center. The research results showed that the industrial robot ensures the positioning of the operating tool pertaining to specified trajectory and required accuracy.

4 Modeling temperature setting

To produce highly-qualified "grown" items it is necessary not only to provide the required penetrating beam trajectory, but also to consider the electron beam energy intensity and velocity, which, in its turn, depends on the thermal-physical properties of used sources.

To produce an industrial item non-vacuum printers should have such electron beam parameters that would optimally further the melting of additive metal powder during the production process itself. In this case, the behaviour of additive material in melting, temperature range of penetrating electron beam and crystallization velocity of used material should be examined.

The mathematical heat transfer model of porous medium and its phase transitions could be written as [10, 11]:

$$\Psi(T)\frac{\partial T}{\partial t} = k(\varepsilon_V, \varepsilon_\alpha)\nabla^2 T + U(q)$$

where Ψ – dimensionless effective heat capacity, including latent heat release of phase transition; T – absolute temperature; t – time; k – temperature conductivity; ε_V and ε_α – porosity characteristics of powder-coated layer; U – thermal intensity of volume source, related to beam power .

Geometric computational region is depicted in Figure 4.

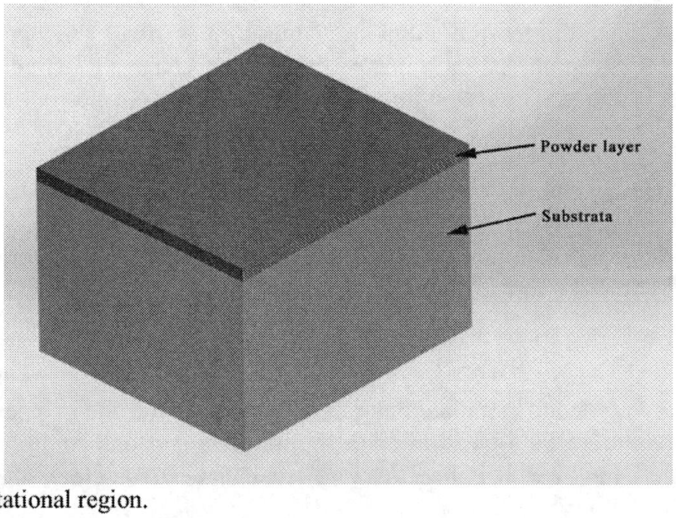

Figure 4. Computational region.

Described fuse layer is titanium carbide powder (TiC) including such properties as [12]:
Thermal capacity in solid phase, J/(g K):
$C_p(T)$ = (-1.608e-18 T^6 + 1.923e-14 T^5 – 9.185e-11 T^4 + 2.234e-07 T^3 – 0.0002895 T^2 + 0.1953 T -3.611)/59.911.
Thermal capacity in liquid phase, J/(g K): $C_{p\_liq} = 1.0476$.

Melting temperature, K: $T_{liq} = 3290$.
Latent heat of melting J/kg : $L_{1\text{-}2} = 1.1873$.
Density, g/cm³ : $\rho = 4.93$.

Powder-coated layer thickness is 0.2 mm. Substrata is of stainless steel, thickness ranging from 1 to 7mm.

Control sintering by electron beam exposure includes the following parameters: 1) actual emission power; 2) beam velocity.

Emission rate depends on the following factors: melt temperature of material powder, thermal conductivity, size and shape of powder particles so as to prevent overheating and further melt boiling. Generally, distribution of beam power is:

$$J(x,y) = F(x)F(y)W\,N\,, \quad (7)$$

$$F(r) = \frac{1}{\sqrt{2\pi\sigma}}\exp\left(\frac{-r^2}{2\sigma^2}\right)$$

$$N = \left(\sqrt{2\pi\sigma}\right)^2$$

$$W = J_0\,V$$

$$J_0 = \frac{I}{2\sigma^2\pi}$$

where, σ– effective beam range, m; I – electron beam current ; V – voltage, 110 kV; x, y – Cartesian coordinates (cross-shaped X/Y axis), the normal of which corresponds to electron beam symmetry axis.

Figure 5 presents the distribution of beam power in the described projector.

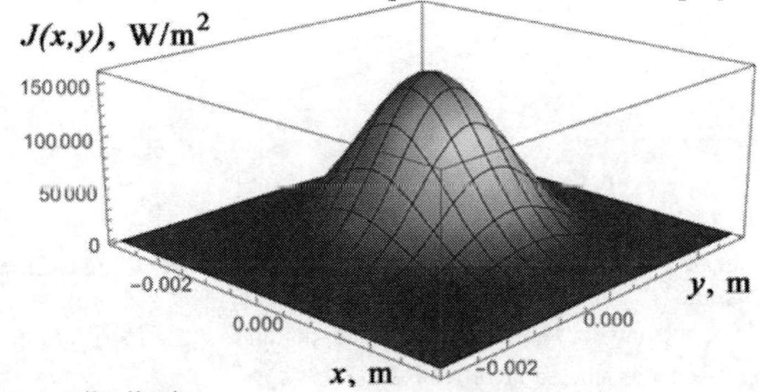

Figure 5. Beam power distribution.

Beam penetration coefficient is characterized by the depth penetration into the powder-coated layer and depends on the powder porosity (shape and size of particles) and radiation surface absorption of powder particles.

Beam velocity is based on the melt temperature: beam velocity decreases for refractory materials and, reversely, increases for low-melting materials.

Modeling electron beam velocity trajectory (based on the results stated in section 3) it is necessary to apply correction for (7):

$$J(x,y,t) = F[x + f_x(t)]F[y + f_y(t)]W\,N\,,$$

where, $f_x(t)$ and $f_y(t)$ – dynamic functions, defining the changing position of electron beam symmetry axis relative to the origin of coordinates along x and y axes, respectively.

Defined equation including initial-boundary conditions is solved numerically, i.e. computer-calculated. In calculating the partial differential equation finite element method is used [8, 13–15]. Applying FDM (finite difference method) in problem-solving of solid bodies is represented as a sum of nodes. Substituting partial differential coefficients by corresponding finite differences the system of linear algebraic equations (SLAE) is derived. Visualized distribution is obtained on the basis of I=15 mA and time t = 0.5 s.

Figures 6 and 7 show that provided parameters and time of electron beam exposure ensure the melting of the powder-coated layer and upper region of substrata (backing).

Figure 6. Fields of phase state: black – «liquid», gray– «solid».

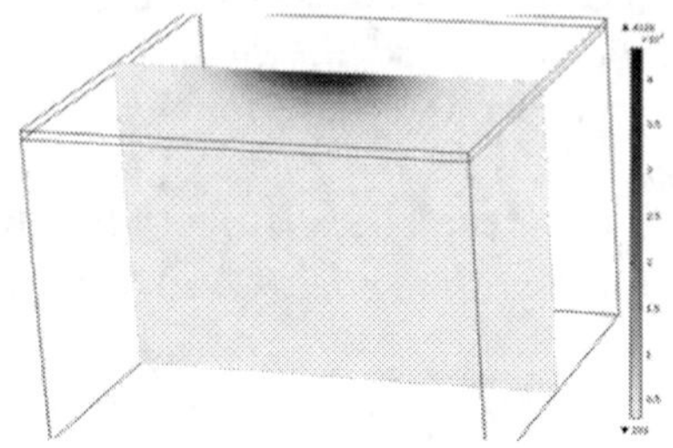

Figure 7. Temperature pattern.

Providing current range of beam within 7.5 -9.5 mA the functional dependency is obtained, reflecting the required electron beam exposure time for melting the powder-coated layer thickness of 0.2 mm and excluding boiling (Figure 8).

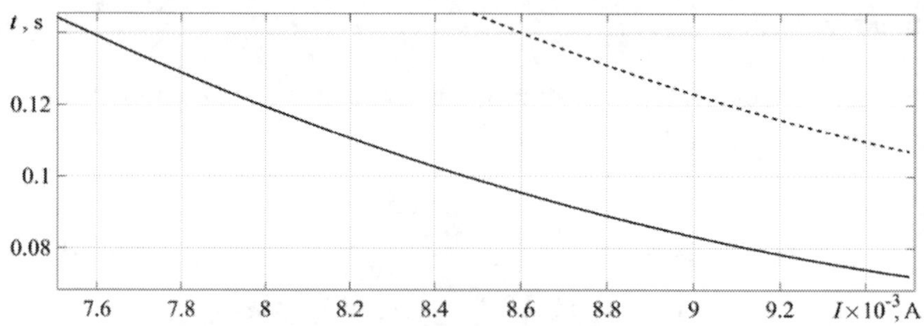

Figure 8. Electron beam current- beam exposure time dependency, providing the melting of the material: continuous line $t'_m(I)$ – for points on the powder-coated layer surface; dotted line $t''_m(I)$ – for points on the separation interface of powder-coated layer / substrata

Based on the modeling results the analytical expression for functional dependency is written (Figure 8), at $I \in [0{,}0075;0{,}0095]$ A:

$$t'_m(I) = 9429I^2 - 196{,}5I + 1{,}088\,,\quad t''_m(I) = 1{,}2\cdot10^4 I^2 - 254I + 1{,}437\,.$$

5 Conclusion

The research revealed the thermal processes within the item during non-vacuum electron beam welding and automated robot control positioning the electron beam in the process itself. The most important control parameters, i.e. beam velocity (scanning) and beam intensity were determined. It was identified that the specific welding regimes depend on the beam penetration coefficient, which, in its turn, depends on the preparation method of powder-coated layer. The modeling results revealed those required parameters of emission power which would provide the melting of additive material excluding boiling.

Acknowledgment

The following research has been financially supported by Ministry of Education and Science of the Russian Federation, Agreement № 14.577.21.0018 (RFMEFI57714X0018).

References

[1] A. Townsenda , N. Seninb, L. Blunta , R.K. Leachb, J.S. Taylord, Precision Engineering **46**, 34 (2016) doi: 10.1016/j.precisioneng.2016.06.001
[2] I. Gibson, D.W. Rosen, B. Stucker, *Additive Manufacturing Technologies: 3D Printing, Rapid Prototyping, and Direct Digital Manufacturing* (Springer, New York, 2014)
[3] A. Aksenov, S. Kornilov, M. Motorin, N. Rempe, D. Shashev , S. Shidlovskiy , E+E **5-6,** 129 (2016)
[4] Y. Zeng, W. Tian, W. Liao, Robotics and Computer-Integrated Manufacturing **42**, 113 (2016) doi: 10.1016/j.rcim.2016.05.011
[5] R. Lia, Y. Zhao, Measurement **88**, 113 (2016) doi: 10.1016/j.measurement.2016.02.038
[6] V.I. Gushenets, E.M. Oks, G.Yu. Yushkov, N.G. Rempe, Laser and Particle Beams **21**, 123 (2003) doi: 10.1017/S0263034603212027
[7] T. Hassel, N. Rempe, S. Kornilov, A. Beniyash, Welding and Cutting **11**, 122 (2012)
[8] S.Yu. Kornilov, N.G. Rempe, S.V. Shidlovskiy, Technical Physics **61**, 841 (2016) doi: 10.1134/S1063784216060116
[9] T. Messaya, R. Ordóñeza, E. Marcilb, Robotics and Computer-Integrated Manufacturing **37**, 33 (2016) doi: 10.1016/j.rcim.2015.06.003
[10] M. Krivilev, E. Kharanzhevski, G. Gordeev, V. Ankudinov, Large-scale Systems Control **31**, 299 (2010)
[11] M.St. Węglowskia, S. Błachaa, A. Phillipsb, Vacuum **130**, 72 (2016) doi: 10.1016/j.vacuum.2016.05.004
[12] M.W. Jr. Chase, *NIST-JANAF Thermochemical Tables* (American Inst. of Physics, New York, 1998)
[13] Y. Liu, Ch. Tian, T. Liu, F. Kong, MATEC Web of Conferences **34**, 02014 (2015) doi: 10.1051/matecconf/20153402014
[14] F. Klocke, D. Lung, H. Puls, Procedia CIRP **8**, 240 (2013) doi: 10.1016/j.procir.2013.06.096
[15] S.V. Panin, V.I. Syryamkin, A.I. Glukhikh, Avtometriya (4), 79 (2003)

INVESTIGATION OF RELATIONSHIP FOR INHOMOGENEOUS SPATIAL DISTRIBUTION OF THE UPPER ATMOSPHERE GLOW WITH THE GEOLOGICAL STRUCTURE OF THE CRUST

Yuriy K. Krakovetsky, *Valentin* Krakovetsky, *Yuriy Y.* Krakovetsky*, *Valeriy* Parnachev, *Lev* Popov

National Research Tomsk State University, 634050, Tomsk, Russia

Abstract. Authors propose a new method for studying of the physical properties of the Earth's crust, based on the effect of interaction of electromagnetic parameters of the magnetosphere, the ionosphere (the upper atmosphere), the atmosphere and the lithosphere. The research has shown that the spatial distribution of the night glow of the upper atmosphere in the zone of high and mid-latitudes is fragmented, sectional inhomogeneity is localized, stable, regardless of the magnitude of the geomagnetic disturbances, and are tied to ground coordinates.

1 Introduction

Nowadays one of the most promising areas of geological and geophysical studies is the remote sensing of the earth's crust structures. This is a very powerful tool to obtain such goals as the research of structural horizons of lithosphere and hydrosphere, localization of energetically active areas in the Earth's crust and hydrosphere, estimation of structure of tension-deformation conditions of various horizons of the Earth's crust. It can also help people with detection of destructions of the Earth's crust and earthquake prediction. There are even more opportunities: Earth crust based local objects zone-circular structure study, mineral deposit prognostication and search, exploration drilling design and others; differentiation degree increase for ecological pollution of atmosphere, water environment, and the Earth's surface; localization of pathogenic areas, land-survey tasks accomplishment. Due to importance of this method many scientists dedicated their work to research in this area, ex. Bedini E. [1], Khan S.D., Mahmood K. [2], Sabins F. [3]

According to the observations made in Norilsk, uneven distribution of auroras in the region was found. It is characterized by the presence of the local isolated minima and maxima, spatially and statistically coincident with the geological structures of the Earth's crust, including the structures of various geological structure - as the host an ore deposit (Norilsk and Talnakh ore nodes), and oil and gas fields [4, 5].

Glow inhomogeneities in the middle latitudes were experimentally registered in 1992 on the fields of the Middle Vasyugan. In the P-345 borehole area, the profile length

* Corresponding author: ukkrakov@yandex.ru

© The Authors, published by EDP Sciences. This is an open access article distributed under the terms of the Creative Commons Attribution License 4.0 (http://creativecommons.org/licenses/by/4.0/).

of 150 km of 56 ° N was covered using a scanning photometer. The results of geophysical data on the profile imposed on the seismic data on the roof of the productive Priobye Bazhenov Formation of Middle and top of the Paleozoic basement showed satisfactory agreement.

2 Equipment and methods

According to the research on our study, the new method was developed for remote ionospheric-telluric profiling, based on a comparison of the spatial distribution of inhomogeneities in the glow of the ionosphere with the geological structure of the crust. The first step is to photograph the atmosphere at a wide-angled camera situated on the ground. After that, it is needed to form the contours luminescence intensity map with reference to the ground coordinates, which is going to be superimposed on the geological basis (Figure 1).

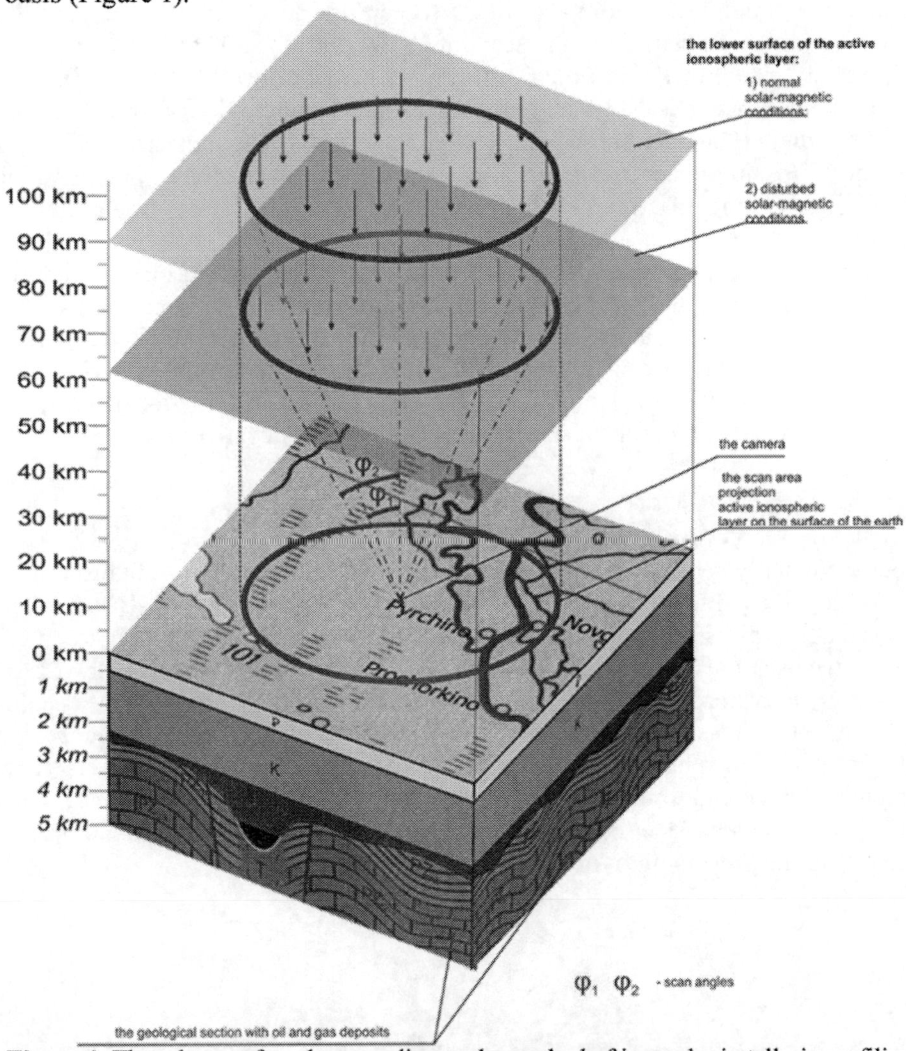

Figure 1. The scheme of works according to the method of ionospheric-telluric profiling.

Formerly photometry ionosphere emission was carried out on a black and white film and the processing was carried out manually by the operator, frame by frame. Now, with the advent of modern photographic and computer equipment processing capabilities greatly enhanced.

The inhomogeneities in the section of the upper atmosphere and the daytime sky in the middle latitudes can be detected via modern photomatrixes. However, the interpretation of this data is strongly hampered by powerful interferences. Overcoming of these problems requires the use of sophisticated mathematical libraries (such as spline interpolation) and serious computing power multithreading.

The use of modern matrix allows you to record the differences in the spectrum and the uneven brightness in increments of up to tens of meters (in terms of ground coordinates) and a resolution on the brightness of up to 0.01–0.001%.

During the process of research we have been creating and constantly improving software that can automatically handle almost an unlimited number of photos, allowing you to quickly gain observation statistics for each shooting point.

The use of a camera with a wide-angle (> 90 °) for the day when the relevant atmospheric and meteorological conditions allows to capture and create the map of the daytime ionosphere glow irregularities in the scale of 1: 100 000 ÷ 1: 200 000.

Many researchers in the work faces the need in photometry of different processes. For this purpose, special equipment is required, realizing various technologies of luminous flux conversion into the digital form for subsequent computer analysis. The most affordable is a CMOS-technology [6], implemented in a large number of professional photographic equipment from different manufacturers. When using such equipment users face a number of serious problems.

All the mass production photographic equipment is focused on getting quality pictures (feature films, documentaries), the most convenient for visual perception. Therefore, a large number of transformations is used for signals derived from the matrix of photosensitive elements to produce a target image in one of the standard RGB-format image. In the process of such a transformation a significant amount of useful information needed for scientific analysis of the latching process can be lost. Here are the negative factors:

2.1 Debayer

Matrix sensors are equally sensitive to all colors in the visible range, so to obtain a color picture they are covered with the mosaic color filter cover (most common Bayer filter, Figure 2) to form a 2×2 pixel cell is composed of red R, blue B and G1, and two green G2 cells. Each pixel receives only ¼ of color information. To get a "standard" tri-color RGB-pixel it is necessary to restore the missing component based on the interpolation of color data from adjacent areas. This process is called debayer, it is based on complex algorithms that are the know-how of the photographic equipment manufacturer, and introduces serious distortions in the raw data read from each individual sensor [7]. In addition, the use of twice the number of "green" sensors (due to the peculiarities of human perception) should also be taken into account in the processing of metering data.

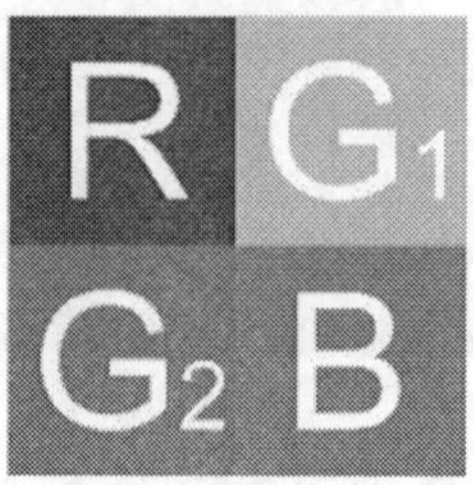

Figure 2. Bayer filter.

2.2 Rounding errors

Standard RGB formats are the 24 or 48-bit (8 and 16 bits per channel, respectively). However, solar cells of matrixes from different vendors often have a different bit. For example, Canon EOS 5D Mark II that widely spread in the market – it's 14-bit [8]. Converting to the so-called "uncompressed» (TIFF or BMP) 24bit format will inevitably cause data loss, significant for the research.

The use of RAW-converters. All professional mass-produced photographic equipment allows to record in a file the unmodified data obtained directly from the image sensor. This is so-called RAW-format. Every producer has his own one. To work with it, you need to use the RAW-converters designed by the manufacturer to significantly expand the possibilities of photographers. Still, it does not allow to explore the source of sensor data "pixel by pixel".

Summarizing the mentioned problems, we can formulate the problem: getting the unchanged data per pixel (color sensor value and the value fixed by the signal) in the form of two-dimensional arrays of data for each separate color coordinates x and y (the starting point - the top left corner of the image). For example, we will use the RAW-file in the format * .CR2, obtained when using the camera Canon EOS 5D Mark II in the maximum available resolution of 21 megapixels.

3 Software

Since all the manufacturers of photographic equipment are hiding algorithms needed to work with their matrix, a large number of commercial and open source software for reading and RAW format data has emerged. Nowadays, most of them are based on the utility dcraw [9]. In our example we will use LibRaw [10] in development, distributed on a license-free basis [11]. Both products are oriented to the needs of professional photographers, but allows you to solve our problem (Figure 3).

```
#include "libraw/libraw.h"

//Reading the file nime via openFileDialog1
openFileDialog1->Filter="Canon RAW|*.CR2";
System::Windows::Forms::DialogResult result = openFileDialog1->ShowDialog();
System::String ^ file_name;
file_name = openFileDialog1->FileName;

// Creating image handler
LibRaw iProcessor;

//Opening the file and reading meta data from it
char* LBR_file_name = (char*)(void*)Marshal::StringToHGlobalAnsi(file_name); iProcessor
.open_file(LBR_file_name);
```

Figure 3. Connecting LibRaw library and opening a file. Learn more about its API (C ++) on the project's official website [10].

3.1 Peculiarities of RAW format

Firstly, for our purposes it is necessary to take into account that 21 megapixels is the total number of sensors, which is made up of 5.25 million pixels of each color Bayer filter (RGGB). Secondly, the recording in RAW-file is carried out line by line. At the same time we do not know (according to the Canon documentation), how these elements are arranged in a conventional 2×2 box from left to right and from top to bottom: RGGB, RGBG or in a different order.

In other words, the odd lines of file will contain duplicate pairs RG1RG1 ... RG1, even - BG2BG2 ... BG2 (there are 4 different versions). The final size of the file (=> our two-dimensional data sets) will be half the original size both vertically and horizontally (if the photo is on the "output", we will have a width 5616x3744, our arrays R, B, G1 and G2 will have a size of [2808, 1872]) (Figure 4).

Third, when working with raw-file, you should keep in mind that it has a "black frame" containing a number of columns and rows with zero values of all pixels. For the definition of "usual size" and "the raw-image size" imgdata.sizes() method is responsible (parameters width and raw_width respectively) [11, 12].

```
// Getting usual and RAW filesize
int CR2_width=iProcessor.imgdata.sizes.width;
int CR2_raw_width=iProcessor.imgdata.sizes.raw_width;
int CR2_height=iProcessor.imgdata.sizes.height;
int CR2_raw_height=iProcessor.imgdata.sizes.raw_height;

// Unpacking the file
iProcessor.unpack();

// Create two-dimensional arrays to store the resulting data
std::vector< std::vector<int> > R(2808, std::vector<int>(1872));
std::vector< std::vector<int> > G(2808, std::vector<int>(1872));
std::vector< std::vector<int> > G2(2808, std::vector<int>(1872));
std::vector< std::vector<int> > B(2808, std::vector<int>(1872));

int CR2_x,CR2_y; // Libraw coordinates
int CR2_position; //unique pixel number for line by line reading
int x,y; // coordinates in converted image

// Line by line reading from converted file and filling the 2-dimensional arrays
for (CR2_y=0; CR2_y<CR2_raw_height;CR2_y++) {
    for (CR2_x=0;CR2_x<CR2_raw_width;CR2_x++) {
    CR2_position=CR2_y*CR2_raw_width+CR2_x;
    // Definig the pixel color.  Possible values are 0-R, 1-G, 2-B, 3-G2
        switch (iProcessor.COLOR(CR2_x,CR2_y)) {
        case 0:
        R[CR2_x/2][CR2_y/2]=iProcessor.imgdata.rawdata.raw_image[CR2_position];
        break;
        case 1:
        G[CR2_x/2][CR2_y/2]=iProcessor.imgdata.rawdata.raw_image[CR2_position];
        break;
        case 2:
        B[CR2_x/2][CR2_y/2]=iProcessor.imgdata.rawdata.raw_image[CR2_position];
        break;
        case 3:
        G2[CR2_x/2][CR2_y/2]=iProcessor.imgdata.rawdata.raw_image[CR2_position];
        break;
        }
    }
}
```

Figure 4. Filling the two-dimensional arrays of R, G1, G2, B.

Currently, the authors are carrying out the experimental work on photometry atmosphere in middle latitudes (Tomsk Region). We obtained reliable data about irregularities of daytime atmosphere glow, reaching the first percents in different color channels (Figure 5).

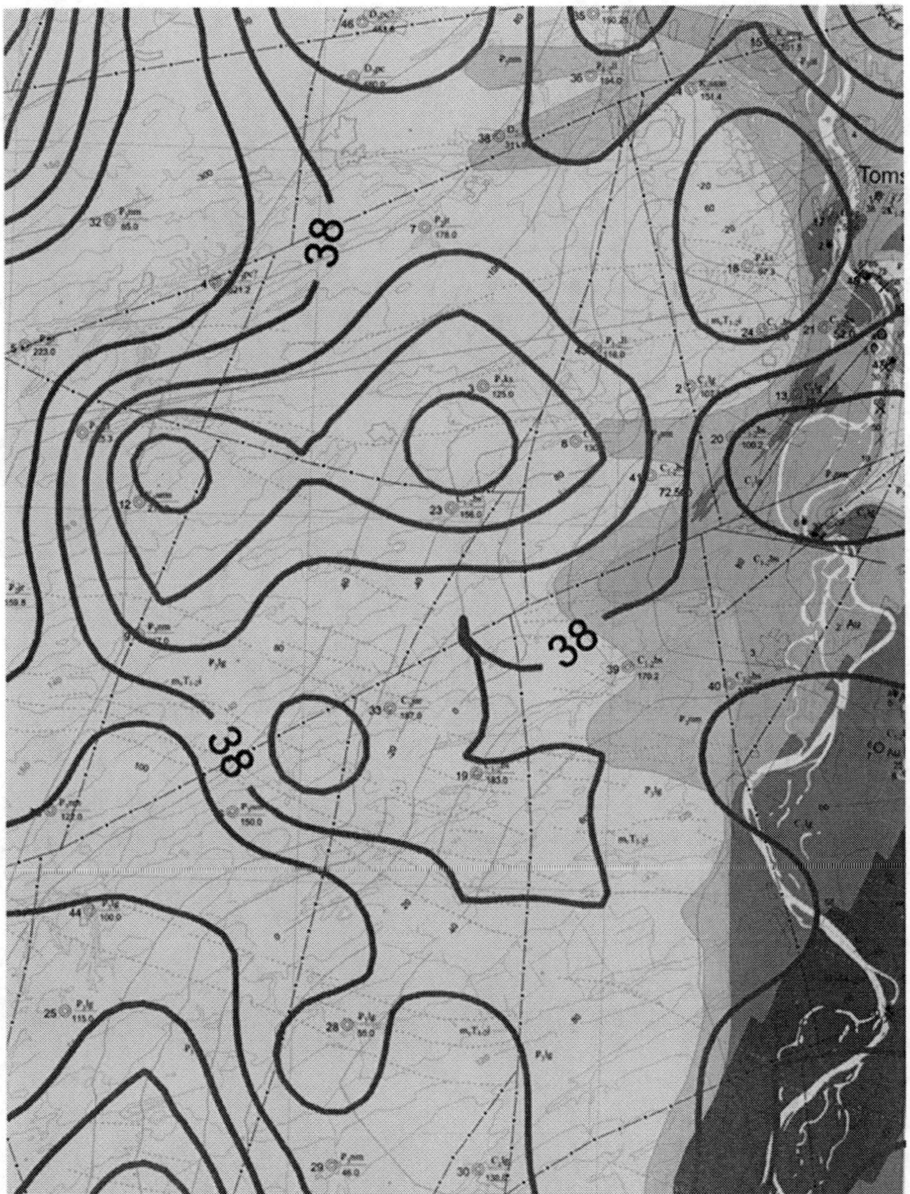

Figure 5. Map of the uneven distribution of the daytime sky illumination, the red (%), Tomsk neighborhood, scale 1: 200 000.

4 Conclusions

1. The new method for remote ionospheric-telluric profiling, based on a comparison of the spatial distribution of inhomogeneities in the glow of the ionosphere with the geological structure of the crust, was invented.
2. A program for photometric data processing was offered.

3. The result is two-dimensional arrays of four color components under a Bayer filter data, are not subjected to any treatment. This information can be used in scientific research, demanding precision measurement at metering.

References

[1] E. M. Bedini, Remote Sensing of Environment **113**, 1208 (2009) doi: 10.1016/j.rse.2009.02.007
[2] S.D. Khan, K. Mahmood, Earth-Science Reviews **89**, 135 (2008) doi: 0.1016/j.earscirev.2008.04.004
[3] F.F. Sabins, *Remote Sensing – Principles and Interpretation* (Waveland Press, Long Grove, IL, 2007)
[4] Y.K. Krakovetzkiy, M.V. Gouhberg, V.A. Pilipenko, L.N. Popov, International Union of Geodesy and Geophysics (IUGG), Abstracts V 1. – P 389 (1987)
[5] L.N. Popov, Yu.K. Krakovetskiy, M.B. Gokhberg, V.A. Pilipenko, Physics of the Earth and Planetary Interiors **57** , 115 (1989) doi: 10.1016/0031-9201(89)90221-5
[6] CMOS (Complementary Metal-Oxide Semiconductor) URL: http://www.engineersgarage.com/articles/what-is-cmos-technology
[7] RGB "Bayer" Color and MicroLenses – Silicon Imaging URL: http://www.siliconimaging.com/RGB%20Bayer.htm
[8] Canon corp., Canon EOS 5D Mark II (2015) URL: http://www.canon.ru/
[9] Dcraw project, the official site of dcraw project. (2016) URL: http://www.cybercom.net/~dcoffin/dcraw/index.html
[10] LibRaw project, The official site (2016) URL: http://www.libraw.su/
[11] LibRaw project, LibRaw documentation (2016) URL: http://www.libraw.su/docs.
[12] LibRaw project, C++ API for LibRaw project (2016) URL: http://www.libraw.su/docs/API-CXX-rus.html

RESEARCH AND DEVELOPMENT OF THE ANTENNA ARRAY FOR GROUND PENETRATING RADAR

Evgeny Balzovsky*, *Yury* Buyanov

National Research Tomsk State University, 634050, Tomsk, Russia

Abstract. The design and results of investigation of the antenna array for ground penetrating radar are presented. The antenna array is intended for measurement of the electrical properties of the road coating and finding of irregularities at monitoring the quality of roads. Antenna elements are made by PCB technology on the plate of FR4. Each antenna element is placed in a reflector formed by aluminum corner. The results of simulation in CST Studio Suite and the results of measurements of the characteristics of the antenna elements, and their mutual influence are presented. The developed antenna array can be used at the ultrawideband pulse excitation, and at a scanning of continuous frequencies in the range of 0.5–2 GHz.

1 Introduction

The ground penetrating radar (GPR) technology is promising for the study of the quality of roads and finding of a latent defects in the asphalt coating during the construction and maintenance of roads to prevent damage to the roadway [1]. Monitoring the status of the roadway portion includes determining the thickness and homogeneity of the top coat layer consisting of a asphalt concrete mix, thickness and humidity of the underlying layers of gravel or crushed stone as well as the detection and localization of cavities [2–4].

Despite the fact that there are a number of technical solutions, as well as the finished commercial devices for sensing of the roadway, a significant obstacle to the widespread introduction of such devices in the practice of diagnosing the state of any roads is low scanning speed while moving the scanner on long stretches of road, as well as the high cost of such devices.

The technologies and algorithms developed for radio-wave imaging systems of objects hidden behind the optically nontransparent obstacles [4] can be applied to create a new inexpensive scanner for inspection of quality of highway coating. Irradiation of the surface layer of asphalt concrete pavement and the underlying layers of pavement by ultrawideband pulses of nanosecond duration provides both the high spatial resolution and the deep penetration of electromagnetic waves due to availability of energy in the low frequency part of the pulse spectrum.

* Corresponding author: tduty5@mail.ru

© The Authors, published by EDP Sciences. This is an open access article distributed under the terms of the Creative Commons Attribution License 4.0 (http://creativecommons.org/licenses/by/4.0/).

For construction of the radar image of the layered structure and the irregularities distribution of the roadway the transceiver of the radar should move along the traffic lane to use the method of synthetic aperture. To increase the width of the scanned lane is advantageous to use switched matrices of transmitting and receiving ultrawideband antennas. Due to the complexity of beam-forming network of the phased antenna array and a high cost of its components it is advantageous to use a switched antenna array. In such matrix the antennas are sequentially connect to the generator or receiver [5], and beamforming algorithm is performed with mathematical processing in a computer.

For the effective radiation of the ultrawideband signals and to record the reflected radiation the antenna should be matched with the feeder in the frequency band of two or more octaves. The pattern of the antenna should have a constant main lobe direction in the same frequency band. Moreover, the antenna should have the smallest possible size if it is used as an element of the array. The antennas that are a combination of electric type and magnetic type radiators [6] satisfy these requirements.

The high spatial resolution can be achieved by the ultrawideband pulses with the spectrum occupying a bandwidth 1–5 GHz. However, the losses in the probed medium increase with increasing of frequency, so to increasing the probing depth it is necessary to use the signals with the spectrum that is shifted toward lower frequencies. The design of the planar ultrawideband antenna with a bandwidth of 0.5–2 GHz is described below. The results of numerical simulation and the measured characteristics are presented.

2 The design of antenna array

The embodiment of the antenna array is shown in Figure 1. The antenna elements 1 are divided into the modules of four antennas and arranged in an aluminum corner 2. The distance between the antennas is equal to 150 mm. The length of one module is equal to 0.6 m. The number of modules depends on the application and can be arbitrary. The number of blocks determines the width of scanned strip of road surface. The antenna array is connected to the switching unit via coaxial connectors 3. The whole array is covered with a plastic housing and mounted on a mobile platform.

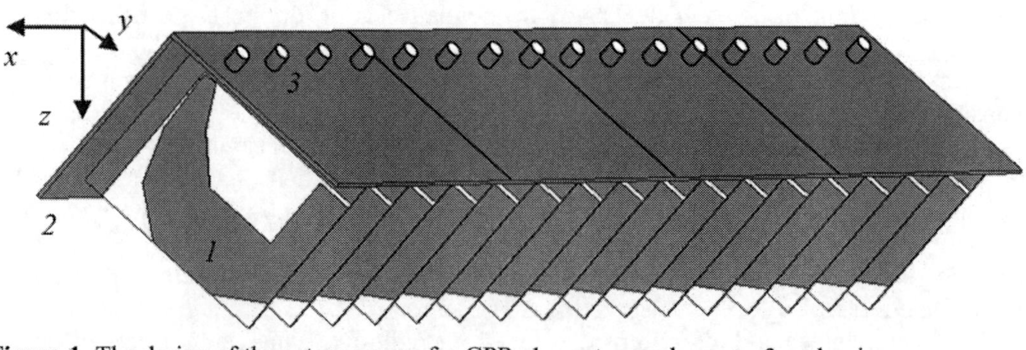

Figure 1. The design of the antenna array for GPR: 1 – antenna elements, 2 – aluminum corner as a reflector, 3 – coaxial connectors.

3 The antenna element

3.1 Design of the antenna element

The developed antenna is made on a plate of FR4 with size of 110×110×1 mm. The electric monopole 1 in Figure 2 is made by PCB technology. The area of the plate 2 that is left free from metal foil represents a radiator of magnetic type. The plate is mounted in the aluminum corner 3 with cross-section of 120×120 mm.

The necessary amplitude-phase relations for the excitation both types of radiators are defined by the sizes of the area 2 and a slot line 4. Tapered slot line 4 performs the function of the impedance transformer. On the back side of the corner the coaxial connector is mounted. The central electrode of the connector is connected to the monopole in the point 5. The direction of main lobe of antenna pattern is indicated by the arrow.

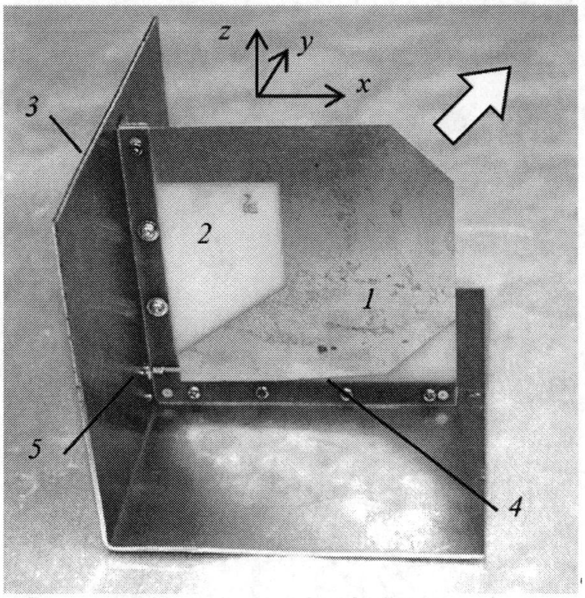

Figure 2. The design of antenna element: 1 – electrical monopole, 2 – magnetic type radiator, 3 – aluminum corner as a reflector, 4 – tapered slot line, 5 – coaxial connector.

3.2 Characteristics of the antenna element

Modeling of antenna element was performed in electromagnetic field simulation software SCT Studio Suite under Educational License. The results of calculation of the voltage stand-wave ratio (VSWR) of a single antenna element in are shown in Figure 3 (curve 1). In the same Figure (curve 2) the VSWR measured by network analyzer Agilent N5230C is shown. There is a good enough agreement between calculated and measured VSWR. The difference of curves may be due to the fact that the losses in the dielectric and the metal had not been taken into account in the model.

The direction of the main lobe of antenna pattern (θ_0, φ_0) lies in the *xz*-plane and is directed at an angle $\theta_0 = 45$ degrees to the *z*-axis over a wide frequency range. The radiated electric field has a linear polarization in the main lobe direction. The plane of polarization coincides with the plane of PCB plate of the antenna. The calculated patterns of the antenna element in *E*-plane and *H*-plane are shown in Figure 4 and Figure 5 (curves 1). The origin (angle = 0) matches the direction of (θ_0, φ_0).

Measurements of antenna patterns were held in the time domain at the excitation of the antenna by bipolar voltage pulses of duration of 0.5 and 1 ns. For each angular position of the antenna under study the pulse waveform at the output of the auxiliary receiving antenna was recorded by the sampling oscilloscope. An angular dependence of the spectral components in the range of 0.5…5 GHz was obtained using Fourier transform. The rectangular temporal window was used for filtering signals reflected from the surrounding environment. The measured patterns of the antenna element at both *E*- and *H*-planes are presented in Figure 4 and Figure 5 (curves 2).

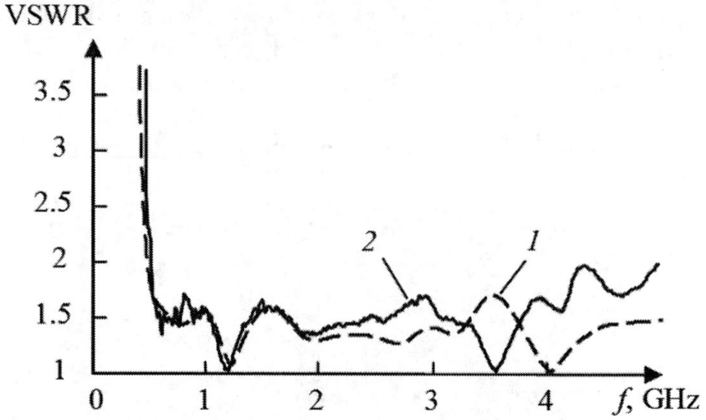

Figure 3. VSWR of the antenna element: 1 – simulations, 2 – measurements.

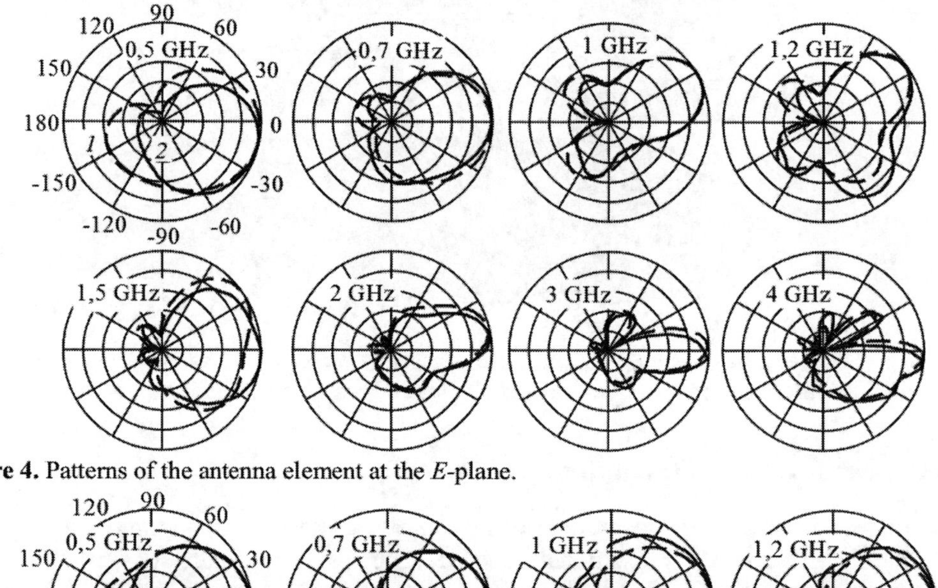

Figure 4. Patterns of the antenna element at the *E*-plane.

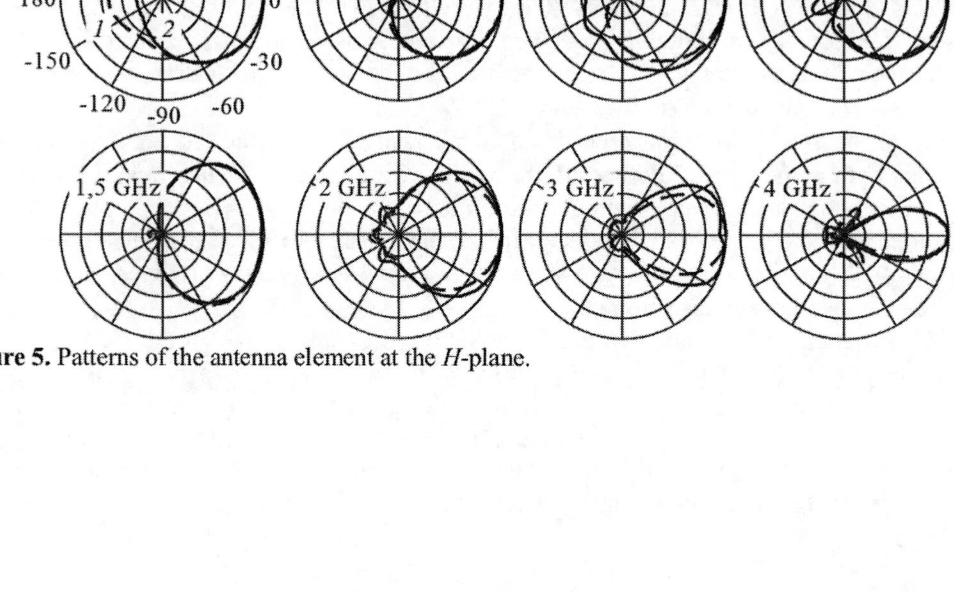

Figure 5. Patterns of the antenna element at the *H*-plane.

DOI: 10.1051/matecconf/20167901036

4 Characteristics of the antenna array

An appearance of one module of the array antenna unit is shown in Figure 6. Since the antenna elements are located in close proximity to each other, it is necessary to consider their mutual influence. To estimate the mutual influence of the elements the measurements of VSWR were performed separately for external elements 1 and 4 in Figure 6 and for inner elements 2 and 3. The measurement results are shown in Figure 7. Change the length of the metal reflector and the presence of neighboring elements for external antennas 1 and 4 results in a slight shift of the lower boundaries of the band matching. At the rest of the frequency band VSWR does not exceed the value of 2. For internal elements (2 and 3 in Figure 6) was found the more significant shift of the lower limit of the bandwidth toward higher frequencies compared with a single antenna (curve 3 in Figure 7). Notwithstanding, the proportion of the reflected power in the frequency band of 0.5…5 GHz is not more than 0.2 of incident power.

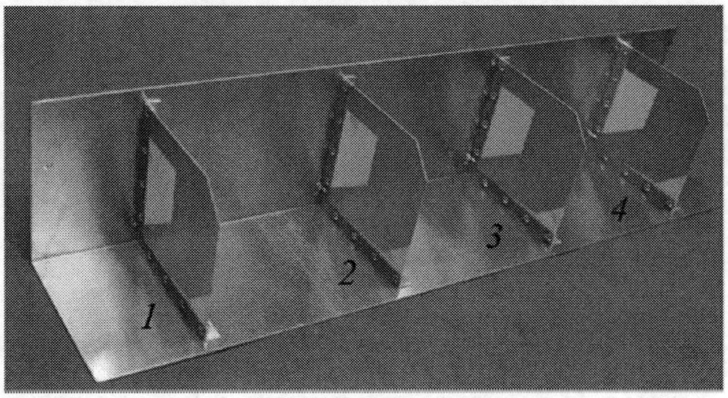

Figure 6. The design of four-element module of the antenna array.

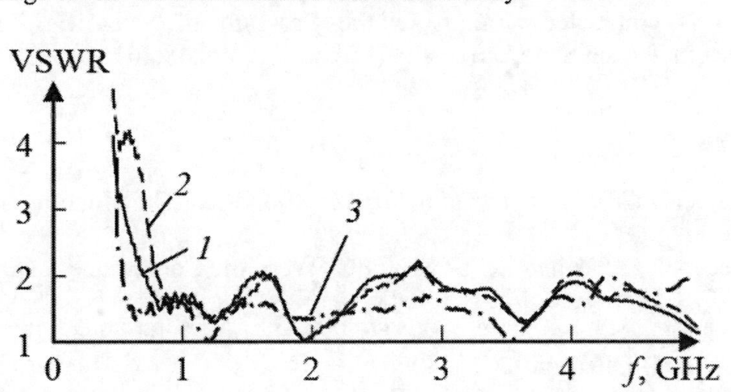

Figure 7. VSWR of antennas in the array. 1 – external antenna elements, 2 – internal antenna elements, 3 – single antenna.

The measured transmission coefficients $|S_{mn}|$ between antenna elements n and m (n, $m = 1…4$) of the array are presented in Figure 8. The adjacent elements (pairs 1–2 and 2–3) have the greatest influence on each other. The maximum of $|S_{21}|$ and $|S_{32}|$ is located below the lower border of frequency band. The mutual influence of the distant elements is smaller than the neighboring. Further, in order to reduce the mutual influence of the antenna element it is assumed to fill partially the space between the antennas by a volume absorbing material.

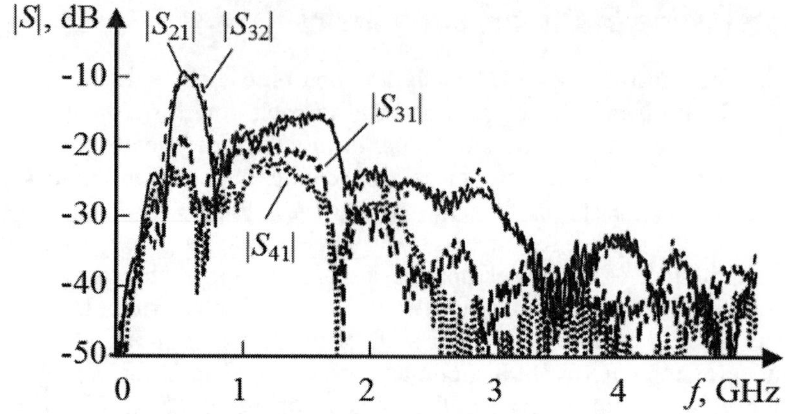

Figure 8. Transmission coefficients between antenna elements in array.

5 Conclusion

The ultrawideband antenna array for use in ground penetrating radar is developed and investigated. The antenna elements are matched with the 50-ohm feeder in the frequency band of 0.5–5 GHz and have a stable direction of main lobe of the pattern. Despite the close proximity of the elements in the array, the level of mutual interference elements is not critical, although it limiting the matching band at low frequencies. The developed array can be excited both by the ultrawideband short pulses and by the continuous scanning frequencies in the range of 0.5…2 GHz.

Acknowledgment

This research is conducted as part of the Program of Scientific Foundation n. a. D.I. Mendeleev in Tomsk State University (Project No. 8.2.48.2015).

References

[1] D. Daniels *Ground-penetrating radar* (London, The Institution of Electrical Engineers, 2004)

[2] Z. Mechbal, A. Khamlichi, MATEC Web of Conferences **16**, 07005 (2014) doi : 10.1051/matecconf/20141607005

[3] Guo M. L., Huang S. P., MATEC Web of Conferences **44**, 02079 (2016) doi : 10.1051/matecconf/20164402079

[4] V. P. Yakubov, S. E. Shipilov, D. Y. Sukhanov, A. V. Klokov *Radiovolnovaya tomografiya: dostijeniya I perspektivi* (Tomsk, NTL, 2014)

[5] R. N. Satarov, I. Y. Kuzmenko, T. R. Muksunov, A. V. Klokov, E. V. Balzovskii, Y. I. Buyanov, S. E. Shipilov, V. P. Yakubov, Rus. Phys. J. **55,** 884 (2013) doi : 10.1007/s11182-013-9896-8

[6] Yu. A. Andreev, Yu. I. Buyanov, V. I. Koshelev, J. Commun. Technol. Electron. **50** , 535 (2005)

MATEC Web of Conferences **79**, 01037 (2016) DOI: 10.1051/matecconf/20167901037
IME&T 2016

USING NI PXI MODULES FOR DIGITAL SIGNAL PROCESSING MICROPROCESSOR TESTING

Vladimir Marfin[1,*], *Pavel* Nekrasov[1], *Andrey* Yanenko[1] and *Andrey* Schepanov[2]

[1]National Research Nuclear University MEPhI (Moscow Engineering Physics Institute), 115409, Moscow, Russia
[2]FSUO "MNIIRIP", 141002, Mytishchi, Russia

Abstract. The article considers the implementation of the external memory interface based on National Instruments modular PXI equipment for environmental testing of digital signal processing (DSP) microprocessors. The block diagram of the developed device pointing out the advantages and disadvantages of this solution is provided. The block diagram of an improved external memory interface is also shown.

1 Introduction

One of the critical steps of the electronic equipment design for extreme environments is the initial evaluation and selection of system's components [1–15], even more so, for functionally complex ICs such as microprocessors and digital signal processors. The most important task here is to ensure the real operation mode of the chip at the highest operating frequency. The market of modern digital signal processing microprocessors is represented by a large number of various architectures and families. This makes the creation of a versatile and flexible system for testing different DSPs an especially relevant problem. The application of NI PXI-7841R reconfigurable I/O module for testing microprocessors is shown in [16]. This solution is based on the implementation of the direct memory access (DMA) controller interface. The main disadvantage of this approach is an irrational choice of the general interface. Not every DSP microprocessor contains a DMA interface, so it is necessary to implement the ability to download an executable program in the chip. Also, DMA controller interface can vary considerably depending on the specific model of the microprocessor under test. The usage of an external memory interface for testing DSP is devoid of these faults.

The article considers the implementation of the external memory interface with the reconfigurable I/O module NI PXI-7841R.

2 Equipment and software

The solution of the problem was implemented in NI LabVIEW 2010 programming environment with installed FPGA software module and modular devices' drivers.

[*] Corresponding author: vamar@spels.ru

© The Authors, published by EDP Sciences. This is an open access article distributed under the terms of the Creative Commons Attribution License 4.0 (http://creativecommons.org/licenses/by/4.0/).

 DOI: 10.1051/matecconf/20167901037

LabVIEW allows one to use visual programming language for mathematical processing of measurements results and generation of visual reports with graphs and statistical information. The external memory interface was developed on the basis of hardware and software system, as described in [16] with the use of such modular devices as:

- NI PXI-7841R module of an analog/digital I/O. This module provides the physical realization of the minimum environment required to complete the work of the DSP processor (timing system, interrupts, initialization, etc.), as well as measurement of logic voltage levels.
- NI PXI-4110 power supply module to set the voltage supply and control IC consumption current.

Modular devices are connected to a personal computer using PXI-1033 chassis.

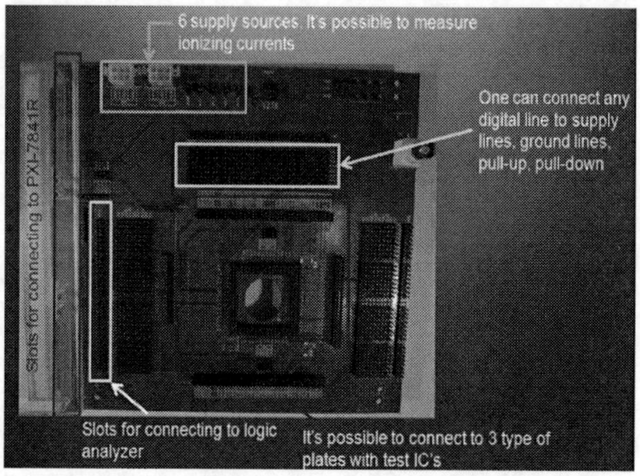

Figure 1. External view of the motherboard printed circuit board.

The main advantage of the hardware-software complex [16] is its versatility, which is achieved mainly by developing motherboard shown in Figure 1, which is connected to used NI PXI modules. Thus, in order to prepare for the testing of a new chip, only adapter board with standardized connectors must be manufactured. This approach significantly reduces the time and effort required to prepare the chip for testing. As an example, Figure 2 shows the adapter board for testing multiprocessing systems.

Figure 2. External view of the adapter printed circuit board with the multiprocessor's system under test.

3 Solution

An external memory interface is needed for connection of external RAM or ROM to a microprocessor. This memory may be used to load executable code into the test microprocessor and for storing intermediate data. In terms of processor performance, test implementation of this interface performs the following tasks:

- storage and loading of executable code;
- receiving and exchanging data with the microprocessor in real-time at full speed;
- high degree of standardization of this interface.

The program of external memory interface controller was created in the Host and FPGA modules. Application code that defines the configuration of the FPGA is developed on the host PC and then loaded into the FPGA. The LabView FPGA contains the most critical hardware blocks that require maximum performance during the interaction with the processor under test. The blocks that are developed in LabView Host do not require execution in real time. This module runs on a PC, so the speed of its implementation depends on the performance of the user's computer and programs priorities. The Host applications tasks include loading bin-file of the test firmware for the microprocessor in the FPGA module memory which is available to be read by the microprocessor and control operation of the entire application, analysis and logging test results.

Thus, to solve this problem it is necessary to implement block RAM on the FPGA Virtex-5 LX30, which is part of the NI PXI-7841R module, as well as physical realization of the external memory interface. Figure 3 shows a block diagram of the initialization memory unit.

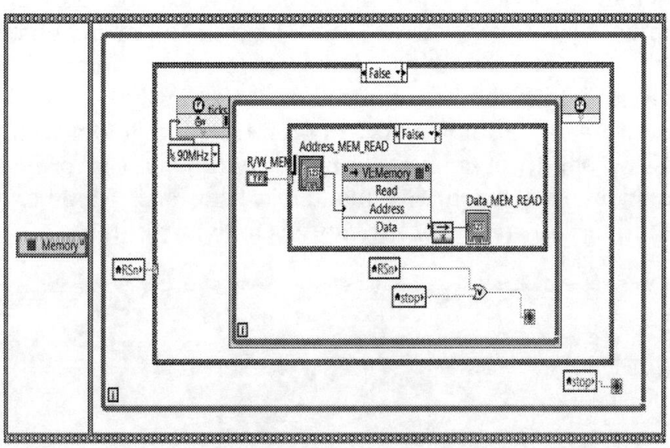

Figure 3. Block diagram of the initialization memory unit on the FPGA.

A block diagram of realization of the external memory interface is shown in Figure 4. The algorithm of this interface is the following sequence of operations:

- expectation of the start gate exchange cycle;
- definition of the type of the exchange cycle (read or write data). Simultaneously, the read address memory put up by a test processor and receiving data in the memory at the specified address with adjustable delay;
- depending on the type of exchange, either setting of information on the data bus (in the case shown in Figure 4), or reading the information and its recording in the RAM memory block occurs.

The advantages of the implemented interface are its high degree of versatility and flexibility, as well as the interface standard. The disadvantage is a relatively low speed. It was found during experiment that a stable work area of the interface is limited to frequencies up to 40 MHz. This limitation is critical if it is impossible to change the frequency of the external memory interface of tested DSP. The analysis revealed that major delays are related to the usage of the Flat Sequence Structure, i.e. uncontrolled and non-fixed time delay in the transition "from frame to frame". It was decided to increase the speed of the interface by developing the improved structure.

Figure 5 is a block diagram of an improved external memory interface.

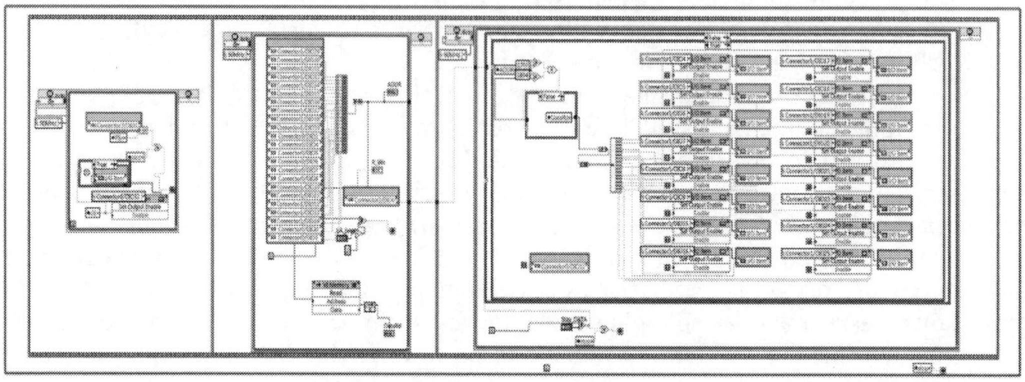

Figure 4. Block diagram of realization of the external memory interface. The read type of exchange.

The main feature of this implementation became the placing of the necessary operations in a single cycle using Time Loop Case Structure (so called State Machine) instead of using Flat Sequence Structure.

The RAM block mode became another implemented feature. Only one access operation to an array of memory cells (read or write) can be carried out at any given time. As a result, it was necessary to stop the operation of the processor in order to read the results of functional control test which the microprocessor stores in the RAM block. To solve this problem, the structure Array of Indicators has been used in parallel with the RAM module. This approach allows one to organize continuous information recording by the processor arbitration and read it from the FPGA module to the Host Module, while the test the microprocessor continues to operate at high speeds in the real time.

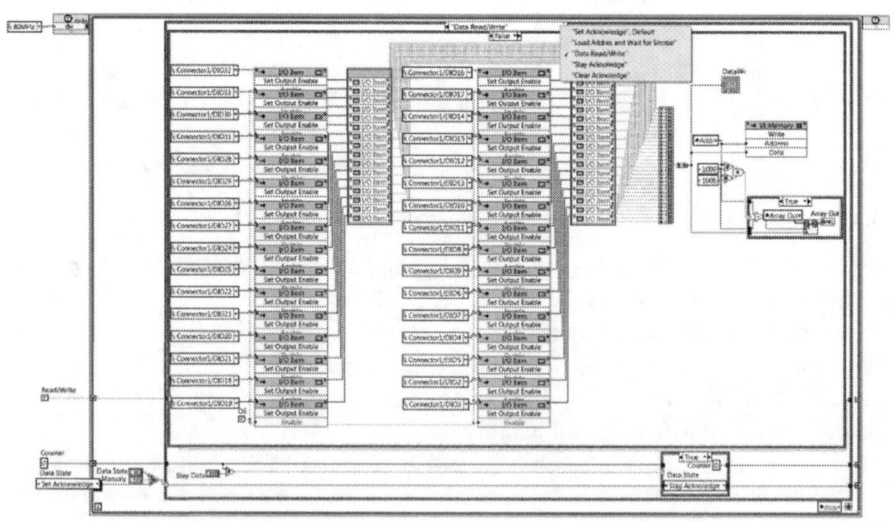

Figure 5. Block diagram of an improved external memory interface. The write type of exchange.

4 Experimental results

To test the efficiency of resulting external memory controller radiation behaviour researches of the DSP processor TMS320C50 (Texas Instrumenrs) were conducted in comparison with the use of DMA controller unit [16]. Functional control was consisted of testing internal memory blocks: 1) at maximum frequency (about 50 MHz), 2) using DMA interface (up to 100 kHz). During the experiment, it was found that functional failure doesn't revealed under exposure to extreme levels using DMA mode while the first method show functional failure that proves the inapplicability DMA mode as the sole means in testing processors under radiation fields.

Analyzing the results, we can conclude that the developed external memory controller provides more critical assessment of the radiation hardness of the DSP processor compared with the previously used DMA interface. This fact proves the significant impact of degradation time characteristics of inner units of DSP processors on the level of radiation hardness that occurs only at extreme frequencies of the scheme and the inability to obtain a critical assessment using a low-speed DMA interface

5 Conclusions

Developed external memory controller interface was successfully used for testing a variety of microprocessors which include general purpose microprocessors (IDT79RC64474 with MIPS architecture, AT697F with the SPARC architecture, Intel 80386, etc.) and DSPs (processor's family by Texas Instruments - TMS320C2x, TMS320C3x, TMS320C5x , TMS320C2xx, TMS320C5000, TMS320C6000 and many others), this unit requires minimal setup that significantly reduces the time of preparation.

Note that this implementation allows us to significantly increase the speed (up to 100 MHz) and doesn't contain uncontrolled time delays. The minimum time interval is equal to the period of the clock signal for the Time Loop cycle. At the same time, this implementation of an external memory interface keeps such advantages as flexibility, standardization and flexibility of the developed external memory interface.

Acknowledgements

The work was performed as part of the grant RFBR № 14-29-09210.

References

[1] A. Y. Nikiforov, P. K. Skorobogatov, A. N. Egorov, D. V. Gromov, Russian Microelectronics **43**(2), 133 (2014) doi: 10.1134/S1063739714020073

[2] M. P. Belova, D. V. Pechenkina, A. Y. Borisov, L. N. Kessarinskiy, D.V. Boychenko, 2015 International Siberian Conference on Control and Communications (SIBCON), 7146995 (2015) doi: 10.1109/SIBCON.2015.7146995

[3] A. G. Petrov, A. L. Vasil'ev, A.V. Ulanova, A. I. Chumakov, A.Y. Nikiforov, Central European Journal of Physics **12** (10), 725 (2014) doi: 10.2478/s11534-014-0503-6

[4] L. N. Kessarinskiy, D. V. Boychenko, A. Y. Nikiforov, Russian Microelectronics **41**(4), 251 (2012) doi: 10.1134/S1063739712040075

[5] I. O. Loskutov, A. B. Karakozov, P. V. Nekrasov, A. Y. Nikiforov, 2015 International Siberian Conference on Control and Communications (SIBCON), 7147128 (2015) doi: 10.1109/SIBCON.2015.7147128

[6] A. O. Akhmetov, D. V. Boychenko, D. V. Bobrovskiy, A. I. Chumakov, O. A. Kalashnikov, A. Y. Nikiforov P. V. Nekrasov, Proceedings of the International Conference on Microelectronics (ICM), 329 (2014) doi : 10.1109/MIEL.2014.6842156
[7] G. V. Chukov, V. V. Elesin, G. N. Nazarova, A. Y. Nikiforov, D. V. Boychenko, V. A. Telets, A. G. Kuznetsov, K. M. Amburkin, IEEE Radiation Effects Data Workshop, 7004589 (2014) doi: 10.1109/REDW.2014.7004589
[8] A. S. Artamonov, A. A. Sangalov, A. Y. Nikiforov, V. A. Telets, D. V. Boychenko, IEEE Radiation Effects Data Workshop, 7004600 (2014) doi : 10.1109/REDW.2014.7004600
[9] A. N. Egorov, A. I. Chumakov, O. B. Mavritsky, A. A. Pechenkin, D. V. Savchenkov, V. A. Telets, A. V. Yanenko, IEEE Radiation Effects Data Workshop, 7004570 (2014) doi: 10.1109/REDW.2014.7004570
[10] I. I. Shvetsov-Shilovskiy, P. V. Nekrasov, A. V. Ulanova, A. A. Smolin, A. V. Sogoyan, 2015 International Siberian Conference on Control and Communications (SIBCON), 7147282 (2015) doi: 10.1109/SIBCON.2015.7147282
[11] A. B. Boruzdina, A. A. Orlov, A. V. Ulanova, N. G. Grigor'ev, A. Y. Nikiforov, 2015 International Siberian Conference on Control and Communications (SIBCON), 7147007 (2015) doi: 10.1109/SIBCON.2015.7147007
[12] D. V. Bobrovskii, O. A. Kalashnikov, P. V. Nekrasov, Russian Microelectronics **41** , 226 (2012) doi: 10.1134/S1063739712040038
[13] P. V. Nekrasov, A. A. Demidov, O. A. Kalashnikov, Instruments and Experimental Techniques **52**, 196 (2009) doi: 10.1134/S0020441209020092
[14] Jin Ming Zhanga and He Zhang, MATEC Web of Conferences **44**, 01030 (2016) doi: 10.1051/matecconf/20164401030
[15] Ning Ma, Shaojun Wang, Syed Mohsin Ali, Xiuhai Cui and Yu Peng, MATEC Web of Conferences **45**, 05001 (2016) doi: 10.1051/matecconf/20164505001
[16] V. A. Marfin, P. V. Nekrasov, O. A. Kalashnikov, K. A. Kagirina, 2015 International Siberian Conference on Control and Communications (SIBCON), 7147142 (2015) doi: 10.1109/SIBCON.2015.7147142

SIMULATION OF SYSTEMS FOR MONITORING, ANALYSIS AND FORECASTING OF THE PRIORITY DIRECTIONS OF SCIENCE AND TECHNOLOGY

Tatyana Abramova[a]

National Research Tomsk State University, 634050, Tomsk, Russia

Abstract. The article presents the automated scheme for monitoring, analysis and forecasting based on the extended conceptual foresight model, which is a system of methods of expert evaluation of the strategic prospects of innovative development, identification of technological breakthroughs that can have maximum influence on economy and society in the medium and long-term outlook.

1 Introduction

Foresight is one of the tools, which allows identifying priority areas of science, technology, innovation and making a concrete action plan for their development. Foresight is a system of methods of expert evaluation of the strategic prospects of innovative development, identification of technological breakthroughs that can have maximum influence on economy and society in the medium and long-term outlook.

Herbert Wells introduced the concept of «Foresight» in 1932. In one of his public speeches, a well-known science fiction writer spoke about the need for «Foresight Professor» – a new scientific specialty aimed at studying future technological breakthroughs and search for areas of their application.

However, as a system of methods, Foresight appeared much later, during the period of 1940-1950. At that time, Foresight has already been considered as a systematic and systemic process enabling to assess future development of a subject area by means of expert evaluations. However, its methodical apparatus was quite limited. American RAND Corporation introduced a number of significant innovations, which gained wide dissemination later.

2 History of Foresight methods

Delphi method, a multi-level expert survey with feedback, was among the first basic Foresight methods [1]. The principle of this method is as follows: at the first stage, a large number of experts (often, several thousands of people) are invited to fill in special

[a] Corresponding author: tanusha-atv@mail.ru

© The Authors, published by EDP Sciences. This is an open access article distributed under the terms of the Creative Commons Attribution License 4.0 (http://creativecommons.org/licenses/by/4.0/).

questionnaires containing questions about the prospects of certain developments, level of their adoption in the country and abroad, etc. Processed questionnaire results, including not only questions but also summarized answers of the previous stage, are directed to the same experts. Respondents have an opportunity to confirm or change their earlier answers in light of the replies of other members of their panel. In practice, the Delphi method is usually carried out in two rounds, although the number of rounds can be greater.

Other countries have used successful US experience of technology Foresight in the middle of the XX century. Based on the Delphi method, «Future Technology in Japan toward the Year 2030» project was implemented in Japan in 1971. The National Institute of Science and Technology Policy of Japan (NISTEP), one of the leading centers in the field of long-term forecasting, conducted the study [2]. Since that time, Foresight at the national level is carried out regularly in this country; currently work on the ninth long-term forecast has been completed.

Since 1980-ies, France, Sweden, Canada, Australia have carried out pioneering work on technological Foresight. During this period, a sharp increase of interest in Foresight studies has been marked. At this time, the US Ministry of Defense and Ministry of Commerce have started the Foresight project aimed to identify «critical technologies»; the Netherlands have implemented Foresight projects under the auspices of the Ministries of Economy, Education and Science; national Foresights in Germany and the UK have drawn a wide public response.

Since 1990ies, prefix «technological» is less frequently used in relation to Foresight. The last two cycles of British Foresights were no longer called «technological»; they reflected broader and more comprehensive vision for the future development. During this period, Foresight has exceeded limits of the study of science and technology; it is widely used for analysis of prospective markets. In recent years, it is applied for prediction of social processes, as well as during the process of formation of the infrastructure of national innovative systems. In fact, it is a complex prediction of development of all sides of society. It is a natural result. A modern innovative model requires developed system of professional training, high educational and cultural level of the population, scientific potential, efficient communications system, and maintenance of a high level of social stability. Today Foresight is focused on assess of prospects for socio-economic development in a global context and ensuring sustainable development in conditions of growing interdependence of the various spheres of society.

In 2000, more than 30 countries have actively used Foresight at the national level. Currently, not only Western Europe, the USA and Japan use this technique, but also a number of developing countries and countries with transition economy - the new EU members, particularly, Hungary, the Czech Republic and Poland. Over the past quarter of the century, individual developed countries and large companies have realized more than one thousand Foresight projects in total.

3 System of Foresight methods

Table 1 shows the main differences between Foresight and forecasting.

Table 1. Comparative analysis of Foresight and Forecasting.

Principle	*Foresight*	*Forecasting*
Involvement	Representatives of the whole society participate	Conducted by scientists
Communication	Communication between participants is important to reach consensus	Communication between participants of forecast is not an important factor
Research horizon	Usually, 10–15 years, sometimes 80-90 years	Horizon is strongly dependent on availability of dynamic data-series. On average, less than that of Foresight
Coordination	Evaluation of development of science and technology are inseparable from economic and social changes	The main focus is on a subject area
Consensus	It determines whether there is a consensus regarding subjects under the study	Presence or lack of consensus is not an important factor
Expertise	Expert methods dominate	Quantitative methods dominate
Alternativeness	Alternative options to achieve the desired image of the future are considered	Basic option is usually considered
Systemacy	Foresight covers not only the subject area, but also its environment	Forecast is made for a specific area
Openness	Initial data, needs, research issues are still opened and can be revised during the course of Foresight. The results are brought to attention of as many people as possible	Initial data, themes and research questions can be determined beforehand

In general, the system of Foresight methods is constantly being developed and improved; great experience for their practical application has been accumulated over the past ten years. Today, a set of approaches covers dozens of techniques - both qualitative (interviews, literature reviews, morphological analysis, «correspondence trees», scenarios, role-plays, etc.) and quantitative (cross-impact analysis, extrapolation, modeling, analysis and forecast of methods indicators, etc.). A number of methods are of synthetic nature. They include abovementioned Delphi method, the road map, the critical technologies as well as multi-criteria analysis, patent analysis, game simulation, etc. A set of methods used in a particular project can be selected based on a variety of factors: time and resource limitations, availability of sufficient number of highly qualified experts, access to information sources and others. However, use of methods that ensure efficient work of experts is a key to success of a project.

Effectiveness of the combined use of different qualitative and quantitative methods has been confirmed. At the same time, it has became obvious that big projects devoted to the choice of technology priorities at the national level require new approaches, which can ensure obtaining of objective assessments based on quantitative analysis of empirical data -

statistical indicators, patent statistics, bibliometric information, etc. In this connection, the idea of the «Foresight triangle» has undergone some changes. I. Miles and R. Popper [3] proposed to add one more vertex – «Evidence» and thus to turn the triangle into the rhombus (Figure 1).

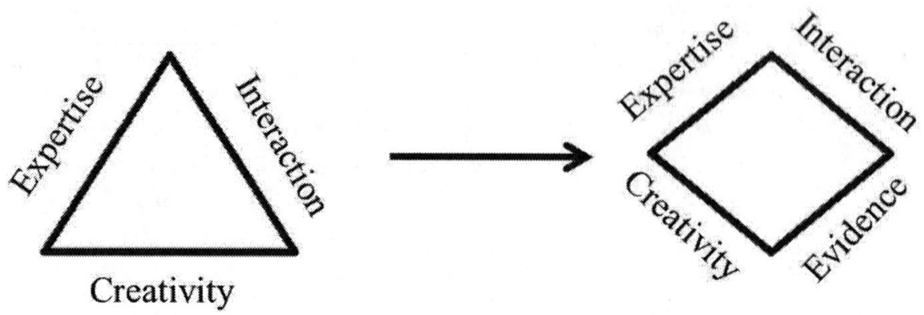

Figure 1. Conceptual transition from the Foresight - triangle to the Foresight – rhombus.

Currently, Foresight includes methods relating to different areas of science: sociology, mathematics, statistics, etc. There are four key categories of Foresight methods [4–9]:

- Expertise. In order to verify results obtained from analysis of primary sources, it is necessary to discuss trends of development of the subject areas with key experts – representatives of science, business, and government organizations.
- Creativity. In order to predict unexpected turns of the future, you must add some creative elements to the analysis. Typical examples of creative methods include wild cards («random factors», «Joker») and weak signals. Essence of the wild cards method is that based on various types of information (scientific and journalistic articles, interviews with experts, etc.) specialists identify events that have a low probability of occurrence, but can have great influence on the further course of development. One of the typical examples of wild cards is an accident at the Fukushima nuclear power plant, which forced the international community to take a fresh look at the prospects of the nuclear energy development.
- Interaction. It is necessary to involve in Foresight process as many sides as possible, including representatives of science, business, government, and to ensure their ongoing dialogue on the future development. If the subject of the research affects wide sectors of society, much attention should be paid to informing people about the Foresight project and its main conclusions. This can be achieved by carrying out various kinds of workshops, panels of citizens, etc.
- Evidence. In order to get reliable Foresight results it is necessary to base on previously published materials and to study best international practice in the subject area. In addition, you can form a pool of the most qualified experts in the subject area based on patent and bibliometric analysis.

Today, Foresight is important for Russia because it is an example of a communication platform that allows wide discussion of actual problems of science and technology, development of innovative strategies with participation of all sides, and coordination of the implementation process.

Foresight makes it possible to provide adequate analysis, monitoring and forecasting development of science and technology. Foresight enables to concentrate innovative resources on the most important scientific and technical problems in the medium and long-term outlook, to prepare proposals on measures of the state scientific, technical and innovative policy. This allows providing authorities for innovations, science and technology of the Russian Federation with current, complete, accurate and regularly updated information in the sector of research and development. Finally, by means of Foresight, it is possible to develop innovative strategies for the industries of the Russian economy, and thereby contribute to strengthening of national competitiveness.

4 Indicators and criteria

Development of indicators for the experts' assessment of technological alternatives as well as selection criteria is a separate subject of study in the work. Now, we can conclude that the content of selection criteria directly depends on the nature of strategic priorities the state government. We can distinguish different «primacy» of strategic national priorities at different levels of government, i.e. the main tasks can include «support of basic science» or «improvement the life quality», «achievement of high rates of economic growth», «development of education and culture», «national defense and security». Ranking of priorities is authority of the legislative branch.

It is advisable to consider the applicability of known criteria of projects evaluation for selection (as a priority) of those technological trends that are being on the stages of development and production of a prototype. Along with this, it is necessary to explore the possibility of primary support not only to developments, which are almost ready for the market launch, but to those researches, which are on the pre-competitive stage of the life cycle, the so-called new and emerging areas of science and technology.

In order to select priority areas and critical technologies, we need to consider «critical» problems taking into account the achieved and achievable level of development of national science and technology. In this case, the «critical problem» is an innovative problem of social and economic sphere, which meets at least one of the following conditions. The problem is critical for many producers or social individuals; or it is a necessary link in the process of manufacturing of a product, which is extremely crucial for economy competitiveness; or it is a necessary link for solving a social problem, which is considered as crucial. Critical issues serve as a motive for the choice of technology.

In this case, participation of representatives of various sectors of society (government, science, industry, business, and social structures) in selection of priorities for scientific and technological development is possible. Use of discussions and surveys within formulation of the so-called «technology roadmaps» or projects of future forecasting (Foresight) seems sensible.

As for the possibility of including social technologies in the lists of priority directions and lists of critical technologies, we can conclude that there are no principle objections for lists to be close to areas such as improving quality of life, support of education and culture. For selection of social technologies, it is necessary to set criteria characterizing possibility to achieve social priorities and socio-economic goals.

We should also note that in conditions of fee-for-scientific and technical information, including patent statistics, and increasing subjectivity of expert assessments, the role of information support for selection of priorities of scientific and technological development is increasing. However, federal statistical data and sampled data are insufficient. In order to achieve completeness, accuracy, relevance of information during implementation of the project on development of lists of priority directions and critical technologies (at any level - federal, regional, sectorial), it is advisable to develop a methodology and to carry out regular comprehensive patent research. This will allow following the patent situation in realization of critical technologies in Russia and abroad, including identification of «patent surge», forecasting of new technological trends, and identification of competitive situation.

Study of the problem of development of a special technology classifier and analysis of opportunities of use of existing national, international, foreign and thematic rubricators and classifiers of scientific and technical information is one more urgent direction for further development of the methodology for selecting priorities of scientific and technological development and critical technologies.

Thus, the system of scientific priorities and critical technologies is not linear. It is a hierarchically constructed multi-level system, which changes dynamically and evolutionary in accordance with the principles of adaptation to the current socio-economic situation of the global and national scale, taking into account trends of global scientific and technological development and anticipatory predictive scenarios of improvement and development of scientific and technological potential of Russia.

All more or less successful projects of recent years are based on integrated approaches. Moreover, we can clearly see a tendency of constant complication of the system of used methods, including application of mathematical methods allowing strengthening Foresight study by scientifically grounded assessments. Such assessments allow making a more detailed scenario forecast based on quantitative evaluations of target indicators and identifying classification rules «If... then» in order to make necessary recommendations on updating the list of critical technologies. We can say that expert opinions and conclusions undergo a kind of test on the possibility to be proved by means of methods of intelligent analysis.

5 Conclusion

Based on the results of the study, we have proposed a conceptual model of conducting expert interviews according to a flexible system of indicators describing the technology.

At the initial stage, it includes a range of quantitative and qualitative indicators outlined by experts. At the next stages, it can be modified by other features depending on socio-economic and political target factors. The developed automated system includes two levels, which are interconnected by analysis and forecast. Automated system, which operates in the iterative mode, is simple and does not need any comments.

In conclusion, we should note that the methodological approach based on the evidence by mathematical methods of analysis develops the technology of Foresight studies and fully integrates in it.

Acknowledgements

The author wishes to thank Tatiana B. Rumyantseva, National Research Tomsk State University, for assistance in the article preparation.

The paper was written as part of the research project No. 8.2.31.2015, carried out with the support of the Program "Research Foundation of Tomsk State University named after D.I. Mendeleev" in 2015 – 2016., grant RFBR No. 16-29-12858.

References

[1] N.C. Dalkey, O. Helmer-Hirschberg, *An experimental application of the Delphi method to the use of experts* RAND Report RM-727-PR (1962)

[2] NISTEP. The 8th science and technology Foresight survey – Delphi analysis. National Institute of Science and Technology Policy, Report PS–14 (2005) URL: http://www.nistep.go.jp/HP_E/researchworks/02_foresight/index.html

[3] R. Popper, *Foresight Methodology*, in L. Georghiou, J. Cassingena, M. Keenan, I. Miles, R. Popper (eds.), *The Handbook of Technology Foresight* (Edward Elgar, Cheltenham, 2008),pp.44

[4] S.V.Gorbachev, V.I. Syryamkin, M.V. Syryamkin, *Intellectual Foresight-Forecast of Scientific and Technological Development of a State* (LAMBERT Academic Publishing, Saarbrucken, 2012)

[5] A.Yu. Apokin, D.R. Belousov, Foresight and STI Governance **3**, 12 (2009)

[6] P.M. Derevyanko, Proceeding of management and economy in creative activity of young scientists (UNECON), 98 (2005)

[7] D.V. Shashev, S.V. Shidlovskiy, V.I. Syriamkin , A.V. Yurchenko, IOP Conference Series: Materials Science and Engineering 81, 012101 (2014) doi: 10.1088/1757-899X/81/1/012101

[8] J.J. Buckley, Fuzzy Sets and Systems **21**, 257, (1987) doi: 10.1016/0165-0114(87)90128-X

[9] A. Kofman, H. H. Aluha, *Introduction of fuzzy set theory in the management of enterprises* (Higher School, Minsk, 1992)

APPLICATION OF AN INTEGRATED AND SELF-CONTAINED ELECTROMAGNETIC ACOUSTIC RECORDER FOR MONITORING THE ROCK MASS STRUCTURE AND DEVELOPMENT OF GEODYNAMIC PROCESSES IN ORE MINES

Anatoly Bespal'ko*, *Ludmila* Yavorovich, *Pavel* Fedotov, *Evgeny* Pomishin and *Maksim* Vedyashkin

Tomsk Polytechnic University, Institute of Non-Destructive Testing, 634050 Tomsk, Russia

Abstract. The paper presents a block diagram of the recorder of electromagnetic and acoustic signals. The recorder provides monitoring of electromagnetic and acoustic signals under changing stress-strain state of heterogeneous materials and rocks. The analog amplifier input sensitivity of electromagnetic signals in 4 channels is 10 μV, that of the electromagnetic signal intensity is 10 μV and that of acoustic signals is 50 μV. The operating frequency range of recording is (1÷100) kHz. The averaging of electromagnetic and acoustic signal amplitudes is performed within 1 or 5 seconds. The data on electromagnetic and acoustic emission materials is recorded in digital format in the internal memory of the recorder. The recorder operates off-line within 7 days. Repeated testing of the recorder in natural conditions of the Tashtagol mine showed that the changes in the amplitude and frequency parameters of electromagnetic signals (EMC) indicate the structural damage in mines made of rocks which differ in their electrical properties. Measurements of the intensity of electromagnetic signals enable monitoring the changes in the stress-strain state of rocks during and after blasting and other geodynamic phenomena.

1 Introduction

Electromagnetic radiation (EMR) occurring in the rocks under mechanical impact has been studied for many years in many countries around the world [1–7]. The key motive of these studies is high potential of EMR to be used for predicting geodynamic events such as earthquakes, rock bursts, landslides and others. Mathematical modeling of mechanoelectrical transformations in dielectrics and the experimental studies showed [8, 9] that monitoring of the electromagnetic emission (EME) and acoustic emission (AE) of heterogeneous materials, including rocks, requires an analog-to-digital instrument to detect and record electromagnetic and acoustic signals in a frequency range of (1÷100) kHz in the internal memory. The input sensitivity of the analog amplifiers of electromagnetic signals is

*Corresponding author: besko48@tpu.ru

© The Authors, published by EDP Sciences. This is an open access article distributed under the terms of the Creative Commons Attribution License 4.0 (http://creativecommons.org/licenses/by/4.0/).

about 10 μV, and that of acoustic signals is not more than 50 μV. Capacitive and inductive sensors are used to receive electromagnetic signals in laboratory and field conditions, and piezoelectric sensors are used to receive acoustic signals. The frequency ranges for electromagnetic emission detection were found in the study of rock block sizes conducted under the guidance of V.N. Oparin, the Corresponding Member of RASб in IM SB RAS [10]. Practical measurements showed the greatest size distribution of *L* blocks in the range of (2÷100) cm. Figure 1 shows the normalized distribution of the rock block sizes obtained through averaging with exposure window of $2\cdot10^{-2}$ m and scan step of 10^{-3} m reported in [10].

The ratio used for the fundamental wave frequency $f = V_з/2L$ at the average speed of sound in rocks $V_з$=3000 m/s indicates that the blocks of these sizes emit mostly electromagnetic signals (EMS) in the above frequency range. The bands of higher amplitude are observed in the frequency spectrum, namely in the range of 12–30 kHz with central frequencies of 15–20 kHz, and in the range of 2–7 kHz. It should be noted that the EMS spectrum contains other frequencies, but they are of lower amplitudes. In addition, the laboratory studies showed that the EMS parameters depend on the characteristics of the excitation acoustic signal. This proves that acoustic signals generated by emerging and propagating cracks have a significant effect on the EMS frequency spectrum, which can also be observed in conditions of natural rock formation. Variations of the amplitude and duration of the excitation acoustic signal, and changed distance of the electromagnetic signal sensor from the radiating surface cause change in the EMS amplitude-frequency characteristics.

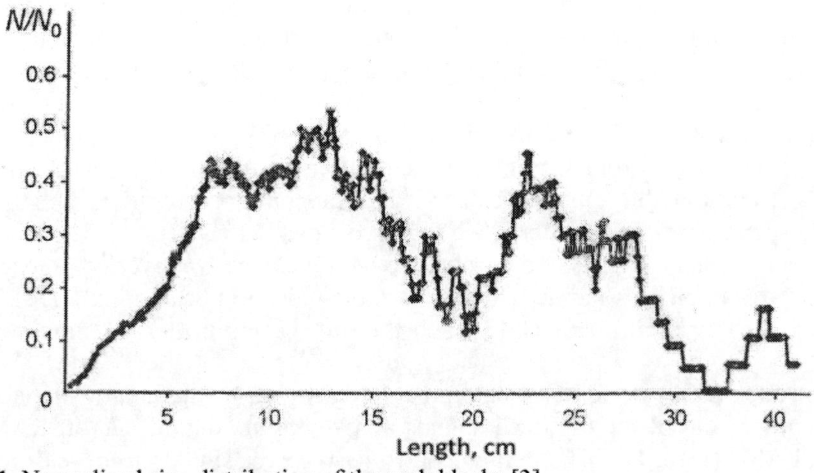

Figure 1. Normalized size distribution of the rock blocks [3].

2 Recorder description

The specified EMS frequency-response characteristics were used to design and manufacture the REMS-1 recorder and to perform its multiple testing in laboratory and field conditions. The REMS-1 recorder transforms electromagnetic signals via the capacity or induction receiver, then it transforms acoustic signals into analog signals with piezoelectric receiver, preprocesses and digitizes the signals, stores the obtained data in the internal memory and transmits the data arrays to PC [11–17]. To decrease the effect of electromagnetic interference, the analogue EMS reception was performed using a differential circuit [18, 19]. To offset from industrial electromagnetic interference, signals below 1 kHz were filtered but not recorded. The obtained values of the averaged EMS and

AS amplitudes were digitized in the wideband of 1÷100 kHz, and the EMS were further filtered in three bands with center frequencies of 2.15 and 100 kHz. In order to reduce the data array and to improve its operational analysis, the EMS and AS amplitudes were averaged in a specified time interval of 1 or 5 seconds. These averaged EMS amplitude values A(t) showed changes in electromagnetic (EME) and acoustic (AE) emissions of heterogeneous materials and rocks under force or acoustic effects the time. These patterns are proportional to the change in the electric E(t) or magnetic H(t) components of the electromagnetic field in the wide-band (F_{WB}) and in the bands with center frequencies of 2 kHz (F_2), 15 kHz (F_{15}) and 100 kHz (F_{100}). During the development of the method the recorder received and recorded acoustic emission of the test objects to increase the reliability of detection of the destructive area formation. AS amplitudes indicate crack propagation in the test object during pre-destruction and destruction processes or acoustic exposure. In case of blasting with known parameters, the electromagnetic acoustic channels were calibrated according to the energy performance of the explosion. In addition, the number of EMS in the given averaging interval was found to change. As a result, the parameters of the change in the intensity of the electromagnetic signal pulse flux could be obtained.

The recorder consists of the units: data-collection unit (DCU); remote capacity (RCR) and induction (RIR) receivers connected in turn to the DCU differential input; acoustic receiver (AR); internal and auxiliary battery power supply (BDA) and power supply unit (PSU) to charge the batteries. The receivers and power supplies are connected to DCU via cable terminals. Figure 2 shows the block diagram of the recorder.

Electromagnetic signals are received by induction or capacity receivers, converted into electrical signals, amplified and fed to the inputs of differential amplifiers (D1, D2) [20, 21]. The signals from the differential amplifiers are the inputs of programmable switching and amplification unit (D3). The amplifier output is connected to three narrowband filters and one broadband filter with a variable gain of (D5÷7) and (D8). The amplified signals are fed to averaging unit (D9÷12) and then to the inputs of ADC microcontroller (D20) [22, 23]. The electrical signals from variable-gain amplifier (D8) are the input of comparator (D14) with a programmable amplitude level. The comparator output is fed to counter (D15), and the signal from the counter is fed to microcontroller (D20).

Piezoelectric acoustic signals are received by the receiver, converted into electrical signals which are amplified and fed to the variable-gain amplifier input (D4). Electric signal D4 is fed to averaging unit (D13), and then to the input of ADC microcontroller (D20).

Microcontroller (D20) performs programmable digitizing of the analog signals in five channels, preprocessing of the digitized signals and stores the digital data in RAM (D18) and PROM FLASH (D16, D17). The microcontroller receives signals from real-time clock (D22) and keyboard signals (S1÷S6) through control logic (D29). Information on the device operation is provided on LCD (H1) [24]. The accumulated data arrays are displayed on PC through serial interfaces without optoisolator RS232 (D24) and optically isolated ones (D25÷27) RS485 (D28) [25, 26]. The recorder is controlled by the programs stored in DCU unit. The operator uses a keyboard to choose the operating modes.

The stabilized voltage of +5V from regulator outputs (D21, D22) is supplied to the components of the electronic circuits, and voltage of +3V from D23 is used for LCD backlighting. The unregulated voltage of +(6÷7) V from the storage battery E_1÷E_6 is supplied to the stabilizer input. The battery is charged through V1 elements via the microcontroller.

The microprocessor data-collecting unit DCU (Figure 1) is controlled by the service program. The remote differential capacity and induction receivers, the piezoelectric sensor

 DOI: 10.1051/matecconf/20167901039

and supplementary external DC power supply, in case of long time operation, are connected through the terminals to the DCU.

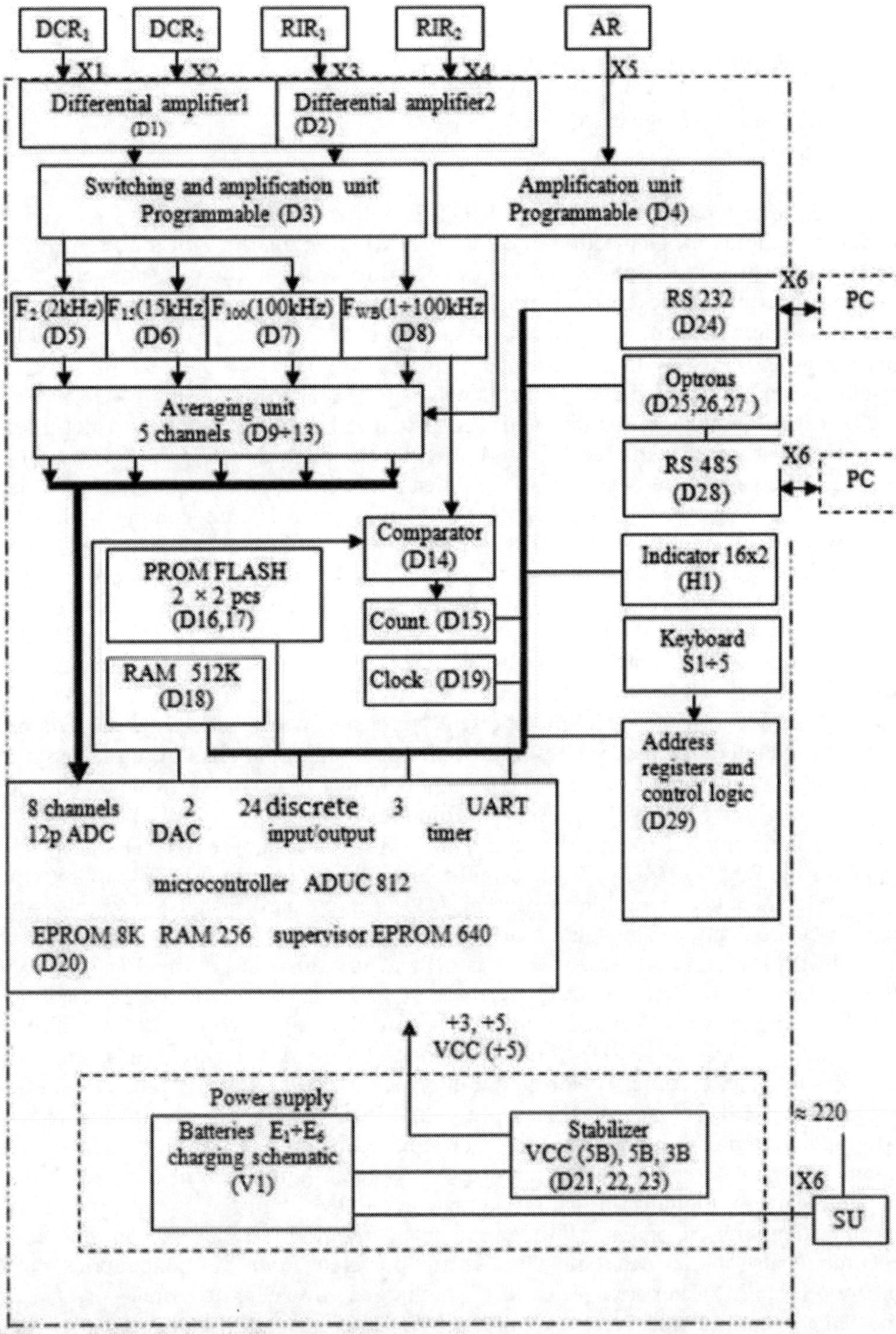

Figure 2. Block circuit diagram of the recorder.

The microprocessor data-collecting unit provides:

- input and pre-processing of analog signals from electromagnetic and piezoelectric acoustic receivers;
- six channel digitizing the analog signals;
- pulse counting;
- accumulation and storage of data of up to 4 MB;
- quick viewing of the information on the LC display;
- the built-in battery operation for at least 40 hours;
- information storage in PC;
- battery charging.

The gain of differential amplifiers (D1, D2) is 4. Switching and amplification unit (D3) provides: programmable connecting to the RCR or RIR; setting the gain in the "menu" (×1, ×4, ×8, ×16), which are displayed on the LCD and recorded in the measurement protocol when the electromagnetic channel amplification gain is set. Three narrow-band filters (D5÷7) are tuned to the band center frequencies of 2 kHz, 15 kHz and 100 kHz, respectively, and are supplied with potentiometers to adjust the gain within 2÷32. The averaging unit [21] supplied with five channels is based on chips of the full-wave rectifier of the average signal (D9÷13) with the time constant τ=1 or 5. Gain unit (D4) amplifies the electric signal of the acoustic receiver and provides the programmed gain (×1, ×2, ×4, ×8, ×16, ×32). The gain value is chosen in the "menu" displayed on the LCD and recorded in the protocol. Comparator (D14) [21] extracts pulse signals and compares the signal amplitude from the amplifier output (D8) (channel without filtration and detection) with the programmed threshold value. If the signal exceeds this value, it generates positive pulse. "Counter" (D15) divides (into 16) a high-frequency signal from the comparator to count the number of pulses entering the microcontroller. Clock (D19) provides real time referencing. Serial interfaces without optoisolator RS232 (D24) and optically isolated interface (D25, D26, D27) RS485 (D28) provide connection to PC. Indicator (H1), a liquid-crystal indicator, has two lines and each line represents sixteen symbols. It displays the data on the parameters and operating modes chosen for unit operation in the form of digits or symbols and controls the input signal values. The keyboard S1÷S4 is used to control and choose the operating mode, and S5 provides power on/off mode.

Control registers and logic D29 selects the synchronization source, D33, 34 are the registers of the RAM address, D36 is a multiplexer control register, D37 is a unit control register, D38 is the decoder of unit selection, D39 is a keyboard case, D8 is optoisolator of external synchronization. The volume of the permanent storage unit PROM FLASH (D16, D17) is 4 MB. It retains data when power is off and stores useful information. The volume of the random access memory RAM (D18) is 512 KB. It records digital information for temporary storage with a maximum microcontroller digitization rate of 200 kHz. Microcontroller ADUC 812 (D20) has 8 channel 12 bit ADC, two DACs, three timer counters, parallel and serial ports for digital information input-output, 256 bytes data RAM, 640 bytes data EPROM and 8 Kbytes program EPROM. The microcontroller provides: program-controlled connection of the RCR or RIR receivers; selection of the channel gain; program-set signal level for the comparator; six channel digitization of the analog signals from the receivers; digital information storage in PROM; digitized signal preprocessing; communication with the operator through the keyboard and LCD; battery charging control. The components of the electronic circuits are powered from E1÷E6 batteries through stabilizers D21÷23. The storage battery is charged with the external power supply connected via controllable current regulator V1. DCU generates supply voltage for the receiver feed circuits, and provides spark protection of these circuits according to GOST 22782.5-78 "intrinsically safe electrical circuit".

DCU software provides: selection of the RCR or RIR, gain factor, averaging time, comparation level; display of the set parameters, operation time, free memory and the values of the received signals on the LCD; mode setting (adjustment, monitoring, transmission of data arrays to the computer, battery charging). During monitoring, information is saved in the form of "pages" in the recorder memory in 256 averaged amplitudes of the F_2, F_{15}, F_{100}, F_{WB}, AE and EMS flux intensities. Each unit contains information on the set EME and AE parameters and start time. The maximum number of the stored monitoring data "pages" is 512. The PC provides a statistical analysis of the collected data. It is a general purpose computer, and it can be installed and used for short time operation in the mine or for permanent remote operation. DCU unit design meets the requirements of its tightness (IR53) and protection against electromagnetic interference and mechanical damage. The unit housing is made of a metal box with sealing on the side of the connectors, and on the keyboard and indicator cover. Unused connectors must be closed off [20]. The front surface of the DCU contains a control panel to set the adjustment mode, to monitor and transmit data arrays to the PC and to charge the batteries. The LCD indicates the settings in the recorder menu in the monitoring mode, transmission of data arrays to the PC, inbuilt battery charging and running EME and AE characteristics. The unit can perform self-testing.

3 Application of the recorder in natural conditions by the example of the Tashtagol Mine

The results of the laboratory studies of the effect of rock contacts on EMS parameters were used to develop a passive method to control SSS changes in the rock mass in the Tashtagol mine.

The observed area was picketed (Figure 3), the sections being spaced at a distance of 3 m. The geology of the sections from 1 to 5 pickets was represented by magnetite. In the area of 5–6 pickets, a dike-formed bulk inclusion was found to contact with skarns on its opposite side.

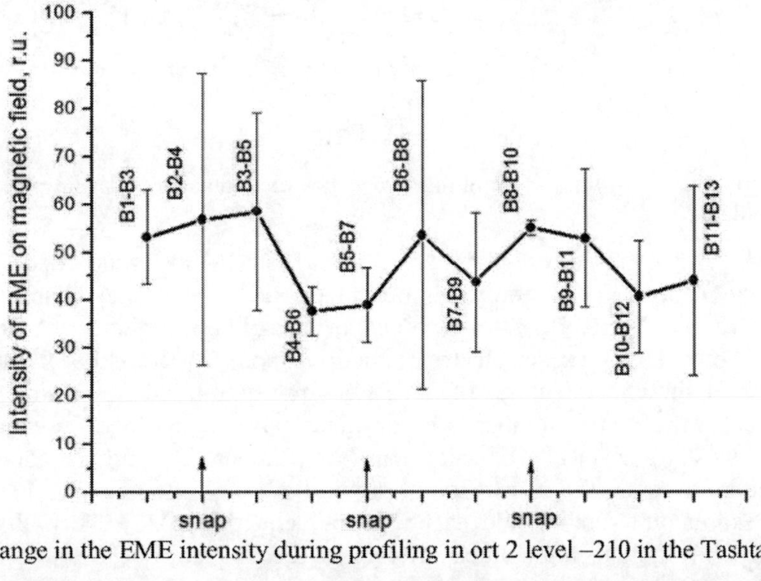

Figure 3. Change in the EME intensity during profiling in ort 2 level –210 in the Tashtagol mine.

In the 9 picket section, a tectonic crack was observed. The studies were conducted by the profiling method using the REMS complex supplied with downhole inductive sensors. The rock pressure geodynamic manifestations detected in this section were followed by a sound effect in the form of clicks. Figure 3 shows the average EME flux intensity in the magnetic component of the electromagnetic field during profiling ort 2 level –210 and the variance in each measurement. The change in the EME flux intensity is detected in the region of ore-dike contact (picket B4–B6), and dike-skarn contact (picket B6–B9). A sharp change in the EME is also observed in the B10 picket section. This area contains tectonic fractures. Earlier studies found that a prediction sign of unstable rock state is not only abnormal change in EME flux intensity, but increase in its dispersion as well.

The studies revealed the greatest variance in EME flux intensity in the above picketed sections. The arrows along the horizontal line show the pickets, where sound effects ("clicks") of rock pressure geodynamic manifestations were recorded. During the EME intensity measurements, significant increase in the electromagnetic radiation intensity dispersion was distinctly recorded before "clicks". After the "click", the EME dispersion sharply reduced to (10–15) %. In addition, it was found that the longer the period of increased EME intensity dispersion, the more powerful the geodynamic effect in the rock.

This confirms the fact that the change in the rock SSS is followed by the change in the electromagnetic field that is manifested by abnormal changes in the EME intensity and its increased dispersion. The same section ort 2 level –210 was taken to study EMS parameters for testing fracturing of rocks of different conductive properties.

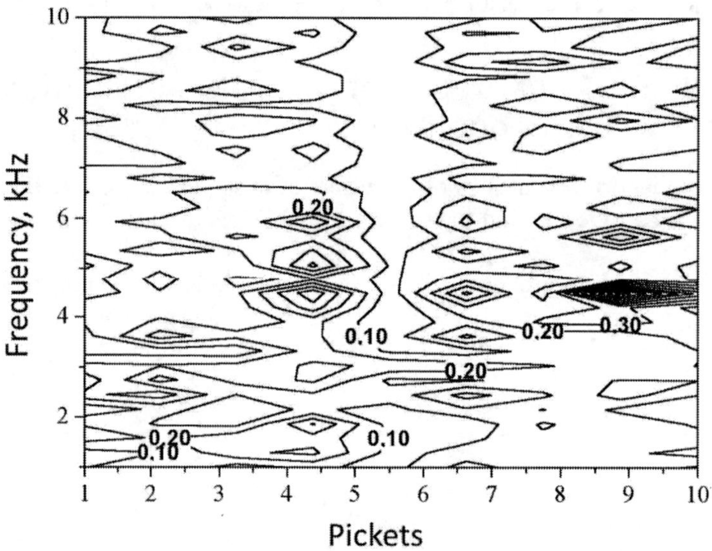

Figure 4. Distribution of isolines of amplitude-frequency electromagnetic signal parameters during subsurface profiling.

The pit edge was exposed to dynamic effects and the EMS was recorded. The recorded electromagnetic responses were analyzed and the amplitude-frequency characteristics were calculated using fast Fourier transform to construct isolines. It should be noted that the dynamic effect on the rock causes electromagnetic responses recorded at frequencies up to 10 kHz. Figure 4 shows the isolines of the EMS amplitude-frequency characteristics. The abscissa indicates the picket numbers, the ordinate shows the EMS frequencies, and the map demonstrates the spectral frequency band amplitudes. The Figure shows a distinct contact boundary between the rock made of ore and dikes on the one side and that formed of dikes and skarns on the other side starting from frequency of 2.5–3 kHz. In the 9 picket section at the point of the tectonic crack emergence, the frequency achieves its maximum of 4 kHz.

The EME observed for contacts of the rocks can be effectively used to monitor the changes in the rock SSS during and after explosion. Figure 5 shows the results of the EME measurement within 5 minutes after blasing of about 300 tons in 11 block between the –210 and –280 m levels in the Tashtagol mine.

Figure 5 shows the electromagnetic emission variations that indicate the response of the rock to blasting and relaxation of its excited state. The EME variations reveal significant difference in the times of the processes occurring in the rock during and after blasting.

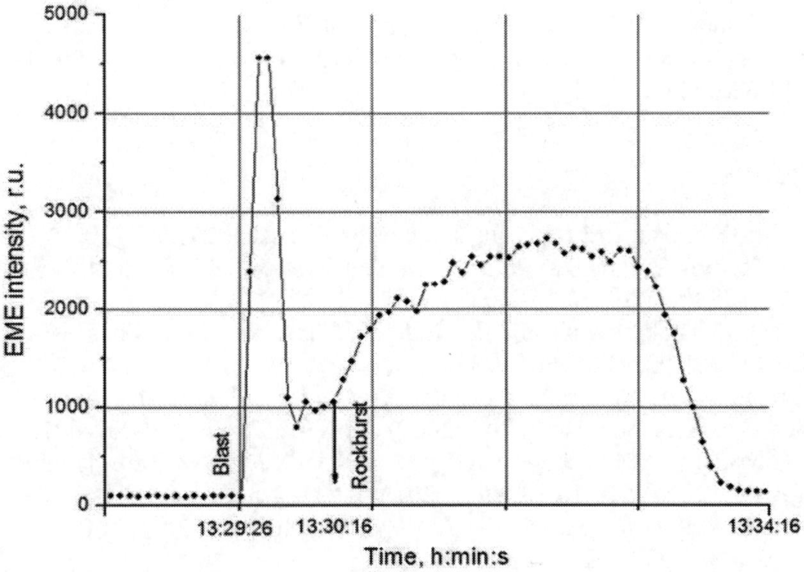

Figure 5. Change in the EME intensity during blasting and subsequent rock burst.

During blasting, the increased EME lasts 15 s. After the rockburst, the rock respond duration is more than 4 minutes, and the seismic station in the Tashtagol mine recorded a great number of geodynamic manifestations of different energy levels in the explosion zone. When the EME intensity dropped to its explosive level and then after 30 minutes, it started to increase up to the values two-fold exceeding the maximum intensity after the rock burst. The increased EME values were observed over 17 hours, and then the intensity dropped to a certain level. At this period of time, the seismic activity in the mine field was observed to decrease.

4 Conclusions

Thus, an analog-to-digital recorder of electromagnetic and acoustic emissions was developed and designed for laboratory and field operation to monitor electromagnetic and acoustic signals under changed stress-strain state of heterogeneous materials and rocks. The analog amplifier input sensitivity of electromagnetic signals in 4 channels is 10 μV, that of the electromagnetic signal intensity is 10 μV, and the acoustic signal sensitivity is 50 μV. The operating frequency range of recording is (1÷100) kHz. The averaging of the electromagnetic and acoustic signal amplitudes is performed within 1 or 5 seconds. The data obtained for electromagnetic and acoustic emissions is recorded in digital format and stored in the recorder internal memory. The recorder battery life time is 7 days.

The field studies in the Tashtagol Mine showed that the changes in the EMS amplitude-frequency parameters can be used to evaluate the structural disturbance in mines formed of rocks which differ in their electrical properties. Measurements of the electromagnetic signals intensity allow monitoring of changes in the rock stress-strain state during and after blasting, and during different geodynamic processes. This monitoring technique can be advantageously used in places where outcropping dikes and veins extend into the area of geodynamics processes in the rock.

Acknowledgments

This work was supported in part by the Ministry of Education and Science of the Russian Federation under the State Research Project and by Russian Foundation for Basic Research Grant No. 14-08-00395.

References

[1] Y. Mori, Y. Obata and J. Sikula, J. Acoustic Emission **27**, 157 (2009)
[2] P. Koktavy, J. Pavelka and J. Sikula, Meas. Sci. Technol. **15**, 973 (2004) doi: 10.1088/0957-0233/15/5/028
[3] V. Frid, A. Rabinovitch and D. Bahat, J.Phys. D: Appl. Phys. **36**, 1620 (2003) doi: 10.1088/0022-3727/36/13/330
[4] A. Pralat, S. Wójtowicz, Acta Geodyn. Geomater. **1,** 111, (2004)
[5] J. Sobotka, Acta Geoph. **57**, 247 (2009) doi: 10.2478/s11600-008-0072-4
[6] H. Kobayashi, K.Horikawa, K.Ogawa, and K.Watanabe, Phil. Trans. R. Soc. **372**, 20130292 (2014) doi: 10.1098/rsta.2013.0292
[7] P. Koktavy, Meas. Sci. Technol. **20**, 015704 (2009) doi: 10.1088/0957-0233/20/1/015704
[8] A.A. Bespal'ko, R.M. Gol'd, L.V. Yavorovich, D.I. Datsko, J. Min. Sci. **39**,12 (2003) doi: 10.1023/B:JOMI.0000008455.54729.30
[9] A.A. Bespal'ko, L.V. Yavorovich, P.I. Fedotov, J. Min. Sci. **43**,472 (2007) doi: 10.1007/s10913-007-0049-8
[10] V.N. Oparin, V.F. Yushkin, A.A. Akinin, E.G. Balmashnova, FTPRPI (5), 34 (1998)
[11] I.P. Golyamin, *Ul'trazvuk malenkaya ehnciklopediya* (Sovetskaya ehnciklopediya, Moskau, 1979) (in Russian)
[12] K.S. Petrov, *Radiomaterialy radiokomponenty i ehlektronika* (Piter, Spb.,2003) (in Russian)
[13] E. Kikuchi, *Ul'trazvukovye preobrazovateli* (Mir, Moscow, 1972) (in Russian)
[14] M.V. Korolyev, Defektoskopiya (4), 65, (1973) (in Russian)
[15] I. Plotnikova, L. Redko, O. Titova, O. Chaykovskaya, IOP Conf. Ser.: Mater. Sci. Eng. **132**, 012023 (2016) doi: 10.1088/1757-899X/132/1/012023
[16] S. Obukhov, I. Plotnikov, M. Kryuchkova, IOP Conf. Ser.: Mater. Sci. Eng. **132**, 012017 (2016) doi: 10.1088/1757-899X/132/1/012017
[17] O.V. Galtseva, S.V. Bordunov, IOP Conf. Ser.: Mater. Sci. Eng. **81**, 012062 (2015) doi: 10.1088/1757-899X/81/1/012062
[18] A.G. Alekseenko, *Osnovy mikroskhemotekhnik* (Yunimedistail, Moscow, 2002) (in Russian)
[19] U. Titce, K. Shenk, *Poluprovodnikovaya skhemotekhnika* (Mir, Moscow,1982) (in Russian)
[20] M.A. Kireev, *Sovremennye zarubezhnye mikroskhemy - usiliteli zvukovoj chastoty spravochnik* (Goryachaya liniya, Moscow, 2004) (in Russian)
[21] *Operacionnye usiliteli i komparatory: spravochnik* (Dodeka-XXI, Moskow, 2002) (in Russian)
[22] Specifikaciya ADuC812 8-kanalnyj precizionnyj 12 razryadnyj ACP so vstroennym mikroprocessorom i FLASH EHRPZU (Spb.: AVTEKS. http://www.autex.spb.ru) (in Russian)

[23] ADuC812: MicroConverter, Multichannel 12-Bit ADC with Embedded Flash MCU Data Sheet (Rev E, 04/2003) URL: http://www.analog.com LCD screen PC1602 URL: http://www.powertipusa.com
[24] Optocouplers 6N137 URL:http://fairchildsemi.com
[25] V.S. Valenko, *Poluprovodnikovye pribory i osnovy skhemotekhniki ehlektronnyh ustrojstv* (Dodeka-XXI, Moskow, 2001) (in Russian)
[26] L. Ruzziconi, S. Lenci , M.I. Younis, MATEC Web of Conferences **16**, 04001 (2014) doi: 10.1051/ matecconf/2014 1604001

MATEC Web of Conferences **79**, 01040 (2016) DOI: 10.1051/matecconf/20167901040
IME&T 2016

RADIOTOMOGRAPHY BASED ON MONOSTATIC INTERFERENCE MEASUREMENTS WITH CONTROLLED OSCILLATOR

Dmitry Sukhanov[a], *Kseniya* Zavyalova

Radiophysics Faculty, Tomsk State University, 634050, Tomsk, Russia

Abstract. The method of three-dimensional tomography based on radioholography measurements with the reference signal transmitted by the transmitter in the near zone and the receiver near zone. We solve the problem of repairing the object signal phase due to the reference signal in the near field in a wide frequency band and the consideration of analytical signals. Here are presented results of experimental studies on application of a tunable YIG (yttrium iron garnet) oscillator in the frequency range from 6.5 to 10.7 GHz for radio tomography of metal objects in air. Holographic principle is applied on the basis of measuring of the interference field amplitude by the detector diode. The interference occurs with the direct wave and waves scattered by the object. To restore the radio images the method of aperture synthesis and extraction of quadrature components at all frequencies sensing are applied. Experimental study on test object shows resolution about 15 mm.

1 Introduction

Tomography at microwave frequencies allows to visualize different electrically inhomogeneous objects with high resolution [1–2]. Currently radio tomography systems are quite expensive and not available to the mass consumer. The most common method to recover the three-dimensional image is a method of aperture synthesizing [3] on the basis of the spatial scan by monostatic radar. Impulse ground penetrating radar (GPR) are widespread to study hidden underground objects and irregularities [1–2]. But the scope of the GPR is not limited, GPR actively used for inspection, monitoring of building structures state, roads and engineering structures. To capture short pulse signals highly stable sampling oscilloscope is required, which significantly increases the cost of the system. Alternative of short-pulse sensing is a sensing using monochromatic signal with scanning frequency. Scanning planar aperture with spatial steps less than a quarter wavelength provides the most complete information about the field scattered by objects. Measuring the amplitude and phase of the scattered waves at different frequencies allows to recover three-dimensional images with range resolution when scanning a wide frequency band and

[a] Corresponding author:sdy@mail.tsu.ru

© The Authors, published by EDP Sciences. This is an open access article distributed under the terms of the Creative Commons Attribution License 4.0 (http://creativecommons.org/licenses/by/4.0/).

without range resolution on probing at the single frequency. To obtain necessary amplitude and phase of the sine and cosine quadratures of the measuring signal. In fact, measurements of a set of quadrature on different frequencies content equivalent information to ultra-wideband short-pulse sensing. However, measurement of quadrature microwave signals is at least technically costly solution as the use of sampling oscilloscopes. Holographic approach is widely used for reconstruction of radio images on single frequencies or for frequency scanning measurements [4–8].

In [9] it was shown that for three-dimensional imaging is sufficient to measure only the cosine signal quadrature. In [9] proposed to use the measurement of the amplitude of the interference field of the direct wave and the wave scattered by objects in the environment. Besides the possibility of visualization of objects hidden underground has been demonstrated [2]. For these measurements, it is possible to apply a scalar network analyzer, but this solution is not applicable for measurements in practice.

In this paper we propose to use a compact microwave YIG oscillator with generated signals in the range from 6.5 to 11 GHz Hewlett-Packard 5086-7343 Oscillator (5.9 – 12.4 GHz) for tomographic measurements. Here is proposed to use the detector diode HSMS - 8101-BLKG. Experiment was conducted on the multi-position sounding of metal object at the distance of 20 cm. The three-dimensional image of the test object has been reconstructed by processing of experimental data. The proposed solution allows to reduce the cost of three-dimensional tomographic systems. The most expensive components of the developed system is a tunable microwave generator.

2 Measurement scheme

It is proposed to consider the scheme of measurements shown in Figure 1.

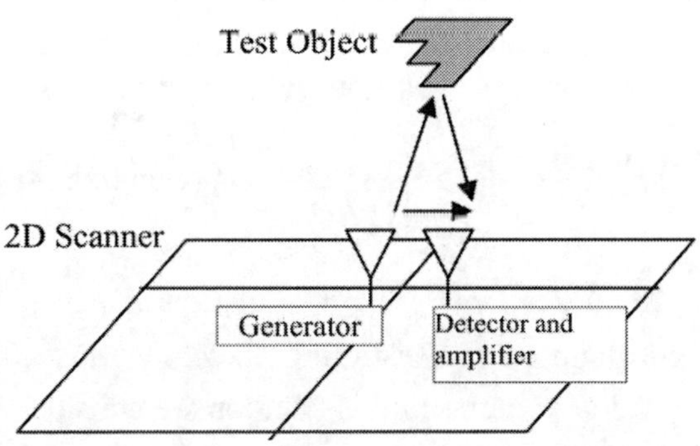

Figure 1. Measurement scheme.

On a two dimensional positioning device has been mounted a unit of the transmitting and receiving antennas (transceiving module). The transmitting antenna is connected to tunable oscillator. Receiving antenna is connected directly to the detector, which is connected to the amplifier. The source antenna irradiates directly receiver antenna and also irradiates the test object.

The scattered from the object wave enters the receiving antenna (object signal), which interferes with the direct signal from the transmitter (reference signal). Assume that the voltage from the detector is proportional to the intensity of the signal in the receiving antenna. Using the interference of object and reference signal the quadrature of signal can be extracted at a current sensing frequency.

We assume that the magnitude of reference signal is much grater than the magnitude of the object signal (at least 10 times). Let us denote $A = |A|\exp(i\alpha)$ – complex magnitude of object signal, $B=|B|\exp(i\beta)$ – complex magnitude of reference signal. Hence the full field in reciewer antenna can be written as the sum: $U=A+B$. Consider that only the intensity $|U|^2$ is measured. The quadrature of object signal has to be extracted from this value. Let us write the value:

$$|U|^2=(A+B)(A^*+B^*)=|A|^2+|B|^2+AB^*+BA^*, \tag{1}$$

and consider that $|A| \ll |B|$, consequently:

$$|U|^2-|B|^2 \approx AB^*+BA^*=|A||B|e^{i(\alpha-\beta)}+|A||B|e^{-i(\alpha-\beta)}=|A||B|2\cos(\alpha-\beta).$$

Denote value

$$C=\frac{|U|^2-|B|^2}{2|B|}e^{i\beta} \tag{2}$$

– reconstructed complex magnitude of object wave in the receiver antenna, where multiplier $\exp(i\beta)$ provides the error of phase reconstruction relatively to reference signal less than $\pm\pi/2$. Assuming that the received signal is analytical applying Hilbert transformation we can recover complete phase of the signals. Hence the measurements are ultra wide band by frequency scanning we can make backward Fourier transformation (to time domain), set to zero negative time part of the signal and then make forward Fourier transformation. This operation allows recovering phase of signals on different frequencies.

The reference signal can be calculated as spherical wave: $B=|B|\exp(ikd)$, where d – distance between the source and receiver antenna, $|B|$ – magnitude of source, which is determined by calibration measurements for current sounding frequency. Besides the reference signal can be written as a solution of the forward task in approximation of the single scattering:

$$A(x,y,\omega)=\iiint_V p(x',y',z')\frac{\exp(2ik|\mathbf{r}-\mathbf{r}'|)}{|\mathbf{r}-\mathbf{r}'|}dx'dy'dz', \tag{3}$$

where ω – sounding frequency; $p(x',y',z')$ – distribution of scattering heterogeneities; $\mathbf{r}'=(x',y',z')$ – current point on the scattering object; $\mathbf{r}=(x,y,0)$ – transceiver module coordinates, V – integration volume. In this solution we consider that receiver and source are combined in a single point.

3 Solution of inverse problem

A three-dimensional distribution of scattering heterogeneities has to be restored from the measured intensity in the measurement plane. To solve the inverse problem we use the method described in [9]. The essence of this method is to restore the three-dimensional spatial spectrum of scattering heterogeneities from the measured field plane-waves spectra at different frequencies, according to the formula:

$$\widetilde{P}(k_x,k_y,k_z)=\widetilde{C}\left(k_x,k_y,\frac{c}{2}\sqrt{k_x^{\,2}+k_y^{\,2}+k_z^{\,2}}\right), \tag{4}$$

where $\widetilde{P}(k_x, k_y, k_z)$ – spatial spectrum of scattering heterogeneities; $\omega = \frac{c}{2}\sqrt{k_x^{\,2} + k_y^{\,2} + k_z^{\,2}}$; $\widetilde{C}(k_x, k_y, \omega)$ – plane wave spectrum of reconstructed field of the object wave in the measurement area at the frequency ω .

Distribution of heterogeneities reconstructed using a three-dimensional Fourier transform:

$$P(x, y, z) = \iiint \widetilde{P}(k_x, k_y, k_z)\exp\left(ik_x x + ik_y y + ik_z z\right)dk_x dk_y dk_z \,. \tag{5}$$

For unambiguous reconstruction of three-dimensional images necessary grid of the frequency with step should be less then $\frac{c}{2R}$, R – distance to the farthest point in the investigated volume. For a larger frequency step with, images of objects will be doubled for different ranges.

4 Experimental studies

To perform experimental studies the laboratory setup has been built. The setup includes: two-dimensional scanner; transceiver module; oscillator YIG Hewlett-Packard 5086-7343 Oscillator (5.9 – 12.4 GHz); detector diode; HSMS – 8101-BLKG. The transceiver module consist of ultra wideband transmitting antenna and ultra wideband receiver antenna (logarithmic patch antenna). The detector diode is connected to low frequency operational amplifier LM 358. The transceiver module scheme is presented on the Figure 2.

Source antenna is located at a distance of 7 cm from the receiving antenna. The flat metal screen mounted between the emitting and receiving antenna allows to decrease and adjust the reference signal level. The excess of the reference signal over the subject signal reached about 10 times by adjusting of the screen position.

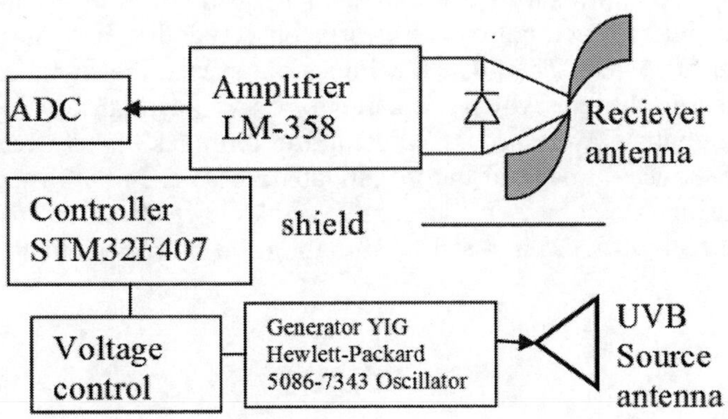

Figure 2. Transciever module scheme.

To start the generator YIG Hewlett-Packard 5086-7343 Oscillator the voltage of 6V is applied to it. The voltage applied to the control coil can be vary from 6 to 11 V to control the frequency. The coil resistance is 50 Ohms. Voltage 6 V provides generation at a frequency of 5.9 GHz, and voltage 12 V provides 1 generation at a frequency of 12.4 GHz. Variation of the voltage on the control coil is controlled by bipolar transistor 2N3055. The transistor base is connected to microcontroller throw optocoupler. By 100mkF capacitor the smoothes of the control voltage is provided. This scheme provides tuning of the frequency generator during the period of 10 ms from 10.7 GHz to 6.5 GHz. Figure 3 shows the frequency of the oscillator as a function of time.

The frequency depends on time nonlinearly, however, there is correspondence between the time and frequency. Therefore the generator frequency can be defined precisely by time. The antenna block is located on the two dimensional scanner, providing positioning region 36 cm on 36 cm with steps of 5 mm. The metal test object has a shape of polygon with the step size of 5 cm. The object has been placed at a distance of 20 cm from the antenna movement plane. Figure 4 shows a photograph of the experimental setup.

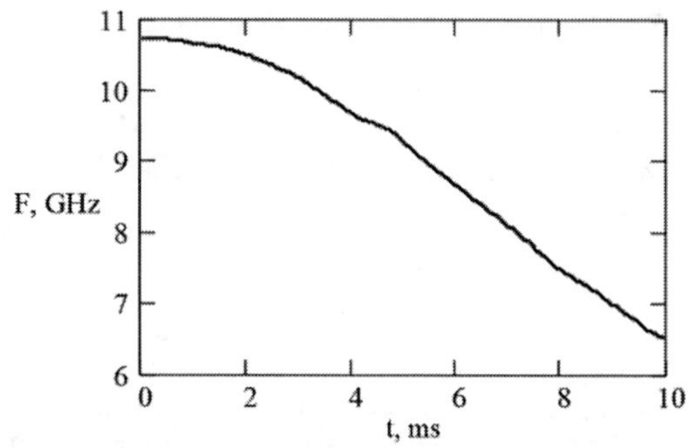

Figure 3. Dependence of oscillator frequency on time during adjustment.

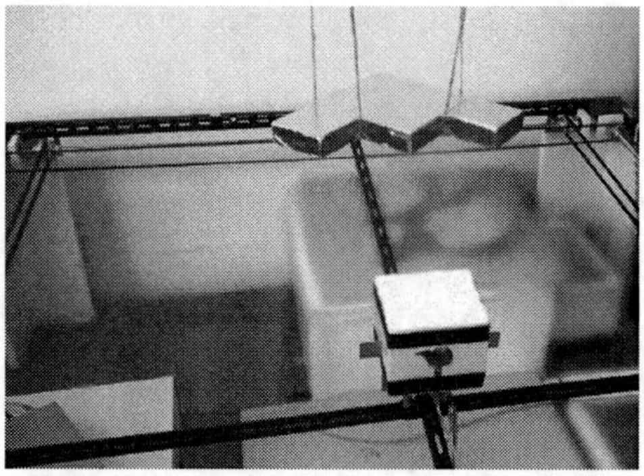

Figure 4. Photography of experimental setup.

Full scanning has been performed on area 36 cm on 36 cm during 10 minutes. As a result of measurements the magnitude of field on 256 frequencies from 10.7 GHz to 6.5 GHz has been acquired. From the measured magnitudes were calculated quadrature signals at different frequencies for the various provisions of the antennas according to the formula (3). Figures 5 and 6 shows a cosine signals quadratures at frequencies of 10 GHz and 6.5 GHz. An interference pattern is clearly observed that indicates the presence of information about the object. The noise level does not exceed the signal level and dynamic range of the measuring system, which is enough to restore quadrature.

Recovered quadratures were processed using the formulas (4 – 5), resulting in a three-dimensional reconstructed image of the investigated volume. In the distance section of 20 cm of 3D image the test object is observed, which is represented in Figure 7. Visual resolution of the reconstructed image by estimating the blurriness of the object edges is about 15 mm.

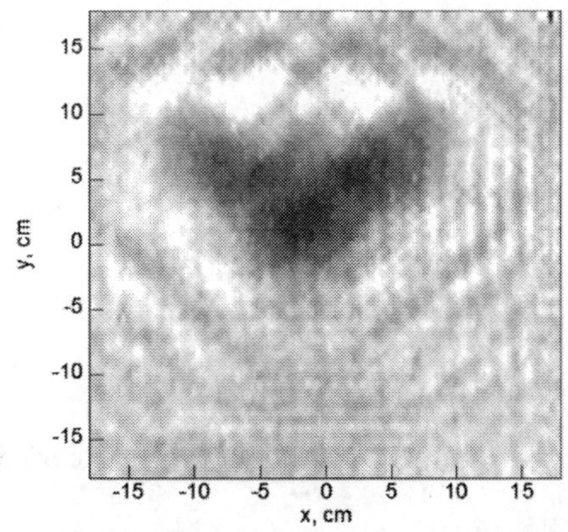

Figure 5. Cosine quadrature on 10 GHz frequency.

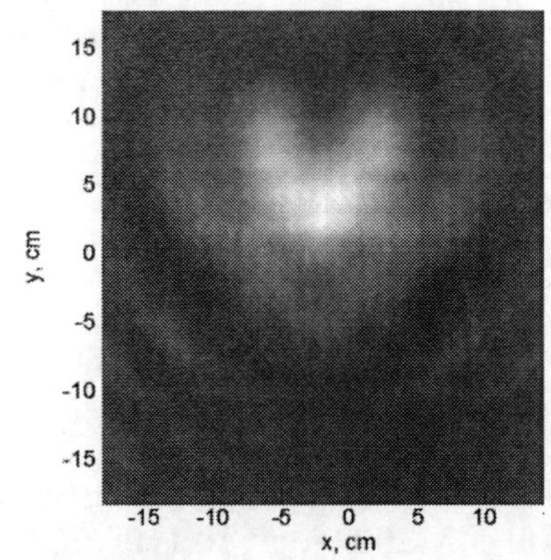

Figure 6.Cosine quadrature on 6.5 GHz frequency.

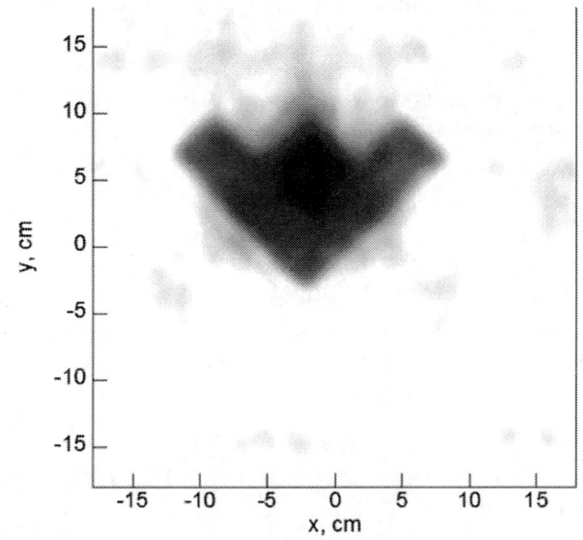

Figure 7.Reconstructed image of the test object.

In addition to the image of the object observed the noise and artifacts. The appearance of the noise associated with outside signals in a reception path circuits. Artifacts caused by the approach of monostatic sounding, but real antennas are not in the same point in space, and are separated by 7 cm. In addition curve of the oscillator frequency from time to time may slightly change according to the transistor 2N3055 temperature. The positioning precision of the two-dimensional scanner is 2 mm also leads to an increase in the noise level. Despite the presence of noise and measurement errors, the image of the test object was restored, which is clearly identified. This result provided by highly stable algorithm for solving of the inverse problem on the basis of the matched filter.

5 Conclusion

In this paper we propose a method of radiowave tomography based on a detector and tunable microwave generator. The result of an experiment of the metal object sensing test at a distance of 20 cm confirmed the presence of an acceptable level of signal-to-noise at all sounding frequencies. The processing of the measured data allowed to restore the image of the test object with a resolution of about 15 mm. Our research demonstrates that the three-dimensional tomography can be created on the basis of low-cost components.

Acknowledgment

The research is supported by Russian Science Foundation project No 16-19-10272.

References

[1] Joaquim Fortuny-Guasch, IEEE Transactions on Geosciences and Remote Sensing **40**, 443 (2002) doi: 10.1109/36.992808
[2] V. P. Yakubov, D. Ya. Sukhanov, Radiophysics and quantum electronics **50**, 299 (2007) doi: 10.1007/s11141-007-0026-8
[3] R.O. Harger, EARSeL Advances in remote sensing **2**, 269 (1993)
[4] E. C. Ngai, 2013 IEEE Antennas and Propagation Society International Symposium (APSURSI), 288 (2013) doi: 10.1109/APS.2013.6710805
[5] A. Popov; I. Prokopovich; V. Kopeikin; D. Edemskij, Proceedings of the 15th International Conference on Ground Penetrating Radar (GPR 2014), 655 (2014) doi: 10.1109/ICGPR.2014.6970507
[6] S.A. Wilson, R.M. Narayanan, M. Rangaswamy, IEEE National Radar Conference – Proceedings **2015**, 925 (2015) doi : 10.1109/RADAR.2015.7131127
[7] Yingzhi Kan, Yongfeng Zhu; Qiang Fu, MATEC Web of Conferences **56**, 02001 (2016) doi: 10.1051/matecconf/20165602001
[8] A. Zhuravlev, S. Ivashov, I. Vasiliev, V. Razevig, 2012 14th International Conference on Ground Penetrating Radar (GPR 2012), 62(2012)
[9] D. Ya. Sukhanov, K. V. Zav'yalova, Technical physics **57**, 819 (2012) doi: 10.1134/S1063784212060229

MATEC Web of Conferences **79**, 01041 (2016) DOI: 10.1051/matecconf/20167901041
IME&T 2016

MEASUREMENT ISSUES OF RADIO FREQUENCY INTEGRATED CIRCUITS WITH DIGITAL CONTROL AT RADIATION TESTING

Konstantin Amburkin*, *George* Chukov, *Vadim* Elesin, *Galina* Nazarova, *Nikolay* Usachev

National Research Nuclear University MEPhI (Moscow Engineering Physics Institute), 115409, Moscow, Russia

Abstract. The results of the development of a portable logical generator/analyzer to be used in an automated hardware and software measurement system for automation of research and testing of TID hardness of digitally controlled RF ICs, are provided. The device has been tested in research of precision parameters of integrated multi-bit RF vector phase shifters.

1 Introduction

In electronic components development today, there is an efficiency-driven trend to combine the functions of an RF transceiver path, digital control and signal processing on a single chip (SoC) or in a single package (SiP). Current and prospective RF transceiver ICs are complex analog-digital functional blocks [1–5]. A comprehensive radiation testing of such RF ICs is possible only if the functions of automated logic control and management have been built in a RF hardware/software measurement system [6].

An example of such an analog-digital block for transceiver modules of RF active phased arrays, is a monolithic IC designed for phase control (6-bit) and amplitude control (5-bit) — phase shifter/attenuator (PSA IC) [7, 8]. Phase and amplitude control is implemented via a parallel or serial digital code. Precision parameters of PSA ICs are determined from results of measuring the frequency dependency of S-parameters in each of the 2048 logical states of the IC. If the parallel control code is specified using mechanical switches (jumpers) and then saved as an S-parameter file, this may require several hours of continuous engineer work, which is unacceptable under the radiation testing [9–11].

To test such ICs, a logical signal generator and logical analyzer had to be introduced into the RF hardware/software measurement system [6]. Commercially available universal logical analyzers cost dozens of thousands of dollars, and their considerable size obstructs their use in radiation testing of RF electronic components in radiation facilities [12–15].

This necessitates developing of an inexpensive portable logical analyzer/generator (PLAN-16M) integrated with a RF hardware/software measurement system and adapted for radiation testing.

* Corresponding author: kmamb@spels.ru

© The Authors, published by EDP Sciences. This is an open access article distributed under the terms of the Creative Commons Attribution License 4.0 (http://creativecommons.org/licenses/by/4.0/).

2 PLAN-16M device description

PLAN-16M has two eight-bit bidirectional ports with variable supply voltage; the device is controlled from a personal computer (PC) via a USB interface.

The functional block diagram of PLAN-16M is shown in Figure 1; it consists of a microcontroller (MCU) enabling connectivity with a PC and managing other IC nodes, bidirectional buffers and a programmable power supply. Two independent external ports allow the device to operate both as a logical analyzer and logical signal generator.

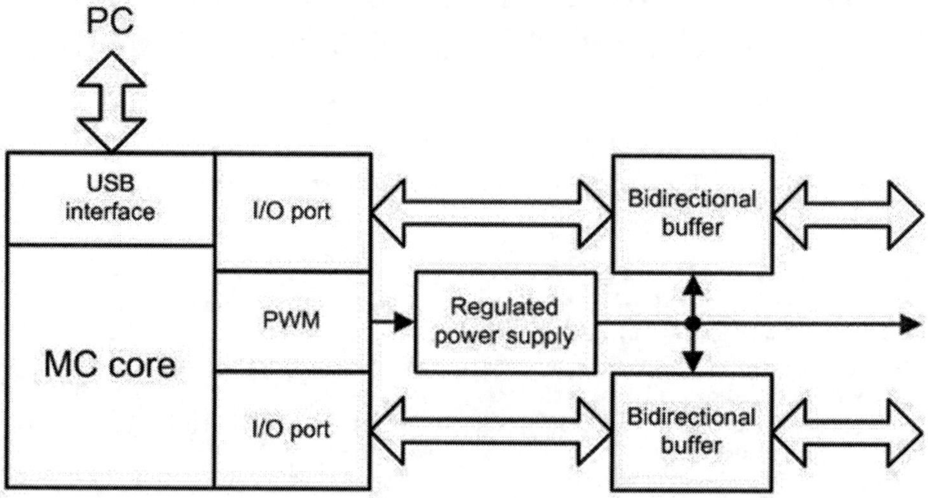

Figure 1. Functional block diagram of PLAN-16M.

The core of the device is the AT90USB162 MCU (Atmel) [16], selected based on the following criteria: availability of a hardware USB interface, required number of *I/O* pins, simple architecture, high availability and low cost. To implement bidirectional ports with variable levels of logic signals, bidirectional buffer ICs 74LVX3245 have been used [17]. The device is fully powered via a USB port. Buffer ICs can be powered both via a built-in programmable source and an external source, to expand the range of levels of logical signals supported. The built-in programmable power supply has been implemented based on the MCU's hardware PWM module and an integrating filter based on the AD826AN operational amplifier (Analog Devices) [18]. Also, the RF IC tested can be powered directly by the built-in programmable PLAN-16M internal power supply. The exterior PLAN-16M is shown in Figure 2. Key features of PLAN-16M are shown in Table 1.

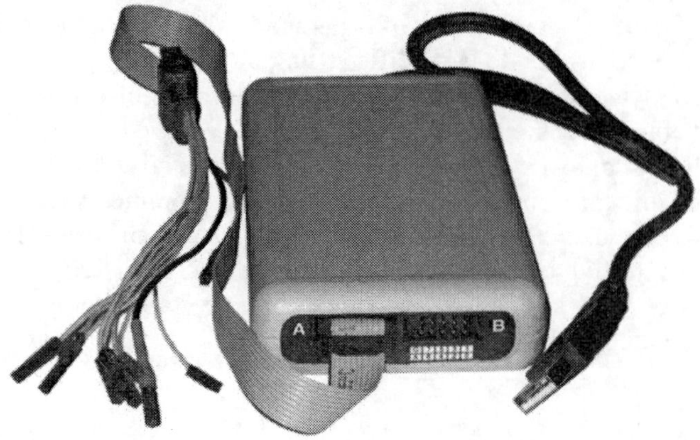

Figure 2. Exterior of PLAN-16M.

Table 1. Key features of PLAN-16M.

Feature	Value
Number and bit of I/O ports	2×8
Supply voltage, V	5
Range of logical signals I/O ports levels, V	1.5… 5
Parallel data setup time code, ms	< 2
Supply Current, mA	60
Package size, mm	134×68×24

When connected to a PC, PLAN-16M is recognized by the system as a standard USB HID-class device with no additional drivers needed to run it. To communicate with the device, a standard library of I/O functions provided by the MCU AT90USB162 manufacturer, is used [19]. To optimize subsequent development of control programs, a specialized library of higher-level functions has been developed. Figure 3 shows a window of the vector circuit analyzer with the program allowing to test PSA ICs both with parallel and serial control.

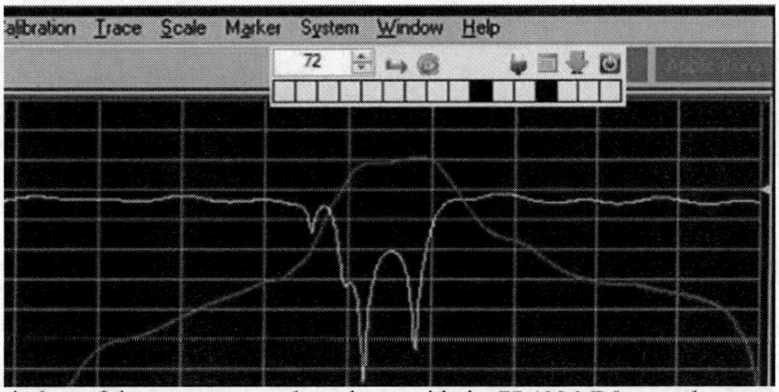

Figure 3. A window of the vector network analyzer with the PLAN 16M control program.

3 Use of the device IN test of parameters of RF phase shifter and attenuator ICs

Testing of multi-bit RF PSA ICs with parallel or serial control has some specifics. First of all, measuring and processing of the phase shifter parameters is a resource intensive process, and also RF test fixture is quite complex.

PSA parameters are classified into primary parameters, i.e. those measured directly, and secondary parameters, which are calculated based on the primary parameters [7, 8]. The primary parameters are S-parameters, the dynamic range parameters (noise Figure and input 1 dB compression point). Secondary parameters are the integral or precision parameters including root mean square (RMS) for the phase error (RMS_{φ}) and for the amplitude error (RMS_A).

Using the device developed, the test routine for multi-bit RF PSA ICs has been automated; the core of this process constitutes measuring of primary parameters and subsequent calculation of secondary PSA ICs parameters. The test routine based on a RF hardware/software measurement system integrated with RF CAD and the universal RF test fixture characterized in a frequency range of 12.5 GHz [20].

The diagram of 2048 states of the amplitude-phase characteristics of the monolithic RF PSA IC including a 6-bit vector phase shifter and a 5-bit attenuator designed by "Progress MRI" with SiGe BiCMOS process [8], is shown in Figure 4. The diagram shows that testing of such ICs, particularly for radiation hardness, is almost non-practicable without the automation of measurement processes (2048 matrices of S-parameters) and mathematical processing of measurement results in real-time (i.e., calculation of RMS_{φ} and RMS_A used as integral criteria parameters).

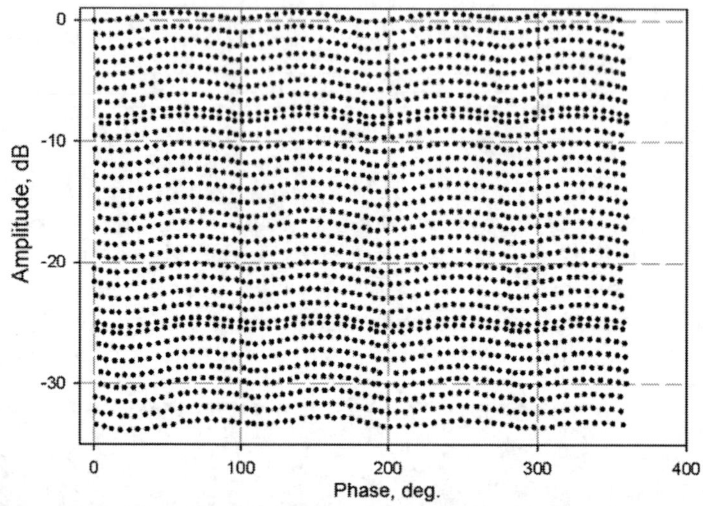

Figure 4. Results of measuring the amplitude-phase characteristics of the PSA IC at a frequency of 5 GHz with PLAN-16M.

4 Optimazation of precision parameters of vector phase shifters with redundant coding

Using PLAN-16M, precision parameters of RF vector phase shifters ICs designed by "Progress MRI" with SiGe BiCMOS process, have been measured [8].

RF vector phase shifters ICs are based on combining of two quadrature signals of a variable amplitude. The input signal is split into quadrature components *I* and *Q* passed through stepped attenuators and then combined back. The degree of attenuation in attenuators determines the amplitude of signals *I* and *Q*, and hence the amplitude and phase of the output signal [8, 21]. For N-bit attenuators, the number of phase shifter states is $2^{2\times N}$; from this set, the basic states are selected, i.e. those which enable phase adjustment at a given step without the amplitude change. Such states are usually selected at the design stage; however the manufacturing tolerance may result in a substantial degradation of phase shifter precision parameters. A way to improve the phase shifter precision parameters is to select the main states after manufacturing.

The procedure used to optimize the phase shifter precision parameters by selection of the initial main operating states based on testing results, consists of two main stages: at the first stage, complex transmission coefficients (S21) are measured in all states of the phase shifter; at the second stage, the initial states are checked and selected based on the given criteria.

The RF hardware/software measurement system [6] has been used at the first stage; to automatically run a complete set of measurements needed for all the phase shifter states, PLAN-16M has been used. Block diagram of the measuring set up is shown in Figure 5. The measurement results have been saved as a file containing complex matrices of S-parameters in the frequency range relevant to each state.

At the second stage, the specialized software searches the main states based on the given criteria which may include RMS_{φ} or RMS_A, amplitude modulation (ΔA), average insertion loss (A_{av})

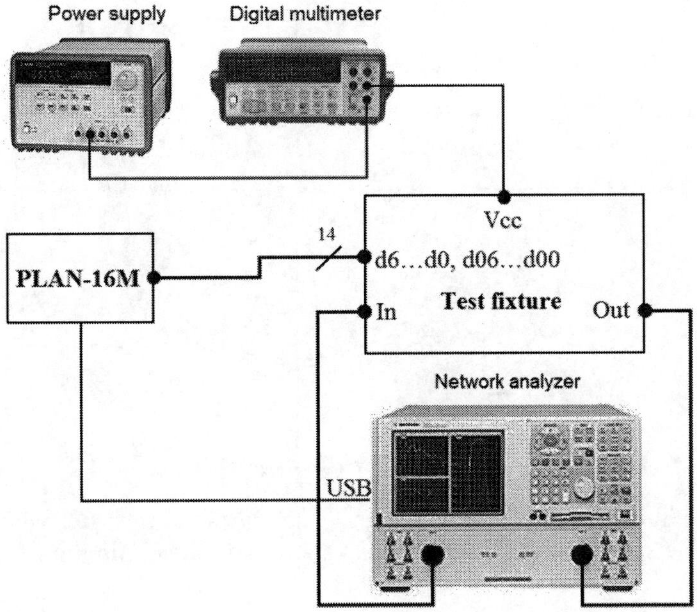

Figure 5. Block diagram of a measuring set up for multi-bit vector phase shifter ICs.

Experimental results obtained for a test 6-bit phase shifter IC operating at 3 GHz and based on 7-bit attenuators (more than 16,000 states total), in the main 64 states according to the specification, have showed unacceptable degradation of its precision parameters (see Table 2).

For the tested phase shifter IC, the optimum states were searched based on the minimum RMS_{φ} criteria, under the conditions that $\Delta A \leq 0.5$ dB and $A_{av} \geq -4$ dB, with the results shown in Figure 6. Phase shifter parameters were measured for the states found; the results are shown in Table 2. For comparison purposes, the Table 2 also shows the parameters of a similar device, a serial HMC647 IC (Hittite). Therefore, using PLAN-16M to automate the measurement of a complete set of complex S-parameters, has substantially enhanced the precision parameters of the test vector phase shifter IC, bringing them to the level of its competitive analogues.

Table 2. Test results for the vector phase shifter IC.

Object	RMS_{φ}, deg.	RMS_A, dB	ΔA, dB	A_{av}, dB
Test phase shifter IC, states according to the specification	3.9	0.8	3.4	-3.7
Test phase shifter IC, states after optimization	0.7	0.14	0.5	-3.9
HMC647, Hittite	1.0	—	1.0	-4.0

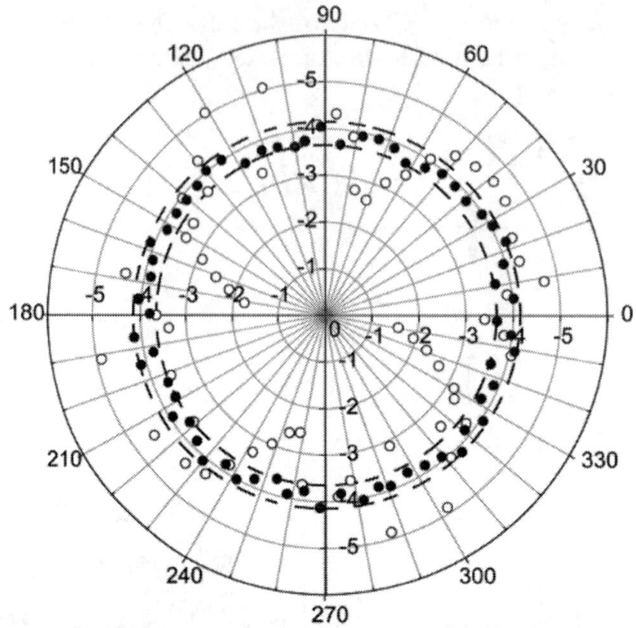

Figure 6. Amplitude-phase diagram of the vector phase shifter IC: white circles – states according to the specification, black circles - states with the minimum RMS_{φ}.

5 Conclusion

Portable logical generator/analyzer PLAN-16M designed for automation of measurement, radiation testing and optimization of digitally controlled RF ICs, has been developed. The device is a new element for the existing RF and MW hardware-software measuring system, adapted to radiation test facility. PLAN-16M has been tested with several domestic and foreign multi-bit phase shifter and attenuator ICs. Testing of such ICs, particularly for radiation hardness, is almost non-practicable without the automation of measurement processes (involving 2048 matrices of S-parameters) and mathematical processing of the measurement results in real-time (i.e., calculating integral criteria parameters, RMS and RMSA). When used within the automated software/hardware measuring system, PLAN-16M provides for recording of the control word and saves the file containing measurement results in less than 1 sec. The process of measuring 2048 states of analog-digital blocks of transceiver modules for RF active phased arrays runs automatically and takes up to 100 minutes, with 70 minutes being the time needed to measure S-parameters, which depends on the measuring equipment capacity. It has been shown that the developed device can help to substantially improve precision parameters of the vector phase shifter and attenuator ICs by selecting optimal states.

Acknowledgments

The authors thank V.V. Repin and I.I. Mukhin from "Progress MRI" (Russia) for the providing the samples and their interest in the work.

This work was partially supported in accordance with agreement between Ministry of Education and Science of the Russian Federation and National Research Nuclear University MEPhI as of November 24, 2014 № 14.578.21.0075 (unique identifier of applied research: RFMEFI57814X0075).

References

[1] J.D. Cressler, *The Silicon Heterostructure Handbook: Materials, Fabrication, Devices, Circuits, and Applications of SiGe and Si Strained_Layer Epitaxy* (CRC Press, London, 2005)

[2] G.N. Nazarova, V.V. Elesin, A.Yu. Nikiforov, A.G. Kuznetsov, N.A. Usachev, G.V. Chukov, CriMiCo'2014, 856 (2014) doi: 10.1109/CRMICO.2014.6959663

[3] V. V. Elesin, G. N. Nazarova, N. A. Usachev, Russian Microelectronics **39**, 134 (2010) doi: 10.1134/S1063739710020083

[4] S. Shi, W. Yang, J. Zhang, Z. Chang, MATEC Web of Conferences **22**, 01005 (2015) doi: 10.1051/MATECCONF/20152201005

[5] X. Liu, MATEC Web of Conferences **44**, 02009 (2016) doi: 10.1051/MATECCONF/20164402009

[6] D.V. Gromov, S.A. Polevich, V.V. Elesin, CriMiCo'2009, 730 (2009)

[7] O.G. Vendik, M.D. Parnes, *Antennas with electrical scanning (introduction to theory)* (Science-Press, Moscow, 2002)

[8] V.V. Elesin, G.N. Nazarova, N.A. Usachev, D.I. Sotskov, G.V. Chukov, V.V. Repin, I.I. Mukhin, Izvestiya vysshikh uchednykh zavedenii. Elektronika (in Russian) **5**, 31 (2012)

[9] O. A. Kalashnikov, V. V. Elesin, D. V. Gromov, CriMiCo'2014, 862 (2014) doi: 10.1109/CRMICO.2014.6959666

[10] V. V. Elesin, G. V. Chukov, D. V. Gromov, V. V. Repin, V. A. Vavilov, Russian Microelectronics **39**, 122 (2010) doi: 10.1134/S1063739710020071

[11] G. V. Chukov, V. V. Elesin, D. V. Boychenko, A. G. Kuznetsov, K. M. Amburkin CriMiCo'2014, 860 (2014) doi: 10.1109/CRMICO.2014.6959665

[12] A.S. Artamonov, A.A. Sangalov, A.Y. Nikiforov, V.A. Telets, D.V. Boychenko, IEEE Radiation Effects Data Workshop **2015-January**, 7004600 (2015) doi: 10.1109/REDW.2014.7004600

[13] G.G. Davydov, A.S. Kolosova, L.N. Kessarinsky, D.V. Boychenko, 2015 International Siberian Conference on Control and Communications (SIBCON), 7147094 (2014) doi: 10.1109/SIBCON.2015.7147094

[14] D.V. Gromov, A.N. Egorov, O.B. Mavritskiy, G.V. Chukov, K.M. Amburkin, CriMiCo'2015, 812 (2015)

[15] A.Y. Nikiforov, O.B. Mavritsky, A.N. Egorov, V.S. Figurov, V.A. Telets, P.K. Skorobogatov, S.A. Polevich, IEEE Radiation Effects Data Workshop, 49 (1996)

[16] Datasheet AT90USB162 IC. URL: www.atmel.com/dyn/resources/prod_documents/doc7707.pdf

[17] Datasheet 74LVX3245 IC URL: www.fairchildsemi.com/ds/74/74LVX3245.pdf

[18] Datasheet AD826 IC URL: www.analog.com/static/imported-files/data_sheets/AD826.pdf

[19] Library of functions for MCU AT90USB162. URL: www.atmel.com/dyn/resources/prod_documents/AVR153.zip

[20] G. V. Chukov, G. N. Nazarova, A. Yu. Nikiforov, CriMiCo'2014, 131 (2014) doi: 10.1109/CRMICO.2014.6959323

[21] Y. Zheng, C.E. Saavedra, IEEE Transactions on Circuits and Systems I: Regular Papers **57**, 752 (2010) doi: 10.1109/TCSI.2009.2025851

MATEC Web of Conferences 79, 01042 (2016) DOI: 10.1051/matecconf/20167901042
IME&T 2016

THE ASSESSMENT OF PROPERTIES OF THE INFORMATION FLOW IN THE SUBSYSTEM «THE MEASURING SENSOR – THE PROCESSING ELEMENT»

Victor Nikolaev[1a], *Anton* Stuchkov[2]

[1]Southwest State University, 405000, Kursk, Russia
[2]Tomsk State University, 634050, Tomsk, Russia

Abstract. Analyzed the scope and types of automated information systems containing subsystems of the "measuring sensor – element processing". On the basis of the Queuing theory a model of evaluating the properties of information flow in the subsystem. The time spent by the sensor data generation and processing time of the output data array to the processing elements described by an arbitrary distribution law. Using little's formula the analytical solution allows to estimate the required level of performance of the subsystem. The research of the obtained solution in the range of parameters of interest. On the basis of General theoretical solutions identified four private practical case, different values of the coefficients of variation of information flows.

1 Introduction

The present stage of the development of the society is characterized by the increasing role of innovative technologies, such as the measuring and information technologies, representing the set of the elements of infrastructure, the subjects of information system which are carrying out measurement, collecting, formation, distribution and use of diverse information in various fields of activity [1–5].

The basic subsystems of the type "the measuring sensor – the processing element" are the most widespread in the automated information systems. Such subsystems are always a component of the automated systems in various fields of activity- science, education, industry, economy, medicine, social sphere [6–10].

In these conditions and the assessment of the properties of information flows in the automated system for the purpose of formation of requirements to parameters of standard elements is important and actual.

[a] Corresponding author: nikovic54@yandex.ru

© The Authors, published by EDP Sciences. This is an open access article distributed under the terms of the Creative Commons Attribution License 4.0 (http://creativecommons.org/licenses/by/4.0/).

2 The theoretical part

The analysis of literature shows that in the all-system plan, the basic subsystems of the type "the measuring sensor – the processing element" consist of the following basic elements [11, 12]:

1. The source (sensor) – the initial elements of receiving of data or information on a state or properties of process, technology or system.
2. The receiver (processing element) – the initial elements which are a part of instruments of measuring and monitoring of information and property space.

Between these elements there is an information flow, the assessment or measurement of properties of which allows to form the requirements to these elements, in particular to the performance level (the capacity of the system) [13, 14].

Generally the initial input data flow from the sensor or measuring system comes to the element or the subsystem of processing. The time spent by the sensor on formation of data is described by unrestricted distribution law $A_j^{(i)}(t)$ and has the intensity

$$\Lambda_j^{(i)} = \frac{dA_j^{(i)}(t)}{(1 - A_j^{(i)}(t))} \tag{1}$$

The time of processing of output data array on the element of processing is distributed under the law $B_j^{(i)}(t)$ with the intensity

$$M_j^{(i)} = \frac{dB_j^{(i)}(t)}{(1 - B_j^{(i)}(t))} \tag{2}$$

Let us also suppose that the variation coefficients of the input flow C_Λ and the time of data array staying at the processing stage C_M is less than unity. As the results of researches on the analysis of temporary characteristics of processing in various automated systems have shown [11–14], such supposition is quite justified.

In the proposed formulation it is necessary, having the set restriction for time of formation of data array T_3, to determine the necessary level of capacity of the subsystem of processing (element) of the information system. Within the stated above and the accepted suppositions the task can be converted to the analysis of the queueing model. In this case the subsystem of processing is considered as single-channel queueing system with unbounded queue to which the unrestricted recurrent flow of applications comes in with the intensity Λ and variation coefficient $0 \le C_\Lambda \le 1$. The intensity of processing also has the unrestricted distribution with average value M and the variation coefficient $0 \le C_M \le 1$. For this case it is not posible to obtain the exact analytical decision, however it is possible to estimate the average queue length, to bound it above and below [14]

$$\begin{aligned} &\frac{\rho^2(C_\Lambda^2 + C_M^2)}{2(1-\rho)} - \frac{\rho(1-C_\Lambda^2)}{2} + \rho \le N_c \le \\ &\le \frac{\rho^2(C_\Lambda^2 + C_M^2)}{2(1-\rho)} + \frac{(1-\rho)(1-C_\Lambda^2)}{2} + \rho \end{aligned} \tag{3}$$

where N_c is the average of the applications which are in the system; $\rho = \Lambda/M < 1$ is the coefficient of the system utilization; C_Λ is the variation coefficient of the input flow of applications; C_M is the processing time variation coefficient.

As at the stage of the macroanalysis we are interested only in the top level of the capacity of the information subsystem providing the set data processing time T_3, then taking the right part (3) and using the Little's equation $N=\Lambda T$, we obtain the top value of time of formation of data array in the system

$$T = \frac{\Lambda (C_\Lambda^2 + C_M^2)}{M^2 2(1-\frac{\Lambda}{M})} + \frac{(1-\frac{\Lambda}{M})(1-C_\Lambda^2)}{2\Lambda} + \frac{1}{M} \cdot \tag{4}$$

having made simple transformations with (4), we obtain the quadratic equation in M

$$M^2(2T\Lambda - 1 + C_\Lambda^2) - M\,2\Lambda(T\Lambda + C_\Lambda^2) + \Lambda^2(1 - C_M^2) = 0 \tag{5}$$

the solution of the equation (5) will be

$$M_{1,2} = \frac{\Lambda(T\Lambda + C_\Lambda^2) \pm \Lambda\sqrt{(T\Lambda + C_\Lambda^2)^2 - (2T\Lambda - 1 + C_\Lambda^2)(1 - C_M^2)}}{2T\Lambda - 1 + C_\Lambda^2} \tag{6}$$

Using the restriction for processing time $T \le T_3$ and taking into account that we need the largest value of M, we obtain

$$M \ge \Lambda \frac{(T_3\Lambda + C_\Lambda^2) + \sqrt{(T_3\Lambda + C_\Lambda^2)^2 - (2T\Lambda - 1 + C_\Lambda^2)(1 - C_M^2)}}{2T_3\Lambda - 1 + C_\Lambda^2} \tag{7}$$

at $0 \le C_\Lambda \le 1$;
$0 \le C_M \le 1$.

The expression (7) allows to determine the necessary capacity of the information subsystem as the stochastic macrosystem at the set parameters Λ, T_3, C_Λ, C_M.

3 The practical part

It is necessary to notice that the majority of the known laws by which the real distributions of an input flow and time of processing are approximated have variation coefficients less than unity. Therefore it is reasonable to select from (7) the special cases which are mostly often found in practice

1 $C_\Lambda = 0$; $C_M = 0$;
$$M \ge \Lambda. \tag{8}$$

2 $C_\Lambda = 1$; $C_M = 0$;
$$M \ge \Lambda \frac{T_3\Lambda + 1 + \sqrt{(T_3\Lambda)^2 + 1}}{2T_3\Lambda}. \tag{9}$$

3 $C_\Lambda = 0$; $C_M = 1$;
$$M \ge \frac{2T_3\Lambda}{2T_3\Lambda - 1}. \tag{10}$$

4 $C_\Lambda = 1$; $C_M = 1$;
$$M \ge \Lambda + \frac{1}{T_3} \tag{11}$$

As it is seen from (8), in case of the determined input flow and constant time of processing the necessary capacity of the information subsystem depends only on the intensity of an input flow and should exceed it. The general time of staying in the system is defined only by the intensity of service as in this case the queues are not formed at the system input.

In other cases the type of dependence is more complex and significantly depends on time T_3, designated for the processing. So at values $\Lambda \approx 1/T_3$ the level of intensity of processing of applications should only insignificantly exceed the value of intensity of the input flow. In case when $\Lambda << \frac{1}{T_3}$, the necessary system performance level should vastly exceed the level of intensity of the input flow. As it is shown by the analysis of expressions (8–11), the maximum value of level of capacity is obtained at the variation coefficients C_Λ=1, C_M=1 and exceeds the value of intensity of the input flow on the value inverse to the set processing time T_3. Therefore further it is reasonable to assess the top level of capacity of system at the information processing phase by the formula (11).

4 Conclusion

A mathematical model of mass service allowing to estimate the average queue length and load factor of the automated subsystem of the "sensor element" for various laws of distribution of internal information flows. The results of the study of this model allowed to identify the four most commonly used in practice, allows to substantiate the requirements of techno-economic characteristics of elements of the automated information-measuring systems at key phases of their life cycle.

Acknowledgments

The paper was written as part of the research project No. 8.2.31.2015, carried out with the support of the Program "Research Foundation of Tomsk State University named after D.I. Mendeleev" in 2015 – 2016, grant RFBR No. 16-29-04388/16.

References

[1] S. Staab, H. Stuckenschmidt, *Semantic Web and Peer-to-Peer, Decentralized Management and Exchange of Knowledge and Information* (Springer Science & Business Media, New York, 2006)
[2] A. J. Bostel, V.K. Sagar, Computing and Control Engineering Journal **7**, 169(1996)
[3] P. Viswanath, V. Anantharam, D. Tse, IEEE Transactions on Information Theory **4**, 1968 (1999)
[4] Y. Zhu, *Multivariable System Identification for Process Control* (Elsevier, The Netherlands, 2011)
[5] J.A. Hanley, A. Negassa, M.D.D.B. Edwardes, J.E. Forrester, American Journal of Epidemiology **157**, 364 (2003) doi: 10.1093/aje/kwf215
[6] Y. Yagi, J.R. Gilbertson, Journal of Telemedicine and Telecare **11**, 109(2005) doi:10.1258/1357633053688705
[7] A.A. Maslak, V.N. Nikolaev, Information and measuring and operating systems **11**, 46 (2013)
[8] O.G. Diev, K.S. Dyagterev, V.N. Nikolaev, High technologies **15**, 21 (2014)
[9] V.N. Nikolaev, A.A.B. Dafalla, L.V. Pleskanos, K.D.A. Kassim, Telecommunications and Radio Engineering **72**, 1219 (2013) doi: 10.1615/TelecomRadEng.v72.i13.80
[10] A.R. Ward, P.W. Glinnj, Queueing Systems **50,** 371 (2005) doi: 10.1007/s11134-005-3282-3
[11] R. Yuan, W. T. Strayer, *Virtual Private Networks: Technologies and Solutions* (Addison-Wesley, Strayer, 2001)
[12] P. G. Harrison, Queueing Systems **41,** 271 (2002) doi: 10.1023/A:1015842221910

[13] T. L. Saaty, *Elements of Queuing theory with applications* (McGraw-Hill, New York, 1961)
[14] D.G. Kendall, Ann. Math. Statist. **24**, 338 (1953)

INFORMATION AND ANALYTICAL WEB MAPPING SYSTEM "MAP OF HEALTH CARE OF KRASNOYARSK REGION"

Alexey Kadochnikov[1,*], *Alexey* Tokarev[1], *Larisa* Chausova[2]

[1]Institute of Computational Modeling SB RAS, 660036, Krasnoyarsk, Russia
[2]National Research Tomsk State University, 634050, Tomsk, Russia

Abstract. Approaches to the modern geoinformation web-systems development and technological features of software implementation are considered, their development trends are discussed. A brief description of the web 2.0 technologies main components is given, the use of which provides the current level of web mapping. Goals and objectives, the main purpose of the formed region's health management informational support system based on mapping web-interface are formulated. A general characteristic of the created software is given, some aspects of its implementation and program architecture are discussed. User interface construction features are discussed using several examples.

1 Introduction

Geographic Information System is a useful tool that aids and assists in health research, health education, planning, monitoring and evaluation of health programmes that are meant to control and eradicate certain life threatening diseases and epidemics [1]. To a large extent, recent years advances in the geographic information systems (GIS) software are related to web technologies. Due to them, an opportunity of access to high resolution satellite images and other data of significant volume appeared, which is displayed on the map and available via web services – city streets photos, operative situation on the roads (traffic jams), weather, emergencies, fires, etc [2]. We can also say that today a new generation of geographic information systems is created using the paradigm of "Web 2.0" technology in addition to traditional mapping applications, whereby more and more significant role is played by end-users, who cooperatively create and update the content of these systems through social networking mechanisms. The considered trends became especially intense in recent years with the growing popularity of mobile devices with geolocation capabilities (smartphones and tablets). At the same time, it should be noted that this area is still being formed and many aspects are not defined and settled yet. A search for optimal solutions goes on in the field of construction and architecture ideology for the distributed GIS software, data access mechanisms, their respective structure and organization, program interface development and data exchange standards, user interface

* Corresponding author: scorant@icm.krasn.ru

© The Authors, published by EDP Sciences. This is an open access article distributed under the terms of the Creative Commons Attribution License 4.0 (http://creativecommons.org/licenses/by/4.0/).

[3, 4]. The existing software analysis in the considered area shows that the geoinformation web-system can be implemented in various ways today. Currently, numerous software developments appear – from simple visualization means for the prepared maps in the browser to complex systems of corporate geospatial information processing in distributed computing environment [5].

In this context, attention should be paid to the fact that a modern trend is to create and implement applied geoinformation web-systems focused on information support for dealing with problems in various areas – governance, environment and natural resources, transport and communications, education, etc. Such a system designed to provide health care control in the region will be considered below [6]. Similar decisions are made today in different countries [7].

2 Mapping web-systems technologies

Information-analytical mapping web-system "Map of health care of Krasnoyarsk region" considered in this article is based on so-called "Web 2.0" technology. This term describes a stage in the Internet development, which involves the use of the following components on the contrary to the old "Web 1.0" technology:

- Web services – platform-independent applications, which are accessed via HTTP protocol, and the data is exchanged in XML or JSON format. As a result, the produced software can use the existing web services instead of performing the required functions again.
- AJAX – an approach to interface building, under which a web page downloads the required data asynchronously without rebooting.
- Web syndication – a mechanism of information dissemination to different pages or websites, usually with the use of RSS.
- Mash-up – a service that completely or partially uses other services as sources of information providing the user with new functionality.
- Tags – keywords that describe the analyzed object or correlate it to a certain category. They are used in blogs.
- Socialization – community creating tools: personal site settings, personal files and photos collections, blogs, etc.

Describing this class of geoinformation and mapping web-systems and services (web-GIS), it should be noted that spatial data processing, distribution and exchange capacity offered by modern web-mapping means generally correspond to the above list of "Web 2.0" features – these are interactive functionally-rich web applications with the content dynamically formed by users. There are a number of such kind systems nowadays [8]. They are adequate to technical possibilities of the Internet - availability, access speed [9].

The use of "Web 2.0" concept in the web-GIS implementation provides an unprecedented level of geospatial technologies democratization. Previously, for example, the formation of cartographic GIS-project data array was performed only by professionals using complex applied systems such as ArcGIS. Today, users upload their own information about road accidents to car navigation systems, draw their maps (OpenStreetMap, Yandex people's map and so on), use mapping services (API Yandex.Maps, Google Maps, etc.) to create personal mapping web interfaces – from a simple web page with an interactive map "How to find us?" to a complex portal with multicriteria spatial queries. A powerful impetus to the web-GIS development and their widespread use was the recent emergence and rapid development of various personal mobile devices with built-in means of satellite positioning and Internet access – netbooks, smartphones, tablets in addition to laptops/ultrabooks [10].

For some functionally rich web-GIS, the term "geoportal" is commonly used. The geoportal is a type of web portal used to access and search for geographic (geospatial) information and related geographic web services (access, editing, and so on). It is one of the spatial data infrastructure key components. The geoportal functionality is closely associated with metadata management subsystem, means of spatial analysis, interactive visualization. Usually, its systemic basis is traditional software and technologies for web content management systems (CMS) [11].

3 Cartographic web-system for health management information support

The classical scheme of data publication in the Internet involves the use of additional software tools for creating graphic images and/or html-reports. A disadvantage of this approach is inability to edit the published data directly, increase of labor costs and publication time, dependence of users on third-party software. In the case of geospatial data web publishing, the situation becomes even more complicated [12, 13].

In the process of web-system "Map of health care of Krasnoyarsk region" designing and developing, a series of practical experiments was conducted, in which different versions of software for publishing geodata on medical institutions were tested. In particular, tools for the formation of interactive thematic mapping information were created without the involvement of specialized geoinformation system software to solve this problem. Instead of professional GIS, data preparing and modifying was used directly through the management subsystem web interface, which is successfully used in today's popular web content management systems [14, 15].

As a result, we have got an efficient and flexible solution that does not require involvement of commercial GIS, which also can be maximally adapted to the problem being solved [16]. The application provides possibilities of distributed multi-user editing of spatial data display style, its dynamic changes based on the values of the corresponding database tables.

In technological terms, the proposed web interface of geospatial data editing is an HTML/JavaScript script, which contacts the PHP handler for additional data and graphic images through AJAX queries. In turn, the PHP handler performs the server side operations, accesses the database and mapping servers and generates data arrays and images, which are then returned to the HTML/JavaScript interface as a response to the asynchronous query. The end user does not see these operations perceiving them as visual editing of the graphic object in the system editor window [17].

The cartographic web interface uses a number of cartographic information forming methods for the user. The first method is map displaying with the use of fragments (tiles). This technology is used by such resources as Google Maps (http://maps.google.com/), Yandex Maps (http://maps.yandex.ru/), Virtual Earth (http://www.bing.com/maps/) and others [18]. The main advantage of this method is the speed of visual information receiving by the user and low server loading during static information displaying. The process of map formation in the client computer consists of several steps using additional program streams, caching, fragments downloading queue, etc. Under this way of map displaying to the user, the map constructing process allows to optimize the downloading process, reduce the web browser load and distribute it more evenly over time. However, when displaying the changing thematic data, this method reduces the speed of the user access to spatial data and increases the server load. To solve this problem, another method of information displaying is used – a single bitmap image is generated by the user's query. Depending on the type of information provided to the user, the system interface uses a combination of these two methods (Figure 1).

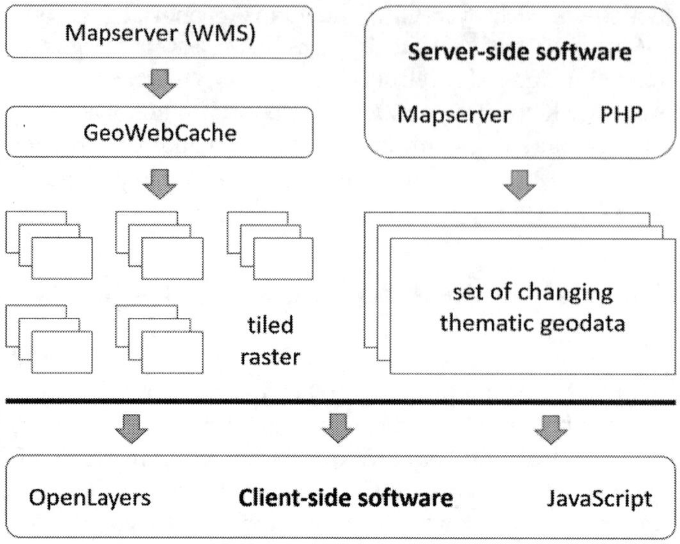

Figure 1. Map formation technology in the client web application.

4 The territorial health resources database

Formation of information-analytical software for the Krasnoyarsk region health management began about 20 years ago. Krasnoyarsk Regional Medical Information and Analytical Center (KMIAC) was established in 2002 performing the functions of integrator in the field of information technologies for the health care system of Krasnoyarsk region; a number of key information systems was developed and introduced into practice.

Within the gradual creation and development of the Krasnoyarsk region system of health and medical insurance, a geoinformation web-system – territorial database of health care resources in the region – is developed since 2008. Its main task is development of an optimal perspective scheme of access to health care for the region residents. Main purposes of the system are:

- professionals and public authorities informing;
- information support for local governments;
- health facilities professionals informing;
- information support for the regional ministry of health employees;
- informing the public about the health care resources location.

This system is a tool to improve the health service quality [19]. The territorial health resources database contains information on all municipal and regional health facilities (over 200 items), health care objects (over 1300 items), about 130 performance indicators for the health care facilities, demographic data, morbidity, a medium-term development plan for the health care institutions network. There are a number of similar system around the world [20–22].

Today, the considered geoinformation web-system is one of the common information space elements of the Krasnoyarsk region health system. All data presented on the map are generated automatically as a result of medical information centralized repository data aggregation. Integrated data can be used to solve various analytical tasks [23].

The system works steadily for more than five years; its maintenance and regular data updating is carried out by KMIAC.

In particular, it is possible to analyze changes in the departmental network of medical institutions – different icons on the map correspond to converted, liquidated and created obstetric centers, clinics and other medical facilities. Any departmental statistics can be presented as a thematic map with automatically generated descriptions (color scales) in the context of municipalities. Data filtering and exporting are also provided.

5 Conclusions

Experience of the developed information-analytical mapping web-system "Krasnoyarsk region health map" practical operation proved the effectiveness of proposed solutions. Professionals and interested users have an opportunity of operative access to the relevant data on the regional health care system. The branch managers and system analysts got a modern tool for the situation analysis and management decisions preparation. Information about the health care resources location is interesting for a wide range of users.

Acknowledgments

The work was supported by grant no. 16-47-240517 from Russian Foundation for Basic Research and Krasnoyarsk Regional Fund of Support of Scientific and Scientific-technical activities.

References

[1] R. Kandwal, P.K. Garg, R.D. Garg, Journal of Biomedical Informatics **42**, 748 (2009) doi: 10.1016/j.jbi.2009.04.008
[2] M. Santini, G. Caccamo, A. Laurenti, S. Noce, R. Valentini, Applied Geography **30**, 394 (2010) doi: 10.1016/j.apgeog.2009.11.003
[3] L.L. Ramírez-Ramírez, Y. R. Gel, M. Thompson, E. de Villa, M. McPherson, Computer Methods and Programs in Biomedicine **110**, 455 (2013) doi: 10.1016/j.cmpb.2013.01.007
[4] A. Koenig, E. Samarasundera, T. Cheng, Public Health **125**, 554 (2011) doi: 10.1016/j.puhe.2011.02.011
[5] O.E. Yakubailik, V.G. Popov, Computational Technologies **6**, 116 (2009)
[6] O.E. Yakubailik, A.A. Kadochnikov, V.G. Popov, A.A. Tokarev, Vestnik of SibGAU **25**, 61 (2009)
[7] Su-su Lei, Yong-tao Gao and Dan-G. Pan, MATEC Web of Conferences **31**, 11003 (2015) doi: 10.1051/matecconf/20153111003
[8] D. Wang, W.C. Ge and Y. G. Wang, MATEC Web of Conferences **44**, 02011 (2016) doi: 10.1051/matecconf/20164402011
[9] A.A. Kadochnikov, Computational Technologies **2**, 70(2007)
[10] Xue Ping Li, Yun An Li, Energy Procedia **16**, 1174 (2012)
[11] A.A. Kadochnikov, V.G. Popov, A.A. Tokarev, O.E. Yakubailik, Journal of Siberian Federal University Engineering & Technologies **4**, 377 (2008)
[12] E. M. Delmelle, H. Zhu, W. Tang, I. Casas, Applied Geography **52**, 144 (2014) doi: 10.1016/j.apgeog.2014.05.007
[13] D. Gkatzoflias, G. Mellios, Z. Samaras, Computers & Geosciences **52**, 21 (2013) doi: 10.1016/j.cageo.2012.10.011
[14] A.G. Matveev, O.E. Yakubailik, Vestnik of SibGAU **2**, 48 (2012)
[15] A. Marano, M. Di Nicolantonio, Procedia Manufacturing **3**, 272 (2015)
[16] O. Yakubailik, A. Kadochnikov, A. Tokarev, 15th International Multidisciplinary Scientific GeoConference (SGEM 2015) **1**, 487 (2015)
[17] A.G. MatveevF, O.E. Yakubailik, Fundamental research **10-15**, 3358 (2013)

[18] T. Kistemann, C. Höser, H. Voigt, Current Opinion in Environmental Sustainability **3**, 506 (2011) doi: 10.1016/j.cosust.2011.10.007
[19] D. A. Lovett, A. J. Poots, Jake T.C. Clements, Stuart A. Green, E. Samarasundera, Derek Bell, Spatial and Spatio-temporal Epidemiology **10**, 67 (2014)
[20] L. Susi, A. K. Mascarenhas, The Journal of the American Dental Association **133**, 636 (2002)
[21] W. Odero, J. Rotich, C.T. Yiannoutsos, T. Ouna, W. M. Tierney, Journal of Biomedical Informatics **40**, 390 (2007) doi: 10.1016/j.jbi.2006.12.007
[22] J.K. Rodgers, K.J. Bergmann, V.L. Salak, D.T. Lackland, V.K. Hinson, Annals of Epidemiology **17**, 723 (2007)
[23] A. Rachmad and D. R. Anamisa, MATEC Web of Conferences **58**, 03005 (2016) doi: 10.1051/matecconf/20165803005

INFORMATION AND COMMUNICATION TECHNOLOGIES IN ENGINEERING EDUCATION

Xavier Maldague[1], *Marina* Kuimova[2*], *Douglas* Burleigh[3], *Sofia* Skvortsova[2]

[1]University Laval, 305 Rue de l'Université, Quebec City, Quebec, Canada
[2]National Research Tomsk Polytechnic University, 634050, Tomsk, Russia
[3]La Jolla Cove Consulting, San Diego CA 92117, USA

Abstract. In the emerging digital era it is difficult to train highly-skilled, competent specialists without the use of information and communication technology (ICT). The use of ICT in education increases the motivation to learn, stimulates cognitive activity and independent work, facilitates information exchange, enables interactive communication between teachers and students, and improves learning outcomes. This paper reviews the literature regarding the use of ICTs in education, explores their advantages and challenges, and surveys first-year students at the Institute of Non-Destructive Testing, National Research Tomsk Polytechnic University to determine their attitude toward ICT in foreign language learning.

1 Introduction

Due to global competition, a university graduate of today must not only possess the specific knowledge and skills needed for his chosen profession, but he must be capable of critical and innovative thinking, and he must be familiar with current software products. Universities strive to use creative and innovative teaching approaches to teach both theoretical and practical knowledge, maximize students' intellectual potential, and train specialists to be able to perform professional tasks and compete in the labour market. In addition to traditional teaching methods and approaches, information and communication technology (ICT) has become a critical tool for engineering training. ICTs include hardware and software, Internet resources, communication technologies, multimedia tools, etc. They expand the learning environment and the range of learning experiences, promote active learning, present the use of the Internet as a critical information resource and offer a wealth of learning online services, courses, and resources.

2 Discussion and research

In an era of global communication, expansion of political, economic and cultural activities, higher education is necessary to meet the needs of the society and introduce innovative

* Corresponding author: mkuimova@mail.ru

© The Authors, published by EDP Sciences. This is an open access article distributed under the terms of the Creative Commons Attribution License 4.0 (http://creativecommons.org/licenses/by/4.0/).

methods and approaches. The inclusion of ICT in higher education creates an open, authentic learning environment, extends educational opportunities, teaches students how to deal with large amounts of information, and provides more autonomy over the learning process.

Computer-based education aims to facilitate the search for educational resources. ICT helps students find educational materials via the Internet and helps teachers create more "learner-centric" instructional activities. In addition, ITCs:

- develop curricula content and provide authentic, relevant information for learning [1–3];
- promote an open, flexible teaching/learning environment [4, 5];
- motivate learning [6];
- enhance students' cognitive performance [7];
- encourage critical thinking and creativity [8];
- improve academic performance and outcomes [9, 10];
- motivate research activity and stimulate open-ended discovery [11];
- open myriad possibilities for interdisciplinary learning [12];
- further exchange of information and intercultural communication [3, 13, 14].

ICT provides many opportunities to use various Internet technologies to solve many educational and pedagogical problems. ICT offers opportunities to:

- work with databases, electronic encyclopedias and e-learning modules;
- use online materials and interactive Web-based textbooks;
- participate in scientific research, creative projects, video conferences and joint activities in the virtual environment;
- communicate quickly at no cost with other students or speakers of the target language around the world;
- create personal Webpages, blogs, interact through social media and mind maps (publish and distribute learner's multimedia information for an international audience).

The use of information technology improves all kinds of cognitive motives, including interest in knowledge, and the contents and process of learning.

ICT involves students in the learning process, promotes the evaluation of their abilities and an increase in mental activity.

ICT enriches the curriculum, complements students' learning experience and makes teaching and learning more interesting and attractive. ICT is widely used at lectures, practical classes, independent work, various types of assessment (formative and summative), work on projects, research and creative activities. It is used as a tool to help accomplish complex tasks and support students' performance of authentic learning tasks. ICT helps students:

- gather data;
- perform on-line exercises and run simulations;
- work at their own pace;
- develop critical and creative abilities (making and editing videos, creating web pages, writing wikis, making presentations, etc.);
- monitor learning process and perform self-assessment tasks;
- take responsibility for their learning and become independent, self-directed learners prepared for lifelong on-going learning.

ICT permits access to online course materials or virtual classrooms and enables blended and distance learning [15]. Even with all it's benefits, ICT is not a substitute for a teacher. Traditional teaching methods and approaches must still be used in conjunction with ICT.

Teachers serve as facilitators; they must be aware of a wide variety of available materials so they can recommend multimedia and other resources effectively, and know how to best deploy them to meet the needs of students [16–20].

The implementation of ITC in foreign language classes provides access to materials and information in the language (one-way communication) and communication with the teacher and other students and native speakers (two-way communication). In addition, it helps:

- promote motivation to foreign language learning;
- provide access to current, up-to-date materials from the country or countries where the target language is spoken;
- offer individual learning pathways and adapt the teaching material to learners' needs;
- provide a means to focus on a specific aspect of language (pronunciation, grammar, vocabulary, etc.);
- provide additional possibilities for independent work aimed at the development of language skills;
- provide a fast feedback to students' answers;
- translate words and phrases from one language into another (electronic translators and dictionaries);
- increase cultural awareness (the Internet affords the means to find out more about history, culture and traditions of countries throughout the world);
- train for professional cross-cultural communication;
- provide access to on-line courses, materials and tests [21, 22].

The use of ICT in teaching is not problem-free. There are challenges and barriers preventing teachers from using ICT:

- inability of universities to provide the physical space for the use of electronic resources;
- poor quality ICT equipment;
- poor quality (slow or intermittent) Internet access;
- teachers may have insufficient computer skills or may lack the ability to use the available technology adequately;
- time-consuming preparation of digital materials and a shortage of class time;
- reduction in personal contact with students;
- students' inability to study independently [23].

Thus, the successful integration of ICT in language learning results in:

- a flexible approach to learning using both computer-based and traditional teaching methods in the classroom;
- the teacher's continuous adjustment to the varying demands of professional competence and knowledge;
- careful and critical selection of Internet resources;
- adaptation to the changing expectations of students;
- teacher's encouragement and support of students.

In our paper we surveyed 43 first-year students of the Institute of Non-Destructive Testing, National Research Tomsk Polytechnic University to determine their attitude towards the integration of ICT in foreign language teaching. The students were given the following questionnaire and asked to provide their views (Table 1).

Table 1. Questionnaire on students' attitude towards ICT in foreign language learning.

№	Question	Agree (%)	Disagree (%)	Don't know (%)
1	I am comfortable using technology in foreign language classes	74.4	18.6	7
2	I think that getting information from ICT is better than using printed materials/textbooks	18.6	60.5	20.9
3	I think that ICT allows me to learn more in the time I have for studies	39.5	25.6	34.9
4	I think that ICT improves my learning	76.8	20.9	2.3
5	I think that the use of ICT allows me to take greater control of my language learning	65.1	20.9	14
6	I use ICT to create my own digital learning resources (Web-pages, blogs, mind maps, etc.)	18.6	51.2	30.2
7	I use ICT to communicate with international peers on topics of personal interest	53.5	41.9	4.6
8	I use ICT to communicate with international peers on education related matters	39.5	55.8	4.7
9	ICT (e.g. the Internet) permits access to current learning materials	79.1	11.6	9.3
10	I would like to use ICT more in language learning	37.2	46.5	16.3

Thus, the survey shows that the students have a positive attitude towards the use of ICT as a tool to enhance language learning. However, they prefer to get information from printed materials rather than from ICT. In addition, students indicated that they find webinars (9.3%) and on-line courses (14%) less interesting and productive than traditional lessons (76.7%) due to the lack of face-to-face contact with their teachers. In fact, such findings are confirmed even in engineering classes where student attention «gets reduced» rapidly if teachers rely too heavily on ICT. So far, a balanced mixed of active teacher presence and ICT was found the best and only «teacher's own experience» combined to her/his «feeling of the class» has been the key of successful ICT implementations [24].

The survey shows that students prefer to communicate with international peers on personal interests rather than on educational issues. The students don't wish to use more ICT in foreign language lessons and this indicates that TPU teachers make an effective, balanced use of ICT to create educational experiences which enrich and extend students' academic abilities and needs

3 Conclusion

Today the ability to independently find, analyze, evaluate, interpret and synthesize information is very important in the educational process. ICT helps to increase the quality of education and has become an indispensable part of the contemporary learning environment.

The integration of ICT into the educational process has great potential for both teachers and students. It helps teachers improve and develop their teaching methods. ICT diversifies students' activities, improves their attention and cognitive motives, increases creative potential and knowledge, expands self-education and self-development, and increases self-regulated learning and independent work. ICT helps bring foreign language and culture to life and make it more tangible. It is worth noting that ICT alone cannot provide a comprehensive basis for language learning. It should be accompanied by teacher's guidelines, traditional teaching methods, conventional textbooks, workbooks, and audio and video materials.

References

[1] B. Shu'aibu, M. Usman, M.S. Saud, Y. Kamin, KJSS **36**, 591 (2015)
[2] D.C.-T. Lo, K. Qian, W. Chen, Proc. FIE, 7344089 (2015) doi: 10.1109/FIE.2015.7344089
[3] P. Resta, T. Laferrière, Educ Inform Tech. **20**, 743 (2015) doi: 10.1007/s10639-015-9419-z
[4] A. Lillo-Bañuls, J.F. Perles-Ribes, R. Fuentes, JTTT **16**, 81 (2016) doi: 10.1080/15313220.2015.1118367
[5] M.V. Kuimova, O.D. Zvekov, iJET **11**, 4 (2016) doi: 10.3991/ijet.v11i04.5430
[6] N. Aghaee, C. Keller, Comput Educ. **94**, 276 (2016) doi: 10.1016/j.compedu.2015.11.006
[7] A. Chaldogeridis, Proceedings of 2015 International Conference on Interactive Mobile Communication Technologies and Learning, IMCL 2015, 7359635 (2015) doi: 10.1109/IMCTL.2015.7359635
[8] G. McMahon, Educ Technol Soc. **12**, 269 (2009)
[9] G.A. Kruchinina, L.I. Tararina, E.E. Sokolova, E.V. Limarova, I.V. Muskhanova, S.M. Arsaliyev, F.U. Bazayeva, N.P. Tagirova, IRMM **6**, 104 (2016)
[10] C.P. Van der Westhuizen, C. Nel, B.W. Richter, Educ Technol Soc. **15**, 190 (2012)
[11] D.K. Saini, L.S. Prakash, H. Gaur, Advances in Intelligent Systems and Computing **409**, 335 (2016) doi: 10.1007/978-981-10-0135-2_33
[12] D. Ktoridou, E. Doukanari, IMCL, 7 (2015) doi: 10.1109/IMCTL.2015.7359605
[13] S. Ribeiro, Lang Intercult Comm **16**, 69 (2016) doi: 10.1080/14708477.2015.1113752
[14] Y. Jiugen, X. Ruonan, C. Li, Proceedings - 2015 7th International Conference on Information Technology in Medicine and Education, ITME 2015, 647 (2015) doi: 10.1109/ITME.2015.85
[15] Y. Pan, MATEC Web of Conferences **44**, 01075 (2016) doi: 10.1051/matecconf/20164401075
[16] M. Kuimova, D. Burleigh, A. Trofimova, MATEC Web of Conferences **48**, 06006 (2016) doi: 10.1051/matecconf/20164806006
[17] A. Uvarov, I. Lezhnina, K. Overchuk, A. Starchak, Proceedings of 2014 International Conference on Mechanical Engineering, Automation and Control Systems, MEACS 2014, 6986935 (2014) doi: 10.1109/MEACS.2014.6986935
[18] N.M. Natalinova, O.V. Galtseva, E.A. Moldovanova, 2016 Third International Conference on Electrical, Electronics, Computer Engineering and their Applications (EECEA), 52 (2016) doi: 10.1109/EECEA.2016.7470765

[19] A.Y. Petrova, O.N. Chaikovskaya, I.V. Plotnikova, Technical Physics **60**, 592 (2015) doi: 10.1134/S1063784215040222
[20] M. Kuimova, A. Kiyanitsyna, A. Truntyagin, SHS Web of Conferences **28**, 01129 (2016) doi:10.1051/shsconf/20162801129
[21] E. Dm. Griaznova, PSR **16**, 4 (2014)
[22] D. Pusca, D.O. Northwood, WTE&TE **14**, 77 (2016)
[23] H. Salehi, Z. Salehi, IJEEEE **2**, 4 (2012)
[24] X. Maldague, ECE dept., U. Laval, private communication (2016)

MATEC Web of Conferences **79**, 01045 (2016) DOI: 10.1051/matecconf/20167901045
IME&T 2016

MODEL OF ARTIFICIAL INTELLIGENCE MEDIUM AND ITS USE IN TREATING DISORDERS OF THE NERVOUS SYSTEM

Vladimir Shumilov[a]

National Research Tomsk State University, 634050, Tomsk, Russia

Abstract. The work is devoted to modeling of the nervous system, the brain. The article considers the mechanism of formation of event traces fixed in the brain in the form of connections between neurons and the effect of these traces on the later passage of signals through the brain. That is, the influence of experience (traces) of previous events on the organism's reaction by forecasting upcoming events. This «forecast» and anticipatory avoidance of dangers based on traces of previous events makes the nervous system, the brain useful for the organism, for its survival and expansion. The author proposes to use a computational model of the brain to study disorders of the nervous system, brain, and make recommendations for their prevention.

1 Introduction

Currently, the number of patients suffering from different diseases of the nervous system increases for a variety of reasons. Therefore, study of diseases of the human nervous system becomes more urgent [1–4]. However, scientists study the problem mainly by means of statistical methods, as well as post-mortem examination of brain pathologies, blood vessels and other organs of the deceased person. Now, there is a great number of different models of nervous systems. Scientists have made careful studies of diverse aspects of the structure and function of nervous systems of various biological organisms. Biochemical processes that provide function of the nervous system are of special attention. Researches investigate behavior of neurons – the main operating units of the nervous system, formation of connections between them, functions of various accessory tissues found in the brain. Neurogenesis - formation of new neurons in the human brain and in the brain of mature animals has been under active study lately, though until recently there has been an opinion that nerve cells of adults do not form, but only die. There is a great number of printed works devoted to various aspects of the processes occurring in the human body. Scientists have achieved great progress in research of various processes in the organism, in the brain. However, the logic of function of the nervous system is lost in numerous studied details and remains unclear. In our opinion, lack of understanding of mechanisms of the nervous system function in general and the brain in particular, is a big obstacle for treating diseases of the nervous system.

[a] Corresponding author: vnshumilov@rambler.ru

© The Authors, published by EDP Sciences. This is an open access article distributed under the terms of the Creative Commons Attribution License 4.0 (http://creativecommons.org/licenses/by/4.0/).

We propose modeling method for study of diseases of the nervous system, which may significantly influence prevention, medical treatment, or at least retardation of development of these diseases. We see this way in use of the computational and electronic models of the nervous system. Within this approach, we represent the brain in the form of a huge electric circuit composed of active units - neural-like units and connections between them.

We have previously proposed principles of function of biological nervous systems and, in particular, the human brain [5,6]. These principles allow us to provide theoretical explanations of all or almost all aspects of operation of the biological nervous system and the brain. We should note that other researchers have also considered the principles and models of the nervous system, similar to those proposed by us ([7–9]). However, they have not suggested any mechanisms of self-formation of connections between neurons, mechanisms of forgetting and retraining. Therefore, they have not implemented their models either as any kind of software or as an electronic device.

Many processes in biological organisms provide function of the brain: mechanical (body movements and mass transfer by blood and lymph), biochemical and electrical. This happens because resources of the organism are limited, so it has to be economical – the same mechanisms are simultaneously involved in various processes. Each subsystem of the body has to perform a variety of functions through interactions with other subsystems in conditions of tight requirement of minimization of resources usage. Thus, blood supplies organs with nutrient, takes away organs' metabolic waste, heats and takes away the heat, protects the organism from foreign bodies, and moves various stimulants within the organism. The organism has to stand cold and hunger, overeating and overheating, poisoning and dehydration, physical overload and microorganisms' attacks in conditions of tight restriction on the organism's resources. Hence, the process of resources saving leads to combing of functions and incredible complication. It would seem that in order to construct a model of the brain it is necessary to consider all various processes running in the brain.

In fact, dealing with any problem, you can ignore many factors and not consider them for the reason that they have no or little influence on the parameters studied in the task. If, however, you cannot ignore the influence of some factors, you can consider the interaction of these factors by means of introduction of processes of equivalent substitutions, which replace the transitional conversion of considered parameters. The method of replacing substitutions is used to solve various problems. It can be used as well for considering information processes in the brain, in the nervous system. Almost any modeling is based on the use of substitutions.

2 Characteristics of the brain

Let us consider our proposed model of the processes occurring in the nervous system and in the brain. For more than 60 years, scientists have undertaken efforts to create artificial intelligence (AI). However, despite the continuing efforts of the specialists and success in particular cases, AI of full value has not yet been created. Despite the low performance of brain's active elements - neurons (the response time of a single neuron is about 0.1 sec.), parallelism of brain structures allows the brain to solve the problem of recognition, to respond to signals from the environment, to manage rapid body movements (running, jumping) for seconds, or even fractions of a second.

Principles of AI's algorithmic and hardware implementation are radically different from the principles of function of the biological brain. Therefore, even if within the current mainstream the AI is created, its study will have low effect for understanding function of the brain, areas of which obviously operate in parallel. Both human and animal brain does not require a big team of highly qualified specialists to provide its activity, but implementation of AI requires it. Moreover, efforts to build AI by traditional ways seem uneconomic to us. Indeed, the possible volume of information in the brain is large (at least 300 terabytes [5]), so processing of information contained in the human brain and its placement in the device with AI, will require thousands of man- years of work of highly qualified specialists.

Therefore, we offer a completely different approach to building a model of the nervous system, based on understanding of mechanisms of function of the nervous system and brain. However, it is necessary to understand the principles of operation of these mechanisms. It is necessary to understand how the nervous system has appeared by a natural way without intervention of supernatural forces in full compliance with the objective laws of nature, and how it has passed the distance from the elementary nervous system to the brain of highly developed animals and humans. In addition, it is necessary to understand how the intellect appears in the brain formed naturally during millions of years of evolution.

Today, the continuing active search of ways for AI creation shows, that there is no common view on the logic of function of the nervous system (NS), on the logic of interaction of neurons (the simplest organisms), and even on the logic of function of the human brain. So far, the brain or the NS looks like a black box, despite the fact that the anatomical structure and biophysical processes in the NS and in the brain are well known [7,8]. Therefore, we will try to find the principles of function of the brain by means of considering of the pre-history of occurrence of the nervous system.

Electrical signals generated by the active elements - neurons when they are excited, are spread in the brain. Each neuron has a great number of inputs – up to 10 000. Neuron is excited when the sum of potentials on its all inputs, that is, the signal on its total input, which acts on the neuron' nucleus, exceeds a threshold level. By means of spending of accumulated nutrients, an excited neuron makes a signal of certain energy (standard capacity and duration), which allows signals to propagate through the branching structures of the brain without rapid attenuation and without increase of the number of signals propagating through the structure of the brain. Too rapid attenuation will not allow signals to reach the addressee - the motor neurons at the output of the NS. Increasing number of signals as they move (with the average rate of reproduction signals greater than one) would cause self-excitation of the brain. Besides other facts, this means that the signals do not occur in the brain, they are only transmitted (transferred, transformed) by brain structures, but are not generated, initiated. The nerve impulses appear only at the NS input under the influence on input brain circuits - receptors of environmental factors (stimuli). Brain neurons can be excited also by direct influence of electric potential.

Muscles or endocrine glands activated by the NS or the brain determine the final reaction of the organism to the received input signals (stimuli). This fact means that the reaction of the organism's NS to a stimulus depends on the path of stimulus signal spread from the input to output of the NS. Therefore, the reaction of the organism depends on the configuration of input signals and structure of the NS (possible signals paths) at the time of signal passing through the NS. On the other hand, it is known that the brain structure changes over time. The number of connections between neurons increases especially rapidly during intensive training [10, 11]. This means that new signal paths are formed in the NS (in the brain) during the training.

So, unlike existing today control units (where information is recorded to a special memory block, read and interpreted by another special block), mechanisms of fixation, storage and interpretation of information (signal propagation) in the biological NS are combined functionally and are not localized in any certain isolated areas of the brain; they are distributed throughout the volume of the brain! Information content of the brain is determined by the structure of the brain, that is, by connections between neurons - the active elements of the NS (between which neurons does communication exist and what is their weight). The number of connections in the brain areas responsible for a certain type of activity increases rapidly during training, acquisition of experience in this type of activity.

What is the mechanism of formation of connections between neurons conducting signals in the NS, in the brain? What is the principle of formation of connections - memory elements? What are the mechanisms of propagation of signals received by the brain, by its structures? Why and how does propagation of signals through the brain structures provide an optimal behavior of the organism adequate to circumstances? Obviously, this mechanism must be extremely simple, provided by a natural process. Simply, there is nowhere to take complex mechanism from. This can be particularly well illustrated by the example of the emerging body, after a few cycles of self-copying, when only a few neurons have appeared in the organism, and there is still no connection between them.

We think that productive approach is based on consideration of the evolution from the simplest organisms up to human beings (and forward?).

3 Functioning logic elements of the brain

As answers to these questions, in [5] we have proposed provisions about principles of function of the brain (NS), which differ from the traditional ones. In accordance with these provisions and the patent based on them [6], the team of scientists of Tomsk State University (TSU) has made research «Development of electronic circuits for creation of a working model of artificial neural-like element». During this work, the specialists have developed the electrical circuits of the model, manufactured and tested the electronic model of an elementary nervous system, consisting of three active neural-like elements (NE), paired connections between them – both existing and potential connections, which do not yet exist in its original state, but which can be formed under appropriate circumstances. The model implemented in the form of an electronic device, consisting of even a small amount of NE, has confirmed the validity of the proposed principles. It has confirmed the possibility of self-creation of connections – event traces (we can say - training and retraining), formation of new stronger connections under the change of circumstances.

The essence of principles of function of the elementary NS (stated in [5]) and the NS's usefulness for the organism can be displayed in several diagrams illustrating the formation of the event traces in the NS, and the impact of these traces on the formation of the NS's reaction to the events that occur later. Electronic model of the elementary biological nervous system based on artificial NE, which was created as a result of research in TSU, has the same operation principles.

Let us consider the diagram of states of two neurons («excited»/ «not excited») when exposed by initially existing input factors (similar to unconditioned reflexes), which are important for the organism controlled by these neurons. The diagram depicts time t (Figure 1).

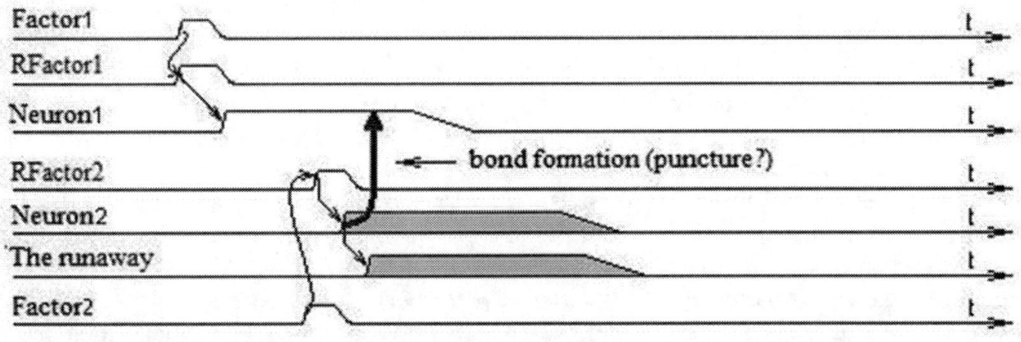

Figure 1. Excitation of neurons 1, 2 as a result of irritation of factors 1, 2 and formation of connection between the excited neurons.

Let Factor 1 precedes Factor 2, and Factor 2 is more dangerous for the organism than Factor 1. When Factors (stimuli) appear, they excite relevant neurons through separate inputs. Besides, appearance of Factor 2 causes unconditional response - the body moves away from Factor 2 so that Factor 2 will stop to influence the organism (its receptors). Neurons may have many inputs, through which can be excited.

What will happen with the organism later, if in the described episode the organism was able to avoid the lethal effects of Factor 2? What will happen if in this episode the event trace was formed, which means approximately simultaneous excitation of neurons N1 and N2? This trace *is a logical connection* between neurons 1 and 2, which is formed *at a moment of simultaneous excitation of neuron.* This connection is realized in real organisms biologically. In principle, in various models it may be formed in any possible way, for example, in mechanical, pneumatic, hydraulic, chemical or galvanic way through potentially possible new inputs. Moreover, at least two options of such connection are possible:

a) output of Neuron 1 (N1) to input of Neuron 2 (N2);

b) output of Neuron 2 (N2) to input of Neuron 1 (N1).

The probabilities of these options are determined by time correlations of stimuli F1, F2 (the order in which they have followed), and the relative position of inputs and outputs of neural-like elements. Let us consider the situation when option a) was realized.

Then, the next time in a similar situation, the events will develop in the following way. In case of excitation of one of already interconnected neurons (say, only N1 by factor F1), immediately after excitation of N1 another neuron N2 will be excited through the event trace formed previously, through existing connection (from output of N1 to input of N2), although harmful factor F2 still does not influence it. And maybe, even will not influence it in this episode. So the diagram given above will be as follows (Figure 2):

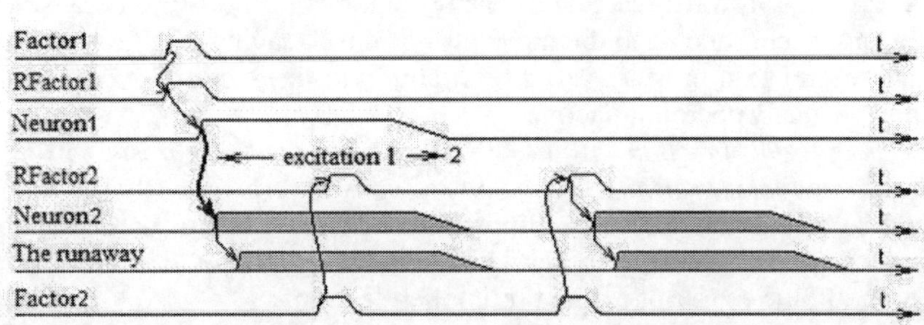

Figure 2. Excitation of Neuron 2 by Factor 1 through the previously self-formed connection N1 – N2.

It means that the presence of the trace (previously formed connection between neurons) enables afterwards the following order of events. As a result of influence of Factor 1, N1 excited through the previously formed connection with N2, immediately excites it. The organism immediately responds to excitation of N2 and «escapes», moves away from Factor 2, although Factor 2 has not yet appeared, and may not appear. If it appears, it will not have a harmful effect on the organism due to «provident», but actually just leading «runaway» from the approaching harmful factor. So the organism with the elementary nervous system of two neurons respond to Factor 2 IN ADVANCE, AHEAD as a result of previously gained experience – formation of the event trace, formation of connection between two excited neurons N1 and N2 (which may be called elementary memory or memory element). As if the organism foresees appearance of F2 due to formation in the previous episode and further existence of communication N1-N2, which is equivalent to memorization of fact of simultaneous action of Factors 1 and 2 in some episode occurred earlier. This leading and, thus, anticipatory to danger (providing provident avoidance of danger) reaction of the organism with the elementary nervous system capable to fix the event traces (to «remember»), will undoubtedly contribute to a more successful survival of the organism and its further expansion in spite of the possible emergence of dangerous Factor 2.

Organisms exist in the material world where events occur due to cause-and-effect relationship of phenomena, so not random, uncoordinated and unconnected events occur in nature, but the chains of causal events. That is, the event A is usually (depending on the surrounding circumstances) followed by the entire sequence of causal events B, C, D, etc. Therefore, the same events- effects usually follow the same events-reasons. For example, smell precedes burning; view and sound of a flying stone precede stroke of the same stone; a flash of lightning (signal speed of 300,000 km /sec) precedes the sound of thunder (speed of 0.33 km /sec). The fact that events are connected in recurring chains provides the body with the ability to detect the approach of a dangerous event and move away from it in advance by detecting some events (probably not dangerous) of this sequence, which precede danger. The leading detection of dangerous events by the organism (forecast) is possible if the previous passage of events chain has left a trace.

Formation of such trace, connection between two neurons is possible under the following conditions. In the natural environment in usual visible macroscopic life, everywhere we can see traces, which have significant effect on the course of events. For example, water flows along the existing paths - furrows, gullies, channels of streams and rivers; rut on the road affects cars movement.

Usually, or at least sometimes, Factor 2 from our example will follow Factor 1 because of the causal event conditionality. However, the organism in advance, even before the arrival of harmful Factor 2, will «predict» appearance of Factor 2 after appearance of Factor 1 by the trace left in the organism during the previous episode. So the organism will provide «foreseeing» forward-reaction to Factor 2 (sometimes in vain, because the harmful Factor 2 may not occur) and «avoid», move away from it, saving its life, continuing later in progeny, and repeating in progeny very useful ability to fix traces of events – to remember pairs of simultaneously operating factors.

This is the useful function of the nervous system of organisms with memory - «foreseeing» (anticipatory reaction) based on previous experience.

It should be noted that we solve almost all challenges facing us upon the model or combination of models (their traces), which we have met earlier as a result of our experience, during the course of explicit or implicit learning.

In order to fix the trace of a random event by forming connections between neurons excited by these events, the inputs and outputs of biological neurons or NE in electronic circuits must form a lattice of possible connections, as shown on Figure 3:

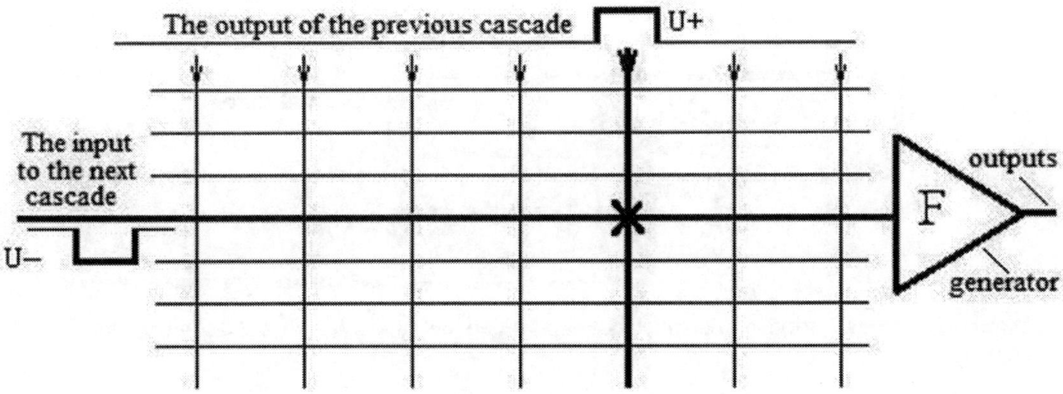

Figure 3. Formation of connections between two random excited NE, outputs and inputs of which are isolated in the initial state.

How can relationship, an event trace, which means simultaneous excitation of two NE, be formed? Figure 4 shows one of the options of fixation, formation of such event. This option of formation of connection between the excited elements has been implemented in the electronic model of the elementary nervous system, created as a result of R & D. This mechanism ensures formation of connection between two simultaneously excited NE.

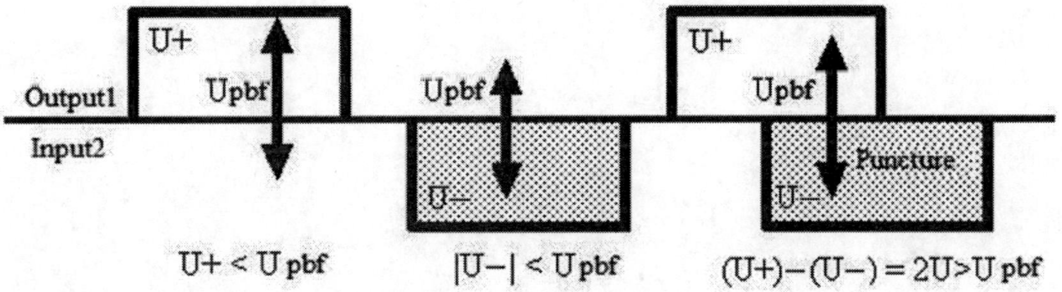

Figure 4. Formation of connection between two simultaneously excited neurons.

Figure 4 illustrates situation when the difference of potentials between two points (Output 1 and Input 2 equal to (U +) – (U-)), at a certain period of time exceeds a certain value U_{pcf} (potential of connection formation). Absolute values of potentials of these two points separately are less than U_{pcf}. So breakdown or another irreversible phenomenon realizing formation of connection – «remembering» (fixation) of an event, can take place only for two simultaneously excited NE, the difference of potentials at inputs and outputs of which exceeds U_{pcf}, and which have mechanical contact (these make formation of connection possible). This simple mechanism provides self-connection between two excited NE, most likely between two biological neurons. Even if a new connection is formed not by means of the breakdown, logical connection can be formed only by means of the described method.

It should be noted that formation of connection with real load characteristics results in the fact that potentials of inputs and outputs of NE cannot remain unchanged throughout the whole period of NE excitation, and they are no more regular rectangular. This is because of the mutual compensation of potentials on output and inputs of excited neurons. Figure 5 shows this fact.

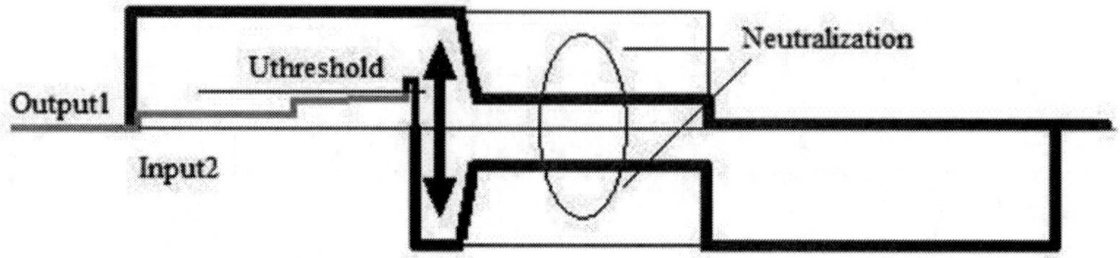

Figure 5. Neutralization of potentials of inputs and outputs of excited NE during formation of connection.

Mutual compensation of potentials on outputs and inputs of two excited NE, shown on Figure 5, provides an opportunity to solve many related problems, both in NS of living organisms, and the NS model implemented based on electronic NE. This model of the elementary NS is based on three electronic neural-like elements, the effect of which is completely transparent and understandable. This simple model shows the possibility of «learning» and «retraining» of the elementary NS implemented as an electrical circuit and acting in accordance with the above-described principles.

As the created electronic model has demonstrated efficiency of the proposed principles of the brain function, it is possible to represent the brain in a form of a very large-scale electric circuit. Figure 6 shows this circuit.

DOI: 10.1051/matecconf/20167901045

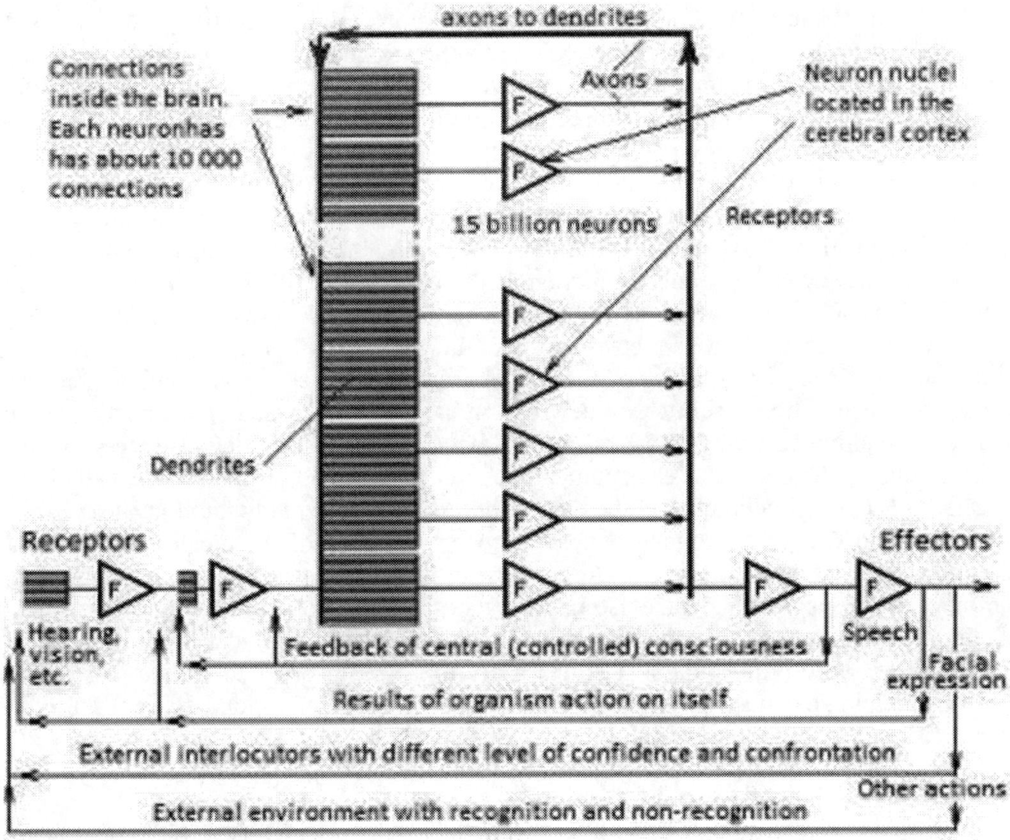

Figure 6. Electric diagram of the human brain.

4 Intellection

Neurons with only local mechanisms cannot determine where signals come to input from. Therefore, all neurons together with their connections act in the same way, no matter where exciting signals come from. Function of each neuron is determined by configuration of connections on its inputs and by configuration of signals coming on these inputs. Neural connections are traces of previous experience of the organism. All neurons being excited, «make their forecast» based on configuration of the current signals (events) and previous experience. Every time these forecasts are single-staged, single-linked. But excitation signals of any neurons in a large electronic circuit of the brain can and act as excitation factors for subsequent neurons, and thus, they create the following links of single-staged forecasts. These single-staged forecasts are combined into long, sometimes intersecting, multi-staged forecasts. Signals passing through such long chains represents thoughts, fragments of thoughts.

With a large number of elements in this circuit, «a thought» being generated by an external factor can move along the circuit for a long time and form new connections between active elements. Configuration of input signals and structure of connections between active elements (neurons), established at the time of passage, that is, the individual «experience» of the scheme define paths of signal passage along the circuit. So the signal originally generated by hunger or splinter when crossing with other chains can turn into a thought about advanced training, a needle, metals, and so on to the structure of the universe through planes and space ships.

Taken into account the huge amount of connections (about $1.5*10^{14}$) and the need to address them (at least 2 bytes per indication of source signal for each possible connection-dendrite in the human brain), it becomes obvious that creation of a device that simulates function of the human brain by direct installation of all these connections is almost impracticable. With so many possible connections it turns out that even with the period of brain life of about 100 years, the average rate of formation of connections between neurons is about 30,000 connections per a second. These connections are formed in the brain as a result of acquisition of experience - training, upbringing, activity in general. Therefore, we propose another way. To simulate function of the brain we should create a device with the same potential for development as the natural human brain. But, as well as the brain, the model in the initial state is not ready for full operation. However, it is ready for development. During the process of the model creation, we set connections, which realize likeness of unconditioned reflexes of newly born babies. Next, based on connections, which realize unconditioned reflexes, we bring up reaction-skills in this device necessary for its comfortable existence in the environment, just as people and animals raise their babies. That is, *to bring up* intelligence in the device, *not to create* the ready intelligence.

In order to strengthen conclusiveness of correspondence of the created models to real biological nervous system, we should confirm in detail correspondence of the models to biological nervous systems in existing living organisms. For confirmation, we need to obtain neuron input/output potential graphs of the simplest biological organisms, such as California sea hare (Aplysiidae). It is necessary to do so in order to ensure that potential graphs of biological nervous systems and of our models correspond to each other. We would like to obtain potential graphs when there is excitation of neurons, especially at the moment of formation of new connections between neurons.

Currently, we are carrying out this work.

Confirmed correspondence of software and electronic models of the biological nervous system will allow more confident studying of the behavior of biological nervous systems on the computational model.

If we are confident that our models reflect processes of biological nervous systems perfectly true, then it will be possible to make recommendations on diagnosing and treatment for biological nervous systems diseases based on the study of models' behavior. Even now, we can give some recommendations concerning ways of medical treatment for diseases. If a person's mechanism of new events memorization has failed but the person remembers rather well events passed a long time ago, we can make the following conclusion. According to our model of the nervous system function - formation of new connections between neurons as memory elements - the memory disorder may be caused by the fact that formation of new connections in the nervous system in the long-term memory area of the brain has stopped for some reason.

The reason could be in reduction of value of excited neurons potential (for example, because of increase of leakage). The value of these potentials becomes insufficient to form new connections in this area [12]. On the other hand, inability to form new connections may be caused by increase of the value of the potential of a new connection formation. For example, by increase of breakdown voltage of membranes separating neighboring neurons.

In both cases, there is need for a certain change of biochemical conditions in the brain, which could change the existing conditions of the brain and restore the ability of brain structures to form connections.

5 Conclusion

The problem of study of diseases and ways of medical treatment is the responsibility of medical researchers. Our task is to provide medical researchers with a convenient tool for studying of various diseases of the nervous system on the computational model, allowing quick change of various parameters that affect the memory model function, and observation the model's reaction on change of these parameters. Currently, we are also carrying out work on creation of the software model of the nervous system with convenient for medical researchers interface.

Together with biologists' work on confirmation of correspondence of processes of biological nervous systems function with our models, it will create a good platform for a humane (non-traumatic intervention in the life of a biological organism), economical, effective (efficient) study of various diseases and disorders of the nervous system, their prevention and medical treatment.

Acknowledgements

The paper was written as part of the research project No. 8.2.31.2015, carried out with the support of the Program "Research Foundation of Tomsk State University named after D.I. Mendeleev" in 2015 – 2016.

References

[1] L. A. Amaducci, W. A. Rocca and B. S. Schoenberg, Neurology **36**, 1497 (1986)
[2] B. E. Tomlinson, G. Blessed and M. Roth, J. Neurological S. **11**, 205 (1970) doi: 10.1016/0022-510X(70)90063-8
[3] F. Matthews and C. Brayne, PLoS Med **2**, 0753 (2005) doi: 10.1371/journal.pmed.0020193
[4] *Dementia UK (*Alzheimer's Society, London, 2007) URL: http://www.alzheimers.org.uk/site/scripts/download_info.php?fileID=2
[5] V.N. Shumilov, *Printsipy funktsionirovanija mozga. Vzgljad inzhenera* (The principles of functioning of the brain. Sight of engineer) (Ekaterinburg, 2008)
[6] V.N. Shumilov, V.N. Voronin , V.I. Solomonov, *The adaptive controller, the base neuro-similar element and a method of organizing the work of such a device* RF Patent № 2475843
[7] A. V. Zhdanov, Science and Life (1), 2000 URL: http://www.nkj.ru/archive/articles/5613/
[8] M. P. Van Den Heuvel, C. J. Stam, R. S. Kahn and H. E. Hulshoff Pol, J. Neuroscience **29**, 7619 (2009) doi: 10.1523/JNEUROSCI.1443-09.2009
[9] Jeff Hawkins, Sandra Blakeslee, *On Intelligence* (Macmillan, New York,2007)
[10] The growth of axons in the developing cerebral cortex of mammals URL: http://scorcher.ru/axiomatics/axiom_show.php?id=305 (15.03.2015)
[11] *Psychophysiology, University textbook*, edited by Y. I. Alexandrov (Piter, SPb,2010)
[12] V. N. Shumilov; V. I. Syryamkin, M. V. Syryamkin, New operational technologies (NEWOT'2015) **1688,** 040007 (2015) doi: 10.1063/1.4936040

MATEC Web of Conferences **79**, 01046 (2016) DOI: 10.1051/matecconf/20167901046
IME&T 2016

DETERMINATION THE AUTHENTICITY AND QUALITY OF THE GOODS

Sergey Minkov, *Andrey* Novikov[1]

National Research Tomsk State University, 634050, Tomsk, Russia

Abstract. Often the question arises, how to assess the quality of the goods. To determine either fake or genuine product, there are many different methods and techniques, but there is no universal method for determining the low-quality products. It is proposed to provide a compact device for use with any mobile device, such as a cell phone, as well as the creation of algorithm and software that will allow the end user to easily use the device for self-determination of the authenticity and quality of the goods, comparing the results of the spectra analysis.

1 Introduction

Food safety is a long-standing problem for consumers. Even if the product was fresh during transportation to the place of sale or at a time when the buyer acquires it, the quality, freshness and safety at the time of use cannot be guaranteed. Thus, a method that will allow the buyer to check the food product in the presence of deviations (violations), immediately prior to use, it will be very convenient for the consumer.

2 Problematic

Non-invasive methods, i.e. methods that are not related to the direct use of the product in the study (testing), as well as possible better suited for use by consumers, as this method provides ease of use and therefore a higher probability of use. Spectroscopic methods for determining the quality of food products meet the requirements to be non-invasive and easy-to-use. But there are two major obstacles to be overcome before the use of such methods consumers.

The first is the need to establish an appropriate spectrometer. Figure 1 shows the structure of a typical spectrometer [1]. Light enters through the entrance slit, is reflected by the mirror 1 at a dispersive element such as a diffraction grating which directs light to a mirror 2. Since the grating disperses light according to its wavelength, each pixel of the CCD-spectrometer detector is lit a certain wavelength. Spectrometers that use this design are complex, require precise alignment, difficult to assemble, and are limited in their ability to miniaturization. The spectrometer is suitable for everyday use by the consumer should be small, durable, easy to use and have features that provide product testing sufficient accuracy and reliability [2].

[1] Corresponding author: unwell@yandex.ru

© The Authors, published by EDP Sciences. This is an open access article distributed under the terms of the Creative Commons Attribution License 4.0 (http://creativecommons.org/licenses/by/4.0/).

The second obstacle is that foods – not simple matter but complex mixtures of substances, which themselves tend to have complex spectra. Thus, to be useful to the consumer, such a recognition system must include means for analyzing complex spectra obtained in real time.

When designing such a spectrometric analysis system should take into account that, for the consumer it would be useful to determine not only the quality and authenticity of the products, but also to apply the device in other areas. For example, such a system can be used in various applications such as the detection of environmental contaminants, detection of explosive and remote chemicals or biological agents, as well as remote diagnostics and monitoring a patient's condition (e.g., blood oxygen levels or glucose levels).

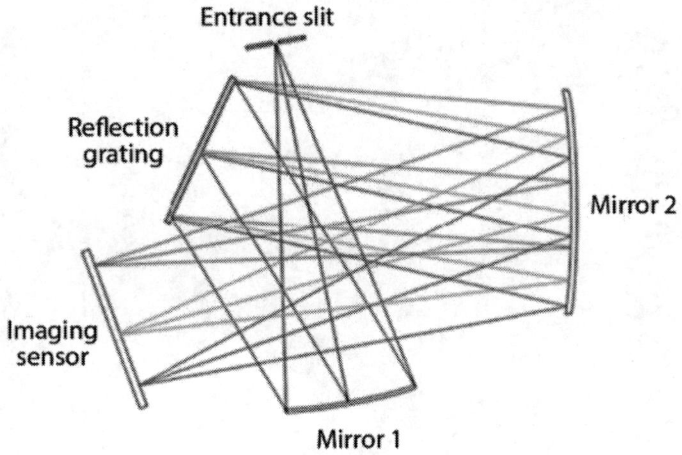

Figure 1. A typical spectrometer.

Compact spectrometer that can be integrated into consumer devices such as cellular phones must be sufficiently durable and have a low cost, to be convenient and practical for spectroscopic measurements performed by the user in a domestic environment. This is a convenient method by which to obtain the spectra (in particular from complex mixtures such as food) and analyze them to meet product quality and authenticity. The compact spectrometer system and its method of operation are designed to meet this need.

3 Principle of operation

The basic principle of a compact spectrometer is as follows (Figure 2). Light enters the optical filter comprising a plurality of sub-filters aligned along the same axis. These sub-filters form a luminous flux on a plane-convex lens (a convex surface facing the optical filter). Several different signals are formed after the lens, which are dispersed along the micro-lens array onto the detector. A micro-lens array arranged so that light is scattered by the detector along an axis parallel to the axis which combines sub-filters.

The detector may be of any suitable type, able to detect light in the required range: from UV to IR radiation, depending on the application.

Using only single filter may restrict the available spectral range. For example, if the angle of incidence greater than 30°, the signal intensity is insufficient due to lens aberrations and this will reduce the efficiency of the detector at large angles [3]. For the angular range of 30° and the optical dichroic filter more than 850 nm spectral range available for the spectrometer is ~ 35 nm. This range is sufficient for use in Raman spectroscopy [4–6]. To solve this problem, an optical filter is used, which consists of a plurality of auxiliary filters, wherein each sub-filter is a dichroic filter with a narrow width in its transmission spectrum wavelength range, and thus, can reach different parts of the optical spectrum. Depending on the number of sub-filters wavelength range available to the spectrometer, can reach hundreds of nm.

In the case of using a plurality of auxiliary filters plano-convex lens overlaps formed on the image plane (one from the sub-filter), and finally the signal received at any particular pixel of the detector is the result of a mixture of different signals collected through the lens. In order to separate the signals from different sub-filters, you must use an array of micro-lenses, which must be placed between the lens and the detector (the convex side to the detector). Such micro-lens arrays are well known and have been used, such as light-field camera [7].

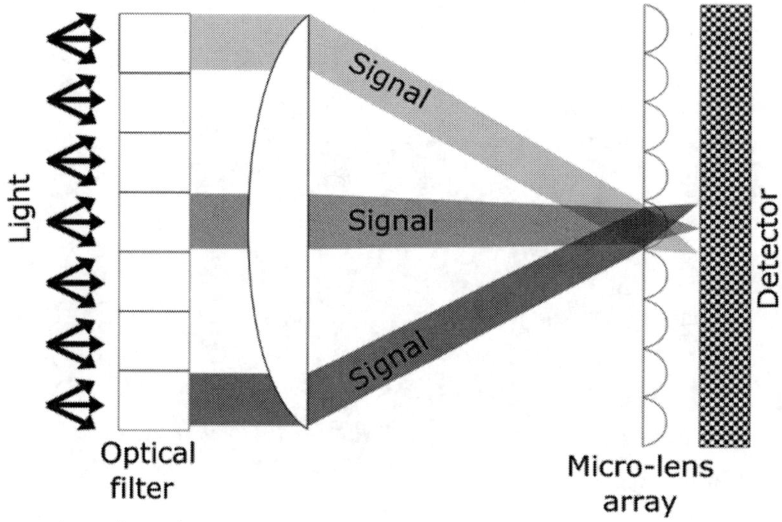

Figure 2. Compact spectrometer.

Their location coincides with the focal plane of the plano-convex lens and the detector plane. Each micro-lens acts as a "super-pixel", which separates the light incident on it from the lens, which in turn, received by a plurality of auxiliary filters. Each micro-lens generates an image at the detector aperture of the optical filter. Such a "micro-image" is a contribution to the signal of each portion (i.e., optical filters) lens aperture.

Sample measurement is performed through the use of the scattered ambient light. In most cases, the natural ambient light is not intense enough to provide a range of high enough quality. Thus, compact spectrometer system includes a light source. The laser or light emitting diode can be used as a light source.

Using the spectrometer is determined the absorption spectrum of the object. Preservation of the spectrum to a database and use it for comparison with newly acquired spectrum will allow to compare objects' characteristics with the characteristics of the reference object (in the database) to assess their authenticity and quality (Figure 3).

The image or sequence of images is captured by the image sensor in the spectrometer. The images are analyzed by a local processor. This step of analysis may include image averaging, compensation of the optical unit aberrations, detector noise reduction by using noise reduction algorithm or image transformation as the original spectrum.

Authenticity and quality are determined by the user independently by comparing the spectra results. For example, the user can read the coefficient spectra overlap greater than 0.8 as evidence that the compared spectra were obtained from identical objects.

The device operates only in conjunction with special software that implements spectral processing and matching algorithm.

Today, mobile technologies cover more and more spheres of human activity. High availability of mobile devices, as well as benefits such as portability and low cost make the need to develop applications for mobile platforms. Therefore availability, ease of use and convenience of the software for the device is best implemented as applications for popular operating systems OS Android and iOS (Table 1).

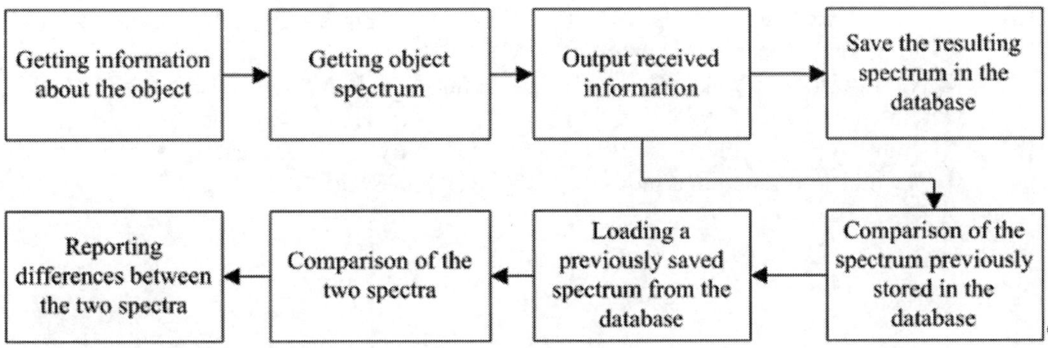

Figure 3. Information Model.

At the initial stage of the application is a native type using optimization design under multiple platforms with a single code base.

Table 1. Worldwide sales of smartphones to end users [8].

OS	2015 Q2, thous. pcs.	Market share in 2015 Q2, %	2014 Q2, thous. pcs.	Market share in 2014 Q2, %
Android	271010	82.2	243484	83.8
iOS	48086	14.6	35345	12.2
Windows Phone	8198	2.5	8095	2.8
BlackBerry	1153	0.3	2044	0.7
Other	1229	0.4	1416.8	0.5
Total	329676.4	100	290384.4	100

4 Conclusion

As a result, an inexpensive reliable spectrometer without dispersion elements and moving parts will be created, as well as software for data analysis. Spectrometer system has no direct optical axis and in the selected embodiments includes no more than four elements. The required accuracy of alignment between the elements is very low, compared with a typical spectrometer. The absence of direct optical axis and the small size of the detector allows to organize the production of spectrometers, which are sufficiently small and economical to fit into devices such as cell phones, but in the other hand have sufficient sensitivity and resolution (typically <10 mm), to obtain spectra of complex mixtures, such as foods.

References

[1] K. Siegbahn, *Alpha-Beta and Gamma-Ray Spectroscopy (*North-Holland Publishing Co., Amsterdam, 1966)

[2] *Opticheskie spectralnye pribory (The optical spectral instruments)* / ed. I.V. Skokov (Mmashinostroenie, Moscow,1984)

[3] D.W. Ward, K. A Nelson, K. J. Webb, New Journal of Physics **7**, 213 (2005) doi: 10.1088/1367-2630/7/1/213
[4] *Osnovy molekulyarnoi spektroskopii (Fundamentals of molecular spectroscopy)*/ ed. Y.A. Pentin, G.M. Kuramshina) (M.: Mir, Moscow, 2008)
[5] Pietro Baraldi, Anna Tinti, MATEC Web of Conferences **36**, 01002 (2015) doi: 10.1051/matecconf/20153601002
[6] D. A. Skoog, F. J.Holler, S. R. Crouch, *Principles of instrumental analysis* (Thomson Brooks/Cole, Australia, 2007)
[7] E.H. Adelson, J.Y.A. Wang, IEEE Trans. on Pattern Analysis and Machine Intelligence **14**, 2 (1992)
[8] IT-industry: the world economy against the IT. URL: http://pressenter.ru/index.pl ?act=PRODUCT&id=2536

MATEC Web of Conferences **79**, 01047 (2016) DOI: 10.1051/matecconf/20167901047
IME&T 2016

RELIEF RESTORATION OF COMPLICATED FORM OBJECTS BY MONOCHROMATIC MICROWAVE RADIATION

Ivan Kuzmenko[1,1], *Vladimir* Yakubov[1], *Timur* Muksunov[2]

[1]Tomsk State University, 634050, Tomsk, Russia
[2] V. D. Kuznetsov Siberian Physical-Technical Institute at Tomsk State University, 634050, Tomsk, Russia

Abstract. Article demonstrates possibility of monochromatic radiation usage for relief restoration. There is a problem with restoration when scanned object is not flat and it is not parallel to the scanning plane. It was discovered that two-dimensional phase distribution could be applied for distance determination. It is reliable way to solve problems listed above. In conclusion offered methods allow monochromatic microwave radiation usage for screening system development.

1 Actuality of problem

Public places' safety assurance is complicated problem which actuality is increasing every year. Terrorism is a large treat for many countries including Russian Federation.

Important infrastructural objects, such as airports, equipped with wide range of technical instruments purposed for human and luggage inspection. Traditional way of search system engineering based on x-ray scanning. But x-ray radiation can be dangerous for human health, especially after multiple exposing. This fact is affirmed by existence of such parameter as number of allowed scans per year. Because of that, already long time researches of alternative ways ago have been started. The alternative scan systems should be as effective as x-ray and safe for human health as well [1–5].

Radiowave scan systems were invented recently. They have image quality at the same level with x-ray introscope and safe for health – allowed number of scans per year is not limited. But existing radiowave scan systems are expensive, costing millions of USD [6, 7].

One of the reasons of such big price is high costing for UHF transceivers with wide frequency range. Adapting of monochromatic radiation is possible way to lower scan system price. Figure 1 presents view of microwave sensor array.

[1] Corresponding author: kuzmenko.ivan.yu@gmail.com

© The Authors, published by EDP Sciences. This is an open access article distributed under the terms of the Creative Commons Attribution License 4.0 (http://creativecommons.org/licenses/by/4.0/).

Figure 1. RSM 2650 sensor array.

Certain results of such researching had already achieved. Figure 2 presents testing object and its radio image which was restored by aperture synthesis method after scanning with sensor array demonstrated above.

a) b)

Figure 2. Testing object (a) and its radioimage (b).

Restored radioimage shows possibility of radiowave scanning system with monochrome sensors. But during the experiment distance of testing object was known before scanning and apparatus system function (ASF) was calculated for exactly that length. In real situation distance can be unknown. In addition to that, objects of scanning, such as human body, usually have non-flat relief.

2 Mathematic modeling

It's necessary to define how will change interferential image when object is not oriented in parallel to scanning plane. For this purpose such situation was modeled with metal pipe as scanning object. Pipe axis is angled by 10° to scanning plane. Interferential images of the pipe in cases of parallel and angled orientation are shown by Figure 3. According to the image, angled orientation causes interferential lines to bend. Restored radioimages are shown by Figure 4. The radioimage of inclined pipe is blurred because of unfocused radiation.

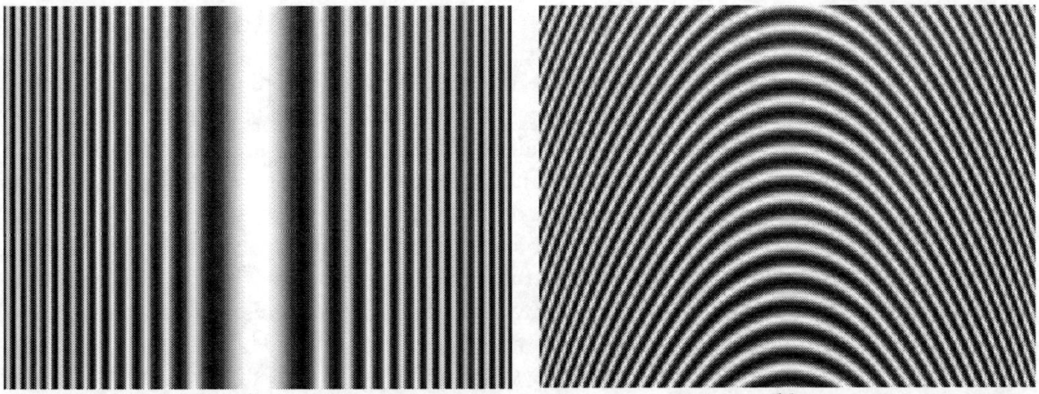

Figure 3. Interferential image of metal pipe with parallel (a) and angled (b) orientation to scanning plane.

Figure 4. Radioimage of metal pipe with parallel (a) and angled (b) orientation to scanning plane.

It will be shown next how changes radioimage in case of ASF for radiation received is calculated for wrong distance.

Figure 5 shows that in this case radioimage quality is considerably reduced. Object positioning is still correct, but its size is considerably bigger than the true one.

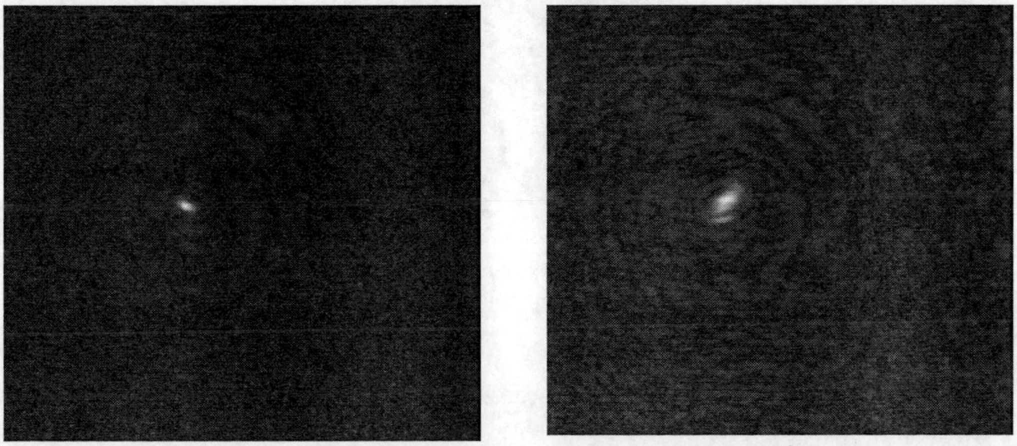

Figure 5. Radioimage in cases of ASF focusing on true (a) and error (b) distance.

DOI: 10.1051/matecconf/20167901047

3 Experimental distance definition and relief restoration

It is known fact that it is possible to define distance by multipositional scanning even in case of monochrome oscillations as scanning signal. Research results of distance definition possibility in case of Doppler sensor array multipositional scanning are presented below.

Figure 6, *a* shows received signal after multipositional scanning of metal pipe at 15 cm distance from scanning plane. For distance definition the signal was focused on distance from 0.1 cm to 30 cm with 0.1 cm step. “Focusing signal on distance” means operation of convolution between ASF calculated for that distance and signal received during scanning. Result is presented at Figure 6, *b*. Image shows that there is distinguishable maximum at 15 cm distance which is corresponding to the object location. This result proofs that monochrome multipositional scanning allows definition of object distance with adequate precision. cm

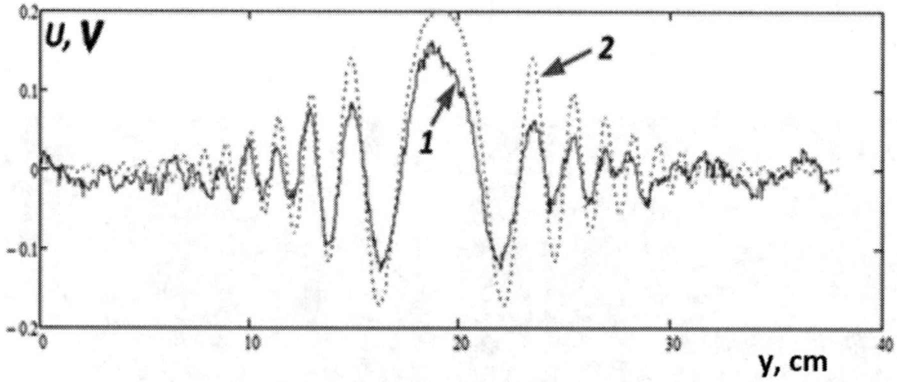

a)

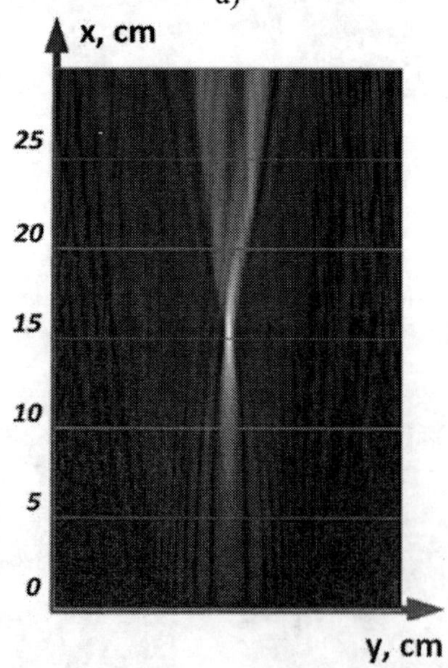

b)

Figure 6. Signal received during multipositional scanning (a) and results of focusing on various distances (b).

Application of radiowave scanner is searching for forbidden items covered on human body under clothing. Relief of human body is not flat, it has its own protrusions and depressions.

DOI: 10.1051/matecconf/20167901047

Applying to the radioimage restoration for such difficult object it means necessity of distance definition for proper radiation focusing.

Serial of experiments was done to solve this problem. One of such experiments consisted of radioimage restoration of two metal plates with 2 mm distance shift between. Plates had size of 35×15 cm and their photograph is presented at Figure 7, *a*. Figure 7, *b* shows result of interferential image focusing on nearest plate distance. But it is not impossible to define which one is nearest by using the image. Focusing on another plate also does not allow correct relief restoration.

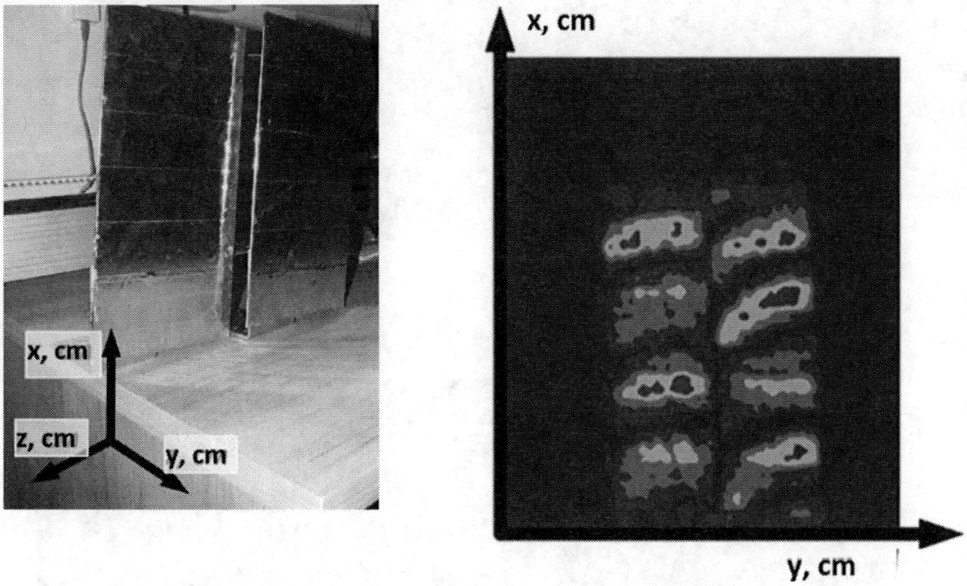

Figure 7. Testing object photograph (a) and it's radioimage (b).

Phase information of radioimage can be used for relief restoration of scanned object. Line 2 at Figure 8 shows dimensional phase distribution at horizontal axis of scanned object. Line 1 shows intensity of the signal. As the graph demonstrates, phase of the signal along the single plane stays on the same level and makes distinguishable leap at the border between plates.

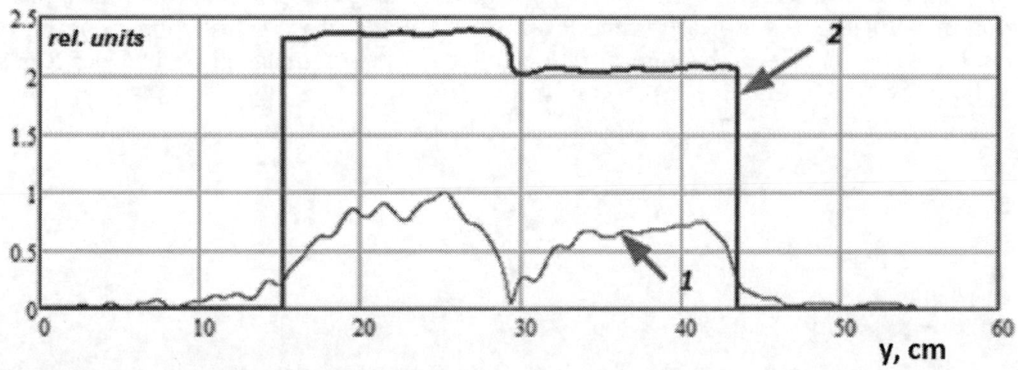

Figure 8. Dimensional phase distribution.

Testing object for next experiment was made of metal plate and dielectric brick glued together. Photograph of the object is shown by Figure 9,*a*. Metal plate is trapezoid with 9 cm and 16 cm bottoms and 25 cm height. Size of dielectric brick is 2×5×9 cm. Interferential image is shown by Figure 9,*b*, restored radioimage is shown by Figure 9, *c*. Form of metal plate is defined and presence of brick is detected.

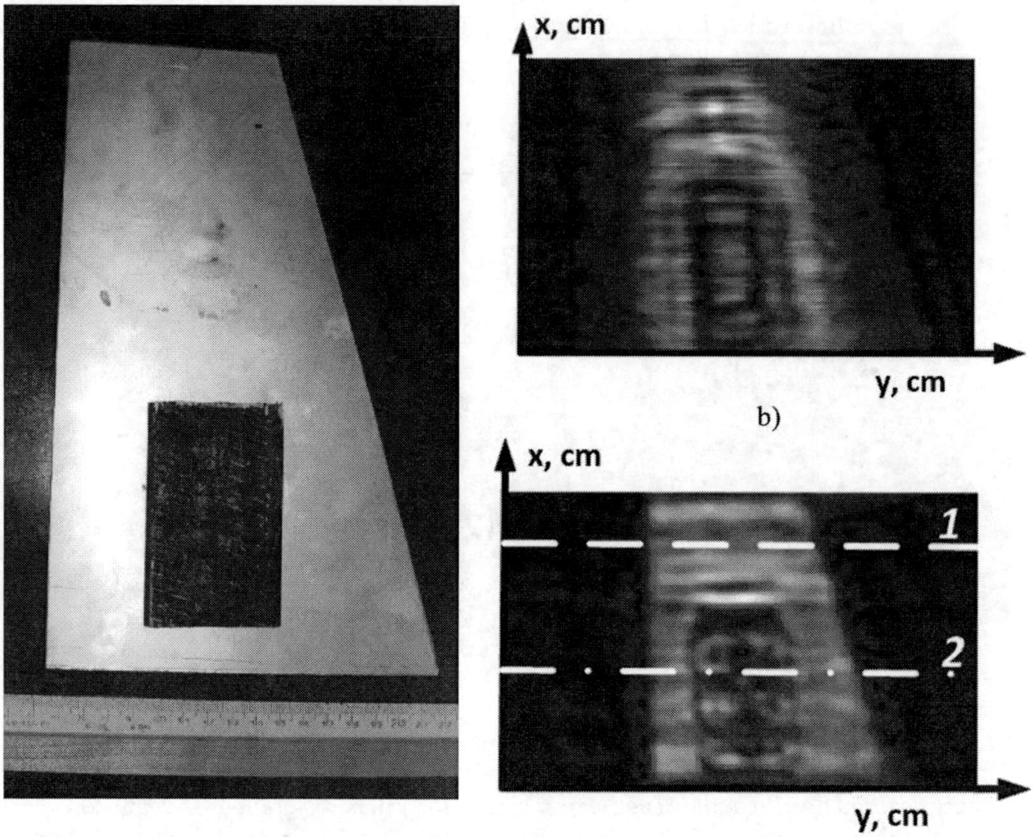

a) c)

Figure 9. Testing object photograph (a), interferential image (b) and radioimage (c).

Two strings of the image were chosen for research. Intensity and phase distribution along line 1 from Figure 9, *c* are shown by Figure 10, *a* graph. As it shows, image phase along metal plate stays at nearly same level. Intensity and phase distribution along line 2 from Figure 9, *c* are shown by Figure 10, *b* graph. As it shows, image phase makes a leap at the dielectric brick position.

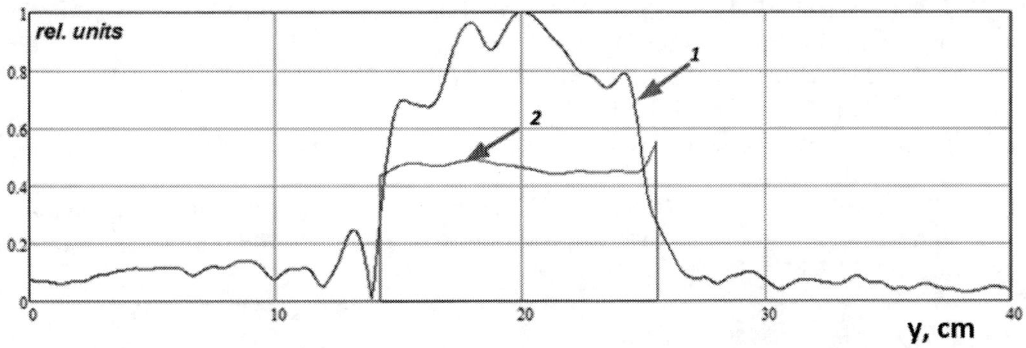

a)

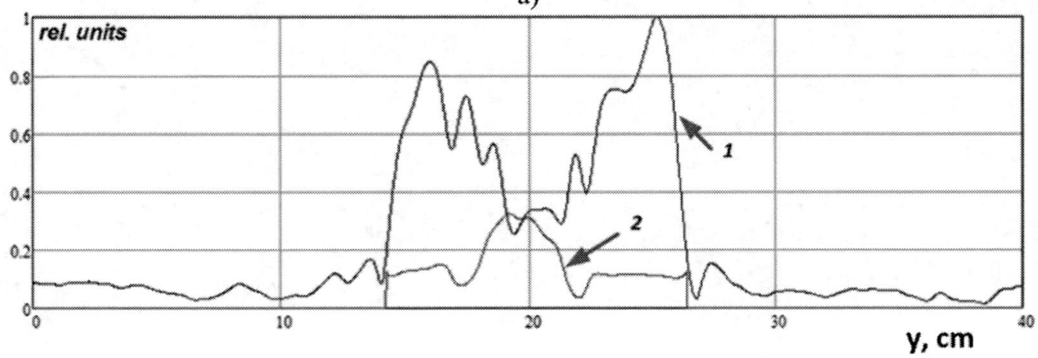

b)

Figure 10. Intensity and phase distribution along metal plate and along metal plate with dielectric brick.

4 Conclusion

Experiments made during research are the proof of relief restoration of non-flat objects by monochrome radiowave radiation. Problem with distance definition can be solved by focusing after multipositional scanning. Phase information of received signal allows restoring relief of scanned object. Proposed methods make possible to use monochrome radiation for radioimage restoration. Using of monochrome sensors as part of scanning systems is possible way to reduce such systems' costs while keeping image quality at the same level.

Acknowledgment

The work is done as part of the State Assignment No. 3.694.2014/K of the Russian Ministry of Education and Science.

References

[1] A. Bennamane, T. Boutkedjirt, MATEC Web of Conferences **16**, 10002 (2014) doi: 10.1051/matecconf/20141610002

[2] R. Appleby, Philosophical Transactions: Mathematical, Physical and Engineering Sciences **362**, 379 (2004) doi: 10.1098/rsta.2003.1323

[3] R. N. Satarov, I. Yu. Kuz'menko, T. R. Muksunov, A. V. Klokov, E. V. Balzovskii, Yu. I. Buyanov, S. É. Shipilov, V. P. Yakubov, Russian Physics Journal **55**, 884 (2013) doi: 10.1007/s11182-013-9896-8

[4] J.E. Bjarnason, T. L. J. Chan, A. W. M. Lee, M. A. Celis, E. R. Brown, Applied Physics Letters **85**, 519 (2004) doi: 10.1063/1.1771814
[5] D.Y. Sukhanov, K.V. Zavyalova, Technical Physics **57**, 819 (2012) doi: 10.1134/S1063784212060229
[6] F.-G. Joaquim, IEEE Transactions on Geosciences and Remote Sensing **40**, 443 (2002) doi: 10.1109/36.992808
[7] J. Groenenboom, A.G. Yarovoy, GPR 2000: Proceedings of the eighth international conference on ground penetrating radar **4084,** 867 (2000) doi: 10.1117/12.383530

TRAINING OF SPECIALISTS IN INFORMATION AND MEASUREMENT INSTRUMENTATION IN THE COMPETENCE DISCOURSE

Svetlana Kvesko[1]*, *Tatyana* Shinn[2]

[1]National Research Tomsk State University, 634050, Tomsk, Russia
[2]Missouri University of Science and Technology, 65401, Rolla, Missouri, USA

Abstract. Refocusing the assessment of the results of educational process from "qualifications" to the concepts of "competence" and "competency" is important in the context for the theoretical and methodological analysis of training in the sphere of information and measurement instrumentation and technologies. The purpose of this paper is to study of the training of today's experts in the field of Information and Measurement Technologies and Instrumentation in the competence discourse. Based on the application of the analysis and interdisciplinary approach methods, the authors concluded the integrative nature of competency-based approach in educational process. The research has revealed the key social competencies and their characteristics.

1 Introduction

Today's socio-cultural context is experiencing the strongest influence of the human creativity. In the technogenic civilization, the rapid advancement of technology and machinery leads to development of new behavioral styles and changes the entire rhythm of living. The society is evolving into a non-linear informational entity.

The inter-dependencies of a human being, society and nature are enormous which may have both positive and negative impact on the process of resolving of various social issues. This, in turn, has direct implications for the informational processes in the society, and as a result, for the process of education. The quality of general education is closely tied to the level of professionals produced by the system. Presently, we witness the growing significance of the application of informational and measurement instruments and technologies for the development of professionalism, professional confidence and image, and therefore, the approach based on the concepts of competency is gaining influence.

2 Method

The shifting focus of the assessment of the education process and of measuring the success of student learning to the concepts of competency dictates the need the for theoretical and

*Corresponding author: svetla_kvesko@mail.ru

© The Authors, published by EDP Sciences. This is an open access article distributed under the terms of the Creative Commons Attribution License 4.0 (http://creativecommons.org/licenses/by/4.0/).

methodological analysis of the professional education programs in the field of informational and measurement instrumentation and technologies. It should be noted, first of all, that the very definition of such approach may have different interpretations.

In our paper, the competency-based approach is to be understood as a cognitive process based on the categories of "competence" and "competency" [1].

The methodological analysis of the problem of professional education in the field of informational and measurement instrumentation and technology assumed the need for a research in order to determine the meaning and the role of competencies in the process of education. The competency-based education is comprised of the methods and procedures related to the development of the educational technology. The meaning of the competency-based approach cannot be understood without the concept of competency as the integral system means of study [2].

The competency approach has a complex hierarchical and multi-level structure with many components and is directly related to the system, structural, functional, problem-focused and genetic approaches to the understanding of the realities and events. Such approach is focused on an individual and his actions. The competency approach relates to the purpose and the outcome of education [3].

3 Main content

The competency approach is very important for the professional education in the field of information and measurement equipment and technology, as it focuses the attention on the important issues and offers integral solution to those. Without a doubt, competency is a requirement the society has for the product of education, however, this concept is somewhat removed from a future professional. Being competent is more than learning the classroom material; it is possession of certain skills, competencies, personal characteristics and qualities.

Administering competency-based education requires a clear definition of competencies and development of valid, reliable assessments. Getting industry input is essential to ensuring the identification of the relevant competencies in the field of informational and measurement instrumentation and technology.

The fundamental competencies in the field of informational and measurement technology are those of the humanities, general professional and occupational. These competencies shape the social position of those being educated, their actions and social roles. In the process of education, the competencies form the guidelines for an individual's future professional activity and for navigating the future.

The competencies stimulate each student to master the means and methods of learning and professional practice, understand the process of assessment and application of information and outcomes. Possession of competencies enables an individual to solve problems, coordinate the work of specialists in multiple spheres of applied informational and measurement technologies and scientific knowledge. The role and specifics of competencies may be analyzed from the perspective of the fundamentals of education in the field of information and measurement technology.

We believe it necessary to stress the importance of the competencies of individual values and meaning for the field of informational technology. These competencies are defined by the moral guidelines in the social and informational world. They mean more than just establishing the level of comprehension, but rather an individual's ability to understand the world while realizing his own contribution to the advancement of the innovations and modern technology. The break-through technologies are highly valued in today's world, but their development is only possible in the context of the value- and meaning-based competencies.

The values- and meaning-based competencies enable individual self-actualization and guide in the choice of life purpose, attitudes and decision-making. These attitudes determine the trajectories of educational and spiritual pursuits. The value-based competencies play a significant role in the formulation of an expert assessment of realities and events, as well as the outcomes of the educational, scholarly, and practical endeavors.

The internal component of the competency (personal competency) is dependent on the individuality and is rooted naturally in the system of individual perceptions of self and one's role in the social context. Personal competency is explicated from an individual's abilities to understand and reflect upon one's role in the society, and to develop life-long goals based on personal values and motivations. The external component of the competency is defined by the individual relationship with the society and social environment. From the moment of choosing a future profession, and individual receives an orientation that determines his actions toward satisfying his aspirations for professionalism.

There is another group of competencies that is cognitive-based. Those are competencies related to educational and learning pursuits and include establishing goals, reflection and self-reflection, logical thinking and methodology. Mastering these competencies requires the use of computer tools and technologies in the educational process. The cognitive-based competencies aid in the refinement of classical forms of education and development of distance-based educational environment.

The special emphasis is being placed on the informational competencies. In the process of acquiring these competencies, an individual develops skills, abilities and knowledge related to the application of information in learning, professional practice and everyday life situations.

The modern means of information and technology require a significant level of knowledge of computers. In everyday life, in professional environment and in the classroom, an individual is faced with the latest technology and equipment, which quickly become outdated and obsolete thus causing a constant stress of catching-up. This may lead to an emotional and sometimes a professional burn-out [4].

Today's digital instruments and technologies in information enable a quick and efficient search for publications and data, aid in their analysis, and streamline interpersonal communication through the use of email, chat, and videoconferencing.

Communicational competencies should also receive our attention. These competencies center around language skills and means of communication. Communicational competency also means the possession of the informational communication skills, knowledge of search engines, social media, video- and audio-communication. The communicational practice within the Internet community plays an important role in the process of mastering these competencies.

Communicational practices are essential components of the cultural dialogues, inter-cultural socialization, as well as of the business interactions. To be successful in communication within the Internet community and related interactions, one must possess linguistic competencies, be skilled at self-presentation, be knowledgeable in the informational technologies and their practical applications [5].

The implementation of communicational practice via informational and measurement instruments requires flexibility in a changing environment and ability to compromise. These skills are highly valuable in the multi-ethnic, multi-lingual and multi-confessional social world.

The competencies discourse outlines the informational context of education. The informational context of education is comprised of the programming tools, software and hardware, information technologies. Computer-mediated instruction facilitates an individualized approach to education of each student. Because individuals learn at a different pace and enter the professional education program with different levels of prior knowledge and experience, this is a fundamental requirement of competency-based education [6].

The analysis of the state of the informational context of education allows us to formulate several key points of today's education and practice in the field of information and education:

1. Today's informational context of education is comprised of multiple components and includes classroom training resources and programming, tests, surveys, graphic and video tools, databases and search engines.
2. Informational context of education has an integral nature, incorporating the wealth of knowledge in the field of informational and measurement equipment, technologies, science, as well as the reference foundation.
3. The distributive quality of the informational context supports the effectiveness and economic efficiency of the technical tools by optimizing the access to the information.
4. The informational context of education is adaptive which enables a flexible use of informational instruments according to the interests and needs of the society.

In the context of the above key points, the informational and measurement field is being regarded as both a part of a general education environment and an autonomic system designed to form an active social and creative personal position that would incorporate the latest informational tools, instruments and technologies [7]. The informational context of education is based upon the whole spectrum of disciplines and subjects and the interactive support factors that are used in the training process.

The term "informational context of education" emerged with the development of the information technology and instruments. Ever since the computer and information technologies infiltrated the education process and became widely used in training, the methodological and administrative support is being designed and regulated for a wide-ranging application of the modern technology.

At implementation of projects both technical, and social there is a question of responsibility of scientists as scientific decisions not always guarantee positive social result. Owing to this fact carrying out social and humanitarian examination is actual. The question of responsibility of designers is treated from the point of view of positive practical application both for all society, and for each certain person, for change of public states. Often responsibility is treated as fault for insufficiently optimum result, for result, which negative in development of social states, for the consequences leading to tragedies, disasters.

4 Conclusion

The global impact of computers and information technology on education has resulted in significant advancement of the technical and didactical resources. As the informational environment evolves, new effective forms of education emerge (distance learning, the organization of independent work in Moodle system, etc.).

Advantages of the modular principle:

1. The informational context of competency-based education is based on the modular principle. The modular approach creates a separate subject within the discipline of information engineering as a unit independent of other subjects. This approach corresponds with the western model of competency-based education in information technology and engineering fields.
2. The advantage of the above principle for the education is in the simplicity of its application.
3. The most important characteristic of competency-based education is that it measures learning rather than time it takes to complete professional training.

There are, however, the disadvantages of using the modular principle:

(1) The contradiction between the integral system of knowledge and the assortment of disconnected from each other individual areas of expertise.

(2) The lack of "the big picture", as the large number of modules/units impedes the effective generalization of the material learnt.

(3) The subjectivity related to personalities of the module creators/authors.

Overall, the concept that incorporates the informational and competency-based context while preserving the traditional system of education, offers a significant contribution toward the development of an innovative model of education [8]. The application of information technology and computer measurement instruments in the process of education leads to the advancement of social networking, while the rapid growth of computer and information technologies and their efficiency facilitates development of the distance education. As the society is evolving into the "society of information", the social and information technologies are impacting the educational system. The problem of changing paradigms in the field of information and measurement instrumentation training must be addressed in the context of general modernization of the process of education. The conducted research illustrates that the information technologies are essential instruments of the educational process.

Acknowledgment

This research is carried out under the scope "Organization and methodic issues of training specialists in Information and Measurement Technologies and Technique. Passes with assistance of a grant of the Russian fund of basic researches №16-07-20218, and the program of improving competitiveness of the TSU (project No. 8.2.31.2015).

References

[1] M. Mayhew, J. Simonoff, W. Baumol, B. Wiesenfeld, M. Klein, Research in higher education **53**, 831 (2012)

[2] A.Yu. Chmykhalo, R.B. Kvesko, T. Shinn. Global Science and Innovation **6**, 85 (2015)

[3] O. Bleikher, S. Kvesko, Applied Mechanics and Materials 770, 651 (2015) doi: 10.4028/www.scientific.net/AMM.770.651

[4] N.V. Sosnin, *Competence-based approach in innovative engineering education* (Izd-vo RO IPK i PRO, Krasnoyarsk, 2006)

[5] J. Cosling & H. Mintzberg. HBR **81,** 54 (2003)

[6] M. Lehmann, P. Christensen, X. Du, M. Thrane. EJEE. **33,** 283 (2008)

[7] S. Gorbachev, M. Syryamkin, V. Syryamkin & E. Vaganova, XVth International Conference Linguistic and Cultural Studies: Traditions and Innovations (LKTI 2015) 206, 382 (2015)

[8] V.I. Syryamkin , E.G. Syryamkina, Procedia - Social and Behavioral Sciences **166**, 471 (2015)

MATEC Web of Conferences 79, 01049 (2016) DOI: 10.1051/matecconf/20167901049
IME&T 2016

METHOD FOR SEPARATION OF ELECTROMAGNETIC RESPONSES OF MECHANOELECTRICAL TRANSFORMATIONS UNDER THE IMPACT OF EXTERNAL NOISE

Anatoly Surzhikov[1], *Anatoly* Bespal'ko[1,*], *Anna* Demikhova[1], *Aleksandr* Bombizov[1,2], *Anton* Loschilov[2]

[1]National Research Tomsk Polytechnic University, 634050, Tomsk, Russia
[2]Tomsk State University of Control Systems and Radioelectronics, 634050, Tomsk, Russia

Abstract. The paper addresses the diagnosis of the presence of defects in concrete, cement-sand and other building and structure mixes using mechanoelectrical transformations. The problem of high levels of electromagnetic noise and interference affecting the correct interpretation of the data to diagnose presence of defects is considered. A version of the electromagnetic response from the unit heterogeneity is proposed, and the assumption of the total electromagnetic signal is suggested. The experiment on acoustic excitation of the sample was conducted and the electromagnetic responses were recorded and filtered using the correlation analysis. The obtained result was compared to the model one to confirm the single response model.

1 Introduction

During production and operation of concrete structures, there is a high demand for their reliability, since it is directly related to life safety. Deterioration of strength characteristics of concrete blocks can be largely due to the emergence of defect-containing areas that may appear during the manufacturing process due to imperfect processes and under the attack of the aggressive environment. In this regard, the development of a rapid diagnostic method suitable for quality control of construction materials both in the laboratory and on site and to analyze the reliability and strength of structures and supporting structures is an important task.

Different approaches to diagnosing of the dielectric material defects have been studied. Ultrasonic methods based on the study of ultrasonic wave propagation in the test product have gained widespread use in inspection of building materials. The easiest way to detect defects and concrete inhomogeneity are methods based on measurement of the speed of sound [1–4]. However, the accuracy of these methods depends on the accuracy of this measurement, which depends on the presence and amount of inhomogeneities. In case of one-sided access to the product, the echo method is commonly used to detect defects and

*Corresponding author: besko48@tpu.ru

© The Authors, published by EDP Sciences. This is an open access article distributed under the terms of the Creative Commons Attribution License 4.0 (http://creativecommons.org/licenses/by/4.0/).

their coordinates of different orientation [5]. The method implies product sounding with short ultrasonic pulses and recording the intensity and the arrival time of the echo signals reflected from discontinuities and inclusions. The methods of nondestructive testing of concrete defects that have been developed in recent years employ Rayleigh waves which are most sensitive to the near-surface cracks oriented vertically to the surface as compared to longitudinal waves [6, 7]. In [8], an approach to assess the real size of the defects using the shadow method of ultrasonic testing is proposed. The authors performed the input of ultrasonic vibrations at different angles using several independent pairs of transducers, thereby eliminating the measurement ambiguity. In the defect-containing region, each converter exhibited the amplitude signal envelopes, and the coordinates of the defect boundaries were recorded. These approaches are based on the principle of elastic (ultrasonic) wave propagation in the medium.

Apart from the existing physical phenomena, mechanoelectrical transformation is used for defects analysis. This is the transformation of the mechanical field energy in structural inhomogeneities or defects in dielectric structures [9]. This approach employs different techniques to conduct the study: short-pulse broadband excitation with subsequent analysis of the recorded responses [10], which increases the signal to noise ratio due to the probing signal energy; excitation by narrowband signals [11–13] and separation of responses from inhomogeneities with filters of high selectivity based on the principles of the test sample linearity.

This approach provides more information on the recorded signals of the mechanoelectrical transformation compared to that obtained by acoustic methods. The main problem encountered is high level of external electromagnetic interference from household and other devices that complicates the analysis and interpretation of responses.

The objective of this research was to search and use approaches to expose the test sample to a narrowband deterministic acoustic signal, and to develop the technique to separate the electromagnetic response during mechanoelectrical transformation in conditions of external electromagnetic interference.

2 Theory

The test sample, assume it is a sand-cement block, is a homogeneous medium with inclusions of empty cavities that represent defects. The equivalent electrical circuit of this sample can be represented as a serial connection of resistance-capacitance circuits. If this sample is exposed to short acoustic pulses, the length of which is less than the time of signal propagation through the sample, it can be considered as a distributed transmission line. In this case, the signal formed is a sum of electromagnetic responses from individual defects and geometric boundaries of the sample. As the sand-cement block is taken as a test sample, the speed of the acoustic wave propagation in the sample will be V=2500 m/s. If the thickness of the test sample is l=0.1 m, the signal path to the opposite boundary of the sample and its return to the exposure source will take $t=2\cdot l/V=1\cdot 0.1/2500=80$ μs. In this regard, it is appropriate to use the probe signal of the duration less than 80 μs to separate and interpret the electromagnetic responses.

For more noise-resistant electromagnetic response, the test sample should be exposed to the acoustic signal in the form of a radio pulse with the carrier frequency f.

As the test sample is a simplified form of the RC-circuits, a short radio pulse induces transient phenomena in each RC-circuit, which can be described by the following formula:

$$A(t)=\begin{cases} A_0\sin[\omega(t-t_0)]\left[1-e^{\frac{(t-t_0)}{\tau}}\right], t<T \\ A_0\sin[\omega(t-t_0)]\left[1-e^{\frac{(t-T)}{\tau}}\right]\left[1-e^{\frac{(T-t_0)}{\tau}}\right], t\geq T \end{cases} \quad (1)$$

where A_0 is the pulse envelope amplitude;

ω is angular frequency ($\omega = 2\pi f$);

t_0 is the pulse signal time;

τ is the time constant of the pulse rise and fall;

T is the time when the pulse envelope reaches its amplitude maximum.

On this assumption, the detected response signal should contain a set of signals:

$$S=\sum_{i=1}^{N} A_i(t) \quad (2)$$

where N is the number of imhomogeneities in the test sample.

Assume that an ideal test sample is free of defects. Initially, N equals to two responses: 1 is the response from the surface through which the sound travels; 2 is the response from the opposite surface of the sample. If a defect is in the middle of the sample, response 3 will be formed.

The difference in time between responses 1 and 2 should be about 80 μs for the sample thickness of 100 mm. Response 3 is to be recorded over 40 μs after response 1. All the responses recorded after 80 ms can be the result of multiple reflections of the acoustic wave that do not bear any useful information on sample defects.

In conditions of external noise effects, most acceptable approach is selective filtration through the correlation analysis of the recorded signal and the model radio pulse. It is known that the duration of the base signal during correlation affects the time and frequency characteristics of the recorded signal. In case of insufficient interference elimination, short signal duration retains time characteristics better. Long signal duration provides the opposite result. In this research, in conditions of high external noise, the base signal and the exposure signal of similar duration are used as a compromise derivative.

3 Experimental results

The studies were performed using the experimental setup, which excites the sample by a series of acoustic pulses and records the electromagnetic responses of mechanoelectrical transformations. Figure 1 shows the block diagram of the experimental setup.

The unit of exposure and measuring (UEaM) consists of a piezoelectric emitter, which induces a series of acoustic pulses of a predetermined shape towards the test sample (test sample SC) and a differential capacitive sensor to receive the electromagnetic response. The differential capacitive sensor has two inputs: measuring and control ones. The measuring input receives a useful signal and electromagnetic interference. The control input is located 12 mm above the measuring one and receives mostly noise but not the signal. Positioning of the UEaM over the test sample is performed with a plotter. The generation and digitization device (GaDD) based on the multi-channel data-acquisition unit NI PXIe-1071/PXI-5412/PXI-5105 generates excitement and digitizes electromagnetic responses. The measuring channel of the analog-to-digital converter is equipped with an auxiliary amplifier and low- and high-frequency Butterworth analog filters to limit the path of the analog bandwidth up to 1 kHz–1 MHz. The digital-to-analog converter output is connected to the input of the high-voltage amplifier (HVA) Tabor Electronics 9200A for 50-fold amplification of the 8 V signal and its transfer to the piezoelectric emitter to excite the sample.

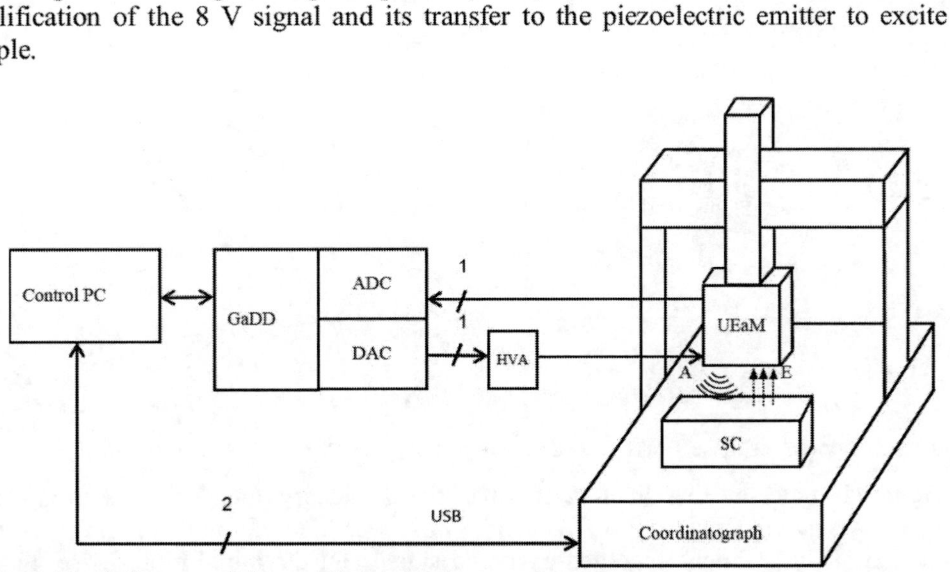

Figure 1. Block diagram of the experimental setup.

A sand-cement sample of 350×200×115 mm with two inclusions located in the center of the sample at a distance of 60 mm from the edge and 150 mm to each other was made for the experiments. The inclusions were foam cubes of 10×10×10 mm and 20×20×20 mm. The diffraction pattern of the test sample is shown in Figure 2.

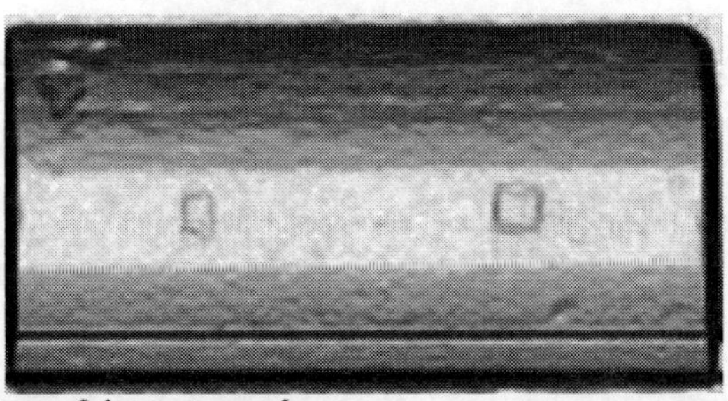

Figure 2. Diffraction pattern of the test sample.

The experimental studies were conducted in several stages.

At the first stage, the electromagnetic responses were measured. Two sample regions (defect-containing and defect-free ones) were chosen to obtain responses. To record the response from the defect, the UEaM was placed above the 20×20×20 mm inclusion using the plotter. To investigate the defect-free region, the UEaM was moved to the area equidistant to both defects. The waveform of the signal to sound the object was a radio pulse with carrier frequency of 122 kHz and duration of 75 μs. The results of multiple exposures of the sample and response recording were averaged. That minimized the stationary electromagnetic interference and noise. Figure 3 shows the averaged recorded responses for 128 measurements, where *A1* is a defect-containing region and *A2* is the region free of defects.

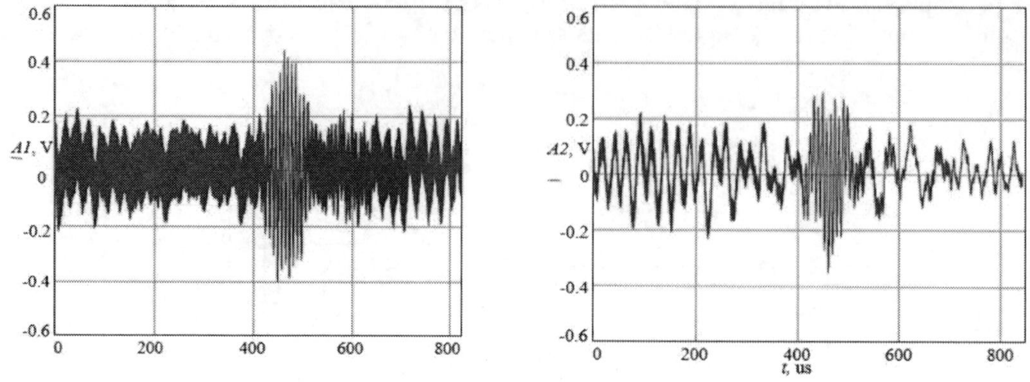

a b

Figure 3. Recorded response 1 (a). Recorded response 2 (b).

Figure 2 indicates that after averaging, the stationary interference and noise are minimal. At the next stage, the response was separated through filtering using the correlation analysis for the recorded averaged signal and the initial radio pulse. Figure 4 shows the electromagnetic responses after the correlation analysis (*A1'* is the defect-containing region; *A2'* is the region free of defects). The values were amplitude normalized to compare the result with those obtained previously.

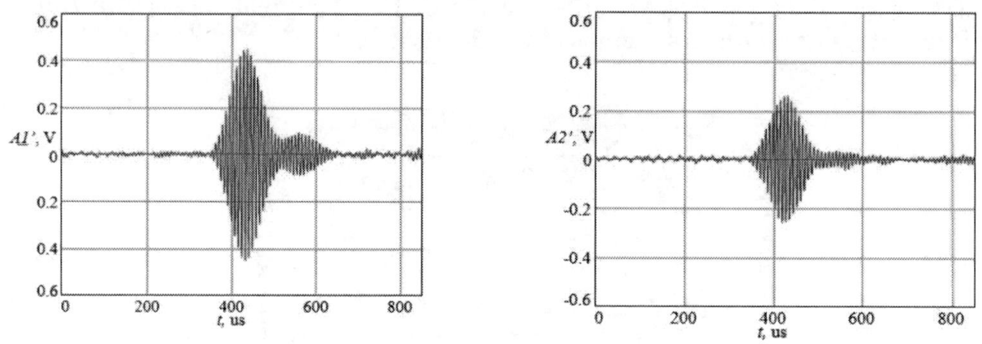

a b

Figure 4. Correlation of response 1 from the defect-containing region and the acoustic signal (a). Correlation of response 1 from the defect-free region and the acoustic signal (b).

After filtering, the obtained signals consisted of the responses from the boundaries of the test sample and the defect (Figure 4, *a*). To verify the results, the response from the defect-free region is shown in Figure 4, *b*.

4 Discussion of the results

During the analysis of the obtained results, the parametrization of the recorded signal was performed using formulas (1) and (2). The pulse time was chosen based on the speed of acoustic wave propagation (2500 m/sec), and the distance the defect (60 mm) and the lower boundary (115 mm). The parameters of the electromagnetic signals obtained for the defect-containing sample region were as follows: for the first response, A_{01}= 0.46 V; t_{01}= 400 μs; f = 122 kHz; $T1$= 475 μs; τ = 34 μs; for the second response, A_{02}= 0.24 V; t_{02}= 450.2 μs; $T2$=525.2 μs; for the third response, A_{03}= 0.12 V; t_{03}= 503.8 μs; $T3$=578.8 μs. For the defect-free region of the sample, the following parameters were obtained: for the first response, A_{02}=0.27 V; t_{02}=400 µs; f=122 kHz; $T1$ = 475 µs; τ=34 µs; for the second response, A_{02}= 0.05 V; t_{02}= 503.8 µs; $T2$=578.8 µs. After model correction, it was found that the response from the defect returned after t=t_{02}–t_{01}=50.2 µs and the depth was 2500×50.2 µs=62.75 mm, while the response reflected from the lower boundary of the sample indicated 129.75 mm depth. The differences in the depth can be attributed to the error of the acoustic wave propagation measurement in the sand-cement block, which may be caused by irregularities of the sample cavities.

The envelopes for signal modules are shown in Figure 5 to compare the results obtained through modelling and experimentally.

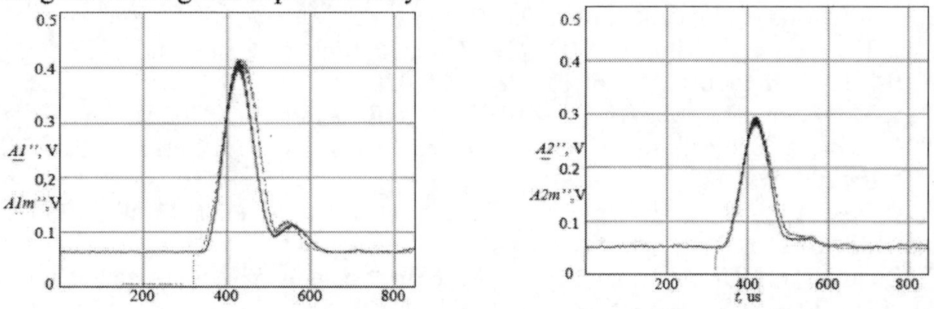

a b

Figure 5. Envelope of the correlation of response 1 with the sounding signal and model signal envelope (a). Envelope of correlation of response 2 with the actuating signal and model signal envelope (b).

The Figures exhibit the coincidence of the model and the experimental results. This confirms the correctness of the model chosen for this experiment.

5 Conclusions

A setup to study the imperfection of concrete, sand-cement mixes and other samples used in the construction industry has been designed. The setup is equipped with digital equipment to process and analyze the measured signals. The analysis of the results obtained in measurement of the electromagnetic response under acoustic effect showed that the correlation analysis allows separation of the useful signal component from electromagnetic interference signals and noise. The designed electromagnetic signal model can be used to test both defect-containing and defect-free regions of the sample. It can also be used to separate responses in conditions of external electromagnetic interference.

Further research will focus on the development of algorithms for parameterization of the recorded signal to decompose it into individual responses and to define the time and amplitude characteristics of these responses. This will allow estimation of the imperfection of the investigated region.

Acknowledgment

This work was supported in part by the Ministry of Education and Science of the Russian Federation under the State Research Project and by RFBR Grant No. 14-08-00395.

References

[1] C. Popovics, K. Komlos, J. Popovics, NDTCE **2**, (1997)
[2] S.V. Bordunov, O.V. Galtseva, J. Phys: Conf. Ser. **671**, 012009 (2016) doi: 10.1088/1742-6596/671/1/012009
[3] E. Vasendina, I. Plotnikova, L. Red'ko, N. Zyablova, IOP Conf. Ser.: Mater. Sci. Eng. **81**, 012070 (2015) doi: 10.1088/1757-899X/81/1/012070
[4] A. Gavrilin, B. Moyzes, O. Zharkevich, J. of Vibroengineering **17**, 3495 (2015)
[5] V.V. Klyuev, *Nondestructive Testing: Handbook* (Mashinostroenie, Moscow, 2004)
[6] D.G. Aggelis, E. Leonidou, T.E. Matikas, Cem. Concr. Compos. **34**, 140 (2012) doi: 10.1016/j.cemconcomp.2011.09.017
[7] K.Z. Kwong, Y.Y. Lim, W.Y.H. Liew, Matec Web of Conferences **47**, 02014 (2016) doi: 10.1051/matecconf/20164702014
[8] V.A. Mosyagin, A.A. Markov. V MIRE NK (in Russian) **4**, 11 (2014)
[9] T.V. Fursa, A.P. Surzhikov, N.N. Khorsov, K.Yu. Osipov, V.A. Zatsepin, Bull. TPU **308**, 10 (2005)
[10] A.A. Bespal'ko, L.V. Yavorovich, P.I. Fedotov, J Min Science **43**, 472 (2007) doi: 10.1007/s10913-007-0049-8
[11] T.V. Fursa, G.E. Utsyn, K.Yu. Osipov, Modern Problems of Science and Education **5**, 145 (2012)
[12] A.P.Surzhikov, O.V. Galtseva, E.A. Vasendina, V.A.Vlasov, E.V. Nikolaev, IOP Conf. Ser.: Mater. Sci. Eng., **110**, 012002 (2016) doi: 10.1088/1757-899X/110/1/012002
[13] V.P. Surzhikov, N.N. Khorsov, P.N. Khorsov, Russ. J. Nondestr. Test. **48** , 85 (2012) doi: 10.1134/s106183091202009x

MATEC Web of Conferences **79**, 01050 (2016) DOI: 10.1051/matecconf/20167901050
IME&T 2016

IMAGE PROCESSING IN INTELLIGENT MEDICAL ROBOTIC SYSTEMS

Dmitriy Shashev[a]

National Research Tomsk State University, 634050, Tomsk, Russia

Abstract. The paper deals with the use of high-performance computing systems with the parallel-operation architecture in intelligent medical systems, such as medical robotic systems, based on a computer vision system, is an automatic control system with the strict requirements, such as high reliability, accuracy and speed of performance. It shows the basic block-diagram of an automatic control system based on a computer vision system. The author considers the possibility of using a reconfigurable computing environment in such systems. The design principles of the reconfigurable computing environment allows to improve a reliability, accuracy and performance of whole system many times. The article contains the brief overview and the theory of the research, demonstrates the use of reconfigurable computing environments for the image pre-processing, namely morphological image processing operations. Present results of the successful simulation of the reconfigurable computing environment and implementation of the morphological image processing operations on the test image in the MATLAB Simulink.

1 Introduction

Nowadays the trend of the development of modern intelligent medical systems (IMS), such as medical robotic systems, assumes the use of a modern equipment (a fully or partially automatic control system). Strict requirements are imposed on those medical systems in terms of quality, reliability, accuracy and speed of performance.

Computer vision systems (CVS) represent a fast-developing field and are widely used in the structure of IMS [1, 2]. In their simplest version they comprise a hardware component in the form of a computing system and a software component in the form of a digital image processing algorithm. The use of high-performance computing systems as part of CVS allows improving qualitative characteristics of its operation, which in turn allows reaching new improved characteristics of the automated process control system as a whole.

The present paper examines the synthesis of reconfigurable digital image processing systems and their use as CVS in IMS. These systems are reconfigurable computing environments for digital image processing tasks.

[a] Corresponding author: dshashev@mail.ru

© The Authors, published by EDP Sciences. This is an open access article distributed under the terms of the Creative Commons Attribution License 4.0 (http://creativecommons.org/licenses/by/4.0/).

2 Structure of control system

In the general case the structure of automated control system with CVS looks as follows (Figure 1).

Figure 1 presents a scheme consisting of the following units:

– *Current image*, produced using any photo or video camera;

– *Object parameters* – parameters, based on the assessment of which the whole system needs to carry out the automation task;

– *Control device* – device that executes the parameter assessment algorithm;

– *Actuating mechanism* – mechanism that on the basis of control device data performs manipulations over the plant;

– *Plant*.

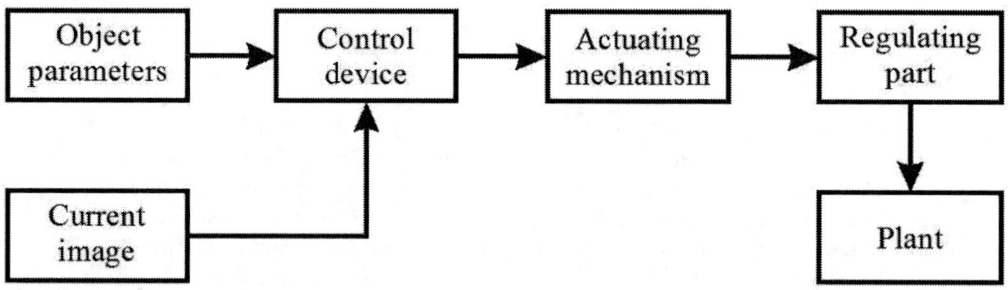

Figure 1. General structure of control system.

Depending on the type of control, in the general case the system control law looks as follows:

$$U(t)=\{A,\ if\ B-\Lambda\cdot B\leq X\leq B+\Lambda\cdot B\} \tag{1}$$

Here $A=[a_1,a_2,...,a_i,...,a_n]$ is the range of control signal values; $B=[b_1,b_2,...,b_i,...,b_n]$ is the range of plant parameter values; $\Lambda=[\lambda_1,\lambda_2,...,\lambda_i,...,\lambda_n]$ is the range of plant parameter identification errors; $X=[x_1,x_2,...,x_i,...,x_n]$ is the range of identified plant parameters.

Therefore, proceeding from the control law described above, qualitative characteristics of control system performance to a large extent depend on the part of its structure that is in charge of identifying plant parameters, i.e. control device that in our case is a CVS as hardware-software complex in charge of executing the preset digital image processing algorithm.

As it was earlier said, the present paper suggests using the reconfigurable digital image processing system.

3 Reconfigurable computing environments

As it was earlier said, reconfigurable digital image processing system is one of the key aspects in performance of the control system demonstrated in Figure 1.

The special feature of the digital image processing system, whose general structure is presented in Figure 2, is the use of *reconfigurable computing environment* (RCE) as computing system.

RCE is a discrete mathematical model of a high-performance computing system consisting of identical and identically interconnected basic elements (*processing elements / PE*), softset for performing any function out of a complete range of logical functions,

 DOI: 10.1051/matecconf/20167901050

memory and any connections with neighbors [3–5]. PE in the RCE used operate on the basis of logical functions of AND, OR, NOT.

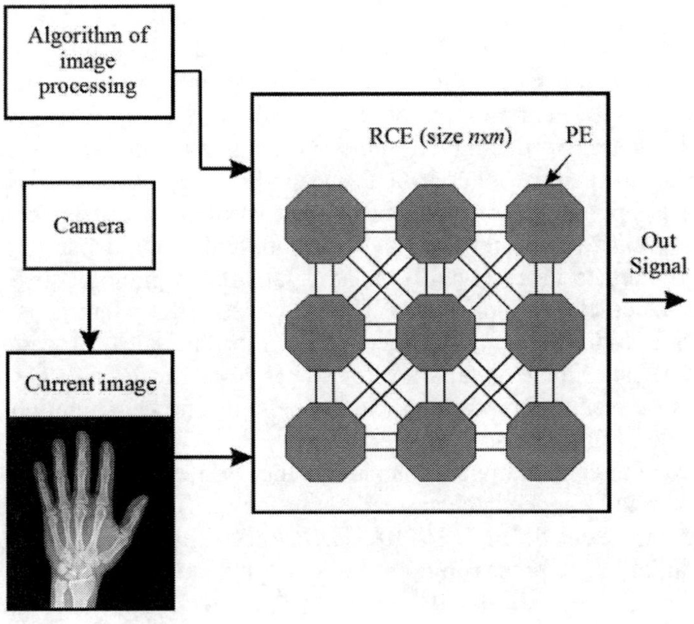

Figure 2. General structure of reconfigurable digital image processing system.

Principles of RCE construction provide it with the following key properties:
– high computing speed;
– reliability, flexibility and structural universality;
– ease of manufacture.

Abstract RCE model is also presented in Figure 2. As one can see, this model reproduces spatial representation of the digital image (matrix with a set of numeric values, where matrix dimensions $n \times m$ are the resolution of the source image and numerical values represent brightness of the corresponding image pixel). Hence RCE size in terms of the number of PE is equal to the resolution of the processed source image.

Therefore, corresponding brightness value of the original image is fed for processing to the input of each PE of RCE used, and then a series of logical operations take place in PE, this series being determined by the image processing algorithm built into RCE operation.

The structure of digital image processing algorithms built into CVS represents a complex set of different operations. The following main stages are usually identified in algorithm operation [6]:
– image pre-processing;
– segmentation;
– highlighting of geometric structure;
– identification of relative structure and semantics.

4 Operations of morphological image processing in RCE

Operations of *morphological image processing* are very often used in the algorithm of digital image processing with CVS at the stages of image pre-processing and segmentation.

The paper suggests a technique for morphological processing of gray scale images adapted to be implemented with RCE.

Mathematical morphology is the instrument of retrieving certain image components that are useful for the understanding and description of object shapes [6, 7]. Morphological methods used for processing of gray scale images are of interest in the context of the present paper.

Using the theory of sets terminology, let us assert that $f(x, y)$ function is called a digital image, if (x, y) represent pairs of whole numbers of Cartesian product Z^2 (Cartesian product of Z multiplied by itself, i.e. multitude of all ordered pairs (z_i, z_j), where z_i, z_j are any whole numbers and f is the transformation that matches brightness value (that belongs to the range of real numbers R) to each pair of coordinates (x, y) [7].

Morphological operations are performed over two images: the one being processed (let us designate it as A) and a special one (let us designate it as B), depending on the operation type and the problem to be resolved. In mathematical morphology such special image is called a structure elements or a primitive. The size of structure element is usually 3×3, 4×4, 5×5 pixels. This is conditioned by the very idea of morphological processing, in the course of which characteristic image details are found. The sought-for detail is described with a primitive, and as a result of morphological processing one can highlight or remove such detail in the whole of the processed image.

There are two fundamental operations in the theory of morphological gray scale image processing: *Dilate* and *Erode*. One can read about the classical mathematical description of such operations in greater detail in [6–10]. Let us determine the primary purpose of these operations in digital image processing:

– for 'dilation' operation, dark details get toned down or disappear completely, processed image becomes brighter than the source one;

– for 'erosion' operation, bright details of the image get toned down.

Coming back to the issue of morphological image processing operations using RCE, there is a need to transform the classical algorithm of such operations into the algorithm that corresponds to RCE requirements described above.

Below the system of functions built into PE operation is presented.

Let us agree that in the following formulas variables a and b are n-bit binary numbers that also represent the values of certain pixels of a gray scale image. In our case 8-bit binary numbers are considered, i.e. 8-bit gray scale images are to be processed.

For the sake of convenience we will designate the value of binary place with a superscript. For instance: $a = 10001000$ – the input 8-bit binary number; $a^8 = 1$ – the value of the eighth binary place of the input binary number a.

Thus, at $n = 8$ and $i = \overline{1,n}$ the execution of the following system of functions is input into PE:

$$
\begin{cases}
R(a,b) = (a^n b^n \vee \overline{a^n \vee b^n}) \cdot (a^{n-1} b^{n-1} \vee \overline{a^{n-1} \vee b^{n-1}}) \cdot \\
\qquad \cdot (a^{n-2} b^{n-2} \vee \overline{a^{n-2} \vee b^{n-2}}) \cdot \ldots \cdot (a^1 b^1 \vee \overline{a^1 \vee b^1}), \\
B(a,b) = a^n \overline{b^n} \vee (a^{n-1} \overline{b^{n-1}} \cdot (a^n b^n \vee \overline{a^n \vee b^n})) \vee (a^{n-2} \overline{b^{n-2}} \cdot (a^n b^n \vee \overline{a^n \vee b^n}) \cdot \\
\qquad \cdot (a^{n-1} b^{n-1} \vee \overline{a^{n-1} \vee b^{n-1}})) \vee \ldots \vee (a^1 \overline{b^1} \cdot (a^n b^n \vee \overline{a^n \vee b^n}) \cdot \\
\qquad \cdot (a^{n-1} b^{n-1} \vee \overline{a^{n-1} \vee b^{n-1}}) \cdot \ldots \cdot (a^2 b^2 \vee \overline{a^2 \vee b^2})), \\
M(a,b) = \overline{a^n} b^n \vee (\overline{a^{n-1}} b^{n-1} \cdot (a^n b^n \vee \overline{a^n \vee b^n})) \vee (\overline{a^{n-2}} b^{n-2} \cdot (a^n b^n \vee \overline{a^n \vee b^n}) \cdot \\
\qquad \cdot (a^{n-1} b^{n-1} \vee \overline{a^{n-1} \vee b^{n-1}})) \vee \ldots \vee (\overline{a^1} b^1 \cdot (a^n b^n \vee \overline{a^n \vee b^n}) \cdot \\
\qquad \cdot (a^{n-1} b^{n-1} \vee \overline{a^{n-1} \vee b^{n-1}}) \cdot \ldots \cdot (a^2 b^2 \vee \overline{a^2 \vee b^2})), \\
D^i(a,b) = (b^i \cdot (R(a,b)) \vee (a^i \cdot (B(a,b)) \vee (b^i \cdot (M(a,b)), \\
\max(a,b) = [D^n(a,b) \ D^{n-1}(a,b) \ \ldots \ D^{n-i}(a,b) \ \ldots \ D^1(a,b)], \\
E^i(a,b) = (b^i \cdot (R(a,b)) \vee (b^i \cdot (B(a,b)) \vee (a^i \cdot (M(a,b)), \\
\min(a,b) = [E^n(a,b) \ E^{n-1}(a,b) \ \ldots \ E^{n-i}(a,b) \ \ldots \ E^1(a,b)], \\
f_1 = \max(\max(\max(\max(y_1, y_2), \max(y_3, y_4)), \max(\max(y_5, y_6), \max(y_7, y_8))), x), \\
f_2 = \min(\min(\min(\min(y_1, y_2), \min(y_3, y_4)), \min(\min(y_5, y_6), \min(y_7, y_8))), x).
\end{cases}
\tag{2}
$$

In the above described system f_1 and f_2 designate 'dilation' and 'erosion' operations respectively.

5 Summary

Modeling of RCE for 'dilation' operation of morphological grayscale image processing, as well as validation of this operation were performed using MATLAB Simulink [11]. Validation was performed by comparing results of RCE-based operations with the same operations available in the Image Processing Toolbox extension of MATLAB suite.

The obtained result of processing the source image (Figure 3, *a*) using the developed RCE (Figure 3, *b*) and the one in MATLAB (Figure 3, *c*) are identical, which indicates the correctness of RCE operation. The result of correct RCE operation for 'erosion' operation was also obtained.

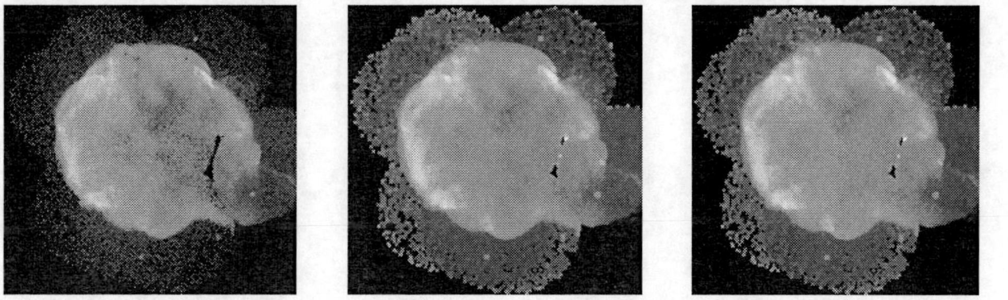

a b c

Figure 3. Results of simulation.

The present research is aimed at studying and developing the automated control systems (ACS), using in IMS, with CVS used in their structure. It was demonstrated that correct approach to selection of CVS to be used affects qualitative characteristics of ACS performance as a whole. In today's world there is a need to develop alternative high-performance computing systems that would allow moving to a new qualitative level of the tasks resolved.

The authors examined and demonstrated the possibility of implementing the RCE model for the tasks of morphological gray scale image processing. RCE can be used to achieve the same results of gray scale image processing, as when conventional methods are used, however processing time is significantly reduced in the first case: 'dilation' and 'erosion' operations are executed over one operating cycle of RCE processing element.

The study and development of RCE for image processing tasks and other classes of problems are the promising lines of research, as far as design concept of such computing environments offers high computational capability compared to computational tools that are currently in use. In turn, there is the possibility of developing a qualitatively new IMS, based on CVS, with a high reliability, accuracy and speed.

Acknowledgment

The reported study was funded by RFBR according to the research project No. 16-37-00082 mol_a.

References

[1] J. F. Giallo II, L.S. Mattos, S. Miller, Z. Neistedt, M. Draelos, G. Lee, E. Grant, CAINE-2012, 225 (2012)
[2] C. H. Shi, MATEC Web of Conferences **44**, 01086 (2016) doi: 10.1051/matecconf/20164401086
[3] S.V. Shidlovskii, J. Comp. Sys. Sci. Inter. **45**, 282 (2006) doi: 10.1134/S1064230706020122
[4] B. Rusyn, M. Kuzio, V. Shmoylov, Auto. Cont. Comp. Sci. **34**, 58 (2000)
[5] L. Shapiro and G. Stockman, *Computer Vision* (Prentice Hall, New Jersey, 2001)
[6] R.C. Gonzalez, R.E. Woods, *Digital Image Processing* (Prentice Hall, New Jersey, 2008)
[7] C. Alcalde, A. Burusco, R. Fuentes-Gonzalez, An. Math. Art. In. **72**, 115 (2014)
[8] ZZ. Zheng and GQ. Zhou, IEEE J. Sel. Top. App. Earth Ob. Re. Sen. **6**, 2338 (2013)
[9] P. Seibold and M. Gerke, VIPIMAGE 2013, 203 (2013)
[10] Z. Fang, M. Yulei, Z. Junpeng, ICCASM 2012, 948 (2012)
[11] R.C. Gonzalez, R.E. Woods, S.L. Eddins, *Digital Image Processing Using MATLAB* (Gatesmark Publishing, Knoxville, 2009)

ADVANTAGES OF NANOSENSORS IN THE DEVELOPMENT OF INTERFACES FOR BIOELECTRIC PROSTHESES

Diana Avdeeva[1*], *Ekaterina* Zakharchenko[1], *Dang Quang* Nguyen[1], *Nikita* Turushev[1], *Aryslan* Abdrakhmanov[1], *Roman* Vorona[1], *Arnold* Popkov[2], *Sergey* Tverdokhlebov[1]

[1]National Research Tomsk Polytechnic University, 634050, Tomsk, Russia
[2]Federal State-Financed Institution "Russian Ilizarov Scientific Center for Restorative Traumatology and Orthopaedics of the RF Ministry of healthcare", 640014, Kurgan, Russia

Abstract. The present research aims to explore the bioelectric activity of muscles using a high-resolution electromyograph and to analyze the prospects of the electromyograph to develop bioelectric patterns for the prosthesis control method based on the data recognition system. The activity of the healthy forearm muscles was investigated during the cyclic activity of fingers in different modes. In addition, the impact of filters on the quality and informativity of myoelectric signals, as well as on the development of bioelectric activity patterns was analyzed. The virtually developed bandpass filters were utilized as experimental filters. The filter impact analysis included the comparison of the signal recorded in the frequency band from 0 to 10000 Hz with the signal filtered in the frequency band from 20 to 500 Hz. The research revealed the advantages of a high-resolution electromyogram for the pattern recognition-based myocontrol.

1 Introduction

Loss of a limb as a result of accident or injury can have a considerable psychological and physical impact. The upper limb amputees experience difficulties with simple thing that do not usually require any efforts. In such cases, prostheses allow people to return to a near normal life: to work, to do sport, to cook, etc.

There are three most common types of hand prosthetic devices: cosmetic, body-powered, and bioelectric prostheses.

Cosmetic hand prostheses are designed to hide the lack of a limb from stares in public. Natural-looking prostheses are focused on the aesthetic appearance, not on the restoration of motor functions. This type of prosthesis has been used to restore a partially amputated hand (Figure 1, *a*) [1].

The first active upper-limb gripping hook prosthesis was developed by Dorrance in 1912. The first version of this prosthesis was limited to gripping and holding objects [2].

* Corresponding author: diana.avdeeva@mail.ru

© The Authors, published by EDP Sciences. This is an open access article distributed under the terms of the Creative Commons Attribution License 4.0 (http://creativecommons.org/licenses/by/4.0/).

The device was a body-powered prosthesis. The executing mechanism of the device was a hook with a pinching mechanism to perform motions by the system of belts and a spring. Movements of the patient's healthy upper arm with one of the expandable prosthesis belts attached were used as control information. The patient had to pull the unclamping belt by a healthy part of the arm to open the clamping mechanism of the prosthesis. The belt pulled back the ends of the clamping mechanism to open the prosthesis. Modern body-powered prostheses function in a similar way. The main disadvantages of the mechanical prosthetic hands are as follows: short service life; limited power grasping; a limited number of degrees of freedom; linkage system can cause discomfort and limit activities; unsightly appearance [1].

The development of science and technology allowed scientists to create prostheses which can be controlled by the registration of bioelectric potentials of a human body. These prostheses are bioelectric (Figure 1, *b*). Bioelectric prostheses perform the function of hand rotation, capturing and holding objects. The main advantage of bioelectric prostheses is a high grasping power and a large number of degrees of freedom. Due to its unique functions, the bioelectric prosthesis allows disabled people to cope with small objects such as a ballpoint pen, fork, spoon, etc. But this type of prosthesis has its own disadvantages: need for maintenance; higher cost and increased weight compared to other types of prostheses; need for the battery charge [1].

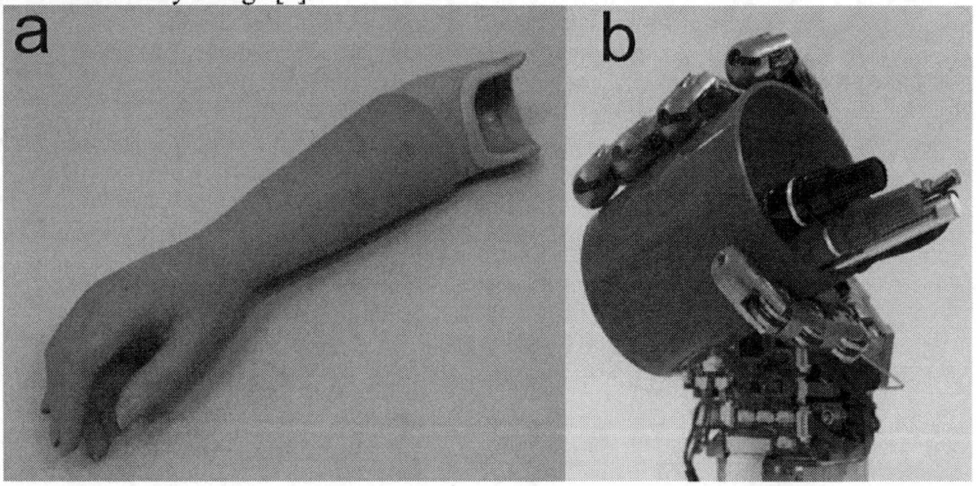

Figure 1. Types of prostheses: a – cosmetic [3], b – bioelectric [4].

Modern bioelectric prostheses are usually controlled by one of the three main control methods: brain-computer interface, neuroelectric, and electromyographic (myoelectric) [5].

The brain-computer interface method uses a subcortical signal as the primary data source. The signal is recorded using the leads implanted into a brain cortex. The obtained data are processed and transmitted to the mechanical part of prosthesis [6]. Prostheses with this type of control are used by people suffering from paralysis of limbs caused by violation of neural connections between the brain and the corresponding organ (tetraplegia, demyelinization of nerve channels). The main disadvantage of the method is its invasiveness, which can cause the infection or injury of the brain during the implantation of electrodes and the prosthesis operation [5].

The neuroelectric method of control uses signals recorded from the active motor nerves of a person. Electrodes can be implanted in nervous tissue (regenerative, an electrode cuff) or placed over the nerves [6]. The main advantages of neuroelectric method are as follows: the possibility of sensitivity restoration by electric stimulation of the sensory nerve fibers and the lack of dependence on the muscles of a missing limb [7]. The disadvantages of the method include the following: invasiveness; low amplitude of the informative signal in comparison with other sources of bioelectric potentials; necessity to design a percutaneous interface; inability to be used in completely paralyzed patients. The main risk of using such prostheses is the possibility of destruction or damage of the nerve and surrounding tissues, and postoperative infections [5].

The electromyographic method uses a myoelectric signal as the primary data source to control prostheses. Myoelectric signal is a signal that is recorded from the excited muscle fibers. This method of prosthesis control consists in registration of the muscle electrical activity signal using surface electrodes. Then the signal is processed by the input amplifier circuits and converted into a digital code. The digital code is analyzed by the microcontroller of the control unit and converted into a command for the prosthesis executing mechanism (Figure 2) [8].

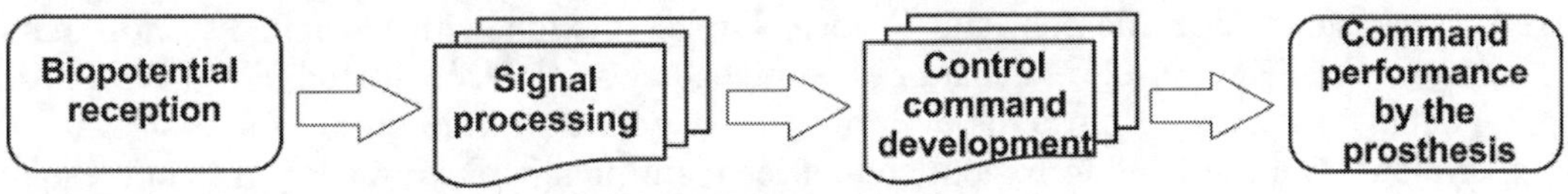

Figure 2. Flow chart of the prosthesis control system.

The main advantages of the electromyographic control method are its non-invasiveness and simplicity of the prosthesis control. The result of the prosthesis operation using this method depends on several factors: the number of informative areas, the arrangement of electrodes on the informative surface, changes in the position of electrodes during operation. The use of prostheses using electromyographic control method requires activity and integrity of the muscle fiber, which is responsible for control of the missing limb [9].

One of the examples of a modern myoelectric prosthesis is a bioelectric prosthesis SmartHand (Figure 1, *b*). This prosthesis was developed by researchers of Lund University in Sweden and Scuola Superiore Sant`Anna in Italy. This device comprises built-in touch sensors on the prosthesis artificial fingers, which are capable of partial restoration of touch perception [4].

Prosthetics falls behind in technological development in spite of the advances in this branch of surgery and organization of prosthesis control. A number of problems in prosthetics remain unsolved [10]. The amputees experience difficulties with significant weight of prostheses; anatomical inequality of the prosthesis and a healthy limb; necessity to make additional efforts which exhaust a patient. All these disadvantages bring discomfort for patients and force them to think about prosthesis. The study of mechanisms of a healthy limb functioning, bioelectric processes in healthy limbs and their interaction will help to determine the appropriate set of bioelectric signal characteristics for the qualitative control of prosthesis.

For example, the determination of specific characteristics of the active muscle myoelectric signal enables organization of the prosthesis control, which uses the patterns of harmonic sets of signal spectrum. This provides the control system with additional functions with minimum number of leads. The organization of this type of control requires myographic investigation of bioelectrical activity of target muscles to control the prosthesis. The quality of the obtained data depends on sensitivity and noise immunity of diagnosing devices.

2 Materials and methods

The research aims to study bioelectric activity of muscles of a healthy limb, namely the forearm, in order to define electrically active zones, the dependence of bioelectric activity of muscles on the geometric position of electrodes in space using a high resolution electromyograph, and the impact of filters on the waveform. The nanosensor-based electromyograph utilized in the research enables measurement of the electrical activity of muscles within the frequency range from 0 to 10,000 Hz with a minimal sensitivity threshold of potential (0.3 μV) without standard filter circuits [5, 6]. A high level of noise immunity is due to nanosensors, which eliminate most adverse effects of electrodes, such as polarization during operation and dryout of contact electrolytes as a result of a long-duration use [11]. The exclusion of filters from the measuring circuit allows us to eliminate the impact of filters on the amplitude and phase characteristics of the measuring signal and to increase the rate of data processing.

The practical part of the research consisted in the registration of myographic potentials from different positions on the forearm during the repetitive isolated finger motions. Three types of repetitive exercises were performed during the conducted experiments: fingers at rest, weak slow finger flexion, and intensive finger flexion with tension. Psychosomatic response of skeletal muscles to changes in ambience was also considered. For this purpose, the patient was asked to close his eyes for a short period of time during the first phase of experiment. The exercises were applied to three main fingers playing a key role in the hand functioning: a thumb, point finger, and middle finger. Bioelectric signals were obtained from the forearm surface using a bipolar configuration of nanosensors with a common reference wire. The sensors for upper-limb prostheses were placed on the electrically active zones, selected according to Vorotnikov's study of dispersal of bioelectric potentials (Figure 3) [9].

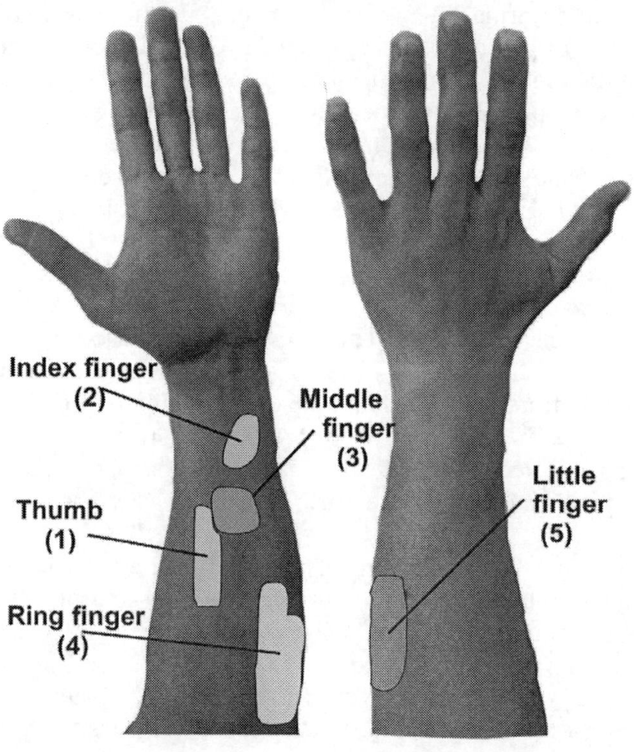

Figure 3. Location of bioelectrical activity zones on the forearm according to S.A. Vorotnikov.

The obtained broadband signals were processed using a bandpass digital filter with cutoff frequencies 20 and 500 Hz, that correspond to the standard range of informative electromyographic signals [12].

3 Results and discussion

The obtained results are considered through the example of bioelectric activity of the point finger at rest and in the state of active finger flexion and extension. Figure 4 shows electromyograms of the rest state (Figure 4, *a*), brain-computer interface (Figure 4, *b*), weak finger activity (Figure 4, *c*), and intensive finger activity (Figure 4, *d*) obtained by a high-resolution electromyograph without filters.

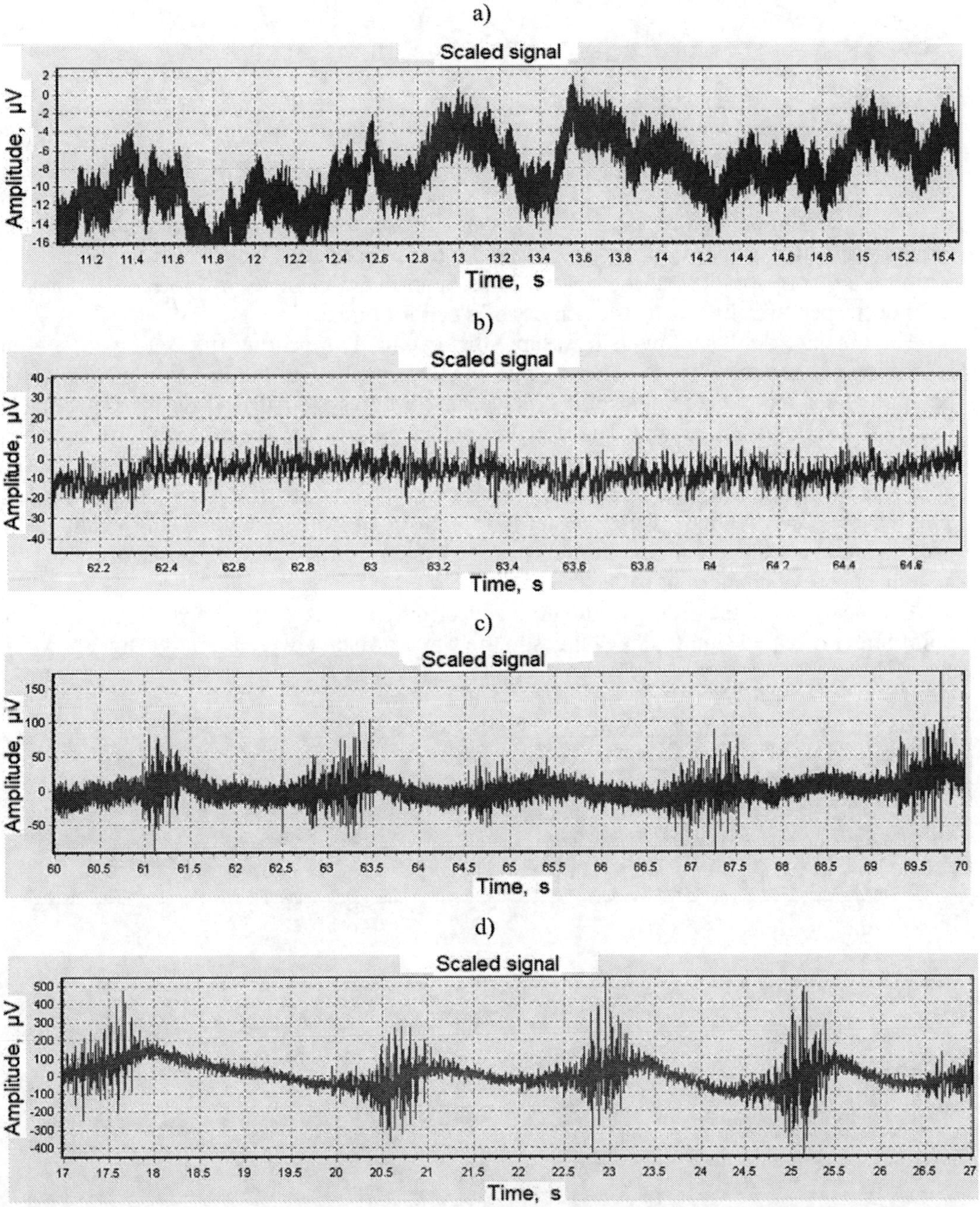

Figure 4. Electromyogram of the point finger activity: a – rest state in the band from 0 to 10,000 Hz; b – brain-computer interface in the band from 0 to 10,000 Hz; c – weak finger flexions in the band from 0 to 10,000 Hz; d – intensive finger flexions in the band from 0 to 10,000 Hz.

The analysis of the rest state electromyogram revealed that at the moment when a patient closed his eyes, the muscle psychosomatic response to changes in visual perception and mental state was registered as a sharp increase of the electromyographic signal direct component on 140 μV (Figure 5).

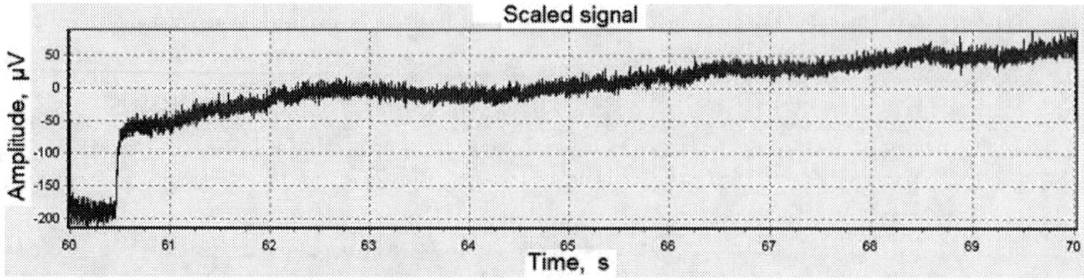

Figure 5. A sharp increase of the direct component in the rest state after closing the eyes.

The analysis of the weak and intensive flexion electromyograms confirmed that bioelectric activity zones of fingers located on the forearm, which were proposed by Vorotnikov, correspond to those in [3]. Thus, significant changes in bioelectric potential level occur periodically during the activity of a certain finger.

The obtained results allow us to assume that cyclically repeating fluctuations of a high-resolution myoelectric signal, which change simultaneously with the finger movements, can be applied for the development of a bioelectric control system based on the detection of potential threshold values and data recognition techniques for specific signal harmonics. For example, low-frequency components contain information not only about the muscle volume alteration, but about the muscle fatigue, the current state of the nervous system, etc. This enables the creation of new systems to control prosthetic devices and evaluate the user's state in a real-time mode. In addition, the systems will be able to perform preventive actions in case of changes in patient condition.

The analysis of the filtered and initial electromyograms of intensive flexions of the point finger revealed significant changes in the waveform and decrease in the signal level.

a)

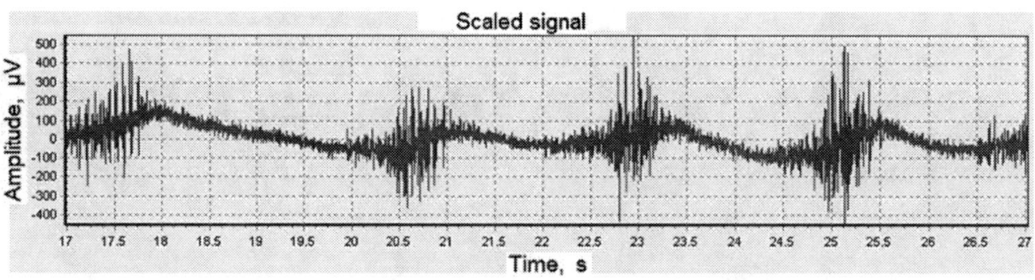

b)

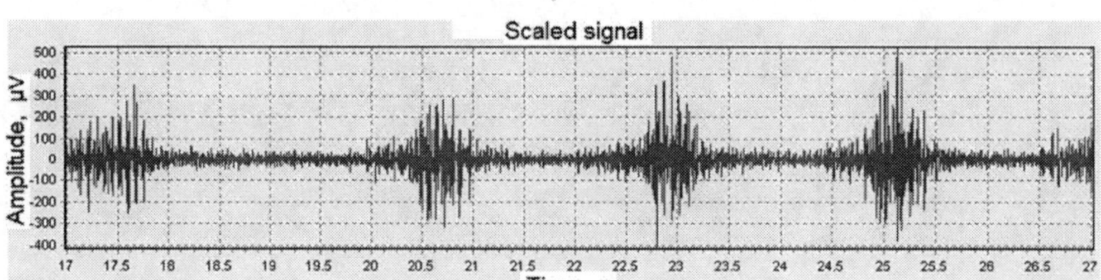

Figure 6. Myographic signals of the point finger activity: a – initial signal; b – filtered signal in the frequency band from 20 to 500 Hz.

Figure 7 shows miographic signals of the thumb, index and middle finger flexions according to S.A. Vorotnikov.

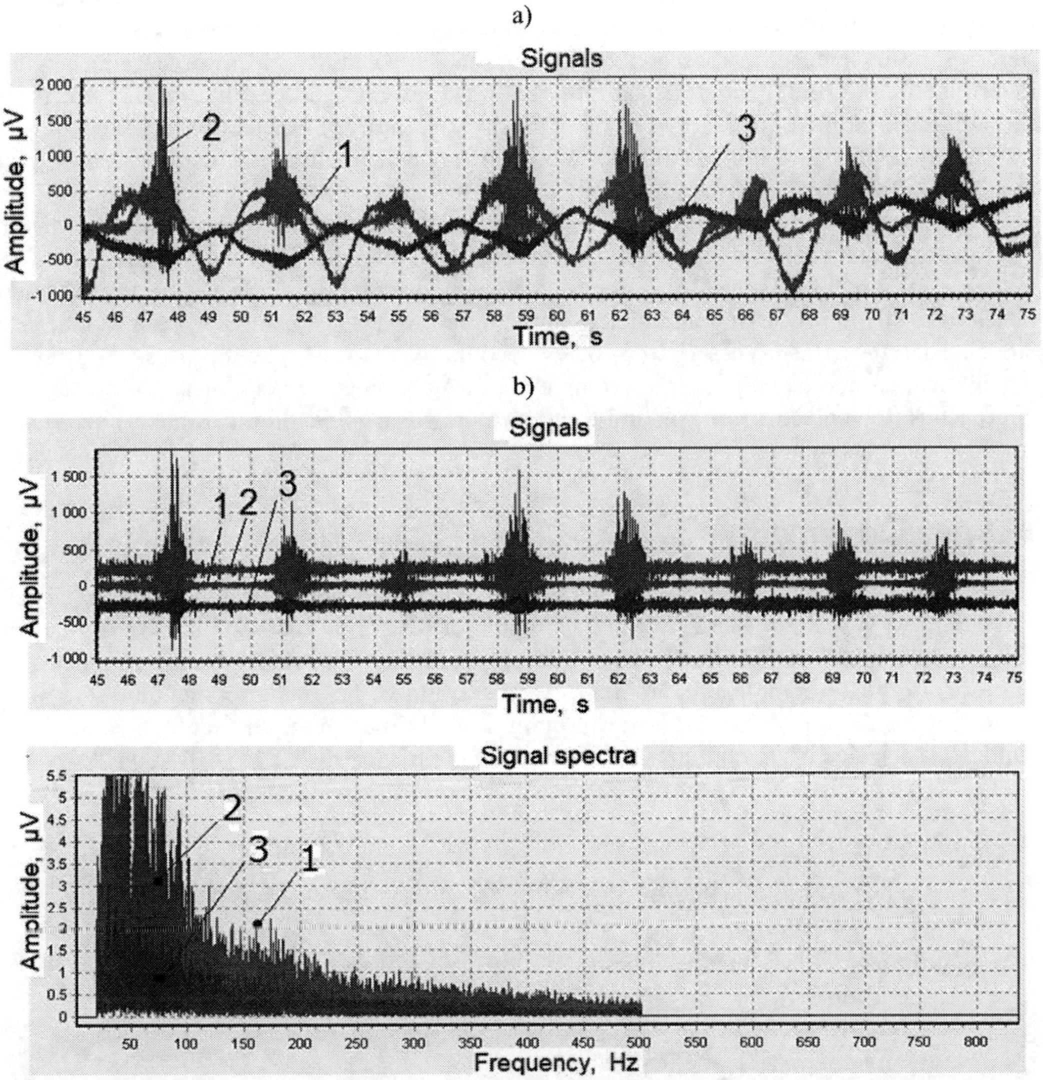

Figure 7. Electromyograms of the finger active flexions: point finger (1), middle finger (2) and a thumb (3): a – in the frequency band from 0 to 10000 Hz; b – in the frequency band from 20 to 500 Hz.

The obtained results showed that filters dramatically distort the waveform, reduce the energy, and remove phase shifts (Figure 7). Since the low-frequency range is not considered, a general picture of electrophysiological processes during muscular contraction cannot be fully evaluated using available modern technologies.

According to the analyzed data, the filter input circuits in bioelectric measuring instruments have a negative impact, thereby reducing the amount of data to be obtained. Modern level of development of the material engineering, computing tools and electrophysiological research potentials exceeds the quality of the current level of myographic research and prosthetic bioelectric control. In addition, the implementation of modern techniques of data processing and obtaining is crucial to provide a general picture of muscle activity.

4 Conclusion

The conducted research confirmed the efficiency of the electrode placement scheme proposed by Vorotnikov for electromyographic study in a wide frequency range using a high-resolution nanosensor-based electromyograph. The obtained electromyograms enabled determination of the patient psychosomatic reaction on changes in visual perception through the example of muscles. The distinguishing features of bioelectric signal enables its utilization for the threshold control in conventional and intelligent prostheses using data recognition patterns.

The input filter circuits limiting the full scale of electromyograph significantly distort the waveform, decrease its energy, and remove phase shifts, thereby reducing the amount of valuable information, especially in the low-frequency range. In addition, the use of filters decelerates signal processing and increases the time response of control system to the input signal. Therefore, it is necessary to find new techniques to measure and process myoelectric signals, which correspond to the current level of technology development, to create new approaches to bioelectric control, and to investigate electrophysiological muscle activity to improve the lives of amputees.

Acknowledgement

The research was financially supported by the Federal Targeted Program "Research and Development in Priority Fields of S&T Complex of Russia in 2014–2020", the Agreement No.14.578.21.0032 dated 05.06.2014 "Development of the experimental sample of a hardware-software complex for noninvasive recording of heart micropotentials in a wide frequency band without filtering and averaging in real time to early detect the symptoms of sudden cardiac death", the unique identifier of the contract: RFMEF157814X0032.

References

[1] Upper extremity prosthetics.
URL:
http://biomed.brown.edu/Courses/BI108/BI108_2003_Groups/Athletic_Prosthetics/UpperLimb.
[2] E.E. Vanderwerker, Inter-Clinic Information Bulletin **15**, 15 (1976)
[3] Forearm external cosmetic prosthesis.
URL: http://www.medicalexpo.com/prod/rslsteeper/product-74956-458446.html
[4] Ch. Cipriani, M. Controzzi, M.Ch. Carrozza, J. of Neuro Engineering and Rehabilitation **8**, 29 (2011) doi: 10.1186/1743-0003-8-29
[5] N.V. Turushev, M.G. Grigoriev, D.K. Avdeeva, Modern equipment and technologies: Proceedings of the XX International scientific-practical conference of students, graduate students and young scientists, 361 (2014)
[6] N.V. Turushev, M.G. Grigoriev, D.K. Avdeeva, 2014 the 4th International Workshop on Computer Science and Engineering - Summer, WCSE 2014, 110947(2014)
[7] C.J. De Luca, J. Med. Eng. Technol. **2**, 57 (1978) doi: 10.3109/03091907809161756
[8] M. Asghari, H. Hu, Biomedical Signal Processing and Control **2**, 275 (2007) doi: 10.1016/j.bspc.2007.07.009
[9] S.A. Vorotnikov, V.S. Strunin, N.A. Vybomov, Caspian Magazine: Management and High Technology (3), 147 (2013)

[10] N. Jiang, S. Dosen, K. Muller, D. Farina, IEEE Signal Processing Mag. **29**(5), 147 (2012) doi: 10.1109/MSP.2012.2203480

[11] M.L. Ivanov, *Development and investigation of the ECG hardware-software complex based on nanosensors for recording of heart micropotentials in real time without averaging and filtering, PhD thesis* (TPU, Tomsk, 2015)

[12] D.R. Safin, I.S. Pilshicov, M.A. Urakseev, V.G. Gusev, Volga region. Technical sciences (2), 88 (2009)

MATEC Web of Conferences **79**, 01052 (2016) DOI: 10.1051/matecconf/20167901052
IME&T 2016

FUZZY SYSTEM OF DISTRIBUTION OF BRAKING FORCES ON THE ENGINES OF A MOBILE ROBOT

Maxim Bobyr[1a], *Vitaliy* Titov[1], *Alexander* Belyaev[2]

[1]Southwest State University, Department of Computer Engineering, 305040, Kursk, Russia
[2]Tomsk State University, 634050, Tomsk, Russia

Abstract. The article presents a fuzzy system of distribution of braking forces on the engines of a mobile robot during its lifting and going down.The block diagram of the system of distribution of braking forces and location of sensors on a mobile robot is given in the paper. Also, fuzzy mathematical model of redistribution of braking forces depending on the conditions of the movement a mobile robot is shown in the article. The result of the simulation of control parameters are presented in the article. The control system of a mobile robot is demonstrated on the example of an autonomous mini-robot on platform Pirate under the control of microprocessor Arduino Mega 2560.

1 Introduction

Today, significant attention is paid to the process of robotics in different industrial spheres [1, 2]. The growth of quantity of robotic systems is increased due to the rapid growth of microprocessor systems. One example of such systems is microprocessor modules Arduino: Uno, Leonardo, Mega et al. As a rule, the control system of a mobile robotincludes a microprocessor control board, sensors which analyze the location and / or the a spatial orientation and the program algorithms of its control.

During the movement of a mobile robot special regimes appear. For example, the movement of a mobile robot along cross-country, in particular its lifting and going down. Thus, during braking the front wheels received greater load than the back ones. There are different systems of force distribution allowing to avoid moments of wheel lockup. If control of a mobile robot uses the concept of Anti-lock braking system (ABS), it successfully controls the unlocking only of the front wheels [3]. Consequently, the turn of a mobile robot is possible during its braking. The researches have shown that the braking forces on the wheels are distributed equally. However, the clutch of the wheels with surfacewhere a mobile robot moves is different. When the concept of Electronic Brake Distribution (EBD) is used force on the wheels is distributed depending upon their speed of rotation [4].Given system consists of three components: a microprocessor control unit, sensors of speed rotation of wheels and actuating mechanisms of control of force distribution on the wheels. It works as follows. A microprocessor unit processes

[a]Corresponding author: maxbobyr@gmail.com

© The Authors, published by EDP Sciences. This is an open access article distributed under the terms of the Creative Commons Attribution License 4.0 (http://creativecommons.org/licenses/by/4.0/).

information from the wheel speed sensors. Based on these data a microprocessor unit concludes which wheels have better clutch with the road surface. After that, EBD redistribute force on the wheels.

When used in mobile robots accelerometers, which are not used in the ABS and EBD system possibilities of distribution braking forceses system are expanded. So accelerometers allow to define the angles of lifting and / or going down during the movement of mobile robots. This allows to make pre-correction in the control program of the movement of a mobile robot, for example, to increase or decrease the speed of rotation of the wheels, and in extreme cases to change the direction of their rotation during going down which has a large angle of declination [5].

At the present time intelligent methods of decision-making control, used in mobile robots are neural networks, evolutionary algorithms, multi-agent systems and adaptive neuro-fuzzy inference system [6–8]. In some cases used the combination of these models. As a rule the selection of one of the aforecited decision-making methods depends on the dynamics of movement of a mobile robot. Thus, it is necessary to take into account the fact that memory of the microcontroller is limited. Therefore, the correct choice of decision-making method also will allow to minimize the program code of control of a mobile robot. The authors propose for increasing efficiency of decision-making control to use adaptive neuro-fuzzy inference system.

2 Structural scheme of a mobile robot

To solve the problem of the movement of a mobile robot along cross-country the authors suggest usingan experimental model assembled on the Pirate 4WD platform shown in Figure 1.

Figure 1. Experimental model of a mobile robot assembled on the platform Pirat 4WD.

The mobile robot consists of a control board – Arduino Mega 2560, two drivers (H-bridge circuit, L293D) for control of four engines of mobile robot, three-axis accelerometer - LIS331DLH. An accelerometer is a part of the IMU-sensor (Inertial Measurement Unit) possessing 10 degrees of freedom. Also IMU-sensor includes a gyroscope, magnetometer (digital compass) and barometer.

In Figure 2 given scheme of the connection H-bridge circuits and an accelerometer to control board of a mobile robot is presented.

 DOI: 10.1051/matecconf/20167901052

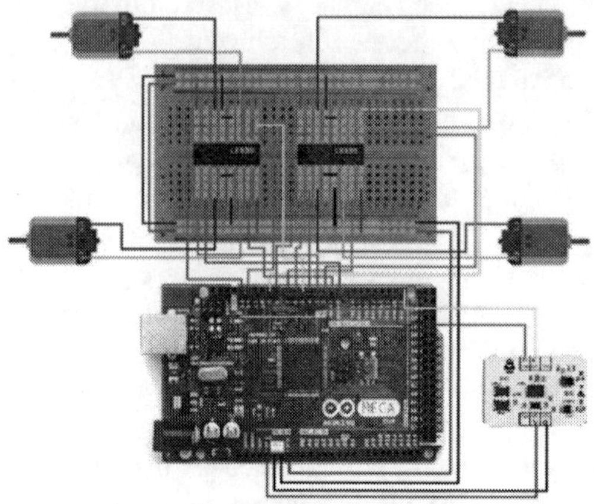

Figure 2. Scheme of the connection H-bridge circuits and an accelerometer to control board.

The concept of interconnection of sensors and a mobile robot with a fuzzy inference system of distribution of braking forces on its wheels is shown in Figure 3.

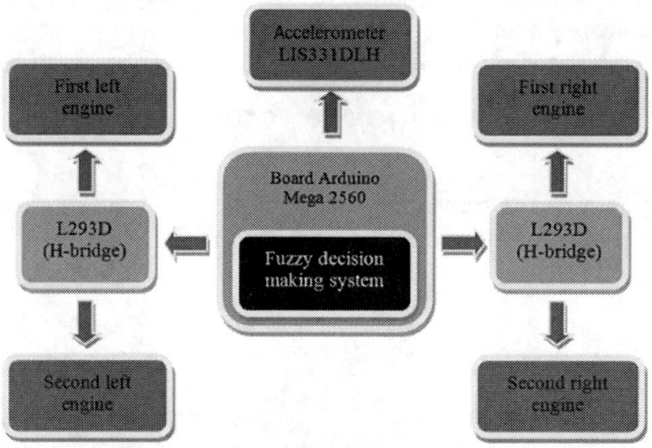

Figure 3. The concept of interconnection of sensors and a mobile robot with fuzzy inference system.

In an experimental model of a mobile robot (Figure 1) are used two DC motors (engines). They work on the basis of pulse-width modulation. Signal strength transmitted to them from the microprocessor unit changes from high (5V or Speed = 255) to low (0V or Speed = 0) levels. Therefore, the control signal transmitted from the microprocessor unit on L293D is in the range from 0 to 255. Let the signal be variable Speed. Thus, the larger value of variable Speed is, the greaterrotation speed of the wheels of a mobile robot is.

Measure of tilt angles during the movement of a mobile robot along cross-country is carried by using three-axis accelerometer. In the initial position of the accelerometer X and Y axes are in the horizontal plane but z-axis is orthogonal to x and $y-$ axis (Figure 4).

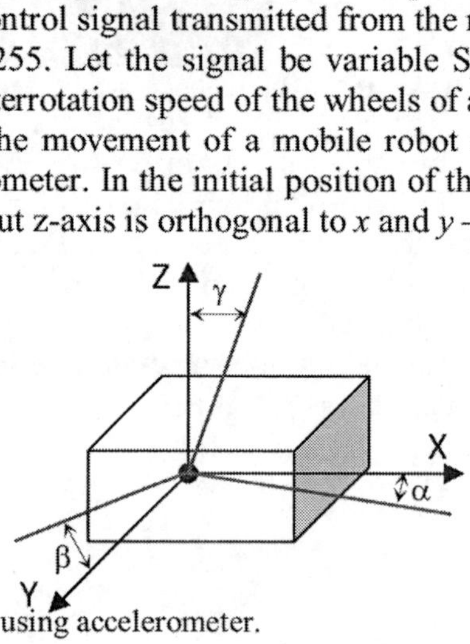

Figure 4. Measuring of tilt angles with using accelerometer.

The values of the tilt angles (Figure 4) are calculated using the following formulas:

$$\alpha = arctan\left(\frac{X}{\sqrt{Y^2+Z^2}}\right) \tag{1}$$

$$\beta = arctan\left(\frac{Y}{\sqrt{X^2+Z^2}}\right) \tag{2}$$

$$\gamma = arctan\left(\frac{Z}{\sqrt{X^2+Y^2}}\right) \tag{3}$$

3 Fuzzy model of distribution of braking forces on the engines of a robot

Formation of braking forces, that is the rotational speed of wheels of a mobile robot, depends on conditions of its movement. Let the speed of rotation of the wheels of a mobile robot (Speed) is output variable. Let the tilt angles defined to formulas (1)–(3) are input variables of the fuzzy system. Further, consider the fuzzy MISO-system having two input variables:

$$X = \{x_1\} + \{x_2\} + \{x_3\};$$
$$Y = \{y_1\} + \{y_2\} + \{y_3\}$$

and one input variable:

$$Speed = \{s_1\} + \{s_2\} + \{s_3\} + \{s_4\} + \{s_5\},$$

wherex_1, …, x_3; y_1,…, $y_3$areterms of first and second input variables of fuzzy sets *X* and *Y*; s_1,…, $s_5$areterms of output variable of fuzzy set *Speed.*

Input and output variables are described by parametric triangular membership functions (MF), which are shown in Figure 5, where μ (x), μ (y), μ (s) are the membership functions of fuzzy sets X, Y, Speed, respectively[9, 10].

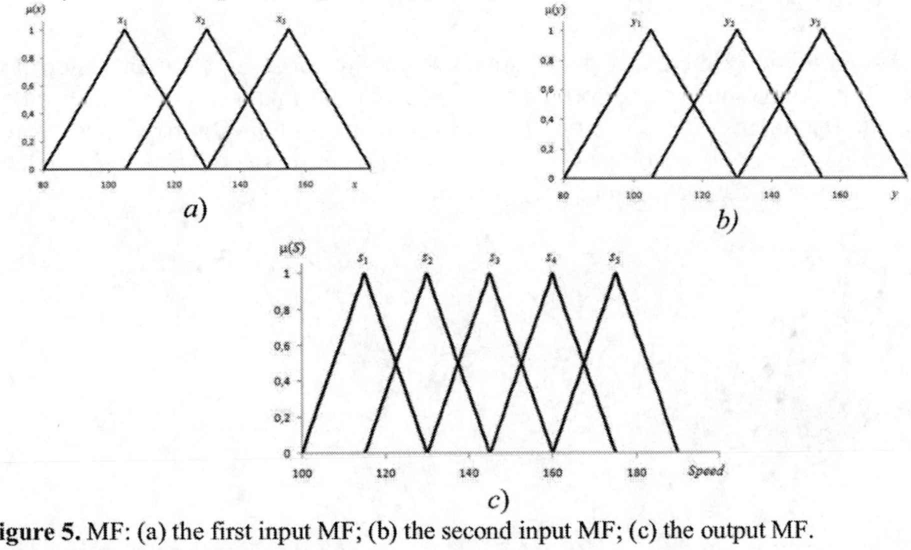

Figure 5. MF: (a) the first input MF; (b) the second input MF; (c) the output MF.

Fuzzy base of knowledge is set by fuzzy rules (FR) and is shown in Table 1.

Table 1. Fuzzy base of knowledge.

FR	If		Then	FR	If		Then	FR	If		Then
FR_1		y_1	s_1	FR_4		y_1	s_2	FR_7		y_1	s_3
FR_2	x_1	y_2	s_2	FR_5	x_2	y_2	s_3	FR_8	x_3	y_2	s_4
FR_3		y_3	s_3	FR_6		y_3	s_4	FR_9		y_3	s_5

Defuzzification is carried on the base of the method of center of gravity or using method of difference areas [11–15].

$$Speed_{CoG} = \frac{\int_{min}^{max} s\mu'(s)ds}{\int_{min}^{max} \mu'(s)ds} \tag{4}$$

where *min*, *max* are limits of integration of fuzzy sets; $\mu'(s)$ is out put fuzzy set after aggregation.

To realize the method of difference areas it's necessary to do the following steps.

Step 1. Determine the total area of the shape output value.

Step 2. Determine the height of degrees of applicability of output terms.

Step 3. Determine areas of every output terms depending on the height values of the degrees of applicability.

Step 4. Determine the total area of the shape with transformed membership functions.

Step 5. Determine the ratio of the total area of the shape of the output variable and the total area of the shape with transformed membership functions.

Step 6. Determine the proportional relationship.

4 Experimental research

For the analysis of fuzzy model of distribution of braking forces is conducted numerical simulation. The simulation was performed with using in the fuzzy inference of hard and soft arithmetic operations of finding the minimum and maximum. During defuzzification were used the models of center of gravity and the difference of areas. After defuzzification surfaces of response of output variable were obtained (Figures 6, 7).

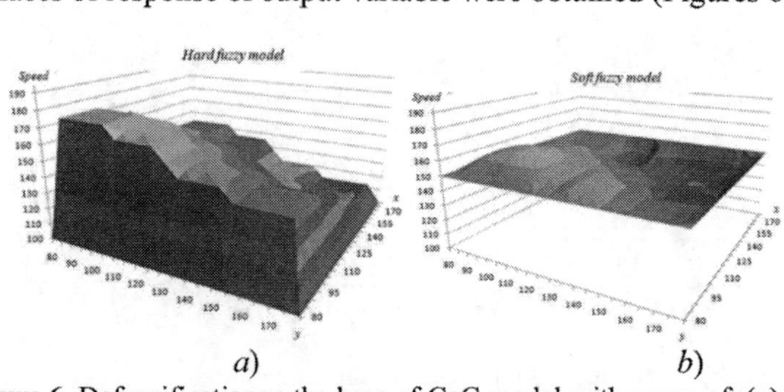

a) *b*)

Figure 6. Defuzzificationon the base of CoG model with usage of: (a) hard arithmetic operations; (b) soft arithmetic operations.

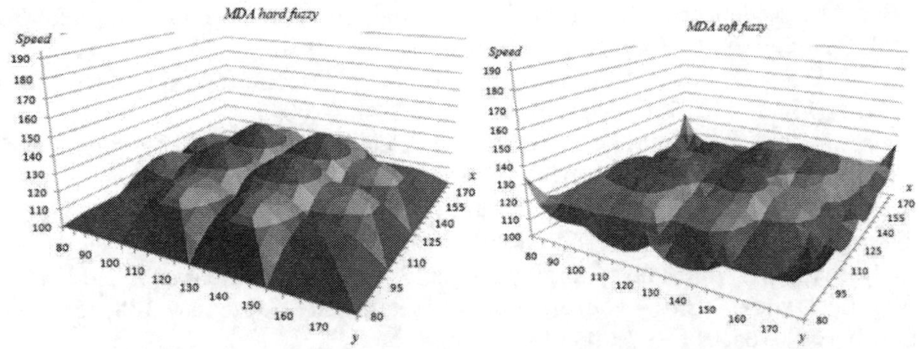

a) *b)*

Figure 7. Defuzzification on the base of MDA with usage: a) hard arithmetic operations; b) soft arithmetic operations.

The simulation showed that during using in the fuzzy inference soft of arithmetic operations the surface of response has more smooth appearance and is additive.

4 Conclusion

The result of research of fuzzy system distribution of braking forces on the engines of a mobile robot showed that to obtain smoother surface of response of the crisp output variable in fuzzy inference it's necessary to use soft arithmetic operations. The presented simulation showed that accuracy of fuzzy systems increased more than 10 times during usage of soft formulas. Proposed in the paper fuzzy control system of a mobile robot's movement allows to increase the speed of the adoption of the control solution more than 15%.

Acknowledgment

The research was done with the support of the RF President's grant for young scientists MD-2983.2015.8, and the program of improving competitiveness of the TSU (project No. 8.2.31.2015).

References

[1] O. Kilincci, S.A. Onal, Expert Systems with Applications **38**,9656 (2011) doi:10.1016/j.eswa.2011.01.159
[2] H.-S. Park, S.-B. Cho, Information Systems **36**, 1124 (2011) doi:10.1016/j.is.2011.04.005
[3] R.-E. Precupa, M.-C. Sabaua, E.M. Petriu, Applied Soft Computing **27**,575(2015) doi:10.1016/j.asoc.2014.07.004
[4] Masao Nakazawa, Osamu Isobe, Sadahiro Takahashi, Yoshito Watanabe, JSAE Review **16**,479 (1995)
[5] P. Staufer, H. Gattringer, Mechatronics **22**, 1043 (2012) doi:10.1016/j.mechatronics.2012.08.009
[6] N. Mendes, P. Neto, Expert Systems with Applications **42**, 8929 (2015)doi:10.1016/j.eswa.2015.07.047
[7] N.-B.Hoanga, H.-J. Kang, Neurocomputing **188**, 12 (2016) doi:10.1016/j.neucom.2015.02.101

[8] D. Bucura, G. Iaccab, M. Gaudesic, G. Squilleroc, A. Tonda, Applied Soft Computing **40,** 416 (2016)doi:10.1016/j.asoc.2015.11.024

[9] M.V. Bobyr, V.S.Titov, A.A. Nasser, Journal of Machinery Manufacture and Reliability **44**,633 (2015) doi: 10.3103/S1052618815070067

[10] I.V. Chernova, S.A. Sumin, M.V. Bobyr, S.P. Seregin, Biomedical Engineering **49**, 263 (2016)doi:10.1007/s10527-016-9545-y

[11] M.V. Bobyr, S.A .Kulabukhov, Vestnik komp'iuternykh i informatsionnykh tekhnologii **9**, 32(2015)doi:10.14489/vkit.2015.09.pp.032-041

[12] M.V. Bobyr, N.A. Milostnaya, Angineering and automation problems **3**, 27(2015)

[13] W.V. Leekwijck, E.E. Kerre, Fuzzy Sets and Systems **108**, 159(1999) doi:10.1016/S0165-0114(97)00337-0

[14] A. Chandramohan , M.V.C.Rao, M.S. Arumugam, Soft Computing **10**, 1047 (2006) doi: 10.1007/s00500-005-0042-6

[15] H.R. Mahdiani, A. Banaiyan, M.H.S. Javadi, S.M. Fakhraie, C. Lucas, Engineering Applications of Artificial Intelligence **26,** 162 (2013) doi: 10.1016/j.engappai.2012.07.001

THE INFLUENCE OF THE RADIATION SOURCE PARAMETERS ON THE ACCURACY OF DIGITAL SPECKLE CORRELATION METHOD

Lin Li[1*], *Fedor* Gubarev[1,2]

[1]National Research Tomsk Polytechnic University, 634050, Tomsk, Russia
[2]V.E. Zuev Institute of Atmospheric Optics, Russian Academy of Sciences, Siberian Branch, 634050, Tomsk, Russia

Abstract. The article presents the impact of the laser radiation divergence and the wavelength on the harmonic distortion detection, as well as the influence of the distance between the optical circuit elements on the harmonic distortion and the sensitivity of amplitude. The research was made by recording the speckle images from the vibrating surface of the object with its digital processing later.

1 Introduction

The method of digital speckle correlation is one of the most common methods for optical remote control and monitoring. The development of nano- and micro-technologies of modern systems brings to the fore the need to study in details this method to obtain the information about objects and processes. As objects of observation can serve machine parts, any mechanisms or living organisms, the sphere of its application is very broad [1–5]. This technique makes possible to receive the pattern, carrying information about the dynamics of the object, by the application of digital image processing systems, comparing the speckle images of the object surface. The method of digital speckle correlation also holds promise for diagnostics in biology and medicine, as it took a significant place in the non-destructive testing of materials [6–9].

The authors [10] consider a number of advantages of this method by comparing it with other methods, in particular with the Doppler method. In their work they also consider the ways of non-contact blood pressure measurement, using digital speckle correlation.

The optical monitoring revealed an interesting feature of the method; it is its ability to read information using the object vibration. The vibration in some processes presents vital information (voice, heartbeat, blood circulation, breathing, etc.), the research in this area presents interest from the point of view of remote monitoring.

The authors [11] examined the optical system, which allows to extract information from vibrating objects located at the distance of 100 m using the method of digital speckle correlation, which demonstrates a great potential of this method. A number of other studies [12–20] also demonstrate the ability to read the information in this way. Nevertheless, the

[*] Corresponding author: wojiaolilin@mail.ru

© The Authors, published by EDP Sciences. This is an open access article distributed under the terms of the Creative Commons Attribution License 4.0 (http://creativecommons.org/licenses/by/4.0/).

problem affecting the quality of the information obtained by this method is not sufficiently investigated.

Therefore, the aim of this work is to study the distortion during the process of decoding the optical signals received by the digital speckle correlation.

2 Experimental techniques

The test object was a cone speaker of the active computer speakers, which are fed from a digital harmonic signal generator SFG-72120 (GW Instek). The diameter of the cone speaker was 7 cm. The voltage amplitude of the loudspeaker was measured by a digital oscilloscope Rigol DS1052E. Unfortunately, we do not know the characteristics of that speaker system (frequency response, phase response), and how well the speaker converts the electrical oscillations into the mechanical ones. But, since we apply the sinusoidal voltage to the speaker, it can be assumed that the oscillation of the speaker cone will also be sinusoidal, as the harmonic oscillations are elementary.

The formation of speckle patterns is carried out by a 5 mW He-Ne laser (632 nm) and a 5 mW semiconductor laser (532 nm) used as a source of a coherent radiation. The beam diameter and the divergence of the He-Ne laser and of the semiconductor laser are 0.8 mm, 1.2 mm and 3 mrad, 2 mrad, respectively. The laser beam was directed to the cone speaker. Speckle patterns were recorded using a high-speed camera Fastec CMOS HiSpec 1 with the lens F0.95 Navitar. The camera was placed at the distance of 50 cm from the loudspeaker. Recorded images were transferred to a computer for further processing. The description of the speckle patterns processing method and the results of its testing are given in [21].

The procedure of investigation, used in this study, is shown in Figure 1. The scheme is used to study the dependence of the quality of the information read from the vibrating object surface. The studies were conducted under various conditions: the diameter of the spot on the surface of the object was varied by changing the distance between the object and the lens; two different radiation sources were used; lenses with different focal lengths, which provided the illumination of the object by the light with a different divergence, were used.

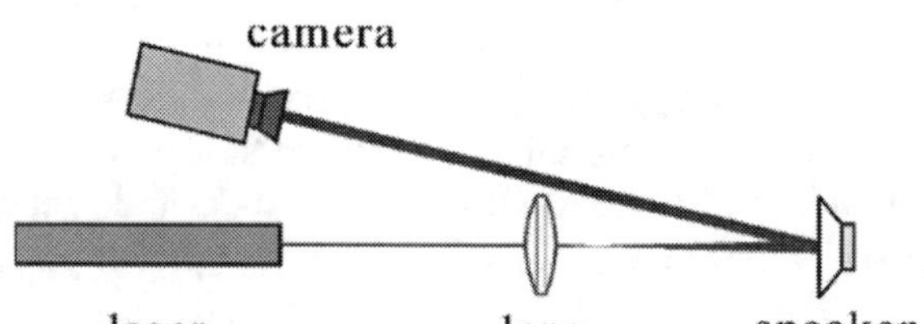

Figure 1. Schematic illustration of the experiment.

3 Results

Due to the fact that the speckle patterns are recorded by a digital camera, i.e. digitized, it can, under certain conditions, make a significant error. Thus, when the speckle size is of the order of the pixel size of the camera sensor or less, the recorded speckle pattern will have a low resolution, which leads to deterioration of the accuracy of oscillation surface decoding. In the case when the size of the speckle is many orders greater than the pixel size, the resolution of the recorded speckle pattern will, of course, be high, however, the quantity of information, contained in the speckle pattern of the same size, will be much less. Therefore, the latter case requires the increase of the speckle pattern size, which leads to the increase of the processing time.

It is well known [22] that the speckle size is directly proportional to the distance between the subject and the camera, and is inversely proportional to the diameter of the laser beam. Figure 2 shows the speckle pattern obtained with different spot size of the laser radiation on the speaker cones. The presented images can demonstrate that by increasing the diameter of the laser spot size the speckle is reduced. The regulation of the laser spot diameter is achieved by moving the focusing lens along the optical axis. However, it should be noted that while the divergence of the laser beam is increased (for the focusing lens, used in the experiment, the divergence of laser radiation was 11 mrad). The choice of such a scheme of the experiment is due to the parallel study of the possibility to apply the simplest scheme of the laser beam diameter varying to detect the objects surface oscillation.

In our next experiment we studied the effect of the laser spot diameter of the object surface on the quality of the process of detection its oscillations. Figure 3 shows the dependence of the sensitivity fluctuation of the amplitude of the decoded signal of the distance between the focusing lens and the speaker cone. The focal distance of the lens was 11 cm. The Figure 3 shows that the movement of the focusing lens from the laser to the speaker leads, firstly, to the decrease of the sensitivity and then to its increase, secondly. The dependence of the sensitivity is expressed in the minimum range distance of 20 to 30 cm (Figure 3). The minimum value of sensitivity in the experiments was about 1 pixel/V.

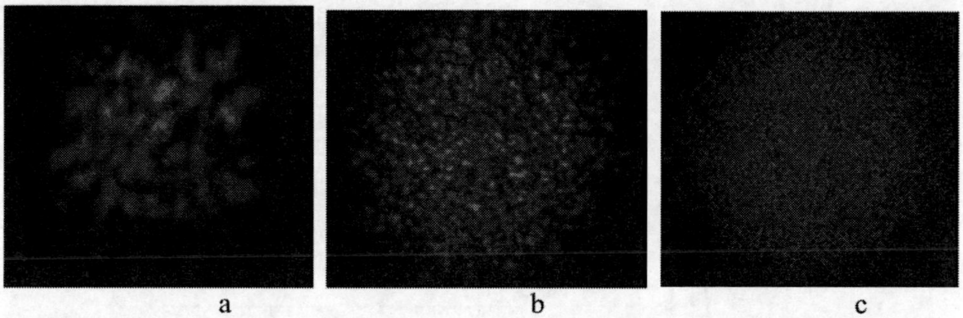

Figure 2. The speckle patterns with different diameter laser spot on the diffuser speaker: a) 11 mm, b) 17 mm, c) 47 mm. The distance between the surfaces of the object and the camera is 50 cm. The camera's exposure time is 1995 ms.

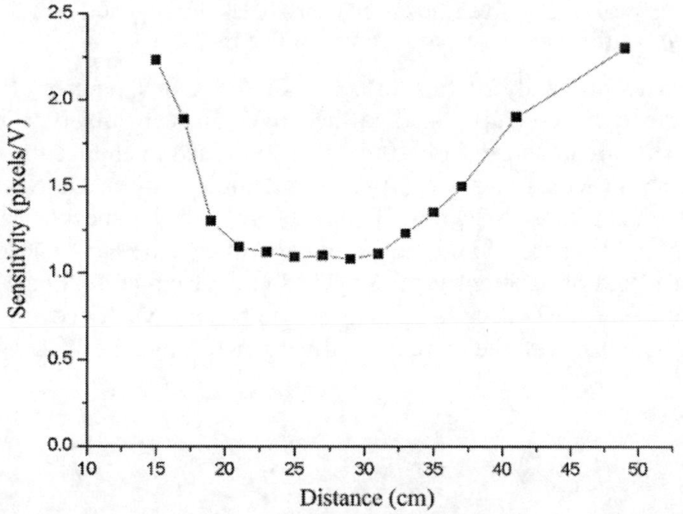

Figure 3. The dependence of sensitivity fluctuation amplitude of the decoded signal of the distance between the focusing lens and the speaker cone. The signal frequency is 50 Hz. The frequency of camera registration is 500 Hz, the camera`s exposure time is 1.995 ms.

It should be noted that the speckle size influences on the optimum size of the image area used for speckle tracking shift. Since the displacement of the lens changes the size of speckles (Figure 1), it is necessary each time to select the optimal size of the image area while moving the lens. The search for optimal image speckle size was based on minimizing the value of the total harmonic distortion (THD) of the decoded signals.

The study also obtained the dependence of THD decoded signal on the sinusoidal voltage amplitude signal applied to the speaker coil, at various distances between the loudspeaker and the focusing lens. The dependence was deducted by averaging the experimental data. The dependence of the THD decoded signal average value on the distance between the focusing lens and the loudspeaker is shown in Figure 4. In addition to the fact that reducing the distance to values close to the focal distance of the lens, led to the sensitivity increase (Figure 3), it also led to a significant increase of the decoded signal THD value (Figure 4). By varying the distance ranging from 15 cm to 47 the THD decoded signal varied in the range from 5% to 15%.

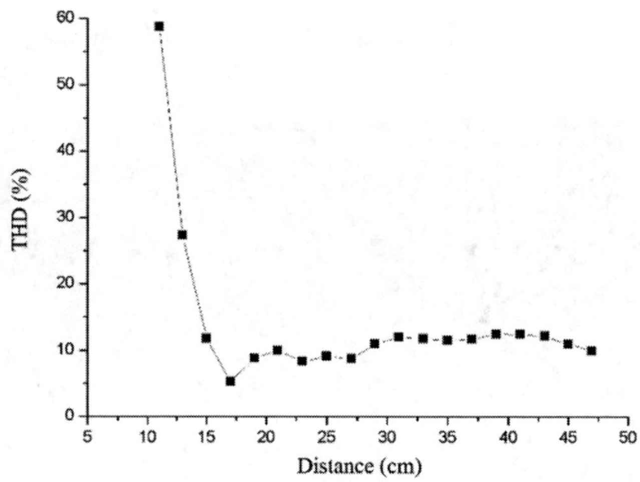

Figure 4. Dependence of the THD decoded signal average value on the distance between the focusing lens and the loudspeaker. The frequency of the sinusoidal signal is 50 Hz. The frequency of camera recording is 500 Hz, the camera`s exposure time is 1.995 ms.

In addition to the study of the influence of the spot diameter on the quality of the received information, this paper studied the effect of beam divergence on the quality of information. The circuit shown in Figure 1 was also used to change the divergence through the use of different lenses and two different radiation sources: He-Ne laser (632 nm) and green semiconductor laser (532 nm). Figure 5 presents the speckle patterns, received by using different light sources. In this series of experiments the He-Ne laser was used without lens and with lenses of different focal length: 8 cm, 11 cm and 50 cm. Figure 6 shows the average value of the THD decoded signal on the beam divergence value using the He-Ne laser. One can see that with the increase of divergence value, the THD value also increases.

a b

Figure 5. The speckle pattern with different sources of radiation: a) He-Ne laser (632 nm), b) a green semiconductor laser (532 nm). The distance between the surfaces of the object and the camera is 50 cm. The camera's exposure time is 1.995 ms.

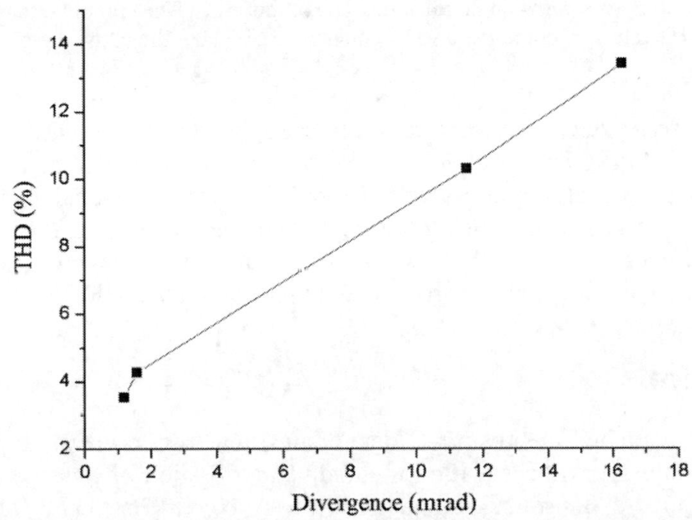

Figure 6. The dependence of the decoded signal average value on the THD values of the beam divergence using the He-Ne laser. The amplitude of the signal voltage is 2 V. The signal frequency is 50 Hz. The frequency of camera registration is 500 Hz, The camera's exposure time is 1.995 ms.

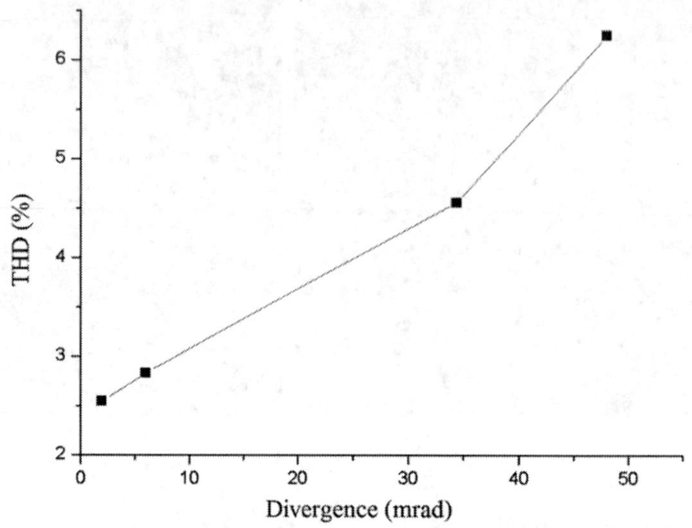

Figure 7. The dependence of the decoded signal average value on the THD value of the beam divergence using the green semiconductor laser. The amplitude of the signal voltage is 2 V. The signal frequency is 50 Hz. The camera registration frequency is 500 Hz. The camera's exposure time is 1.995 ms.

In the next series of experiments the green semiconductor laser was used without a lens and with lens of different focal length: 8 cm, 11 cm and 50 cm, as in the previous experiments. Figure 7 shows the dependence of THD decoded signal average value on the value of beam divergence when using the green semiconductor laser. It also shows that with increasing of divergence values the THD value also increases. However, this value is much smaller if we compare it with the THD average value when using He-Ne laser.

4 Conclusions

In this paper we studied the quality of data acquisition from a vibrating object surface by means of a digital speckle correlation method. The vibration of loudspeaker was used, to which was applied a sinusoidal voltage as the test object. The quality of the information extracting was assessed by comparing the decoded signals with the harmonic ones.

It is shown that the divergence of the beam affects the received information quality. The smaller the divergence of the beam is, the smaller the value of the coefficient of harmonic distortion is, i.e. it provides the better information extraction.

Using a collecting lens to illuminate the object, when the distance between the controlled object and the lens reduces to values, which are close to the focal distance of the lens, on the one hand, it leads to an increase of sensitivity, on the other, to a significant increase of the harmonic distortion values of the decoded signal. Thus by varying the distance between 15 to 47 cm the total harmonic distortion of the decoded signal was changing in the range from 5% to 15% under the experimental conditions.

References

[1] S. Noh, I. Yamaguchi, Jpn. J. Appl. Phys. **31**, L1299 (1992) doi: 10.1143/JJAP.31.L1299

[2] D.J. Chen, F.P. Chiang, Y.S. Tan, H.S. Don, Appl. Opt. **32**, 1839 (1993) doi: 10.1364/AO.32.001839
[3] T. Fricke-Begemann, Appl. Opt. **42**, 6783(2003)
[4] D. Amodio, G.B. Broggiato, F. Campana, G.M. Newaz, Experimental Mechanics **43**,396 (2003) doi: 10.1177/0014485103434004
[5] J. Guanchang, W. Zhen, B. Nikeng, Y. Xuefeng, Opt. and Lasers in Engineering **39**, 457 (2003) doi: 10.1016/S0143-8166(02)00028-3
[6] B.R. Durig, W.H. Peters, M.A. Hammer, Proc. SPIE. **0954**, 438 (1989)
[7] V.V. Tuchin, J. Biomed. Opt. **4**, 106 (1999)
[8] M.A. Vilensky, D.N. Agafonov, D.A. Zimnyakov, V.V. Tuchin, R.A. Zdrazhevskii, Quantum Electron **41**, 324 (2011)
[9] M. Golberg, D. Fixler, A. Shainberg, *et al,* J. Biomed. Opt. **18**, 101310 (2013)
[10] Y. Beiderman, I. Horovitz, N. Burshtein *et al,* J. Biomed. Opt. **15**, 061707 (2010) doi: 10.1117/1.3505008
[11] Z. Zalevsky, Y. Beiderman, I. Margalit, S. Gingold, M. Teicher, V. Mico, J. Garcia, Opt. Express **17**, 21566 (2009) doi: 10.1364/OE.17.021566
[12] W.O. Wong, Opt. and Lasers in Engineering. **28**, 277 (1997)
[13] Sama Molaie, Sohrab Ahmadi, Hadi Seyed Arabi, Indian J. Sci. Res. **3**, 90 (2014)
[14] S. Bianchi, Appl. Opt. **53**, 931 (2014) doi: 10.1364/AO.53.000931
[15] Z. Chen, C. Wang, C. Huang, H. Fu, H. Luo, H. Wang, Opt. Communications **331**, 6 (2014) doi: 10.1016/j.optcom.2014.05.038
[16] Z. Chen, J. Chen, Journal of Vibration and Shock **34**, 121 (2015)
[17] J. Quisberth, Z. Wang, H. Nguyen, Conference Proceedings of the Society for Experimental Mechanics Series **9**, 105 (2016) doi: 10.1007/978-3-319-21765-9_14
[18] P. Vaz, T. Pereira, E. Figueiras, C. Correia, A. Humeau-Heurtier, J. Cardoso, Biomedical Signal Processing and Control. **25**, 188 (2016) doi: 10.1016/j.bspc.2015.11.013
[19] C. Lal, A. Banerjee, N.U. Sujatha, J. Biomed. Opt. **18,** 111419 (2013) doi: 10.1117/1.JBO.18.11.111419
[20] N. Sujatha; A. Banerjee, Proc. SPIE. **9302**, 93021R (2015)
[21] F.A. Gubarev, L. Li, M.S. Klenovskii, A.Ph. Glotov, MATEC Web of Conferences **48**, 04003 (2016) doi: 10.1051/confmatec/20164804003
[22] M. Franson, *Speckle optics* (Mir, Moscow, 1980)

MULTIPARAMETER SYSTEM FOR MONITORING THE STATE OF URBANOSPHERUM BASED ON MULTIVENDOR DEVICES OF DATA COLLECTION

Aleksandr Pilipenko[1] , *Anastasiia* Pilipenko[2]

[1]Orel State university named after I.S. Turgenev, 302026, Orel, Russia
[2]Proton-Electrotex, 302040, Orel, Russia

Abstract. The authors explain the problem of partitioning of the control object on the controlled elements, and give a description of the mathematical apparatus. The authors propose different types of portable devices for collecting information about the state of urbanospherum on the basis of different hardware and software platforms that combine integrated information system. In the work the presented algorithms and block diagrams of data collection devices are based on the controller NI MyRIO, micro-computer Raspberry Pi and Arduino microcontroller system. The authors explain the approach to the optimization of systems of management of urban resources and processes with the using of systems condition monitoring of urbanospherum.

1 Introduction

The city is a multi-parameter system consisting of a large number of resources which are various according to the purpose and the content, and which are integrated by the processes taking place between them. These are industrial enterprises and public organizations, institutions of health and education, utilities and service sector enterprises, construction and energy companies, etc., successful functioning of which depends directly on transportation infrastructure, the condition of linear objects and buildings, from the management of life supporting processes, security environment and, primarily, from the human condition – from biosphere compatibility of nature and socio-technical structures. At present, there is no structured data about the interaction processes between participants in the life of the city, their impact on the urban environment. Information about the objects and the relationships between them are completely different from the point of view of the format, the synchronization in time, the methods of processing, storage and transmission. These factors do not allow timely to track and to provide ecological, technogenic safety of the city and the rapid response of special services to emergency situations.

[1] Corresponding author: a@pilipenko.info

© The Authors, published by EDP Sciences. This is an open access article distributed under the terms of the Creative Commons Attribution License 4.0 (http://creativecommons.org/licenses/by/4.0/).

2 Concept

Markets of public safety in urbanized areas that include fire safety, water safety, public security, traffic safety, environmental and industrial safety were analyzed. It was revealed that the system of city management and urban farms could be described by quantitative and qualitative indicators of biosphere compatibility of different structures. The objective function of the system of city management should strive for minimum cost, minimum time of decision-making and the minimum number of emergency situations. The very same objective function of the management system is multivariable, which depends on many parameters of the urban environment, the collection of which is possible due to monitoring systems. The development of multiparameter systems of monitoring the condition of urban processes and resources and integrated implementation of information systems in all of the management, administration and emergency services will not only create a system of quality management of a city with elements of feedback, but also give the opportunity to assess the work quality of city structures.

The goals of the development of multiparameter monitoring systems of the urbanospherum state are the following:

1) the increase of security and the improvement of the quality of life of the city population;

2) the increase of city management efficiency;

3) the transition from information systems that use different databases to a single integrated information environment that provides automated support of processes of activity of departments of the Ministry of Emergency Situations (MES), authorities, organizations and enterprises of the city, without losing any existing data;

4) the reduction in time-consuming routine operations related to the exchange of information performed at all stages of technological processes;

5) the improvement of the validity of obtained, processed, and stored information used in the process of activity of divisions of the MES, authorities, organizations and enterprises of the city, by centralizing its storage, regulation of use;

6) the reduction of terms and improvement of quality of technological processes, as well as the preparation and registration of documents;

7) technological support for collecting and processing the accounting information, automation of reporting;

8) the creation of a tool of automated support for management decision-making on the basic directions of activity of divisions of bodies of the MES, the authorities, organizations and enterprises of the city.

The criterion of achieving the set goals is the organization of a single information space which ensures the complete implementation of the goals of creating a multiparameter system of monitoring the state of the urbanospherum, the ability of the divisions of the MES, authorities, organizations and enterprises to perform their functions with the least time cost, to improve the quality of life and to provide comprehensive security of the city population.

The urbanospherum refers to some part or parts of the urban environment; or the environment or conditions of existence, similar to the conditions of existence in the urban environment; or the environment, from resource-dependent urban environments [1]. A distinctive feature of the urbanospherum is the presence of a single purpose of its functioning. For example, the urban water supply system is understood by all as a system of single purpose functioning, it is controllable, and thus it is the urbanospherum. However, autonomous regional water supply system, being a part of a citywide water supply system, also has a common goal and also is controllable, that it is the urbanospherum. Several urban spheres can be located on the same territory and use the same resources, the elements of the management system and even be located completely on the same material objects.

The authors propose a multi-parameter system of monitoring the urbanospherum condition which includes:

– Application subsystems are a set of software means and their documentation designed to solve relatively narrow classes of problems in specific subject areas designed to a specific customer, which are closely related to each other for organizational, informational, technological or other grounds.

– Providing subsystems are a set of program-technical means, intended for organization of interaction between different application subsystems among themselves and between the data collection devices.

– Data collection system is a set of software and hardware means for collecting, processing and transmitting information about the state of the urban environment into providing subsystems.

Examples of application subsystems are "System-wide sectoral data bank", "Normative-reference information", "Geoinformation system", "Ecoinformation system", "Monitoring of the city resources", "Monitoring of urban processes", etc. All of these subsystems can be developed in any programming environment on any programming languages, but united by a common database. However, the authors believe that the graphical programming environment LabVIEW from National Instruments is the most appropriate environment for the development of automated systems. The biggest advantage of LabVIEW environment is the large number of developers around the world. Almost all universities that implement training direction 15.03.04 "Automation of technological processes and productions" have LabVIEW in their educational program. A large number of developers lead to a large number of examples, libraries and modules to work with almost any device from the microcontroller system Arduino to the programmable quadcopter Ar Drone.

Let us consider the development of the most common subsystems and their elements. Figure 1 shows a screenshot of a window application subsystem device "Management of the data collection devices" in the mode of recommendations on the placement of data collection devices. The task of this subsystem is to calculate the optimal location of devices for data collection depending on resources and priority of the borders of the urbanospherum.

The use of triangulation [2] reduces the problem of state analysis of city resources and processes to the finite element method, where many smaller urban spheres are controlled by different types of data collection devices, and the recommendation on the necessary controlling actions affecting the urban sphere as a whole is based on these data. The operator selects the number of active data collection devices, which would mean the number of elements in the triangulation. The number of priority resources of the city shall be selected using the unit Radio Button, which is connected to the element Tab Control and Case Structure, on which input field coordinate city resources are located, and this allows displaying and using the selected number of data entry fields. The coordinates of the triangulation points are set manually or filled from the database automatically, then using the API functions the territory map is loaded with the site www.google.ru/maps/, on which the points and the results of triangulation are mapped with which with the help of virtual devices of NI Vision module.

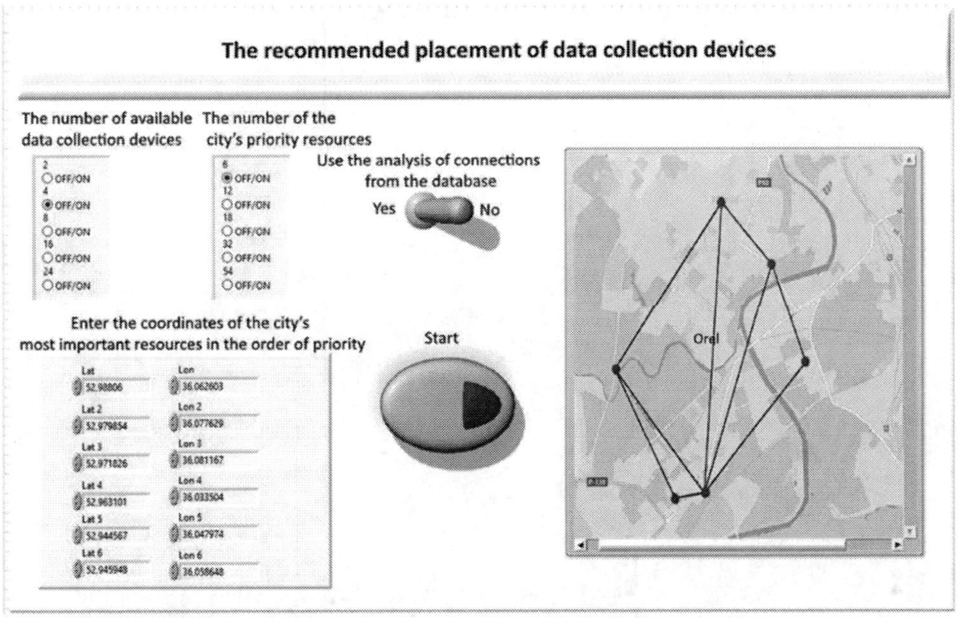

Figure 1. The picture of the window of the application subsystem in the mode of the placement of data collection devices.

The mathematical apparatus for constructing the convex hull of the points (Figure 2) and the optimal triangulation by a greedy algorithm is partly implemented in MathCAD [3, 4], and partly in LabVIEW. The connection of MathCAD and other mathematical software with programming environments are usually implemented through the development of own dll libraries, but the LabVIEW community has already been implemented in Mathcad-LabVIEW Integration Wizard with the developed library MCVI Wizard.llb.

The application subsystem is connected to the MS SQL database, which contains information about major businesses, emergency services, administrative structures and the processes taking place between them. The function of linkage analysis allows to refine the triangulation and to build it not with the aim of obtaining the minimum sum of the lengths of edges, but with the aim of connecting resources among themselves according to links laid down in the database. The flowchart of the subprogramme responsible for the interaction with the database using the NI LabVIEW Base Connectivity Toolkit Data [5] is presented in Figure 3.

Data collection devices can be completely different both at the level of hardware, and at the level of programming languages and environments. Let us consider several types of data collection devices:

1. Portable devices for collecting information on the state of the personal urbanospherum.
2. Stationary devices for collecting data about an urban environment that is installed on public transport and pillars. The point of data collection will have an extensive array of sensors and components for photo-video-recording.

```
VEKPR(A,B) := | 1
              | Z ← A_1·B_2 − A_2·B_1
              | 1
              | Z
```

```
VEKPR(ZAWCSORT^⟨1⟩, ZAWCSORT^⟨2⟩) = 13.474
```

```
VECOBOL(coord, wypobol) := | 1
                           | for i ∈ 1.. rows(wypobol)
                           |    | Q^⟨i⟩ ← coord^⟨(wypobol_i)⟩
                           |    | 1
                           | 1
                           | Q
```

```
WYP(thk, Tk) := | 1
                | NUMMER_thk ← 0
                | ZAWCSORT ← csort(Tk, 2)
                | RM ← ZAWCSORT^T
                | TT ← RM^⟨1⟩
                | i ← 1
                | NUMMER_i ← 1
                | OBL_i ← 1
                | 1
                | while 1 = 1
                |    | Sh ← 0
                |    | for k ∈ 1.. thk
                |    |    | if NUMMER_k = 0
                |    |    |    | 1
                |    |    |    | PRG ← (RM^⟨k⟩ − TT)
                |    |    |    | for j ∈ 1.. thk
                |    |    |    |    | 1
                |    |    |    |    | if NUMMER_j = 0
                |    |    |    |    |    | PRT ← (RM^⟨j⟩ − TT)
                |    |    |    |    |    | if VEKPR(PRG, PRT) ≥ 0
                |    |    |    |    |    |    | Sh ← 1
                |    |    |    |    |    |    | 1
                |    |    |    |    |    | otherwise
                |    |    |    |    |    |    | Sh ← 0
                |    |    |    |    |    |    | break
                |    |    |    |    |    |    | 1
                |    |    |    |    |    | 1
                |    |    |    |    | 1
                |    |    |    | if Sh = 1
                |    |    |    |    | ko ← k
                |    |    |    |    | break
                |    |    |    |    | 1
                |    |    |    | 1
                |    |    | 1
                |    | 1
                |    | break if ko = 1
                |    | NUMMER_ko ← 1
                |    | TT ← RM^⟨ko⟩
                |    | i ← i + 1
                |    | OBL_i ← ko
                |    | NUMMER_1 ← 0 if i > 1
                |    | 1
                | 1
                | OBL
```

Figure 2. The construction of the convex hull of the set of points for triangulation in MathCAD.

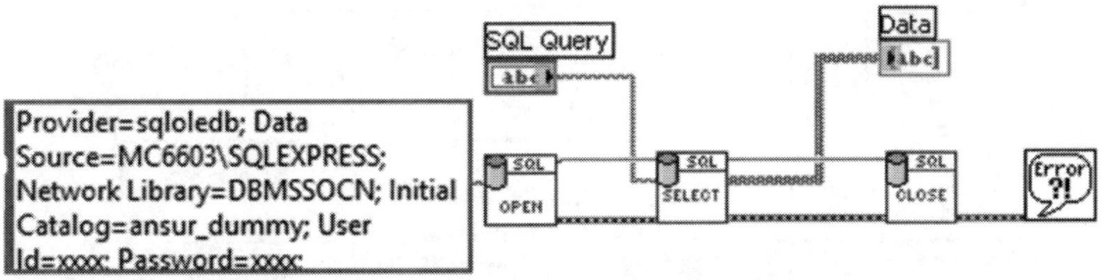

Figure 3. The flowchart of the subprogram responsible for the interaction with the database.

The personal urbanospherum, being a special case of the urbanospherum, retains all its properties. Everything that a particular person living in the urbanized area communicates with can apply to this concept. The development of the wearable device for collecting information about the human condition is performed on two different systems and programming languages:

1. The laboratory version of the device [6]. The device is based on a chip Arduino Lilypad on the base of the microcontroller AVR by connecting the chipset and sensors to it. The programming language is C++, the programming environment is Arduino IDE. The small size of electronic components allows implementing the system in the form of "the smart clothing". The link with other system modules is organized by the wireless channel of the Zigbee data.

2. The casual version of the device. The information gathering device is based on a ready product Samsung Gear S, which has a completely open operating system Tizen, which gives the ability to change even the core of the system. The sensor pulse, accelerometer, gyroscope, GPS, microphone, in a stylish designer case with optimized power consumption. Communication interfaces are GPRS, Bluetooth and Wi-Fi, the programming is supported in JavaScript

The devices for gathering information about the human environment are also presented in two variations:

1. A stationary device that is installed on public transport and pillars. Information gathering device about the external environment is on the basis of NI MyRIO [6], which is based on a Cortex ARM 9 and has a built-in accelerometer, gyroscope, Wi-Fi and plenty of ports for connecting sensors and cameras. The programming environment from the manufacturer NI LabVIEW is a set of libraries and modules allowing to solve tasks of any complexity including the collection and processing of images.

2. The mobile version of the device is based on a compact micro-computer Raspberry Pi 2, which also runs on the Cortex ARM 9 and has a USB. Inputs, GPIO ports and a connector for connecting the camera. Connecting an additional extension board of GPIO ports allowed to increase their number from 17 to 30 and to add analog inputs. The device weight with all connected sensors is not more than 300 grams. The operating system of the micro-computer is Linux and Windows, the recommended programming language is Python.

Figure 4 shows a conceptual diagram of data collection devices reflecting connections between hardware components of data collection devices. The connection between data collection devices and application subsystems is implemented through wireless and wired technologies, and it doesn't have to be popular, but energy-consuming data transfer channels GPRS and Wi-Fi [7]. Almost all devices work with all receivers/transmitters Zigbee, which allow transmitting information at a distance of from 10 to 1500 meters with high energy efficiency and with the support the cryptographic protection. The algorithm of the work of the subprogramme of a message receiver over a wireless channel Zigbee is shown in Figure 5.

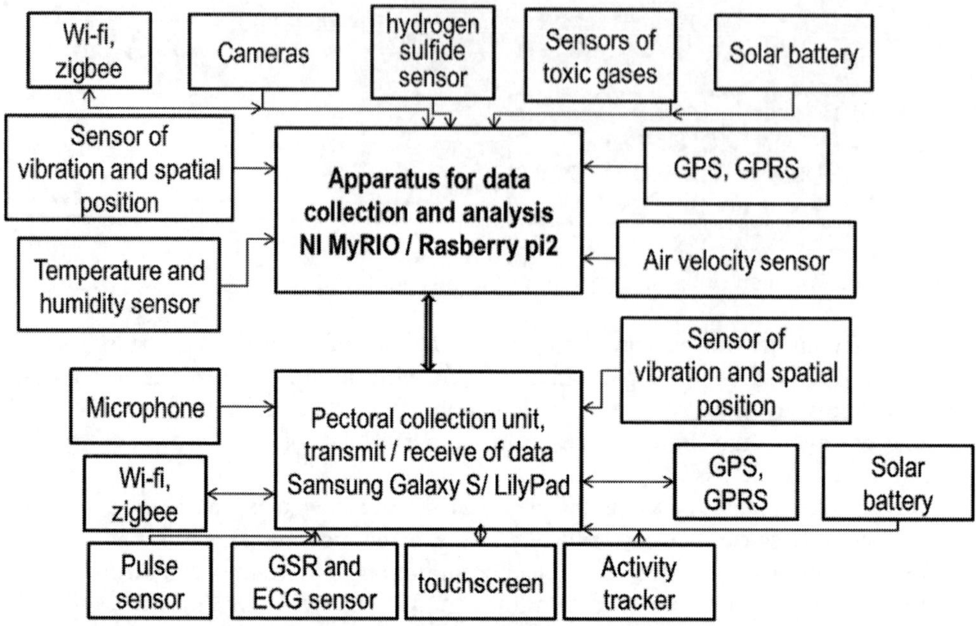

Figure 4. Conceptual diagram of the data collection devices.

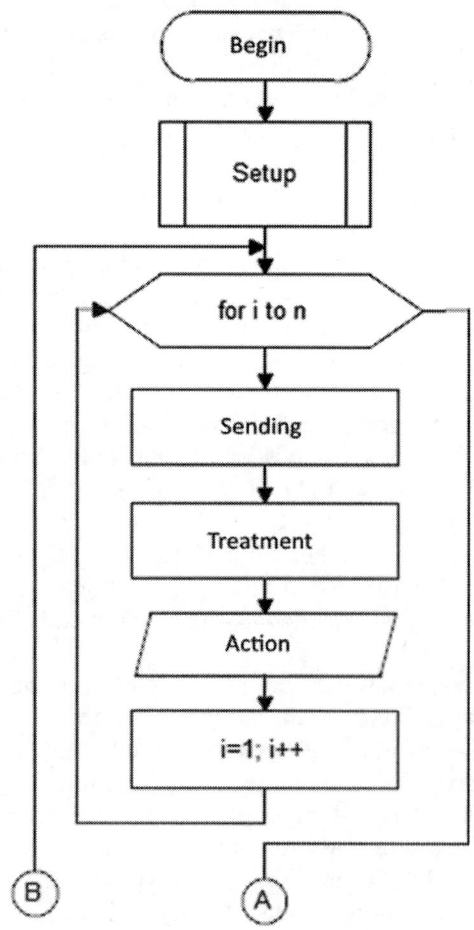

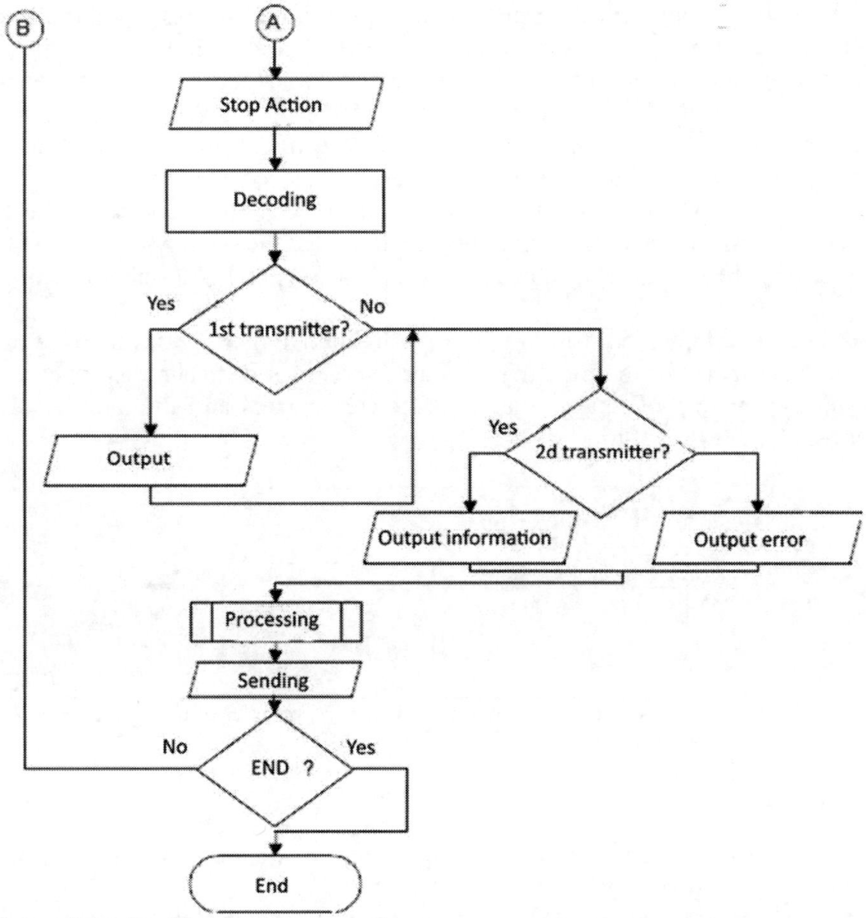

Figure 5. The receiver algorithm.

Stationary devices have great performance and accuracy of measurements, and also allow the connection of devices of photo and video- fixing (cameras) to collect and to display operational data on incidents. The programme structure of the computer vision is presented in Figure 6.

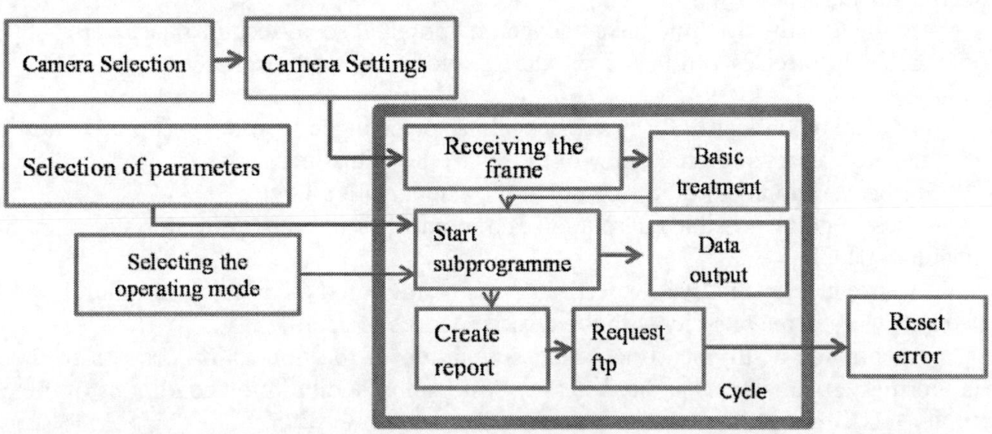

Figure 6. Structural scheme of the system under development.

The algorithms for collecting photo-video data, initial processing and their transfer to the FTP server have been developed and tested (Figure 7) [8]. The developed subsystems allow to determine fires and ignitions based on the algorithms for the analysis of the illumination intensity, and to determine the speed of the vehicle, the entering vehicle registration, the analysis of speed and density of traffic flows, the definition of a traffic collision based on the algorithms of the detection and comparison of images. The whole effect is that the system deducts a few frames out of each other, and when the same code is subtracted from the code of the pixel, 0 is obtained that the code system of colors is equal to zero, i.e. if there is a moving object, in the array of pixels, the photos will be non-zero pixels.

To determine the speed it is enough to write a code that divides the size of the road in front of the camera for the time that the car overcomes it. It is rather easy to define time knowing the number of images that have nonzero pixels and the number of images the camera shoots per unit of time.

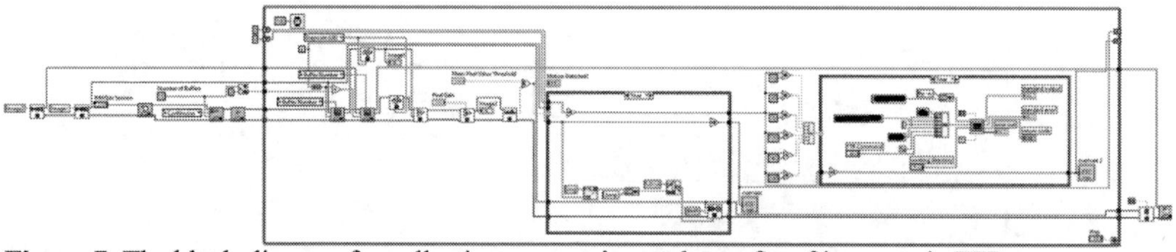

Figure 7. The block-diagram for collecting, processing and transfer of images via FTP.

3 Conclusion

The developed multi-parameter monitoring system of the state of the urbanospherum is a modular system that allows:

– to work out application subsystems for specific structural divisions;

– to develop and to implement providing subsystems to combine existing information systems;

– to obtain systematic and current data on the state of urban processes;

– to develop methods for qualitative and quantitative analysis of the security environment of the urbanized territories;

– to obtain quantitative estimates of the resources and processes of the urban environment;

– to identify the dynamic links between natural and socio-technical structures of the city that have the greatest influence on the development of biosphere compatibility of city safety;

– to display to the user the processes taking place in a city under specified conditions, to establish the reasons of occurrence of those or other situations.

Further development of this complex of systems will allow:

– to predict the possible consequences of control actions and prevent risks to the life and health of people;

– to produce qualitative (expert) analysis of the interaction between natural and socio-technical structures of a city with the desired degree of detail;

– to provide a sufficiently detailed simulation of the interaction process of the MES, authorities, organizations and businesses, residents of a city in the conditions of the current time;

– to implement the automated decision-making support into the management of urban resources and processes.

As a result of theoretical studies the mathematical model of construction of a convex hull and triangulation, which is now used to split the town in parts, and will continue to be used in the implementation of projects related to GLONASS and persons' identification. In the course of performing applied research we developed a device that raised the interest of many industrial enterprises – a device analyzing the state of the operator and his personal urbanospherum, and it is a portable individual unit to carry out monitoring of the movement of staff on the workshop floor, deviations in person's psychophysiological state and deviations in the parameters of aero-gas-dynamic control. Such systems allow not only to reduce the response time to an emergency situation, but also to optimize production processes and enterprise management system as a whole. The developed mathematical model of triangulation will be checked and analyzed by using the device of collection and analysis of data about urbanospherum. All developments are carried out with the support of the Grant of the President of the Russian Federation and regularly registered in Rospatent and published in peer-reviewed journals included in the list recommended by the Higher Attestation Commission.

References

[1] V.G. Abashin, A.V. Pilipenko, Proceedings of the International Conference on Intelligent Information Systems (IIS2013) E **9**, 196 (2013)

[2] D.-T. Lee, B. J. Schachter, International Journal of Computer & Information Sciences **9**, 219 (1980) doi: 10.1007/BF00977785

[3] *Visualization program for solving the problem of finding a triangulation: the registration certificate of the software for the computer.* The Russian Federation. No. 2016611364; declared 03.12.15; published 01.02.16

[4] *The program of debugging the problem of finding a triangulation by the method of conjugate faces of barrier and penalty functions: The registration certificate of the software for the computer.* The Russian Federation. No. 2016610841; declared 03.12.15; published 20.01.16

[5] K. Elmansouri, R. Latif, B. Nassiri, F. M. R. Maoulainine, Journal of medical systems **38** , (2014) doi: 10.1007/s10916-014-0039-8

[6] R. Lynn, A. Chen, S. Locks, C. Nath, T. Kurfess, IFIP Advances in Information and Communication Technology **459**, 27 (2015) doi: 10.1007/978-3-319-22756-6_4

[7] *Software for automation of the premises on the Arduino platform: the registration certificate of the software for the computer.* The Russian Federation. No. 2015662105; declared 28.09.2015; published 17.11.2015

[8] *Multi-function control system based on machine vision: the registration certificate of the software for the computer.* The Russian Federation. No. 2013613901; declared 04.03.2013; published 18.04.2013

METHOD OF CREATION OF "CORE-GIS-SEISMIC ATTRIBUTES" DEPENDENCES WITH USE OF TRAINABLE NEURAL NETWORKS

Denis Gafurov[a] [1], *Oleg* Gafurov [2]

[1] Krasnoyarsk Research Institute of Geology and Raw Materials, 660049, Krasnoyarsk, Russia
[2] International Laboratory of Vision Systems, Tomsk State University, 634050, Tomsk, Russia

Abstract. The study describes methodological techniques and results of geophysical well logging and seismic data interpretation by means of trainable neural networks. Objects of research are wells and seismic materials of Talakan field. The article also presents forecast of construction and reservoir properties of Osa horizon. The paper gives an example of creation of geological (lithological -facial) model of the field based on developed methodical techniques of complex interpretation of geologic-geophysical data by trainable neural network. The constructed lithological -facial model allows specifying a geological structure of the field. The developed methodical techniques and the trained neural networks may be applied to adjacent sites for research of carbonate horizons.

1 Introduction

Dependence establishment between a complex of geophysical investigations (GIS) in a prospecting well and the actual material in rock samples form (core) is complex and relevant objective of geological exploration. Coring along wellbore in a production range covers only a small part of well, in relation to intact samples extraction difficulties and, as a rule, a small sinking with coring opened by well in production range of the geological section.

The standard technique of GIS data interpretation allows to calculate total porosity factor (f_p) of breed's and component structure of a collector along wellbore. Calculations were implemented by means of linear petrophysical equations system solution, according to the established model of solid matrix. For the studied field (Talakan oil-gas condensate field) the model of solid matrix consists of dolomite, clay, salt, chalk-stone and anhydrite.

The combined equation was solved by using a system of methods: neutron gamma logging (NGL), gamma-ray density logging (GDL), acoustic logging (AL). Other methods of the GIS are auxiliary at standard technique application for identification of porosity factor and component structure of collector.

Auxiliary complex of the Talakan field consists of such GIS methods as density logging (DL), micro lateral resistivity logging (MRL), well diameter logging (WDL).

[a] Corresponding author: gafur35@mail.ru

© The Authors, published by EDP Sciences. This is an open access article distributed under the terms of the Creative Commons Attribution License 4.0 (http://creativecommons.org/licenses/by/4.0/).

In general, the technique based on lithological model of solid matrix creation by a system of NGL,GDL, AL GIS methods, significantly underreports porosity factor (f_p) in a carbonate collector because caverns are neglected in spite of the type of producing Osa horizon collector, which is characterized as complex cavern porous-fissured collector.

Unlike the standard technique, the benefit of mathematical tool of neural networks is possibility of all input data usage, in our case, GIS data (including the methods, for which linear dependences between values of logging and collector properties are absent and have difficult interpreted nonlinear nature). Compared to a standard technique, use of all GIS data allows carrying out the forecast of the physical parameters, characterizing collector along well trunk. As a result, it raises the essentially greater number of data to describe a geological section in well [1–3].

This paper offers the technique of geological field model creation based on establishment of standard joints between analytical researches of core material, GIS data and dynamic attributes of the seismic wave field by means of neural networks (core – GIS – seismic attributes).

Combination of geological -geophysical data was carried out by mathematical tool of neural networks implemented by means of "NeuroInformGeo" intellectual geographic information system, developed in "InformGeoService" LLC (O. M. Gafurov). "InformGeoService" LLC in concern with "Krasnoyarskgeophysics" CJSC (A.A. Kontorovich, D. O. Gafurov, etc.) created forecasting methods development of geological section and evaluation of Osa horizon collectors (Talakan field).

During the period from 2004 to 2015, the technique has gained further development in relation to various petroleum districts based on "InformGeoService' LLC ' (O. M. Gafurov) and "RN-Krasnoyarskpetroleum R&D Institute" LLC (A.K. Bitner, A. A. Kontorovich, D.O.Gafurov). In 2011, certificates and the Russian Federation patent were received [4].

2 Core material analysis

Analytical researches of core material and assessment of Osa horizon breeds reservoir properties have been performed by analyzing the well materials data. Wells are located in different parts of the field. Talakan field core material was selected in rather small amounts; less than 10% of wells with a representative core recovery (>70%) were drilled in the range of the productive Osa horizon.

After conducting analytical core material researches, as a result, we allocated nine lithological types of the breeds, which are characterized by various reservoir properties (Table 1). According to analytical researches results, wells test results, hydrodynamic logging (HDL), etc., boundary values for carbonate collectors have been determined. For Osa horizon carbonate collectors of the Talakan field: porosity factor f_p >6%, clayiness index I_{cl} <10%.

Table 1. Reservoir properties of lithological types of the Osa horizon breeds.

Lithology	Reservoir characteristics, %	Comments
Salt;	$K_p < 1$	Not collector
Clay material;	$K_p < 2$, $K_{sh} > 25$	Not collector
Salinized dolomites, calcareous, sulphated, clayed;	$K_p < 1$, $K_{sh} = 10$- 12	Not collector
Clay dolomit, chalk-stone dolomits;	$K_p = 2 \ldots 6$, $K_{sh} < 10$	Not collector
Anhydrite – lowclayed dolomit;	$K_p < 2$, $K_{sh} < 10$	Not collector
Lowclayed - bituminous sulphated dolomit;	$K_p = 6 \ldots 10$	Collector

Organogenic porous dolomit with trace impurity of terrigenous materials, sulphates and salts	K_p= 10…15	Collector
Organogenic highly porous dolomit.	$K_p \geq 15$	Collector

The analysis shows that reservoir properties of the allocated lithological types are rather well sustained on the area. Well-logging curves and the allocated lithotypes allow carrying out the forecast of collecting properties of breeds along wells trunk, which did not provide with core material.

3 Establishment of core-GIS dependence, forecasting of lithological types of breeds along well trunk

The review of domestic and foreign [5–10] studies devoted to interpretation of the geological – geophysical data by neural networks, indicates researchers tendency to focus attention on the choice of an optimum mathematical training method of neural network, at preservation of the simplified triplex architecture. The present work offers quite different approach. The main idea is creation of training boundary conditions (criteria) under which the dynamic mechanism of a neural network architecture adaptation goes to sophistication and increasing the quantity of training examples. Figure 1 presents the scheme of neural networks application implemented in the present work.

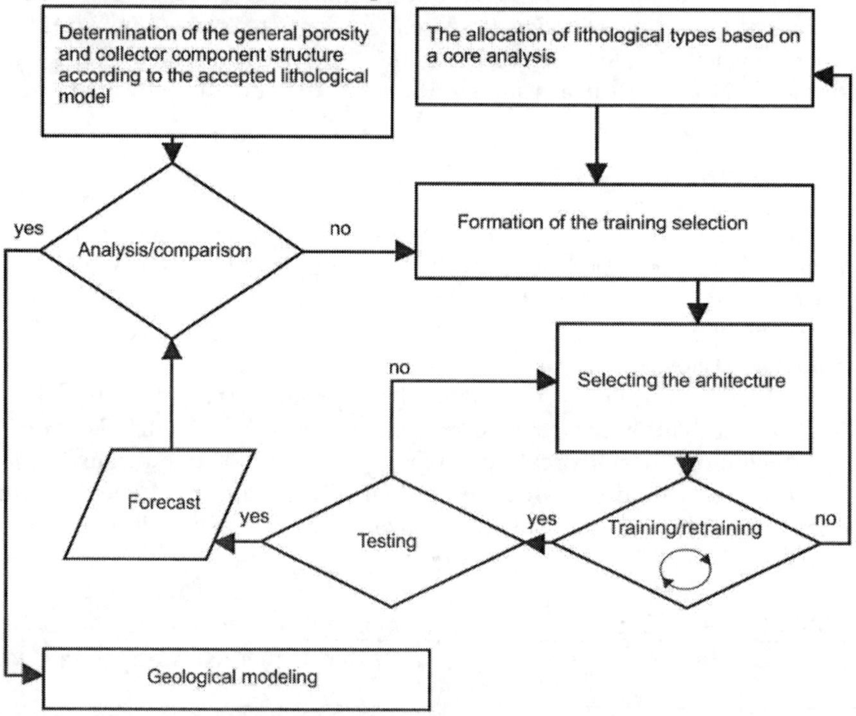

Figure 1. Scheme of the method of GIS data interpretation by trainable neural networks.

The full range of GIS used in neural network configuration includes the following system of methods: GL, NGL, WDL, AL, DL, MRL, and GDL. Additional studies consisted of hydrodynamic logging (HDL) and formation testing of log cable (TLC).

Forecast technique of lithological breeds types and effective thickness of the Osa horizon by GIS, which is set on intervals of lithological breeds types on a core, includes the following set of procedures [11–12]:

- selection of initial elementary architecture of a neural network;
- forming of the training selection, representing comparison of lithological breeds types intervals (classes) to discrete counting of the readings taken from GIS diagrams;
- choice of the boundary value, determining forecast quality in %;
- training of a neural network (Fletcher-Reeves conjugate gradient method) and forecasting;
- check of forecast results on the reference data which have entered to training selection;
- check of forecast results on the examination data, which are not included in the training selection.

Successful result means that the task is solved, negative result means that system gives a value of the maximum forecast probability and sophistication of network architecture happens automatically. In other words, the quantity of layers and neurons increases and calculation of expected parameters repeats. This process is non-infinite. After a number of iterations, network architecture sophistication process stops and the best result, received in the previous iterations, is considered as the optimal solution.

Figure 2 presents three examination wells (wells provided with the full GIS complex and representative core recovery). Examination wells have not entered to the training selection, where the neural network shows allocation and forecasting results.

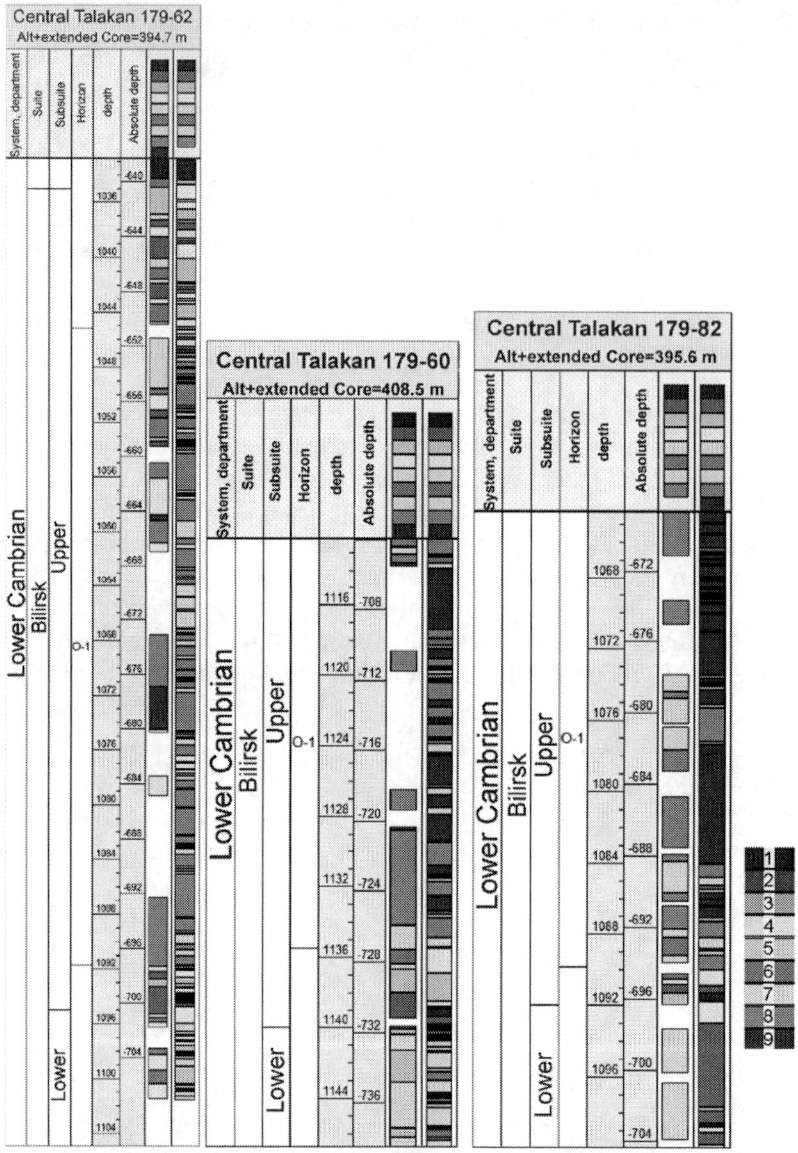

Figure 2. The tested wells (in the left column contains the allocated lithotypes on a core, in the right – expected values of lithological types along well trunk). A color scale of compliance to nine allocated classes of lithological breeds types. <u>***Not collector:***</u> 1*) salts 2) clay material;3) salinized dolomites, calcareous, sulphated, clayed; 4) clay dolomit, chalk-stone dolomits; 5) anhydrite dolomit calcareous (nonporous) highly clayed; 6) anhydrite – lowclayed dolomit.* <u>***Collector:***</u> *7) lowclayed - bituminous sulphated dolomit; 8) Organogenic porous dolomit with trace impurity of terrigenous materials, sulphates and salts; 9) Organogenic dolomit highly porous porosity factor.*

Figure 3 gives comparison of application results between standard technique of GIS interpretation and neural networks application with the most characterized core information to an examination well. There are dependences on the productive Osa horizon for 160 discrete reports of examination well. Dependences given in Figure 3 testify to higher quality of the forecast mathematical tool of neural networks.

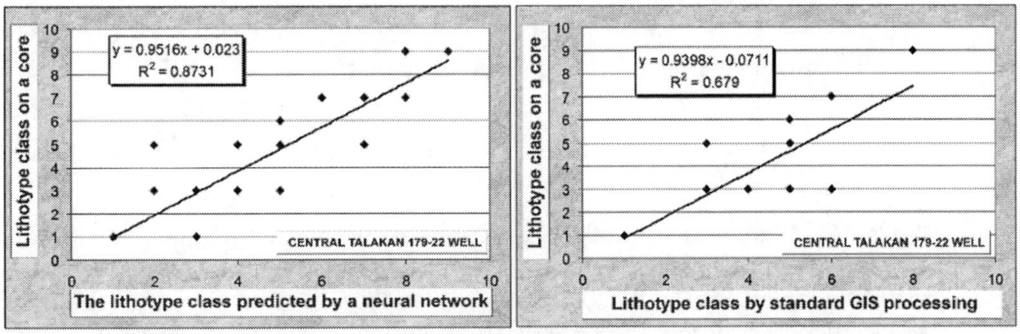

Figure 3. Comparison of GIS data interpretation techniques.

4 Establishment of GIS –seismic wave field attributes dependence, forecasting of effective thickness of the productive horizon

Seismic material during adaptation of technique presented by 2D irregular network of the seismic profiles, fulfilled on the Talakan field. Based on results of complex interpretation of analytical data, GIS materials and seismicity it was supposed to perform map development of the effective flooding space thickness of the Osa horizon with further creation of facial zonation scheme. The purpose is profound understanding of field geology and distribution of layer physical properties on the area.

Technique implementation in "NeuroInformGeo" geographic information system looks as follows:

1. Networks of N_a dynamic areal attributes with common starting point are coming to input.
2. In order to form etalon, wells N_s are used, in which tests the GIS have been carried out. For areal interpretation, all wells are getting to given or preselected area, and being in close proximity to a profile or in a band of confidential probability from a profile. Wells break into classes of the predicted geological sign.
3. For every i^{th} well, where i=1,…, N_s, based on geophysical attributes, $D_i\{x,y\}$ variety of a set of n_i points lying in a radius R_i with the center in i^{th} well. The choice of a confidential interval radius depends on geological conditions and well environment parameters. Whereby, all points, which have got to confidential space near one well carry identical information capacity.
4. The standard formation in the vector form with the average values of geophysical parameters and scales, which provides informative contribution assessment of each parameter. Points in a circle used for registration of an image or classes. Formation of vectors with coordinates $\{x,y\} \in D_i$ and geophysical parameters values implemented by using a set of the calculated coordinates near the wells $D_i\{x,y\}$. At the same time, for each jth point from the $D_i\{x,y\}$ variety, vector V_j^i (with length of N_a +1) is formed. The components of vector are values of $v_{j1},\ldots,$ v_{jN_a} geophysical parameters attributes and class number k, to which i^{th} well is related: $V_j^i = (v_{j1},\ldots, v_{jN_a}, k)$. Thus, sample constructed on the set of points

$D_i\{x,y\}$, relating to i^{th} well, is the set of vectors $U^i_j=\{V^i_1, V^i_2,\ldots, V^i_{n_i}\}$. Sample constructed on all chosen wells is formed by the set:

$$U=\bigcup_{i=1}^{N_s} U_i = \{U_1,\ldots,U_{N_s}\} \quad (1)$$

5. Wells, which are in a strip of probability belief in the first iteration, are dividing into two parts. One part of wells are used as etalon sample, another part forms test sample in a random way for forecast quality testing. In other words, the set of U is divided into two subsets U_{train} and U_{test}, forming training and test samples.
6. Multilayered neural network with N_a input and one output neuron, accepting OUT $\in$ value to the allocated k class value is forming by a method of neural networks training technique based on back propagation of error algorithm. Neural network is training on a set U_{train}. We select the architecture of a network empirically, and it provides the minimal training error.
7. Importance indicators χ_{N_a} are defined for each N_a of input parameters. This type of indicators characterizes informative contribution of known parameter (attribute) to results. The neural network calculates a gradient of assessment function on input signals and the trained values of a network. Thus, following formula determines the importance indicator N_a of the parameter in solving *q-o* case:

$$\chi^q_{N_a} = \left|\frac{\partial H_q}{\partial w_{N_a}}(w_{N_a} - w^*_{N_a})\right| \quad (2)$$

where, in *q*-o case solving, importance indicator shows how much *q*-o case solving assessment function value can be affected. If running value of w_{N_a} parameter changes to next designated value $w^*_{N_a}$ for N_a parameter. Final parameter of signification N_a is calculated as general mean:

$$\chi_{N_a} = \frac{1}{n}\sum_{q=1}^{n}\chi^q_{N_a}, \quad (3)$$

where n is number of cases. Thus, calculated value of signification parameter for N_a parameter assumes, in linear fit, absolute value of assessment function variation at input parameter (signal) moving off from the network. Non-informative parameters are moving off from the network and neural network is retraining.

8. Training control handled by comparison of predicted values variety with a set of assigned classes value in testing wells U_{test}. In the set of value U_{test} network generalization error is calculated by divergence network answers with known answers. If results are not satisfactory, training repeats. By cross validation technique the set U is divided again into two subsets U_{train} и U_{test}, which forming training and testing selections in the ratio *4/5* and *1/5*. Training and testing are implemented by forecast correlation factor calculation with testing sample. Farther, by all possible methods we divide initial sample multistage with saving of forecasting classes intervals and with network architecture sophistication. In result of iterations, true average error of neural networks extension is calculated. In case

of unsatisfactory results, next decisions have to be implemented: rechecking of initial geological geophysical data, another value forecasting possibility, which also characterize investigated environment.

9. In case of satisfactory problem solving, forecast is start. The program divides all multidimensional attribute space into classes of fixture or similarity to one or other standard and in output forms the map of standards for areal interpretation or forms a section with allocation of complexes (classes) to temporary or deep model.

As a result of training and testing, the set of seismic parameters has been determined. This set has allowed to perform with rather high degree of reliability effective thickness and facial situations forecast in Osa reservoir forming on Talakan area. As an example, Figure 4 gives effective thickness map of Osa horizon in Talakan field.

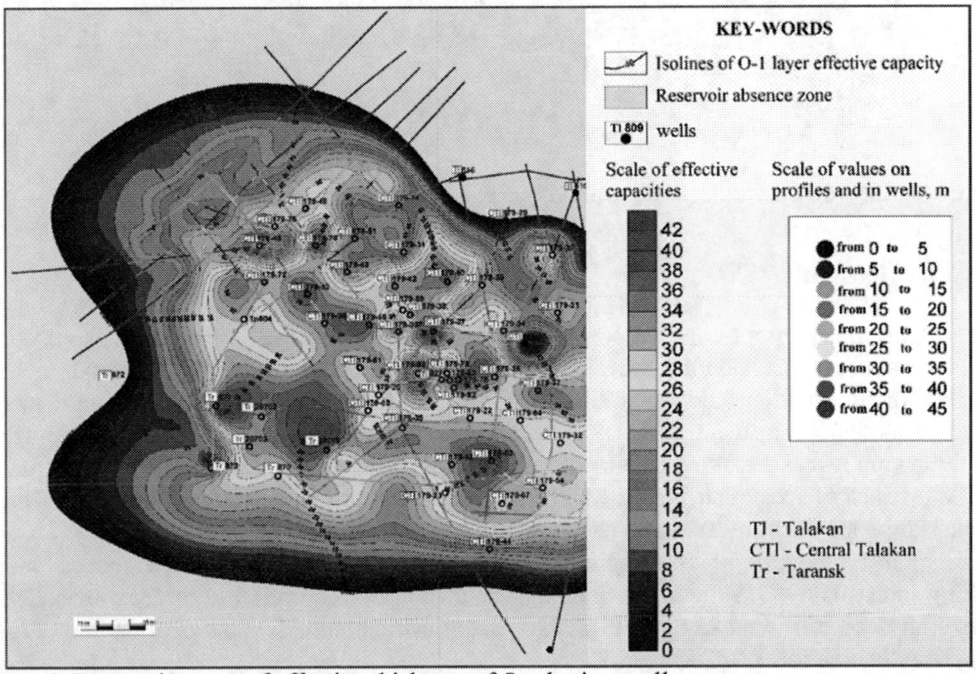

Figure 4. Forecasting map of effective thickness of Osa horizon collector.

5 Lithological facial field model creation

We constructed the lithological facial scheme of the Talakan field within checking results of coherence forecast to geological ideas about object. Scheme construction based on detail lithology splitting of subsuite drift by «NeuroInformGeo» geological information system, by forecasting of effective thickness and core analysis results. Figure 5 presents selected facial conditions of sedimentation in lithology facial scheme. Distribution of conditions on the field area (laterals) and in a vertical sense of upper Berilsk subsuite also has shown in scheme.

Complex analysis of geological –geophysical material shows that, collector properties of Osa horizon in Talakan field depends on paleogeography of sedimentation area:

- ➢ the best collectors zones of Osa horizon connected with organogenic dolomites, which formed in central and peripherical parts of organogenic structure;
- ➢ mostly, organogenic saline dolomits characterized by transient effective thicknesses. Their formation carried out in lagoon conditions with low water exchange.

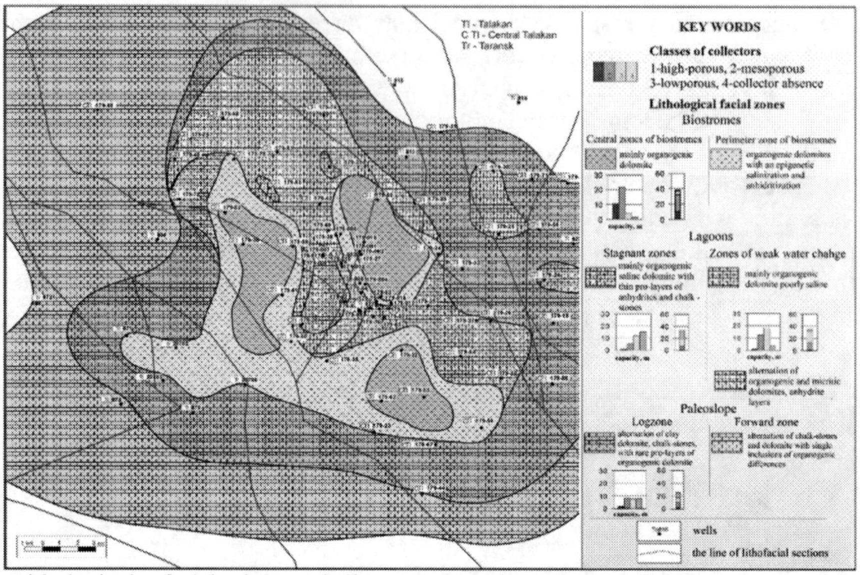

Figure 5. Lithological – facial scheme of oil, gas and condensed steam field.

6 Conclusion

One of the most important and complex tasks, which faces geologists and geophysicists carrying out researches on the Siberian platform, is reduced to collectors quality forecast of the carbonate Vendian and the Cambrian horizons, in which amount of oil and gas are concentrated. The problem of small volume of core selecting is particularly acute at GIS materials analysis stage for the purpose of lithological-petrophysical wellcuts creation and at the forecast of areal distribution of a collector, which can be carried out only according to subaerial geophysical measurements.

Standard recovery of lithological-petrophysical wellcuts determines success of the testing intervals by GIS materials choosing. It means that the reliable forecast of collectors distribution on the area can form a reliable basis for the planned wells location and diggings solid model creation.

The executed researches shows that neural networks application in case of analytical data integration, GIS materials and seismic exploration allows solving the following tasks: problems of geological section forecasting and collectors quality evaluation of both by wells, and on the area. Developed methodical techniques with some calibration can be used at the stages of investigation and additional exploration of oil and gas fields in the carbonate horizons of already found fields. In addition, taking into account statistical data at a search, these techniques can be used at the territories, which are poorly studied by drilling [12].

Acknowledgment

The paper was written as part of the research project No. 8.2.31.2015, carried out with the support of the Program "Research Foundation of Tomsk State University named after D.I. Mendeleev" in 2015 - 2016, grant RFBR No. 16-29-04388/16.

References

[1] W.S. McCulloch, W. Pitts, *Logical calculus of the ideas immanent in neurons activity* (Publishing House of Foreign Literature, Moscow,1956)

[2] Cheng-Dang Zhou and Xi-Ling Wu, Proceedings of the Asia Pacific Oil and Gas Conference, 357 (1993)

[3] N. Silpngarmlers, T. Ertekin, Proceedings - SPE Annual Technical Conference and Exhibition, 3201(2002)

[4] O. M. Gafurov, D. O. Gafurov, M. V. Pankov, N. B. Krasilnikova, A. O. Gafurov, A. K. Bitner, *A method of operational wells deposit places determination in case of hydrocarbons fields development* Patent Russian Federation, MPK G01V1/00 (2006.01), G01V11/00 (2006.01)

[5] V.G. Tsaregorodtsev, Lecture notes in computer science **3606**, 186 (2005)

[6] Y. Deng, Q. Chen, J. Wang, Proceedings of the International Oil and Gas Conference and Exhibition in China (IOGCEC),317 (2000) doi: 10.2118/64652-MS

[7] Y. Du, W.W. Weiss, J. Xu, D. Li, Proceedings - SPE Annual Technical Conference and Exhibition, 3449 (2003) doi: 10.2118/84445-MS

[8] I.S. Agbon, J.C. Araque, SPE Hydrocarbon Economics and Evaluation Symposium, 82014(2003) doi: 10.2118/82014-MS

[9] M. Nikravesh, A.R. Kovscek, A.R. Murer, T.W. Patzek, Proceedings - SPE Annual Western Regional Meeting, 681 (1996)

[10] S. Mohaghegh, R. Arefi, S. Ameri, Proceedings – SPE Annual Technical Conference and Exhibition, 28394 (1994)

[11] D.O. Gafurov, Bulletin of the Tomsk Polytechnic University **309,** (32) 2006

[12] D.O. Gafurov, O. M. Gafurov, V.A. Kontorovich, Technology of seismic exploration (4), 85 (2014)

GEOINFORMATION SUPPORT AND WEB TECHNOLOGIES FOR PROBLEMS OF HYDROBIOLOGICAL MONITORING OF YENISEI RIVER

Anna Andrianova[1,2], *Nicolay* Shaparev[1,3], *Oleg* Yakubailik[1,*]

[1]Institute of Computational Modeling SB RAS, 660036, Krasnoyarsk, Russia
[2]Institute of Ecology of Fishery Reservoirs, 660097, Krasnoyarsk, Russia
[3]National Research Tomsk State University, 634050, Tomsk, Russia

Abstract. Results of long-term hydrobiological monitoring for the Yenisei river section from the Krasnoyarsk hydroelectric power station dam to the village Zotino (Turukhansk district of the Krasnoyarsk region) are considered. Researching of zoobenthos showed significant changes of its structure after the Krasnoyarsk HPP dam construction. In particular, the proportion of amphipods (crustaceans) in the zoobenthos biomass increased by 10 times. In the expeditionary studies, seven species of amphipods were identified, but it should be noted that only two species develop and clearly dominate among them. Data obtained during the hydrobiological research formed the basis of geospatial database developed as a result of the work. The geospatial database with the observation results is available in the ICM SB RAS geoportal. The geoportal services provide data visualization in the form of interactive thematic maps, information about the objects on the map as "tooltips", information downloading in tabular form, direct access to the data through the web mapping services. Information and analytical support for the monitoring system based on the considered approach can simplify access to the needed information, provide new opportunities for analysis and modeling, promote the development of cross-disciplinary researches.

1 Introduction

The Yenisei is the main water artery of Siberia, which is among the seven largest rivers of the world and the most abounding river of our country. Over many years, an extensive array of diverse biotic and abiotic data is accumulated on the Yenisei ecosystem that is a valuable material for analysis. Using of new data processing techniques such as geoinformation and cartographic modeling will provide additional information on the features of aquatic organisms spatial distribution and help to search for relationships with various environment factors.

* Corresponding author: oleg@icm.krasn.ru

© The Authors, published by EDP Sciences. This is an open access article distributed under the terms of the Creative Commons Attribution License 4.0 (http://creativecommons.org/licenses/by/4.0/).

Usage of modern geoinformation system technologies and the Internet provides an effective solution of all problems related to the geospatial data submission, processing and analysis. Creation of the geospatial database with the results of field researches and implementation of geoinformation web-system for information and analytical support of the hydrobiological monitoring form the basis for interdisciplinary studies [1, 2].

2 Hydrobiological monitoring

It is known that fundamental change in the hydrological, hydrochemical and hydrobiological regimes happened in the Yenisei since the commissioning of the Krasnoyarsk hydroelectric power station. Consequences of the regulation can be traced at a distance of more than 1.5 thousand kilometers from the hydrounit point. Environmental transformations in aquatic communities require in-depth study and long-term monitoring both to assess the environmental damage and to develop scientific bases of strategy to overcome the negative effects and restore the biological potential of the Yenisei ecosystem. Currently, hydrobiological monitoring system is almost absent in the Yenisei.

Integrated monitoring of aquatic ecosystems is impossible without researches of such an important biological unit as the zoobenthos community. Zoobenthos consists of invertebrates (such as leeches, mollusks, crustaceans, worms, insect larvae, etc.), which live in the water bodies on the bottom surface and inside it. The role of zoobenthos in aquatic ecosystems is hard to overestimate. First, the community of benthic invertebrates is a link in the trophic chain and a food object for most Siberian fish species (sturgeon, whitefish, grayling, carp, etc.). Second, zoobenthos organisms contribute to the natural self-purification of natural waters as they become active mineralizers of organic substances and water biofilters in their life. Third, they are reliable biological indicators for the ecological state of different water bodies as there is a clear link between the zoobenthos indices and the content of pollutants in the bottom layers of water and sediments. Structural changes of benthic communities and species diversity reduction occur under the influence of pollution.

One of the main features of zoobenthos and its individual representatives is the density (the ratio of the organisms number to the square unit), which is expressed by the abundance (ind./m^2) and biomass (g/m^2).

Regulation of the Yenisei led to global changes in the zoobenthos of the river, first of all – to the spread of Baikal origin amphipods (crustaceans) above the Angara estuary. At the same time, their share in the Yenisei zoobenthos biomass increased by 10 times and they occupied a dominant position [3, 4]. It is known that crustaceans are among the most active aquatic organisms self-dispersing outside the natural habitat. In addition, in order to enrich the food supply for fish, the Baikal amphipods were successfully settled in many ponds and lakes of the USSR [5]. Invasions of these organisms lead to significant changes in the recipient water bodies, in particular, to a decrease in species diversity and stability of aquatic ecosystems [6].

The problem of environmental consequences assessment for the introduction of alien organisms in aquatic ecosystems is extremely important. The related issue of risk assessment for the aquatic organisms invasions is equally important. Consequences of the amphipods introduction, which belong to polluting species, are poorly studied and require long-term researches [7, 8].

3 Geoinformation web-system

It is believed that the first cartographic web application is the Map Viewer program created in 1993 by the Palo Alto Research Center (PARC) of the Xerox company, which allowed the users to interactively send requests from the browser to the server and receive fragments of maps in GIF format. This application and its functional concept became the ancestor for the majority of web mapping systems later versions. Roughly, since 1998 large companies - manufacturers of GIS software entered the process of these systems creation: ESRI, Intergraph and others began to develop commercial applications for web mapping. Since the same moment of time, the most successful non-commercial open-source project - MapServer started its history, the development of which began at the University of Minnesota. In subsequent years and nowadays, there is an intensive formation of the web-cartography market. Numerous software developments appear - from simple visualization tools of prepared maps in the browser to complex distributed systems of corporate geo-information processing [9–11].

The resulting concept of mapping web-application development involves the creation of a program complex, the execution of which is carried out simultaneously on the server computer and the client computer (multi-tier architecture), as well as formation of a spatial data set in the format of popular GIS or using geospatial databases [12]. The initial geodata preparation for web applications was carried out outside the web-GIS using the desktop GIS – QGIS [13].

The generated geospatial database content included the results of research expeditions – data on the zoobenthos numbers and biomass in the Yenisei (from the Krasnoyarsk hydroelectric power station dam to the Angara river estuary), its distribution by species, banks of the Yenisei, depth, soil type, presence of higher aquatic vegetation.

Distinctive characteristics of the geoinformation web-system considered in this paper for the hydrobiological monitoring tasks solution are the following features [14]:

• integration of mapping web-application to the web content management system, its user access controls, administration and interface configuration, formation of the web portal information blocks;

• user interface improvement: creation of map and geodata controls in the style of traditional desktop GIS – floating panels with button instruments, interactive treelike drop-down menus with the map layer lists, contextual setting of data display properties, etc.;

• enhanced support for the geospatial data exchange between the mapping web-application elements and other systems based on open technology standards - web services;

• used geospatial data kits design as directories with the appropriate metadata, creation of independent software for navigation and search for geoinformation in these directories. In the literature a lot of suchlike integrated geospatial data processing systems can be found, for example [15, 16].

User interface of the considered systems is made as a web resource with two access modes – user and administrative ones. The user interface is made for navigation and search for published information resources. The administrative interface allows editing of various data to create new information resources and manage their publication [17].

The development is based on PostgreSQL 8.x database with PostGIS module to store the spatial data, PHP 5.x script language to implement the application business logic. Formation of user and program interfaces was carried out with the help of many open-source libraries and technologies – Ext2js, TinyMCE, HTML_MetaForm, Smarty, PEAR, Zend Framework and others.

4 Hydrobiological studies results

The materials are a part of the complex work "Determination of forage resources production potential for the Yenisei river fish at the section from the Krasnoyarsk hydroelectric power station dam to the village Zotino", which was started back in the 2000s. The primary role in the fish forage resources of the considered area belongs to the benthic fauna. One of the main objectives was to evaluate changes in the river benthic fauna after the regulation.

According to physical and geographical conditions, the structure of the river valley and bed, the water regime, the Yenisei is divided into the upper (source - the Angara estuary), middle (the Angara – the Lower Tunguska estuary) and lower (the Lower Tunguska – the Yenisei estuary) areas. The studies covered a section of about 780 km long in the middle of the Yenisei River: 325 km – in the mountainous Upper Yenisei (from the Krasnoyarsk hydroelectric power station dam to the Angara estuary) and 456 km – in the semi-mountainous Middle Yenisei (from the Angara estuary to the village Zotino). During carrying out the field work for the selection of zoobenthos samples, 10 sections were selected in the specified part of the Yenisei.

Screening and treatment of the benthic samples were carried out according to standard hydrobiological methods [18]. In each section, the soil sampling was carried out at both banks at two depth (D) gradations (less than 1 m and 1 to 2 m) using a circular Dulkeit's scraper (capture area 1/9 m^2) designed specially for the Yenisei in the 1930s. During the hydrobiological material collecting, watercourse depth, water temperature, flow rate, soil type and degree of the watercourse overgrowing with macrophytes (aquatic vegetation) were determined.

During the period of studying, about 300 samples of zoobenthos were collected and processed in the Yenisei and its tributaries. All results were statistically processed using Microsoft Excel and Statistica 9 software. In Figure 1, aggregate data is presented in the form of arithmetic averages, average errors, confidence intervals. Student's t-criterion was used for the data comparison with a probability of 95% (p = 0.05).

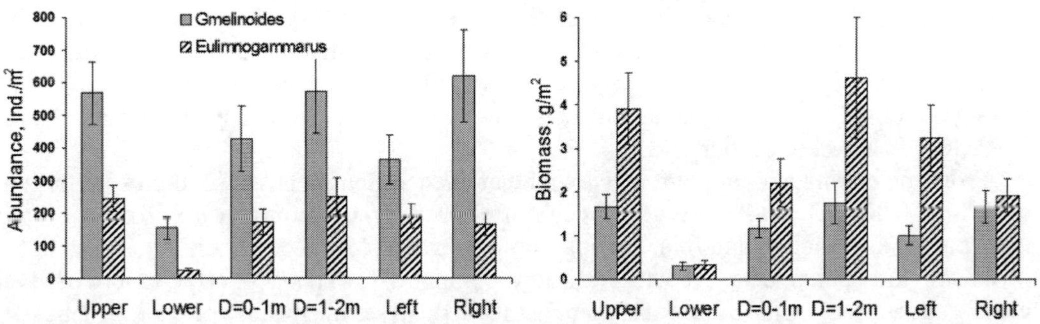

Figure 1. Spatial distribution of the Baikal amphipods in the Yenisei.

To form the geospatial database, materials on the spatial distribution of amphipods were used, which currently dominate among the Yenisei zoobenthos communities. In the studied area, seven species of amphipods were identified, but only two species, which are endemic to Baikal, massively develop and clearly dominate among them: *Eulimnogammarus viridis* Dybowsky, 1874 and *Gmelinoides fasciatus* Stebbing, 1899.

The spatial dynamics of the amphipods quantity and biomass in the Yenisei (Figure 1) is characterized by inhomogeneous distribution, but a clear decline trend is obvious in the transition from the upper to the lower section (i.e., above and below the Angara).

It was revealed that the number of *Gmelinoides* exceeds the number of *Eulimnogammarus* more than twice in the upper section and more than five times in the lower section. It is known that the *Gmelinoides* amphipod is very ecologically plastic. He is hardy to significant eutrophication, resistant to a number of pollutants, can adapt to a range of adverse factors and thus is one of the first species to colonize habitats contaminated with industrial waste [19].

It is believed that the right bank of the Yenisei is more productive due to its more extensive river network. *Gmelinoides* abundance and biomass are really slightly higher at the right bank; for *Eulimnogammarus*, this trend is not noted. In addition, a slight tendency to increase the quantity and biomass of both amphipods species with depth is revealed.

5 Data presentation with GIS

The first step in the web-GIS creation for hydrobiological monitoring was the formation of a geospatial data set for the field research results registration in the GIS and visualization of the results [20]. In particular, the work on the sailing sheets georeference to satellite images was carried out by locally-affine transformations (at least 15 points for each sheet), layers of vector data with the fairway, kilometer markers, watercourse alignment, etc. were created. To visualize various data on the fairway and river banks (left/right), a series of specialized polygonal and linear layers was also formed.

In the second step, conversion algorithms were developed for the calculation and generalization of the initial hydrobiological monitoring data array into the geopositioning values system taking into account the used data aggregation. As a result of the software performance, thematic geodata layers are formed, visualization of which is made by regular means of the ICM SB RAS geoportal.

The considered geoinformation web-system is built according to so-called service-oriented architecture – it can be regarded as a set of interrelated software components for spatial data operations – its import/export, cataloging, visualizing, creating, processing, distribution, etc [21, 22]. The technological basis is a set of program interface libraries; they provide access to the functions and context of web pages mapping elements, thematic data visualization. The main characteristics of the system are high degree of the user interface interactivity, very fast regeneration of images made of small caching raster fragments (tiles), almost instant response of the move (shift) and map scale change tools, extensive settings of visualization and geodata management. Support of geodata exchange standards – KML, GML, WMS, etc. is provided [23].

From the organizational point of view, the created system is based on the ICM SB RAS geoportal. The geoportal is a type of web portal used to access and search for geographic (geospatial) information and related geographic web services (access, editing, and so on). It is one of the spatial data infrastructure key components. The geoportal functionality is closely associated with metadata management subsystem, means of spatial analysis, interactive visualization [24]. A popular CMS system - Drupal web content management system – was selected as a basis for the content publication.

6 Conclusions

It should be noted that the experience of various research and applied systems design and development shows high efficiency of using geoinformation web-systems and services as the foundation for information and analytical resources and geospatial data integration.

Modular architecture of the system, the use of standard web services to exchange data between these modules provide rapid adaptation of existing software to customer's requirements, replication of individual components, their complementarity. In turn, registration of created resources – spatial data in the centralized geoportal directory provides the ability to re-use them. This service-oriented approach based on active introduction of web technologies in the applied information systems is increasingly used today.

Acknowledgments

The work was supported by grant no. 16-47-240517 from Russian Foundation for Basic Research and Krasnoyarsk Regional Fund of Support of Scientific and Scientific-technical activities.

References

[1] L. Berrahou, N. Lalande, E. Serrano, G. Molla, L. Berti-Équille, S. Bimonte, S. Bringay, F. Cernesson, C. Grac, D. Ienco, F. Le Ber, M. Teisseire, Computers & Geosciences **85**, 126 (2015) doi: 10.1016/j.cageo.2015.09.012
[2] Li Li, B. Zheng, L. Liu, Procedia Environmental Sciences **2**, 1510 (2010) doi: 10.1016/j.proenv.2010.10.164
[3] A.V. Andrianova, Tomsk State University Journal of Biology **1**, 74 (2013)
[4] M.I. Gladyshev, A.V. Moskvicheva, Doklady Biological Sciences **383**, 138 (2002)
[5] D.V. Matafonov, R.M. Kamaltynov, M.T. Itigilova, Sibirskiy ekologicheskiy zhurnal **5**, 595 (2006)
[6] N.A. Berezina, Hydrobiologia **590** (1), 15 (2007) doi: 10.1007/s10750-007-0753-z
[7] R. Liu, T. Xie, Q. Wang, H. Li, Procedia Environmental Sciences **2**, 1307 (2010) doi: 10.1016/j.proenv.2010.10.141
[8] J. Graham, G. Newman, C. Jarnevich, R. Shory, T. J. Stohlgren, Ecological Informatics **2**, 177 (2007) doi: 10.1016/j.ecoinf.2007.03.006
[9] O.E. Yakubailik, V.G. Popov, Computational Technologies **6**, 116 (2009)
[10] A. Vyas, S. John, MATEC Web of Conferences **57**, 05002 (2016) doi: 10.1051/matecconf/20165705002
[11] X. Jin, D. Guo, S. Wang, X. Zhao, Y. Song, A. Liu, MATEC Web of Conferences **22**, .04027 (2015) doi: 10.1051/matecconf/20152204027
[12] M.K. Yetik, M. Yüceer, R. Berber, E. Karadurmuş, IFAC Proceedings Volumes (IFAC-PapersOnline) **7**,798 (2009)
[13] A.A. Kadochnikov, V.G. Popov, A.A. Tokarev, O.E. Yakubailik, Journal of Siberian Federal University Engineering & Technologies **4**, 377 (2008)
[14] P. Yue, M. Zhang, Z. Tan, Environmental Modelling & Software **69**, 128 (2015) doi: 10.1016/j.envsoft.2015.03.017
[15] X.-l. Xie, Z. Long, W.-Q. Liao, X.-J. Guo, MATEC Web of Conferences **56**, 01013 (2016) doi: 10.1051/matecconf/20165601013
[16] C. P. Korose, R.A. Locke, C. S. Blakley, C. H. Carman, Energy Procedia **63**, 3945 (2014) doi: 10.1016/j.egypro.2014.11.424
[17] O.E. Yakubailik, Vestnik of SibGAU **1**, 40 (2010)
[18] E. Comino, M. Bottero, S. Pomarico, M. Rosso, Land use policy **36**, 381 (2014) doi: 10.1016/j.landusepol.2013.09.006
[19] M.A. Barbashova, S.A. Malavin, E.A. Kurashov, Russian Journal of Biological Invasions **4**, 219 (2013) doi : 10.1134/S2075111713040036
[20] Q.-G. Liu, Y. Chen, J.-L. Li, L.-Q. Chen, Ecological Modelling **203**, 279 (2007) doi: 10.1016/j.ecolmodel.2006.11.028
[21] Z. Qun, Y. Yujin, K. Yuena, Physics Procedia **25**, 2235 (2012) doi: 10.1016/j.phpro.2012.03.376
[22] S. Borza, V. Petrescu, Process Safety and Environmental Protection **101**, 9 (2016) doi: 10.1016/j.psep.2016.01.002
[23] O.E. Yakubailik, A.A.Gosteva, M.G. Erunova, A.A. Kadochnikov, A.G. Matveev, A.S. Pyataev, A.V. Tokarev , Vestnik of KemSU **52**, 136 (2012)

[24] Nina M Kelly, Karin Tuxen, Computers, Environment and Urban Systems **27**, 527 (2003)

SPECIAL ASPECTS OF THE DEVELOPMENT OF HEALTH INFORMATION SYSTEMS AS ONE OF THE AREAS OF EXPERTISE OF UNIVERSITY STUDENTS MAJORING IN PROFESSIONAL INSTRUMENT ENGINEERING

Elena Vaganova*

National Research Tomsk State University, 634050, Tomsk, Russia

Abstract. The article describes distinguishing features of development and production economics of health information systems. Characteristics of medicine as a subject area from the point of view of developers are presented. Basics of medical informatics and the special aspects of development of medical products are proposed to be one of the areas of expertise of university students majoring in professional instrument engineering.

1 Introduction

Health information systems (HIS) – a set of software, hardware, and data and knowledge bases for automation of health care processes in medical and prophylactic institutions. The market of health (healthcare, hospital) information systems (HIS) was initiated in 1960s, when they were used primarily for hospital inventory. Nowadays HIS are included in the integration of all clinical, financial and administrative processes. Modern HIS consist of multiple applications, meeting the needs of different departments in a hospital. Hospitals using HIS have access to a fast and reliable data, including patient records (demographics, gender, age and others) [1–3].

2 Special aspects of the development of health information systems

Medicine has made a significant progress in the diagnosis and treatment of many diseases, involving more and more technically advanced tools and methods. HIS meets the needs of the medical and management personnel in the arrangement of information used for improvement of healthcare and decision-making [4]. Besides the primary objective a corporate HIS carries out the following tasks:

- common information space, intended for immediate access to information;
- improving the quality of medical records;

* Corresponding author: hailun@mail.ru

© The Authors, published by EDP Sciences. This is an open access article distributed under the terms of the Creative Commons Attribution License 4.0 (http://creativecommons.org/licenses/by/4.0/).

- control of health care quality, and as a result, reduction of probability of medical errors;
- increasing the transparency of a medical institution;
- analysis of the economic aspects of health care;
- deducing the time of examination and treatment.

HIS of a medical center provides a single information space, where doctors, nurses, hospital administrators work closely with each other. The control process is performed my means of multimedia data, including texts, images, numerical values, and so on, providing comprehensive solutions of control problems. Complexity and ambiguity of the subject area include responsibility for other people's health. Thus main feature of medical informatics is each HIS should be developed precisely in terms of conditions in which it will be operated.

The developers of health information systems have to deal with a constantly changing subject area. The most important sources of these changes are:

- development of social and economic spheres;
- development of medical science;
- the influence of information technologies on patients' behavior (they become more informed) and health management in general.

Before developing a HIS a number of important issues must be considered:

- Detailed analysis of the structure of the organization and collaborative mechanisms between the participants of internal and external processes.
- Planning software and hardware tools and resources necessary for the development and operation of the system.
- Detailed calculation of the resources necessary for the development and maintenance of the system.
- Information awareness regarding the software systems used previously.

The object of medicine is a human being. The subjects of the medicine are etiology, pathogenesis, diagnostics, treatment, and prevention of diseases. The complexity of medicine as a field of human activity is reflected in management of health care institutions and in the development and operation of HIS.

Medicine has made a significant progress in the diagnosis and treatment of many diseases, involving more and more technically advanced tools and methods. All the complexity of medicine includes such issues as management of medical institutions and the of development and implementation of medical information systems. The developers of health information systems have to deal with a constantly changing subject area. The most important sources of these changes are:

– development of social and economic spheres;
– development of medical science;
– the influence of information technologies on patients' behavior (they become more informed) and health management in general.

Another special feature of health information is privacy. In Russia the rights of citizens for confidentiality of health encounters are protected by Federal laws regarding health care [5].

3 Understanding principles of the development of health information systems as one of the educational competencies

Creation of complex software products, including HIS, by large groups of experts requires efficient production management, planning of costs and deadlines. The developers of complex software systems, as a rule, do not know the basics of economic issues of industrial production of complex technical products, and managers do not understand completely properties of objects developed, as well as special features of technological processes. This complicates the implementation process as well. Thus, basic knowledge of the specifics of the development of medical products should be considered as one of the competencies of future engineers.

Competency-based approach, applied in basic educational programs of engineering majors, must take into account constant changes in the labor market. Occupational activities of graduates of basic educational programs majoring in engineering consist of research and development, design and engineering, production and technology, and management. Research and development activity can include basic principles of the development of health information systems and the characteristics of medical informatics. For one of the competencies of production and technological activity of design engineers include the understanding of the principles of the development of health information systems and the characteristics of medical informatics [6].

4 Conclusion

In higher education of the Russian Federation a lot has been done in recent years to implement competency-based approach to studies (knowledge, skills). The number of educational programs for undergraduate and graduate students majoring in engineering is constantly growing.

Russia is a comparatively young participant in the market discovering the achievements of heath informatics, reproduction and restoration of human resources on the basis of new technologies.

Due to the fact that in recent years, in Russia a growing number of health information systems for automation of health care institutions have been developed and implemented [7,8], basics of medical informatics, including specifics of the development of medical devices should be included in educational programs of engineering majors.

Acknowledgement

The study was performed within the framework of the Agreement № 14.579.21.0019 (unique identifier for the project - RFMEFI57914X0019) on the «Development of the device for daily monitoring of the fetus and the mother during pregnancy by controlling the parameters of the cardiovascular system based on the acoustic data», signed between the R&D company «Diagnostika+» LLC and the Ministry of Education and Science of the Russian Federation, and the program of improving competitiveness of the TSU (project No. 8.2.31.2015).

References

[1] R. Haux, International journal of medical informatics **79**, 599 (2010)

[2] N. S. Fleming, S. D. Culler, R. Mccorkle, E. R. Becker, D. J. Ballard, Health Affairs **30**,481 (2011)

[3] R. Hillestad, J. Bigelow, A. Bower, F. Girosi, R.Meili, R. Scoville and R. Taylor, Health Affairs **24**,1103 (2005)

[4] M. Ngafeeson, *Healthcare Information Systems: Opportunities and Challenges* (IGI Global, Hershey, 2014)

[5] Ya. I. Guliev, I. F. Gulieva, E. V. Rjumina, O. A. Foht, Je. F. Tavlybaev, A. Ju. Programmnye sistemy: teorija i prilozhenija **4**, 3 (2012)

[6] S.L. Eremina M.D. Gladkova Tomsk State University Journal of Economics **3** , 187 (2013)

[7] E. V. Vaganova M. V. Syryamkin, AIP Conference Proceedings **1688,** 040012 (2015)

[8] V.I. Syryamkin, A.S. Bureev, D.S.Zhdanov, A.V. Osipov, S.A. Klestov, Biomedical Engineering **46**, 112 (2012) doi: 10.1007/s10527-012-9281-x

MANUFACTURING THE TECHNOLOGY OF XENON-CONTAINING DRINKING WATER AND ITS INFLUENCE ON SOME PSYCHOPHYSIOLOGICAL CHARACTERISTICS OF MAN

Irina Verkhovskaya[a], *Evgeny* Petrov, *Alexander* Verkhovskiy

National Research Tomsk State University, 634050, Tomsk, Russia

Abstract. This study is aimed to describe the technology of making xenon-containing drinking water and its impact on health and psychophysiological characteristics of 26 volunteers after regular admission. Testing conducted after 2 months of daily use such water showed significant improvement of emotional state and health of volunteers. This technology allows obtaining a large amount of drinking water saturated with xenon, which can be used in therapeutic and prophylactic purposes.

1 Introduction

The therapeutic capabilities of xenon are well known and have been used successfully [1, 2]. Mechanisms of action of xenon on the body associated with its effect on receptors [3–5] and the state of membranes [6–9]. Xenon is an amphiphilic substance, but its solubility in oils is 20 times higher than in water [10]. As a result xenon may accumulate in any hydrophobic structures, for example, in the middle of a cell membrane bilayers [11]. Therefore, it is expectable the impact of xenon on the vital organs, including the CNS. Now xenon is used in the form of inhalations, which results in decrease in fatigue, improve sleep, prevent the development of stress, and improve the adaptability of the organism [12]. Using our own technology of xenon dissolving in a variety of liquid funds and products, and with questionnaire test, we decided to see how overall psychophysiological profile of adult men may change after daily drinking water with xenon. Our results showed an improvement in psychosomatic characteristics that we associate with a regular intake of drinking water with xenon.

[a] Corresponding author: alverkhovniy@mail.ru

© The Authors, published by EDP Sciences. This is an open access article distributed under the terms of the Creative Commons Attribution License 4.0 (http://creativecommons.org/licenses/by/4.0/).

2 Materials and methods

For effective administration of xenon to the drinking water, we have developed a technology using a high speed stirrer in a sealed vessel volume 10l. Flowsheet is shown in Figure 1.

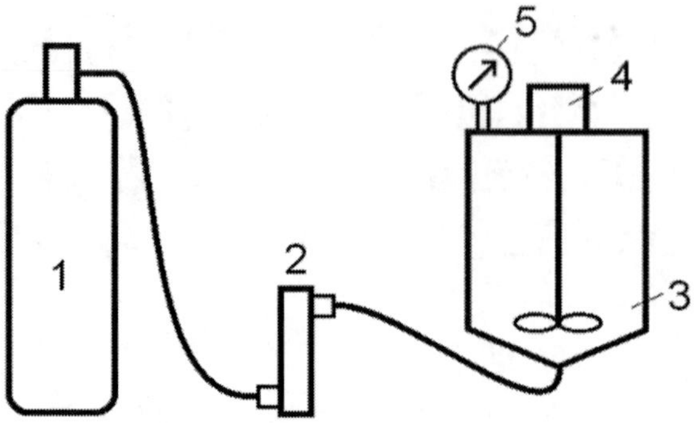

Figure 1. Block diagram of the apparatus for producing the xenon-containing drinking water: 1 – cylinder with xenon, 2 – flowmeter, 3 – water tank, 4 – high speed mixer, 5 – manometer.

The amount of dissolved xenon monitored by the change in pressure inside the container. Xenon (> 99.999%, Tianjin Dongchuangrixing Technology Co. Ltd, China) was fed through a 5 mm tube into water tank through the flowmeter. Xenon concentration measured by gas chromatography was 10.7 vol.%. The work was done within two months in the summer time. The study involved 26 male workers aged 21–49 years on a voluntary basis after receiving an informed consent from each volunteer. All participants were given every day 3 bottles of drinking water saturated with xenon volume of 330 ml. Water with xenon prepared by our own technology. Xenon concentration in water was 10.7% (vol). To determine the concentration of xenon gas chromatography method used. To evaluate the psycho-physiological state is a questionnaire, which are commonly used in medical examination. To reply to a questionnaire proposed 3 options: None; Occasionally; Have / Noticed. The results were evaluated by comparing the responses to the questionnaire before and after 2 months of drinking daily water with xenon. The data were processed statistically (Microcal Origin, USA).

3 Results and discussion

Results of testing before and after 2 month of consuming xenon-saturated water are presented in Table 1.

Table 1. Results of testing.

Question	Responses before receiving water saturated with xenon			Responses after receiving water saturated with xenon		
	No	Occasionally	Have / Noticed	No	Occasionally	Have / Noticed
Do you have headaches or dizziness?	9	7	10	17	5	4
Have you had faintings?	20	6	0	22	4	0
Do you have digestive	4	8(diar	8	11	6(diarr	4

problems: bloating, constipation, and diarrhea?		rhea, constipation)	(constipation) 6 (diarrhea)		hea, constipation)	(constipation) 5 (diarrhea)
Do you have breathing difficulties: rapid breathing, feeling short of breath?	13	9	4	15	9	2
Do you experience palpitations, heart sinking?	11	12	3	10	12	4
Do you experience excessive sweating?	16	4	6	19	4	3
Do you notice decreased performance, increased fatigue?	9	6	9	19	3	5
Do you have sleep disorders?	7	8	9	18	5	3
Do you notice blurred vision?	20	3	3	20	3	3
Do you notice a hearing loss, ringing in the ears?	15	7	4	22	3	1
Do you have a sense of inner tension or worrying?	23	3	0	23	3	0
Do you have disturbing thoughts, anxiety, fear?	16	6	4	17	4	5
Do you look after your appearance?	19	2	5	15	2	7
Do you enjoy communicating with friends, colleagues?	7	4	15	4	3	19

Analysis of the responses showed that after 2 months of daily drinking of xenon-containing water the shift of subjective psychophysiological indicators take place. It was noticed a general decrease of anxiety and physiological disorders (Figure 2). Almost all workers (25 of 26) would like to continue to drink water with xenon because felt an improvement in their overall condition.

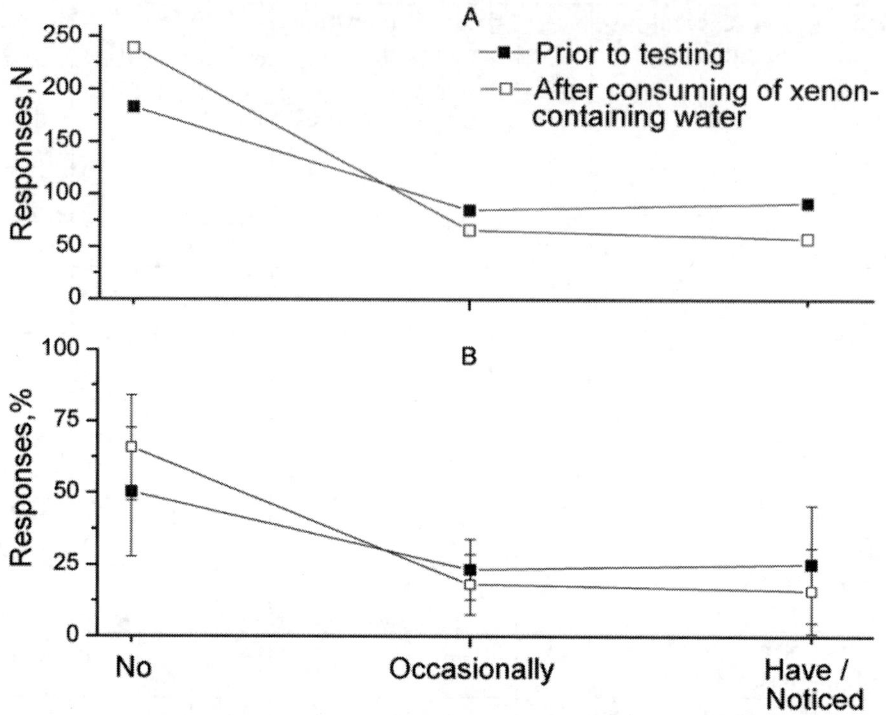

Figure 2. Effect of xenon-containing drinking water physiological indicators of all volunteers.

There is a decrease amount "disturbing" (92 to 58) and an increase of "positive" responses (from 183 to 239) (A, in absolute values) and, respectively, from 25% to 16% and from 50% to 65% (B as a percentage of all volunteers). Thus, when answering the question "Do you have headaches or dizziness?" answered "no" 9 volunteers of 26 (34.6%) prior to testing and 18 of 26 (65.4%) after 2 months of consuming xenon-containing water. Data are presented as Mean ± SD.

The results showed that regular use of drinking water with xenon has a positive effect on the general state of the organism, similar to the effects of inhalation of xenon use during therapeutic procedures. In studies with inhalation of xenon and nitrous oxide, conducted on men-volunteers, it has shown a positive sedative effect of xenon, a more pronounced compared with nitrous oxide [13].

4 Conclusions

Our results suggest a new, alternative method of administering xenon to the human body as part of the drinking water. Thus, we propose the use of drinking water with xenon as an additional procedure between xenon inhalation. This study is the first attempt to quantify the effect of xenon in the composition of drinking water on human psycho-emotional performance. In the future we plan to use advanced methods of questioning involving a large number of volunteers.

Acknowledgment

The paper was written as part of the research project No. 8.2.31.2015, carried out with the support of the Program "Research Foundation of Tomsk State University named after D.I. Mendeleev" in 2015 – 2016.

References

[1] B.D. Jordan, E.L. Wright, Xenon as an anesthetic agent., AANA J. **78,** 387 (2010)

[2] B. Lachmann, S. Armbruster, W. Schairer, M. Landstra, A. Trouwborst, G.J. Van Daal, *et al.,* The Lancet **335**, 1413 (1990) doi: 10.1016/0140-6736(90)91444-F

[3] S.L.M. de Sousa, R. Dickinson, W.R. Lieb, N.P. Franks, Anesthesiology **92**, 1055 (2000)

[4] P.-L. Chau, Br. J. Pharmacol **161** 288 (2010) doi:10.1111/j.1476-5381.2010.00891.x.

[5] T. Yamakura, R.A. Harris, Anesthesiology **93**, 1095 (2000) doi:10.1097/00000542-200010000-00034

[6] R.D. Booker, A.K. Sum, Biochim. Biophys. Acta - Biomembr **1828**, 1347 (2013) doi:10.1016/j.bbamem.2013.01.016

[7] R.S. Cantor, Biochemistry **36**, 2339 (1997) doi:10.1021/bi9627323

[8] M. Weinrich, D.L. Worcester, J. Phys. Chem. B. **117**, (2013) 16141 doi:10.1021/jp411261g

[9] E. Yamamoto, T. Akimoto, H. Shimizu, Y. Hirano, M. Yasui, K. Yasuoka, J. Phys. Chem. B. **116**, 8989 (2012) doi:10.1021/jp303330c

[10] J.H. Lawrence, W.F. Loomis, C.A. Tobias, F.H. Turpin, J. Physiol. **105**, 197 (1946)
doi: 10.1113/jphysiol.1946.sp004164

[11] L.M. Stimson, I. Vattulainen, T. Róg, M. Karttunen, Cell. Mol. Biol. Lett. **10**, 563 (2005)

[12] N. V. Stryapko, T.G. Sazontova, V.I. Potievskaya, A.A. Khairullina, I.B. Vdovina, A.N. Kulikov, *et al.*, Gen. Reanimatol **10**, 50 (2014) doi:10.15360/1813-9779-2014-2-50-56

[13] A. Bedi, C. McCarroll, J.M. Murray, M.A. Stevenson, J.P.H. Fee, Anaesthesia. **57** , 233 (2002) doi:10.1046/j.0003-2409.2001.02455.x

NI PXI-BASED AUTOMATED MEASUREMENT SYSTEM FOR DIGITAL ASICs VERIFICATION

Georgiy Sorokoumov[a], *Dmitriy* Bobrovskiy, *Oleg* Kalashnikov, *Anastasiya* Ulanova and *Yuliya* Moskovskaya

National Research Nuclear University MEPhI (Moscow Engineering Physics Institute), 115409, Moscow, Russia

Abstract. The paper describes a structure of the automated measuring system used to control digital ASICs electrical and functional parameters. The automated measuring system is based on National Instruments PXI modules. The PXI-7954R module is the most significant module in the system. Hardware and software operations of the measuring system are discussed in the paper. The measuring system is based on test vectors for digital ASICs functional verification.

1 Introduction

Modern digital ASICs are an optimal solution for unique high-reliability systems with a broad class of applications: data collecting and processing, encoding / decoding and transmitting information over a long distance and others. The possibility of implementation of unique projects combining diverse digital blocks within single IC chip is an important reason behind a developers' choice of hardware digital ASICs. ASICs are widely used in design of electronic systems for operation in a spacecraft, for systems controlling technical processes in reactors of nuclear power plants and other high-reliability systems [1–4].

In modern digital ASICs, up to 200 user I/O lines and significant internal resources for the construction of various types of digital circuits are available for developers. A significant part of the equipment developers don't disclose information about electrical schematic or the firmware diagrams. Developers provide only an order card containing a composition of test vectors. The composition of test vectors describes all possible logic states of the digital ASICs firmware. Also composition of test vectors regulates points in which electrical parameters such as supply current, leakage current, high and low levels output voltage are controlled. In this case, the only possible way to conduct functional checks of a digital ASIC is to use a measuring system based on the operation by order cards. The important task is to control functional and electrical parameters of ASICs at different stages of equipment development. The capability of the measuring system to be used with different firmware is required for an easy modernization. The suggested technical solution allows us to minimize the measuring system setup time for controlling digital ASICs with different firmware.

[a] Corresponding author: gssor@spels.ru

© The Authors, published by EDP Sciences. This is an open access article distributed under the terms of the Creative Commons Attribution License 4.0 (http://creativecommons.org/licenses/by/4.0/).

2 Equipment

Standard PXI modules (National Instruments) have been applied in order to solve this problem. PXI-7954R module is the main system component. NI PXI-7954R module is based on Virtex-5 LX110 FPGA chip. NI PXI-7954R module is a multifunctional reconfigurable I/O module and has the following features: 132 single-ended or 66 differential digital I/O lines, 128 MB onboard memory (DRAM) and direct memory access (DMA) to transfer data.

Standard composition of a measuring system is based on NI PXI modules: two DC power supplies modules PXI-4110; FPGA module PXI-7941R (when ASIC I/O lines are more than 132, another FPGA module PXI-7954R can be added). All measuring system modules are combined in PXI-1033 chassis.

The following tasks were solved using NI PXI-7954R module:

- storing test vectors sequence in DRAM and directing all I/O lines in the same vectors to enable signal discharge;
- storing test results for each unload vector in the current test frame;
- transferring test results information to PC.

3 Software

In order to load test vectors in PXI-7954R DRAM and to implement operation control algorithms and I/O, LabVIEW environment has been used. LabVIEW graphical programming allows user to build a virtual instrument (VI) logic algorithm, quickly developing data transmission algorithms from NI modules devices to a HOST PC, generating visual reports about testing in graphs, to analyze text documents that contain basic information for the measuring system, and generate text and graphical log reports.

The digital ASIC's order card is a text document containing a composition of test vectors that determines the condition of all I/O of the IC. The completeness of an order card may be different. Usually the composition of test vectors included in the order card allows us to check all possible modes and logic states of the test ASIC firmware. Figure1 shows an order card example. Adopted unique designation – string of characters «*» informs about the need to control the electrical parameters of test's sample ASIC in a given vector of order card. In some cases, the point of control of electrical parameters is given in the text description to an order card which is issued by the ASIC's manufacture.

Symbols «H», «L», «X», «Z» were used as a universal notation which indicates the status of ASIC's output lines – a logic high level, logic low level, the state of line is not important, high-impedance state. Characters «0», «1» indicate low and high input logic levels of the digital ASIC. The beginning and the end of the vector may be indicated by different symbols. In the example, this symbols are «R1 std» (beginning) and «;» (end). In the software, settings of the measuring system have the ability to define characters for the beginning and the end used in a given order card for the vectors.

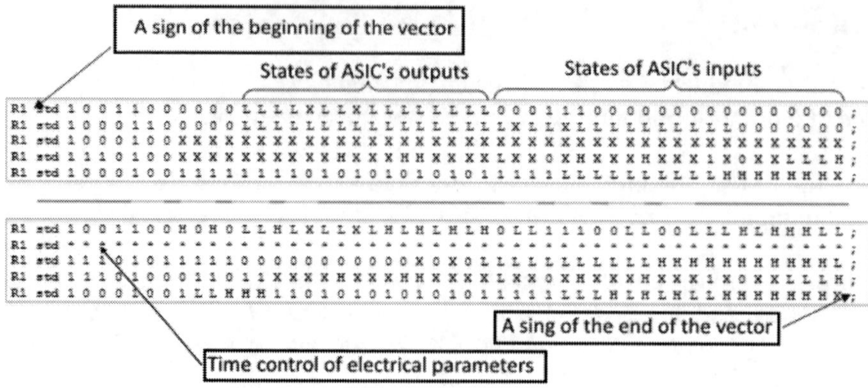

Figure 1. Example of typical order card.

Measurement system's software consists of two parts. The first part is the user part (HOST). The second part is the hardware based on FPGA module and DC power modules. The HOST is running on PC. The HOST provides a user interface and manages the hardware part of the measuring system. The HOST is also responsible for the accumulation of statistics about failures in a sample chip functional test. Types of faults are described in [5–11]. It's recording the results of parametric and functional control.

The main functions performed by the different parts of the software are following:

<u>HOST part</u>

1. Configuration of the measuring system. The user selects the number of involved FPGA PXI-7954R modules (1 or 2). The number of modules is based on the number of user I/O lines in ASIC.
2. Loading file with the order card. In the automatic transmission mode determined the direction of each I/O line in each vector.
3. Formed array of states I/O lines, transmission direction and control mask for each vector.
4. Formed array is loaded into the FPGA PXI-7941R module DRAM.

<u>FPGA part</u>

5. At the limit operating frequency, data is unloaded from DRAM. The data stream discharged from the module memory contains three vectors: data, direction and mask.
6. Data in direction array customize each I/O line of PXI-7954R module. Data array is used to set the values of the input ASIC's lines. Data array also has reference information on the output ASIC's lines. Figure 2 shows fragment of the code generating the output data which is based on data and direction arrays.
7. Next, the state of ASIC's I/O lines is read. Formed response array is based on the mask array. Then the formed data array is compared with the reference data array. The number of differences is determined. The response data array is then stored in FPGA module's memory and can be read by HOST for the detailed analysis of any failures.

Figure 3 shows the front panel of the developed HOST VI. Boot time and reading of the order card's vectors controlling the ASIC's currents are shown.

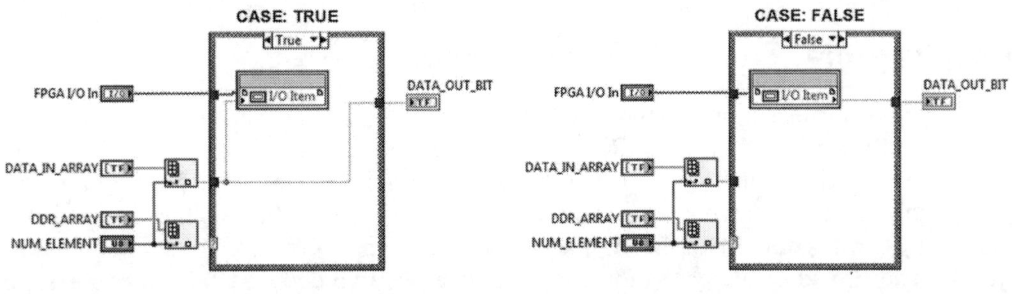

Figure 2. Fragment of the program code. Setting up, set and reading the state ASIC's line.

The number of vectors downloaded in the FPGA module's memory is limited by the memory volume available in the FPGA module (for the PXI-7954R module – 128 MB DRAM). If the volume of the order card exceeds available memory, the data from the order card divided into blocks of suitable size. Cyclic repetition of the operations described above allows us to perform complete verification of the test according to the ASIC's order card to limit frequency of enumeration test vectors with pauses to load the next part of the vectors into the FPGA module internal memory.

The software provides the logging results of the functional test (number of failures) and extended diagnostic information in text files. The diagnostic information contains the number of failure vectors, the read sate output lines of ASIC, and the reference value of ASIC's output lines. The measuring system controls electrical parameters of the ASIC in order card breakpoints. Currents of the test sample ASIC are monitored using built-in PXI-4110 module ammeters.

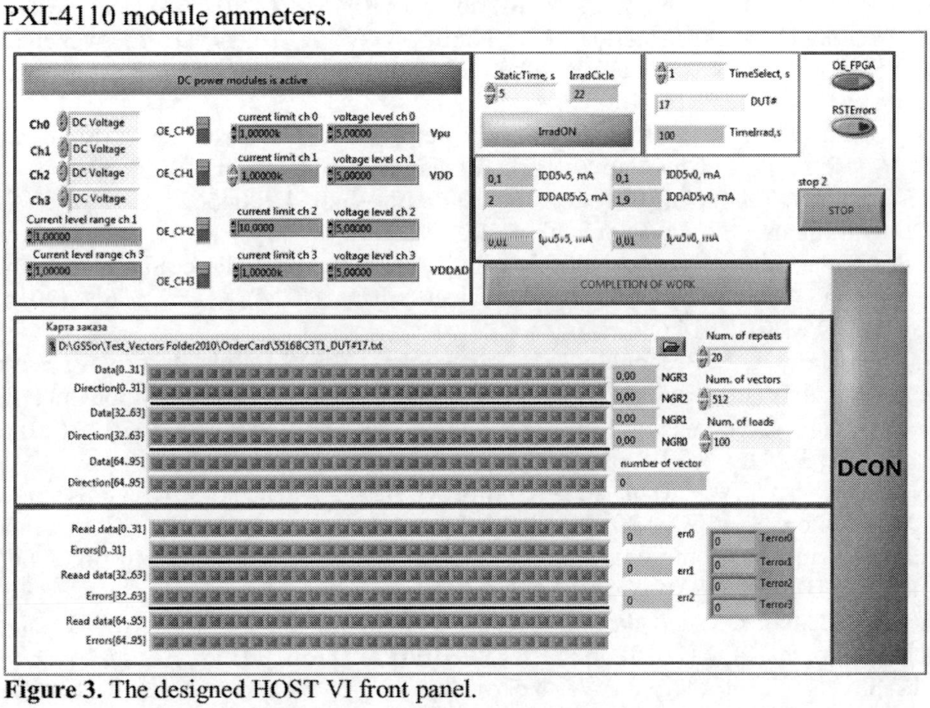

Figure 3. The designed HOST VI front panel.

4 Implementation and future prospects

The developed measuring system has been successfully tested in JSC «Specialized Electronic Systems» (ENPO SPELS) and Institute of Extreme Applied Electronics of National Research Nuclear University MEPhI. The principle of modularity of equipment, as well as the ability to easily zoom code, allows us to increase the number of used input-output lines proportionally to the number of lines used in FPGA modules of the measuring system. The choice of the base FPGA module is a positive decision for the entire system. FPGA module allows testing of integrated circuits not only using the order card, but, as described in [12–20], using the approach of building an FPGA module full test environment study of ASIC's firmware.

Acknowledgements

The work was performed as a part of the grant RFBR № 14-29-09131.

References

[1] A. I. Chumakov, A. V. Afonin, V. A. Polunin, Russian Microelectronics **42**, 424 (2013) doi: 10.1134/S1063739713070032

[2] O.A. Kalashnikov, A.Y. Nikiforov, Proceedings of the International Conference on Microelectronics (ICM), 6842189 (2014) doi: 10.1109/MIEL.2014.6842189

[3] A.S. Artamonov, A.A. Sangalov, A.Y. Nikiforov, V. A. Telets, D.V. Boychenko, IEEE Radiation Effects Data Workshop **2015-January**, 7004600 (2014) doi: 10.1109/REDW.2014.7004600

[4] L.N. Kessarinskiy, D.V. Boychenko, A.G Petrov, P.V. Nekrasov, A.V. Sogoyan, V.S. Anashin, P.A. Chubunov, IEEE Radiation Effects Data Workshop **2015-January**, 7004562 (2014) doi: 10.1109/REDW.2014.7004562

[5] A.O. Akhmetov, D.V. Boychenko, D. V. Bobrovskiy, A. I. Chumakov, O. A. Kalashnikov, A.Y. Nikiforov, P. V. Nekrasov, Proceedings of the International Conference on Microelectronics (ICM), 6842156 (2014) doi: 10.1109/MIEL.2014.6842156

[6] A.I. Chumakov, A. A. Pechenkin, D.V. Savchenkov, A.S. Tararaksin, A. L. Vasil'ev, A. V. Yanenko, Proceedings of the European Conference on Radiation and its Effects on Components and Systems (RADECS), 6131420 (2011) doi: 10.1109/RADECS.2011.6131420

[7] G.N. Nazarova, V.V. Elesin, A.Y. Nikiforov, A. G. Kuznetsov, N. A. Usachev, G. V. Chukov, CriMiCo 2014 - 2014 24th International Crimean Conference Microwave and Telecommunication Technology Conference Proceedings, 6959663 (2014) doi: 10.1109/CRMICO.2014.6959663

[8] D.V. Boychenko, O. A. Kalashnikov, A.B. Karakozov, A. Y. Nikiforov, Russian Microelectronics **44**, 1 (2015) doi: 10.1134/S1063739715010035

[9] I. I. Shvetsov-Shilovskiy, P.V. Nekrasov, A. V. Ulanova, A. A. Smolin, A. V. Sogoyan, Proceedings of the International Siberian Conference on Control and Communications (SIBCON), 7147282 (2015) doi: 10.1109/SIBCON.2015.7147282

[10] A.G. Petrov, A.L. Vasil'ev, A.V. Ulanova, A I. Chumakov, A.Y. Nikiforov, Central European Journal of Physics **12**, 725 (2014) doi: 10.2478/s11534-014-0503-6

[11] D. V. Bobrovskii, O. A. Kalashnikov, P. V. Nekrasov, Russian Microelectronics **41**, 226 (2012) doi: 10.1134/S1063739712040038

[12] P. V. Nekrasov, A. A. Demidov, O. A. Kalashnikov, Instruments and Experimental Techniques **52**, 196 (2009), doi: 10.1134/S0020441209020092
[13] A. O. Akhmetov, A. V. Yanenko, A. I. Bazhan, Proceedings of the European Conference on Radiation and its Effects on Components and Systems (RADECS), 6937364 (2013) doi: 10.1109/RADECS.2013.6937364
[14] A.B. Boruzdina, A. V. Ulanova, A.G. Petrov, V. A. Telets, P.Reviriego and J.A. Maestro, Proceedings of the European Conference on Radiation and its Effects on Components and Systems (RADECS), 6937393 (2013) doi: 10.1109/RADECS.2013.69373693
[15] A.B. Karakozov, O.V. Korneev, P.V. Nekrasov, M. N. Sokolov and D. A. Zagryadsky, Proceedings of the European Conference on Radiation and its Effects on Components and Systems (RADECS), 6937407 (2013) doi: 10.1109/RADECS.2013.6937407
[16] A.V. Demidova, A.Y. Borisov, L. N. Kessarinskiy, D.V. Boychenko, Proceedings of the International Siberian Conference on Control and Communications (SIBCON), 7147036 (2015) doi: 10.1109/SIBCON.2015.7147036
[17] I.O. Loskutov, A.B. Karakozov, P.V. Nekrasov, A.Y. Nikiforov, Proceedings of the International Siberian Conference on Control and Communications (SIBCON), 7147128 (2015) doi: 10.1109/SIBCON.2015.7147128
[18] V.A .Marfin, P.V. Nekrasov, O. A. Kalashnikov, K. A. Kagirina, Proceedings of the International Siberian Conference on Control and Communications (SIBCON), 7147142 (2015) doi: 10.1109/SIBCON.2015.7147142
[19] A. B. Boruzdina, A. A. Orlov, A. V. Ulanova, N. G. Grigor'ev, A.Y. Nikiforov, Proceedings of the International Siberian Conference on Control and Communications (SIBCON), 7147007 (2015) doi: 10.1109/SIBCON.2015.7147007
[20] M.E. Cherniak , A.V. Ulanova, A.Y. Nikiforov, Proceedings of the International Conference on Microelectronics (ICM), 6842119 (2014) doi: 10.1109/MIEL.2014.6842119

MATEC Web of Conferences 79, 01060 (2016) DOI: 10.1051/matecconf/20167901060
IME&T 2016

USE OF CREEPING SPEED MODE IN DISCRETE CONTROL AND MEASURING EQUIPMENT WITH DIGITAL-PROGRAM CONTROL

Lubov Payuk*, *Nataliya* Voronina, *Sergey* Korepanov, *Olga* Galtseva, *Nataliya* Natalinova

Tomsk Polytechnic University, 634050, Tomsk, Russia

Abstract. The creeping speed mode is studied with calculation of the resultant vector of the flux linkage ψ_0 of the double-fed machine. It is shown that use of the double-fed machine as the actuating motor in the creeping speed mode increases the working off accuracy of the motion law. A mathematical model of the double-fed machine to study various modes of the CNC machine including the creeping speed mode is developed. The method to estimate the load parameters of the double-fed machine is presented. Practical implication of the study is confirmed by the results obtained in the evaluation of the influence of load parameters on the dynamic indicators of actuating electromotor.

1 Introduction

Currently, any measuring equipment incorporates the systems of the fixed current parameters, comparison with a reference element and compensation of the measuring error [1]. In these systems, modern CNC machines are used; many engines with relatively small power are involved in their work, the basic modes of operation of these engines are "unnatural" [2]. These modes include frequent reversal, braking by various ways (dynamic, double-current, capacitor-dynamic, etc.), getting the creeping speed according to the schemes for the joint feeding, step and torque modes, etc. [3–9].

The creeping speed mode is described in detail below. The feature of the creeping speed mode is an operation at the speed of the tool movement from 2 to 10 mm/min [3]. This mode is widely used in the CNC machines (for processing and plasma cutting), aboard mechanisms of antenna rotation, detection systems for mobile and ground equipment, fixation systems of outdoor surveillance, textile industry (warp sizing machine), and discrete measurement systems with CNC (two-dimensional electromagnetic displacement sensors) [10–12]. In practice, this type of the tool movement is often implemented due to the presence of linear asynchronous engines (LAE), actuating motor and ball-and-screw unit or electromagnetic frame.

* Corresponding author: lubapa81@mail.ru

© The Authors, published by EDP Sciences. This is an open access article distributed under the terms of the Creative Commons Attribution License 4.0 (http://creativecommons.org/licenses/by/4.0/).

2 Research method

The method to develop the creeping speed in the double-fed machine (DFM) at the different inclusions of magnetic fields is presented. The analysis of influence of the load parameters on the dynamic indicators of actuating electromotor is carried out.

The double-fed machine is referred to as the controlled (full controlled) machine of alternating current; voltages are supplied to its stator and rotor clips. These voltages are composed of two components: voltage of compensation and voltage (law) of control. This type of electrical machines is known to be universal; it has high power characteristics. This electrical machine is implemented as a micromachine; it is a decisive factor for measurement systems [13, 14].

The creeping speed mode (Figure 1) is implemented in the feeding of stator and rotor windings of the DFM by sinusoidal voltage of different frequencies. In the first case, the difference between frequencies is 1 Hz; in the second case, it is 40 Hz at the initial phase of the supply voltage $\varphi=\pi/2$.

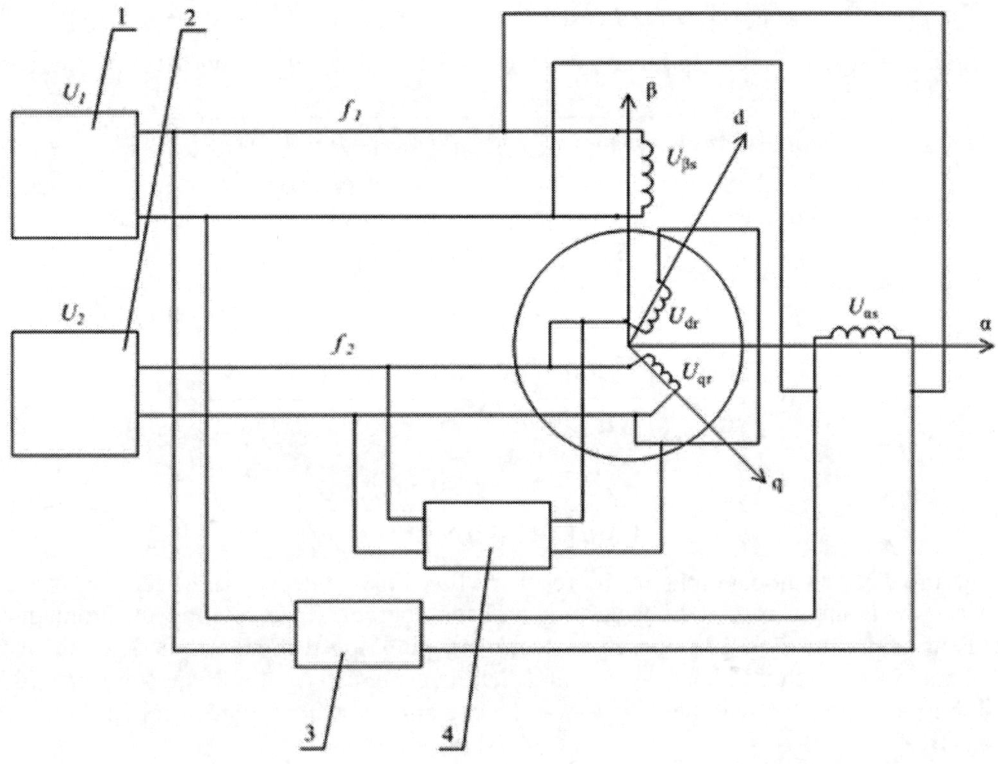

Figure 1. Feeding scheme of the double-fed machine in realization of the creeping speed mode: 1, 2 are the sources; 3, 4 are the phase shifters.

The following equations are true for operation of the DFM in the creeping speed mode: $\omega_1=\omega_2=\omega$, and $\omega_3=\omega_4=\omega`$ if $\alpha=\beta=\gamma=0$, $\varphi=\pi/2$.

Equations (1) for the flux linkages of stator and rotor windings in the transformed coordinate system with due regard to the accepted assumptions and the mode of the machine operation are

$$\psi_{\alpha s} = \psi_{ms} \, sin(\omega t);$$
$$\psi_{\beta r} = \psi_{ms} \, sin(\omega t);$$
$$\psi_{\alpha r} = \psi_{mr} \, sin(\omega` t) cos\,\chi + \psi_{mr} \, sin\left(\omega` t + \frac{\pi}{2}\right) sin\,\chi; \qquad (1)$$
$$\psi_{\beta r} = -\psi_{mr} \, sin(\omega` t) sin\,\chi + \psi_{mr} \, sin\left(\omega` t + \frac{\pi}{2}\right) cos\,\chi.$$

Equations (2) for the squares of the real and imaginary parts of the generalized flux linkage vector are

$$\psi_\alpha^2 = \left(\psi_{ms} \sin(\omega t) + \left(\psi_{mr} \sin(\omega`t)\cos\chi + \psi_{mr}\cos(\omega`t)\sin\chi\right)\right)^2 =$$
$$= \left(\psi_{ms}\left(\sin(\omega t) + \mu(\sin(\omega`t)\cos\chi + \cos(\omega`t)\sin\chi)\right)\right)^2 = (\psi_{ms}(\sin(\omega t) + \mu\sin(\omega`t + \chi)))^2,$$
$$\psi_\beta^2 = \left(\psi_{ms} \sin(\omega t) - \left(\psi_{mr} \sin(\omega`t)\sin\chi + \psi_{mr}\cos(\omega`t)\cos\chi\right)\right)^2 = \qquad (2)$$
$$= \left(\psi_{ms}\left(\sin(\omega t) - \mu(\sin(\omega`t)\sin\chi + \cos(\omega`t)\cos\chi)\right)\right)^2 = (\psi_{ms}(\sin(\omega t) - \mu\cos(\omega`t - \chi)))^2.$$

Equations (3) for the amplitude value of the flux linkage vector in the creeping speed mode are

$$\psi = \psi_{ms}\sqrt{(sin(\omega t) + \mu\, sin(\omega`t + \chi))^2 + (sin(\omega t) - \mu\, cos(\omega`t - \chi))^2}. \qquad (3)$$

Equations (2) for the movement of the spatial vector of flux linkage in the air gap for the creeping speed mode (3) are

$$\chi_0 = \mathrm{arctg}\left(\frac{\psi_{ms}\sin(\omega t) - \psi_{mr}\sin(\omega`t)\sin\chi + \psi_{mr}\cos(\omega`t)\cos\chi}{\psi_{ms}\sin(\omega t) + \psi_{mr}\sin(\omega`t)\cos\chi + \psi_{mr}\cos(\omega`t)\sin\chi}\right) =$$
$$= \mathrm{arctg}\left(\frac{\sin(\omega t) - \mu(\sin(\omega`t)\sin\chi + \cos(\omega`t)\cos\chi)}{\sin(\omega t) + \mu(\sin(\omega`t)\cos\chi + \cos(\omega`t)\sin\chi)}\right) = \qquad (4)$$
$$= \mathrm{arctg}\left(\frac{\sin(\omega t) - \mu\sin(\omega`t + \chi)}{\sin(\omega t) + \mu\cos(\omega`t - \chi)}\right).$$

Figure 2 shows hodographs of the resultant flux linkage vector at the feeding of stator and rotor windings from the current and voltage sources. Two rotating electromagnetic fields arise during the DFM operation in the creeping speed mode for hodograph of the resultant vector of flux linkage ψ_0: the first field is generated by the frequency f_1=f_2=50 Hz and the second one is generated by f_3=f_4=51 Hz from the primary and secondary elements respectively.

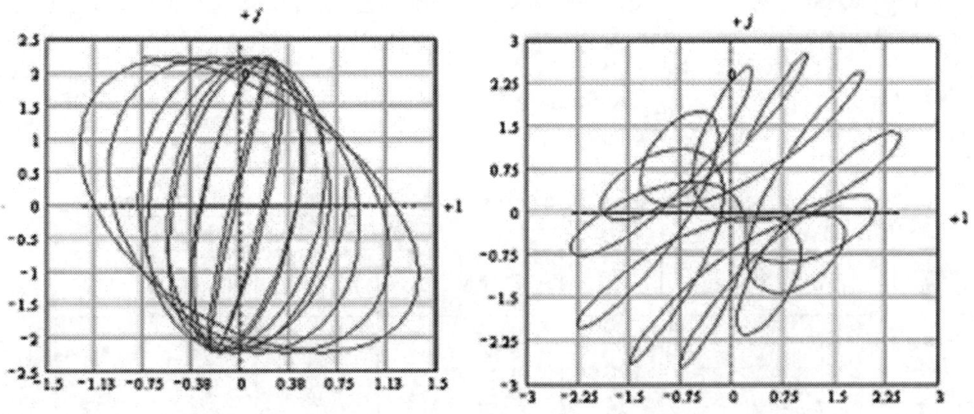

(a) (b)

Figure 2. Hodographs of vectors ψ_0 at the feeding of stator and rotor windings from the current sourse (a) and from the voltage sourse (b) at $\omega_1=\omega_2=1$; $\omega_3=\omega_4=1.02$.

Consider two variants of interaction of the magnetic fields in the DFM in the creeping speed mode:

1. It is necessary to set the following signals of the control system at the consonant interaction of electromagnetic fields: $\gamma_1=\gamma_2=1$; $\gamma_3=\gamma_4=1.225$; $\omega_1=\omega_2=\omega$; $\omega_3=\omega_4=\omega`$; $\alpha=\beta=\gamma=0$, $\varphi=\pi/2$. The voltages on the clips of stator and rotor windings in transformed axes α, β, 0 are in the following form:

$$\begin{aligned} U_{\alpha s} &= U_{ms}\sin(\omega t); \\ U_{\beta s} &= U_{ms}\sin(\omega t); \\ U_{\alpha r} &= U_{mr}\sin(\omega`t)\cos\chi + U_{mr}\sin(\omega`t+\phi)\sin\chi; \\ U_{\beta r} &= -U_{mr}\sin(\omega`t)\sin\chi + U_{mr}\sin(\omega`t+\phi)\cos\chi. \end{aligned} \tag{5}$$

2. In the case of the opposite interaction of electromagnetic fields the angular frequency of rotation and the initial phases on the axes are set similarly to the above case $\gamma_1=\gamma_2=1$; $\gamma_3=1.225$; $\gamma_4=-1.225$. Equations for the voltages on the clips of the stator and rotor windings are similar to the system (5) with the only difference: the sign "+" before the second summand is replaced for the sign "–" in the second part of the two equations.

A mathematical model of DFM is given relative to the currents of stator and rotor in the axes α, β, 0 in the iterative form at zero initial conditions. It is presented in the system (6). DFM was based on the asynchronous engine with phase rotor 4AK160S8UZ with protection degree of IP44, and its asynchronous rotational speed is n=750 rpm with power P_2 =5.5 [kW].

This model allows investigating transients in the DFM in various modes of its operation. Table 1 shows the parameters of actuating motor 4AK160S8UZ.

Table 1. Parameters of actuating motor 4AK160S8UZ.

Energy indicators			**Mechanical characteristic**			**Equivalent circuit parameters**				
cos φ	p	η	m_m	S_{nom}	S_{crit}	X_μ	R'_1	X'_1	R'_2	X'_2
[pu]	[-]	[%]	[pu]	[%]	[%]	[pu]	[pu]	[pu]	[pu]	[pu]
0.7	4	80	0.5	6.4	9	1.6	0.06	0.11	0.094	0.175

Gabriel Kron's equations are the basis for the DFM mathematical model. Gabriel Kron's equations are the system of differential equations of voltages and motions with due regard to the accepted assumptions and describe the energy transformations in the electric machine. It is also necessary to calculate the base values and coefficients of the model, which are shown in Tables 2 and 3.

Table 2. Estimated base values of parameters of actuating motor.

Main base values				**Auxiliary base values**						
U_b	I_b	P_b	ω_b	t_b	Z_b	L_b	$M_{mech.\,b}$	$L_{mech.\,b}$	$R_{mech.\,b}$	$C_{mech.\,b}$
[V]	[A]	[W]	[rad/s]	[s]	[Om]	[H]	[W·s/rad]	[W·s³/rad³]	[W·s²/rad²]	[W·s/rad]
311.1	18.2	8.5	314.3	0.0032	17.07	0.054	108.297	0.00439	1.38	433.19

Table 3. Coefficients of mathematical model [pu].

σ	α_S	α'_s	α_r	α'_r	k_S	k_r
0.158	3.702	0.222	3.756	0.213	0.935	0.901

Mathematical model of DFM is of the form [pu]:

$$\begin{pmatrix} t_{j+1} \\ i_{S\alpha_{j+1}} \\ i_{S\beta_{j+1}} \\ i_{R\alpha_{j+1}} \\ i_{R\beta_{j+1}} \\ \omega_{j+1} \\ \chi_{j+1} \end{pmatrix} = \begin{cases} t_j + dt \\ i_{S\alpha_j} + dt \cdot \begin{cases} \alpha_S U_1 \sin(\omega'_1 t_j + \alpha) - \alpha_S K_R \left(U_3 \sin(\omega'_1 t_j + \gamma) \cdot \cos\chi_j \ldots + U_4 \sin(\omega'_2 \cdot t_j + \varphi) \cdot \sin\chi\right. \\ -i_{S\beta_j}\omega_j \dfrac{K_S \cdot K_R}{\sigma} - i_{S\alpha_j}\alpha'_S - i_{R\beta_j}\omega_j \dfrac{K_S}{\sigma} + i_{R\alpha_j}\alpha'_R K_S \end{cases} \\ i_{S\beta_j} + dt \cdot \begin{cases} \alpha_S U_2 \sin(\omega'_1 t_j + \beta) - \alpha_S K_R \left(-U_3 \sin(\omega'_1 \cdot t_j + \gamma) \cdot \sin\chi_j \ldots + U_4 \sin(\omega'_2 \cdot t_j + \varphi) \cdot \mathrm{co}\right. \\ -i_{S\alpha_j}\omega_j \dfrac{K_S K_R}{\sigma} - i_{S\beta_j}\alpha'_S + i_{R\alpha_j}\omega_j \dfrac{K_S}{\sigma} + i_{R\beta_j}\alpha'_R K_S \end{cases} \\ i_{R\alpha_j} + dt \cdot \begin{cases} \alpha_R \left(U_3 \sin(\omega'_1 t_j + \gamma) \cdot \cos\chi_j \ldots + U_4 \sin(\omega'_2 t_j + \varphi) \cdot \sin\chi_j\right) \ldots - K_R \alpha_S U_1 \sin(\omega'_1 t_j + \\ +\alpha'_S K_R i_{S\alpha_j} - i_{S\beta_j}\omega_j \dfrac{K_R}{\sigma} + i_{R\beta_j}\omega_j \dfrac{1}{\sigma} - i_{R\alpha_j}\alpha'_R \end{cases} \\ i_{R\beta_j} + dt \cdot \begin{cases} \alpha_R \cdot \left(-U_3 \sin(\omega'_1 t_j + \gamma) \cdot \sin\chi_j \ldots + U_4 \sin(\omega'_2 t_j + \varphi) \cdot \cos\chi_j\right) \ldots - K_R \alpha_S U_2 \sin(\omega'_2 t \\ -\dfrac{K_R}{\sigma} \cdot \omega_j \cdot i_{S\alpha_j} - \alpha'_S \cdot K_R \cdot i_{S\beta_j} - i_{R\alpha_j} \cdot \omega_j \cdot \dfrac{1}{\sigma} - i_{R\beta_j} \cdot \alpha'_R \end{cases} \\ \omega_j + dt \cdot \left(\dfrac{1}{J}\left(-L_m\left(i_{S\beta_j} i_{R\alpha_j} - i_{S\alpha_j} i_{R\beta_j}\right) - R_g \omega_j - C_m \chi_j - M_c\right)\right) \\ \chi_j + dt \cdot \omega_j \end{cases} \tag{6}$$

where R_g is a coefficient of damping torque of load; C_m is a coefficient of position torque of load; J is a total torque of inertia of engine; M_c is a torque of static load on the shaft of engine; χ is a law of motion of the movable element.

3 Results and discussion

Figure 3 shows results of mathematical modelling of the DFM (6). As an example, the dependencies of the DFM electromagnetic torque in the starting and creeping speed modes are presented.

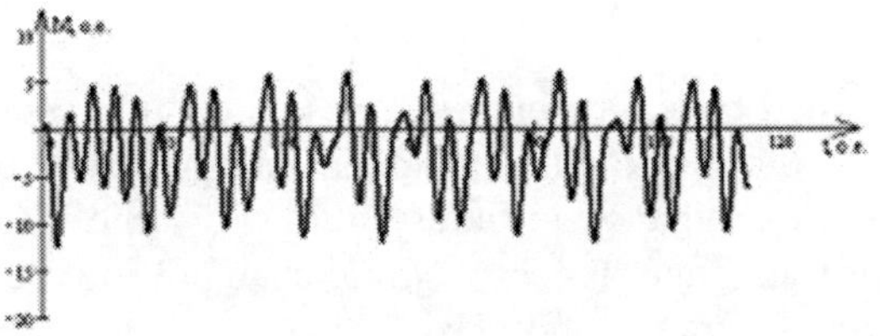

(a)

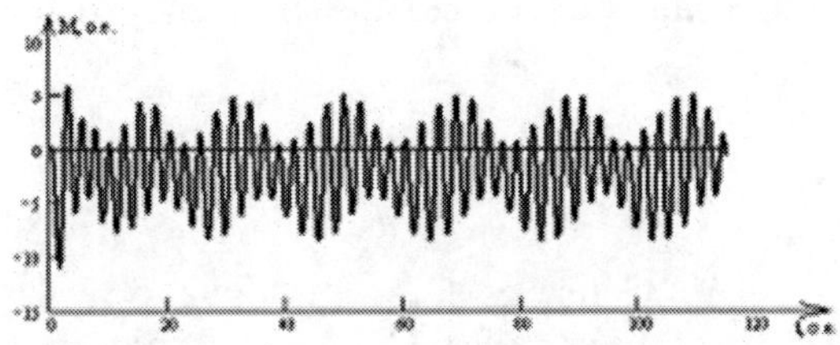

(b)

Figure 3. Dependence of electromagnetic moment on the time at the frequency difference of 1 Hz (a) and of 40 Hz (b).

The dependencies were obtained by the modelling; they are adequate and the model is sustainable. The disadvantage is inability to investigate the model in case of simultaneous changes in the load parameters.

The influence of load parameters on the dynamic indicators of actuating motor is important as the additional disturbances and measurement errors lead to incorrect measurement results. The load torque of inertia J_n (from 80 to 2500 pu) is selected as a changeable load parameter.

Figure 4 shows the results of investigation of dynamic indicators: time of transition process τ_{tp} and shock values of electromagnetic torque M_{sh} on load torque of inertia J_n.

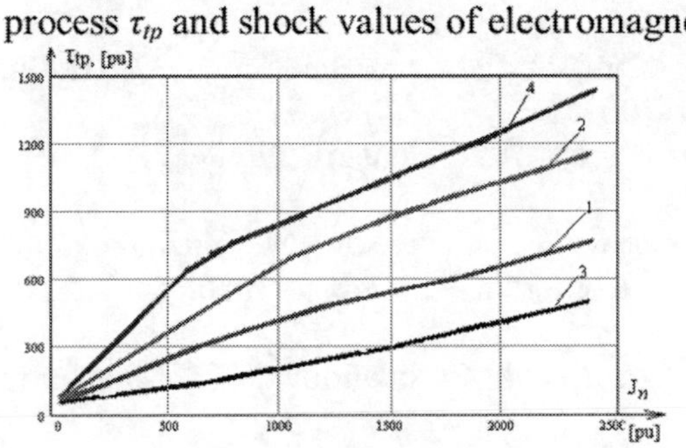

(a)

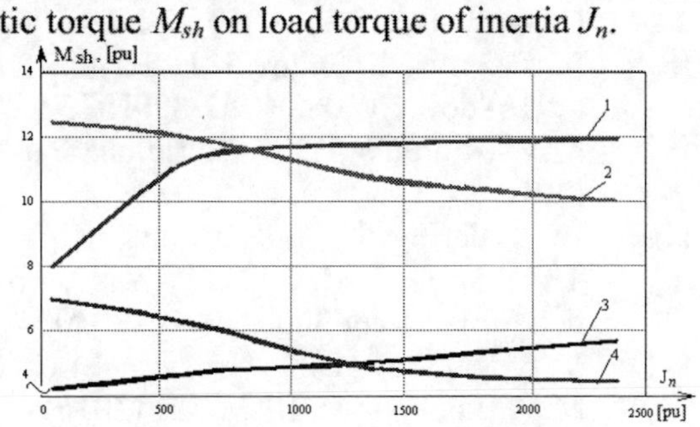

(b)

Figure 4. Dependence of time of transition process (a) and shock values of electromagnetic torque (b) on inertia load: consonant interaction of electromagnetic fields at the frequency difference of 1 Hz (curve 1) and of 40 Hz (curve 3); opposite interaction of electromagnetic fields of 1 Hz (curve 2) and of 40 Hz (curve 4).

In the creeping speed mode, the change in time of the transition process on load torque of inertia $\tau_{tp}=f(J_n)$ is observed to increase; it is typical of special modes of electromechanical energy converters based on the DFM.

In the case of the consonant interaction of electromagnetic fields, the shock values of electromagnetic torque M_{sh} decrease with the increase in the inertia torque of load J_n (Figure 4, *b*, curves 1 and 3), and these values increase in the case of the opposite interaction of electromagnetic fields (Figure 4, *b*, curves 2 and 4).

4 Conclusion

1. The change of the load torque of inertia J_n leads to the increase in time of transition process τ_{tp} and decrease in the shock values of electromagnetic torque M_{sh} in the case of the consonant interaction of electromagnetic fields. With the same load parameters, the DFM has the maximal time of transition process at the consonant interaction of electromagnetic fields and at the frequency difference of 40 Hz.
2. The practical application of the DFM in discrete-measuring systems of the CNC machines is possible with some limitations: the frequency difference must not exceed 10 Hz.

References

[1] N.M. Natalinova, O.V. Galtseva, E.A. Moldovanova, 2016 Third International Conference on Electrical, Electronics, Computer Engineering and their Applications (EECEA), 52 (2016) doi: 10.1109/EECEA.2016.7470765
[2] M.H. Fisher, *Magnetic amplifier creeping speed control* Patent US 2663833A, (1953)
[3] A.V. Aristov, N.A. Voronina, *Device for control two-phase asynchronous motor in the intermittent movement mode* Patent RF 2009125766/22 (2009)
[4] L.A. Payuk, N.A. Voronina, O.V. Galtseva, J. Phys.: Conf. Ser. **671**, 012044 (2016) doi: 10.1088/1742-6596/671/1/012044
[5] N.S. Starikova, V.V. Redko, G.V. Vavilova, J. Phys.: Conf. Ser. **671**, 012056 (2016) doi: 10.1088/1742-6596/671/1/012056
[6] Y.A. Chursin, E.M. Fedorov, Optics&Laser Technol. **67**, 86 (2015) doi: 10.1016/j.optlastec.2014.09.017
[7] A.P. Surzhikov, T.S. Frangulyan, S.A. Ghyngazov, Lysenko, E.N., J. Therm. Anal. Calorim. **102** , 883 (2010) doi: 10.1007/s10973-010-0912-8
[8] L. Kozlova, E. Bolovin, L. Payuk, IOP Conf. Ser.: Mater. Sci. Eng. **132**, 012005 (2016) doi: 10.1088/1757-899X/132/1/012005
[9] M.V. Kuimova, D.D. Burleigh, Yu.Yu. Arnst, A.E. Sentsov, Ponte **72**, 127 (2016)
[10] R. Kodermyatov, M. Ivanov, M. Yuzhakov, V. Kuznetsov, M. Yuzhakova, E. Timofeeva, MATEC Web of Conferences **48**, 05004 (2016) doi: 10.1051/confmatec/20164805004
[11] V.Y. Kazakov, D.K. Avdeeva, M.G. Grigoriev, N.M. Natalinova, I.V. Maksimov, M.V. Balahonova, BLM **7**, 1 (2015)
[12] O.V. Galtseva, S.V. Bordunov, N.M. Natalinova, S.V. Mazikov, IOP Conf. Ser.: Mater. Sci. Eng. **132**, 012003 (2016) doi: 10.1088/1757-899X/132/1/012003
[13] L. Cheng, X.J. Ye, D. R. Sun, Y.Y. Ye, Y. Jin, Appl. Mech. Mat. **416-417**, 652 (2013) doi: 10.4028/www.scientific.net/AMM.416-417.652
[14] A.Y. Petrova, O.N. Chaikovskaya, I.V. Plotnikova, Tech. Phys. J. **60**, 592 (2015) doi: 10.1134/S1063784215040222

MATEC Web of Conferences 79, 01061 (2016) DOI: 10.1051/matecconf/20167901061
IME&T 2016

BUILDING MODERN VIBRATION DIAGNOSTICS SYSTEMS BASED ON THE FREQUENCY-TIME TRANSFORMATIONS OF A MEASURED SIGNAL

Vasikh Yasoveev, *Marat* Akhmetshin[a]

Ufa State Aviation Technical University (USATU), 450008, Ufa, Russia

Abstract. Basic methods of analysis of vibration transducers signals were reviewed. Continuous wavelet transform, being a time-frequency transform, was found to be an advanced mathematical tool for analysis of vibration signals. Experimental studies revealed obvious changes in the continuous wavelet transform spectrum depending on the existing defects. A method for detection and identification of technological violations based on the analysis of CWT spectrum components and normalized correlation coefficient was suggested. In accordance with the suggested method software for vibration diagnostics was developed.

1 Introduction

At present industrial testing and technological equipment and systems have reached a high level. It affects not only the cost and complexity of the equipment itself but also the ability to be qualitatively repaired without costly and labor-consuming operations.

Vibration is the most informative indicator in terms of health assessment of virtually any class of equipment. However, vibration testing requires that issues of selection of the necessary technical devices should be seriously addressed, as well as revealing diagnostic indicators for each specific class of equipment.

2 Problem statement

Development of effective systems for non-destructive and non-disassembling monitoring of technological violations and item's defects is a task of exceptional importance, and the possibility to quickly and adequately identify incorrect modes and deviations of the item's parameters from normalized ones is a priority trend.

3 State of the art and relevance

Systems for monitoring and control of vibration testing [1] used nowadays to process a measured signal generally employ a spectral analysis method involving direct Fourier transform (FT), which has a number of limitations and disadvantages. FT does not provide

[a] Corresponding author: ahmaratyan@rambler.ru

© The Authors, published by EDP Sciences. This is an open access article distributed under the terms of the Creative Commons Attribution License 4.0 (http://creativecommons.org/licenses/by/4.0/).

localization of the analyzed signal in a time area, since the eigenfunction of the Fourier transform is a sine wave that is determined over the whole time interval and, consequently, FT does not take into account changing of the analyzed signal frequency over time.

Wavelet analysis is a special type of linear transformation of signals and physical data [2, 3]. The basis of eigenfunctions employed by signals decomposition has many specific properties and possibilities. Wavelet functions of the basis allow focusing attention on local peculiarities of the analyzed processes that cannot be revealed by traditional Fourier and Laplace transforms. For instance, fields of various physical parameters of natural environments belong to such processes in geophysics. This applies especially to fields of temperature, pressure, and seismic path profiles, as well as other physical values.

Wavelet transform of a one-dimensional signal [4] is a representation of this signal as a generalized Fourier series or integral by a combination of the basis functions

$$\psi_{ab}(t) = \frac{1}{\sqrt{a}}\psi\left(\frac{t-b}{a}\right), \tag{1}$$

constructed from mother (prototype) wavelet $\psi(t)$ which has certain properties due to time shift operations (b) and time scale change (a). Multiplication factor $1/\sqrt{a}$ provides independence of these functions norm from scale value a.

As parameter a increases, the frequency resolution increases and the time resolution decreases, and vice versa, as parameter a decreases, the frequency resolution decreases and the time resolution increases. Therefore, WT is an adaptive transform. Flexible time-frequency window extracts both low frequency and high frequency signal characteristics equally well. This WT feature is a big advantage for analysis of signal local properties.

4 Methods

It was suggested to employ the mathematical apparatus of continuous wavelet transform (CWT) to diagnose vibration parameters. In order to reveal CWT spectrum behavior, an experimental study was performed [5], and models of main intrinsic defects and technological violations were defined and analyzed.

By processing and continuous wavelet transforming of a data array obtained during experimental testing with the help of an application operating under MatLab 7.11.0, signals skeletons and CWT results in the form of 3D surfaces were obtained.

During analysis of the skeletons and CWT results' surfaces it was found that manifestation of each analyzed defect type depends on the vibration frequency.

Frequencies for every defect type at which the defect manifests were identified, as well as the CWT spectrum components that are informative for each defect type. These components generally have areas of local maxima.

Figures 1 and 2 demonstrate the difference between the signal skeleton under normal mode and the one when defect No.1 emerges at the testing frequency of 500 Hz. Thus, when the defect is developed up to 50% of the maximum value, it manifests at 100 Hz and 500 Hz; and for testing frequency of 500 Hz the informative spectrum components are a=8..25 wherein the emergence of local maxima is observed.

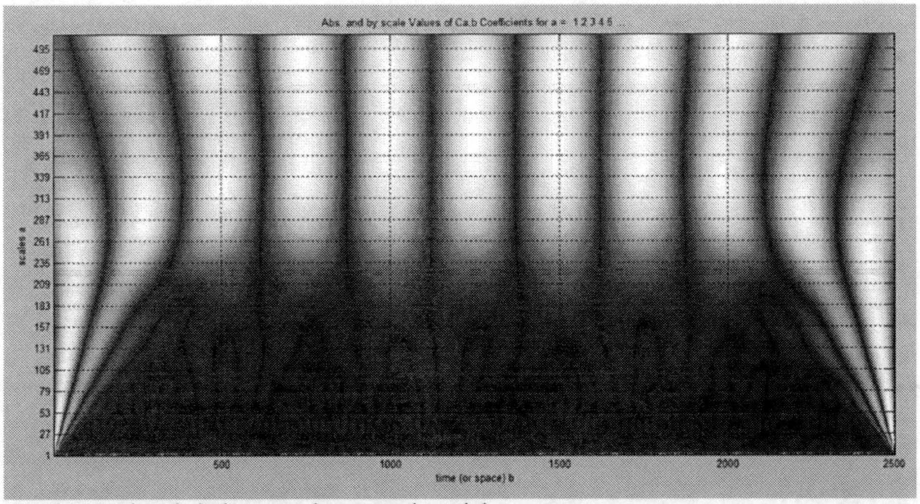

Figure 1. 500 Hz signal skeleton under normal model.

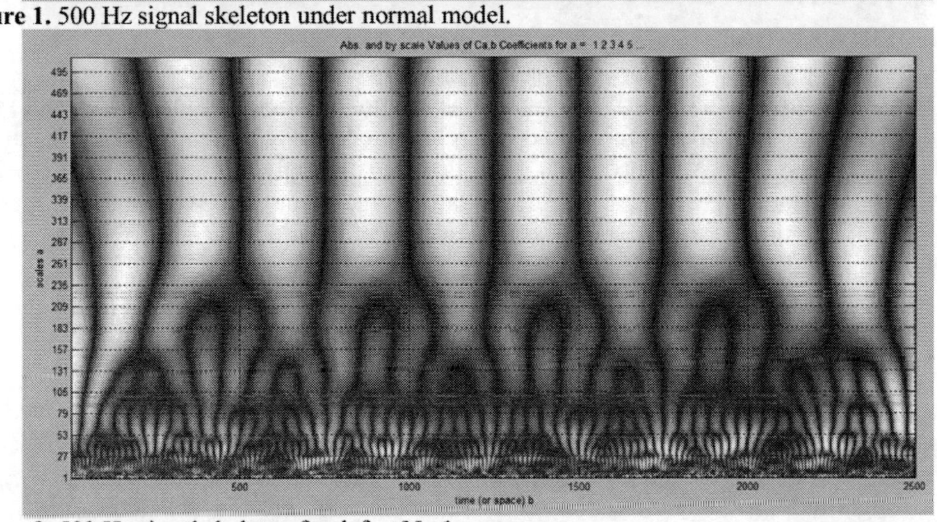

Figure 2. 500 Hz signal skeleton for defect No.1.

Figure 3 shows the 3D surface model of CWT results for the signal frequency of 1 kHz. It is clearly seen that a=150..200 for the carrier frequency. Emergence of defect No. 2 is characterized by emergence of high frequency components of the wavelet spectrum when a<80.

For each defect type the autocorrelation function (AF) and intercorrelation function (IF) were defined. During evaluation of the autocorrelation function of the analyzed signals it was determined that the maximal normalized correlation coefficient is approximately equal to 1 when the relative shift equals zero. Therefore, we can claim that the analyzed signals have such property as ergodicity, and it is sufficient to perform analysis using values obtained from one period of testing. The minimum signal sampling duration for analysis is 0.02 s.

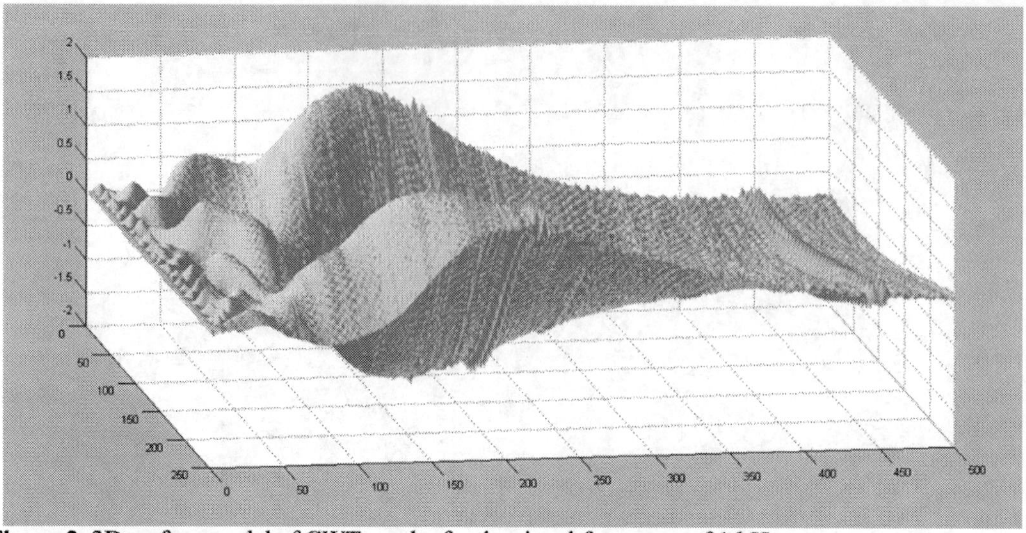

Figure 3. 3D surface model of CWT results for the signal frequency of 1 kHz.

During correlation analysis of informative spectrum components for each defect type it was determined that maximal normalized correlation coefficient equals 0.75. It means that defect modes (modes in which defects were detected) can always be identified correctly. There is no correlation between the defects as they manifest on different CWT spectrum components. Figure 5 shows graph of intercorrelation function of informative CWT spectrum component for defect No. 1.

5 Discussion and application of results

During analysis of the informative CWT spectrum components, rules for diagnostics and criteria for determination and identification of the defects based on evaluation of local maxima values and maximal normalized correlation coefficient were defined.

An algorithm for identification and determination of the existing defects and technological violations was developed in accordance with the rules for diagnostics. Figure 4 shows a typical cycle of the algorithm.

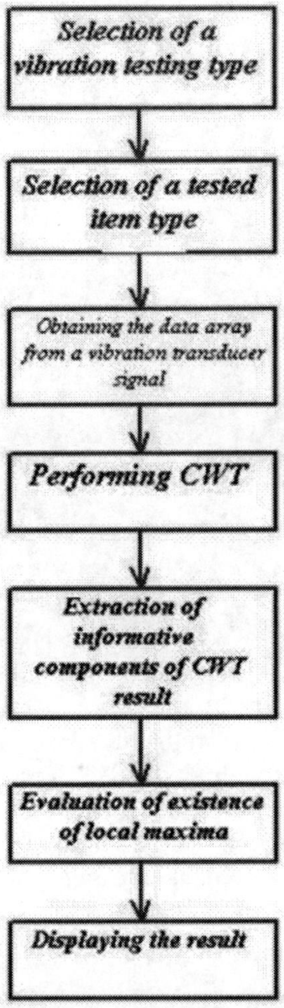

Figure 4. Cycle of the algorithm for identification and determination of the existing defects and technological violations.

This algorithm is the basis of the developed application software whose interface is illustrated in Figure 5.

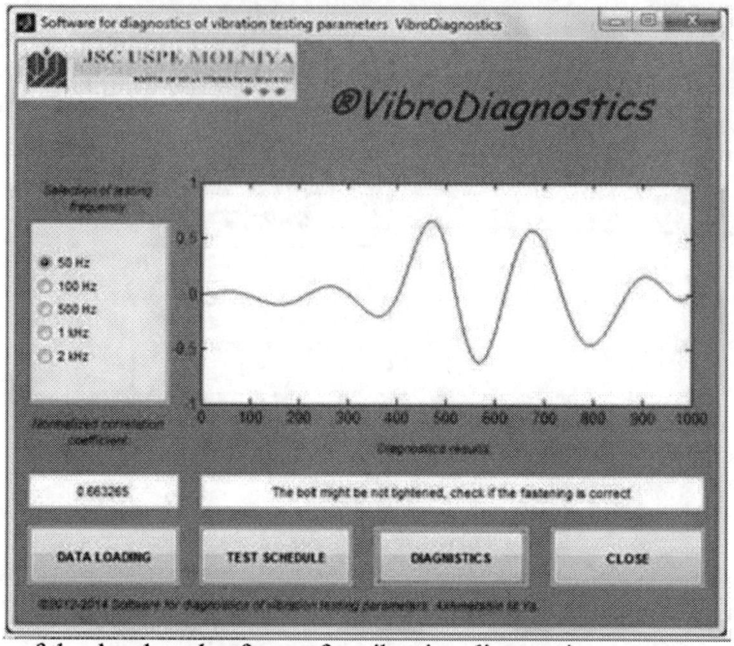

Figure 5. Interface of the developed software for vibration diagnostics.

The software makes it possible to load the .mat or .xls files of the analyzed signals from a computer hard disc, look through the loaded signals graphs, as well as the interconnection function graph. The software interface has round buttons to select testing frequency. Clicking DIAGNOSTICS button starts the main program the operation of which is followed by displaying the context message on diagnostics results and correlation coefficient value, as well as the intercorrelation function graph.

6 Conclusions

Following the study of the existing methods to analyze measured signals in modern vibration diagnostics systems, methods have been developed enabling to analyze signals and their changes not only in terms of their frequency, but also in a time scale. An opportunity has emerged to assess the effects of ambient conditions on the structure of the analyzed signals, in particular, vibration signals.

In the course of the experimental study, CWT spectrum changes have been revealed and information components of CWT spectrum corresponding to each defect type have been defined. The correlation analysis of the studied signals has shown that each defect type can be correctly identified.

Rules for diagnostics and criteria for determination and identification of the defects based on evaluation of local maxima values and maximal normalized correlation coefficient, which became the basis of the vibration diagnostics application algorithm, have been defined.

Analysis of the signals of a time area makes it possible to predict and to early detect defects still barely manifested, which has a large practical significance when developing modern vibration diagnostics systems. The developed method will also be useful for the development of highly effective neural network algorithms for processing not only vibration signals but also any other complex measured signals requiring consequent frequency-time analysis.

References

[1] L. Debnath, International Journal of Mathematical Education in Science and Technology **29**, 677 (1998) doi: 10.1080/0020739980290504
[2] C. Chiann, P. A. Morettin, Journal of Nonparametric Statistics **10** , 1 (1999)
[3] M. C. Hu, I. S. Tsai, The Journal of The Textile Institute **91**, 420 (2000)
[4] L. Shao, X. Lin, X. Shao, Applied Spectroscopy Reviews **37**, 429 (2002) doi: 10.1081/ASR-120016391
[5] M. Ya. Akhmetshin, V. Kh. Yasoveev, Pribory (8), 33 (2014)

INTERNET AS AN INSTRUMENT TO TRANSMIT THEORETICAL KNOWLEDGE

Svetlana Kvesko[1]*, *Svetlana* Kvesko[2], *Nataliya* Kabanova[3], *Daria* Shamrova[4]

[1]National Research Tomsk State University, 634050, Tomsk, Russia
[2]Lyceum № 10, Krasnoyarsk, 660113, Russia
[3]National Research Tomsk Polytechnic University, 634050, Tomsk, Russia
[4]Michigan State University, 49254, Michigan, USA

Abstract. The problem of transmitting theoretical knowledge and the role of the Internet in it require the solution due to the existing modernization of theoretical knowledge transmission process. The objective of this research is to define the role of the Internet in transmitting theoretical knowledge as it is the extremely important resource of the modern society. According to the carried out analysis of the problem and its solution the information technology is not only the mean that accumulates the volumes of knowledge, but also the tool of its social use, forms of social activity by way of social and information technology. As a result, using method of the methodological analysis in combination with competency-based approach we revealed that the Internet as a diachronic way of transmitting knowledge and experience is characterized by a polyagentity and interdisciplinarity.

1 Introduction

The history of society is the history of transmitting theoretical and practical knowledge and experience from a person to a person, from a generation to a generation, from one epoch to another. This is the history of transmitting society's social memory (sociocode). Knowledge and experience transmission can be of various methods [1]:

1. Transmission is carried out either by means of address communication, during communicative practice, or through the address communication having operational character. The activity of people as individuals in social environment of the cohabitation and joint activity is coordinated during such communication. This kind of communication is called synchronous. Transmission of both empirical and theoretical knowledge is carried out by means of synchronous communication.
2. Information and experience can be transferred trough time, i.e. from generation to generation, from epoch to epoch, here is transmission of knowledge and experience occurs. Such way is called diachronic.

These two ways differ from each other in character of feedback communication. At synchronous transmitting knowledge, information is known to both subjects of educational process, while at diachronic communication, the information is known to one subject. The

*Corresponding author: svetla_kvesko@mail.ru

© The Authors, published by EDP Sciences. This is an open access article distributed under the terms of the Creative Commons Attribution License 4.0 (http://creativecommons.org/licenses/by/4.0/).

knowledge is generally connected with transmission, with diachronic communication namely. Generally, theoretical knowledge is connected with transmission.

Both synchronous, and diachronic ways of transmitting knowledge use language, sign reality, as the main information storage and transmission devices. We should note that here also artificial languages are used, which that is connected with use of information and measuring equipment and technology.

2 Method

The aim of this research is to define the role of the Internet in transmitting theoretical knowledge as an extremely significant resource of modern society, as not a mean for accumulating the scope of knowledge only, but the instrument of its social use, as a form of social activity in social and information technology. The method of the methodological analysis in accordance with a competence-based approach is applied to detect necessary characteristics of transmitting theoretical knowledge.

3 Main content

Transmitting knowledge is the unity of objective and subjective knowledge expressed through language: either natural, or artificial, and through writing. The objective knowledge is kept in various material forms, including computer memory, books etc. We deal with an object in information stored in objective knowledge, for example, in computer memory. The object has not been interpreted and transformed in speech. In this case objective knowledge is a part of the social activity presented in signs. This objective knowledge is theoretical knowledge which is objective according to its content.

However, degree of adequacy of percepting information by means of transmission (diachronic communication) depends on a subject and personal form of theoretical knowledge and its cover. The subjective knowledge represents language model of the situation which is subjective and individual and personal expression of language of the broadcast theoretical knowledge [2]. This knowledge is a result of systematization of statements, theses, provisions, the points of view.

Adequacy and extent of closeness to the objective content of theoretical knowledge depends on experience, a variety of relationships of a personality with culture of society, intelligent potential of an individual. And as a result, methods of interpretation and formalization get importance. Interpretation broadens the sense-conceptual field of information, but formalization promotes developing forms of transmitting theoretical knowledge.

Transmission of theoretical knowledge is carried out when a neutrality is, and there is no any personal influence, exact reflection. The language picture of the world is the picture of the world reflecting the world by means of languages both natural, and artificial. The last one is characteristic of the Internet.

The Internet as the instrument of transmitting theoretical knowledge is based on use of technology of communication practice when the knowledge is addressed to a great number of addressees.

Transmitting theoretical knowledge is connected with a defined type of society. Pre-information epoch societies paid much attention to the teacher/lecturer, imparting knowledge to pupils/trainee. In information society information technology including the Internet, have impact on various forms of activity in society and, certainly, on transmitting theoretical knowledge. As a result, theoretical knowledge transforms to information resource. Its storage, processing and transmission to the subject of informative process is provided with information technology which allow to obtain, process and transmit great deal of information at high speed [3].

Computer storage devices give an opportunity to form a synchronous way of transmitting knowledge in addition to a diachronic way. Consequently, the information system of theoretical knowledge is being formulated, which is able to coordinate the process transmitting the sum of knowledge in connection with address communication [4].

The unity of communication (synchronous way) and transmission (diachronic way) causes feedback, i.e. provides correction of knowledge transmission programs. As the essence of transmission is transmission of theoretical knowledge from one part having information to the other part not having it, therefore, the Internet becomes the instrument of transmission.

Theoretical knowledge transmission via the Internet is based on communication technologies, on the principles of synchronism, results in increase of erudition and education of a person, extent of modern society intellectualization.

Theoretical knowledge transmission under the mass use of the Internet has no clear boundaries, and is not controlled practically, and consequently it leads to polyagent (multilateral) and interdisciplinary use. The absence of information selection criteria and its variety complicate the use of interdisciplinary approach and a polyagentity in assimilating theoretical knowledge received via the Internet.

We can claim that the universal and conceptual way of a person's inclusion to social reality and bank of theoretical knowledge via the Internet is an essential elements of using computer technologies in information space by a person for work.

In the 21st century information transmission becomes not a simply important and essential resource. This leads to updating computer technologies and the Internet as the instrument of transmitting information. As a result, expansion of using electronic resource brings about the emergence of various developed, multilateral (polyagent) social networks, which are characterized by existence of interdisciplinary communications. All these factors cause forms of people social activity, such as social and humanitarian technologies together with computer.

Theoretical knowledge, being abstract and deep on its essence, explains the phenomena, processes, and events. It is widely used by a person to familiarize in social and natural environment, to forecast and draw up a road map.

Due to the accelerated information and computer technologies development, their intensification, the innovative forms of electronic learning occurs, which assume training people via the Internet. Such method of studying allows to higher intellectualization of modern society, increasing intellectual potential of citizens [5]. However, great deal of information, various assessments of theoretical objectives, the giving uncontrolled standpoints complicate formulating a clear picture of event, making the selection of significant theoretical knowledge difficult.

The method of transmitting theoretical knowledge via Internet transmission provides disregarding qualities, skills and opportunities of a particular personality. Such knowledge transmission influence formulating world outlook, lines o thoughts, value system. Such theoretical knowledge transmission is informal as it tends to a certain convention, blurring, and multilateral interpretation of the gained knowledge [6]. On the Internet it is possible to receive any knowledge by means of a discourse. Searching and developing new ways of transmission had lead to informal methods of processing, understanding, assimilating greater content of theoretical knowledge.

The solution of the problem of theoretical knowledge transmission, the question of speed and volumes of the imparted theoretical knowledge via the Internet is important for revealing specifics and features of computer and information technology development influence on processes in society and in the nature, on language and culture, on education and upbringing.

Increase in speed of transmitting knowledge is caused by formalization of communication language that leads to creation of universal language [7].

4 Conclusion

Having studied the essence of the Internet as instrument of transmitting theoretical knowledge we received the following results:

1. We revealed that transmitting knowledge is a unity of both objective and subjective knowledge.
2. Theoretical knowledge transmission is carried out when there is neutrality, and there is no any personal influence, precise reflection. f
3. The Internet as the instrument of transmitting theoretical knowledge is based on the use of communication practice technology when knowledge is addressed to a great number of addresses.
4. It has been proved that computer data storages give an opportunity to form a synchronous way of transmitting knowledge in addition to a diachronic way. As a result, the information system of theoretical knowledge is formulated, which is able to coordinate the process of transmitting the sum of knowledge in connection with address communication.
5. It has been founded that the unity of communication (a synchronous way) and transmission (a diachronic way) causes feedback, i.e. provides correction of knowledge transmission programs.
6. The essence of transmission is definitely the transmitting of theoretical knowledge from a having information part, to the part which has not it; the Internet here is the instrument of transmission.
7. We have revealed that transmitting theoretical knowledge in the context of the Internet mass use has no distinct limits, and it is basically not controllable. Consequently it leads to polyagent (multilateral) and interdisciplinary use.
8. We have established that due to the lack of information criteria selection, and owing to its variety the use of interdisciplinary approach and a polyagentity in digestion of theoretical knowledge received from the Internet is complicated.
9. It is arguable that the universal and conceptual way of a person's inclusion to the social reality and theoretical knowledge bank via the Internet is an essential element of using computer technology by a person to work in information space.
10. We emphasize that there is an updating of computer technology and the Internet as an instrument of transmitting information. As a result, the widening usage of an electronic resource brings into existence an extremely developed, various, multilateral (polyagent) social networks which are also characterized by the existence of interdisciplinary communications.
11. We have proved that the emergence of the innovative forms of electronic learning assuming people training via the Internet results from the accelerated development of information and computer technologies and their intensification.

According to the carried-out methodological analysis based on polyagent and interdisciplinary approaches we can draw conclusions:

Information technology is the cornerstone of modern economy. This new economy is based on knowledge. The technology of generating knowledge makes essence of productivity in modern economy.

Transmission of theoretical knowledge, qualitatively developing. as a way of knowledge transmission, is capable to give an impulse to changes in economy, culture, education and in other spheres of life.

Information technologies represent a new form of social activity. They act as disambiguation of a contradiction between growth of volume of knowledge and social use of information technology.

The diachronic way of transmitting theoretical knowledge, which is closely connected with the Internet as the transmission instrument, led to emergence of electronic learning, i.e. the learning based on using the Internet.

Language is used in transmitting theoretical knowledge as the main social sign reality. It should be noted that there is a problem of creation of universal language.

Improvement of process of understanding of the broadcast knowledge acts as an urgent task in a solution of the problem of transfer and processing of theoretical knowledge by means of use of the Internet as the tool.

The problem of transmitting theoretical knowledge and the role of the Internet in it require the solution due to the existing modernization of the process of transmitting theoretical knowledge. As it appears from the undertaken study the information technology is the instrument of social use of theoretical knowledge transmission. Therefore we may state that the Internet represents a diachronic way of transmitting knowledge.

Acknowledgment

This research is carried out under the scope "Organization and methodic issues of training specialists in Information and Measurement Technologies and Technique". Passes with assistance of a grant of the Russian fund of basic researches №16-07-20218, and the program of improving competitiveness of the TSU (project No. 8.2.31.2015).

References

[1] V.L. Anoshkina, S.V. Rezvanov, *Education. Innovation. Future. Methodological and sociocultural problems* (Publishing house of RO IPK and PRO,Rostov-on-Don, 2001)

[2] M. Porter. Harvard Business Review **76** , 77 (1998).

[3] N. Kachalov, A. Kornienko, R. Kvesko, Yu. Nikitina, S. Kvesko, Zh. Bukharina, Procedia Social and Behavioral Sciences **206**, 459 (2015) doi: 10.1016/j.sbspro.2015.10.083

[4] I.B. Ardashkin, R.B. Kvesko, O.M. Ardashkina IFOST-2013 **2**, 692 (2013)

[5] M. Lehmann, P. Christensen, X. Du & M. Thrane. EJEE. **33** , 283 (2008)

[6] H. Etzkowitz, L. Leydesdorff, Research Policy **29** , 243 (2000)

[7] V.I. Syryamkin, E.G. Syryamkina, Procedia Social and Behavioral Sciences **166**, 471 (2015)

DIGITAL TECHNOLOGIES IN PROVIDING DEVELOPMENT OF ALGORITHMS SURGICAL TREATMENT OF SUPRAVENTRICULAR ARRHYTHMIAS

Vladimir Evtushenko[1,3a], *Konstantin* Smyshlyaev[3], *Alexander* Bykov[1], *Yury* Kistenev[1], *Elena* Pavlyukova[3], *Vadim* Katkov[3], *Igor* Kurlov[3], *Yana* Anfinogenova[2,3] and *Alexey* Evtushenko[1,3]

[1]National Research Tomsk State University, 634050, Tomsk, Russia
[2]National Research Tomsk Polytechnic University, 634050, Tomsk, Russia
[3]Federal State Budgetary Scientific Institution "Research Institute for Cardiology", 634012, Tomsk, Russia

Abstract The aim of the study was the development and clinical application of patient selection algorithm for surgical treatment of long-lasting persistent atrial fibrillation. The study included 235 patients with acquired heart disease and coronary artery disease, which in the period from 1999 to 2015 performed surgical treatment of long-term persistent atrial fibrillation (RF "MAZE III" procedure) in conjunction with the correction of the underlying heart disease. The patients were divided into 2 groups according to the method of operation: the group 1 – 135 patients (76 women and 59 men) who have applied an integrated approach to surgery for atrial fibrillation, including penetrating method of RF effects on atrial myocardium and the study of the function of the sinus node before and after the operation (these patients were operated on from 2008 to 2015). The group 2 – 100 patients (62 women and 38 men) with a "classical" method of monopolar RF "MAZE III", which the sinus node function was not studied. We used the combined (epi- and endocardial) method of RF «MAZE». This algorithm is decreasing of possible permanent pacemaker postoperatively. The initial sinus node function in these patients, measured using the original method, the basic line of this algorithm was taken. The results showed that use this algorithm for selection of patients allows significantly reduce the possibility of pacemaker implantation in the postoperative period.

1 Introduction

Improving the efficiency of radiofrequency (RF) ablation "MAZE III" due to sinus rhythm (grade 4 on a scale of Santa Crus) [1] is an actual problem in modern surgical arrhythmology.

[a] Corresponding author: evtushenko.vladimir@gmail.com

© The Authors, published by EDP Sciences. This is an open access article distributed under the terms of the Creative Commons Attribution License 4.0 (http://creativecommons.org/licenses/by/4.0/).

The drug treatment of AF is usually ineffective and reduced to the control of ventricular rate and prevention of thromboembolic complications. Measures aimed at the preservation of sinus rhythm are effective only in paroxysmal AF when the main mechanism of its development is a trigger. When myocardial substrate for maintenance of AF is forming, and the micro re-entry is developing, drug therapy becomes ineffective [2–4]. Thus, the development of surgical treatment of persistent and prolonged persistent AF becomes especially over other methods.

At the same time, the effectiveness of carefully performed radiofrequency procedure "MAZE III" in some cases significantly reduced due to postoperative sinus node dysfunction. Therefore, even a well-executed surgery does not lead to a significant improvement in patient well-being. In addition, these patients are in need of long-term temporary pacing and prolonged presence in the intensive care unit, and in some cases they need the pacemaker implantation.

In a number of scientific papers was shown [3–8], that the long persistent AF reduces the likelihood of recovery of sinus rhythm after the procedure, "MAZE III", however, does not provide a method to estimate the initial functional state of the sinus node during AF. Atrial fibrillation makes excessive load on atrial cardiomyocytes, during episodes of atrial fibrillation. This leads to the activation of neurohumoral mechanisms (mainly cardiac angiotensin II), which trigger mechanism "structural remodeling", leading to dilatation, hypertrophy, structural disorganization of myofibrils fibrosis and atrial myocardium, that escalate the processes of "electrical remodeling" and supports AF [4, 9, 10]. Thus, atrial fibrillation itself maintains and exacerbates the atrial fibrillation, leading to cell degeneration of sinoatrial node.

Thus, the preservation of the sinoatrial node function in patients during atrial radiofrequency fragmentation scheme "MAZE III" is one of the leading problems.

In the literature, forecasting methods the recovery of sinoatrial node function after surgical treatment of AF with cardiopulmonary bypass has not been found. The close similar is a method for predicting the successful restoration of sinus rhythm after the procedure, "MAZE III", associated with the preoperative analysis of electrocardiograms and standard chest leads [3, 11, 12]. The method is a f-wave analysis on the electrocardiogram. Restoration of normal sinus rhythm is possible if the amplitude of the f-waves of more than 0.1 mV.

The method disadvantage is it does not provide direct information about the preoperative electrophysiological condition of sinoatrial node. This one allows to predict only the restoration of normal atrial contractile activity. It cannot give a reliable forecast for the restoration of normal sinoatrial node function after the procedure, "MAZE III". In addition, an important disadvantage is the fact that there is no clear algorithm for the selection of patients for atrial radiofrequency fragmentation scheme "MAZE III", is based on the prevention of sinoatrial node dysfunction.

The aim of study is prognosis normal sinus rhythm recovery RF procedure "MAZE III", performed with cardiopulmonary bypass.

2 Materials and methods

The study included 235 patients with acquired heart disease and coronary artery disease, which in the period from 1999 to 2016 performed surgical treatment of long-term persistent atrial fibrillation (RF "MAZE III" procedure) in conjunction with the correction of the underlying heart disease. The patients were divided into 2 groups according to the method of influence: the first (primary) group –135 patients (76 women and 59 men) who have applied an integrated approach to surgery for atrial fibrillation, including penetrating method of RF effects on atrial myocardium and the study of the function of the sinus node before and after the surgery (these patients were operated on from 2008 to 2015). The second group – 100 patients (62 women and 38 men) with a "classical" method of monopolar RF "MAZE III", which the sinus node function was not studied (operated from 1999 to 2008.) [13]. We used the combined (epi- and endocardial) method of RF «MAZE» in both groups. Clinical characteristics and operations by groups are shown in Table (Table 1).

All patients before surgery were classified as II–III functional class (FC) according to the classification of New-York Heart Association (NYHA). Heart diseases were diagnosed on the basis of anamnesis, results of physical and instrumental studies. All patients underwent preoperative echocardioscopy with the definition of the morphology of valvular lesions and the degree of hemodynamic transvalvular blood flow (gradient, valve orifice area, effective area and regurgitant volume, the ratio of the effective area of the valve opening in the patient's body surface area). The etiology of heart diseases and hemodynamic types listed in the table (Table 1). Both groups of patients are identical, the differences between them in terms of preoperative statistically insignificant. Beta-blockers received 221 patients (94%), angiotensin converting enzyme (ACE) inhibitors and diuretics obtained all patients (Table 2). Some patients received prehospital digoxin and amiodarone. 5 days prior to surgery patients were canceled beta-blockers, amiodarone and digoxin. According to the test with a 6-minute walk 221 patient was classified to III functional class (FC) of heart failure (according to NYHA), 14 patients – to the class II of NYHA. Also, all patients recorded 12-lead ECG, according to testimony – Daily Holter ECG monitoring. All patients older than 40 years (men) and 45 years (women) before surgery performed coronary angiography. Sampling solid, inclusion criteria for patients were the presence long persistent atrial fibrillation, resistance to drug therapy, combined heart defects requiring surgical correction. Exclusion criteria were the presence of severe comorbidities, multiple organ dysfunction syndrome, intolerance of iodine preparations, patient reluctance to participate in the study.

In the first group, all patients underwent electrophysiological study (EPS) of the sinoatrial node function by proposed technology. We used diagnostic complex "Elcart-M" ("Electropulse", Tomsk). Electrophysiological study protocol included a portion of the right atrium stimulation in order to assess the automatic function of the sinoatrial node (the definition of recovery time of sinoatrial node function and correct recovery time of sinoatrial node function) and becomes overdrive stimulation of the right atrium to the development of AV block grade 2 ("Wenckebach point"). Electrophysiological study of preoperative sinoatrial node function performed intraoperatively before the heart structure separation and beginning of cardiopulmonary bypass. This was done to minimize the external impact on the sinoatrial node tissue. The study was carried out as follows: intraoperative transesophageal performed ultrasound of the heart and in the absence of thrombosis of heart cavities cardioversion was performed. In the case of sinus rhythm after 10 minutes was performed electrophysiological study of the function of the sinoatrial node. According to the results of preoperative EPS Group 1 patients were divided into 2 subgroups: Subroup 1 with normal sinus automatism (118 patients); Subroup 2 with signs of sinoatrial node dysfunction (17 patients). After the main phase of the operation and remove the cannula lung machine from the cavities of the heart and great vessels, in case of restoration of sinus rhythm, control performed EPS.

Table 1. Clinical characteristics of patients.

	Group 1	Group 2
The total number of patients, n	135	100
Male / female ratio	0.78	0.61
Mean age, years of	53.5±8.8	50.8±9.8
FC (NYHA)	2.3±1.1	2.4±0.9
The right atrium diameter, mm	62.5±5.7	63.2±7.9
The left atrium diameter, mm	50.3±6.12	53.2±7.9
The left atrium volume, ml	225.1±3.6	223.4±4.8
Pulmonary artery systolic pressure, mmHg	48.5±3.6	49.6±10.4
The mean lasting of AF, years	4.3±2.4	4.6±3.1
Diseases which are complicated by AF		
Congenital heart diseases	10	14
acquired heart diseases	112	83
Ischemic heart disease	8	3
The etiology of heart diseases		
Rheumatism	74	71
Degenerative aortic valve calcification	3	1
Connective tissue dysplasia	36	7
Congenital septal defect	10	16
Dilated cardiomyopathy	4	2
Operations		
Mitral valve replacement	42	35
MV replacement and tricuspid valve reconstruction	15	34
Mitral and aortic valves replacement	10	3
Mitral and aortic valves replacement, tricuspid valve reconstruction	-	3
Mitral valve reconstruction	48	4
Aortic valve replacement	2	1
ASD closing	10	16
ASD closing+CABG	-	1
CABG	4	3
Mitral valve replacement+CABG	4	-
Left atrioplasty	28	11

Statistical processing of the data was performed using the software package SPSS 15.0 for Windows (SPSS Software Products). Compliance testing samples of normal distribution was performed Shapiro-Wilk test criterion (Shapiro-Wilk). Description of data was carried out using mean and standard deviation – M ± StD. The Student's t-test was used for assess the reliability of inter-group differences. Paired Student's t-test was used for assess the reliability of intragroup differences (dynamics of indicators pre-surgery within one group). The significance level "p" was taken to be 0.05, respectively, the confidence level (p) = 0.95.

Table 2. Drug therapy before surgery.

The group of drugs	Group 1 (n, %)	Group 2 (n, %)
Beta-blockers	129 (90,6%)	92 (92%)
ACE inhibitors	135 (100%)	99 (99%)
Diuretics	135 (100%)	100 (100%)
Digoxin	22 (16 %)	31 (31%)
Amiodarone	42 (31%)	12 (12%)

 DOI: 10.1051/matecconf/20167901063

3 Results

As a result of the study, patients in Group 1 in the Subgroup 1 mean recovery time of sinoatrial node function and correct recovery time of sinoatrial node function initially was 1033.7 ± 101.2 ms and 242.3 ± 40.1 ms respectively. All patients of Subgroup 1 had postoperative sufficient rate sinus rhythm. During the EPS after the intervention, the mean recovery time of sinoatrial node function and correct recovery time of sinoatrial node function in this subgroup was 1020.3 ± 203.1 ms and 305 ± 60.4 ms, respectively (Figure 1, 2).

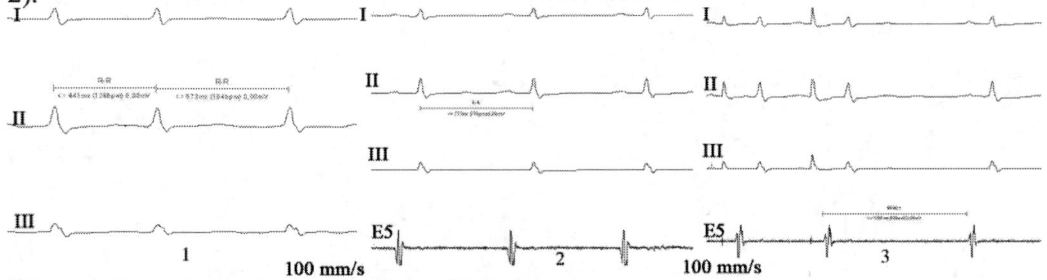

Figure 1. Patient's electrograms (subgroup 1) before surgery:
1) the original electrogram; I, II, III – standard leads; 2) the electrogram after cardioversion and sinus rhythm; I, II, III – standard leads; E5 – the sinoatrial node electrogram; 3) frequent stimulation of the right atrium in order to assess the automatic function of the sinoatrial node (recovery time of sinoatrial node function); I, II, III – standard leads; E5 – the sinoatrial node electrogram.

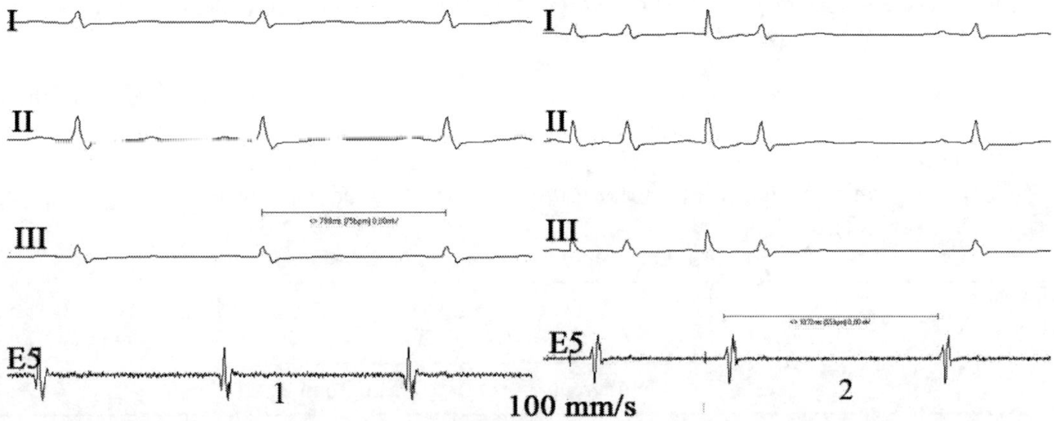

Figure 2. Patient's electrograms (subgroup 1) before surgery:
1)the original electrogram; I, II, III – standard leads; 2) frequent stimulation of the right atrium in order to assess the automatic function of the sinoatrial node (recovery time of sinoatrial node function); I, II, III – standard leads; E5 – the sinoatrial node electrogram.

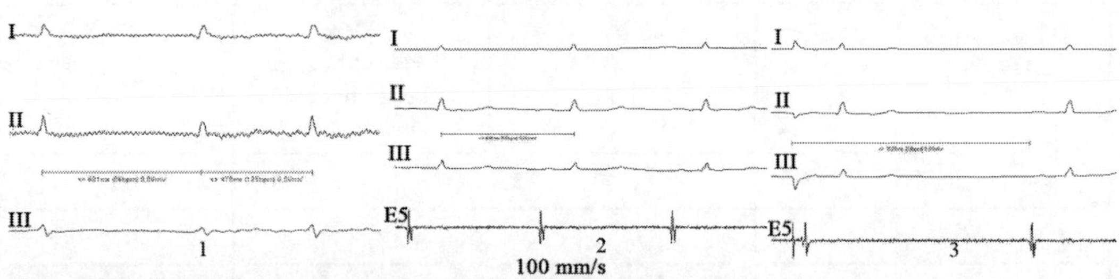

Figure 3. Patient's electrograms (subgroup 1) before surgery:
1) the original electrogram; I, II, III – standard leads; 2) the electrogram after cardioversion and sinus rhythm; I, II, III - standard leads; E5 – the sinoatrial node electrogram; 3) requent stimulation of the right atrium in order to assess the automatic function of the sinoatrial node (recovery time of sinoatrial node function); I, II, III – standard leads; E5 – the sinoatrial node electrogram.

 DOI: 10.1051/matecconf/20167901063

The recovery time of sinoatrial node function and correct recovery time of sinoatrial node function in patients of subgroup 2 initially had mean of 1995 ± 30.1 ms and 971.5 ± 102.4 ms, respectively (Figure 3).

In this subgroup of patients after surgery was observed sinoatrial node dysfunction and required temporary pacing mode AAI in up to 14 days (Figure 4). In 12 (71%) patients in the Subgroup 2 for the early postoperative period AF returned. 5 (29%) patients in this subgroup had pacemaker implantation. Thus, the total number of patients with irreversible sinoatrial node dysfunction was 17 people (12% of the patients of the group) (0 – 1 grading scale Santa Crus [1]).

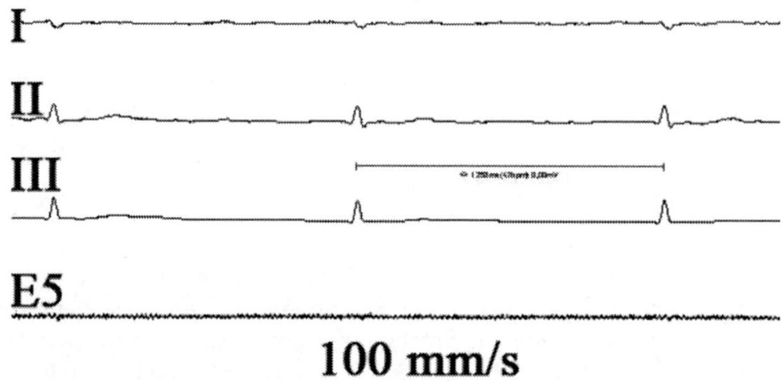

Figure 4. The electrogram of the patient from subgroup 2 after surgery; I, II, III – standard leads; E5 – the sinoatrial node electrogram (no activity).

Patients of Subgroup 1 did not require a temporary pacemaker, on average, more than 3.4±2.1 days. All of them had sinus rhythm sufficient rate.

Patients in group 2 had a significantly higher percentage of ineffective treatments "MAZE III", which consists in returning the AF and the need for pacemaker implantation ($p<0.05$) (Table 3).

Table 3. Early and long-term results by groups.

	Group 1 (N= 135)		**Group 2 (N=100)**
	Subgroup 1	Subgroup 2	
Sinoatrial node dysfunction (in terms of up to 3.4 ± 2.1 days)	56 (47%) (41% of the total number of patients in the group)	17 (100%) (12% of the total number of patients in the group)	86 (86%)
Sinoatrial node dysfunction (in terms of > 14 days)	0	13 (76%) (9.6% of the total number of patients in the group)	22 (22 %)
ECMO	3 (2.2%)		-
Mortality	2 (1.5%)		2 (2%)
Sinus rhythm in discharge	125 (93%)		78(78%)
Long-terms results			

A) Sinus rhythm without antiarrhythmic drugs	120 (88%)	51 (51%)
B) Sinus rhythm with antiarrhythmic drugs	126 (93%)	68 (68%)
Pacemaker	5 (3.7%)	10 (10%)
Mortality	4 (2.9%) (non-cardiac causes)	5 (5%) (1 patient died of hemorrhagic stroke)
Thromboembolic complications	No	No
Non-effective	3.2%	19%

In all patients, treatment of AF which has been found to be effective, there remains an active systole left and right atrium (A wave amplitude greater than 0.8 m/s). The A-wave is either absent or its amplitude <0.5 m/s in patients with sinoatrial node dysfunction.

4 Discussion of the results

Development of sinoatrial node dysfunction in patients with long-standing persistent atrial fibrillation is one of the reasons for the ineffectiveness of surgical treatment. This complication is not life-threatening in itself, but requires a long-term presence of the patient in intensive care, time spent pacing for a long time, and in some cases, the implantation of a permanent pacemaker [8, 14]. The sinoatrial node dysfunction develops as a result described by many researchers expressed the defeat of the host tissue and arteries, and its supply perinodal structures [5, 6, 15–17]. As pointed out by a number of researchers, sinoatrial node dysfunction provoked chronic AF, which in most cases is in these patients lengthening the time intraatrial and shortening of atrial refractoriness [3].

Possible causes of the sinoatrial node dysfunction after radiofrequency "MAZE" to "open" the heart are:

1. The mechanical damage of the sinoatrial node arteries during access to the mitral valve.
2. The thermal damage to the sinoatrial node during radiofrequency exposure on the myocardium.
3. The initial sinoatrial node dysfunction.

The first reason can be solved choice the atriotomy to the mitral valve [18, 19]. When selecting the atriotomy approach to the mitral valve is necessary to consider the type of myocardial perfusion. In all cases, preferred the left atriotomy approach is not leading to damage of the arteries supplying the sinus node, but it is in some cases (at small sizes of left atria) provides visualization of mitral valve worse than the Guiraudon approach. When there is the right or balanced types of heart circulation, superior approach by Guiraudon is preferred. The transseptal approach to the mitral valve through right atria leads to damage of the arteries feeding the components of the conduction system in 100% cases [19].

To prevent thermal and wave damage to the sinus node is proposed and applied in clinic technique penetrating effects on atrial myocardium. Features of this method lies in the fact that the original design of the monopolar electrode allows for more effective impact on the transmural atrial myocardium in a shorter period of time, which reduces the area of damage and prevents involvement in the zone of influence elements of the conduction system of the heart [20].

However, to evaluate the functional state of the sinus before surgery during the phase transition was not possible. The literature found messages on technology research sinoatrial node functions in patients with long-term persistent AF. To solve this problem, we developed an original method of intraoperative evaluation sinoatrial node functions in these patients. It allows electrophysiological study with the definition of recovery time of sinoatrial node, correct the recovery time of sinus node function and "Wenckebach point" (becoming more frequent stimulation of the right atrium to the development of AV block degree 2).

Evaluation of sinoatrial node function using the technology developed before the procedure "MAZE III", correlates with its function in the late postoperative periods. Thus, the proposed technology allows to evaluate the function of the sinus node and conduction system of the heart in patients with long-standing persistent AF.

The findings suggest that while initially the compromised function of the sinoatrial node should not be expected in the postoperative effects of surgical treatment of AF in these patients. Moreover, given the initial severity of the condition of these patients, in some cases is unreasonable increase in the time of cardiopulmonary bypass to prevent complications, such as haemolysis, air embolism and systemic inflammatory response. Thus, to determine the indications for the procedure of "MAZE III" expedient to EPS on technology developed before the intervention with the assessment of the functional state of the sinoatrial node. When it is appropriate we need to assess the initial severity of the patient's condition and the effect on it of atrial fibrillation [21–25]. If AF has no pronounced negative effect on the hemodynamics of the patient, it makes sense to restrict the atrial appendages off by their resection and suturing, in addition to the main stage of the operation to reduce the risk of thromboembolic complications.

Thus, based on the above-described method for evaluating the function of sinus node selection algorithm can provide patients with long persistent AF for surgical treatment.

5 Conclusion

The proposed method for evaluating sinus node function in patients with long-standing persistent atrial fibrillation can be used to predict the successful restoration of sinus rhythm after the procedure, "MAZE III". A set of activities aimed at preserving the function of the sinus node, is able to increase the effectiveness of surgical treatment of atrial fibrillation. If there is the original sinus node dysfunction in patients, the surgical treatment of long-term persistent AF is not very promising in terms of the sinus rhythm restoration.

Acknowledgements

The study was supported by a grant from the Russian Science Foundation (№ 15-15-10016), and the program of improving competitiveness of the TSU (project No. 8.2.31.2015).

References

[1] J. Q. Melo, J. Neves, P. Adragao, R. Ribeiras, M. M. Ferreira, L. Bruges, M. Canada, T. Ramos, European Journal of Cardio-Thoracic Surgery **12,** 739 (1997) doi: 10.1016/S1010-7940(97)00252-2

[2] S. Benussi, A. Galanti, V. Zerbi, Y. A. Privitera, I. Iafelice and O. Alfieri, J. Thorac. Cardiovasc. Surg. **139,** 1131 (2010) doi: 10.1016/j.jtcvs.2009.12.039

[3] H.-Y. Chang, Y.-J. Lin, L.-W. Lo, S.-L. Chang, Y.-F. Hu, C.-H. Li, T.-F. Chao, W.-H. Yin and S.-A. Chen, Europace **15,** 205 (2013) doi: 10.1093/europace/eus219
[4] S. Fareh, C. Villemaire and S. Nattel, Circulation **98,** 2202 (1998)
[5] A. Elvan, K. Wylie and D. P. Zipes, Circulation **94**(11), 2953 (1996)
[6] J. Gomes and P. Kang, Circulation, **63,** 80 (1981)
[7] J. M. S. Lee and J. M. Kalman, Europace **15,** 161 (2013) doi: 10.1093/europace/eus223
[8] G. Peretto, A. Durante, L.R. Limite and D. Cianflone, Cardiology Research and Practice **2014,** 615987 (2014) doi: 10.1155/2014/615987
[9] S. Saksena, A. Prakash, R. B. Krol and A. Shankar, Am. J. Cardiol. **84,** 880 (1999) doi: 10.1016/S0002-9149(99)00459-2
[10] A. W. Teh, P. M. Kistler, G. Lee, C. Medi, P. M. Heck, S. J. Spence, P. B. Sparks, J. B. Morton and J. M. Kalman, J. Cardiovasc Electrophysiol **23,** 232 (2012)
[11] A. S. Nemkov and A. V. Sydorchuk, Bulletin of Surgery **168**, 21 (2009)
[12] H. G. Je, J. W. Lee, S. H. Jung, S. J. Choo, H. Song, S. C. Yun and C. H. Chung, European journal of cardio-thoracic surgery **36,** 272 (2009) doi: 10.1016/j.ejcts.2009.02.058
[13] A. V. Evtushenko, M. B. Knyazev, V. M. Shipulin, S. V. Popov, I. V. Antonchenko, V. O. Kiselev, I. V. Ponomarenko, V. V. Evtushenko and M. V. Krivov, Siberian Medical Journal **2,** 79 (2005)
[14] M. Pasic, M. Musci, H. Siniawski, O. Grauhan, B. Edelmann, T. Tedoriya, Y. Weng and R. Hetzer, J. Thorac. Cardiovasc. Surg. **118,** 287 (1999) doi: 10.1016/S0022-5223(99)70219-9
[15] S. Saksena, I. Giorgberidze, R. Mehra, M. Hill, A. Prakash, R. B Krol and P. Mathew, Am. J. Cardiol. **83,** 187 (1999) doi: 10.1016/S0002-9149(98)00822-4
[16] Y. H. Yeh, B. Burstein, X. Y. Qi, M. Sakabe, D. Chartier, P. Comtois, Z. Wang, C. T. Kuo and S. Nattel, Circulation **119,** 1576 (2009)
[17] S. Nath and D. E. Haines, Prog. Cardiovasc. Dis. **37**(4), 185 (1995) doi: 10.1016/S0033-0620(05)80006-4
[18] A. V. Evtushenko, V. V. Evtushenko, K. A. Petlin, E. M. Belenkova and I. V. Antonchenko, Herald of arrhythmology **69,** 5 (2012)
[19] V. V. Evtushenko, A. V. Evtushenko, K. A. Petlin, V. M. Gulyaev, E. M. Belenkova and V. M. Shipulin, Clinical Physiology of circulation **4,** 39 (2008)
[20] A. V. Evtushenko, V. V. Evtushenko, K. A. Petlin, E. M. Belenkova, E. K. Knyazeva, I. V. Antonchenko, V. Kh. Vaizov, V. M. Shipulin and S. V. Popov, Herald of arrhythmology, **48,** 15 (2007)
[21] D. K. Swanson, W. J. Smith, T. Ibrahim and A. S. Wechsler, Innovations: Technology and Techniques in Cardiothoracic and Vascular Surgery **6**(4), 276 (2011) doi: 10.1097/IMI.0b013e31822b4d22
[22] M. La Meir, S. Gelsomino, F. Lucà, R. Lorussoc, G. F. Gensinic, L. Pisond, F. Wellensb and J. Maessena, Interact. Cardiovasc. Thorac. Surg. **14**, 445 (2012) doi: 10.1093/icvts/ivr142
[23] W. Saliba and O. M. Wazni, Clin Cardiol. **34**, 12 (2011) doi: 10.1002/clc.20826
[24] C. Wenning, P. S. Lange, C. Schülke, A. Vrachimis, G. Mönnig, O. Schober, L. Eckardt and M. Schäfers, EJNMMI Res. **3**, 81 (2013) doi: 10.1186/2191-219X-3-81
[25] G. Nasso, R. Bonifazi, A. Del Prete, G. Del Prete, V. Lopriore, F. Bartolomucci, A. M. Calafiore and G. Speziale, J. Thorac. Cardiovasc. Surg. **142**, e41 (2011) doi: 10.1016/j.jtcvs.2011.04.009

MATEC Web of Conferences 79, 01064 (2016) DOI: 10.1051/matecconf/20167901064
IME&T 2016

NEAR-FIELD INTERACTION OF CLOSED CELLS FOR METAMATERIAL CREATION

Aleksandr Mironchev*, *Aleksandr* Gorst

National Research Tomsk State University, 634050, Tomsk, Russia

Abstract. This article presents the results of numerical and computer modeling of the flat closed conductor with different variants of arrangement. The interaction of the conductors is examined and the results of active and reactive part of the Poynting vector for each structure is presented. According to the results the model with identical parameters for each element was built and examined for the presence of metamaterial properties.

1 Intorduction

Back in the late 1960s, V.G. Veselago predicted the unique properties of materials with a negative real part of the permittivity and permeability [1]. Later, a special interest was aroused by the works of D. Pendry of the possibility of developing "superlenses" allowing overcoming the diffraction limit [2].

Previously, the authors of this work presented the possibility of metamaterial creation for the radio range as an artificial composite medium made up of a certain oriented linear and circular conductors [3, 4]. Such a medium may have a negative refractive index. In the theoretical consideration of the works above stated the approach of not interacting elements has been introduced. For more complete analysis when creation of the structure with the properties of the metamaterial it is necessary to consider the influence of elements on each other at the incidence of a plane wave [5].

2 Modelling the interaction of confined conductors

The problem of the circular conductors' structure arrangement is sufficiently capacious. It is necessary to select geometric dimensions of the elements, their orientation, and the distance between them. To estimate the interaction between the individual elements of the array we will consider Poynting vector along the axis of distribution of the incident plane wave [6, 7].

To consider the active (Re(P)) and reactive (Im(P)) parts of the Poynting vector it was provided a modeling using CST Microwave studio software package. This product allows considering different characteristics of the electric (E) and magnetic (H) fields.

* Corresponding author: mironchev42@mail.ru

© The Authors, published by EDP Sciences. This is an open access article distributed under the terms of the Creative Commons Attribution License 4.0 (http://creativecommons.org/licenses/by/4.0/).

For the first model a closed conductor with the outer radius – 4 mm, inner radius – 2 mm, conductor thickness – 1mm was chosen. The ring was in free space i.e. with permittivity (ε) and permeability (μ) equal to unity. The plane electromagnetic wave falls to the conductor, the vector of the electric component was directed perpendicular to the axis of the ring. The estimation was performed at the frequency $F = 1 \div 13$ GHz.

To calculate the active and reactive parts of the Poynting vector it is required to estimate three components of the fields E and H, such as amplitude ($|A|$) and phase (φ), for the further calculation of the complex field defined by the formula:

$$B = A \cdot \exp(-i\phi) \qquad (1)$$

where A is amplitude of the field, φ is phase.

Calculating the necessary characteristics and specifying the electric and magnetic fields of each projection on the coordinate axes, it becomes possible to calculate the Poynting vector by the following formula:

$$\vec{P} = \frac{1}{2}\left[\vec{E} \times \vec{H}^*\right] \qquad (2)$$

where $\vec{E}$ is a vector of the electric field, $\vec{H}$ is a magnetic field vector.

Figure 1 shows the real part of the Poynting vector along the direction of distribution of the incident wave in the plane of the ring. The diagram is made for three frequencies 7 ÷ 9 GHz where the maximum effect of the incident wave interaction with the circular conductor is observed. Normalization to the value of the Poynting vector modulus of the primary wave is made. The boundary of the ring is indicated by dotted lines.

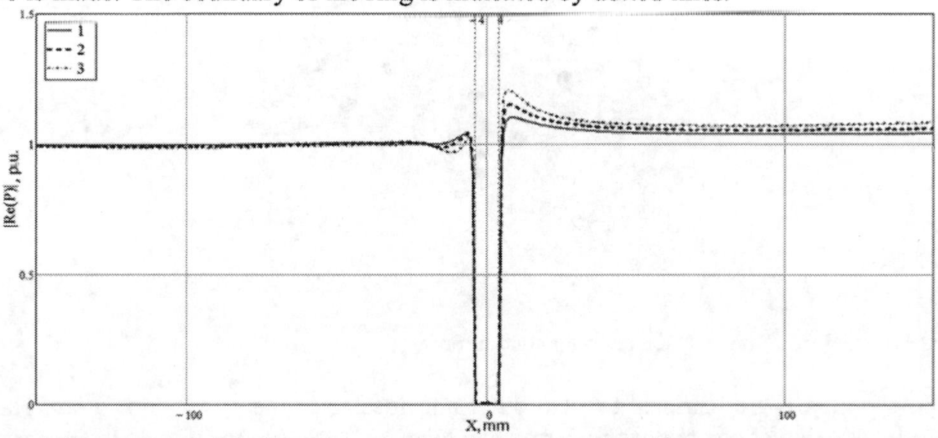

Figure 1. The real part of the Poynting vector at the plane wave incidence to the circular conductor for three frequencies: 1– for 7 GHz, 2 – for 8 GHz to 8, 3 – for 9 GHz.

The diagram shows that at the incidence of the plane wave to the closed conductor it acts as a secondary source of electromagnetic waves. This is due to vectoring of currents on the ring. As it can be seen, the radiation is both against the direction of the incident wave distribution and co-directed with it. But it can be seen that in front of the ring the secondary field decreases rapidly as the incident wave dominates. For the conductor the effect of increasing of the Poynting vector active part is observed that is associated with the addition of the incident and the generated waves. It can also be noted that the real part of the Poynting vector (further Re(P) for simplicity) at a frequency of 9 GHz has a maximum index. This occurs due to the approach to the resonant frequency.

DOI: 10.1051/matecconf/20167901064

When considering the imaginary part of the Poynting vector (further Im(P) for simplicity) presented in Figure 2 it can be seen that in front of the ring there is a field of the incident wave interference with the wave generated by the ring. At the distances of about 15 mm from the ring in both sides the amplitude is much higher; this is due to the near zone of radiation of the wave resulting from the generation of the ring.

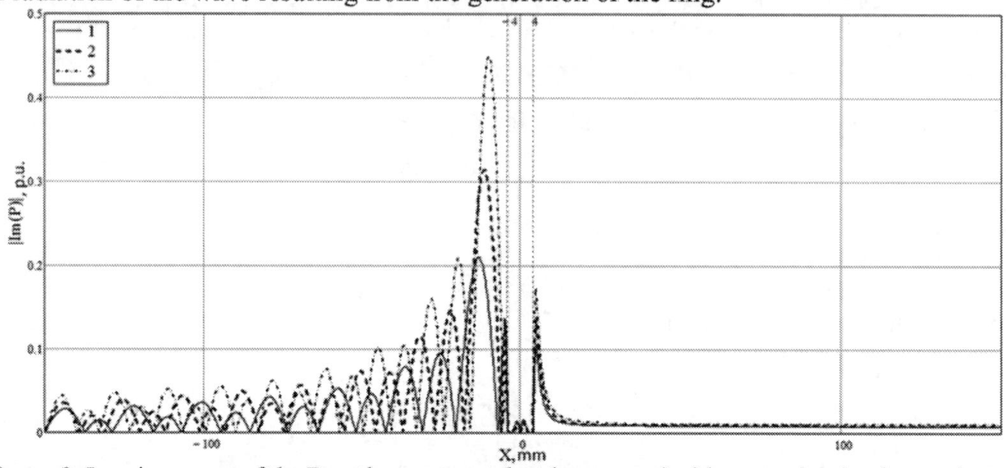

Figure 2. Imaginary part of the Poynting vector at the plane wave incidence to the circular conductor for three frequencies: 1 – for 7 GHz, 2 – for 8 GHz, 3 – for 9 GHz.

Now let`s consider Re(P) and Im(P) for three closed conductors (rings) being axially (along the axis OZ) as shown in Figure 3.

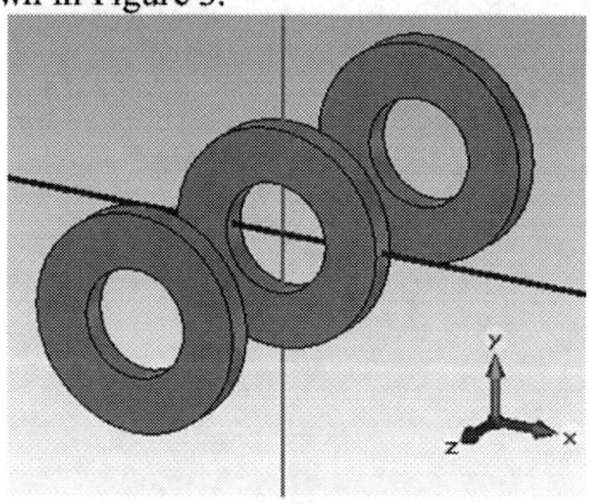

Figure 3. Model from three closed co-axial conductors.

The conductors are identical to each other and located at the distance of 5mm. Having the results of calculations and using formulas (1) and (2), we get Re(P) and Im(P) for three closed conductors. Figure 4 shows the comparison diagrams of the active component of the Poynting vector at the frequency of 9 GHz of one ring with three rings.

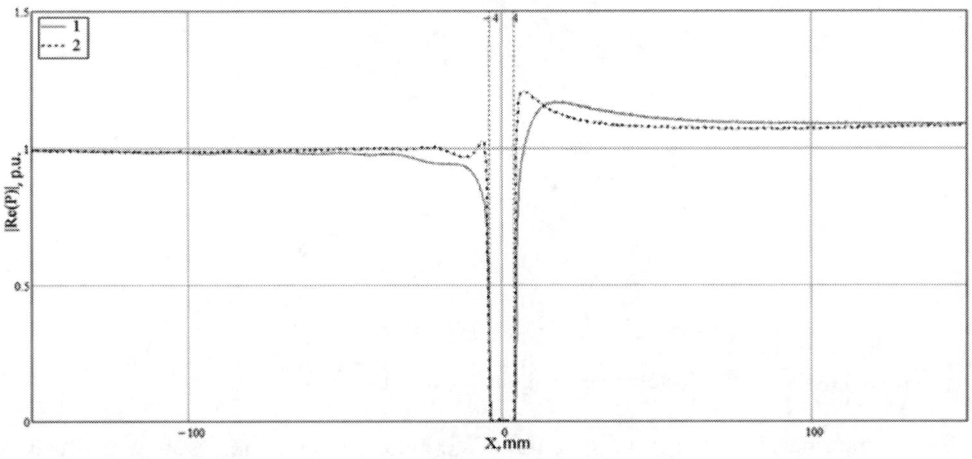

Figure 4. Real part of the Poynting vector at the frequency of 9 GHz: 1 – for three rings, 2 – for one ring.

Figure 4 shows that the real part of the Poynting both for one ring and for the three rings is almost identical, except for some changes nearby the ring. The diagram clearly shows that the increase of the amplitude out of the conductors is happened not in a sharp jump but smoothly due to the influence of the neighboring rings.

Figure 5 demonstrates the dependence of the reactive part of the Poynting vector on the distance. In front of the conductors the behavior of the imaginary part is similar with one ring; amplitude increase is due to the presence of the rings being on both sides of the center one. But in the near field out of the rings the reactive part is significantly less. At the other frequencies at a given location of the rings the changes in the real and imaginary components of the Poynting vector are insignificant. Maximum interaction is observed at the frequency of 9 GHz.

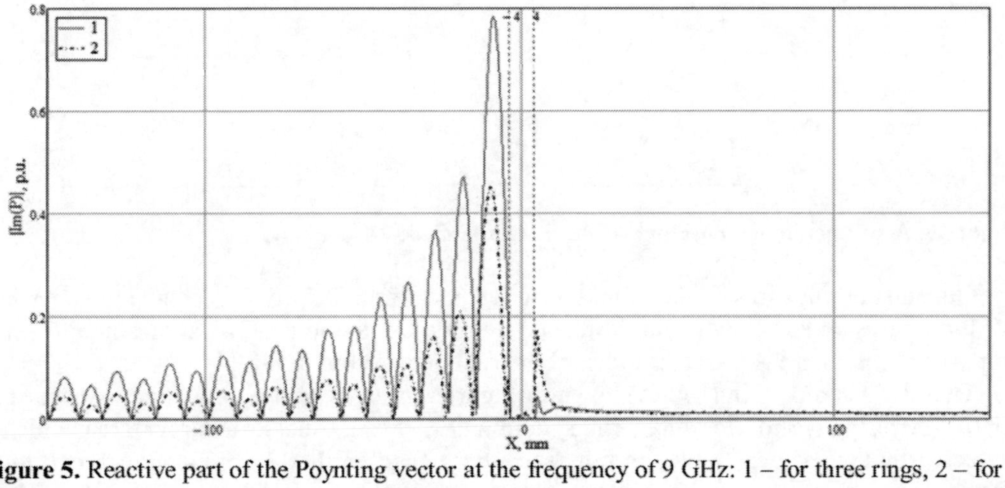

Figure 5. Reactive part of the Poynting vector at the frequency of 9 GHz: 1 – for three rings, 2 – for one ring.

Further, in Figure 6 there is a model for the case of the location of additional rings above and below (along the OY axis). E vector of the plane wave is directed along the axis OY.

DOI: 10.1051/matecconf/20167901064

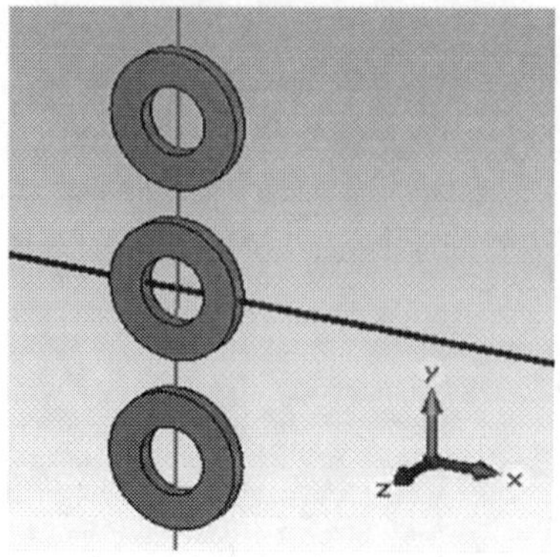

Figure 6. Model of closed conductors located along the axis OY.

The distance between the rings is 2 mm. We also compare the result with Re(P) and Im(P) for one ring; the results are given at the frequency of 9 GHz as in other frequencies the diagrams character is similar and maximum interaction is observed exactly at this frequency.

Diagrams for the active component of the Poynting vector are shown in Figure 7. We can see that the amplitude of Re(P) in the case with 3 rings before the conductors up to a distance of 8 cm from them is equal to unity. With the approach to the rings the decrease of the amplitude associated with the reflected wave and near field generated by the conductors is observed.

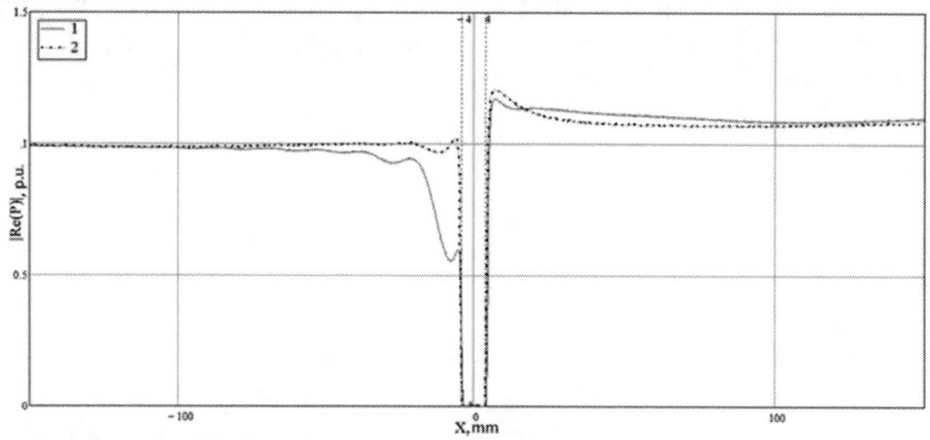

Figure 7. Active part of the Poynting vector: 1 – three rings, 2 – one ring.

Outside the rings it is evident that the active part of the Poynting vector in the near field for three rings ranks below in amplitude to one ring, but in the far field at distances from 15 mm to 115 mm there is a greater index of the real part for three rings.

Then let`s compare Im(P). As shown in Figure 8 in front of the rings the imaginary part of the Poynting vector for three rings dominates, this is due to a larger area which is reflected by the incident wave, and the secondary field emitted by the conductors. Out of the elements an increase in the reactive part of the Poynting vector in the case of three rings can be noticed. Moreover, the decrease of Im(P) for three closed conductors is slower than in the case with one ring, and compared with the reaction field index for one ring at the distance of 30 mm.

DOI: 10.1051/matecconf/20167901064

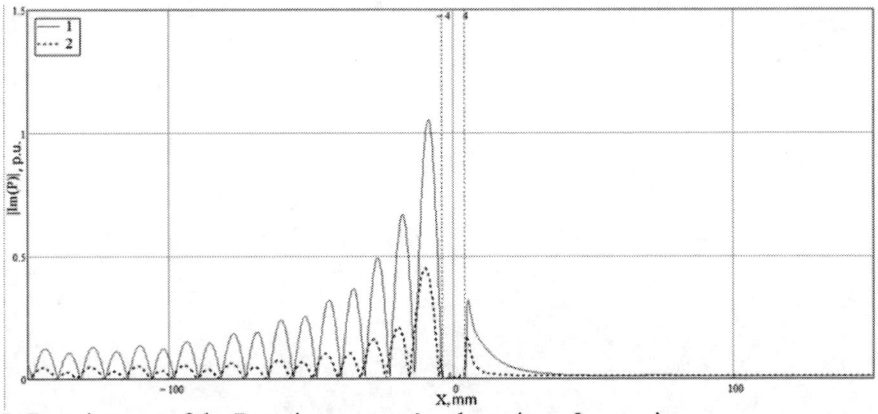

Figure 8. Reactive part of the Poynting vector: 1 – three rings, 2– one ring.

Another possible case is shown in Figure 9, the construction of a model from three closed conductors, it is their location along the axis of distribution of the incident wave (along the axis OX).

Figure 9. The model of closed conductors along the axis OX.

The distance between the rings as in the previous case is 2 mm. As before, the interaction of three rings will be compared with one ring. The results are given at 9 GHz as in the previous cases.

From Figure 10 it can be seen a clear predominance of the active component of the Poynting vector for three rings, which indicates a sufficiently strong interaction of the rings facing each other, but with increasing of the distance from the rings the amplitude goes up to one level. However, in front of the elements the active field is comparable.

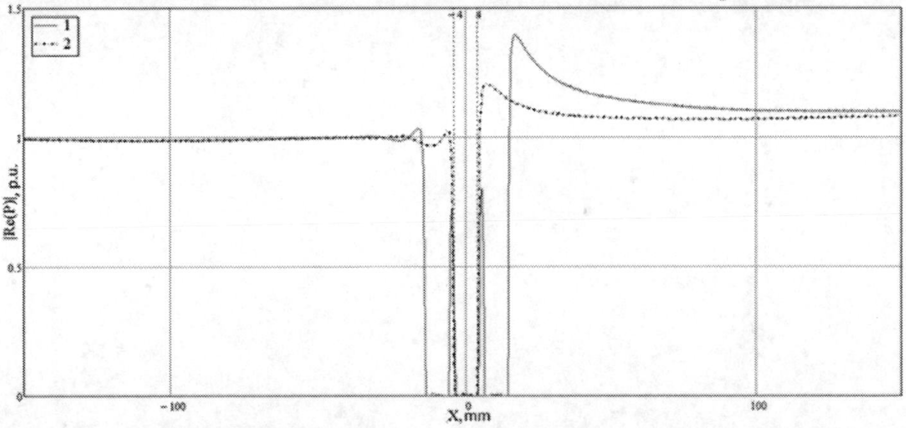

Figure 10. Active part of the Poynting vector, 1 – three rings, 2 – one ring.

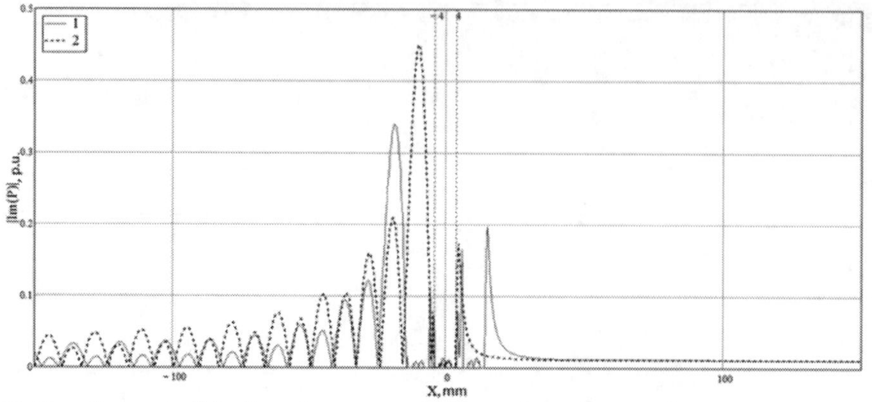

Figure 11. Reactive part of the Poynting vector: 1 – three rings, 2 – one ring.

The diagrams shown in Figure 11 illustrate the comparison of the reaction field in the case of the incidence of the plane wave to one ring element and three identical ring conductors arranged one behind another. It can be seen that for one conductor significantly higher indexes of the imaginary part of the Poynting vector to the elements than for three rings are observed. Out of the conductors the pattern is comparable both in magnitude and the rate of decay.

3 Modeling the structure with metamaterial properties

These results allow us to estimate the interaction of the ring elements with each other and to help in the modeling of structures in order to achieve the properties of metamaterials.

Based on all the above mentioned studies the lattice from the circular conductors with identical parameters was modeled. The lattice is consisted of four layers and had 324 elements. Because the metamaterial is characterized by a negative refractive index or a negative phase incursion in the layer, so there was this parameter.

In Figure 12 there is a diagram of the phase incursion for the lattice of the closed ring conductors at the frequency of 9 GHz. It can be seen that there is a small negative phase incursion at passing the lattice. This incursion may be counted into the effective refractive index. At this frequency its real part is Re(n) = 0.344. At the frequency of 7 GHz it was Re(n) = - 0.993, which can be explained by a stronger interaction of elements on each other, and slightly larger near field for each of them.

DOI: 10.1051/matecconf/20167901064

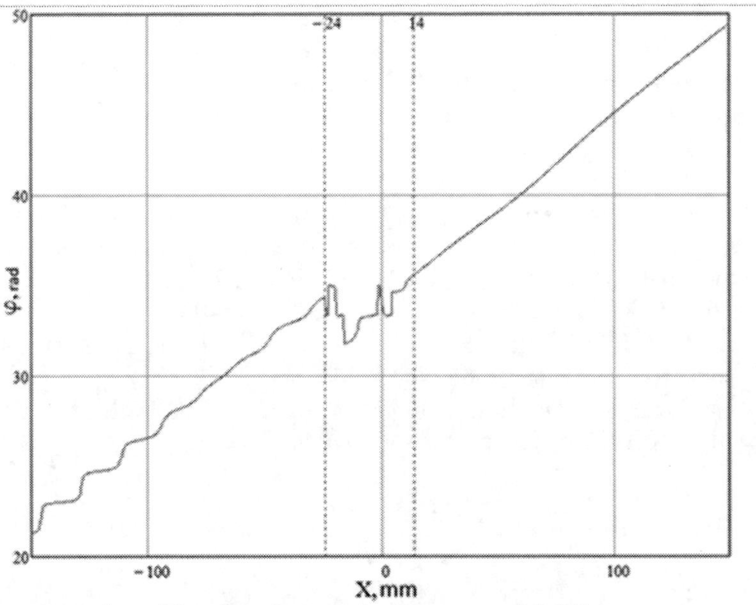

Figure 12. Phase incursion of the plane wave at the frequency of 9 GHz.

4 Conclusion

In view of the above stated diagrams it can be concluded that the maximum increase of the active part of the Poynting vector out of the elements can be reached at the location of the rings one after another (along the axis OX). Similar effect, but slightly smaller in size, can be reached at the location of the rings on top of each other (along the axis OY), minimal effect is observed at the location of the rings on the sides (along the axis OZ). The real part of the Poynting vector in front of the conductors is significantly reduced only in case if the rings are located above each other (along the axis OY), while in the model with rings being one after another (along the axis OX) there is a slight increase of the active field in front of the elements. In the case with the rings located on the sides (along the axis OZ) there is a decrease in the level before the conductors, but it happens quite smoothly. As for the reactive component of the field, in front of the elements it is maximum in the case of location of the rings over each other (along the axis OY), and it is minimum at the location of the rings one after another (along the axis OX). Out of the conductors the imaginary part of the Poynting vector is as maximum at the location of the rings on top of each other (along the axis OY), and minimum in the case of the rings location on the sides (along the axis OZ). At location of the rings one after another (along the axis OX) the decrease of the reactive field out of the layer is faster than at the location of the rings one above the other (along the axis OY).

Thus, by virtue of the dative analysis of the near-field interaction of the closed circular conductors with each other we were able to build a lattice model with the properties of the metamaterial.

Acknowledgment

This research carried out in 2016 was supported by «Russian Science Foundation» grant (No. 16-19-10272).

References

[1] V.G. Veselago, Sov. Phys. Usp. **10**, 509 (1968)
[2] J. B. Pendry, Physical review letters **85**, 3966 (2000) doi: 10.1103/PhysRevLett.85.3966
[3] V.P. Yakubov, V.P. Belichenko, V.V. Fisanov *Fundamentals of electrodynamics radiation and its interaction with the material* (Publishing NTL, Tomsk, 2010)
[4] V.P. Yakubov, A.S. Mironchev, A.G. Andreytsov, I.O. Ponomareva, Russian physics journal **53** , 895 (2011) doi: 10.1007/s11182-011-9507-5
[5] J.B. Pendry, D.R. Smith, Physics Today **57**, 37 (2004)
[6] I.V. Lindell, A.H. Sihvola , Electromagnetic Waves and Applications **19**, 861 (2005) doi: 10.1163/156939305775468741
[7] Dan Xiaa, Dianyou Song and Zhiyong Wang, MATEC Web of Conferences **61**, 4 (2016) doi: 10.1051/matecconf/20166102004

 DOI: 10.1051/matecconf/20167901065

A SYNERGISTIC EFFECT IN THE MEASUREMENT OF NEURO-FUZZY SYSTEM

Sergey Gorbachev[a], *Nadezhda* Gorbacheva, *Sergey* Koynov

National research Tomsk state University, 634050, Tomsk, Russia

Abstract. We consider a new type of hybrid neuro-fuzzy system based on fuzzy and neural computing in hierarchical sequential structure, the total effect exceeds the effect of each component separately. The proposed system can be applied to multi-criteria analysis, automatic classification on signs and obtain evidence-based estimates of the efficiency of scientific and technical solutions and technologies, engineering and robotics. An example of a neuro-fuzzy system measuring the intensity of the emotions of a robot, with the extraction of diagnostic decision rules "If ... then".

1 Introduction

The concept of soft computing (soft computing) was first mentioned in the work of Zadeh (Lotfi A. Zadeh) in 1994 [1]. Soft computing - is a complex computer methodology based on fuzzy logic, evolutionary computation, Neurocomputing and probabilistic calculations, with the later inclusion of chaotic systems, trust networks and learning theory sections. Components do not compete, but they create a synergistic effect. The guiding principle of soft computing - is accounting inaccuracies, uncertainty, partial truth, and approximation to achieve robustness, low cost solutions that better fit with reality. Four constituents of soft computing include:

- — Fuzzy logic - approximate calculations, granulation information on the calculation of the words.
- — Neurocomputing - learning, adaptation, classification, system modeling and identification.
- — Evolutionary computation - synthesis, configuration and optimization using systematic random search and evolution.
- — Probability calculations - uncertainty management, trust networks, chaotic systems, prediction.

Traditional computer calculations (hard computing) and the measurement is too precise for the real world. There are two classes of problems for soft computing and measurement: First, there are problems, for which complete information can not be obtained, and, secondly, the problem is not defined adequately.

Each individual component of soft computing has a number of internal problems, mutual compensation and which creates a synergistic effect. Neural networks solve the problem of approximation or classification, and setting the balance is due to the training.

[a] Corresponding author: gsv@mail.tsu.ru

© The Authors, published by EDP Sciences. This is an open access article distributed under the terms of the Creative Commons Attribution License 4.0 (http://creativecommons.org/licenses/by/4.0/).

But the knowledge distributed across many affairs and results of the neural network seem inexplicable to the user, that is needed explication of knowledge. Fuzzy inference systems, on the contrary, have explicit knowledge in the form of products and make it easy to build explanations protocol. However, this clarity is achieved by a long and creative in nature prior knowledge extraction process and subsequent debugging set of rules. The inability to extract knowledge from data, settings membership functions during training, automatic reduction of the rules - the problem of inference systems. Application of genetic algorithms requires a prior study of the problem in order to select the model of probability, the formulation of the objective function. Selection soft computing individual components as a base typically defines the overall system architecture of a hybrid (a hybrid system in this case it refers to the "soft" system).

In [2] it is noted that in dealing with classification tasks to measure the effectiveness of multidimensional objects in recent years there has been an increase in the number of so-called hybrid algorithms, in which two or more soft computing technology integrated (fuzzy logic, neural networks, genetic algorithms) to improve the overall performance algorithms. Fuzzy Neural Network - a fuzzy inference system by the methods of neural networks. The conclusion is based on the apparatus of fuzzy logic, but the parameters of membership functions are configured using neural networks learning algorithms. Therefore, for the selection of the parameters of such networks use the method of back propagation, initially proposed for training of multilayer perceptron. For this fuzzy control module is in the form of the multilayer network, consisting usually of four layers: the layer of input variables fuzzyfication, the layer of activation conditions aggregation, the layer of fuzzy rules aggregation and the output layer (Figure 1).

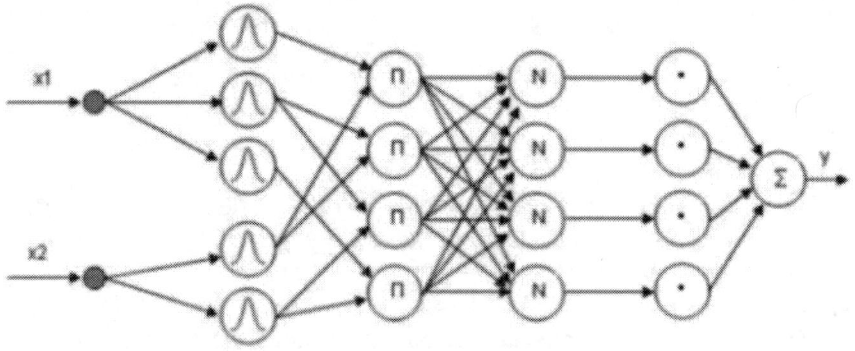

Figure 1. An example of a hybrid neuro-fuzzy.

Currently classifiers built using fuzzy neural networks, expert systems, and other methods, allow to obtain sufficiently accurate results, however, lead to the construction parametrically complicated, cumbersome models with a large number of neurons, and, consequently, to increase the amount of memory and time the network. The analysis of the literature reveals a contradiction, which is the lack of complete solutions for the construction of simple and compact classifier providing a compromise between accuracy and interpretability of classification rules.

2 Purpose and Method

The purpose of this research is to develop algorithms for constructing a hybrid neuro-fuzzy classifier for analyzing and measuring the intensity of the emotion of the robot, characterized by a set of disparate parameters, which provides not only a high classification accuracy, but also consists of a compact and interpreted a set of classification rules.

3 Results and discussion

In the domestic and foreign literature describes the use of decision trees as a powerful evolutionary methodology for solving problems of classification and regression [3]. The following simple example (on the IBM SPSS Modeler Infocenter Site) shows a decision tree for making a car purchase (Figure 2).

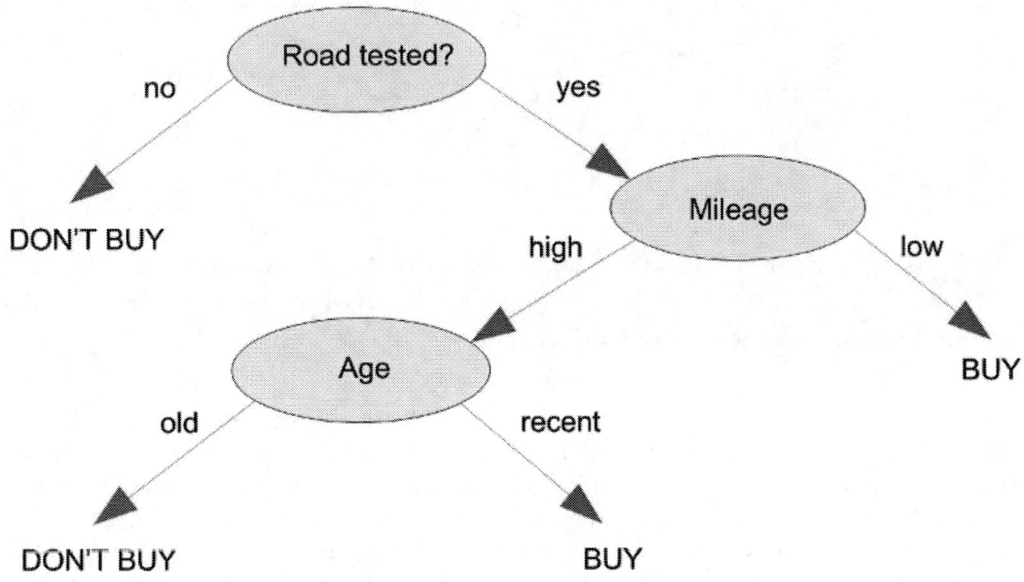

Figure 2. A decision tree for making a car purchase.

In Figure 2 shows the classification tree iris. Classification is into three classes and passes in the parameters: length\thickness of the sepal (*SepalLen, SepalWid*) and length\thickness of the petal (*PetalLen, PetalWid*). As you can see, each node should belong to the class (depending on which items got more in this node), the number got there of observations N and the number of each class. Also not the leafy tops, is the condition of the transition is one from a child. Respectively, for these conditions and breaks the sample. As a result, this tree almost perfectly (6 out of 150 incorrectly) klassificeret the original data on which it was trained.

If the monitoring condition is true, then the transition on the left edge, if false - on the right. Typically, each node includes examining one independent variable. Sometimes in the tree node, two independent variables are compared to each other or determined by some function of one or several variables.

If a variable that is checked at the node that takes categorical values, each possible value corresponds to a branch emerging from a node of the tree. If the value of the variable is a number, verified by more or less than this value a constant. Sometimes the area of numeric values is divided into intervals (the test value falls in an interval). The leaves of the trees correspond to the values of the dependent variable, i.e. classes.

A common disadvantage of the development of traditional decision trees is the requirement of certainty of the input data, which is achieved by applying weighted average values of input parameters that may result in significantly displaced resulting point estimates. It is also obvious that the requirement of determinacy of input data is an unjustified simplification of reality, as real complex objects of analysis are characterized by many uncertainties: uncertainty of initial data, the uncertainty of the external environment, the uncertainties associated with the nature, options, and model implementation or the object's state.

To improve the accuracy of classification and decision-making, the author developed the algorithm for constructing neuro-fuzzy decision tree having the property of adapting parameters using neural network learning. As input data we use a fuzzy set with the membership function of the following form (Figure 3).

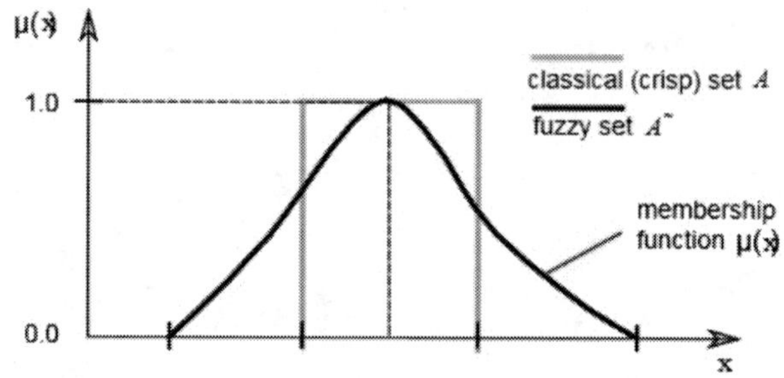

Figure 3. The membership function of fuzzy set *A*.

For any set *X*, a membership function on *X* represent fuzzy subsets of *X*. The membership function which represents a fuzzy set $A^{\sim}$ is usually denoted by μ_A. For an element *x* of *X* the value $\mu_A\,(x)$ is called the membership degree of *x* in the fuzzy set $A^{\sim}$. The membership degree $\mu_A\,(x)$ quantifies the grade of membership of the element *x* to the fuzzy set $A^{\sim}$. The value *0* means that *x* is not a member of the fuzzy set; the value *1* means that *x* is fully a member of the fuzzy set. The values between *0* and *1* characterize fuzzy members, which belong to the fuzzy set only partially.

In the forward loop fuzzy decision trees are built based on the fuzzy ID3 algorithm [3]. Algorithm The Iterative Dichotomizer 3" (ID3) is one of the earliest algorithms for learning decision trees (1985), using recursive partitioning subsets in the nodes of the tree on one of the selected attributes, and is widely used in problems of decision-making. In General it is as follows – starts with the root of the tree, which contains all examples of the training set. To separate it selects one of the attributes, and for each the values it accepts the construction branch, and you create a child node in which distributed its containing record.

The procedure is repeated recursively until then, until the knots will not be only examples of one class, then they will be declared the leaves and the branching stops. The most problematic step here is the choice of which attribute should be split. The classic ID3 algorithm uses a criterion increase information or decrease in entropy.

In the loop feedback parameters [4, 5] of fuzzy decision trees proposed to adapt based on the stochastic gradient algorithm by traversing back from leaf to root nodes. With this strategy in the course of adaptation parameters, the hierarchical structure of fuzzy decision trees intact, and the neural network back-propagation algorithm directly on the structure of fuzzy decision trees improves the accuracy of training without compromising interpretability.

Note that in order to reduce input information and presenting it in understandable to interpret the form of the algorithm was developed gazifikatsii input data based on the Kohonen self-organizing maps, the Value of fuzzy input parameters, described in linguistic terms, can be represented functionally by a set of membership functions. For the given set

 DOI: 10.1051/matecconf/20167901065

of functions, we find several new fuzzy sets, which are considered as the result of clustering the original data for describing a plurality of membership functions (Figure 4).

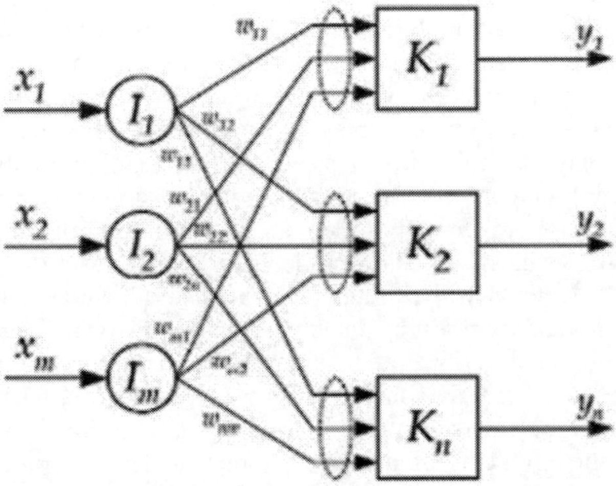

Figure 4. The algorithm fuzzy- input data by Kohonen self-organizing maps .

4 Insights

On the basis of the conducted research we can draw the following conclusions. To solve the problem of multidimensional measuring the intensity of the emotions generated by the emotional agent for the robot to vliyaniyem diverse input parameters, as a neuro-fuzzy system can use not the classical 4-layer network (Figure 1), and neuro-fuzzy decision tree structure which is constructed by the method of self-organization by fuzzy ID3 algorithm and for tuning of parameters (membership functions and identification of sheets) are used to reverse the cycles of iteration neural network method with back propagation mistakes. This hybrid neuro-fuzzy system based on a combination of methods of "sequential hierarchy" (decision trees) + fuzzy logic + neural networks". Hence, in weakly structured environments multi-criteria analysis of the appropriate joint use of these methods for modelling of emotional agent robot, is used to support decision-making in robotic systems, as a compensation for the frailties of the individual component of the methodology of soft computing expands the possibilities for building intelligent systems with incomplete input information. Such hybrid neuro-fuzzy system have the most powerful cognitive capacities (modelling sensation, perception, recognition, learning and remembering patterns with the purpose of revealing of knowledge from data). They find a substantially wider field of application than other methods of synthesis of fuzzy sets and neural networks. This is because such systems enable the best use of the "strong" side of a fuzzy system interpretability of knowledge) and neural networks (the ability to learn on the data).

5 Conclusion

In the course of the study of neuro-fuzzy systems were obtained the following results:

1. The developed method of synthesis of neuro-fuzzy classifier based on decision tree, neural network includes adapting parameters of the constructed decision tree. The model focuses on the input vector data does not depend on the mutual ratios of the elements of the

input vectors does not use the learning procedure used in neuro-fuzzy models with the teacher, and eliminates the problem of choice of their structure.

2. Determined that the proposed neural network method for adapting parameters of a decision tree based on the backpropagation of the signal from the leaves of the tree to the root, in contrast to the known optimization problems, improves the classification accuracy of fuzzy decision trees a fuzzy system with direct distribution of signal) without prejudice to the interpretation, preserving the hierarchical structure of the tree.

3. In contrast to the known method of synthesis of neuro-fuzzy networks the training set the method of display the rules in its structure [6], leading to the construction of a parametrically complex, cumbersome model with a large number of neurons, and, consequently, to increase the amount of memory and uptime of the network, the method of synthesis of neuro-fuzzy classifier based on decision tree allows to synthesize models with a small number of elements with a high level of generalization, with the possibility of linguistic interpretation of the result by displaying diagnostic decision rules "If ... then".

In the General case described neuro-fuzzy system allows automated to solve the difficult problem of multi-criteria analysis and measurement of analysis objects having an arbitrary shape, with the intersection of classes is characterized by high accuracy and efficiency for solving problems of decision support under uncertainty, the low training time, the reliability against noise and distortion of information.

Acknowledgments

The paper was written as part of the research project No. 8.2.31.2015, carried out with the support of the Program "Research Foundation of Tomsk State University named after D.I. Mendeleev" in 2015 – 2016., grant RFBR No. 16-29-04388\16.

References

[1] L.A. Zadeh, Com. of the ACM **37**, 77 (1994) doi: 10.1145/175247.175255

[2] S.V. Gorbachev, V.I. Syryamkin, *Neuro-fuzzy techniques in intelligent systems of processing and analysis of multidimensional information* (Tomsk state University, Tomsk, 2014)

[3] L. Breiman, J.H. Friedman, R.A. Olshen, C.J. Stone, *Classification and regression trees* (Wadsworth & Brooks/Cole Advanced Books & Software, Monterey, 1984)

[4] D.V. Shashev, S.V. Shidlovskiy, Optoelectronics, Instrumentation and Data Processing **51**, 227 (2015) doi: 10.3103/S8756699015030036

[5] D.V. Shashev, S.V. Shidlovskiy, V.I. Syriamkin , A.V. Yurchenko, IOP Conference Series: Materials Science and Engineering **81**, 012101 (2014) doi: 10.1088/1757-899X/81/1/012101

[6] N. Yarushkina, Int. J. General Systems **30** , 71 (2001)

E-GOVERNMENT IMPLEMENTATION

Phonepaseuth Solinthone[1,2a], *Tatyana* Rumyantseva[1]

[1]National Research Tomsk State University, 634050, Tomsk, Russia
[2]Ministry of Science and Technology, 2279, Lao PDR

Abstract. Information and communication technologies (ICTs) are playing an increasingly vital role in the daily lives of people, revolutionizing work and leisure and changing the rules of doing business. In the realm of government, ICT applications are promising to enhance the delivery of public goods and services to citizens not only by improving the process and management of government, but also by redefining the traditional concepts of citizenship and democracy. The ICTs' effect on societies are both far-reaching and uneven. On the one hand, ICT is fueling the transition from industrial-based economies to knowledge-based societies. On the other hand, ICT still has little or no impact in the lives of people in many countries. This wide disparity in the impact of ICT around the world today underscores the uneven progress of economic development. It also highlights the critical role of government in the information age. The goal of this article is to provide a brief overview of E-government studies and their evaluation, to clarify the major issues, which can make E-government to be more effective, transparent and responsive. The article considers the significance of contemporary E-government activity.

1 What is e-government?

Definitions of E-government range from "the use of information technology to free movement of information to overcome the physical bounds of traditional paper and physical based systems" to "the use of technology to enhance the access to and delivery of government services to benefit citizens, business partners and employees." The common theme behind these definitions is that E-government involves the automation or computerization of existing paper-based procedures that will prompt new styles of leadership, new ways of debating and deciding strategies, new ways of transacting business, new ways of listening to citizens and communities, and new ways of organizing and delivering information [1].

Ultimately, E-government aims to enhance access to and delivery of government services to benefit citizens. More important, it aims to help strengthen government's drive toward effective governance and increased transparency to better managing country's social and economic resources for development.

The key to E-government is establishment of a long-term, organization-wide strategy to constant improvement of operations with the view of fulfilling citizen needs by

[a] Corresponding author: phonepaseuth.slt@gmail.com

© The Authors, published by EDP Sciences. This is an open access article distributed under the terms of the Creative Commons Attribution License 4.0 (http://creativecommons.org/licenses/by/4.0/).

transforming internal operations such as staffing, technology, processes and workflow management.

1.1 What are the types of E-government transactions?

E-government services focus on four main customers: citizens, business community, government employees, and government agencies. E-government aims to make interaction with citizens, businesses, government employees, government agencies and other governments more convenient, friendly, transparent, inexpensive and effective.

In an E-government system, individuals are able to initiate a request for a particular government service and then receive that government service through the Internet or some computerized mechanism. In some cases, the government service is delivered through one government office, instead of many. In other cases, a government transaction is completed without direct in-person contact with a government employee.

1.2 What are the specific types of services types of E-government?

There are four types of E-government services:

- Government-to-Citizen (G2C).
- Government-to-Business (G2B).
- Government-to-Employee (G2E).
- Government-to-Government (G2G).

1.2.1 Government-to-citizen

G2C includes information dissemination to the public, basic citizen services such as license renewals, ordering of birth/death/marriage certificates and filing of income taxes, as well as citizen assistance for such basic services as education, health care, hospital information, libraries, and the like.

Case 1. Singapore's E-citizen Portal: A Case Study in G2C Transactions

Through Singapore's E-citizen portal, Singaporeans are able to access about 1,600 E-services pertaining to business, health, education, recreation, employment, and family. Of this, 1,300 E-services are completely transacted by citizens with government online. The E-citizen portal is divided into categories based on the real-life needs of every individual, with every single ministry and statutory board providing E-services through the same portal. Singaporeans thus have one-stop access to government services; they are spared having to navigate through the bureaucratic jungle. A few of the popular E-services are the following: submitting application forms for purchase of apartments, searching for school information, employment search, career development, and voter registration. As of June 2002, about 77% of public services deemed feasible for E-delivery were enabled for online delivery [2].

1.2.2 Government-to-business

G2B transactions include various services exchanged between government and the business community, including dissemination of policies, memos, rules and regulations. Business services offered include obtaining current business information, downloading application forms, renewing licenses, registering businesses, obtaining permits, and payment of taxes. The services offered through G2B transactions also assist in business development, specifically the development of small and medium enterprises (SME). Simplifying application procedures that would facilitate the approval process for SME requests would encourage business development.

On a higher level, G2B services include E-procurement, online government-supplier exchange for the purchase of goods and services by government. Typically, E-procurement Web sites allow qualified and registered users to look for buyers or sellers of goods and services. Depending on the approach, buyers or sellers may specify prices or invite bids. E-procurement makes the bidding process transparent and enables smaller businesses to bid for big government procurement projects. The system also helps government generate bigger savings, as costs from middlemen are shaved off and purchasing agents' overhead is reduced.

1.2.3 Government-to-employee

G2E services encompass G2C services as well as specialized services that cover only government employees, such as the provision of human resource training and development that improve the bureaucracy's day-to-day functions and dealings with citizens.

1.2.4 Government-to-government

G2G services take place at two levels: at the local or domestic level and at the international level. G2G services are transactions between the central/national and local governments, and between department-level and attached agencies and bureaus. At the same time, G2G services are transactions between governments, and can be used as an instrument of international relations and diplomacy.

2 The goals of E-government

2.1 What are the goals of E-government?

The Working Group on E-government in the developing world has identified five broad categories of goals commonly pursued for E-government. E-government is a means to accomplish these broader social goals, goals that move beyond mere efficiency of government processes to that of overall reform and development. The goals are not listed in any particular order of importance, as each country must determine its priorities in E-government.

1. Creating a better business environment.
2. Customers online, not in line.
3. Strengthening good governance and broadening public participation.
4. Improving the productivity and efficiency of government agencies.
5. Improving the quality of life for disadvantaged communities.

3 E-government and human development

3.1 How do ICTs facilitate good governance?

ICT is an enabler of efficient and effective function of government. In turn, more efficient government function enables improved and better governance.

The Cristal Web site was specifically created to fulfill the mandate of a law that requires that the state makes available "to whatever institution or interested person" the following information related to the administration of public funds:

- Execution of budgets, to the lowest level of disaggregation.
- Purchase orders and public contracts.
- Financial and employment data concerning permanent and contracted staff.
- An account of the public debt, including terms, guarantees, interest costs, etc.; outstanding tax and customs obligations of Argentine companies and people.
- Regulations governing the provision of public services.
- All information necessary for the community control of social expenditures.

A primary goal of the Cristal program is to create a better-informed citizenry that can provide control over political representatives more effectively. While the content of the Web site is directed to all citizens, journalists are particularly important audience of the site, as newspapers and television enable much wider dissemination of its contents.

3.2 What are the implications of equality of access to government information and services?

The implementation of E-government facilitates citizen participation in governance by increasing access channels to government. It broadens opportunities for citizen participation, opening new channels of communication between constituents and their representatives and bringing marginal groups (i.e., women, physically challenged, indigenous peoples) into mainstream participatory channels.

However, as E-government initiatives have the potential to democratize the delivery of basic services and "level" the effects of development, these same initiatives can also further citizens distance from government and even deepen existing disenfranchisement. Policy-makers, in trying to achieve development goals through E-government, should consider projects that would deliver the most benefits to the broadest number of people.

3.3 What is digital democracy?

Digital democracy is a term used when the use of information and communications technology enhances citizen participation in the democratic process. It is the computerization of political discourse, policy-making and the political process with the end in view of increasing, enhancing, and deepening citizen participation in the policy- and decision-making processes of government through a spectrum of activities: electoral campaigns, voting, consultation and participation in the policy process, public opinion polling, and communication exchange between elected officials and constituents.

While opportunities are increasing for citizens to be more actively involved in the policy-making and decision-making process, much of government decision-making is still hidden from the wider public. ICT integration in government processes facilitates greater openness, transparency, and accountability. As more information is given to the public, greater citizen engagement in the overall policy process is considered as necessary as greater accountability for public officials.

Case 2. Improving Service Delivery To Women Through ICT: The Women's Services Portal in British Columbia, Canada

To improve service delivery to women, the Government of British Columbia has created a portal dedicated to delivering services for women. The women's services site [3] is embedded within the British Columbia government portal called "BC Connects" [4].

This Web site provides information, government assistance and training to women in general, as well as to aboriginal, immigrant and minority women living in the province of

British Columbia. The services include application for business loans and childcare subsidy, employment preparation training, job matching, health information services, and counseling and legal aid.

4 Advantages and disadvantages and the challenges of E-government implementation

4.1 Advantages of E-government

Advantages and benefits of E-government implementation are the same for both developed and developing countries. However, E-government applications have many benefits for citizens, business and government entities. E-government applications allow people, businesses, and government sectors to access to available government information 24 hours a day, 7 days a week, which improves the quality of these services. Implementation of E-government can reduce cost and levels of organizational processes by streamlining and re-organizing operating procedures. Moreover, use of E-government systems can improve the performance of government agencies because they will be able to deliver the public service effectively and efficiently for all customers. In addition, E-government has great benefits regarding economizing and improving of governments' service operations, including efficiency, reduced transactional costs, increase the transparency and increased services for citizens. Furthermore, we can identify E-government benefits as follows:

- Reduction of customers' and organizations' time, effort and costs.
- Improvement of service delivery and citizens' satisfaction.
- Increase of users' ICT skills, internet knowledge and computer usage.
- Creation of new business and work opportunities.

In addition, E-government implementation has many advantages such as:

- Improvement of efficiency of government agencies in data processing.
- Improvement of services through better understanding of users' requirements, thus aiming for seamless online services.
- Sharing information and ideas between all government agencies and department to build one mega database.
- Assisting in reaching government's economic policy objectives by promoting productivity gains inherent in ICT and e-commerce.
- Improvement of transparency, accuracy and facilitating information transfer between government and customers.
- Assisting in building trust between governments and citizens.
- Essential factor in good governance by using internet-based strategies to involve citizens in the policy process, illustrating government transparency and accountability.

To conclude, it is clear that implementation of E-government saves resources, effort and money but it can also extensively increase service quality levels and reduce time spent in government departments. [5]

4.2 Disadvantages of E-government

In spite of some advantages gathered from successful E-government implementation, there are also some disadvantages. Some of these disadvantages are briefly discussed in this paper below and include lack of equality in public access to the Internet, lack of trust and cyber-crime, hyper- surveillance, false sense of transparency and accountability and costly infrastructure.

4.2.1 Lack of equality in public access to the Internet

Studies have shown that there is potential for reduction in the usability of government online due to factors such as the access to Internet technology and usability of services and the ability to access to computers (MSG, 2008), plus literacy of the users and the ability to use the computer.

There are users who are illiterate (do not know how to read and write), who would need assistance. An example would be the elderly (senior citizens). Generally, senior citizens do not have much education and they would have to approach a customer service officer for assistance. Bwalya et. al. echoes the same sentiments that some of the disadvantages of E-government implementation could be the likelihood of excluding citizens and businesses that have no access to technology due to the digital divide (e-exclusion).

4.2.2 Lack of trust and cyber crime

Even though the level of confidence in the security offered by government web sites are high, the public are still concerned over security, fear of spam from providing email addresses, and government retention of transaction or interaction history. Similarly, Varros posits this too that, despite the efforts of government agencies to ensure the safety of citizen's personal data, E-governance websites are still liable to attack from hackers. Personal data can be exposed and there is less trust to how the information is kept secure and whose hands it lands on. This is also pointed out by Bwalya et. al. (2012), that one of the disadvantages could be generally a lack of engagement due to the anticipated low levels of trust by citizens of the E-government platforms.

4.2.3 Hyper surveillance

Sharma, Bao and Qian [6] hints that even though developing countries attempt to improve public services through E-government implementation, they also turn to increase control over people through E-government. Augmented contact between government and its citizens is bi- directional (it goes both ways).

Once E-government starts to develop and become more sophisticated, citizens will be forced to interact electronically with the government on a larger scale. This could potentially lead to a lack of privacy for civilians as their government obtains more and more information on them. There are very real concerns about turning over much information to the government by the citizens or businesses. While the government may see like a benevolent organization, it is possible this may not be true in the future or that outside governments/or parties could use this information to harm the citizens of this country and the businesses.

4.2.4 False sense of transparency and accountability

Those against E-government argue that online governmental transparency is dubious because the governments themselves maintain it. Information can be added or removed from the public eye. To this day, very few organizations monitor and provide accountability for these modifications. Majozi (2008) also highlights that hidden agendas of government groups that could influence and bias public opinions are among disadvantages of E-government implementation.

4.2.5 Costly Infrastructure

According to Vassos [7], an efficient E-government system requires all citizens or at least the vast majority to have access to the Internet. Therefore, Internet-enabled devices, hardware such as routers, and connection infrastructure are essential to connect to government websites.

Additionally, public sector agencies need advanced servers and security systems to cope with vast amounts of information and fire walls for complex cyber threats. All these requirements constitute a costly investment, far beyond the reach of less developed economies.

Ngulube [8] concurs to this that one of the disadvantages of E-government implementation is technological costs. The costs include infrastructure development, interoperability of the technologies, permanent availability and preservation, education and training of operating and using the technology, cost structures and benchmarking. The cost cannot only be disadvantageous to the government, but also to the citizens and the businesses who cannot afford.

4.3 The Challenges of E-government

4.3.1 Who pays for E-government?

Like any government infrastructure project, E-government can be done in phases and the costs of implementation will depend on current infrastructure availability, supplier and user capabilities, and mode of service delivery (whether through the Internet or through telephone hotlines and one-stop shops). The more complicated and sophisticated the kind of services the government wants to offer, the more expensive it is. [9]

Governments should focus on small, self-financing or outsourced projects. Because E-government projects must be financially sustainable, there must be a revenue/ cost-reduction model in place from the beginning. Smaller projects with a clear revenue-generation strategy and minimal initial investment are the most likely to be sustainable over the long term. For instance, web sites are one of the easiest and cheapest ways to achieve high impact E-government with a minimum of investment.

E-government projects are, more often than not, long-term endeavors, requiring large capital infusion in software, hardware, infrastructure and training. A viable financing plan should not only pay for the immediate needs to jumpstart E-government; it must also consider its long-term financing options for the sustainability of the project.

There are various business models for funding E-government projects, and the private sector plays a critical role in these. Under partnership arrangements, the private sector builds, finances and operates public infrastructure such as roads and airports, recovering costs through user charges. Various financing schemes exist from soft and development assistance loans from donor/multilateral aid agencies to partnerships and outsourcing deals with private third party vendors under special financing schemes (e.g., the Build-Operate-Transfer or BOT scheme) that can minimize the initial cost to government.

BOT and its variants are usually the favored financing models/arrangements for government projects that require large and immediate financing from the private sector. Under BOT, the private sector designs, finances, builds, and operates the facility over the life of the contract. At the end of this period, ownership reverts to the government. A variation of this is the Build-Transfer-Operate (BTO) model, under which title transfers to the government when construction is completed. Finally, with Build-Own-Operate (BOO) arrangements, the private sector retains permanent ownership and operates the facility on contract.

Cooperation, rather than competition, with the private sector can facilitate effective E-government. Government can encourage private sector investment by complementing and supporting private sector efforts rather than duplicating them. The key to E-government is to improve citizen access to service delivery, not expand the role of government. Government should not attempt to create products, services where public-private partnerships or private service providers can adequately provide these products, and services more efficiently and effectively.

4.3.2 How do you get the wider public to use E-government services?

Any E-government policy must consider a citizen-centered approach. This means that E-government should be an end-user or demand-driven service. However, many citizens do not use e-government for several reasons, among these unfamiliarity with ICT, lack of access, lack of training, and concerns about privacy and security of information. While E-government may provide ease and convenience in the delivery of public services, and offer innovative government services, none of these will prompt citizen use unless the concerns mentioned above are resolved.

4.3.3 Why are security and protection of privacy important?

Security generally refers to the protection of information system assets and control of access to information. Security policies and strategies are context- and information-specific.

Privacy refers to the right for information attributed to an individual (also called "nominal information") to be treated with an appropriate level of protection. Information privacy protection laws are often put in place to regulate this.

Protecting the privacy of citizens and assuring them that their personal information will not be compromised is critical in E-government because this is the key to user trust. Without this assurance, no one will use E-government services.

5 An overview of the 2014 united nations E-government survey

The Department of Economic and Social Affairs produce the United Nations E-government Survey every two years. It is the only report in the world that assesses the E-government development status of the 193 United Nations Member States. It serves as a tool for decision-makers to identify their areas of strength and challenges in E-government and to guide E-government policies and strategies. The publication also highlights emerging E-government trends, issues and innovative practices, as well as challenges and opportunities of E-government development. Each chapter provides an analysis of the Survey's data, as well as highlights strategies, challenges and opportunities. The Survey is intended for government officials, academics, intergovernmental institutions, civil society organizations, the private sector and citizens at large. Table 1 shows the E-government Development Index (EGDI) of the year 2014. [10]

Table 1. The E-government Development Index (EGDI) 2014. Top ten EGDI Rank with Russia Federation, China, Lao PDR .

Rank	Countries List	EGDI Percentage
1	Republic of Korea	0.946
2	Australia	0.910
3	Singapore	0.908

4	France	0.894
5	Netherlands	0.890
6	Japan	0.887
7	United States of America	0.875
8	United Kingdom	0.870
9	New Zealand	0.864
10	Finland	0.845
27	Russian Federation	0.730
70	China	0.545
152	Lao People's Democratic Republic	0.266

5.1. Case study of E-government of Korea

▪ Korea has actively pursued E-government as a crucial means to make its government more competitive, by leveraging the world's best information and communications technology including broadband Internet.
▪ After laying the groundwork for E-government, including the National Basic Information System (NBIS) computer networks in the 1980s and streamlining of applicable laws and institutions in the 1990s, the Korean government made the implementation of E-government a major national agenda for the 2000s. It has concentrated on 11 major tasks for E-government (2001~2002) and 31 major tasks for the E-government roadmap (2003~2007). As a result, E-government has become firmly established in all areas of Korean government.
▪ Korean E-government has produced visible results: both efficiency and transparency of administrative work have been significantly improved; administrative civil services have been greatly enhanced; and opportunities for people to participate in the policy-making process have been expanded.
▪ Accordingly, the effectiveness of the E-government of Korea is widely acknowledged by the international community and various E-government systems are being exported to foreign countries. The 2010 UN Global E government Survey shows that Korea is ranked first among all the member countries, given the highest possible scores in the categories of Online Service Index and the E-participation Index.
▪ Korea is now promoting E-government that is focusing on utilization and convergence by consolidating services to maximize the convenience of users and implementing a seamless digital cooperation system connecting government departments and agencies, in order to improve the overall quality [11].

5.1.1. Achievements

A. Improvement of efficiency and transparency of administrative work

- Use of electronic documents has become standard practice, and most administrative business such as personnel management, finance, and procurement are being handled electronically. This greatly and innovatively enhances the overall efficiency of government administration.
- All central administrative organizations have introduced a standardized business process system (On-nara) to record all decision-making procedures of the government, increasing transparency of administration significantly.

B. Provision of people and company-focused administrative services

- The age of civil service processing at home has been fully ushered in by implementing integrated online civil service processing channels and enabling notification, filing, and payment of taxes via the Internet.
- Various corporate activities are supported efficiently by means of a single window for corporate support and processing of logistics, customs clearance, and trading online.

C. Strengthening of communications with the people about government policies

- People can now participate in the government decision-making process more easily through a single window linked to all administrative organizations that provides comprehensive civil services and receives public suggestions.
- People can request and check a wide range of administrative information online easily. It is possible to check national records anytime, anywhere.

D. Increased efficiency of information resource management

- The government's integrated computing center comprehensively manages all information systems of the government, improving its capability to respond to system errors and security threats.
- Introduction of Enterprise Architecture (EA), which acts as a comprehensive informatization blueprint enables systematic management of E-government.

5.1.2. E-government Best Practices

- Electronic Procurement Service (www.g2b.go.kr).
- Electronic Customs Clearance Service (portal.customs.go.kr).
- Comprehensive Tax Services (www.hometax.go.kr).
- Internet Civil Services (www.egov.go.kr).
- Patent Service (www.kiporo.go.kr).
- e-People: Online Petition & Discussion Portal (www.epeople.go.kr).
- Single Window for Business Support Services (www.g4b.go.kr).
- On-nara Business Process System (BPS).
- Shared Use of Administrative Information (www.share.go.kr).
- National Computing & Information Agency (NCIA).

6 Conclusion of world E-government

Public administration, being the cornerstone of governments' work is essential for improving peoples' lives. As illustrated in this chapter, amidst the economic, social and environmental challenges, E-government is continuing to play an important role in enabling the delivery of quality public services that meet citizen needs and goals by transforming how the public sector works.

Countries must decide the level and extent of their E-government initiatives based on their specific national development context.

In addition to effective planning and deployment of E-services, governments may consider enhancing their ICT infrastructure and raise the level of human capital, including improvement of the ICT literacy of citizens, to make use of the new technologies to realize the full benefits of online and mobile services. This should go hand in hand with capacity development of leadership in E-government and public servants as facilitators of online public services.

These countries are ahead, because as the world economy begins to recover from one of the worst economic crises in decades, information and communication technologies (ICT) is bound to play an increasingly prominent role as a key enabler of renewed and sustainable growth, given that it has become an essential element of the infrastructure underpinning competitive economies. ICT will continue spreading its revolutionary power to modernize economies and societies and improve living conditions and opportunities around the world.

ICT performance will remain crucial not only for developed countries for sustaining and enhancing their innovation potential and long-term competitiveness, but also for middle-income and developing countries in fostering structural transformations, increasing efficiency as well as reducing the digital, economic, and social divides within their territories and more advanced economies. If ICT plays a central role in ensuring economic sustainability, it can and must play an equally central role in promoting environmental and social sustainability, both as an industry and as a key element of enabling infrastructure. In mind opinion, I believe sustainability, in all its components, should be a priority for policymakers, businesses, and civil society alike to foster a more inclusive and crisis-resilient world. ICT and the ICT industry must now rise to increase information of their countries, improve the quality of education, increase knowledge for their people, developments of economic, social, political, culture and improve management systems by using ICT.

Acknowledgements

The paper was written as part of the research project No. 8.2.31.2015, carried out with the support of the Program "Research Foundation of Tomsk State University named after D.I. Mendeleev" in 2015 – 2016., grant RFBR No. 16-29-12858.

References

[1] E-government wikibooks URL: https://en.wikibooks.org/wiki/E-government/
[2] Case 1. Singapore's e-citizen Portal URL: http://www.ecitizen.gov.sg
[3] Case 2. The women's services site URL: http://www.cserv.gov.bc.ca/womens_services/links/index.htm
[4] The British Columbia government portal site URL: http://www.gov.bc.ca
[5] S.R. Joseph, International journal of technology marketing 5(9), 18 (2015)
[6] G. Sharma, X. Bao, W. Qian, International Journal of Advanced Research in Computer Science and Software Engineering **2**, 19 (2012)
[7] T. Vassos,Advantages and disadvantages of E-governance URL: Available at http://www.ehow.com/info_8473642_advanta ges-disadvantages egovernance.html
[8] P. Ngulube, The Nature and Accessibility of E-government in Sub Saharan Africa. International Review of Information Ethics **7**,1 (2007)
[9] M. Alshehri, S. Drew, Implementation of e-Government: advantages and challenges URL: http://www98.griffith.edu.au/dspace/bitstream/handle/10072/40620/72631_1.pdf?sequence=1
[10] UN E-government Survey 2014 URL:https://publicadministration.un.org/egovkb/portals/egovkb/documents/un/2014-survey/e-gov_complete_survey-2014.pdf
[11] Case Study of E-government of Korea URL: http://unpan1.un.org/intradoc/groups/public/documents/UNGC/UNPAN043625.pdf

MATEC Web of Conferences 79, 01067 (2016) DOI: 10.1051/matecconf/20167901067
IME&T 2016

AUTOMATIC FIRE DETECTION SYSTEM BASED ON CONTOUR ANALYSIS VIDEO IMAGES

Irina Egoshina[1], *Dmitry* Titov[2,a], *Anton* Stuchkov[3]

[1] Volga State University of Technology,424000, the Republic of Mari El, Yoshkar-Ola, Russia
[2] Southwest State University, 405000, Kursk, Russia
[3] Tomsk State University, 634050, Tomsk, Russia

Abstract. A general structure of an automatic fire detection system and the results of a study video analysis algorithms for detecting based on contour analysis. Given contour extraction processes objects in the image and make decisions about the discovery of fire or the absence thereof. Experimental researches on the video images of the forest area done with the use of the developed structure of the automatic fire detection system.

1 Introduction

The efficiency of the fire problem depends from the early detection of the fire place. Usually space monitoring and aviation are using for the decision of this actual problem. Every method has its benefits and weaknesses, but only the automatization of the work of the territory monitoring system, which should determine and automatically recognize the fire place in the early beginning, and then information of the necessary services can solve the problem in the area of fire safety. That's why the development of the automatic fire detection system based on modern intellectual technologies is a very important task nowadays.

The development of modern technologies of the technical vision allows determining the fire on digital images in the automotive way. Such systems with the use of different sensors get the data about the environmental condition and analyze the fire probability by the determination of such dangerous visual objects as smoke or fire without the operator work and inform him just after getting the final decision about real danger [1, 2].

2 Structure of the automatic system and methods of determination of dangerous fire factors

Developed system is made for determination of the fire and its' fireplace in the real-time mode. The existing info communication framework of the data transfer can be used for the system functioning.

Enlarged structure of the automation system consists of these functional blocks (Figure 1):

[a] Corresponding author: amazing2004@inbox.ru

© The Authors, published by EDP Sciences. This is an open access article distributed under the terms of the Creative Commons Attribution License 4.0 (http://creativecommons.org/licenses/by/4.0/).

- system of sensors;
- software block;
- fire alarm system;
- PC operator.

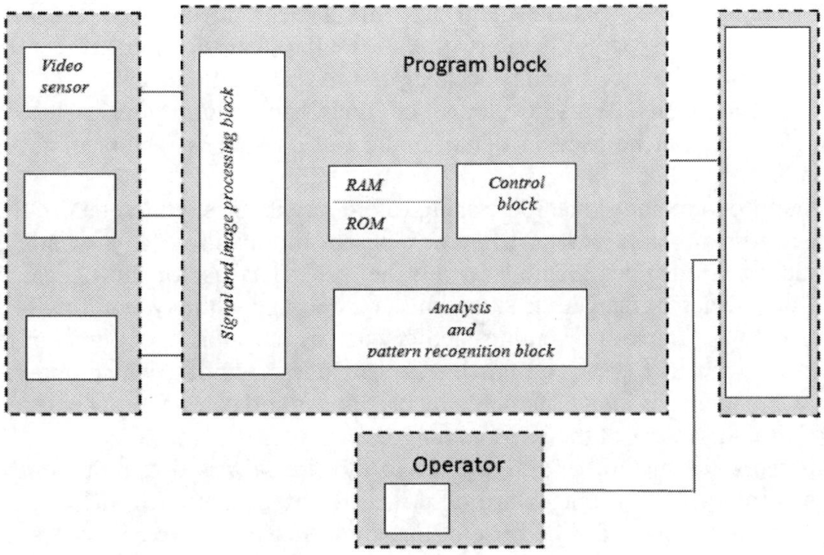

Figure 1. Enlarged structure of the automatic fire detection system.

The system of sensors, which includes video sensors, infrared sensors, ultraviolet sensors, controls such parameters as the increasing of the air temperature, appearance of the open fire, detection of certain changes in the video image. Because of the simultaneous use of the ultraviolet and infrared channels the amount of fireplaces, their location and fire intensity can be detected. The use of three types of determination channels and of the special algorithm of the signal and image processing will decrease the possibility of false operations and more effective control the situation in the zone of the system work [3].

Software block has such functions as the preliminary signal and image processing from the sensors; archiving of the video image; storage of the database for determination of the fire objects; analysis and fire detection. Control block allows to get the final decision of the start of the fire on the intellectual level, i.e. the control signal will go to the fire alarm system after analyzing of several object parameters.

The task of the operator in this process is to react timely on the system signals.

Basic fire factors which can be registered by the video sensor and preventing factors are described down [4,5]:

- Open flame differs highly from the backlight by the intensity and frequency of the electromagnetic radiation. Sometimes the flame can be determined by the video sensor, based on the special algorithm of the signal processing. The advantage is the possibility to control of big open grounds. The additional advantage is the possibility to fix the exact place of the fire in the automotive mode. Preventing factors are noises from the sun, changes of the time of the day, rainfalls, different movements of people in the frame, but there is no problem to fix this factors. The technical difficulty can cause during the creation of the special video sensor with a wide dynamic range.

- A column of smoke on the open ground can be easily determined by the strong change of the image contrast on the big size of the image. In this case same problems appear as in the previous one.
- Indoor smoke: firstly smoke appears in the upper part of the space in the very beginning of the fire. The main preventing factor in this case can be insufficient contrast in the space in the dart time of the day, but this problem can be solved by the use of "reference signals" or the use of the infrared sensors.

Well known methods and developed on their basis algorithms of the video signal processing, which can be used in determination of dangerous factors of the fire, can be divided in 3 groups [6–8].

The first group includes methods, using references images, gotten before the dangerous situation in normal conditions in different types of illumination (day, evening, rain, etc.). The second group includes methods, using the exact database of the typical video image blocks of such different dangerous situations as the form and the type of the flame, intensity of the smoke and different preventing factors, such as the solar flare, lights of moving cars, and different types of flares. The third group includes methods, which analyze static and dynamic components of image elements by brightness and color components, and also with the use of special sensors of the infrared range.

In this work the method and algorithm of the fire factors determination based on the contour analysis of video images are described. Contour approach allows not to analyze inside points of the image and because of that to decrease the size of processed data by the movement from the function with two variables to the function of the one variable. The consequence of this is the possibility of work of the system in the real-time mode.

3 Algorithm of the image analysis for the fire detection based on the contour analysis

Described automatic system includes the records of video images of the controlled surface and contour standards of the image objects. The result of work of the signal and image processing block is the basic analysis of the image characteristics (brightness, contrast and others, which depend from the peculiarities of the shooting and parameters of the equipment) and their preliminary processing [9–12].

Analysis and pattern recognition block by the gotten video images of objects can react in a such way:

- Detection of the object contour on the image.
- Formation of the scalar product of the recognizing contour with the standard contour.
- Comparison of the real part of the scalar product with the threshold value.
- Getting the final decision about the fire determination.

Generally, the algorithm of detection is as follows. We have standard contour G of the image object of the controlled surface before dangerous situation and determined contour **N** of the object in the current image. Standard can be formed depends from the weather situation, illumination, etc.

In this case the algorithm of detection will look:

$$\mathrm{Re}(\mathbf{N}, G) \geq \sigma^2 \ln \Lambda_0 + 0{,}5\|G\|^2 = U_0, \tag{1}$$

where Λ_0 is the threshold value, U_0 is the threshold, which depends from the energy of the standards and noise contours and risks.

Thus, for getting the final decision about the fire detection in the exact place the object contour **N** should be determined and scalar product of the determined contour **N** with the standard contour **G** should be done. Then the real part of the scalar product is comparing with the threshold value, depends from the energy difference, noise dispersion, risks [13].

Researches in the fire detection were made on the example of video images of the forest area (Figures 2–4). The object contour was determined on the image (the upper level of the forest area) and then was compared with the standard contour of the exact area (before the dangerous situation).

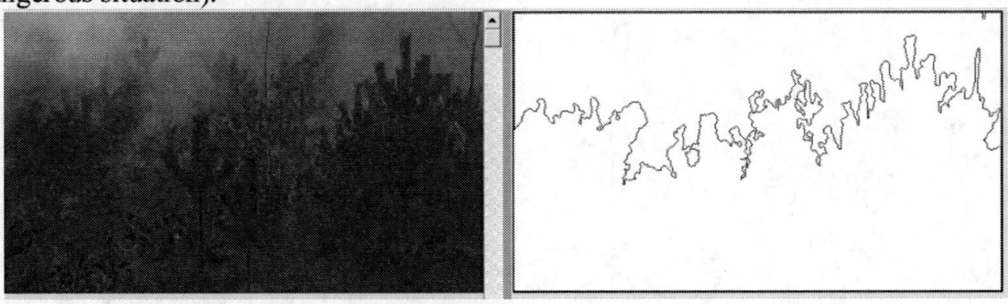

Figure 2. Video image and the formed standard contour of the object G (upper level of the forest area).

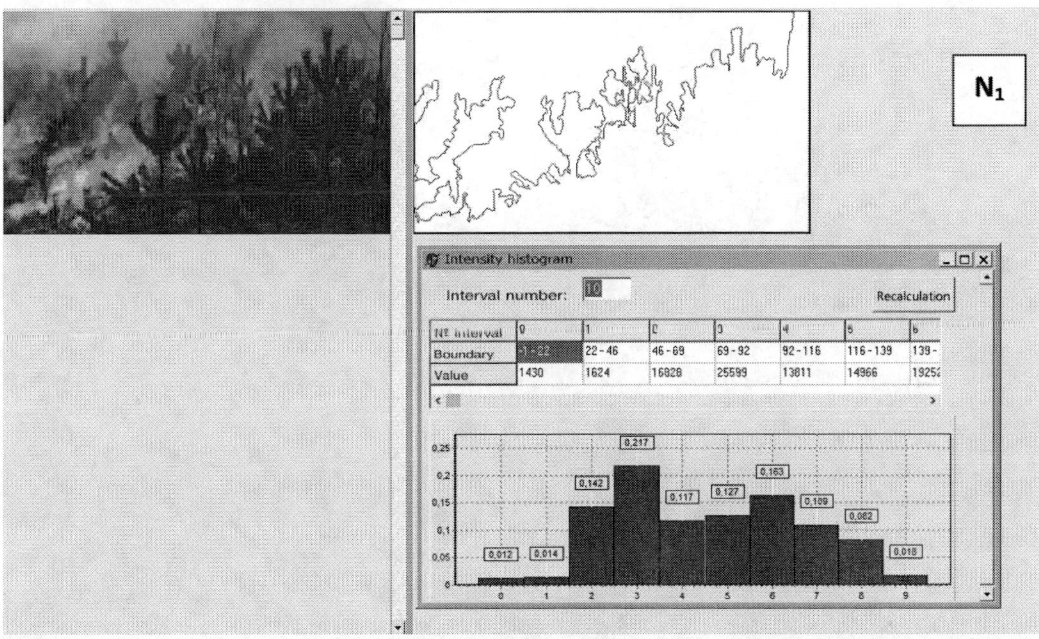

Figure 3. Video image of the forest area before the appearance of the dangerous situation on the exact are and it's processing (contour N_1 is determined).

DOI: 10.1051/matecconf/20167901067

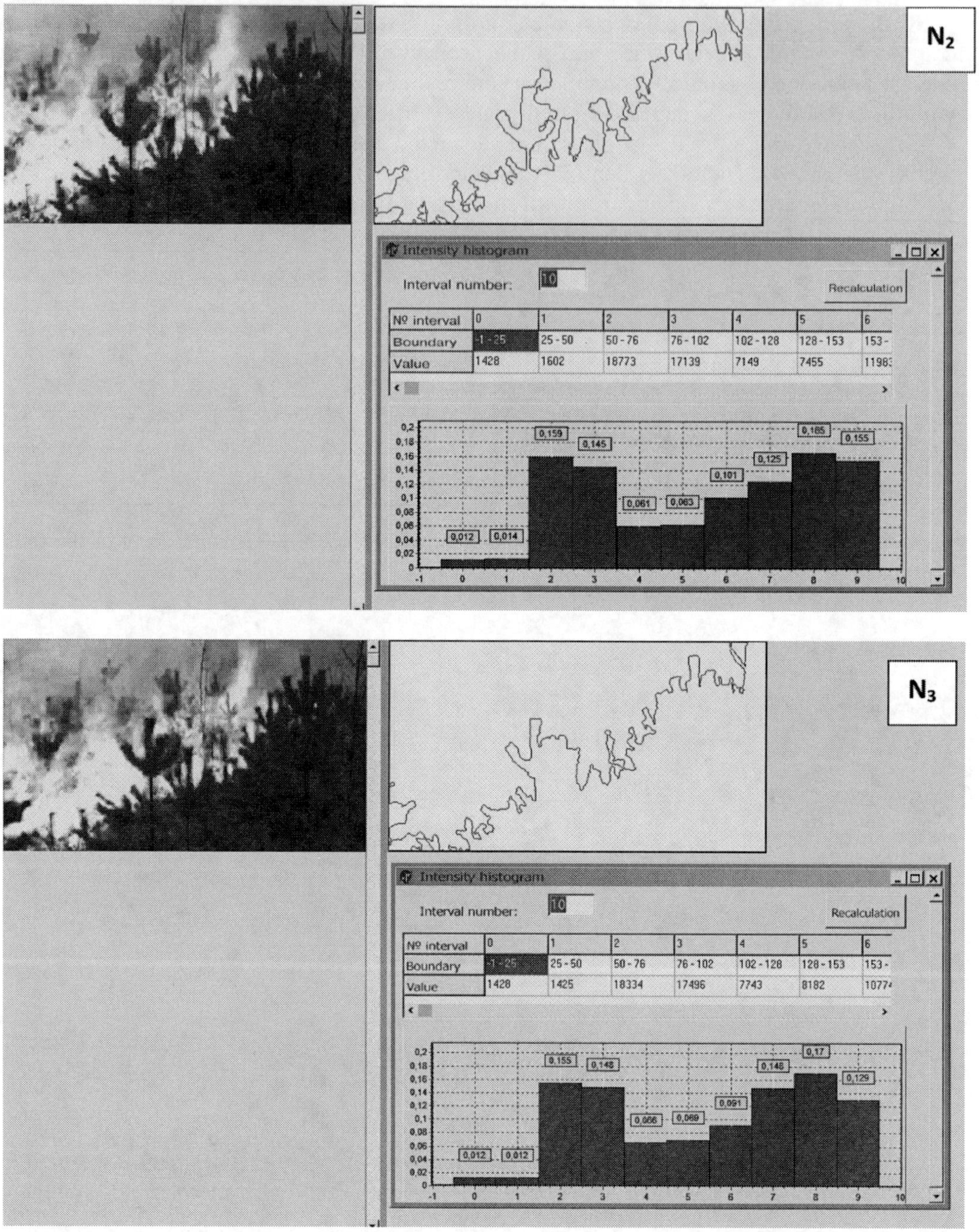

Figure 4. Video images of the forest area during the fire and their processing with interval 5 seconds (contours N_2, N_3 are determined).

4 Conclusion

Experimental researches, which were made on the video image of the forest area, shown that during the fire appearance the object contour is changing during the all dangerous situation. The results of the developed algorithm show that in the image with determined contour $\mathbf{N_1}$ the final decision about the fire absence is done, in images with determined contours $\mathbf{N_2}$, $\mathbf{N_3}$ the final decision about the fire appearance is done.

Thus, contour analysis can be shown as the mathematical model, which provides the way for decision of the problem of the image processing and the private problem of the fire appearance and dangerous situations on video images.

Acknowledgements

The paper was written as part of the research project No. 8.2.31.2015, carried out with the support of the Program "Research Foundation of Tomsk State University named after D.I. Mendeleev" in 2015 – 2016, grant RFBR No. 16-29-04388/16.

References

[1] A. Kuznetsov, V. Musalimov, A. Saenko, K. Trambitskiy, Journal of Instrumental Engineering **55** , 51 (2012)
[2] M.V. Rukin, Safety systems (4), 61 (2009)
[3] C. Koulas, Proceedings of SPIE - The International Society for Optical Engineering **7298**, 72983Q (2009) doi: 10.1117/12.818987
[4] I Y. Furman, A. Krevetskiy, A. Peredreev, A. Rozhentsov, R. Hafizov, I. Egoshina, A. Leushin, *Introduction in the contour analysis and it's applications for the image and signal processing* (Fizmalit, Moscow, 2003)
[5] H. Kennedy, Opt. Eng. V. **30** , 1771 (1999)
[6] N. Koshchavtsev, Proceedings of SPIE - The International Society for Optical Engineering **4369**, 81 (2001) doi: 10.1117/12.445278
[7] N. Topolskiy, A. Chlenov, T. Butcinskaya, M. Zemljanukhin, International workshop "Built Heritage: Fire Loss to Historic Buildings" (2004)
[8] E. Wixon, H. Rocky, Method and apparatus for determing ambient conditions from an image sequence, such as fog, haze or shadows, US patent no. 6037976 (2010)
[9] T. Yamagishi, M. Kishimoto, Fire detection system utilizing relationship of correspondence with regard to image overlap, US patent no. 5926280 (1999)
[10] R. Borsh, *Imaging fire detector* WO patent no. 01/67415 (2001)
[11] D.V. Shashev, S.V. Shidlovskiy, V.I. Syriamkin , A.V. Yurchenko, IOP Conference Series: Materials Science and Engineering **81**, 012101 (2014) doi: 10.1088/1757-899X/81/1/012101
[12] D.V. Shashev, S.V. Shidlovskiy, Optoelectronics, Instrumentation and Data Processing **51**, 227 (2015) doi: 10.3103/S8756699015030036
[13] M. Eisman, Proceedings of SPIE – The International Society for Optical Engineering **5783**, 71 (2005) doi: 10.1117/12.606718

AUTOMATED DATA PROCESSING SYSTEM FOR OSCILLOSCOPE MEASUREMENTS DURING AN ENVIRONMENTAL TEST

Sergey Shmakov[1,*], *Maxim* Cherniak[2], *Anna* Boruzdina[2], *Peter* Skorobogatov[2], *Alexander* Nikiforov[2]

[1] JSC "Specialized Electronic Systems", 115409, Moscow, Russia
[2] National Research Nuclear University MEPhI (Moscow Engineering Physics Institute), 115409, Moscow, Russia

Abstract. The paper discusses the automation of the processing of images obtained with an oscilloscope using National Instruments LabVIEW software. The mathematical background is presented as well as standard hardware communication interface. The block diagram of the electrical signal acquisition and processing system and the front panel with an example of processed waveforms is shown in the paper.

1 Introduction

The calibration process is an integral part of an environmental test using laser techniques to model the destabilizing effects on a device [1–3]. The main objective of the calibration process is a determination of the calibration factor. The calibration factor is determined by comparison of the dependences of the reaction of the device under the external influence and a laser source in pulse mode. One of the problems encountered during the work with external influence sources is interference signals. Hence, there is a need for software processing of the information received from the measurement equipment [4, 5]. In particular, the challenge is to create the software allowing to determine the integral from the waveform (the area under the curve), since the integral of the interference is significantly smaller than that of the useful signal. It is also necessary to automate the process of determining the minimum and maximum of the test signal during the laser testing. Ease data collection and processing based on existing test systems are required as well. It is necessary to ensure the ability to save the information in a convenient form for further work.

2 Hardware and software

Operating mode of the chip during the environmental test is set using a PXI platform based on the NI PXI–1033 chassis. This platform includes NI PXI-7951R FPGA module with NI 6581 I/O adapter, NI PXI-4110 module as a voltage supply, and PXI-4070 digital multimeter for control of the current consumption.

* Corresponding author: sbshmak@spels.ru

© The Authors, published by EDP Sciences. This is an open access article distributed under the terms of the Creative Commons Attribution License 4.0 (http://creativecommons.org/licenses/by/4.0/).

The software for interfacing Tektronix TDS 2024B oscilloscope with a computer and further data processing was implemented in NI LabVIEW programming environment with NI-VISA module.

3 Description of the solution

Block diagram of the system for collecting and processing electrical signals is shown in Figure 1.

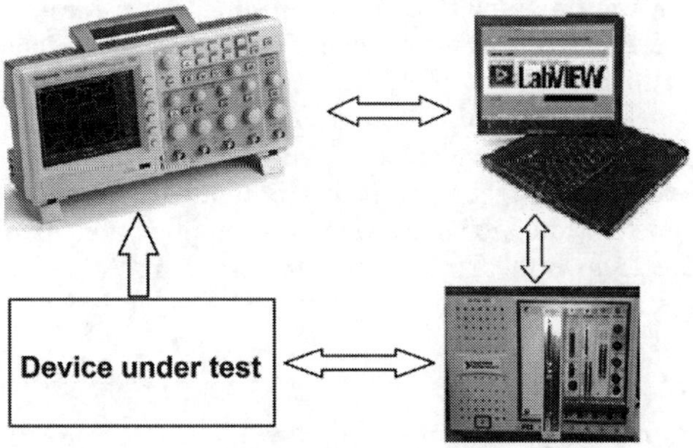

Figure 1. Block diagram of the system for collecting and processing electrical signals.

During the experiment, PXI-7951 FPGA module with NI 6581 adapter and PXI-4110 supply module set the control signals and bias of the DUT respectively. Oscilloscope measured the parameters of the output signals.

Test signals are inputted to an analog oscilloscope and a one-dimensional array of digital data is transmitted to a computer using USB interface.

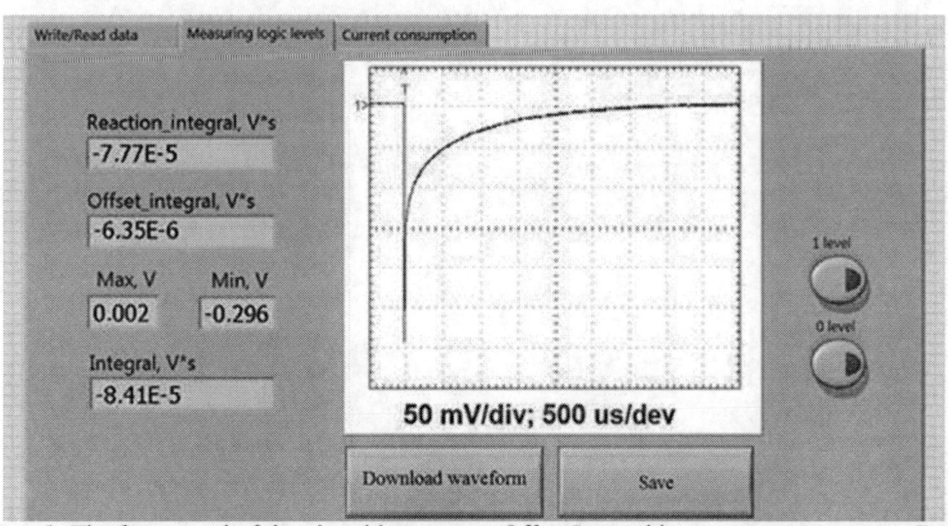

Figure 2. The front panel of the virtual instrument. Offset Integral is a constant component, Reaction Integral is the total integral, Integral is the total integral with DC component subtracted, Min – the minimum value of the signal on the waveform, Max - the maximum value of the signal on the waveform.

The array has a capacity of 2500 elements and contains the signal digitized by the oscilloscope in ADC units. Each element of the array contains the result of digitizing by 8

DOI: 10.1051/matecconf/20167901068

bit ADC in a form of an integer number. Values fall in a range from -127 to 128. To determine the transformation coefficient of ADC conversion units the expression (1) is applied in Volts. N – the number of cell divisions the oscilloscope grid; D – vertical division of the oscilloscope

$$K = N \cdot D/256 \quad (1)$$

The minimum and the maximum value are determined by the oscilloscope and obtained by the program from Meas1 and Meas 2 registers.

Integral calculation program performs the summation of all the data from an array of post-trigger, which were multiplied by the corresponding values of the scale factor sweep oscilloscope. If there is a constant component available in the signal, it will be taken into account in the final result (2).

$$Integral = Reaction\_integral - Offset\_integral \quad (2)$$

The constant component is calculated by the method indicated above, but 250 array elements are taken before trigger point, that is, until the arrival of the signal, which is multiplied by a constant equal to N/250 where N– number of elements in the array after the trigger. The Figure 2 shows the front panel of the program.

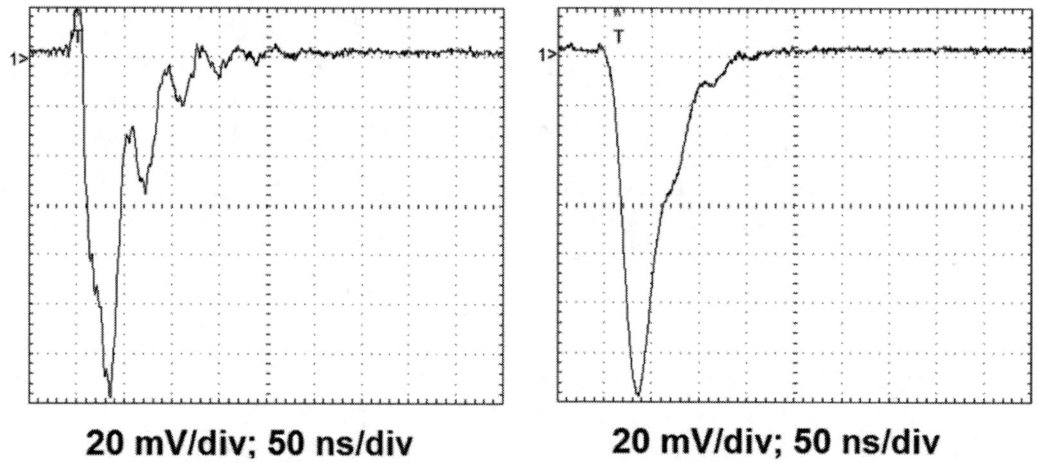

Figure 3. Typical reaction during calibration: *a* – External influence; *b* – Laser light source.

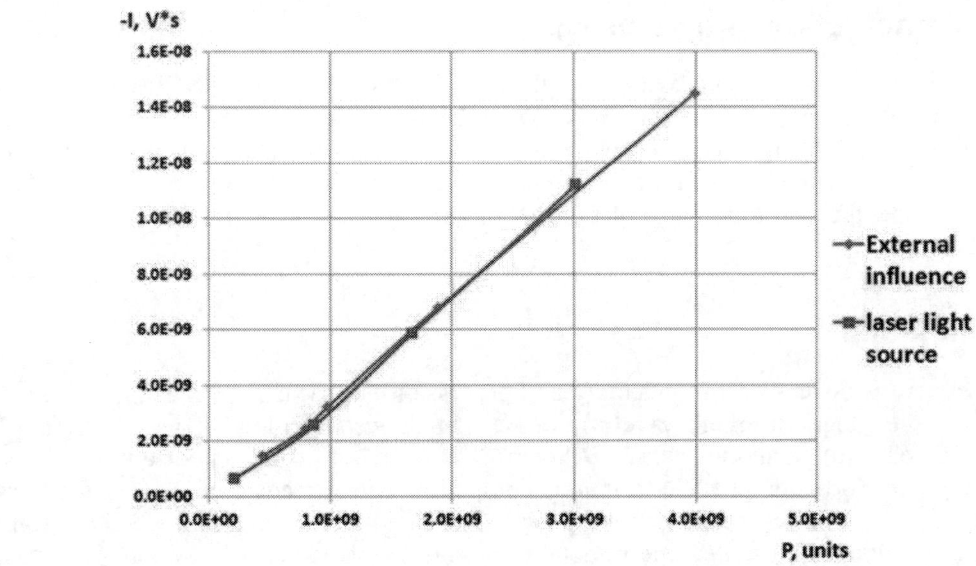

Figure 4. Combined charts based on the calibration coefficient obtained using the developed software.

The program allows the waveform to be saved as a png file (Figure 3), and, as seen on Figure 2, can be easily integrated as a functional unit of the test system [6–12], and allows us to work with the oscilloscope remotely, if necessary [13].

The possibility of remotely work is provided by sending to the oscilloscope commands such as CH<x>:SCAle <NR3> (set scale-division value), MEASU:MEAS1:VALue? (check the scale-division value), HORizontal:MAIn:POSition? (check the position of the start of the trigger), HORizontal:MAIn:POSition <NR3> (set the position of the beginning of the trigger) via NI-VISA driver.

Using the developed software, graphs shown in Figure 4 were obtained and matched using the calibration factor. During an environmental test, this software is useful for monitoring parameters such as output voltage drop at logical zero and logical one (Figure 5), the pulse response of current consumption.

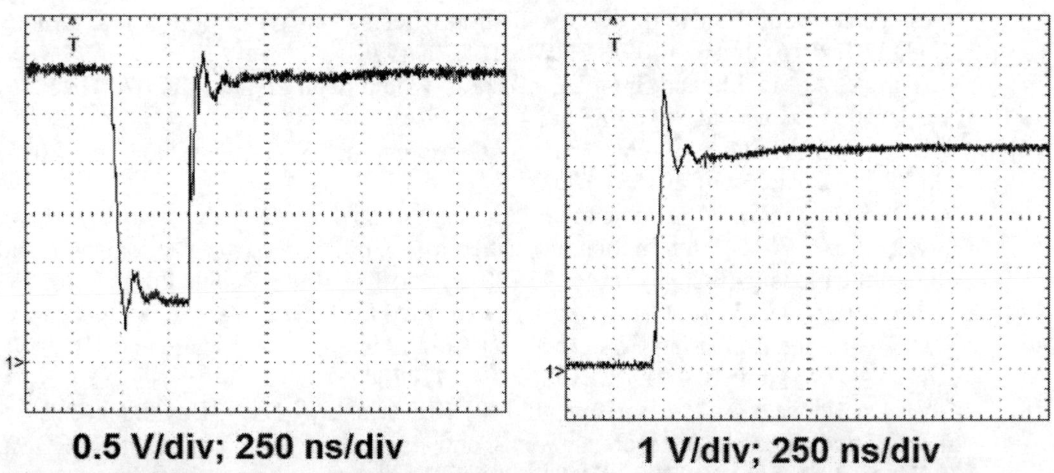

Figure 5. Typical reaction: *a* – Waveform obtained to monitor the output high logic level; *b* – Waveform obtained for low logic level control (switch from the low logic level to a high logic level).

4 Implementation and its prospects

The developed software is successfully implemented and used in JSC "ENGOs SPELS" as a signal processing system with Tektronix TDS 2024B oscilloscope.

Improvement of software for measurement dynamic characteristics is planned, such as data access time (time elapsed since an enable signal and the output lines to establish relevant data), the rise time and fall time of the signal, control of the frequency of the output signals.

5 Conclusions

Thus, system is developed for collecting and processing electrical signals and allowing to determine the integral from the waveform of the signal under examination and to determine the maximum and minimum value. Developed software gives the opportunity to save measured waveforms in .png file format, is compatible with already existing test facilities and can be used as a separate functional block. One should notice that developed software allows controlling the oscilloscope remotely; this fact is of particularly importance when working on test facilities.

Acknowledgments

The work performed under grant agreement between the Russian Ministry of Education and the National Research Nuclear University MEPhI on November 24, 2014 № 14.578.21.0075 (unique identifier RFMEFI57814X0075 for Applied Scientific Research).

References

[1] P. K. Skorobogatov, A. Y. Nikiforov, A.N. Egorov, Russian Microelectronics **44**, 8 (2015) doi: 10.1134/S1063739715010084

[2] A.Y. Nikiforov, P. K. Skorobogatov, Russian Microelectronics **35**, 138 (2006) doi: 10.1134/S1063739706030024

[3] A.N. Egorov, A.I. Chumakov, O.B. Mavritsky, A. A.Pechenkin, D. V. Savchenkov, V. A. Telets, A. V.Yanenko, IEEE Radiation Effects Data Workshop **2015-January** , 7004570 (2016) doi:10.1109/REDW.2014.7004570

[4] W. Fan, Y. An, D. Liu and S. Guan, MATEC Web of Conferences **40,** 07002 (2016) doi: 10.1051/matecconf/20164007002

[5] X.Tan, MATEC Web of Conferences **31,** 09003 (2015) doi: 10.1051/matecconf/20153109003

[6] M. P. Belova, D. V. Pechenkina, A. Y. Borisov, L. N. Kessarinskiy, D.V. Boychenko, 2015 International Siberian Conference on Control and Communications (SIBCON), 7146995 (2015) doi: 10.1109/SIBCON.2015.7146995

[7] A. B. Boruzdina, A. A. Orlov, A. V. Ulanova, N. G. Grigor'ev, A.Y. Nikiforov, 2015 International Siberian Conference on Control and Communications (SIBCON), 7147007 (2015) doi: 10.1109/SIBCON.2015.7147007

[8] A.V. Demidova, A. Y. Borisov, L. N. Kessarinskiy, D.V. Boychenko, 2015 International Siberian Conference on Control and Communications (SIBCON), 7147036 (2015) doi: 10.1109/SIBCON.2015.7147036

[9] M.E. Cherniak , A.V. Ulanova, A.Y. Nikiforov, Proceedings of the International Conference on Microelectronics (ICM), 6842119 (2014) doi: 10.1109/MIEL.2014.6842119

[10] I. I. Shvetsov-Shilovskiy, P.V. Nekrasov, A.V. Ulanova, A. A.Smolin, A. V. Sogoyan, 2015 International Siberian Conference on Control and Communications (SIBCON), 7147282 (2015) doi: 10.1109/SIBCON.2015.7147282

[11] I. O. Loskutov, A. B. Karakozov, P.V. Nekrasov, A.Y. Nikiforov, 2015 International Siberian Conference on Control and Communications (SIBCON) , 7147128 (2015) doi: 10.1109/SIBCON.2015.7147128

[12] G. G. Davydov, A. S. Kolosova, L. N. Kessarinskiy, D.V. Boychenko, 2015 International Siberian Conference on Control and Communications (SIBCON), 7147094 (2015) doi: 10.1109/SIBCON.2015.7147094

[13] A.S. Artamonov, A.A. Sangalov, A.Y. Nikiforov, V. A. Telets, D.V. Boychenko, IEEE Radiation Effects Data Workshop, 7004600 (2014) doi: 10.1109/REDW.2014.7004600

MATEC Web of Conferences 79, 01069 (2016) DOI: 10.1051/matecconf/20167901069
IME&T 2016

REDUCING THE IMPACT OF ELECTROCONDUCTIVITY AND THE GAP BETWEEN THE PIPE AND THE TRANSDUCER AT MEASURING THICKNESS OF ELECTROCONDUCTIVE PIPE WALLS USING THE EDDY-CURRENT METHOD

Evgeny Yakimov * and *Daniil* Ustyugov

Tomsk Polytechnic University, Institute of Non-Destructive Testing, 634050, Tomsk, Russia

Abstract. The paper describes a dual-frequency method for reducing the impact of changes in the gap size between the eddy-current transducer and the pipe, as well as the pipe electrical conductivity on the eddy-current thickness gauge readings. A block-diagram of the dual-frequency eddy-current thickness gauge is proposed for light-alloy drill pipes. The amplitude and signal phase dependencies on the wall thickness in the range from 6 to 17 mm and the gap in the range from 0 to 13.5 mm were studied, the results are presented. The digital signal processing algorithms based on the piecewise-linear approximation of low-frequency and high-frequency signal phase dependencies on the wall thickness are proposed. It is shown that the proposed correction algorithms can reduce the error caused by variations of electrical conductivity and the gap between the transducer and the pipe.

1 Introduction. Measuring the thickness of pipe walls

Light-alloy drill pipes of TB and TBP types made of aluminum alloy D16 are manufactured in accordance with the Russian standard GOST 23786-79 [1]. The pipes are used to perform the workover and the structure, exploration and production drilling. The pipe walls are worn during in-service use, and therefore the wall thickness of the main pipe section must be controlled according to the Russian standard. With due regard to the assortment, the nominal wall thickness of the main pipe section can vary from 7.5 to 13 mm. The limit thickness deviation shall not exceed ± (0.4–0.5) mm for pipes of high accuracy and ± (0.7–1.7) mm for pipes of normal accuracy. A non-destructive testing method is not considered. The measuring instruments must have the accuracy of ± 0.1 mm.

* Corresponding author: shishkovka@mail.ru

© The Authors, published by EDP Sciences. This is an open access article distributed under the terms of the Creative Commons Attribution License 4.0 (http://creativecommons.org/licenses/by/4.0/).

Ultrasonic devices provide high accuracy of measurement, and therefore are used to measure the thickness of metal objects. However, a good acoustic contact (through the liquid layer) is required, that decreases the speed of testing.

The advantages of the eddy-current testing method are high-performance control and sensitivity, and the possibility of non-contact measurement. These advantages enable the method automation and provide a continuous control along the entire length of the pipe. However, signals of the eddy-current transducer considerably depend on the material electrical conductivity and the gap between the transducer and the pipe.

In the previous paper [2], it was shown that a dual-frequency eddy-current method can reduce the dependence of thickness gauge readings on the gap size. Consider the fact that the method is capable of reducing the impact of changes in the pipe electrical conductivity on the thickness measurement results. With this, the block diagram of the gauge (Figure 1) comprising the surface eddy-current transducer (ECT), generator (G), amplitude and phase detectors (APD), data acquisition board with ADC, and personal computer (PC) can be used.

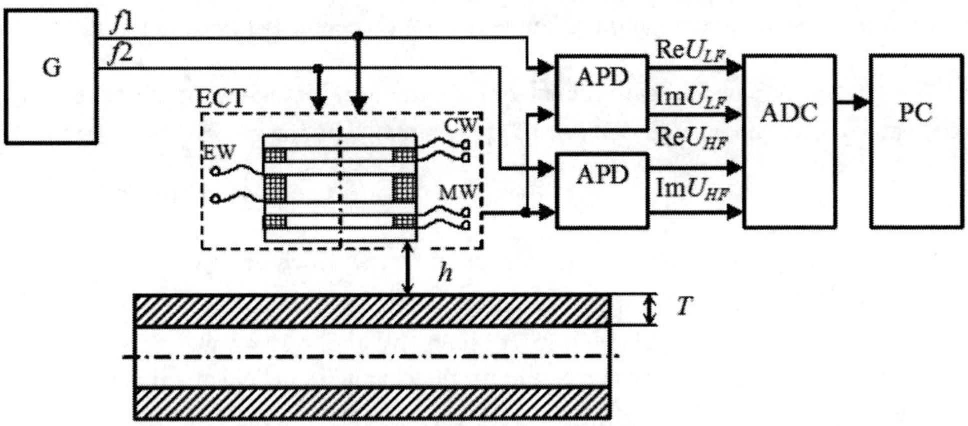

Figure 1. Block diagram of the thickness gauge based on the dual-frequency eddy-current method.

The eddy-current transducer comprises the excitation winding (EW), measurement winding (MW) and compensation winding (CW). The signal is fed from the generator to the transducer EW, which is the sum of two harmonic signals. The first frequency *f*1 is principal and is intended to measure the wall thickness *T*. The frequency must be up to several hundred hertz and therefore is referred to as low frequency (LF). The second frequency *f*2 is selected so that the signals of this frequency slightly depend on the pipe wall thickness and at the same time change with variations of the pipe electrical conductivity and the gap *h* between the pipe and the transducer. These properties appear in the frequency of several kilohertz, so the frequency is called high frequency (HF).

During the measurement, signals from the eddy-current transducer arrive at the amplitude and phase detectors, which output voltages are proportional to the amplitudes of imaginary and real components of the low-frequency signal *f*1 and high-frequency signal *f*2: $\mathrm{Re}U_{LF}$, $\mathrm{Im}U_{LF}$, $\mathrm{Re}U_{HF}$, $\mathrm{Im}U_{HF}$.

Then, signals arrive at the data acquisition board with ADC, which performs analog-digital conversion of signals and transfer them to the PC.

The PC records and processes the data.

The values of the ECT signals are used to calculate the amplitude A_{HF} and phase φ_{HF} of the high-frequency signal (*f*2), and the phase φ_{LF} of the low-frequency signal (*f*1):

$$A_{HF} = \sqrt{(ReU_{HF})^2 + (ImU_{HF})^2}\,, \tag{1}$$

$$\varphi_{HF} = arctg(\mathrm{Im}U_{HF} / \mathrm{Re}U_{HF})\,, \tag{2}$$

$$\varphi_{LF} = arctg(\mathrm{Im}U_{LF} / \mathrm{Re}U_{LF})\,. \tag{3}$$

The value of the high-frequency signal amplitude A_{HF} is taken to calculate the gap h between the ECT and the pipe surface:

$$h = Kh \cdot \left[cg_0 + cg_1 \cdot e^{cg_2 \cdot amp} + cg_3 \cdot e^{cg_4 \cdot amp}\right], \tag{4}$$

where cg_k is constant coefficients, $amp = A_{HF} / A_{HF0}$ is the relative amplitude of the high-frequency signal, A_{HF0} is the amplitude of the high-frequency signal for zero gap, and Kh is a correction factor used for non-zero gap calibration.

The approximation error in accordance with expression (4) does not exceed 0.08 mm [2].

The dependence of the wall thickness T on the low-frequency signal phase φ_{LF} for a fixed value of the gap h (Figure 2) can be approximated by the piecewise-linear function

$$T = K_{LF}(h, jtx) \cdot \left[\varphi_{LF} + \Delta\varphi_{LF(tx)} + \Delta\varphi_{LF} + Kc \cdot \varphi_{LF}\right] + B_{LF}(h, jtx), \tag{5}$$

where $K_{LF}(h, jtx)$ and $B_{LF}(h, jtx)$ are the coefficients, which depend on the gap h and the number jtx of the site, where the thickness depends on the phase φ_{LF}; $\Delta\varphi_{LF}$ and Kc are the coefficients used to calibrate the readings by the two known values of the pipe thickness; $\Delta\varphi_{LF(tx)}$ is the phase correction for the change in the electrical conductivity.

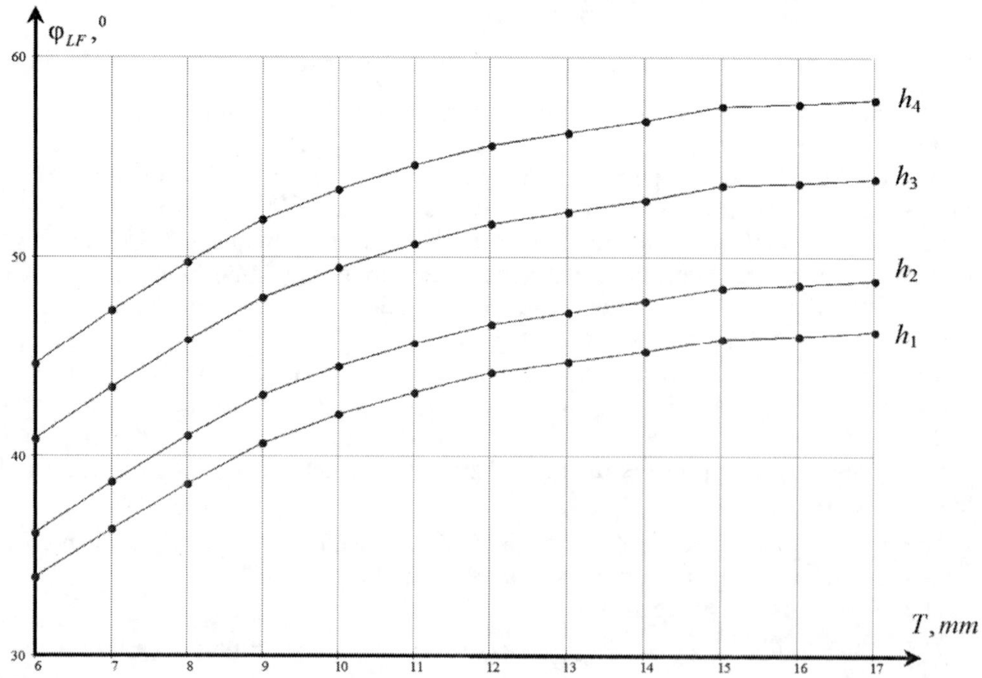

Figure 2. Dependence of the low-frequency signal phase φ_{LF} on the pipe wall thickness T for the fixed values of the gap h.

The correction is calculated based on the assumption that the conductivity is temperature dependent [3–10]. However, the electrical conductivity can vary due to the different chemical composition of the tube material.

The phase dependence on the temperature can be approximated by the linear function

$$\Delta\varphi_{LF(tx)} = -Kt_{LF} \cdot (tx - t0), \tag{6}$$

where Kt_{LF} is the coefficient of proportionality, tx is the pipe temperature, and $t0$ is the temperature at which the thickness gauge was calibrated.

The pipe temperature is calculated from the difference of the measured high-frequency signal phase $\varphi_{HF(tx)}$ and the phase $\varphi_{HF(t0)}$ measured at temperature $t0$,

$$tx = t0 - Kt_{HF} \cdot \lfloor \varphi_{HF(tx)} - \varphi_{HF(t0)} \rfloor, \tag{7}$$

where Kt_{HF} is the coefficient of proportionality.

That is, the preliminary evaluation of the thickness is made with no regard to the conductivity correction, and then the expected value of the high-frequency signal phase $\varphi_{HF(t0)}$ and the phase difference $\lfloor \varphi_{HF(tx)} - \varphi_{HF(t0)} \rfloor$ are calculated. The change in the high-frequency signal phase against the gap h and the thickness of the tube T are taken into account (Figure 3). A piecewise-linear model of dependence is used

$$\varphi_{HF(t0)} = T \cdot \lfloor Kk_{HF(jhx)} \cdot h + Bk_{HF(jhx)} \rfloor + \lfloor Kb_{HF(jhx)} \cdot h + Bb_{HF(jhx)} \rfloor, \tag{8}$$

where $Kk_{HF(jhx)}$, $Bk_{HF(jhx)}$, $Kb_{HF(jhx)}$ and $Bb_{HF(jhx)}$ are the coefficients of piecewise-linear approximation of the dependence of the high-frequency signal phase on the gap.

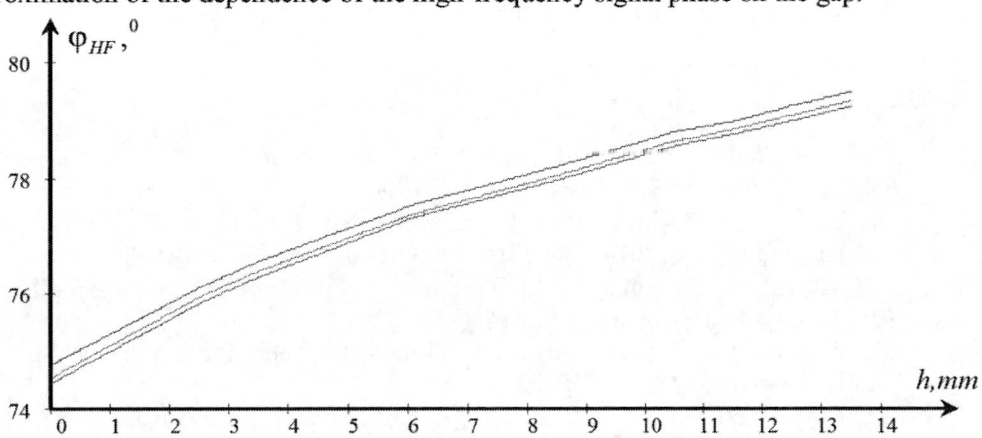

Figure 3. Dependence of the high-frequency signal phase φ_{HF} on the gap h for the fixed pipe wall thickness T.

2 Thickness gauge calibration

To calibrate the readings when measuring a zero gap, the ECT is placed directly on the pipe and the high-frequency signal amplitude A_{HF0} is saved in the memory.

For the non-zero gap calibration, the ECT is placed on the plate of known thickness and the correction factor Kh used in equation (4) is calculated

$$Kh = h2 / h. \tag{9}$$

The thickness calibration is performed in two steps.

At the first step, the ECT is installed on the sample pipe with a known thickness value $T1$, and the ECT signals are saved in the memory. Next, the ECT is installed on the sample pipe with a known value $T2$, and the ECT signals are saved in the memory. The thickness of sample 1 should be less than the thickness of sample 2.

The current values of the correction factors $\Delta\varphi_{LF(i)}$ and $Kc_{(i)}$ used in equation (5) to correct the gauge readings are calculated by the system of equations (10)

$$\left.\begin{array}{r} Tx1 = K_{LF(1)}(h, jtx)\cdot\left[\varphi_{LF(1)} + \Delta\varphi_{LF(i-1)} + Kc_{(i-1)}\cdot\varphi_{LF(1)}\right] + B_{LF(1)}(h, jtx) \\ T1 = K_{LF(1)}(h, jtx)\cdot\left[\varphi_{LF(1)} + \Delta\varphi_{LF(i)} + Kc_{(i)}\cdot\varphi_{LF(1)}\right] + B_{LF(1)}(h, jtx) \\ Tx2 = K_{LF(2)}(h, jtx)\cdot\left[\varphi_{LF(2)} + \Delta\varphi_{LF(i-1)} + Kc_{(i-1)}\cdot\varphi_{LF(2)}\right] + B_{LF(2)}(h, jtx) \\ T2 = K_{LF(2)}(h, jtx)\cdot\left[\varphi_{LF(2)} + \Delta\varphi_{LF(i)} + Kc_{(i)}\cdot\varphi_{LF(2)}\right] + B_{LF(2)}(h, jtx) \end{array}\right\}, \tag{10}$$

where $\Delta\varphi_{LF(i-1)}$ and $Kc_{(i-1)}$ are the coefficients used in the previous calibration; $Tx1$ and $Tx2$ are the thickness gauge readings with the previous correction factor values; $T1$ and $T2$ are the values of the thickness of sample pipes.

3 Conclusion

The experiments showed that the dual-frequency eddy-current method used to measure the wall thickness of the electrically conductive pipes according to the described algorithms reduces the impact of the gap between the pipe and the eddy-current transducer, and that of the changes in the pipe electrical conductivity on the measurement results.

References

[1] *GOST 23786-79. Drill pipes made of aluminum alloys. Technical conditions* (Publishing House of Standards, Moscow, 1990)

[2] E.V. Yakimov, A.E. Goldstein, V.F. Bulgakov, Yu.V. Alkhimov, V.Yu. Belyankov, Test. Diagn. (11), 14 (2014) DOI: 10.14489/td.2014.011.pp.014-018

[3] Z.-G. Zhang, H.-X. Feng, W.-H. Zhao, MATEC Web of Conferencesb **40**, 04001 (2016) doi : 10.1051/matecconf/20164004001

[4] A.Y. Petrova, O.N. Chaikovskaya, I.V. Plotnikova, Tech. Phys. J. **60**, 592 (2015) doi: 10.1134/S1063784215040222

[5] A.A. Bespalko, L.V. Yavorovich, P.I. Fedotov, Russ. J. Nondestr. Test. **47**, 680 (2011) doi: 10.1134/S1061830911100068

[6] O.V. Galtseva, S.V. Bordunov, N.M. Natalinova, S.V. Mazikov, IOP Conf. Ser.: Mater. Sci. Eng. **132**, 012003 (2016) doi: 10.1088/1757-899X/132/1/012003

[7] A.P. Surzhikov, T.S. Frangulyan, S.A. Ghyngazov, N.N. Koval, Nucl. Instrum. Methods Phys. Res., Sect. B **267**, 1072 (2009) doi: 10.1016/j.nimb.2009.01.144

[8] R. Kodermyatov, M. Ivanov, M. Yuzhakov, V. Kuznetsov, M. Yuzhakova, E. Timofeeva, MATEC Web of Conferences **48**, 05004 (2016) doi: 10.1051/confmatec/20164805004

[9] A. Gavrilin, B. Moyzes, O. Zharkevich, Journal of Vibroengineering **17**, 3495 (2015)

[10] O.V. Galtseva, S.V. Bordunov, S.N. Torgaev, IOP Conf. Ser.: Mater. Sci. Eng. **110**, 012094 (2016) DOI: 10.1088/1757-899X/110/1/012094

MATEC Web of Conferences **79**, 01070 (2016)
DOI: 10.1051/matecconf/20167901070
IME&T 2016

THE RADIATION FOCUSING BY THE ARRAY OF WAVEGUIDE BASED ELEMENTS WITH CONTROLLABLE REFLECTION

Aleksandr Eremeev[1], *Vladimir* Antipov

National Research Tomsk State University, the Department of Radiophysics, 634050, Tomsk, Russia

Abstract. The authors propose a solution of the problem of radiovision using the reflective array, each element of which can change the reflection coefficient under the action of external control voltage. The focusing abilities of flat reflection array of monochromatic radiation were studied to solve the problem of radiovision. The array element based on waveguide with a controlled reflection coefficient was developed. The phase shift switching is 180°.

1 Introduction

Modern radiovision systems are applied for numerous purposes: to control the quality of materials, buildings and constructions as well as for medical applications. The development of computing technologies provided the application of the radiotomography techniques for the remote non-destructive testing and the diagnostics of the internal structure of radio-semitransparent media and the shape recovery of radio opaque objects. The task of radiotomography is the transformation of the data obtained from the multiangle scanning of target objects into the 3D visualization [1]. The electronically and electromechanically scanned arrays are developed to obtain the multiangle projections of the wave field. However, the design and development of such arrays are complicated and expensive task, as it requires the development and control of numerous microwave transceiver channels.

The development of the electronically scanned arrays widely prevails nowadays over the electromecanically scanned ones. Together with conventional phased arrays, the arrays of elements with controllable phase of reflection are extensively applied. The reflection elements are usually based on ferrite phase shifters, micromechanical systems or varactor diodes [2].

The phase switching of the array reflection coefficients enables the simulation of the Fresnel zone plate that provides the focusing of feeder radiation at a certain point. The waves from two neighboring Fresnel zones are counter-phased. The electronic scanning can be performed in a certain angle sector with the change in geometrical arrangement of zone rings [3–4]. In this case, the only transceiver at fixed position can be used.

[1]Corresponding author: aleksandr_eremeev_93@mail.ru

© The Authors, published by EDP Sciences. This is an open access article distributed under the terms of the Creative Commons Attribution License 4.0 (http://creativecommons.org/licenses/by/4.0/).

DOI: 10.1051/matecconf/20167901070

2 Formulation of the problem

In this research, the authors propose a solution of the problem of radiovision using the reflective array, each element of which can change the reflection coefficient by the external voltage control. The detailed task scheme presented in Figure 1. A transmitted spherical wave be propagated in a free space defined in the Cartesian coordinate system. The transmitter (1) is placed at Z axis and at the h height from the reflective array (2). The test object (3) is placed above the array.

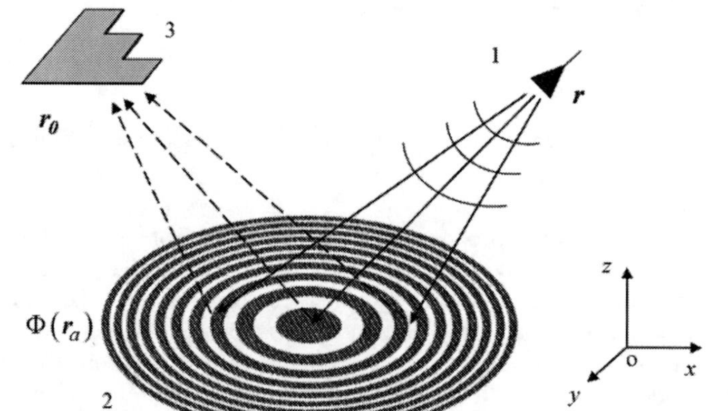

Figure 1. Task scheme.

Set the focusing point coordinates as (x_0, y_0, z_0). Then, the Fresnel zone boundaries are defined by the equation:

$$\sqrt{z_0^2+(x-x_0)^2+(y-y_0)^2}+\sqrt{h^2+x^2+y^2}=n\frac{\lambda}{2} \quad (1)$$

where n is the number of the Fresnel zone boundary, λ is the wavelength of the sounding signal.

Previously, two problems were solved [5]. The equation for reflection coefficient distribution was obtained, which provided the radiation focusing at the certain point. Also, the equation describing the field at the receiver point was obtained as well. The using of the 2D reflective array with controlled reflection coefficient to focus the monochromatic radiation for radiotomographic application were analysed. The numerical simulations with a test object approved the recovery of the test object shape.

3 Waveguide element of the reflection array

A waveguide was used as an element of the reflection array. The scheme of the waveguide array element and its pilot prototype are presented in Figure 2.

 DOI: 10.1051/matecconf/20167901070

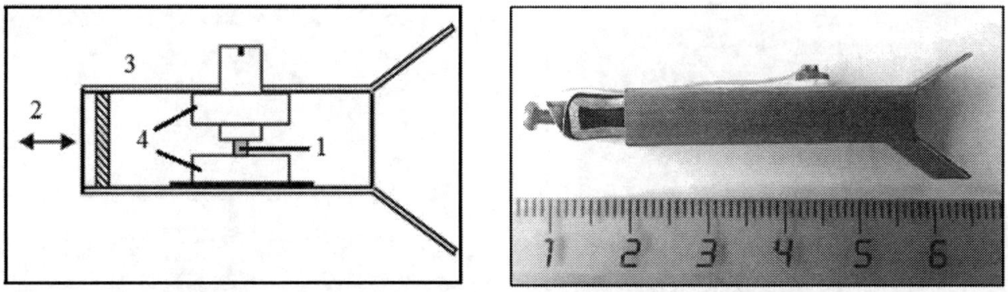

Figure 2. Array element structure (1– semiconductor diode capacitor, 2– shorting plunger, 3– waveguide, 4 – impedance transformer) and its external view.

The varactor diode was installed into the waveguide section 11×5.5 mm in size. The waveguide impedance could be switched from low to high by altering the varactor capacity through the voltage adjustment. In that case, the installed varactor diode could either shunt the waveguide or transmit the propagated wave to the shortening plunger. In the latter, the phase shift to be set as 180° by the shorting plunger position.

To match the waveguide, its narrow walls were cut to make the horn-shaped aperture with bent wide waveguide walls. That technique reduces the reflection as well as increases the effective area of the array. The impedance transformer was built of two symmetrically inserted conductive discs, one of which was insulated from a waveguide wall with a dielectric pad. The varactor diode was soldered to the end of the screw which was extended through the non-insulated disc to the contact with the insulated one. The control voltage was applied to the insulated disc. The waveguide element is simply constructed that makes possible its volume production as well as provides its cost competitiveness with printed phase shifters.

4 The numerical and experimental results

The array element experimentally demonstrated the reproducibility of characteristics. The graph illustrating the 180° phase shift of the reflection coefficient is presented in Figure 3.

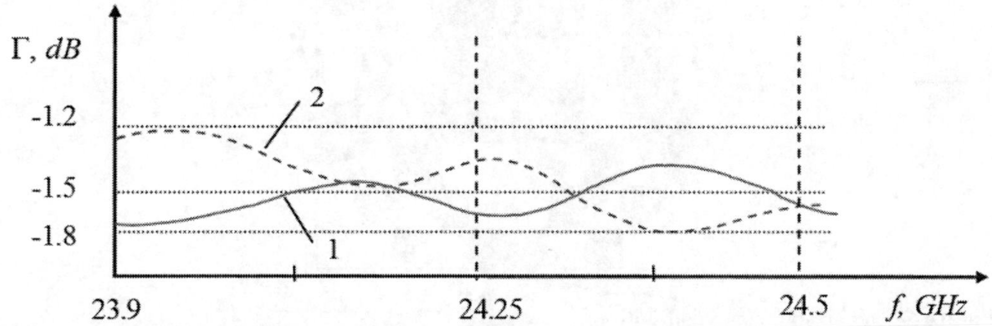

Figure 3. Reflection coefficient in the waveguide element: 1 — control voltage applied to the varactor diode, 2 – no voltage applied.

The full line graph (1) corresponds to the signal reflected when the shift voltage was applied to the varactor diode. The dashed line (2) illustrates the signal without the voltage shift. The array structure required the special holders, with 10 waveguide elements in each holder. The holders are set one above the other (Figure 4, *a*).

DOI: 10.1051/matecconf/20167901070

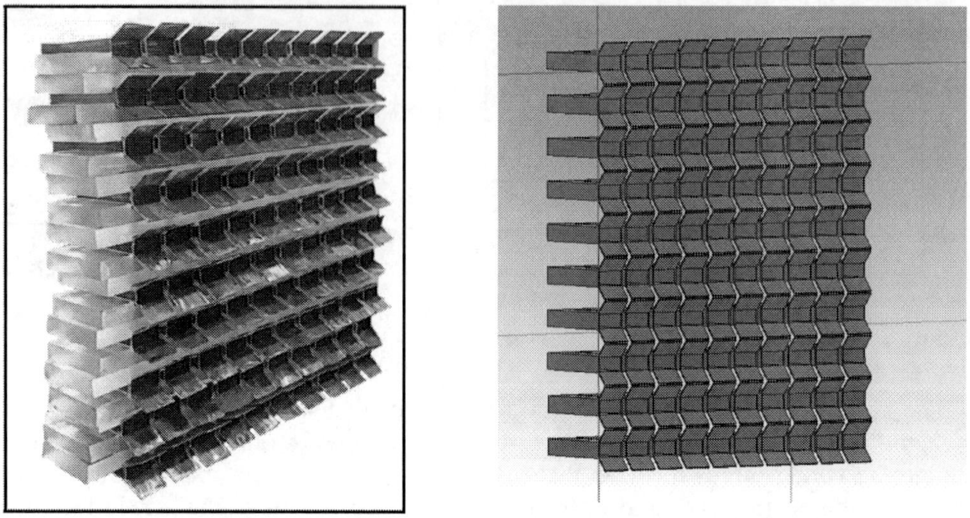

a b

Figure 4. Reflection array sample in the size of 10x10 elements, and its numerical model.

To develop the experimental sample of the reflection array, the waveguide element and the array were simulated in CST Microwave Studio. Array model is presented in Figure 4,*b*.

The distribution of the reflection coefficient $\Phi(\boldsymbol{r}_\mathrm{a})$ was predefined, which was calculated by the algorithms [5] corresponding to the array size. The calculations were performed for the frequency of 24 GHz. The distribution of the reflection coefficient, which provides the focusing at the coordinates $(0,0,10)$ is presented in Figure 5.

Figure 5. The Fresnel zone distribution.

The field focused with the waveguide-based reflection array is presented in Figure 6. It was calculated in CST Microwave Studio.

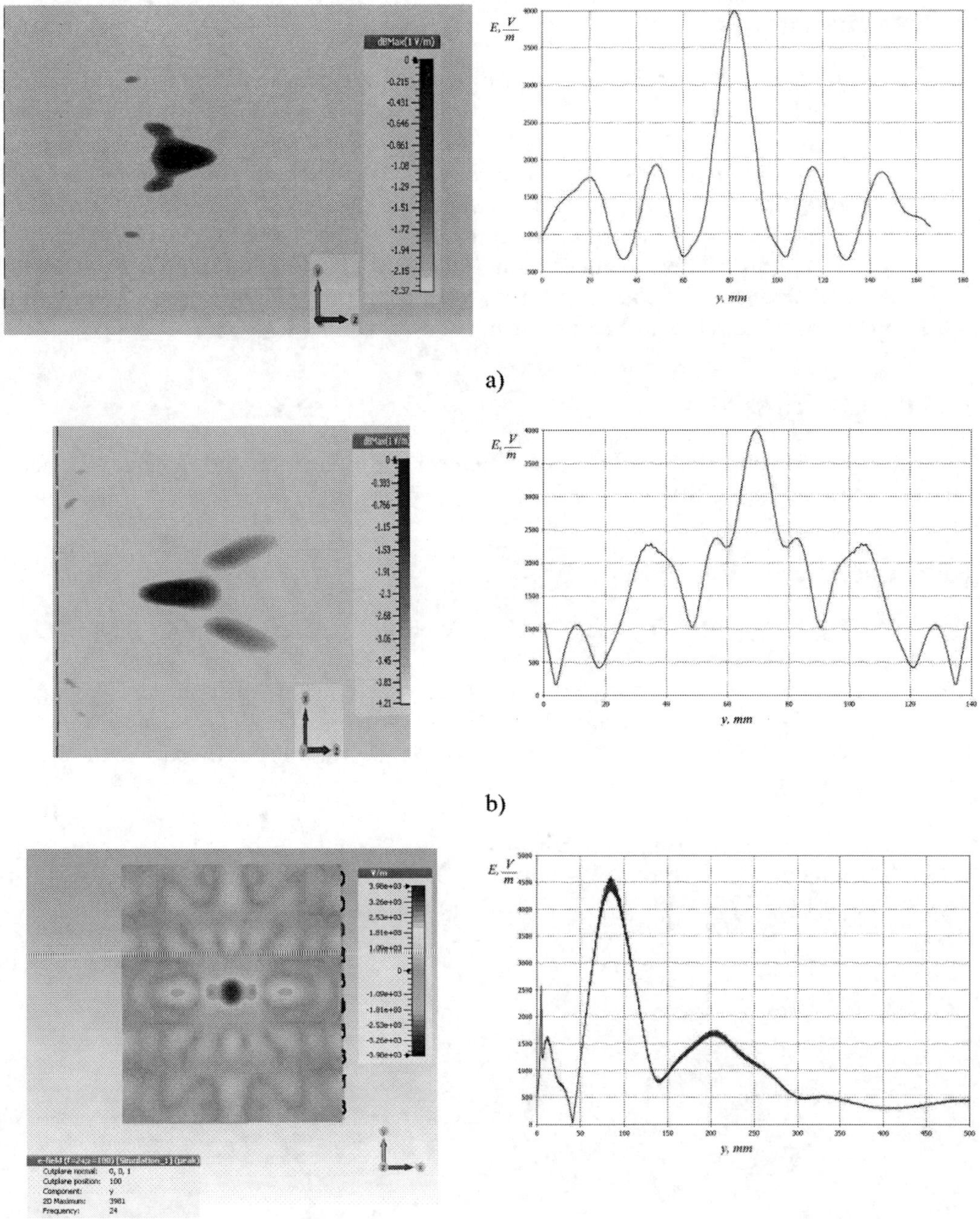

Figure 6. Radiation focusing with the reflection array.

The field distribution in the *yz* plane is presented in Figure 6, *a*. Figure 6, *b* presents the field distribution in the *xz* plane. Figure 6, *c* shows the field distribution in the *xy* plane as well as the field strength variation depending on the distance from the array. It is indicated that the field was focused at a certain point $(0,0,10)$ which the distribution of the reflection coefficient corresponded to (Figure 5).

5 Conclusion

So, the capability of the flat reflection array to focus a monochromatic radiation at a certain point were studied. The element of the array based on waveguide with a controlled reflection coefficient was developed. The controlled phase shift is 180°. The array model based on waveguide elements was simulated in CST Microwave Studio. These simulations proved the monochromatic radiation focusing with the array made up of waveguide elements. A hardware and software complex to control the array is being currently developed. The proposed development of the reflection tomographic setup is considered to be a Russian analogue of the equipment from other countries. The proposed structure of the reflection array provides the high-rate focusing at a certain point.

Acknowledgment

The research was supported by the Russian Ministry of Education and Science as a part of the state order No 3.694.2014/K.

References

[1] Y. Kan, Y. Zhu, Q. Fu, MATEC Web of Conferences **56**, 02001 (2016) doi: 10.1051/matecconf/20165602001
[2] S. V. Hum, J. Perruisseau-Carrier, IEEE Transactions on Antennas and Propagation **62**, 183 (2014) doi: 10.1109/TAP.2013.2287296
[3] G. Leon, L. Herran, M. Munoz, F. Las-Heras, Y. Hao, Prog. Electromagn. Res. **54**, 125 (2014)
[4] M. Hajian, G. Vree, L. Ligthart, IEEE Antennas Propag. Mag. **45**, 13 (2013) doi: 10.1109/MAP.2003.1252806
[5] S. Shipilov, A. Eremeev, V. Yakubov, J. Phys.: Conf. Ser. **671**, 012052 (2016) doi: 10.1088/1742-6596/671/1/012052

MATEC Web of Conferences **79**, 01071 (2016) DOI: 10.1051/matecconf/20167901071
IME&T 2016

THE AUTOMATED SYSTEM FOR PARAMETRIC AND FUNCTIONAL CONTROL OF THE LOGIC ICS

George Davydov, *Anna* Kolosova*, *Adeliya* Sharipova, *Dmitriy* Boychenko

National Research Nuclear University MEPhI (Moscow Engineering Physics Institute), 115409, Moscow, Russia

Abstract. The article discusses implementation of the automated system based on the use of National Instruments PXI platform and designed for parametric and functional control of the logic ICs. The features of specialized software and typical results, obtained using the developed system, are observed.

1 Introduction

Logic ICs are widely used in all complex systems and devices. The main advantage of logic elements, in comparison with other digital chips, is their high performance (low delay times), as well as low supply power. During the certification tests it is often necessary to develop special software for each type of IC separately [1, 2]. The universal automated system developing is an urgent task for unification of the radiation test for different types of logic ICs.The work is dedicated to the development of an automated system for parametric and functional control of the logic ICs.

2 Equipment and software

The automated system has been developed for this research is based on National Instruments PXI platform [3] and consists of chassis with modular measurement units, as well as specialized software.

The following hardware was used in the developed system: NI PXIe-1073 (chassis), NI PXI-4110 (power supply), NI-6229 (digital and analog data acquisition), NI-5114 (oscilloscope), NI-2529 (relay switches), NI-6723 (digital and analog output), PXIe-6556 (high-speed digital input/output with PPMU and 200MHz bandwidth).

The specialized software was designed using the NI LabVIEW for manage the modular devices.

* Corresponding author: askol@spels.ru

© The Authors, published by EDP Sciences. This is an open access article distributed under the terms of the Creative Commons Attribution License 4.0 (http://creativecommons.org/licenses/by/4.0/).

3 Solution

The system is designed as the real-time monitor of the main parameters of ICs before, during and after irradiation. During the influence, the device under test (DUT) is operating in one of the preset electric modes (static / dynamic mode, low power mode, "shutdown" mode, etc) [4]. The most typical parameters of DUT are shown in Table 1.

Table 1. Typical parameters for logic ICs.

Parameter	Range	
	min	max
Power Supply current I_{CC}	≤1nA	≥100 mA
Input current I_I	≤1nA	≥100 mA
High output voltage V_{OH}	-1V	Ucc (0…6V)
Low output voltage V_{OL}	-1V	Ucc (0…6V)
Functional control	Static	≤100MHz

Any test means the different regimes of DUT for measurement of each its parameter. Modes and parameters are regulated in DUT's datasheet, which usually include typical and specific parameters of IC.

The system software allows saving measurement results in useful and friendly format.

Figure 1 illustrates the control panel of the developed system.

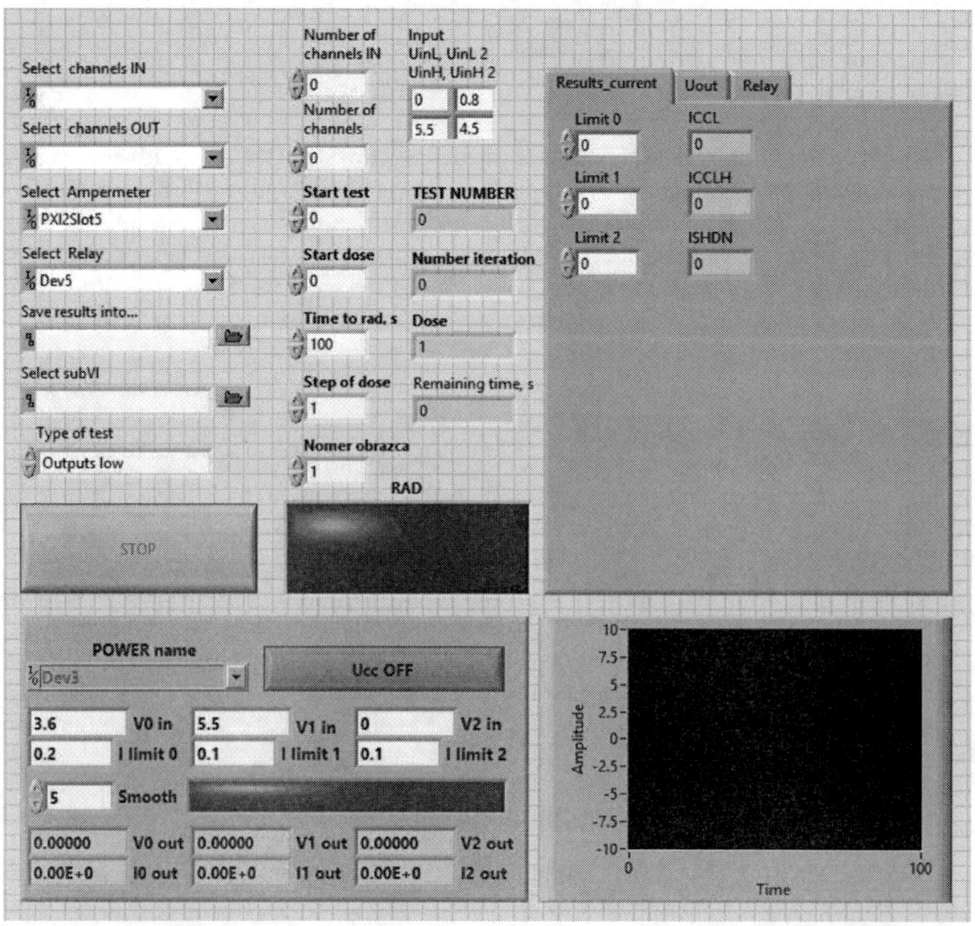

Figure 1. The control panel of the developed system.

DOI: 10.1051/matecconf/20167901071

The specialized software contains several independent subprograms operating in parallel and communicating through the local variables:

- power source control subprogram sets the output voltage and maximum output current and performs continuous measurement of the actual values of output voltages and currents on the 3 channels, automatically switching the current measurement ranges on 2 channels and quickly disconnecting the power circuits in the case of latchup effect.
- NI-5105 – allows to monitoring the functioning of the chip charts in real time.
- Subprogram of the current measurements can automatically switch between all available current ranges of DMM PXI-4071 and carry out the current measurements of less than 1 nA at a frequency down to 5 Hz (with down to 500 measurements in each point).
- The switch module subprogram ensures the correct relay switching and protects against switching collision.
- Subprogram of test algorithm determines the sequence of the test procedures, the sequence of parametric control, as well as the time of start / stop of the source of influence.
- NI-6723, NI-6229 and NI-6556 – the digital and analog data processing subprogram allows setting / reading of analog and digital signals.
- DUT mode, actual values and data timing are managed by the main algorithm subprogram.
- Subprogram of source of exposure control allows to initiate or stop the impact on chip. View subprogram determined by the type of exposure source.

4 The typical results of tests performed by the developed system

The tests results of the two IC studies, using the developed system, are listed below. The DUTs were SN74AHC14D and SN74LVC14ADR – 6 independent Schmitt-Trigger Inverters.

The AHC series (Advanced High-Speed CMOS – an advanced high speed CMOS structure) improves performance in low-speed input signal power in the presence of significant interference; provide additional protection against static discharge circuit and overload outlets and have a relatively large range of supply voltage [5].

The LVD series (Low Voltage CMOS) is specifically designed to work with low-voltage and low power consumption [6]. The CMOS LVC family provides devices with supply voltage in the range from 3.3 V to 5 V.

The radiation test was performed by using the Cs137 isotopic source "Gamma-Panorama MEPhI". During the experiment the parameters are contain in Table 1 were controlled.

The truth table for DUT is usually described in its datasheet.

It was found that the critical functional mode of both ICs is the regime providing low output static voltage. TID dependences of the supply current of both of the tested ICs are shown in Figure 2. The change of the other monitoring parameters with TID was insignificant.

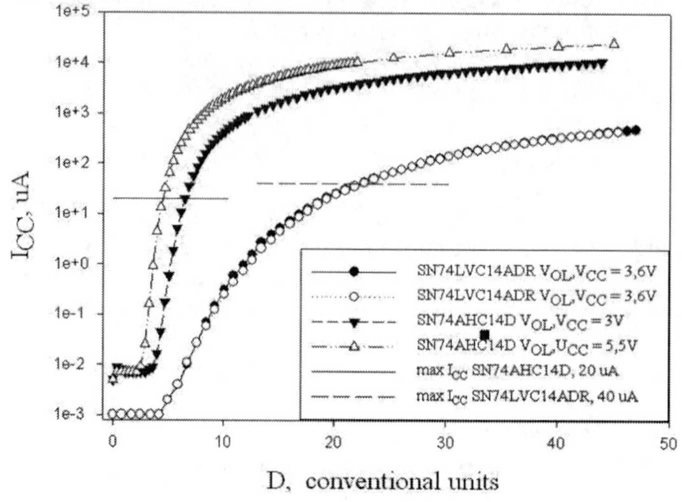

Figure 2. TID dependence of the supply current for tested ICs in various regimes.

The levels of parametrical failure have been determined for tested ICs. LVC devices possess more stable radiation behavior in comparison with AHC devices. Probably, such difference may be caused by presence of guard output structure in AHC series; that structures may be not optimized for TID influence.

Research in a range of temperatures was conducted using the developed system. The object of research was the logical IC - 6 logic elements "NOT" made by bipolar technology. Studies were carried out at ambient temperatures of 25 ° C ± 10 ° C, 125 ° C and -60 ° C.

During operation it was found that the chips have a high radiation resistance. Figure 3 shows TID dependence of the supply current increment for tested ICs in various temperatures.

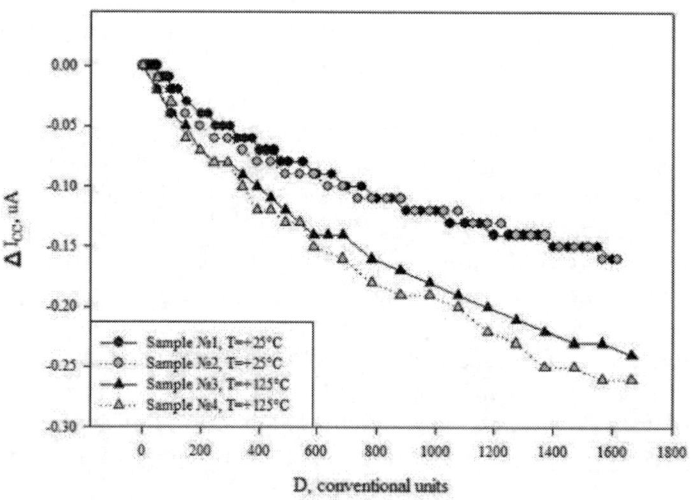

Figure 3. TID dependence of the supply current increment for tested ICs in various temperatures.

5 Conclusions

PXI platform equipment with LabVIEW development system is an optimal solution for engineering hardware and software control systems like described in the article.

The developed system allows to carry out research at other environmental conditions and facilities (pulsed electron accelerators, electron LINACs, X-ray testers etc.)

The developed automated system has successfully implemented at JSC "ENPO SPELS", Moscow, Russia, and was used to test more than 20 IC types. The plan for further work in this direction is to expand the developed system on other IC types and to implement the elements of this system into existing systems for parametric control of electronic devices [7, 8].

Acknowledgements

The paper is a part of the research done within the government assignment given by the ministry of education and science of Russian Federation (ref. number RFMEFI57414X0115).

References

[1] A.Y. Nikiforov, D.V. Boychenko, V.A. Telets, Book of abstracts RAD-2014, 300 (2014)

[2] D.V. Boychenko, O.A. Kalashnikov, A.Y. Nikiforov, A.V. Ulanova, D.V. Bobrovsky, P.V. Nekrasov, Facta Universitatis Series: Electronics and Energetics **28** , 153 (2015)

[3] PXI Platform, National Instruments Corporation URL: http://sine.ni.com/np/app/main/p/ap/global/lang/ru/pg/1/sn/n24:PXI-FSLASH-CompactPCI/

[4] G. G. Davydov, A. S. Kolosova, L. N. Kessarinskiy, D.V. Boychenko, 2015 International Siberian Conference on Control and Communications (SIBCON), 7147094 (2015) doi : 10.1109/SIBCON.2015.7147094

[5] AHC/AHCT Designer's Guide, TI, 2000 URL: http://www.ti.com/lit/ml/scla013d/scla013d.pdf.

[6] LVC Designer's Guide, TI, 1996 URL: http://www.ti.com/lit/ml/scba010/scba010.pdf.

[7] M. Z. Malik, A. Farooq, A. Ali, G. Chen, MATEC Web of Conferences **40,** 07001 (2016) doi : 10.1051/matecconf/20164007001

[8] W. Song, W. Zhang, MATEC Web of Conferences **22,** 03001 doi : 10.1051/matecconf/20152203001

FUSION OF THRESHOLDING RULES DURING WAVELET-BASED NOISY IMAGE COMPRESSION

Yury Bekhtin[1,a], *Alexander* Lupachev[2], *Dmitry* Titov[3], *Vladimir* Siryamkin[4,5]

[1] Ryazan State Radio Engineering University, 390005, Ryazan, Russia
[2] National Research University "Moscow power engineering Institute", 111250, Moscow, Russia
[3] Southwest State University, 405000, Kursk, Russia
[4] Tomsk State University, 634050, Tomsk, Russia
[5] Tomsk Polytechnic University, 634050, Tomsk, Russia

Abstract. The new method for combining semisoft thresholding rules during wavelet-based data compression of images with multiplicative noise is suggested. The method chooses the best thresholding rule and the threshold value using the proposed criteria which provide the best non-linear approximations and take into consideration errors of quantization. The results of computer modeling have shown that the suggested method provides relatively good image quality after restoration in the sense of some criteria such as PSNR, SSIM, etc.

1 Introduction

Images which are registered from different sources and which have to be transferred or archived can be distorted by specific noises having a multiplicative character. In many cases, video sources and data links add their own noises during image formation and transferring. For example, we can discuss about synthetic aperture radar (SAR) images with multiplicative noise known as speckle, infrared devices with fixed pattern noise (FPN) like unstable photo element's voltage sensitivity, etc. There are a lot of methods and algorithms invented in last decades trying to effectively filter and/or compress noisy images. In the most cases, the tasks of filtering and compression are decided separately. Therefore, filtering and compression algorithms are not harmonized, that leads to new distortions and artifacts onto the images restored after compression. Moreover, being theoretically focused on an additive Gaussian noise model, any filtering algorithm inevitably leads to unsatisfactory results.

The good idea is to combine filtering and compression within one and same procedures of noisy image processing. The main decisions which we have chosen for this overview are relatively old and can be grouped into three sets. The first set of methods presented by works [1, 2] was developed using the Rissanen's principle of minimal description length (MDL) [3]. Here, the final compression ratio is determined by the achieved maximal quality of the decompressed image. The second set of methods presented by works [4, 5]

[a] Corresponding author: yuri.bekhtin@yandex.ru

© The Authors, published by EDP Sciences. This is an open access article distributed under the terms of the Creative Commons Attribution License 4.0 (http://creativecommons.org/licenses/by/4.0/).

exploits the idea of so-called optimal operation point (OOP) which is the compression ratio or bit rate (BR) measured in bit per pixel (bpp), when the maximal peak-to-signal noise ratio (PSNR) is achieved. Existence of OOP can be explained by existence of so-called "dead zone", an interval of quantization near zero, size of which is changed in dependency on the given bit rate for any codec. Again, OOP provides the desired quality of compressed images but does not guarantee the necessary bit rate for transmission through channels with fixed bandwidths. The third set of methods presented by our works [6, 7] and other works is based on forming an image of high quality under the given bit rate (quota of bits).

The most of mentioned works uses wavelet transformation which allows consolidating filtering and compression by thresholding techniques. Usually, so-called fast discrete wavelet transform (FWT) [8, 9] is the base of many suggested methods and algorithms. This wavelet transform in the form of dyadic decomposition has been applied to image processing by S.Mallat [8]. The image is decomposed into several high-frequency pseudo images (subbands) containing wavelet coefficients of details in horizontal, vertical and diagonal orientations with increasing scale. Different thresholding rules are the bases of wavelet filtering methods [10, 11]. There are some popular shrinkage policies: semisoft, non-negative and *n*-degree garrote and hyperbole thresholding [10]. Nevertheless, semisoft thresholding where hard, soft and Vidacovic thresholding rules are related is the most popular in practice because the one parameter (the threshold value) is needed to calculate only. Otherwise, non-negative garrote rule [10], for instance, requires two parameters to be estimated that gives an additional optimization problem. Concentrating on semisoft thresholding rules in this paper, we can note that the effectiveness of filtering for any thresholding rule depends on different image features such as texture, type and intensity of noise, spatial and radiometric resolutions, etc. In other words, each of thresholding rules can win in the contest for the best quality of the filtered image. Moreover, the situation becomes unpredictable when the noised image is undergone by wavelet-based compression that gives errors of so-called non-linear approximation and quantization [12]. Therefore, this paper comprises an attempt to develop the theoretical base to fuse different semisoft thresholding rules during data compression of noisy images that provides the fully automatic scheme of choosing the best thresholding rule and the corresponding threshold value.

This paper consists of five sections. After this introduction (Section 1), Section 2 describes the problem definition. Section 3 contains the obtained criteria for different thresholding rules and discusses what to expect from the fusion of the thresholding rules. The results of computer modeling and comparison are represented in Section 4. Conclusion remarks are reflected in Section 5.

2 Problem definition

Let us assume that the compression is subjected to the observed signal Y, which is the projection of the unknown original image X distorted by multiplicative noise Z with unity mean and variance σ_z^2:

$$Y = XZ. \tag{1}$$

The noise Z may have different probability density function (pdf), for instance, an exponential pdf for speckle in SAR images; normally distributed for FPN in infrared sensors, etc.

Multiscale analysis provides a decomposition of a noisy signal (1) in the form of FWT for a given number of levels Q. Because the wavelet transform is the result of successive convolutions, we define an operator $W^{[j]}$ that generates the wavelet coefficients at each level j, $j=1,...,Q$. Then, the wavelet decomposition of a noisy signal with a fixed basis may be represented as follows:

$$W_Y = W^{[j]}Y = W^{[j]}XZ = W^{[j]}X + W^{[j]}X(Z-1)$$
$$= W^{[j]}X + W^{[j]}XZ_c = W_X + W_\xi, \quad (2)$$

where $W_X = W^{[j]}X$, $W_\xi = W^{[j]}(XZ_c)$ are centered and uncorrelated random processes

$$E[W_X] = 0,\ E[W_\xi] = 0,$$
$$E[W_X W_\xi] = W^{[j]}W^{[j]}E[Y]E[Z_c] = 0.$$

Therefore, in wavelet domain, the multiplicative model (1) becomes the additive one (2).

Any wavelet coefficient can be represented as

$$w_{Y_i}^{[j]} = w_{X_i}^{[j]} + w_{\xi_i}^{[i]},\quad i = 1,...,I^{(j)},\ j = 1,...,Q, \quad (3)$$

here $I^{(j)}$ is the quantity of wavelet coefficients at each level j, $j = 1,...,Q$, and the noise in wavelet coefficients can be shown as:

$$w_{\xi_i}^{[j]} = \sigma_{w_\xi}^{[j]} \cdot e_i,$$

where $|e_i| \le 1$, $\forall i \in I^{(j)}$, $\sigma_{w_\xi}^{[j]} > 0$, $j = 1,...,Q$, and here, we consider conditionally that e_i is Gaussian noise with zero mean, and $\sigma_{w_\xi}^{[j]2}$ stands for its variances which are unknown, $j = 1,...,Q$. Therefore, we complicated the model for noise which is used for computer modeling and reflects heterogeneous features of an image in FWT subbands. There are some works where different hypotheses about noisy wavelet coefficients distribution are suggested. The generalized Gaussian distribution has been very well adopted in imagery on wavelets (see, e.g., [13]). Here, we consider that the high-pass filtering leads to an approximately symmetrical form of the distribution for wavelet coefficients. Hence, the properties of multiplicative noise in spatial domain are changed in such manner that we can use the additive white noise model in wavelet domain. This is our lonely presumption which has no the strong mathematical background.

Thus, the problem is to perform data compression of a noisy signal (1) by encoding the wavelet coefficients (3) in such way that we could improve and estimate a quality of the processed noised image. It can be done by use of mean square error (MSE) criterion in the form of the Euclidean norm which can be written as follows

$$E\left\{\left\|W_X - W_{\hat{X}}\right\|^2\right\} \to \min, \quad (4)$$

where $W_{\hat{X}}$ is the wavelet coefficients of the restored (after compression) image.

It is because the original signal is distorted, then the coefficients $W_{\hat{X}}$ should be considered as estimators of the wavelet coefficients for the original signal X after processing of the noisy wavelet coefficients W_Y. It is shown in the papers [8, 9] that the error of restoration (4), which is computed in the wavelet domain, is equivalent to the averaged square of the norm of the error in the spatial domain:

$$E\left\{\left\|X - \hat{X}\right\|^2\right\} = E\left\{\left\|W_X - W_{\hat{X}}\right\|^2\right\} \quad (5)$$

Let us suppose that the noisy signal (1) is represented by I discrete samples (pixels); in the case of FWT, it gives the same quantity of the wavelet coefficients I. Then, after compression, we have M so-called significant wavelet coefficients only ($M < I$) which are undergone by quantization before statistical encoding. If we ignore the possible small losses of the signal energy during encoding, then the estimators of the wavelet coefficients can be represented as follows:

$$w_{\hat{X}_k} = \begin{cases} \langle w_{Y_k} \rangle, \text{ if } k \le M, \\ 0, \text{ if } k > M+1, \end{cases} \quad \forall k \in [1,...,I], \tag{6}$$

where the brackets $\langle \cdot \rangle$ denote the operation of uniform quantization (*uq*). To calculate the total error caused by the approximation and quantization, we first consider the sum of squares of the errors in estimating the wavelet coefficients:

$$\begin{gathered} \sum_{k=1}^{I}(w_{X_k} - w_{\hat{X}_k})^2 = \sum_{k=M+1}^{I} w_{X_k}^2 + \sum_{k=1}^{M}(w_{X_k} - w_{Y_k} - \sigma_{uq\,k})^2 = \\ \sum_{k=M+1}^{I}(w_{Y_k} - w_{\xi_k})^2 + \sum_{k=1}^{M}(w_{\xi_k} + \sigma_{uqk})^2 = \sum_{k=M+1}^{I} w_{Y_k}^2 + \sum_{k=M+1}^{I} w_{\xi_k}^2 + \sum_{k=1}^{M} w_{\xi_k}^2 + \sum_{k=1}^{M} \sigma_{uqk}^2 = \\ \sum_{k=1}^{I} w_{Y_k}^2 - \sum_{k=1}^{M} w_{Y_k}^2 + \sum_{k=1}^{I} w_{\xi_k}^2 + \sum_{k=1}^{M} \sigma_{uqk}^2 . \end{gathered} \tag{7}$$

For the sake of notation simplicity, we enter the next terms:

$\sigma^2 = \frac{1}{I}\sum_{i=1}^{I} w_{Y_i}^2$ is the variance calculated for all of wavelet coefficients;

$\tilde{\sigma}^2 = \frac{1}{M}\sum_{i=1}^{M} w_{Y_i}^2$ is the variance of the thresholded (significant) wavelet coefficients, $M / I = \varepsilon$;

$\sigma_{W_\xi}^2 = \frac{1}{I}\sum_{i=1}^{I} w_{\xi_i}^2$ is the variance of wavelet-coefficients for noise with zero mean;

$\sigma_{uq}^2 = \frac{1}{M}\sum_{i=1}^{M} \sigma_{uq\ i}^2$ is the quantization variance for M significant wavelet-coefficients.

Then, the formula for the MSE criterion can be written as follows

$$E\left\{\left\|W_X - W_{\hat{X}}\right\|^2\right\} = \frac{1}{I}\sum_{i=1}^{I}(w_{X_i} - \hat{w}_{X_i})^2 = \sigma^2 - \varepsilon\tilde{\sigma}^2 + \sigma_{W_\xi}^2 + \varepsilon\sigma_{uq}^2 , \tag{8}$$

It follows from equality 8 that minimizing the value $E\left\{\left\|W_X - W_{\hat{X}}\right\|^2\right\}$ is equivalent to maximizing the value $\varepsilon(\tilde{\sigma}^2 - \sigma_{uq}^2)$. In order to obtain the best MSE, it is necessary to provide the maximal variance of the significant wavelet coefficients (because it gives the good signal approximation [8]) and the minimal variance of quantization errors. These requirements are contradictory, because the larger M of the significant wavelet coefficients increases the total quantization error simultaneously. On the other hand, we can control both these components in equality 8 by different methods including thresholding techniques.

Therefore, our problem can be described in the terms of the approximation theory where any thresholding rule can be considered as a procedure of non-linear approximation of the noised signal (image) when the maximal value $\varepsilon(\tilde{\sigma}^2 - \sigma_{uq}^2)$ is achieved. Having different thresholding rules, we have a nice opportunity to choose the best rule because we obtain different signal approximations. Therefore, the problem converges to the search of the generalized method suitable to estimate the effectiveness of any thresholding rule. Our solution placed below has been derived for so-called semisoft thresholding rules which are described in Section 3. For computer modeling and comparison, the quality of the restored after compression image is estimated by several criteria such as peak signal-to-noise ratio (PSNR) and structural similarity index (SSIM).

3 The fusion of thresholding rules

The main idea of our method is to iteratively refine a non-linear approximation of the noised image using so-called coherent structures which are the components of the image and which are strongly correlated with a chosen wavelet basis [15]. Having a variance σ_z^2, noise Z in Y can be determined as components of the decomposed image w_{ξ_k}, $k = 1,\ldots,I$, that have no a strong correlation with the given basis. If I is relatively large then there is the probability up to 1 that [8, 14]

$$\frac{\max_{1\le k\le I}\left|w_{\xi_k}\right|}{\|Z\|} \le \frac{\sqrt{2\ln I}\,\sigma_Z}{\sqrt{I}\,\sigma_Z} = \frac{\sqrt{2\ln I}}{\sqrt{I}} = \rho_I \tag{9}$$

for any orthonormal wavelet basis. The parameter ρ_I can be considered as the supreme estimator of the correlation coefficient for correlation between Gaussian noise and any given wavelet basis. It follows from equality 9 that this estimator does not depend on noise variance. It has been experimentally proven by our experiments [6] that the theoretical value of ρ_I is also the highest asymptote of the correlation coefficients for correlation between Daubechies wavelets, symmlets, some biorthogonal wavelets, on the one side, and, on the other side, noises having different pdfs and their combinations, e.g., log-normal, exponential, Gamma-distributed, etc.

It is known from the theory of wavelets [8, 14] that the nonlinear approximation of any signal decomposed by wavelet transform is referred as inverse wavelet transform W^1 for the first (significant) M wavelet coefficients after their sorting

$$\hat{X} = \sum_{k=1}^{M} W^{-1}\left\{w_{Y_k}\right\}. \tag{10}$$

where

$$\left|w_{Y_k}\right| \ge \left|w_{Y_{k+1}}\right|, \qquad \forall k \in [1,\ldots,I]. \tag{11}$$

The residual image Y_M can be computed as the difference between the input image and the pseudo image obtained by means of the inverse wavelet transform for the coherent structures

$$Y_M = Y - \sum_{k=1}^{M} W^{-1}\left\{w_{Y_k}\right\} = \sum_{k=M+1}^{I} W^{-1}\left\{w_{Y_k}\right\}, \tag{12}$$

The residual image Y_M can not be recognized as a noise if the following inequality

$$\rho^2(Y_M) = \frac{\left|w_{Y_M}\right|^2}{\sum_{k=M+1}^{I}\left|w_{Y_k}\right|^2} > \rho_{I-M}^2 \tag{13}$$

is true. Not taking into consideration the quantization errors, the estimator $\hat{X}$ of the original image is the sum of M coherent structures (10).

Hard thresholding. The equality 13 shows that pursuit of the coherent structures looks like hard thresholding (see Figure 1, *a*) of the wavelet coefficients using the threshold [8, 10, 14]

$$\tau = \rho_{I-M}\sqrt{\sum_{k=M+1}^{I}\left|w_{Y_k}\right|^2}. \tag{14}$$

Therefore, the threshold value τ is automatically set by any change in the quantity M of the significant wavelet coefficients. Taking into consideration equality 8 and the following conclusion from Section 2, we can derive the method of the noised image processing based on hard thresholding:

$$f(M) = \varepsilon(\tilde{\sigma}^2 - \sigma_{uq}^2) \to \max_M \tag{15}$$

$$\forall \left|w_{Y_i}\right|^2 \geq \rho_{I-M}^2 \sum_{i=M+1}^{I} \left|w_{Y_i}\right|^2 = \rho_{I-M}^2 I\left(\sigma^2 - \varepsilon\tilde{\sigma}^2\right). \tag{16}$$

Let the wavelet-coefficients of the noised image be arranged according to equality (11). Then, the proposed algorithm using the matching pursuit can be represented as follows [6].

1. Set $M = I - 1$ (initializing).
2. Compute the cumulative sum of wavelet-coefficient squares and, according to (14) calculate the threshold value for the given correlation coefficient ρ_{I-M}.
3. Compute the new (corrected) value M of the significant wavelet coefficients from the inequality (16).
4. Check: is the maximal value of the cost function (15) achieved?
5. If the cost function (15) has the maximal value then calculations are finished. Otherwise, go to the step 2.

It is possible to use different rules to finish calculations when the maximum of the cost function (15) is achieved, for example

$$f(M^{(n)}) - f(M^{(n-1)}) \leq \delta,$$

where δ is the reasonable error of calculations showing the accuracy of the algorithm; $M^{(n)}$ is the estimator of the quantity of the significant wavelet coefficients obtained at the iteration with number n, n=2, 3, ...

Because the algorithm presented by equalities 15 and 16 is based on the hard thresholding rule, then the estimators of the processed wavelet coefficients are automatically obtained as $w_{\hat{X}_i} = \langle w_{Y_i} \rangle$, $i = 1,\ldots,M$.

<u>*Soft thresholding*</u>. During soft thresholding (see Figure 1,*b*), the wavelet coefficients are also set to zero if their values are below the threshold, but simultaneously, there is a reduction of the details (the wavelet coefficients of high-frequency subbands of the FWT) to the threshold value $\tau = \left|w_{Y_M}\right|$ [8, 10, 11]:

$$w_{\hat{X}_k} = \begin{cases} \text{sign}(w_{Y_k})\left(\left|w_{Y_k}\right| - \left|w_{Y_M}\right|\right), & \text{if } \left|w_{Y_k}\right| \geq \left|w_{Y_M}\right|, \\ 0, \text{ if } \left|w_{Y_k}\right| < \left|w_{Y_M}\right|. \end{cases} \tag{17}$$

In practice, the processing scheme in the form (17) is applied to the details only; while the wavelet coefficients of the approximation remain unchanged. Therefore, we need to analyze the three components.

$$\frac{1}{I}\sum_{i=1}^{I}\left(w_{X_i} - w_{\hat{X}_i}\right)^2 = \frac{1}{I}\sum_{i=1}^{M}\left(w_{X_i} - w_{\hat{X}_i}\right)^2 + \frac{1}{I}\sum_{i=M+1}^{I} w_{X_i}^2 = \tag{18}$$

$$\frac{1}{I}\sum_{i=1}^{t}\left(w_{X_i} - w_{\hat{X}_i}\right)^2 + \frac{1}{I}\sum_{i=t+1}^{M}\left(w_{X_i} - w_{\hat{X}_i}\right)^2 + \frac{1}{I}\sum_{i=M+1}^{I} w_{X_i}^2,$$

where t is the number of the first-ordered wavelet coefficients corresponding the approximation. The third (last) term in equality 18 represents the error which is inevitably appeared due to the thresholding of the wavelet coefficients. We obtain for the first two summands in equality 18:

$$\frac{1}{I}\sum_{i=1}^{t}\left(w_{X_i}-w_{\hat{X}_i}\right)^2+\frac{1}{I}\sum_{i=t+1}^{M}\left(w_{X_i}-w_{\hat{X}_i}\right)^2=\frac{1}{I}\sum_{i=1}^{t}\left(w_{Y_i}-w_{\xi_i}-\left\langle w_{Y_i}\right\rangle\right)^2+$$
$$\frac{1}{I}\sum_{i=t+1}^{M}\left(w_{Y_i}-w_{\xi_i}-\left\langle\left|w_{Y_i}\right|-\left|w_{Y_M}\right|\right\rangle\mathrm{sign}(w_{Y_i})\right)^2=\frac{t}{I}\sigma_{\kappa в}^2+\frac{t}{I}\sigma_{W_\xi}^2+ \tag{19}$$
$$\frac{1}{I}\sum_{i=t+1}^{M}\left(\left|w_{Y_i}\right|-\left\langle\left|w_{Y_i}\right|-\left|w_{Y_M}\right|\right\rangle\right)^2+\frac{M-t}{I}\sigma_{W_\xi}^2\sim\varepsilon\sigma_{\mathrm{uq}}^2+\varepsilon\sigma_{W_\xi}^2+(\varepsilon-\theta)\left|w_{Y_M}\right|^2,$$

where $\theta=\frac{t}{I}$ is the coefficient showing how many wavelet coefficients of the approximation are in the total quantity of the wavelet coefficients. We assumed during our derivation of equality 19 that $\frac{1}{I}\sum_{i=t+1}^{M}\left|w_{Y_M}\right|\sigma_{\mathrm{uq}\,i}\approx 0$ because the numbers of quantization errors $\left|\sigma_{uq\,i}\right|\le\frac{\Delta}{2}$ with different signs are equiprobably ($\overline{\sigma}_{\mathrm{uq}}=0$) and $w_{Y_i}=\mathrm{sign}(w_{Y_i})\left|w_{Y_i}\right|$ (Δ is an interval of quantization). Finally, the equality 8 is modified for the case of soft thresholding:

$$E\left\{\left\|W_X-W_{\hat{X}}\right\|^2\right\}=\sigma^2-\varepsilon\tilde{\sigma}^2+\varepsilon\sigma_{\mathrm{uq}}^2+\sigma_{W_\xi}^2+(\varepsilon-\theta)\left|w_{Y_M}\right|^2. \tag{20}$$

Hence, we need to also modify the method of processing based on equality 15 and equality 16:

$$f(M)=\varepsilon(\tilde{\sigma}^2-\sigma_{\mathrm{uq}}^2)-(\varepsilon-\theta)\rho_{I-M}^2 I(\sigma^2-\varepsilon\tilde{\sigma}^2)\to\max_M \tag{21}$$

$$\forall\left(\left|w_{Y_i}\right|-\left|w_{Y_M}\right|\right)^2\ge\rho_{I-M}^2\sum_{i=M+1}^{I}\left(\left|w_{Y_i}\right|-\left|w_{Y_M}\right|\right)^2. \tag{22}$$

The algorithm corresponding to equalities 21 and 22 contains the same steps with some modifications as it is for the case of hard thresholding. The estimators of the processed wavelet coefficients can be calculated as follows:

$$w_{\hat{X}_i}=\begin{cases}\left\langle w_{Y_i}\right\rangle, & i=1,\ldots,t,\\ \mathrm{sign}(w_{Y_i})\left\langle\left|w_{Y_i}\right|-\rho_{I-M}\sqrt{I(\sigma^2-\varepsilon\tilde{\sigma}^2)}\right\rangle, & i=t+1,\ldots M,\\ 0, & i=M+1,\ldots,I.\end{cases} \tag{23}$$

<u>*Vidacovic thresholding*</u>. The thresholding function derived by Vidacovic [10] comprises some "medium" variant of semisoft thresholding as it can be seen from Figure 1, *c*:

$$w_{\hat{X}_k}=\begin{cases}\mathrm{sign}(w_{Y_k})\left\langle\sqrt{\left|w_{Y_k}\right|^2-\left|w_{Y_M}\right|^2}\right\rangle, & \text{if } \left|w_{Y_k}\right|\ge\left|w_{Y_M}\right|,\\ 0, & \text{if } \left|w_{Y_k}\right|<\left|w_{Y_M}\right|.\end{cases} \tag{24}$$

We obtain after some algebra:

$$E\left\{\left\|W_X-W_{\hat{X}}\right\|^2\right\}=\sigma^2-\varepsilon\tilde{\sigma}^2+\varepsilon\sigma_{uq}^2+\sigma_{W_\xi}^2+\frac{1}{I}\sum_{k=t+1}^{M}\left(\left|w_{Y_k}\right|-\sqrt{w_{Y_k}^2-\left|w_{Y_M}\right|^2}\right)^2 \tag{25}$$

The corresponding algorithm for noisy image processing can be also described by two equations:

$$f(M)=\varepsilon(\tilde{\sigma}^2-\sigma_{\mathrm{uq}}^2)-\frac{1}{I}\sum_{i=t+1}^{M}\left(\left|w_{Y_i}\right|-\sqrt{w_{Y_i}^2-\rho_{I-M}^2 I(\sigma^2-\varepsilon\tilde{\sigma}^2)}\right)^2\to\max_M \tag{26}$$

$$\forall\left(\left|w_{Y_i}\right|^2-\left|w_{Y_M}\right|^2\right)\ge\rho_{I-M}^2\sum_{i=M+1}^{I}\left(\left|w_{Y_i}\right|^2-\left|w_{Y_M}\right|^2\right) \tag{27}$$

The estimators of the processed wavelet coefficients can be found as follows:

$$w_{\hat{X}_i} = \begin{cases} \langle w_{Y_i} \rangle, & i = 1,\dots,t, \\ \operatorname{sign}(w_{Y_i}) \left\langle \sqrt{\left| w_{Y_i} \right|^2 - \rho_{I-M}^2 I(\sigma^2 - \varepsilon\tilde{\sigma}^2)} \right\rangle, & i = t+1,\dots,M, \\ 0, \quad i = M+1,\dots,I. \end{cases} \quad (28)$$

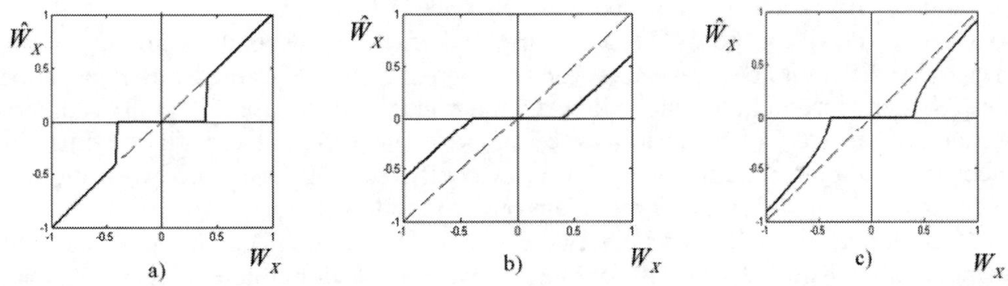

Figure 1. Thresholding rules: hard (a), soft (b), and Vidacovic (c).

Fusion of thresholding rules. In order to fuse different thresholding rules, we need to perform some formalization. Suppose that there is a set of q thresholding algorithms $\{A_1, A_2, \mathrm{K}, A_q\}$, which can search for coherent structures in the chosen wavelet basis. Let the results of executing the corresponding algorithms be the values of the objective functions $\{f_i(M_i),\}_{i=1}^{q}$ and the set of the corresponding values $\{M_1, M_2, \mathrm{K}, M_q\}$. Then, to select the best coherent structures, we must use the criterion:

$$M_{opt} = M(A_{i^*}) = \arg\max_{i=1,\dots,q} \{f_i(M_i)\}. \quad (29)$$

Therefore, equality 29 determines the best thresholding algorithm A_{i^*} (or/and the corresponding thresholding rule).

The obtained wavelet coefficients are sent to any wavelet coder where they are statistically encoded and transferred. The restored image can be considered as the sum of the coherent structures (in the form of equality 10) which are restored by the corresponding decoder on the side of recipient or from the database (in the case of archiving).

Our expectations. In the case of data compression of noisy images, we have too many factors influencing on the final result. Therefore, we need to determine the strategy of using the suggested method which can answer, by our opinion, on the following two important questions:

1) What, in average, have we to expect from the fusion of thresholding rules?
2) How "to tune" the suggested method for practical applications, when, for example, bandwidths of data links are limited?

In order to obtain an answer for the first question, we should remind that there were a lot of theoretical works devoting to wavelet shrinkage. In last decades, Mallat, Coifman, DeVore, Donoho, Johnston and their collaborators explored statistical optimality and risks of wavelet shrinkage for cases of signal processing, approximation, and statistics (see, e.g., [8, 10, 15, 16]).

These works help to find out the high estimators of risks for any thresholding rules; however, it does not allow us to compare or choose the best thresholding rule in the given cases.

On the other side, there were a lot of works containing the results of practical applying of the semisoft thresholdiong rules to different fields of imagery. Unfortunately, in those works, comparison is limited by types of images and noises (see, e.g. [17, 18]). If we take this way for proving effectiveness of the suggested method, then we can put in the difficult situation of endless comparing by pairs, for example, hard and soft thresholding under different bitrates, etc.

Nevertheless, here, we try to give our (maybe, "too engineering") answer for the first question; our thoughts are not controversial to the main theoretical results.

When we consider a noisy image, then the behavior of the ordered wavelet coefficients (equality 11) is changed; we observe a difference between two curves for small wavelet coefficients (details of the FWT) (see Figure 2). Differences depend on intensity of noise. It means that different noise variances lead to different threshold values; however, it does not mean that any thresholding rule will be a winner in the given case. Figure 3 (in the form of sketch for the case of a strong noise) shows some different situations which can be encountered during modification of wavelet coefficients in accordance with thresholding rules (14), (17), and (24). These situations have the following explanations.

If the input image (1) is relatively clear (noise is too small) then we observe that the quantity of undistorted wavelet coefficients with small amplitudes (details) is relatively large; hence, we can choose the small threshold value $\left|w_{Y_M}\right|$ to set these coefficients to zero for compression. It is obviously, that the hard threshodilng rule has to win because no modification is done in the residual wavelet coefficients. In the case of noisy image, the situation is not so obvious because the dynamic rate of wavelet coefficients related to details is increased due to noise; hence, the threshold value should be increased. If we register the case of a strong noise, then, we consider, the soft trhesholding rule should be a winner because the residual distorted wavelet coefficients are changed (smoothed). It can be noted from Figure 3, where "soft" modification sets the part of the significant wavelet coefficients (the residual details) to be close to "an ideal", high frequency noise-free coefficients. Also, as it seen from Figure 3, in general, the Vidacovic thresholding rule takes an intermediate position between soft and hard threshoding rules. It means that the Vidacovic thresholding rule modifies wavelet coefficients related to middle frequencies of the signal. Hence, we cannot predict all possible results.

Therefore, we should go to the second question because error of quantization depends also on available bit rate for data links or the given quota of bits for image archiving. The problem is that the interval of uniform quantization depends on both the given bit rate and the dynamic rate of the residual wavelet coefficients:

$$\Delta = 2\left(\left|w_{Y_1}\right| - \left|w_{Y_M}\right|\right)\cdot 2^{-R_C/M}, \tag{30}$$

where R_C is the given budget of bits; $\left|w_{Y_1}\right|$ is the first coefficient from the ordered wavelet coefficients having the maximal value. In the cases of low bitrates ($R_C < I$), there are possibilities to tune a full scheme of noisy image compression if to determine so-called optimal operation point (OOP) [4, 5] of a coder. It has also been experimentally proved by the author for a case of multiplicative noise that any coder has such bit rate (i.e. OOP) where the maximal peak signal-to-noise ratio (PSNR) is achieved.

Therefore, our experiments have to show that the suggested method chooses different thresholding rules and OOPs in dependency on noise variance and image textures.

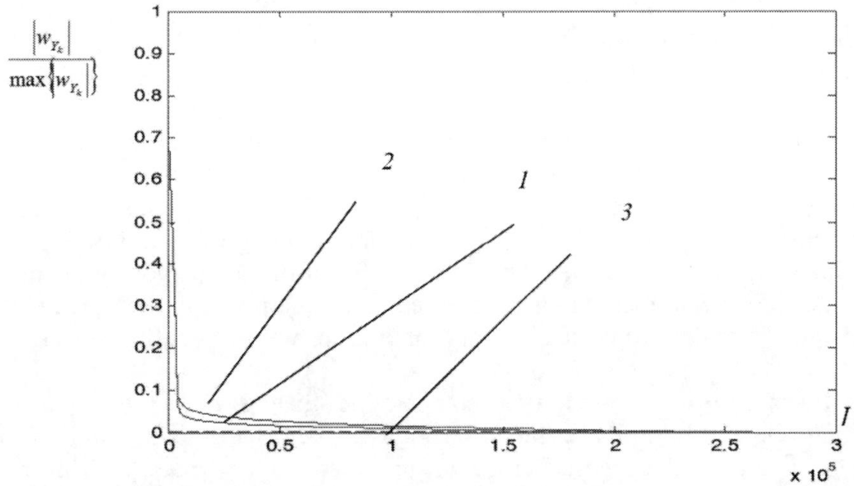

Figure 2. The curves of the ordered wavelet coefficients of the test mage sized 512×512 with added noise: $\sigma^{[j]}_{w_\xi}$ =23.02 (1), 39.18 (2), and 0 (3); $j=1,\ldots,Q$, Q=3.

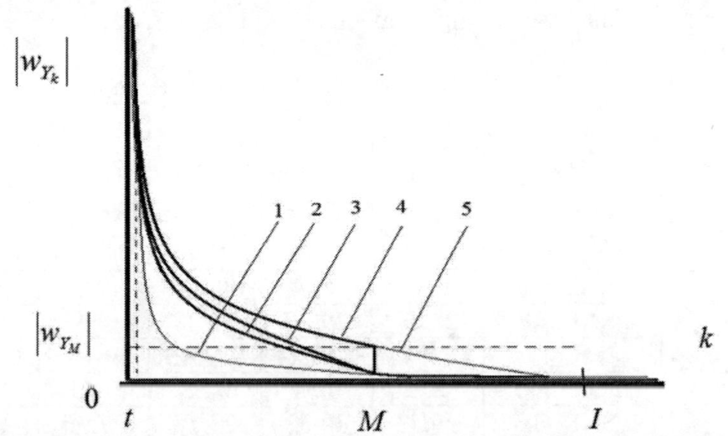

Figure 3. The sketch showing curves of the ordered wavelet coefficients for noise-free image (1), after soft (2), Vidacovic (3), hard (4) thresholding, and for an image distorted by a strong noise (5).

4 Results of modeling

For our experiments, we formed the library of test images containing 25 reference images which have been taken from the library described in [20]. The initial formats all of images were converted into the gray scale format with 8 bpp. Moreover, then, all of the converted images were divided into two groups with homogeneous and heterogeneous textures. The type of texture was determined by use of so-called variation coefficients which can serve as indicators of homogeneity within the given window [17]. Image is considered to be homogeneous if the quantity of points (pixels) from the set of homogeneous points is more than the reasonable threshold. We set this threshold value to 60 %.

We used the FWT of three levels (Q=3) based on the wavelet CDF 9.7 (Cohen-Daubechies-Feauveau). To simulate a multiplicative noise, the random number generators were applied to model speckle-noise with exponential pdf and unity mean. The variance of noise was a subject to change during experiments. In order to get a completely full processing of noised images, we used SPIHT (Set Partitioning In Hierarchical Trees) coder software [19] which has been modified to regulate the dead zone in dependency of the results obtained by the chosen thresholding rule (equality 30). The bit rate value R_C / I was changed within low bitrates (up to 1 bpp) with the step 0.01 bpp in order to find out the OOP.

The results of computer modeling are represented in Table 1. The table contains results in the form of different characteristics such as PSNR, SSIM [21] and OOP. The results obtained by the suggested method (i.e., the chosen thresholding rule) are marked. We compared the obtained results with "not-chosen" thresholding rules and SURE and Oracle thresholding rules under the same OOP. The SURE thresholding algorithm is described in [8]. "An ideal denoising method" was represented by the principle of "Oracle" when the threshold value is calculated if the original (non-noised) wavelet coefficients are known [8, 16].

Some distorted and restored after compression images with the obtained optimal thresholding rule and bit rate are shown on Figures 4–11. One can see from Table 1 and Figures 4–11 that the suggested method provides better image enhancement both numerically and visually.

In spite that there are other thresholding rules, we do not place here results of comparison with these rules because it is difficult to find out any statistical lows during pair comparing. Moreover, we understand that our method cannot be a winner in many cases if to compare to powerful and very complex methods as [1, 2, 6, 7, 13, 14, 22].

Table 1 The results of comparison.

Image		Homogeneous			Heterogeneous		
Thresholding rule	$\sigma_{w_\xi}^{[J]}$	17.43	28.56	57.76	13.82	38.23	54.46
Hard thresholding	PSNR, dB	**32.41**	27.88	22.32	**30.34**	25.88	20.44
	SSIM	**0.9834**	0.7366	0.5832	**0.9234**	0.6646	0.4832
	OOP, bpp	**0.52**	0.34	0.23	**0.43**	0.26	0.14
Soft thresholding	PSNR, dB	28.67	29.27	**26.38**	26.45	27.72	**24.82**
	SSIM	0.7845	0.8734	**0.6745**	0.6445	0.6941	**0.6126**
	OOP, bpp	0.52	0.34	**0.23**	0.43	0.26	**0.14**
Vidacovic thresholding	PSNR, dB	29.73	**30.32**	23.75	27.35	**28.24**	21.23
	SSIM	0.8341	**0.9231**	0.6154	0.6843	**0.8137**	0.5247
	OOP, bpp	0.52	**0.34**	0.23	0.43	**0.26**	0.14
SURE	PSNR, dB	30.36	28.21	24.21	28.64	26.93	22.84
	SSIM	0.8874	0.8122	0.6327	0.8341	0.7122	0.5878
	OOP, bpp	0.52	0.34	0.23	0.43	0.26	0.14
Oracle	PSNR, dB	32.68	31.32	27.12	31.81	29.12	25.22
	SSIM	0.9882	0.9521	0.6823	0.9327	0.8716	0.6445
	OOP, bpp	0.52	0.34	0.23	0.43	0.26	0.14

DOI: 10.1051/matecconf/20167901072

Figure 4. Original (noise-free) image.

Figure 5. Distorted image with $\sigma_{w_\xi}^{[J]}$ =18.4.

Figure 6. Distorted image with $\sigma_{w_\xi}^{[J]}$ =27.8.

Figure 7. Distorted image with $\sigma_{w_\xi}^{[J]}$ =42.3.

DOI: 10.1051/matecconf/20167901072

Figure 8. Restored image from Figure 5 (hard thresholding, OOP 0.43 bpp).

Figure 9. Restored image from Figure 6 (Vidacovic thresholding, OOP 0.27 bpp).

Figure 10. Restored image from Figure7 (soft thresholding, OOP 0.12 bpp).

Figure 11. Restored image from Figure7 by SURE, OOP 0.12 bpp.

5 Conclusion

The problem of noisy image compression is very complex; and, certainly, the suggested method does not pretended to decide related tasks in full. The main aim of this paper is to demonstrate that there are possibilities to build more flexible schemes of noisy image processing. The suggested method can be expanded by means including other thresholding rules, e.g. non-negative and *n*-degree garrote. However, it will require using a multi-criteria problem definition.

Because the suggested method exploits the idea of pursuit the coherent structures, than it leads to relatively high computational and timing expenditures. We estimate the complexity of our method as $O(3I\log_4 I)$ operations taking into consideration that we have three thresholding rules and choosing the quantity M of the significant wavelet coefficients is similar to search of the best path during wavelet package processing. In spite, we consider that there are some possibilities to decrease these expenditures due to using so-called pipe-line and unrolling loop methods of programming to seek the maxima of cost functions (15), (21), and (26) in parallel. It allows us to promote this method for onboard mage processing using FPGA implementations.

Acknowledgment

The paper was written as part of the research project No. 8.2.31.2015, carried out with the support of the Program "Research Foundation of Tomsk State University named after D.I. Mendeleev" in 2015 – 2016, grant RFBR No. 16-29-04388/16.

References

[1] S. Chang, B. Yu, and M. Vetterli, IEEE Trans. Image Process. **9**, 1532 (2000)
[2] N. Saito, Simultaneous noise suppression and signal compression using a library of orthonormal bases and the minimum description length criterion in Wavelets in Geophysics, Academic Press, New York, 299, (1994).
[3] J. Rissanen, IEEE Trans. on Info. Theory. **46** 2537 (2000) doi: 10.1109/18.887861
[4] O. Al-Snaykh and R. Mercereau, IEEE Trans. Image Process. **7**, 1641 (1998) doi: 10.1109/83.730376
[5] R. Öktem and K. Egiazarian, Electron. Lett. **35**, 1830 (1999) doi: 10.1049/el:19991261
[6] Y. Bekhtin and A. Bryantsev, Proc. SPIE **8179**, (2011) doi: 10.1117/12.897902
[7] Y. Bekhtin, Joint adaptive wavelet thresholding and bit allocation for data compression of noisy images, 2007, http://ticsp.cs.tut.fi/images/a/a1/Cr1006.pdf (2015)
[8] S. Mallat, A wavelet tour of signal processing (Academic Press, 1999)
[9] M. Shensha, IEEE Trans. on Signal Proc. Papers **40**, 2464 (1992) doi: 10.1109/78.157290
[10] B. Vidakovic , Statistical Modeling by Wavelets, John Wiley & Sons (1999)
[11] A. Bruce and H. Gao, Proc. of SPIE **2569**, 270 (1995)
[12] A. Chambolle, R. DeVore, N-Y. Lee *et al.*, IEEE Trans. Image Processing **7**, 319 (1998) doi: 10.1109/83.661182
[13] F. Argenti, T. Bianchi, L. Alparone, IEEE Trans. on Image Processing. **15**, 3385 (2006) doi: 10.1109/TIP.2006.881970
[14] S. Mallat and Z. Zhang, IEEE Transactions on Signal Processing **41**, 3397 (1993) doi: 10.1109/78.258082
[15] R. DeVore, B. Jawerth, B. Lucier, IEEE Trans. on Info. Theory. **38**, 719 (1992) doi: 10.1109/18.119733
[16] D. Donoho, IEEE Trans. on Info. Theory **41**, 613 (1995) doi: 10.1109/18.382009
[17] E. Hervet, R.. Fjortoft, P. Marthon *et al.*, SPIE **3497**, 43 (1998) doi: 10.1117/12.331361
[18] M. Mittal, V. Singh, R. Krishnan, ISSPA 1999–Proc. of the Fifth International Symposium on Signal Processing and Applications **2**, 781 (1999) doi: 10.1109/ISSPA.1999.815788
[19] A. Said, W.A. Pearlman, IEEE Trans Circ. and Syst. Video Tech. **6**, 243 (1996) doi: 10.1109/76.499834

[20] N. Ponomarenko, V. Lukin, A. Zelensky, K. Egiazarian, M. Carli, F. Battisti, Advances of Modern Radioelectronics **10,** 30 (2009)

[21] Z. Wang, A. Bovik, H. Sheikh, E. Simoncelli, IEEE Transactions on Image Proc. **13**, 600 (2004)doi: 10.1109/TIP.2003.819861

[22] D.V. Shashev, S.V. Shidlovskiy, Optoelectronics, Instrumentation and Data Processing **51**, 227 (2015) doi: 10.3103/S8756699015030036

MATEC Web of Conferences **79**, 01073 (2016) DOI: 10.1051/matecconf/20167901073
IME&T 2016

RADIOMETRIC RECEIVER FOR PASSIVE MICROWAVE IMAGING SYSTEM

Anton Ubaichin[1*], *Alexander* Filatov[2], *Egor* Alexeev[2], *Gregory* Zhuk[2], *Maxim* Anishin[2], *Stanislav* Gazitov[2], *Sergey* Tarasov[2], *Boris* Utkin[2]

[1]Tomsk Polytechnic University, Institute of Non-Destructive Testing, 634050, Tomsk, Russia
[2]Tomsk State University of Control Systems and Radioelectronics, 634050, Tomsk, Russia

Abstract. The paper describes a new method to develop a zero-type *W*-band radiometric receiver. The block diagram and operating algorithm of the digital control system are presented. The main advantages of the zero method in passive radio physical investigations in comparison with classical engineering solutions are shown.

1 Introduction

Passive thermal radio imaging systems are currently being in demand. An object can be detected by radio thermal features without active radio emission. The passive thermal radar system operates regardless of the weather and time of day as compared to the equipment based on infrared methods. Therefore, these systems can be used in navigation in conditions of restricted visibility (aviation security), for radiolocation of unmanned aviation vehicles, in environmental monitoring, for medical applications and etc. Over the last years, much attention has been paid to the radio systems used for imaging of objects hidden behind not optically transparent media due to increased terrorism threat [1].

2 Study objectives

Over the last 40 years, the radio imaging system has become sufficiently advanced due to improved microwave devices: antennas with low loss and high directivity, special processing algorithms in real time and etc. However, no significant techniques were found to improve the design of radiometric receivers. The radiometric receivers currently used are based on total power or modulation principle proposed by Dick's in 1946 [2]. It is well-known, that the receiver gain fluctuation has a major impact on sensitivity of these receivers. The gain fluctuation has a maximum effect near the zero frequency and its spectral density decreases by 1/f law (similar to flicker noise). Increase in the averaging time in this type of radiometric receivers is a good approach to improve the fluctuation sensitivity. As a result, the impact of low frequency gain fluctuations within the correlation time at different frequencies becomes less significant [3].Therefore, in most radiometric systems, the decreased measurement time inevitably leads to sensitivity degradation.

*Corresponding author:ubaychin@tpu.ru

© The Authors, published by EDP Sciences. This is an open access article distributed under the terms of the Creative Commons Attribution License 4.0 (http://creativecommons.org/licenses/by/4.0/).

DOI: 10.1051/matecconf/20167901073

3 Theory

The use of zero-type radiometric receiver decreases the averaging time and maintains the sensitivity at nominal level due to insignificant effect (theoretically null) of gain fluctuations on measurement results. However, the use of the zero measurement method in the radio imaging system is constrained due to the following reasons. The zero-type radiometer is a particular type of the modulation radiometer. It employs symmetry amplitude pulse modulations of both the fiducial noise source and antenna noise similar to that in the modulation radiometer. The difference is that the fiducial noise source is adjustable and it is tuned to the antenna signal through the feedback loop. In this case, the linearity of the radiometric system depends on the linearity of the fiducial noise source feedback loop, rather than that of the receiver. This makes the system more complicated and causes feedback loop fault and system delay [4]. In addition, it increases the requirements to the fiducial source linearity. The modified zero measurement method used in the radiometric receiver, which allows increase in the measurement rate without fluctuation sensitivity loss, is described in [5]. The modification implies the use of two types of synchronous processing pulse modulations, width and amplitude ones.

4 Experimental researches

Figure 1 shows the block diagram of the radiometric receiver with 3-mm wavelength used in the passive imaging system based on the modified zero-type method. It decreases the measurement time with given both sensitivity and long time stability, and eventually, it improves the image frequency rate.

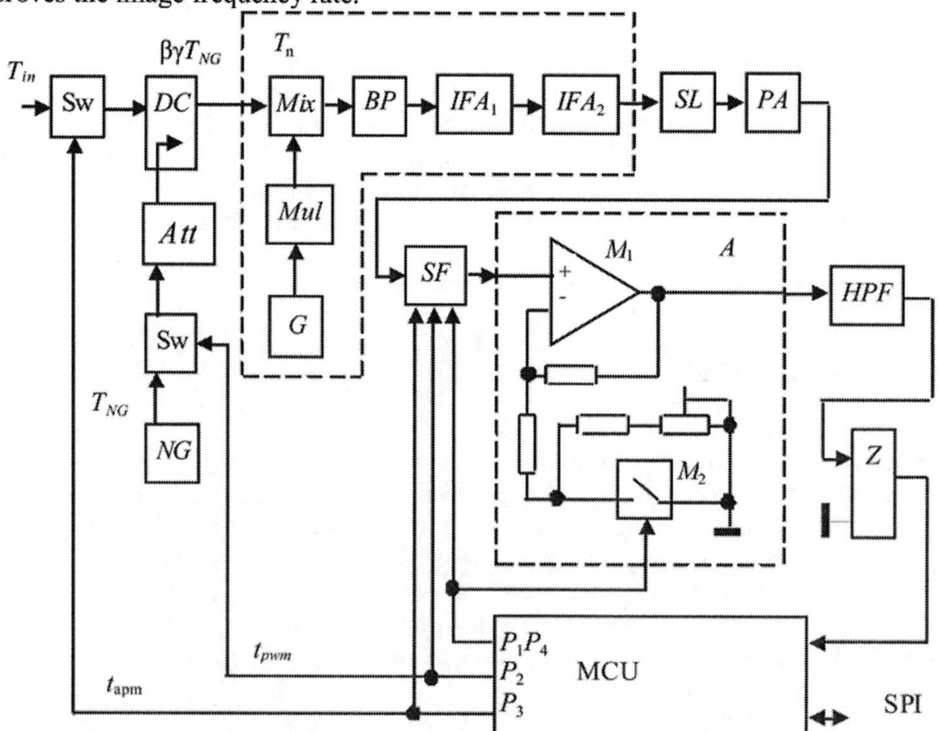

Figure 1. Block diagram of the radiometric receiver.

A directional coupler DC is placed at the receiver input and the pulse-width signal of noise generator NG modulated by microwave switch Sw_2 is the input of the directional coupler. The level of the signal is adjusted by attenuator Att during the radiometric system calibration. The amplitude modulation is processed by microwave switch Sw_1. The receiver designed in the heterodyne circuit consists of generator G with frequency of 45.5 GHz, frequency multiplier (doubler) M and mixer Mix. Thus, the 91 GHz signal is fed to the mixer. The output signal of the mixer passes through the band pass filter with band pass 2...4 GHz to the input of two serially connected intermediate frequency amplifiers IFA_1 and IFA_2 with the total gain of about 61dB. Thus, the operating bandwidth of the receiver is 93...95 GHz. A square law detector SLD is placed at the output of the second intermediate frequency amplifier. The output signal of the detector is equal to the square of the input voltage, hence, to the input power. The square law detector based on the Schottky diode is designed according to the "diode is the current to voltage converter" circuit. Its sensitivity is about 10uW/V. The post-detector section consists of preamplifier PA, synchronous low pass filter SF and invert amplifier A with discrete adjustable gain. The next components are conventional modules for zero-type receiver realization: high pass filter HPF, which excludes the direct current, and analog zero-crossing comparator Z.

The operating principle of the radiometric receiver is as follows. A pulse signal for amplitude modulation from the microcontroller output to the control Sw_1 input is a periodic serial pulse with duration t_{apm} and duty cycle 2. The signal of the noise generator in the noise adding channel is pulse-width modulated by Sw_2 that is controlled by t_{pwm}. Thus, there are three noise signal levels at the receiver input:

-while pulses t_{apm} and t_{pwm}, signal A equal to $\beta\gamma T_{NG} + T_N'$;

-while only t_{apm} pulse, signal B equal to T_N;

-without both t_{apm} and t_{pwm}, signal C equal to $T_{in}+T_n$

where T_{in} is the noise temperature at the system input, T_{NG} is the noise temperature of the generator fiducial noise, T_n is the amplifier noise, β is the coupling factor of the directional coupler, γ is loss in the attenuator. Signal T_n' is the subject of careful consideration. This signal is the input of the mixer while Sw_1 is off, and it is the sum of three signals: $T_n' = T_n + T_nR + T_{ninv}$. The first term T_n is the noise temperature of the receiver. The second term is the same signal, but reflected with R coefficient by Sw_1 while it is off. The third term T_{ninv} is the result of interference of the receiver noises and these noises reflected from Sw1. One way to reduce this interference is fulfillment of the condition $s \geq C_0/2\sqrt{\varepsilon df}$, where s is the electrical length of the waveguide between the Sw_1 output and the mixer input, C_0 is the speed of light in vacuum, df is the receiver bandwidth and ε is the electric permittivity. The gain of the amplifier A is adjusted during calibration by variation of the dividing coefficient when M_2 is off (see Figure 1).

$$\alpha T_n' = T_n. \tag{1}$$

As noted above, the operating principle of the radiometric receiver is based on the modified zero measurement method. According to this modification, the antenna noise temperature is determined by the duration of the pulse-width modulated signal, when zero balance is established, by the following expression [5]:

$$t_{pwm} = (C - B)t_{apm} / (A - B), \tag{2}$$

where$A$, B and C are the mean of the signal levels considered above. After substitution of A, B and C in (2) and considering (1), we obtain

$$t_{pwm} = T_{in}\, t_{apm} / \beta\gamma T_{NG}\,, \tag{3}$$

The last formula (3) implies that the antenna signal is determined indirectly by the duration of the pulse-width modulation, and for this, only one fiducial noise source is used.

 DOI: 10.1051/matecconf/20167901073

The new 3-mm range radiometric receiver was designed for theory, block diagram and operating algorithm confirmation. The calibration is a one of the experimental research stages of the designed radiometric receiver. The calibration is carried out by classic technique, and consists in alternately connection of afiducial noise sources to the receiver input. Due to the complexity of the fiducial noise source for millimeter wave range, external sources, for example, clear sky, special prepared absorptive materials and etc are used for two point calibration. To implementation of the calibration were used two external noise sources – "cold" fiducial is the sky in clear weather, "hot" fiducial is special prepared absorptive Carbone-based material, which was pre-placed in the volume with temperature control for a long period of time, sufficient to uniform heating of absorptive material.

The series of experimental research is carried out after the calibration. To execute this, the radiometric receiver is placed into the volume with temperature control, able to set negative and positive temperature, operating in the cyclic mode between two target values of the temperatures. The input of the receiver with antenna is aimed to the "hot" fiducial, which placed outside the temperature control volume, thus, in the receiver input is the signal with noise power close to 300 K. The internal temperature of the temperature control volume is changing with specified speed to determine the temperature stability of measurement results. During the cyclic temperature changing the receiver operated in three modes – full power, modulation and zero. At the changes of the temperature from +25°C to –5°C, the error corresponds to full power mode is 12.7%, modulation mode is 6.1% and zero mode is 1.9%.

Figure 2 shows the dependence of output radiometer signals operation in three modes from the time, when the temperature is cyclic changed from +25°C to +10°C.

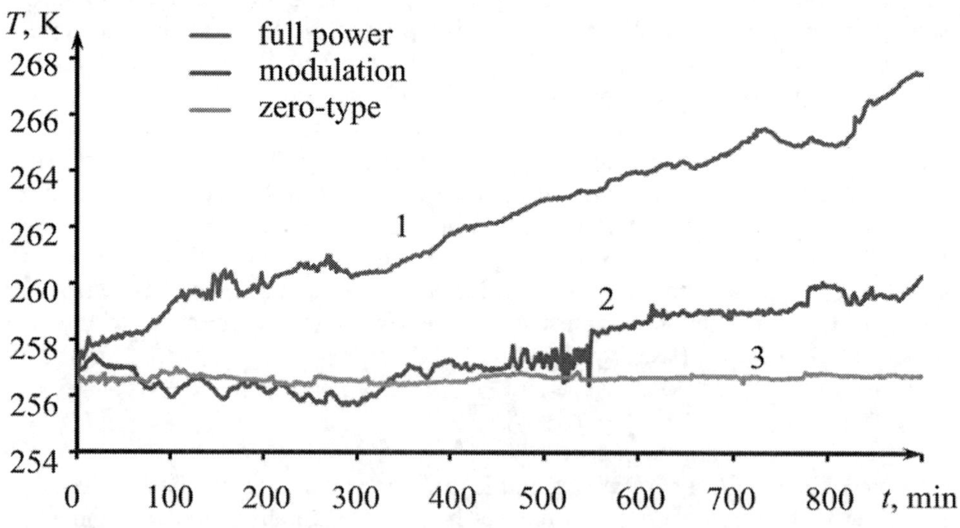

Figure 2. The influence of cyclic temperature changing to the output signals of radiometer in full power (1), modulation (2) and zero (3) modes.

5 Discussion

As should be expected, a change of operating temperature is main influence to the measurement accuracy. As shown in conducted experiments, the radiometric receiver has a best long time and temperature stability in the zero operating mode, it is fully corresponds to researches conducted earlier [5]. General analysis of Figure2 shows the some down shifting of the noise temperature scale. It is caused by error in the determination of the noise temperature of the "cold" fiducial or loss in medium between antenna and "hot" fiducial during the experimental research. These factors have no significant influence to the measurement results due to normalization of the dynamic range.

6 Conclusion

The proposed technical solution significantly simplifies the receiver input high-frequency section. It becomes similar to that of the classical modulation radiometer based on the differential measurement method. The modulation radiometer is used to develop systems for various applications due to simple technical realization of the input high-frequency section and decreased receiver noise temperature effect within the measurement interval. In spite of this, the effect of the drift and fluctuation of the receiver gain on measurement results is great, because the gain is a slope of the radiometer transfer characteristics [5]. In the described zero-type microwave radiometer based on the zero measurement method, the effect of gain fluctuation is minimized, and the slope is determined by a high precision resistive divider. Thus, the circuits are simplified and measurement accuracy is achieved.

Acknowledgments

The reported study was partially supported by RFBR, research project No. 15-07-04971A.

References

[1] A. Camps, J. Tarongi, Rem. Sens. **2**, 1826 (2010)doi:10.3390/rs2071826
[2] M. Kaisti, M. Altti, T. Poutanen, Remote Sens. **8**, 85 (2016), doi:10.3390/rs8020085
[3] X. Lu, J. Li, F. Tao, P. Duan, MATEC Web of Conferences **44**, 02084 (2016) doi:10.1051/matecconf/20164402084
[4] A. Bespalko, A. Surzhikov, L.Yavorovich, P. Fedotov, Russ. J. Nondestr. Test. **48**, 221 (2012) doi: 10.1134/S1061830912040043
[5] A. Filatov, A. Ubaichin,Meas. Tech. **54**, 1160 (2012) doi:10.1007/s11018-012-9865-6

DEVELOPMENT OF TELEMEDICINE SYSTEM

Denis Kuznetsov[a]

National Research Tomsk State University, 634050, Tomsk, Russia

Abstract. The tendency of computer technologies implementation to all areas of life is obvious and indisputable. Informatization of such area as medicine, gives huge advantage to health care service. For example, use of telemedicine technologies will allow making professional medical care to the removed patient. The telemedicine is distance medical care of patients by means of computers, the Internet and other communication technologies. Nowadays, telemedicine is developed subarea of health services delivery and medical education system. It is the reliable and available instrument for the improvement of healthcare systems. In the remote regions of Russia forming of medical services is rather difficult, but last years, Russian organizations show interest in implementation of telemedical services. Telemedicine promotes cooperation between the medical organizations, allows organizing effective system of medical personnel training. Priority tasks of telemedicine development are following: standardization of the applied medical, computer and telecommunication technologies, development of information resources. Approach "telemedicine" is rather new, especially for Russia, so the problem of healthcare services improvement is a priority task of our society and one of the main aspects in policy of the Russian Federation. In this paper, we raise the issue of algorithms development for carrying out the medical consultations in telemedicine system

1 Introduction

All types of information: official, confidential and descriptive – are stored in the database (D), with which the system of telemedical consultations needs to interact. The module of the database provides interpretation of authorized user requests, with next transmission to database, data acceptance, encoding and interpretation of results [1, 2]. The module of the database interacts with the desktop module, through safety system, receiving from it requests of users. Also, it exchanges data with bases of all system via the module of network interaction which is responsible for authorization of users, encoding and protection of confidential information in a D.

[a] Corresponding author: novorostomsk@gmail.com

© The Authors, published by EDP Sciences. This is an open access article distributed under the terms of the Creative Commons Attribution License 4.0 (http://creativecommons.org/licenses/by/4.0/).

DOI: 10.1051/matecconf/20167901074

2 Database Module

Functions of the D module, according to solvable tasks, are divided into two classes: for operation with the database and for dataful operation. The first are necessary for link establishment with PostGreSQL, for transmissions of requests, acceptances of responses. The second – for provision of data to other system modules in a necessary look and a format [4].

Figure 1 gives the algorithm of D module operation. Connection setup with database happens at the program start stage during initialization of system resources. The Delphi component, which is used for operation with different databases – ADOConnection of TADOConnection classes, is created. After component creation, initialization string assigns to the component, also other necessary properties are set and, at last, connection opens. Further, initialization string involves authentication and service data, which is necessary for connection with the database. If connection with a D was not set for any reason, the module under abnormal condition finishes a program runtime [5].

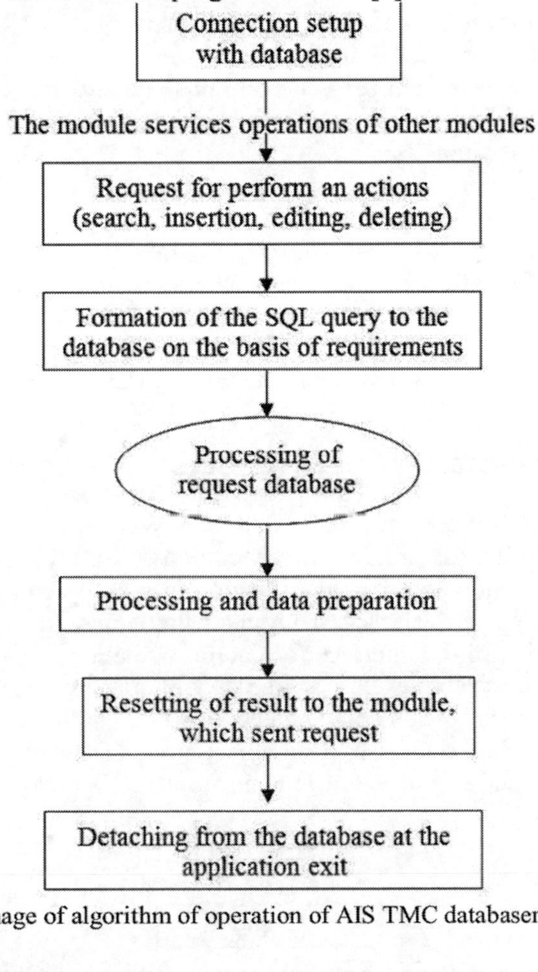

Figure 1. Schematic image of algorithm of operation of AIS TMC databasemodule.

3 Execution of SQL queries

Execution of SQL queries to the database is carried out by means of ADOQuery delphi-component of TADOQuery class. Depending on type of request there are different functions of their execution. As follows, the SQL requests, implying availability of the answer from the database in the form of any data (requests like SELECT). Requests are carried out by function, which returns the ADOQuery component as result of the accomplishment, allowing to conduct, further work with the data obtained from a D. On the contrary, requests as INSERT, UPDATE, DELETE do not imply return from a D of any data (except confirmation about successful accomplishment of request) therefore the ADOQuery component is destroyed after accomplishment of function in order to avoid memory leakages [6, 7].

Upon the end of operation of application, the ADOConnection component finishes connection with a D, and then is destroyed.

Both components: ADOConnection and ADOQuery – are a part of the ADODB module of the Delphi CodeGear 2009 language.

In order that the AIS TMC system interacted with D in unitary way, we use the mechanism of automatic formation of SQL queries to D. Special object - Delphi class was developed. Object ensures functioning of basic functions with the database – TDB Unit or "Database object". All information units of system: users, patients, teleconsultations, etc. – are based on this object [9].

The database object is the central class for all elements of AIS TMC system. So, we will describe hierarchical structure of D objects.

The class TWorkUnit is a main base class. Immediate ancestor of all executive classes of objects descriptions system:

– TWUTMUsers (class of users);
– TWU_TM_GroupsUsers (class of users groups);
– TWUTMPatients (class of patients).

For example, users of AIS TMC have different characteristics: login, password, identification data. Information on these characteristics and their description are set in TWU TM Users class, for patients – respectively in the TWU TM Patients class, etc. Copies of these classes are the working tool of system. They service data and interactions, store information, control a status of a D and the tables [8].

4 Schematically hierarchy of D objects

All derived classes of TWorkUnit, for example TWU TM Users, use the universal algorithms of TWorkUnit class, implement generation of SQL queries to the database for formation of tables, selection and data modification in a D, based on structure description of objects, defined in derived classes. They also implement data synchronization, data transformation, formation of the interface for dataful operation, activation of encoding and decoding functions.The executive classes of descriptions are extensions of TWorkUnit class, which contain all information about objects.

In turn, TWorkUnit class comprises a subobject – the TDB Unit class in which all tools for work with SQL inquiries and data are concentrated. Figure 2 presents schematically hierarchy of classes [10].

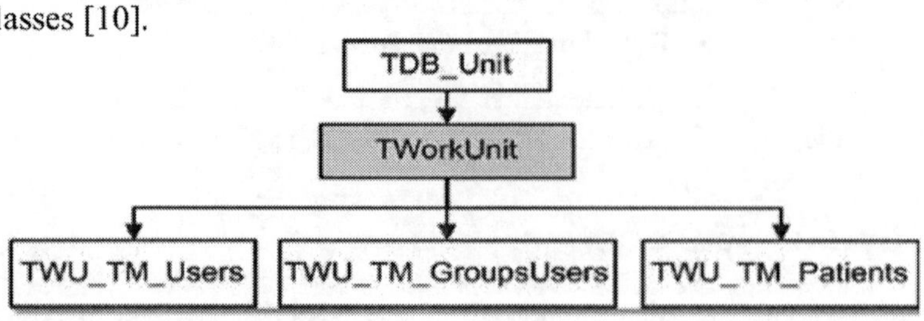

Figure 2. Schematically hierarchy of D objects of AIS TMC.

Objects of TWUTMUsers, TWUTMGroupsUsers, etc. are representation of entities of users of AIS TMC , patients, etc in Delphi (program implementation of objects). Copies of

TWU TM Users class in the AIS TMC system formally correspond to patients (in other words, each patient is a copy of derived class (patients with parent class the Homo sapiens). Thus, real patients (as well as other information units, for example, users or groups) correspond in the program to copies of the TWU TM Users class which, in turn as a data set correspond to record set in D tables.

All derived classes of TWorkUnit work by the general principle.Iin Delphi selected variable corresponds to each characteristic of a class, in the database – a field or a set of fields of the table. The characteristic has the data format (number, a line, a set of values, the image, etc.), and also a role which defines semantic loading of the characteristic. For example, the surname of the practicing doctor – the user of AIS TMC – is important only for other users of system, and login and the indicator of this doctor – key characteristics, they are processed essentially differently [12].

The description of object is created from characteristics. Based on this description, TDB Unit creates SQL queries to the database. In case of initialization of derived classes copies of TWorkUnit class, function of check of D integrity checking and creation of tables is executed. Figure 3 provides algorithm of functioning.

At a stage of system parameters preparation we select resources under auxiliary variable and set primary value . Then we begin search of the tables, serviced by this class. If the table doesn't exist – we start SQL query formation by table creation [11]. Otherwise check begins whether all fields of the table corresponding to attributes are present at it. Thus, the database always is in a status necessary for us, and database changes dynamically are traced and improve, and we exclude situations when the SQL query to this table comes to the end with an error because of its incorrect structure.

We creat SELECT section of SQL query in the same way. SELECT section is used for data sampling.

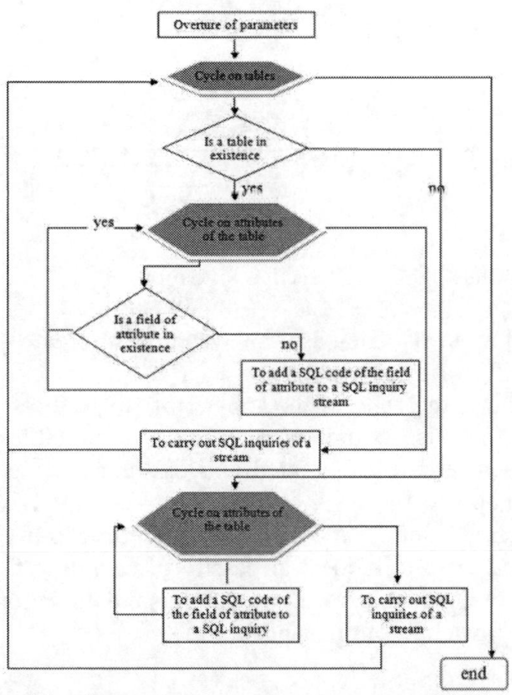

Figure 3. Control-flow chart of D table control.

Formally SELECT section of SQL query looks as follows: "SELECT list of the FROM fields list of the tables WHERE conditions".

The system collects each section in turn, based on characteristics of attributes, – each attribute belongs to the table, possesses a certain data type, and also can be connected somehow to other attribute from other table. In addition, during the search we specified additional conditions which need to be considered [3].

The system automatically creates request.

Schematically the algorithm is provided in Figure 4.

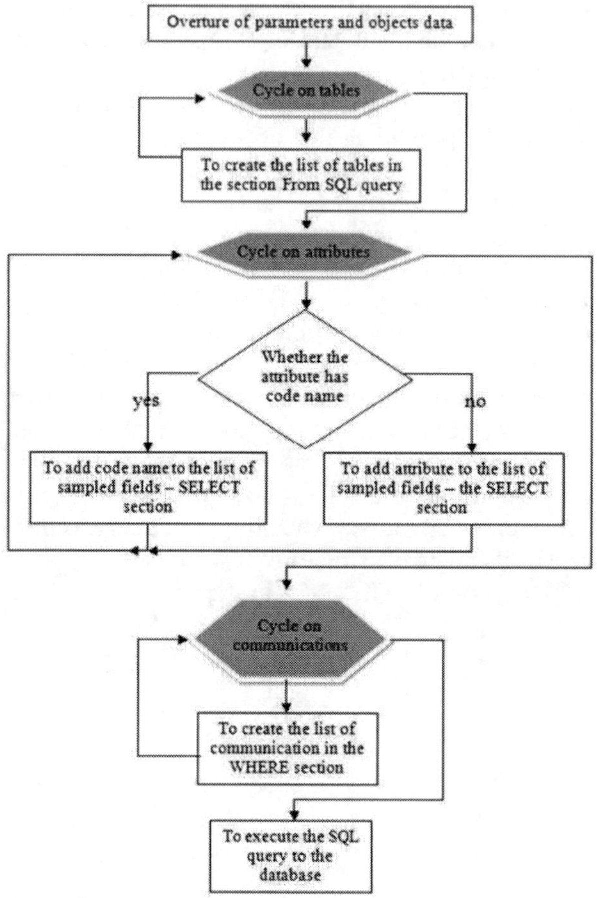

Figure 4. Control-flow chart of SELECT section formation.

For a data recording in the database the system also creates the SQL query depending on the task, the INSERT or UPDATE type.

During the recording, the copy of the "object of Delphi" class contains data, which are required to be added to basis. As incremental fields of one tables can be keys at others, for tables requests of an insertion are created and executed sequentially. Thus, attributes of each table get over [12].

At this time system, leaning on a role and sometimes on type, defines how data are stored and how to add them to request, necessity of encoding them, etc. In completion,we executed SQL queryfor each table, result of which is the indicator of new record.

Figure 5 presents flowchart of this algorithm.

DOI: 10.1051/matecconf/20167901074

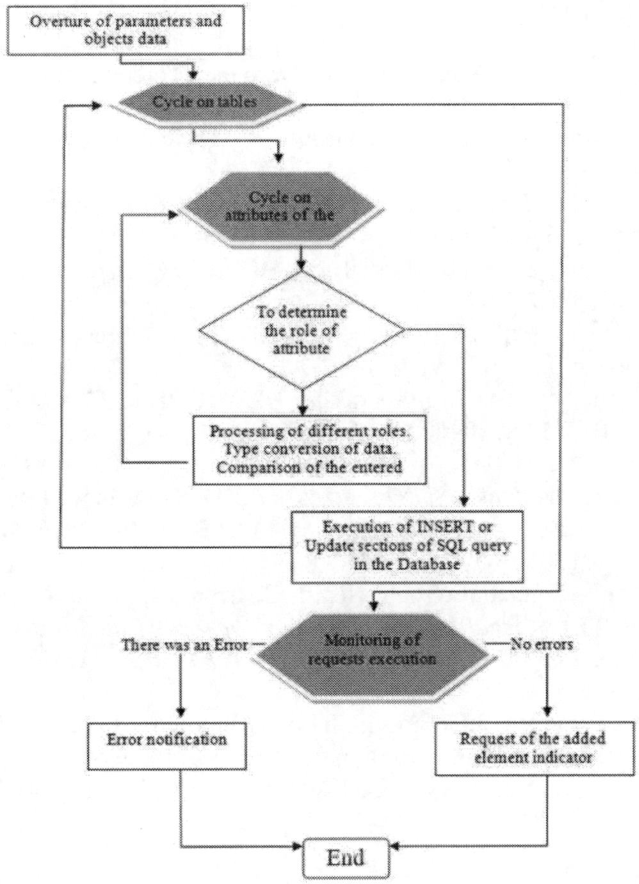

Figure 5. Control-flow chart of insert and editing records in D.

5 Conclusion

Thus, the system itself controls processes of operation with the database independently, provides the flexibility in construction and changing of D structure, which allows for developer quicker program creation. For program formation we developed different modules: the module of network interaction, module of security policy, the module of an exchange of video information, the module of control demonstration tasks, the module of desktop, but it will be stated in the following materials.

The developed telemedical system integrates all types of healthcare institutions – the central and regional regulations, the central, regional and region clinics and hospitals, medical academies and institutes, archives and libraries and other treatment-and-prophylactic establishments.

Acknowledgements

The paper was written as part of the research project No. 8.2.31.2015, carried out with the support of the Program "Research Foundation of Tomsk State University named after D.I. Mendeleev" in 2015 – 2016.

References

[1] D.A. Person, J.S. Hedson, K.J. Gunawardane, Telemedicine Journal and e-Health **9**, 95 (2003) doi: 10.1089/153056203321593642
[2] A. Szot *et al.*, International Journal of Medical Informatics **73**, 65 (2004) doi: 10.1016/j.ijmedinf.2003.10.002
[3] V. Patterson *et al.*, Journal of Telemedicine and Telecare **7**, 52(2001) doi: 10.1258/1357633011936714
[4] I. Qaddoumi *et al.*, Pediatric Blood & Cancer **48**, 39 (2007) doi: 10.1002/pbc.21085
[5] P.V. Pradeep *et al.*,World Journal of Surgery **31**, 1665 (2007) doi: 10.1007/s00268-007-9108-1
[6] D. N. Kuznetsov, V. I. Syryamkin, NEW OPERATIONAL TECHNOLOGIES (NEWOT'2015) 1688, 040004 (2015) doi:10.1063/1.4936037
[7] E.M. Strehle, N. Shabde, Archives of Disease in Childhood **91**, 956(2006)
[8] I Matveev, E Siemens, A Yurchenko and D Kuznetsov, IOP Conference Series: Materials Science and Engineering **132,** 012010 (2016) doi: 10.1088/1757-899X/132/1/012010
[9] WHO Group Consultation on Health Telematics, *A health telematics policy in support of WHO's Health-for-all strategy for global health development : report of the WHO Group Consultation on Health Telematics* (World Health Organization, Geneva,1997)
[10] D. N. Kuznetsov, V. I. Syryamkin, Innovation theory-2015, 453 (2015)
[11] W. Einthoven, Archives Internationales de Physiologie (4), 132 (1906)
[12] D. N. Kuznetsov, V. I. Syryamkin, Telecommunications (8), 44(2015)

MATEC Web of Conferences **79**, 01075 (2016) DOI: 10.1051/matecconf/20167901075
IME&T 2016

DESIGN OF NEURO-FUZZY DECISION TREES

Tatyana Abramova[a]

National Research Tomsk State University, 634050, Tomsk, Russia

Abstract. In order to improve accuracy of fuzzy decision trees classification we propose a procedure of parameters adaptation by means of neural network training. In the direct cycle, fuzzy decision trees are based on the algorithm of fuzzy ID3; in the feedback cycle, parameters of fuzzy decision trees are adapted based on stochastic gradient algorithm by traverse to the root nodes back from the leaves. Using this strategy, the hierarchical structure of the fuzzy decision trees remains fixed.

1 Introduction

Domestic and foreign literature describes the use of decision trees as a powerful evolutionary methodology for solving problems of classification and regression [1–5]. Being a DATA MINING tool (detection of hidden knowledge from data), decision trees are used for search and retrieval of interpretable classification rules, which are clear for humans. We should note that many packages for intellectual data analysis already contain methods for constructing decision trees; they are the perfect tool for decision support systems.

2 Principles of constructing decision trees

Decision tree is a tree, the leaves of which contain values of the target function, and the remaining nodes contain branch conditions, which determine the edge to go (for example, "Sex is male"). If the condition is true for this observation, transition is made along the left edge, if it is false – along the right. Usually, each node includes checking of one independent variable. Sometimes two independent variables are compared with each other in the tree node, or a function of one or more variables is determined.

If a variable that is checked in the node takes categorial values, then the branch emanating from the tree node corresponds to each possible value. If the value of the variable is a number, it is checked to see if the value is greater or less than a constant. Sometimes, numerical range is divided into intervals (to check if the value hits one of the intervals).

The leaves of the trees correspond to the values of the dependent variable, i.e. classes. Figure 1 shows the iris classification tree. Classification contains three classes (marked with red, blue and green on the Figure 1) and has the parameters: length \ width of sepals (SepalLen, SepalWid) and length \ width of petals (PetalLen, PetalWid).

[a] Corresponding author: tanusha-atv@mail.ru

© The Authors, published by EDP Sciences. This is an open access article distributed under the terms of the Creative Commons Attribution License 4.0 (http://creativecommons.org/licenses/by/4.0/).

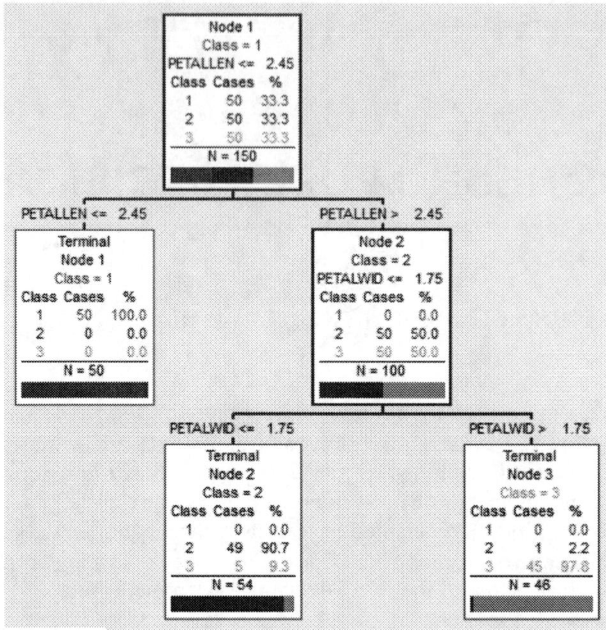

Figure1. Iris classification tree.

As we can see, each node contains class belonging (depending on the fact, what elements have hit this node in a greater number), number of observations N, and number of each class. Not leafy tops also contain a transition condition – to one of the subsidiaries. Sample is divided according to these conditions. As a result, the tree has classified initial data (exactly initial data, those on which it was trained) almost perfectly (6 out of 150 are not correct).

3 The main methods, which use decision trees

Classification and regression trees (CART) was the first method invented in 1983 by four well-known scientists in the field of data analysis: Leo Breiman, Jerome Friedman, Richard Olshen and Stone (Table 1) [2].

The essence of the algorithm is in ordinary construction of a decision tree [6]. On the first iteration, we build all possible (in a discrete sense) hyper planes, which would divide our space into two. Number of observations in each sub-space of different classes is counted for each subdividing. As a result, we select subdividing, which has maximally allocated observation of one of the classes in one of the sub-spaces. Accordingly, this subdividing is the root of our decision tree, and two subdivisions will be the leaves on this iteration.

On the next iterations, we take the worst leave (in the sense of number of observations of different classes) and conduct the same operation on its subdividing. As a result, the leave becomes a node with some subdividing and two leaves.

We continue to do so until we reach the limit on the number of nodes, or the overall error (the number of misclassified observations over the tree) fails to improve from one iteration to another. However, the resulting tree will be "retrained" (will be made-to-the training sample) and, accordingly, will not give normal results on other data. In order to avoid "retraining", it is possible to use test samples (or cross-validation) and, accordingly, to make reverse analysis (so-called, pruning), when the tree is reduced depending on the result on the test sample [7].

This is a relatively simple algorithm, which results in one decision tree. It is convenient for the primary data analysis, for example, to check presence of relationships between variables.

Random Forest is a method invented after CART by one of the four scientists - Leo Breiman in co-authorship with Adele Cutler [3]. The method is based on the use of committee (ensemble) of decision trees.

The essence of the algorithm is that random sampling of variables is made on each iteration, and then, constructing decision trees starts on this new sample. Besides, "bagging" takes place – sampling of random two-thirds of observations for training, and the remaining one-third is used to evaluate the results. This operation is done hundreds or thousands times. The resulting model will be the result of "voting" of a tree set obtained during simulation.

Stochastic Gradient Boosting is a data analysis method introduced by Jerome Friedman [4] in 1999. It represents solution of the regression problem (which can include classification) by constructing committee (ensemble) of "weak" predictive decision trees.

On the first iteration, a decision tree limited by number of nodes is constructed. Then the difference is counted between the value predicted by the resulting tree and multiplied by learnrate ("weakness" coefficient of each tree) and the unknown variable on this step. The next iteration is based on this difference. This continues until the result improves. It means that on every step we try to correct the mistakes of the previous tree. However, it is better to use check data (not involved in the simulation), because retraining is possible on the training data.

Table 1. Comparative analysis of methods, which use decision trees.

Method	*Advantages*	*Disadvantages*
CART	Rapid construction of the model. Simple interpretation (because of the model simplicity, you can easily draw a tree and observe all tree nodes).	It often comes to a local decision (for example, a hyperplane has been selected on the first step, which maximally divides the space on this step; but it will not lead to the optimal solution).
Random Forest	High quality of results, especially for data with large number of variables and small number of observations. Possibility to parallelize. No test sample is required.	Each tree is huge; as a result, obtained model is huge. Long period of model construction in order to achieve good results. Complex model interpretation (it is difficult to interpret hundreds or thousands of huge trees).
Stochastic Gradient Boosting	High quality of results, especially for data with large number of observations and small number of variables. Rather small model size (compared with the previous method), since each tree is limited to specified dimensions.	Test sample (or cross-validation) is required. Inability to parallelize well. Relatively weak resistance to erroneous data and retraining. Complex model interpretation (as in Random forest).

DOI: 10.1051/matecconf/20167901075

Common disadvantage of constructing traditional decision trees is requirement for certainty of the input data, which is achieved by applying the average values of the input parameters of the analyzed technology. This can lead to receiving significantly shifted point estimates of project performance indicators. It is also clear that the requirement for determinancy of input data is unjustified simplification of reality, because any technology is characterized by many uncertainties: uncertainty of input data, uncertainty of external environment, uncertainty associated with the nature, options and a model of project realization, uncertainty of requirements for technology effectiveness. Uncertainty factors determine technology risk, i.e. a danger of resources loss, revenue deficiency, or additional costs.

4 Neuro-fuzzy trees

In order to improve the accuracy of classification, the author suggests using neuro-fuzzy decision trees, which have property to adapt parameters by means of neural network training. In the direct cycle, fuzzy decision trees are based on fuzzy ID3 algorithm [5]. The feedback cycle, parameters of fuzzy decision trees are adapted based on stochastic gradient algorithm by traverse to the root nodes back from the leaves.

As initial data, we use the so-called triangular fuzzy numbers with a membership function of the following type (Figure 2).

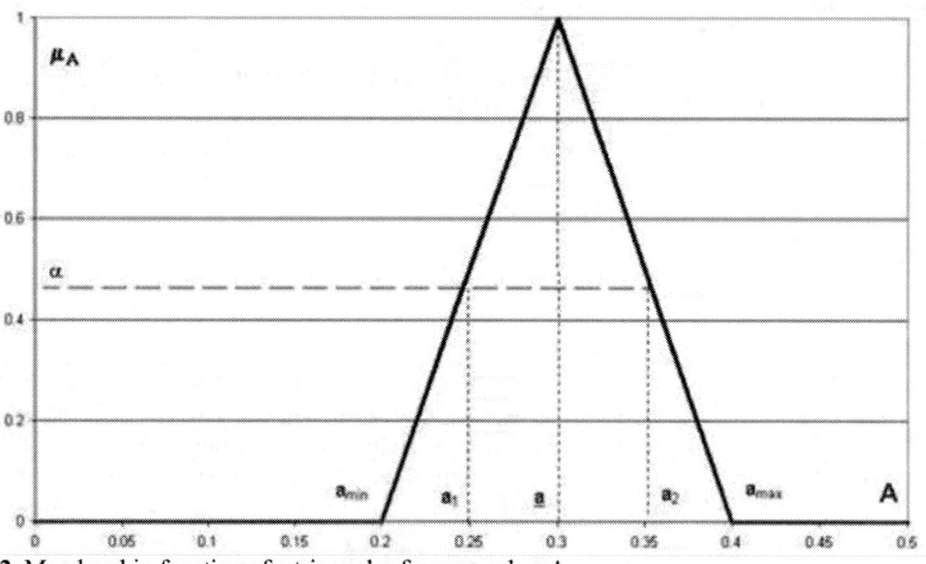

Figure 2. Membership function of a triangular fuzzy number A.

These numbers model the following statement: "Parameter **A** is approximately equal to α and is clearly in the range $[a_{min}, a_{max}]$".

In general, a fuzzy number means a fuzzy subset of a universal set of real numbers, which have a normal and convex membership function. This description allows experts to take parameter interval $[a_{min}, a_{max}]$ as input information and most expected value of α, then the appropriate triangular number A = $(a_{min}, \alpha, a_{max})$ is built. Selection of three important points of initial data is rather common in investment analysis. Often subjective probabilities of realization of appropriate ("pessimistic", "normal" and "optimistic") initial data scenarios are compared with these points. Further, we will call parameters $(a_{min}, \alpha, a_{max})$ as *significant points* of triangular fuzzy number **A**.

We should note that attributes of technological innovative projects are classified as subjective or objective. Subjective attributes include quality features such as technical level, advantages of the enterprise, innovative risk, project management; we will estimate them by linguistic values presented by fuzzy numbers based on expert interviews.

Objective (quantitative) attributes include plans of investment costs, etc. These quantitative features are reduced to a common scale in order to provide compatibility with linguistic values of subjective features. Let us consider typical descriptions of the technology projects attributes (Table 2).

Table 2. Technology projects attributes.

№	Investment costs of the project (*10 mil. rubles), $\times 1$	Technical level {1,2,3},$\times 2$	Advantages of the enterprise , $\times 3$	Innovative risk, $\times 4$	Level of the project management,$\times 5$
1.	0.82	0.0; 0.1; 0.3	average	low	low
2.	0.81	0.0; 0.1; 0.9	average	average	average
3.	0.78	0.0; 0.3; 0.5	good	high	low
4.	1.00	1.0; 0.9; 0.0	fairly poor	very low	very high
5.	0.97	1.0; 0.8; 0.6	poor	average	average
6.	0.80	0.0; 0.2; 0.9	very poor	low	average
7.	0.96	0.0;0.4; 0.9	very poor	average	very high
8.	0.78	0.0; 0.1; 0.2	very good	average	very low
9.	0.98	0.0; 0.3; 0.8	fairly good	average	average
10.	0.78	1.0; 0.7; 0.4	good	average	fairly average
11.	0.98	0.0; 0.2; 0.5	poor	low	very low
12.	0.81	0.0; 0.3; 0.9	good	average	high

The first attribute (investment costs of the project) is numerical. The third and the fifth attributes (advantages of the enterprise and level of the project management) are described in linguistic terms such as good, very poor, etc. The value of the second attribute is defined on a fuzzy set {1, 2, 3}. Innovative risk based on fuzzy classification is in linguistic terms.

Fuzzification involves conversion of attributes' numerical values to linguistic terms in order to reduce information and present it in a human-understandable form convenient for decision-making. One of the ways to determine membership functions of these linguistic variables is expert opinion or human perception. In order to automate this procedure, you can use statistical methods and fuzzy clustering based on self-organizing neural network training. Let us consider the second method.

Let there is a data set X, which must be conversed to k linguistic variables T_j, $j=1,2,...,k$. For simplicity, assume that T_j function has a form of triangulation:

$$T_1(x)=\begin{cases} 1 & , \quad x \le a_1 \\ (a_2-x)/(a_2-a_1), & a_1 < x < a_2 \\ 0 & , \quad x \ge a_2 \end{cases} \tag{1}$$

$$T_j(x)=\begin{cases} 0 & , \quad x \geq a_{j+1} \\ (a_{j+1}-x)/(a_{j+1}-a_j), & a_j \leq x < a_{j+1} \\ (x-a_j)/(a_j-a_{j-1}) & , \quad a_{j-1} < x < a_j \\ 0 & , \quad x \leq a_{j-1} \end{cases} \tag{2}$$

$$T_k(x)=\begin{cases} 1 & , \quad x \geq a_k \\ (x-a_{k-1})/(a_k-a_{k-1}), & a_{k-1} < x < a_k \\ 0 & , \quad x \leq a_{k-1} \end{cases} \tag{3}$$

Parameters, which are to be defined for each attribute, form k centers $\{a_1, a_2,..., a_k\}$. A neural network algorithm – Kohonen's self-organizing maps - is an effective method for determination of these centers [3].

Let us consider a numerical attribute of the project - investment costs for a group of examples in Table 2. By Kohonen's self-organizing maps, we define: $a_1= 0.68$, $a_2= 0.76$, $a_3= 0.82$.

Then the membership functions for variable x of one of linguistic variables T_j, (j = 1, 2, 3) are described as follows:

$$T_1(x)=\begin{cases} 1 & , \quad x \leq 0.68 \\ (0.76-x)/0.08, & 0.68 < x < 0.76 \\ 0 & , \quad x \geq 0.76 \end{cases} \tag{4}$$

$$T_2(x)=\begin{cases} 0 & , \quad x \geq 0.82 \\ (0.82-x)/0.06, & 0.76 < x < 0.82 \\ (x-0.68)/0.08, & 0.68 < x < 0.76 \\ 0 & , \quad x \leq 0.68 \end{cases} \tag{5}$$

$$T_3(x)=\begin{cases} 1 & , \quad x \geq 0.82 \\ (x-0.76)/0.06, & 0.76 < x < 0.82 \\ 0 & , \quad x \leq 0.76 \end{cases} \tag{6}$$

Obviously, these linguistic terms can be described as «low», «average» and «high». The second column of Table 3 shows the degree of proximity of the attribute «investment costs» to these three membership functions.

For description of linguistic and corresponding numerical values, we assume that the membership functions of this linguistic term are known.

The measure of similarity between linguistic terms can be determined by their functions as follows:

f(A,B) = *0.5*{S(A,B)* + *S(B,A)}* where *S(A,B)* and *S(B,A)* represent degree of sub-multiplicity *A* in *B* and *B* in *A* respectively (sub-multiplicity as a degree of membership of one set to another). Here, sub-multiplicity *A* in *B* is defined as $S(A,B) = M(A \cap B)/ M(A)$, where *M* is the sum of degrees of membership of conversion of a fuzzy set to the final state.

Using the described function *f* we can calculate the degree of membership of each of two linguistic terms «Advantages of the enterprise».

The value of fuzzy attributes, for example, the attribute «technical level», can be functionally represented by a set of membership functions (Table 3). For a given set of functions, we find some new fuzzy sets, which are considered as a result of clustering of initial data for description of membership functions of set.

Table 3. Fuzzy sets of technology projects data after the neural network training.

№	Investment costs of the project (*10 mil.			Technical level {1,2,3},×2			Advantages of the enterprise ,×3			Innovative risk, ×4			Level of the project management,×5		
	low	avg.	high	low	avg.	high	good	avg.	poor	good	avg.	poor	high	avg.	low
1	0.12	0.86	0.02	0.46	1.00	0.46	0.28	0.46	0.96	0.68	1.00	0.28	1.00	0.58	0.36
2	0.00	0.92	0.08	0.56	1.00	0.56	0.38	0.97	0.42	0.36	0.56	0.92	0.54	1.00	0.58
3	0.96	0.00	0.02	0.35	0.68	1.00	0.36	0.63	0.92	0.96	0.26	0.38	1.00	0.57	0.38
4	0.00	0.00	1.00	0.92	0.78	0.36	0.85	0.13	0.08	0.88	1.00	0.32	0.18	0.22	0.76
5	0.11	0.06	0.83	1.00	0.58	0.35	0.94	0.58	0.36	0.28	0.36	1.00	0.56	1.00	0.58
6	0.12	0.31	0.57	0.88	0.42	0.24	0.36	0.95	0.48	0.98	0.38	0.27	0.56	0.38	0.56
7	0.00	0.00	1.00	0.38	0.64	0.98	0.45	0.95	0.48	0.92	0.46	0.56	0.88	1.00	0.28
8	1.00	0.00	0.00	1.00	0.57	0.32	0.26	0.42	0.98	0.26	1.00	0.38	0.57	0.38	0.59
9	0.00	0.00	1.00	0.34	0.55	1.00	0.96	0.39	0.28	0.18	0.68	0.98	0.82	0.68	0.24
10	0.08	0.86	0.06	0.56	1.00	0.56	0.44	0.29	0.95	1.00	0.45	0.56	0.26	1.00	0.56
11	0.00	0.00	1.00	0.32	0.98	0.28	0.68	0.45	0.68	0.12	0.56	0.32	0.43	0.68	0.46
12	0.46	0.54	0.00	0.96	0.56	0.25	0.98	0.27	0.36	0.78	0.56	0.68	1.00	0.36	0.68

5 Conclusion

Thus, the developed method of constructing neuro-fuzzy decision trees allows you to get rid of weighted estimations of the input data and has a property of neural network adaptation of the parameters based on stochastic gradient algorithm by traverse to the root nodes back from the leaves. In the direct cycle, fuzzy decision trees are constructed based on fuzzy ID3 algorithm. In the feedback cycle, parameters of fuzzy decision trees are adapted. Because of this strategy, the hierarchical structure of the fuzzy decision tree remains fixed.

In conclusion, we note that the proposed approach of use of back-propagation algorithm directly on the structure of fuzzy decision trees improves the accuracy of their training without damage to interpretability.

Acknowledgements

The author wishes to thank Tatiana B. Rumyantseva, National Research Tomsk State University, for assistance in the article preparation.

The paper was written as part of the research project No. 8.2.31.2015, carried out with the support of the Program "Research Foundation of Tomsk State University named after D.I. Mendeleev" in 2015 – 2016., grant RFBR No. 16-29-12858.

References

[1] S.V. Gorbachev, V.I. Syryamkin, S.A. Koynov, *Intelligent system of strategic business planning with fuzzy multiple estimation of effectiveness and risks* (LAMBERT Academic Publishing, Saarbrucken, 2012)

[2] S.V. Gorbachev, V.I. Syryamkin, I.B. Rudakov, *Recognition of complex constructed oil and gas pools based on neuro-fuzzy portraits* (LAMBERT Academic Publishing, Saarbrucken, 2013)

[3] S.V. Gorbachev, V.I. Syryamkin, *Neuro-fuzzy methods in intelligent systems of processing and analysis of multivariable information* (Publishing House of Tomsk State University, Tomsk, 2014)

[4] S.V. Gorbachev, V.I.Syryamkin, M.V. Syryamkin, *Intellectual Foresight-Forecast of Scientific and Technological Development of a State* (LAMBERT Academic Publishing, Saarbrucken, 2012)

[5] C. Z. Janikow, Proceedings of the Sixth International Symposium on AI, 360 (1993)

[6] D.V. Shashev, S.V. Shidlovskiy, V.I. Syriamkin , Yurchenko A.V. IOP Conference Series: Materials Science and Engineering **81**, 012101 (2014) doi: 10.1088/1757-899X/81/1/012101

[7] S.V. Panin, I.V. Shakirov, V.I. Syryamkin, A.A. Svetlakov, Avtometriya (1), 37 (2003)

A COMPARISON OF DATA COMPRESSION METHODS FOR SOLVING PROBLEMS OF TEMPERATURE MONITORING

Hussein Mogahed[1], *Alexey* Yakunin[a2], *Larisa* Suchkova[2]

[1]Minia University, 61111, Minia, Egypt
[2]Altai State Technical University, 656038, Barnaul, Russia

Abstract. The paper considers the problem of large processing huge amounts of data for temperature monitoring of man-made and natural objects associated with the lack of data compression efficiency in real time when they are transferred and stored in the presence of anomalies in the information signal in the form of sudden changes and outlier. The solutions of existing methods were described and new approaches were proposed. The results of experimental comparison of proposed and known solutions are included.

1 General definition of the problem

Nowadays, remote control and monitoring systems are widely used in modern technologies of automated production and scientific research [1].

One of the main demands in the monitoring systems, especially, for environmental monitoring and energy accounting is the ability to reproduce the recorded data at sufficiently large time intervals, up to several years. In some cases, the dynamics of registered processes require that the sampling period does not exceed 5–30 seconds. Considering that the number of different types of stored data may reach tens or even hundreds and requires huge amount of storing volume.

Therefore, in most technological and environmental monitoring systems, all data are usually exposed to preliminary on-line processing, one of its objectives is converting the original data in order to minimize the amount of physical memory required to store them. To make such changes in the monitoring system, it requires a dedicated storage subsystem. This subsystem must guarantee storing the source data with sufficient accuracy for later reconstruction, and a partial or full treatment to simplify and improve the efficiency of subsequent sampling and analysis of the monitoring results. This is necessary for using the stored data to address new problems related to the study of the controlled objects, also, to use this data for further improvement of processing algorithms.

One of the major requirements for storage systems is to minimize the total volume of stored data, which is achieved by using various methods of compression. Another, no less

[a] Corresponding author: yakunin@agtu.secna.ru

© The Authors, published by EDP Sciences. This is an open access article distributed under the terms of the Creative Commons Attribution License 4.0 (http://creativecommons.org/licenses/by/4.0/).

important requirement, is to minimize the time access to the data. These requirements are particularly relevant when the amount of stored data becomes too large. For example, in studies of atmospheric turbulence, the temperature data sampling time can be 5–30s at a resolution of about 0.05 ° C and a number of sensors 10 or more [2]. Almost the same time and amplitude resolutions are needed to control the emergency situations in residential and office buildings, however the total number of temperature sensors in this case can reach several hundred [3–7]. It is easy to verify that the uncompressed storage in this case may require about 315 MB per month (100 sensor * 8 bytes (time) * 2 bytes (measured value) * 365 days * 24 hours * 120 sample / hour for one sensor at 30 – second sampling rate). As far as, the number of sensors in complex systems can be significantly bigger, and the storage period may be several years, it is clear that to solve the problem of fast compression and decompression of large data sets in real-time, the traditional approaches, for example in [8–17], may be not helpful. This is explained by the fact that purely statistical methods are universal, as a rule, the block processing of the input data, whereas the dictionary methods of compression are based on the adaptation of some models without taking into account the real properties of the recorded data stream.

From this it follows that it is the most reasonable to add the known universal methods of compression with specially developed methods of stream processing and transformation of the data allowing to carry out most effectively preliminary compression of information. As it will be shown further, such decision allows refusing the subsequent application of the algorithms of compression using resource-intensive statistical strategy. This is especially relevant in implementing the monitoring systems based on the Internet - technologies and microcontrollers with low power consumption and, therefore, low data processing capacity.

Another important issue relating to storing data is optimization of data structure and extent of their preliminary processing. Indeed, the demand for various kinds of data is determined by their relevance and pragmatic value, which in turn depends on the nature of the problem to be solved by an information system. Therefore, in some cases it is advisable to keep the same data with different degree of post-processing and therefore detailed. A typical example of this approach is widely used in SCADA – systems.

Nature of preliminary data processing in many respects is defined also by their original quality. Thus, in some cases, may require compensation of systematic errors due to the inaccuracy of the sensor calibration and the effect on it of various influencing variables.

To eliminate the influence of random errors due to internal noise and external equipment, as well as to minimize the dynamic distortion of the signal caused by the inertia of the primary device, typically use different filtering methods. It follows that the lossless data compression is not always the best solution and always need a reasonable tradeoff between signal quality and quantity of information loss during processing and controlled process granularity.

The features of the pre-processing subsystem and data storage optimization in monitoring systems can be summarized as:

- optimization of storage data structures and methods of presentation specific controlled processes;
- the development of compression methods as the source, and pre-processing, "purified" data: both with and without the use of filtration to reduce random noise.

Let us continue briefly on the consideration of the use of both known and new approaches to improve the efficiency of data storage subsystems and results compare the effectiveness of the use of certain techniques for solving problems of climate, environmental and technical monitoring.

2 Compression methods based on optimization of the data structure

To reduce the amount of stored data, classical compression techniques can be used, for example, the approximation of the observed data using linear or parabolic function, Fourier series or the applications described in [18, 19].

The method, described in [20], an approach that allows no loss of accuracy to minimize the amount of data stored on the server simultaneously with reducing time access to database. The basis of this approach was the principle of optimization of the type of fields used for recording samples and to record values of time corresponding to these points. Its foundation was laid such an important feature of the monitoring of the recorded raw data, like the uniformity of the sample and the low accuracy of the measurement of controlled physical quantities limited in most practical applications, the value of reduced error is no more than 0.5…1%.

Therefore, optimization was proposed to use for the storage of regularly updated data fields with only fixed-length integer without sign (in the MySQL database and a number of other databases such fields have the option "unsigned" [21–24]). Since the actual data samples can be both positive and negative, and maybe have a fractional part, it is proposed to carry out the normalization of the record with the formula [20]:

$$y = C * x + y_0. \tag{1}$$

where x – observed data, C– some multiplier, and y_0 – shift.

Since many databases provide rich types of fields that differ in the number of bytes, always it is possible to pick up a field type with minimum amount of bytes that suitable for recording the normalized data. For example, to record temperature values with a resolution of 0.0625 ° C in the range from -55 to + 125C ° (that is provided, for example, when using digital DS18B20 sensors), two bytes is sufficient instead to keep it as a real number with a field of 4 bytes. Moreover, it will be possible to store data in this range, even with a resolution of up to 0.005 ° C, if selecting C = 200 and y_0=1000. Such resolution for temperature recording can be obtained by averaging several measurements. In addition, even for the storage of such sizes as daily and monthly consumption of heat, water and electricity for most objects with sufficient commercial and operational accounting accuracy, two bytes will be quite enough.

In addition, it is possible to reduce the size of the field due to the transition from the store values collected since the start (reset) to the values in the system, is accumulated in a relatively small intervals, just as in a sigma-delta ADC and video compression systems. With this campaign sometimes is sufficient to store the data in fields with sizes up to 1 byte.

The next step, which allows to move at least a two-fold reduction in the volume of stored data, is the transition from the field for storing the date and time (according to the standard ANSI / ISO SQL 92 is a field of type DATE, TIME, DATETIME, and TIMESTAMP [22]), to unsigned integer fields . The practice of long-term application of this approach has shown that it is expedient to transform this first convert the time to UTC – format that eliminates the problem of time shifts due to irregular daylight savings clock of administrative time and possible change of time zones, the local time for a specific area. Then to view data for a specific period of time in a particular locality translation UTC – time to local time can be done either with the use of stored in a separate database table on time shifts and standard database functions and CGI – handler, or by writing codes for conversion functions.

When determining the amount allotted under the temporary reference value of the field must be guided by the minimum necessary time data sampling period, and the time required to store these data. In the base of the form are recorded values.

$$k_i = (t_i - t_o) / \Delta t, \quad (2)$$

where t_i – observation time, t_o – system start time, and Δt – sampling interval. For example, the following Table 1 shows that for most practical applications of this approach can store the temporary sampling in field of 1 to 3 bytes instead of 8 bytes allocated for storing the date and time in most databases. Since the lifespan of software without modification makes about 3–7 years, in the table in bold fields that provide the most effective combination of terms of data storage, field width, and the data sampling interval.

Table 1. The dependence of the maximum possible storage time, years of the sample period and the size of the field.

Number of bytes	Data sampling interval									
	1c	10c	30c	1 min	5 min	10 min	30 min	1 hour	1 day	1 month
1									0.7	21
2						1.3	**4**	**7.5**	180	
3	0.53	**5.3**	**10.6**	32	160	133				
4	136									

As seen in this table, the most appropriate field is 2 bytes and sampling intervals 1 hour. This eliminates from the database table, which records the measurement results, a separate field for storing temporary labels, since, as already mentioned, the majority of controlled variables also require normalizing after its storage for two-byte field. Furthermore, if the structure of the incoming data clearly fixed and the data is received regularly, it is possible to remove the record timestamps and the identifiers of controlled parameters. Assuming that the size of the field for storing the identifier of the controlled parameter is one byte, and fields to store time stamps and the data itself - two bytes, then the transition to the fixed structure of the measurement results will reduce the total recording length of 5 bytes to 2 bytes, i.e. reduce the amount of data stored in 2.5 times. At the same time, this approach eliminates the need for indexing and data tables without increasing the time to search for the information required, since in this case, the right place in the table is calculated in advance, and is set in a typical option OFFSET SQL - SELECT command.

However, in real systems, failures and loss of receiving information from certain channels are always possible. Therefore, to improve the storage reliability and the possibility of recovery even in case of partial destruction of a file, it is offered to enter a tag for identifying bad or missing data and transmitting time label of feature using records that are not used for recording the monitored parameters values. The most convenient for this is using a combination of all zeros or all ones if the data is stored as unsigned integers. This does not violate the normalization mechanism and practically does not affect the dynamic range of possible changes in the controlled variable.

Additional data compression can be achieved if use bitwise processing when the values of several samples of the same or different controlled parameters are packed into one field in the database table. However, such a solution is difficult to standardize, since the nature of the packaging bit will always depend on the particular set of stored values and their dynamic range. In addition, for each case will have to write the code for the compress, and to decompress the data, which is not very convenient.

A much more effective in terms of data compression, and improve the reliability of storage and provides faster access method based on their grouping into separate clusters for some common characteristics. Among these, first of all, are the equal intervals of samples, as well as the probability of a simultaneous demand. As a result, a common database will be a separate set of tables, each of which will be kept homogeneous sampling time, location, the destination, the measurement units or other status information.

The specificity of the majority of the monitoring systems, and first of all, interactive systems is the need for the best possible real-time access to current information, while access to information for prior periods is not so urgent, and may require more time. To implement this functionality, the database should provide a buffer tables with small volume for data with a small sample interval. Simultaneously, the current data is written into the buffer table and large blocks are overwritten in the main table. This approach minimizes both the access time and the time of recording data.

Consider further compression methods based on the account of features of controlled processes, as the presence of a strong correlation between adjacent samples of the controlled parameter.

2.1 Difference schemes compression

In practice, very often situations that some time for a series of samples of controlled magnitude does not change its consecutive values or these changes happen actually at the noise level, complicating the analysis of the observed process. Examples of such signals observed in conducting studies to assess the outdoor air temperature influence on the temperature in the controlled room, shown in Figure 1.

In this Figure, each step-jump signal corresponds to the temperature change by 0.0625 ° C. It would seem to get rid of such jumps generated by random noise is quite easily by smoothing the signal using digital filters. However, it is known that any low-pass filter increases the lag effect of the system [25–27].

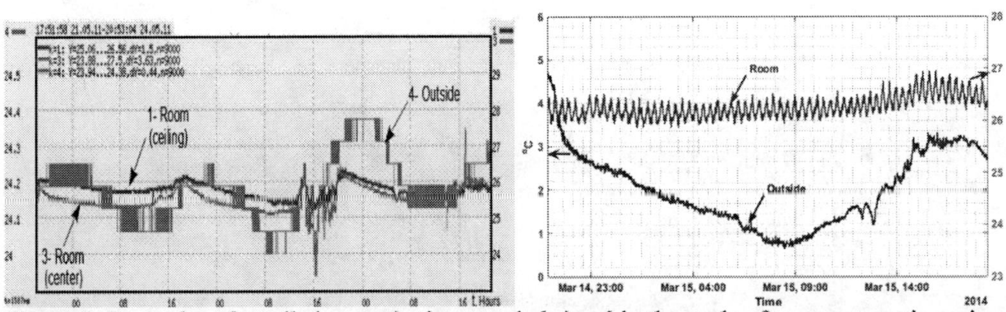

Figure 1. Examples of small changes in the recorded signal in the study of temperatures in various points of the controlled room and the outside air.

If the monitored parameter changes over time rather smoothly, has a clearly expressed trend line, and the difference between the parameter values between adjacent samples is much smaller than the range of its possible changes, a high compression ratio is provided by difference schemes using run-length encoding algorithm (RLE, run-length encoding) [28–30]. Compression scheme using RLE commonly used for streaming video and audio compression. However, unlike the multimedia processing systems, monitoring systems has a higher demands regarding reliability of information storage. Therefore, in the unacceptable situation, when failure in one count information can lead to distortion of the whole data block.

The essence of the proposed modifications RLE – compression methods is that the difference is not taken between adjacent samples (e.g., between image frames), but between the current count and a fixed time interval reference value. Due to this, failure of one reading will not affect, in any way, the reliability of data recording in the other readings.

Since, the algorithms with partial loss of data provide the best compression ratio and, on the other hand, increase the signal quality by smoothing the random noise. In [31] was carried out an experimental assessment of level of such noise. The initial data were taken of temperature indoors and outdoors, such as those shown in Figure 1. To determine the noise level of the temperature sensor was placed in a box made of foam with a wall thickness of 2 cm thermostat. Sampling interval was 30 sec with a total of 2030 samples, the first of 30 samples of them were used for the averaging and discarded from the end. To find the average (reference) values of the sensor, signal was smoothed by moving average that has a rectangular window with a width of 10 samples. At the same time, the noise was defined as the difference between the original and the smoothed signal. The latter was taken as it or rounded up to the nearest integer. For greater clarity, the signal was reckoned not in terms of temperature, but in its levels of quantization. For temperature, the digital sensor DS18B20 used in the experiment has quantization equaled 0.0625 ° C.

Histograms of the probability distribution of observed deviations for different selections of the reference signal are shown in Figure 2.

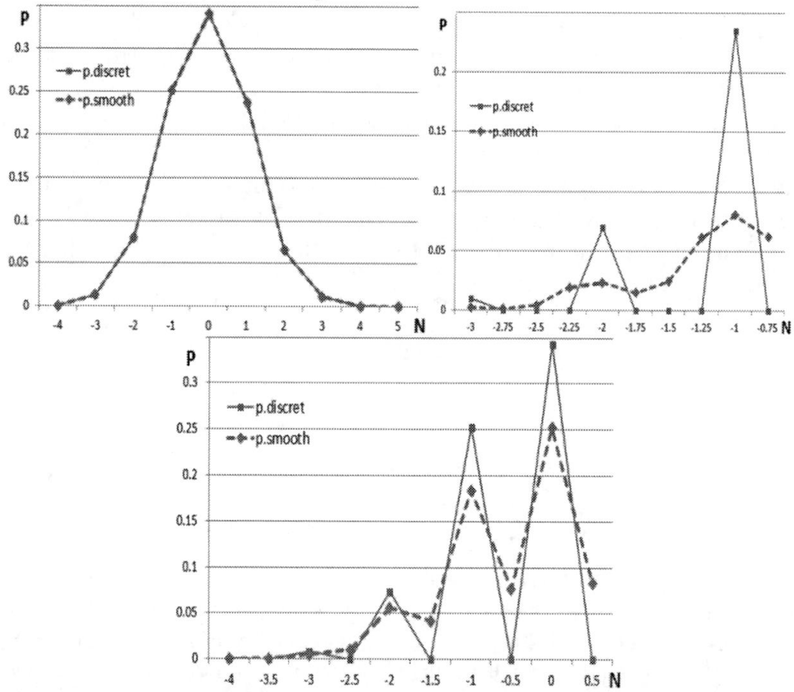

Figure 2. Probability Histogram of the temperature deviations (in N quantization levels) for various values of the partition intervals ΔN:
a) ΔN = 1; b) ΔN = 0.5; c) ΔN = 0.25. The deviations were calculated relative to the mean of the smoothed (p.smooth) or medium sampled (p.discret) of signal level.

As can be seen from this Figure, the sensor noise is pronounced discrete nature, and their function is well approximated by a normal distribution law. It can be assumed that the noise caused by the maximum deviation of the signal from its true value lies within three quantization levels, and noise STD is approximately one level of quantization. It follows that the modified RLE compression method should also provide a recovery error signal in the same range, i.e. within one - three quantization levels, without significant loss of information data.

To estimate efficiency of the proposed methods of data compression, a series of experiments based on the developed information and measuring system of technical and environmental monitoring described in [2, 32] was carried out.

2.2 Experimental Procedure for the comparison of various compression algorithms as applied to the temperature monitoring

The essence of the conducted field experiments was as follows. For processing, two temperature signals of 8200 samples (slightly less than 3.5 days) inside and outside building were used.

One were a relatively smooth in nature, while the other contained anomalies in the form of a sufficiently rapid temperature changes caused by both natural factors (wind, cloud) and anthropogenic (airing, the inclusion of a heater). Initially, all of the samples were subjected to two-fold smoothed moving average with a width of rectangular window were 2 samples and the first 4 samples are discarded. Simultaneously, controlled the maximum deviation between the original and the smoothed signal Δx does not exceed 1…3 quantization levels. This treatment allows, on the one hand, minimizing random noise, and on the other - to keep all the little details of the changes in the signal level.

Since the RLE method works with integer values, all samples of the smoothed time series were rounded to the nearest integer values.

During compression, each signal was divided into equal time intervals. For each interval, polynomial coefficients of polynomial approximation were calculated. When approximating trigonometric functions of the coefficients of the Fourier series were determined using a fast algorithm "butterfly". For the possibilities of its application the number of the samples in the interval was selected multiple 2^n.

After finding the coefficients of the approximating functions for each interval, the difference between these functions and the original smoothed time series was calculated and the difference signal standard deviation, maximum deviation and coefficient of determination were calculated.

The described sequence of operations, since splitting a temporary row data into intervals, repeated for various extent of intervals (i.e. for various number of samples inside an interval) until the maximum value of the deviation did not exceed 2-3 quantization levels.

At the final stage of the experiment, the compression ratio was estimated for packed data; it always remains the same for the reference methods and is determined by the expression:

$$K_{compress} = 0.25 \cdot K_{samples} / K_{coef}. \qquad (3)$$

where $K_{samples}$ – the number of samples within interval; K_{coef} – the number of coefficients in the approximation function. The presence in the formula of the constant coefficient 0.25 is because, unlike the original integer valued sample values that occupy at most 1-byte, the coefficients of the approximating functions are real type and for storing their values already require 4 bytes for each one.

It is important to note that, unlike the proposed compression method, the approximating functions did not store the samples in the database, but the real coefficients of the approximation polynomial.

3 Experimental results

Experimental results of applying different compression methods for compressing the data summarized in Table 2, where $K_{compress}$ – the compression ratio determined from the expression (3); STD – standard deviation of the error; Δx – the maximum modulus of the difference between the approximated and original signals, $K_{samples}$ – the number of samples within in the interval, K_{coef} – the number of real coefficients in the approximated function, K_{rep} – the number of repetitions of the new values of controlled quantities.

 DOI: 10.1051/matecconf/20167901076

Table 2. The experimental results for temperature series compression.

Type of approximation function	K_{coef}	$K_{samples}$	Δx	STD	$K_{compess}$
Temperature fluctuations in the room, without anomalies					
Constant	1	6	0.29	0.59	1.50
Linear function	2	20	0.29	0.71	2.50
3rd polynomial	4	50	0.29	0.79	3.13
6th polynomial	7	70	0.29	0.56	2.50
9th polynomial	10	300	0.29	0.32	7.50
8 harmonics Fourier series	18	2880	0.29	0.79	40,00
RLE1	K_{rep}=3	-	0.29	0.49	77.59
Temperature fluctuations in the room, with anomalies					
Constant	1	4	0.29	0.59	1.00
Linear function	2	10	0.29	0.48	1.25
3rd polynomial	4	25	0.29	0.55	1.56
6th polynomial	7	35	0.29	0.53	1.25
9th polynomial	10	300	0.29	0.31	7.50
8 harmonics Fourier series	18	2880	0.29	0.80	40,00
RLE1	K_{rep}=3	-	0.29	0.83	67.84
Outdoor temperature without anomalies					
Constant	1	4	0.29	0.44	1.00
Linear function	2	8	0.29	0.36	1.00
3rd polynomial	4	20	0.29	0.48	1.25
6th polynomial	7	35	0.29	0.53	1.25
9th polynomial	10	300	0.19	0.52	7.50
8 harmonics Fourier series	18	1000	0.19	0.45	13.89
RLE1	K_{rep}=5	-	0.19	0.97	68.61
Outdoor temperature with anomalies					
Constant	1	2	0.29	0.22	0.50
Linear function	2	5	0.29	0.12	0.63
3rd polynomial	4	15	0.29	0.21	0.94
6th polynomial	7	20	0.29	0.22	0.71
9th polynomial	10	300	0.34	0.88	7.50
8 harmonics Fourier series	18	1000	0.34	0.58	13.89
RLE1	K_{rep}=5	-	0.34	0.95	61.65

As an example, Figure 3 shows a portion of a temperature series and its error of reconstruction using various approximation functions. It shows that the reconstruction error for anomaly areas (in this case – outliers) increases dramatically. This leads to reduce the number of samples within the approximation interval. Typical plots of the relation between the interval size and the approximation error STD, as well as, the maximum error of approximation are shown in Figure 4.

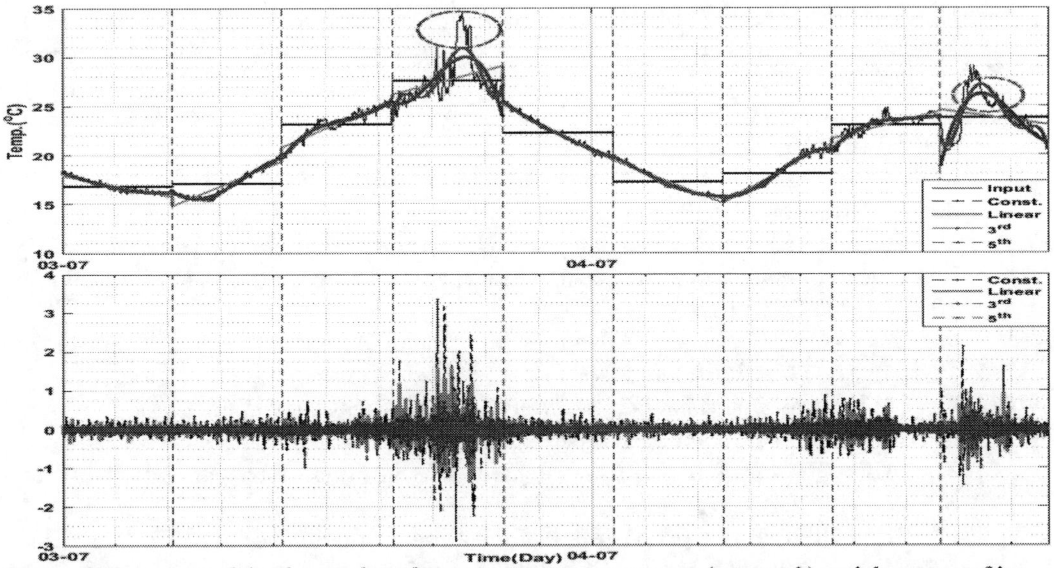

Figure 3. Fragment of the time series of temperature measurements (top graph) and the errors of its approximations using different approximation functions.

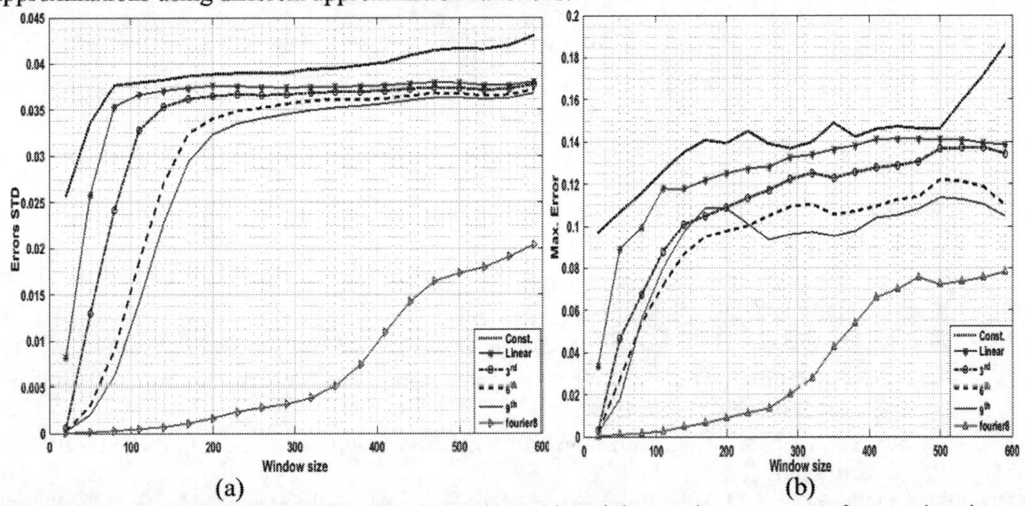

Figure 4. Dependency of approximation errors STD (a) and the maximum errors of approximations (b) on the number used to approximate $\boldsymbol{K}_{\text{samples}}$ samples.

4 Conclusion

As the performed experiments showed, the proposed method for compression has clear advantages with respect to approximation methods, especially in signals with anomalies. In addition, it has less computational complexity, since all the operations are performed in integer arithmetic and it uses input data directly from the ADC output. However, if the accuracy is not sufficient to reproduce the input data, it is better to use high order polynomial approximation or Fourier series for input data that has not anomalies.

References

[1] M. Hirayama, MATEC Web of Conferences **59**, 01002 (2016) doi: 10.1051/matecconf/20165901002

[2] H. M. Hussein, R. V Kuntz, L. I. Suchkova, A. G. Yakunin, Sci. J. of the theor and applied researches (Phys.) (1), 210 (2013) URL: http://izvestia.asu.ru/2013/1-1/phys/ TheNewsOfASU-2013-1-1-phys-13.pdf.
[3] Y. Li, H. Zhao, J. Fan, MATEC Web of Conferences **22**, 04008 (2015) doi: 10.1051/matecconf/20152204008
[4] M. F. Othman and K. Shazali, Procedia Eng. **41**, 1204 (2012) doi: 10.1016/j.proeng.2012.07.302
[5] M. Haefke, S. C. Mukhopadhyay, H. Ewald, Conference Record - IEEE Instrumentation and Measurement Technology Conference, 5944154 (2011) doi: 10.1109/IMTC.2011.5944154
[6] G. Ohring , A. Gruber, Advances in Space Research **28**, 207 (2001) doi: 10.1016/S0273-1177(01)00350-7
[7] G. B. Wiersma, *Environmental Monitoring* (CRC Press, New York, 2004)
[8] M. Malyutov, B. Ryabko, J. Astola, *Compression-Based Methods of Statistical Analysis and Prediction of Time Series.* (Springer International Publishing, Switzerland, 2016)
[9] C. Zhen, B. Ren, Int. Conf. on Computational Intelligence and Software Eng., 5366670 (2009) doi: 10.1109/CISE.2009.5366670
[10] M. B. Lin, Y. Y. Chang, IEEE Trans. on VLSI Systems **17**, 1297 (2009) doi: 10.1109/TVLSI.2008.2003512
[11] I. Suarjaya, IJACSA **3**(8), 14 (2012)
[12] P. Lindstrom, M. Isenburg, IEEE Trans. on Visualization and Computer Graphics **12** , 1245 (2006) doi: 10.1109/TVCG.2006.143
[13] D. Salomon, G Motta, *Handbook of data compression* (Springer-Verlag, London, 2010)
[14] N. Kimura, S. Latifi, ITCC '05 **2**, 8 (2005)
[15] G. Graefe, L. D. Shapiro, Proc. 1991 Symposium on Applied Computing, (1991)
[16] J. Ziv, A. Lempel, IEEE Trans. on Information Theory **23**, 337 (1977) doi: 10.1109/TIT.1977.1055714
[17] H. M. Hussein, A. G. Yakunin, Polzunovsky vestnik (2), 65 (2013) [in Russian]
[18] H. M. Hussein, A. G. Yakunin, XI inter. scientific-practical Conf. "Science and technology: step into the future-2015", 73 (2015)
[19] K. Sayood, S. Bhanja, N. Ranganathan, *Lossless Compression Handbook* (Academic Press, Cambridge, 2003).
[20] H. M. Hussein, L. I. Suchkova, M. A. Yakunin, Polzunovsky Almanac, AltSTU (2), 48 (2012). http://elib.altstu.ru/elib/books/Files/pa2012_2/pdf/048Hussein.pdf. [in Russian]
[21] MySQL 5.0 Reference Manual, URL: https://docs.oracle.com/cd/E19078-01/mysql/mysql-refman-5.0/
[22] V. Vaswani, *MySQL(TM): The Complete Reference* (McGraw-Hill, New York, 2004)
[23] C. Vicknair, D. Wilkins, Y. Chen, ACM-SE '12, 176 (2012) doi: 10.1145/2184512.2184554
[24] B. D. Blansit, J. of Elec. Resources in Medical Libraries **3**, 135 (2006) doi: 10.1300/J383v03n03_10
[25] R. Losada, *Digital Filters with MATLAB* (MathWorks, Inc., Natick, 2008).
[26] S. W. Smith, The Scientist and Engineer's Guide to Digital Signal Proc. (1), 261(1997)
[27] G. Taubin, IEEE Fifth Inter. Conf.on Computer Vision, 852 (1995)
[28] S. W. Smith, The Scientist and Engineer's Guide to Digital Signal Processing, 481 (1997)

[29] B. Žalik, N. Lukač, Signal Processing: Image Communication **29**, 96 (2014) doi: 10.1016/j.image.2013.09.002
[30] D. Salomon, *Data Compression: The Complete Reference* (Springer Science & Business Media, Berlin, 2004).
[31] H. M. Hussein, Proc. 16th Inter. Conf. "Measuring, monitoring, information", AltSTU **1**, 11 (2015). URL: http://mca.altstu.ru/download/proceedings2015.zip
[32] Altai State Technical University, Monitoring system, [Online] URL: http://abc.altstu.ru

VIBRA-SENSORS PXI-BASED TEST COMPLEX CONTROLLED BY LABVIEW

Alexey Borisov[a], *Leonid* Kessarinskiy, *Dmitry* Boychenko

National Research Nuclear University MEPhI (Moscow Engineering Physics Institute), 115409, Moscow, Russia

Abstract. The article devoted to construction of an automated test complex for parametric and functional control of vibra-sensors ICs (such as integrated accelerometers and MEMS-generators). The test complex based on PXI-standard measurement devices and controlled by LabVIEW soft. The complex's characteristics was confirmed by calibration service and been certified.

1 Introduction

Vibration-sensitive devices such as accelerometers and MEMS-generators, are widely used in industrial and space systems with external factors hardness requirement. Consequently it required to provide laboratory hardness tests with functional and parametric samples control of such sensors [1–4].

Most of the available today on the market test solutions (EMIC, Sentek Dynamics, Thermotron, TIRA) are designed for vibration tests without additional external factors. They are quite bulky, multipurpose devices, and unsuitable for external factors test. Besides standard vibration testers are used for rather big and heavy device-under-test, for example packaging box of ICs, and are excessive for single chip. The aim of creation custom PXI-based test complex was to provide vibration-sensitive devices tests for microchips or micromodules [5–10] weighing typically less than 100 grams. So there was a need to create test complex answered to next requirements:

- Mobility.
- The possibility to set the vibration to a wide range of frequency and vibroacceleration.
- The possibility of automated controlling of frequency and vibroacceleration.
- The possibility to integration into the existing test complexes.

The work is devoted to present a custom designed automated test complex for parametric and functional control of vibration-sensitive devices during the radiation testing.

2 Hardware and software

First step was to define an executive part of the complex, includes a source of vibration and a power amplifier for this source, and then to convert a software commands to executive hardware vibration.

a Corresponding author: aybor@spels.ru

© The Authors, published by EDP Sciences. This is an open access article distributed under the terms of the Creative Commons Attribution License 4.0 (http://creativecommons.org/licenses/by/4.0/).

An executive hardware action was based on an electrodynamics vibration table with a requirement mobility rate and amplitude.

The photo of a designed executive part is shown on Figure 1.

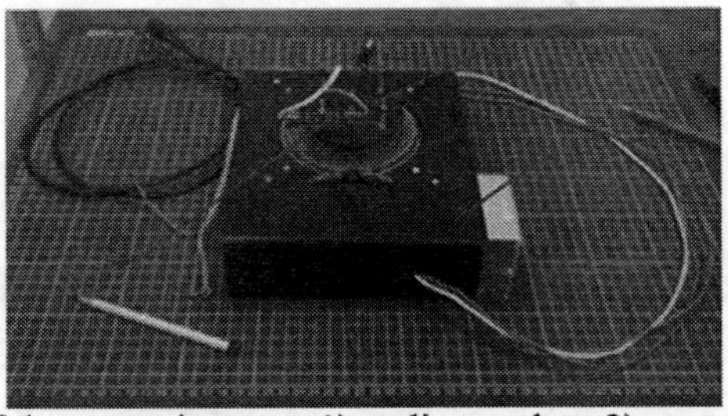

Figure 1. The appearance of the executive part: 1) audio speaker, 2) power amplifier, 3) accelerometer.

The main parts of the vibro-table are described below.

1) Audio speaker

The audio speaker with large diameter of a membrane and 80 watts of electrical power was chosen as the basis of executive part of a complex. The audio speaker has wide bandwidth that allows for a wide range of vibration frequency.

2) Power amplifier

The audio speaker supplied from the audio amplifier corresponding by electrical power. High electrical power and diameter of the speaker membrane provides a wide range of vibroacceleration. Small size of the executive part provides mobility.

3) Accelerometer

Full automation of parametric and functional control process is possible by reference through high speed accelerometer with analog output.

3 Solution

1) Hardware part

Represented executive part was certified in the prescribed manner. Achieved charachteristics of the vibration load is shown in Table 1.

Table 1. Characteristics of the executive part of complex.

Name of parameter	Minimum	Maximum
Vibro acceleration	0.05g	1.5g
Frequency	5 Hz	1.5 kHz

The next step of the complex development was solving the problem of integration one to existing external factors hardness test complexes based on the National Instruments PXI-standard measurement devices. Another words it needed to developed PXI-based controller with LabVIEW soft.

Functional generator PXI-5402 used as a setting device for form and amplitude of the oscillation, a vibroacceleration setting. High accuracy of vibroacceleration defining was achieved by sufficient precision of output amplitude.

Output voltage amplitude of these accelerometer measured by PXI-5114 oscilloscope. The software calculate vibroacceleration and thereafter and adjust the output amplitude of PXI-5402 functional generator.

2) Software part

The system's software was created NI LabVIEW. Front panel of electrodynamic vibration table control program is shown on Figure 2.

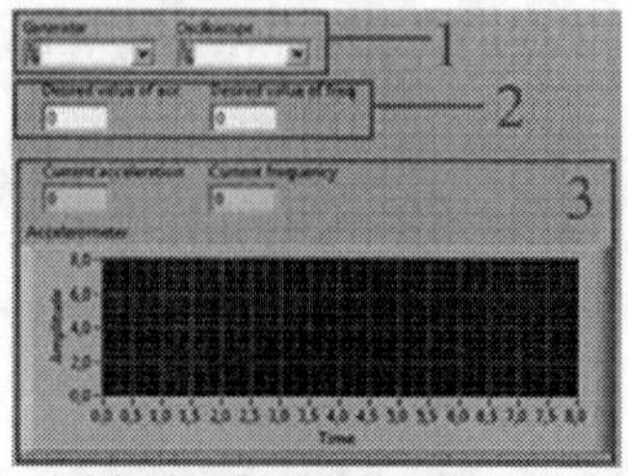

Figure 2. Front panel of control VI: 1) block of initialization, 2) block of desired parameters setting, 3) block of reached values.

To use the program must enter the desired frequency and acceleration.

Measured acceleration sensor, or any other device under test can be mass comparable to the mass of the movable part vibrating table, it can change the resonant frequency of the system. The same effect on the resonance frequency is the presence in the device own resonant frequencies, if present massive moving parts in its composition. Unpredictable change in the resonant frequency may result in a sharp increase in acceleration at this point, which may be critical for some devices and cause their destruction. Thus, by measuring the transfer characteristics of the object and just when moving from one frequency to another, there is a need for a smooth change of the oscillation frequency by adjusting the amplitude of a continuous acceleration in order to maintain a predetermined level. The main parts of the software are shown in Figures 3 and 4.

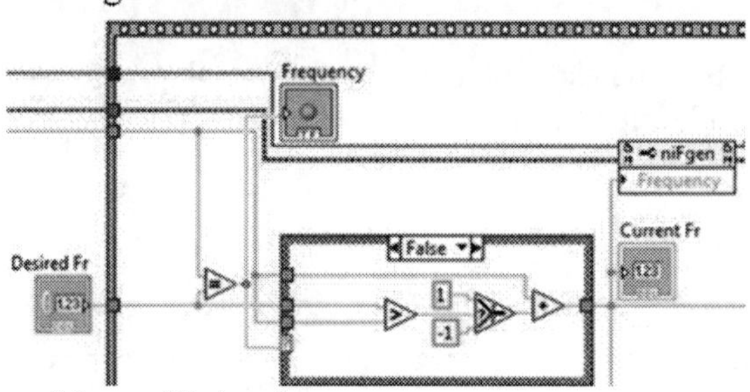

Figure 3. Smooth change of the oscillation frequency.

The first step in the algorithm is the frequency change in the desired direction. The current frequency is compared to the entered in the appropriate box on the front panel of the value of the desired frequency, if these values do not match, is given to the frequency change command in the direction desired by a small amount. Figure 4 corresponds to the minimum change in frequency of 1 Hz, but, depending on the task, the equipment allows to change the frequency in increments of up to 1 mHz. After the change of vibroacceleration occurs check for compliance specified in the appropriate box on the front panel. Measured current value of the acceleration by reading from the accelerometer readings in a few oscillation periods, subtracting from the resulting waveform DC component (acceleration of gravity), and calculate the actual maximum acceleration on the basis of the oscillation amplitude and the sensitivity of the accelerometer in this case is 0.66 V/g.

If the measured acceleration measurements fit within acceptable boundaries, the algorithm proceeds to further tuning frequency. If the measured value is out of range, then the value of the gain generator modified to decrease or increase, then the measurement is repeated and the change in gain continues until the desired acceleration values, and then repeats step frequency control.

Thus, the application of this algorithm allows you to change the frequency and acceleration values smoothly within a wide range and, thus, does not exceed the permissible limits on acceleration [12, 13].

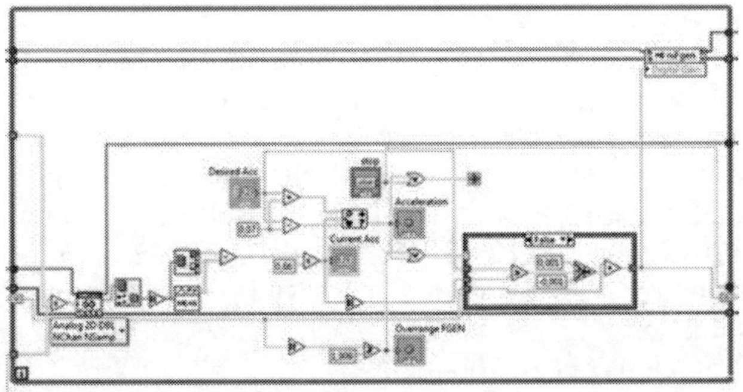

Figure 4. Smooth change of the vibroacceleration.

4 Implementation and its prospects

The developed automated complex allows automatically measured all of the criterial parameters such as frequency response function of a vibration-sensitive devices during the radiation tests at other environmental conditions and facilities (pulsed electron accelerators, electron LINACs, X-ray testers etc.) [11].
The developed automated system has successfully implemented at JSC "ENPO SPELS", Moscow, Russia.

Acknowledgment

The paper is a part of the research done within the government assignment RFMEFI57414X0115 given by the ministry of education and science of Russian Federation.

References

[1] A. S. Artamonov, D. V. Boychenko, A.Y. Nikiforov, A. V. Sogoyan, N. A. Shelepin, V. A. Telets, M. G. Tverskoy, Proceedings of the European Conference on Radiation and its Effects on Components and Systems (RADECS) **1**, 296 (2002)

[2] V. V. Belyakov, V. S. Pershenkov, G. I. Zebrev, A. V. Sogoyan, A. I. Chumakov, A.Y. Nikiforov, P. K. Skorobogatov, Mikroelektronika **32**, 31 (2003)

[3] V. V. Belyakov, A. I. Chumakov, A.Y. Nikiforov, V. S. Pershenkov, P. K. Skorobogatov, A. V. Sogoyan, Mikroelektronika **32**, 134 (2003)

[4] V. V. Belyakov, V. S. Pershenkov, G. I. Zebrev, A. V. Sogoyan, A. I. Chumakov, A.Y. Nikiforov, P. K. Skorobogatov, Russian Microelectronics **32**, 25 (2003) doi: 10.1023/A:1021809802818

[5] D. V. Boychenko, A.Y. Nikiforov, P. K. Skorobogatov, A. V. Sogoyan, Proceedings of the European Conference on Radiation and its Effects on Components and Systems (RADECS), 5205522 (2007) doi: 10.1109/RADECS.2007.5205522

[6] A. V. Grebenkina, D.V. Boychenko, A.Y. Nikiforov, V. A. Telets, V. V. Amelichev, V.A.Kharitonov, Proceedings of the International Conference on Microelectronics (ICM), 133 (2014) doi: 10.1109/MIEL.2014.6842103

[7] A. B. Boruzdina, A. A. Orlov, A. V. Ulanova, N. G. Grigor'ev, A.Y. Nikiforov, Proceedings of the International Siberian Conference on Control and Communications (SIBCON), 7147007 (2015) doi: 10.1109/SIBCON.2015.7147007

[8] I.O.Loskutov, A. B. Karakozov, P.V. Nekrasov, A.Yu. Nikiforov, Proceedings of the International Siberian Conference on Control and Communications (SIBCON), 7147128 (2015) doi: 10.1109/SIBCON.2015.7147128

[9] I.I. Shvetsov-Shilovskiy, P.V. Nekrasov, A.V. Ulanova, A.A. Smolin, A. V. Sogoyan, Proceedings of the International Siberian Conference on Control and Communications (SIBCON), 7147282 (2015) doi: 10.1109/SIBCON.2015.7147282

[10] V. A.Marfin, P.V. Nekrasov, O. A. Kalashnikov, K. A. Kagirina, Proceedings of the International Siberian Conference on Control and Communications (SIBCON), 7147142 (2015) doi: 10.1109/SIBCON.2015.7147142

[11] A.S. Artamonov, A.A. Sangalov, A.Y. Nikiforov, V. A. Telets, D.V. Boychenko, IEEE Radiation Effects Data Workshop **2015-January**, 7004600 (2014) doi: 10.1109/REDW.2014.7004600

[12] D. Hartono, D. Halim, A. Widodo and G.W. Roberts, MATEC Web of Conferences, **65**, 1003 (2016) doi: 10.1051/matecconf/20166501003

[13] Zhi Zhou, MATEC Web of Conferences **65**, 5001 (2016) doi: 10.1051/matecconf/20165905001

MOBILE COMPLEX FOR RAPID DIAGNOSIS OF THE TECHNOLOGICAL SYSTEM ELEMENTS

Alexey Gavrilin*, *Boris* Moyzes, *Alexzander* Cherkasov, *Kirill* Mel'nov, *Xiaoliang* Zhang

Tomsk Polytechnic University, Institute of Cybernetics, 634050, Tomsk, Russia

Abstract. The article shows the up-to-dateness of the new informing and measuring tools and technologies development. It is reviewed the mobile complex for runtime diagnostics of technological system "machine-tool-instrument-detail". It was found that the use of the complex allows to identify the frequency area in which the appearance of resonance of the technological system elements is possible, and thus to draw a conclusion on the technical state of the diagnosed object. It is concluded that there is the prospects for the use of the above mentioned mobile complex for vibration diagnostics.

1 Introduction

The vibration that occurs in the production equipment, leads not only to a positive effect, useful for various technological tasks [1–8], but also negative effect. In terms of metal working vibration can have a significant effect on the accuracy of the equipment operation of (accuracy and surface roughness), the reliability of its operation, the protection level of installed in close proximity the precision manufacturing equipment, the safety of the personnel on the consequences of possible emergency situations.

The development of scientific and technological progress and the requirements for the processingaccuracy has set the development trends of metal working equipment design, the main ones are:

- increased productivity through increased automation and intensification of the cutting modes (high-speed cutting);
- improved accuracy of processing [9], including the vibration damping [10] and the choice of optimal operating regime [10];
- improved reliability equipment operation due to on-time diagnostics [9], including vibration diagnostics to determine vibration activity of the technological system elements which negatively affect the working accuracy of workpiece and tool life.

The development of the metal-working equipment designs led to the further development of information and measurement technology [11–14] for vibration diagnostics of the technological system "machine-tool-tool-instrument-piece".

Diagnostic facilities can be both mobile and stationary, but the stationary ones cannot be used for fast diagnostics of metalworking equipment. It should be noted that there is a wide range of parts in appliances.

* Corresponding author: tom-gawral@list.ru

© The Authors, published by EDP Sciences. This is an open access article distributed under the terms of the Creative Commons Attribution License 4.0 (http://creativecommons.org/licenses/by/4.0/).

There are known the designs of the mobile diagnostic complexes of foreign and domestic origin as well as their application techniques [11–14]. These complexes show good results, but they are complicated to operate (instruction and special skills are required) that determined the necessity to develop an up-to-date, compact, reliable and easy-to-use diagnostic system for vibration diagnostics.

2 Mobile complex for rapid diagnosis of the technological system elements

The scientists of Tomsk Polytechnic University in collaboration with the company "VITEK Siberia" (Tomsk) has developed and tested the diagnostic system [10], consisting of three modules (Figure 1):

- piezoelectric sensors;
- vibration and measurement unit;
- a laptop with the operating software.

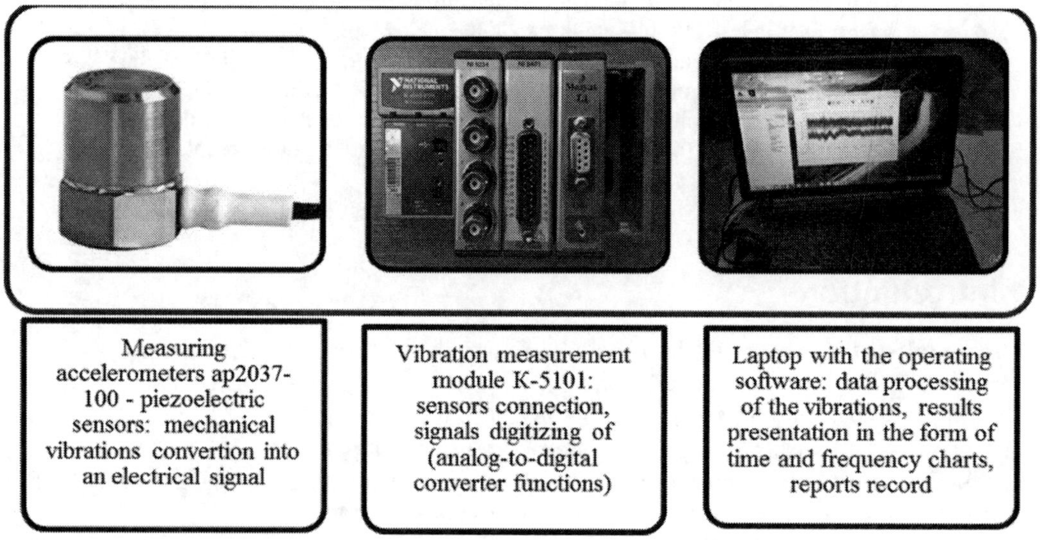

Figure 1. Modules of the diagnostic complex.

It was required for the diagnostic complex to be competitive on the market which determined modification of the software in the follow terms:

- presentation of the diagnostic results – the applicability of the first-fifth orders filters of the low and high frequency as well as band and-reject, which allows to adapt to different conditions faster (Figure 2) are added;
- signal processing – frequency band of the amplitude spectrum selection for the vibration pattern analysis (Figure 3) was simplified;
- channel information selection – up to four or more.

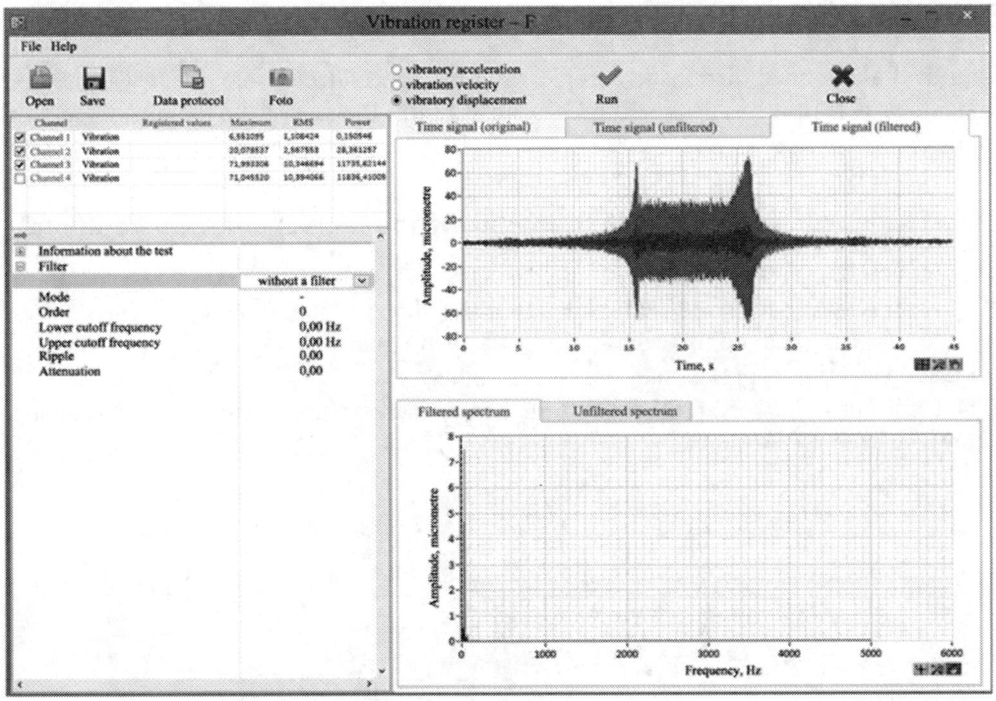

Figure 2. Software window fragment of "Vibration register – F": left – channel information selection and filtering parameters settings; right – signal waveform by three channels when accelerating to 2000 rev/min, dwelling and braking and its spectrum.

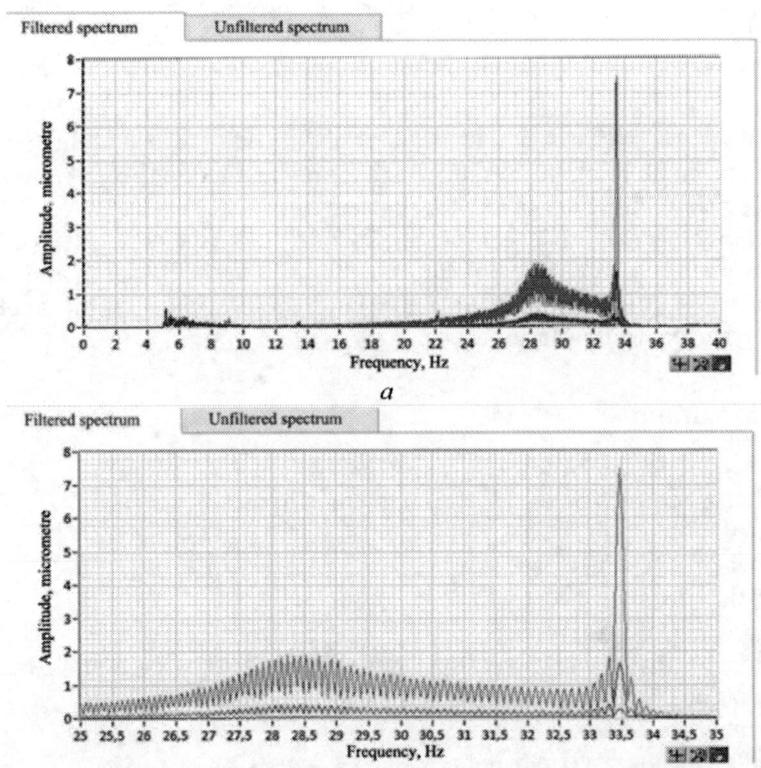

a

b

Figure 3. Software window fragments of "Vibration register –F" for different scales: *a* – 1 Hz; *b* – 0.5 Hz.

3 The diagnostics procedure by the mobile diagnostic complex

Testing of the upgraded diagnostic complex and the procedure development was assigned to students as part of the educational process to simplify the use of the diagnostic complex while testing.

3.1 The diagnostic complex setting (basic diagnostic scheme development)

The complex setting includes the follow steps:

- the selection of the diagnosing unit and fixing the vibration directions;
- the selection of bonding technique and a sensing point;
- the connection complex modules to each other

The screwing lathe 16K20 model was chosen for researching. Sensors layout is shown in Figure 4 (magnet fastening).

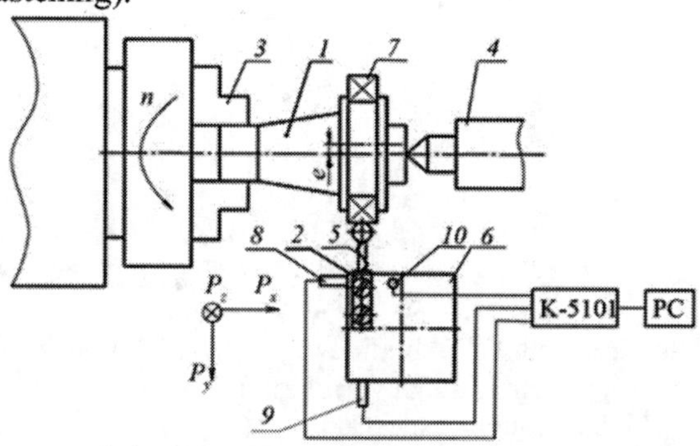

Figure 4. The basic diagram of the complex setting: *1* – wobbler shaft; *2* – holder; *3* – three-jaw chuck; *4* –tail block; *5* –elastic element; *6* – toolholder; *7* – bearing part; *8–10*–accelerometers; K-5101 – vibration measurement module; PC – a laptop; P_x, P_y, P_z – operative directions of cutting force components.

Sensors are placed on the toolholder (due to the inability to install them directly on the spindle of the machine-tool), the axes *X*, *Y* and *Z* are the directions of vibration registration, P_x, P_y, P_z – operative directions of exciting cutting force components.

Excitation was performed during rotating spindle due to the eccentricity *e* by the elastic element *5* (by spring- actuated piston plunger), mounted in the holder *2*, and fastened to the toolholder 6. The elastic element *5*, acting through the bearing *7* on the shaft *1*, loads the spindle by the harmonically varying force.

3.2 Complex setting

To set the complex the following steps are required:

- to identify the sensor channels on the computer (Figures 2, 4):
  - channel 1 (blue) - the axis *X*;
  - channel 2 (red) – the axis *Z*;
  - channel 3 (green) – the axis *Y*;
- to configure the software – select:
  - registered values: vibration displacement (selected), vibration velocity, vibration acceleration;
  - tool: cutter (selected), the miller;
  - the filter type and its parameters etc.
- to configure the software – select:

- registered values: vibration displacement (selected), vibration velocity, vibration acceleration;
- tool: cutter (selected), the miller;
- the filter type and its parameters etc.

3.3 The modes selection and the test operation plan development

As a power impact source the spring-loaded plunger rod has been chosen (Figure 4). Spindle rotation speed is 2000 rev/min (maximum).

Experimental plan: to register the vibration patterns during acceleration, dwelling and breaking.

3.4 Starting the diagnosing object and registering the results

Vibrations pattern obtained during the experiment is shown in Figure 2.

3.5 Report making

The test results are generated automatically in Microsoft Word format with the time and spectral diagrams.

3.6 The data analysis and recommended guideline

Special software allows to choose a specific time range (Figure 5, *a*) for reviewing and analyzing and the corresponding signal spectrum (Figure 5, *b*), namely a sharply increased amplitude fluctuations zone.

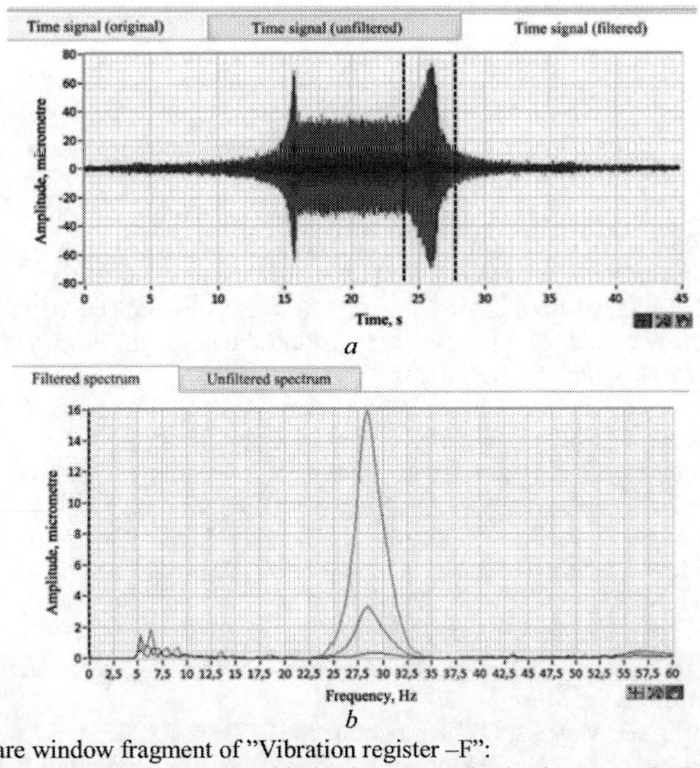

Figure 5. Software window fragment of "Vibration register –F":
a – waveform through the three channels with the time span selection between 24-28 s; *b* – signal spectrum.

Vibration pattern (Figure 5) shows the presence of oscillation amplitude jump at a frequency of 33.45 Hz, which is close to the disturbing frequency at the spindle:

$$f_{df} = \frac{n_{sp}}{60} = \frac{2000}{60} = 33.3 \text{ Hz},$$

where n_{sp} – spindle speed, rev/min.

4 Conclusion

In addition it is possible to determine the frequencies close to resonance, according to the obtained values of the vibratory displacement (Figure 2) as well as the crest factor value can be determined, as the ratio of maximum amplitude value of time signal to RMS (Table 1).

Table 1. Measurement results of vibratory displacement, RMS and crest factor calculation.

Channel	Maximum	RMS	Crest factor
1	6,55	1,11	5,9
2	20,08	2,57	7,81
3	71,99	10,35	6,96
4	71,05	10,39	6,84

Increased values of cross-factor (above 1.41) indicate the presence of impact in the drive elements, which requires thorough diagnostic.

Thus, the mobile complex for fast diagnosis of the technological system elements with improved special software allows to optimize the signal processing in the terms of filtration and frequency analysis, and the research carried out by students, has demonstrated how simple to operate the complex.

References

[1] A.I. Nizhegorodov, A.N. Gavrilin, B.B. Moyzes, Key Eng. Mater. **685**, 320 (2015) doi: 10.4028/www.scientific.net/KEM.685.320

[2] A.I. Nizhegorodov, A.N. Gavrilin, B.B. Moyzes, J. Phys: Conf. Ser. **671**, 012037 (2016) doi: 10.1088/1742-6596/671/1/012037

[3] R.M. Mustafina, I.A. Plotnikov, I.V. Plotnikova, O.N. Tchaikovskaya, J. Phys: Conf. Ser. **671**, 012046 (2016) doi: 10.1088/1742-6596/671/1/012046

[4] A.Y. Petrova, O.N. Chaikovskaya, I.V. Plotnikova, Tech. Phys. J. **60**, 592 (2015) doi: 10.1134/S1063784215040222

[5] A.P. Surzhikov, E.N. Lysenko, V.A. Vlasov, A.V. Malyshev, E.V. Nikolaev, RUSS PHYS J. **56**, 681 (2013) doi: 10.1007/s11182-013-0085-6

[6] O.V. Vasileva, Y.N. Isaev, A.A. Budko, A.I. Filkov, J. Phys.: Conf. Ser. **552**, 012026 (2014) doi: 10.1088/1742-6596/552/1/012026

[7] I. Plotnikova, L. Redko, O. Titova, O. Chaykovskaya, IOP Conf. Ser.: Mater. Sci. Eng. **132**, 012023 (2016) doi : 10.1088/1757-899X/132/1/012023

[8] S. Obukhov, I. Plotnikov, M. Kryuchkova, IOP Conf. Ser.: Mater. Sci. Eng. **132**, 012017 (2016) doi : 10.1088/1757-899X/132/1/012017

[9] A.M. Dalsky, *Reference-book of technologist-mechanical engineer* (Mashinostroenie, Moscow, 2003)

[10] A. Gavrilin, B. Moyzes, O. Zharkevich, JVE **17**, 3495 (2015)

[11] C. Grigoriev, M. Kozochkin, F. Sabirov, V. Sinopalnikov, Vestnik MSTU STANKIN **4**, 27 (2010)

[12] M. Kozochkin, A. Privalov, Instrum. Syst.: Monit., Control and Diagnostics **2**, 18 (2013)
[13] M. Kozochkin, F. Sabirov, Vestnik MSTU STANKIN **3**, 14 (2008)
[14] D. Hartono, D. Halim, A. Widodo, G.W. Roberts, MATEC Web of Conferences **65**, 01003 (2016) doi: 10.1051/matecconf/20166501003

POSITIONING FOR PEOPLE BEHIND BARRIERS IN REAL TIME WITH UWB PULSE SENSING

Sergey Shipilov [1], *Rail* Satarov, *Ivan* Fedyanin, *Evgeny* Balzovsky, *Vladimir*Yakubov

National Research Tomsk State University, 634050, Tomsk, Russia

Abstract. This paper features the technology for identifying the location of real people behind the barriers in real time using the UWB sensing. The opportunity to obtain the radar images of moving and not moving people was presented in this paper. The oscillation rate of a human chest can be defined by the continuous recording the not moving person. The point of this technique is in calculation of the differential signal in successive periods of time. The range of application to be extended considering the particular conditions. The research results can be used in intelligence service as well as for searching people trapped under the rubble.

1 Introduction

Terrorism and counter terrorism policy as well as its impact on everyday life are of high relevance nowadays as there is always a high chance of hostage-taking or destruction of the engineering facilities. In both cases, every second counts for human lives. Therefore, the equipment for remote accurate detecting people in real time behind the barriers are very important.

The use of X-rays is preferred for that purpose because of high resolution and penetration. However, its hazard for living beings and the one-side access to the test object prevent X-rays from the application in the searching and detecting people behind barriers. Therefore, the radiowave tomography techniques are the most reasonable choice.

The ultra-wideband (UWB) radiation provides the high longitudinal spatial resolution, which is particularly important to detect hidden objects behind the dielectric barriers. The radiation of this type provides the sufficient penetration into a barrier to create a radiotomogram of the test area. This process is based on tomography of multi-angle remote measurements of scattered radio wave radiation. Radiowave tomography is anyway based on ray-focusing effect, which enables inverse transformation of wave projections of test objects and propagation medium. It appears that almost all methods which are available or just in progress now, may be considered under this aspect. The multiple effects (scattering and diffraction) of interacting the wave fields with propagation medium inhomogeneities can be considerably reduced by using of spatiotemporal radiation focusing.

The developed algorithms of creating the tomogram should be resistant to noise and interference, which significantly impact the received signal in the one-side sounding of the objects behind the dielectric barriers[1, 2]. The developed algorithms should also provide

[1]Corresponding author: s.shipilov@gmail.com

© The Authors, published by EDP Sciences. This is an open access article distributed under the terms of the Creative Commons Attribution License 4.0 (http://creativecommons.org/licenses/by/4.0/).

the adequate operating speed when the requirements to the measuring and recording equipment as well as to the computer hardware are not high. There is no doubt about the principle opportunity to build the radar equipment to detect people behind barriers. That was proven by researches published [1, 2] and by the authors as well [3–5].

2 Inverse focusing problem

The algorithm of radiotomography imaging is based on the focusing of the reflected UWB signals. The main idea of the data processing is referred to the technology of the synthetic aperture radar with the focusing. All signals, recorded at different transmitting-receiving points $\boldsymbol{\rho}$, are summed in phase for each predefined focusing point $\mathbf{r}$ with the delay correction:

$$F(\mathbf{r}) \equiv \sum_j S\left(\boldsymbol{\rho}_j, t = 2\frac{|\mathbf{r} - \boldsymbol{\rho}_j|}{c}\right).$$

The transmitting receiving points can be randomly located. In approximation of the single scattering, when $S_0(t)$ is transmitted signal and the received one is

$$S(\boldsymbol{\rho}_j, t) = k_0^2 \iiint_V S_0\left(t - 2\frac{|\mathbf{r}_1 - \boldsymbol{\rho}_j|}{c}\right)\frac{\Delta\varepsilon(\boldsymbol{\rho}_1, z_1)}{\left(4\pi|\mathbf{r}_1 - \boldsymbol{\rho}_j|\right)^2}\left(d^3\mathbf{r}_1\right),$$

the following equation can be written

$$F(\mathbf{r}) = \iiint_V \Delta\varepsilon(\boldsymbol{\rho}_1, z_1) Q(\boldsymbol{\rho}_1, z_1; \mathbf{r})\left(d^3\mathbf{r}_1\right),$$

where

$$Q(\boldsymbol{\rho}_1, z_1; \mathbf{r}) = \sum_j S_0(t = 2\frac{|\mathbf{r} - \boldsymbol{\rho}_j|}{c} - 2\frac{|\mathbf{r}_1 - \boldsymbol{\rho}_j|}{c})\frac{k_0^2}{\left(4\pi|\mathbf{r}_1 - \boldsymbol{\rho}_j|\right)^2}$$

works as the hardware function of the system. With the data sufficiency, this function approximates to the «blurred» δ function which provides the identification of the focusing results with the $\Delta\varepsilon$ permittivity distribution:

$$F(\mathbf{r}) \sim \Delta\varepsilon(\boldsymbol{\rho}, z).$$

As previously stated, the task of the sensing behind barriers is reduced to the detection of moving objects in some cases. Some research proposed the differential – residual technique for the detection of moving people based on filtering the signals from the stationary objects. The point of this technique is in calculation of the differential signal in successive periods of time.

$$dS(t, T_i) = S(t, T_i) - S(t, T_{i-1}),$$

where $dS(t, T_i)$ is the filtered signal, $S(t, T_i)$ is the recorded signal, $S(t, T_{i-1})$ is the signal received in the previous time period. In this case, the signals reflected from the stationary background objects located in the test area are equal in different time periods and filtered by subtraction. The signal difference dS, obtained from the moving object is other than zero and can be used for the radar imaging the moving objects.

However, that approach didn't consider the speed of the sounded object. When the object doesn't move, the difference of signals from neighboring frames is close to zero. This results in disappearance of the information spot in the radar image. The authors proposed the modified algorithm accounting the object speed to determine the optimal sampling rate that maximizes the SNR for the differential signal. So, the previous equation was modified to

$$dS(t, T_i) = S(t, T_i) - S(t, T_{i-N}),$$

where N is the number of the frame that provides the maximal intensity of the differential signal.

When a person doesn't move, their presence behind the barrier can be detected by breathing movements of their chest or abdomen. Study the Figure 1. The S_2 signal reflected from the chest during the exhalation differs from the S_1 signal reflected during the inhalation with the delay $T = 2\Delta x / c$, where Δx is the chest shift when inhalation/exhalation.

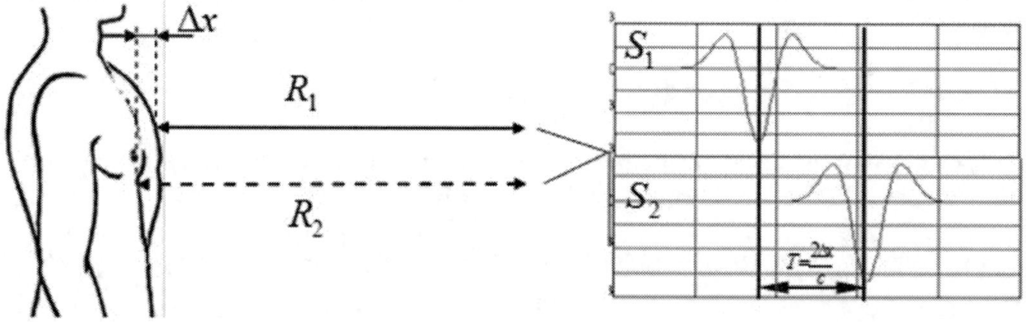

Figure 1. Record of human breath oscillation.

The data obtained in UWB sensing to be converted to distinguish the human breath oscillations. Let's consider the reflected signal spectra. The complex amplitudes of the spectrum signals received in sequential time periods agree up to the $\exp(i\omega T)$ factor:

$$S_2(\omega) = S_1(\omega)\exp(i\omega T)$$

When multiplying the $SS_i(\omega)$ spectra of the received signals by the $S_1(\omega)^*$ complex conjugated spectrum, we obtain:

$$S_i(\omega)S_1(\omega)^* = |S_i||S_1|\exp\left(i[\varphi_i - \varphi_1]\right) = |S_i||S_1|\exp\left(i\omega T\right).$$

If we separate the phase from the equation obtained, then

$$T_i = \frac{\arg\left(S_i(\omega)S_1(\omega)^*\right)}{\omega}, \quad \omega = 2\pi f,$$

where f is the significant frequency picked out from the received signal spectrum by the predefined level.

3 Experimental results

To obtain multiangle projections of the sounded area, an UWB clocked array of 8 receivers and 1 transmitter was used. The receiving antennas were polled sequentially in clocking mode. The arrangement of the elements in the array is presented in Figure 2, *a*. The array element [2] is presented in Figure 2, *b*. It was a printed antenna with a screen in form of the metal angle. The developed antenna is a plate of the foil-laminated glass fiber 110×110×1 mm in size. The arrow points the direction of maximal radiation. The antenna is designed for the radiation of UWB pulses with duration of 0.3… 0.4 ns.

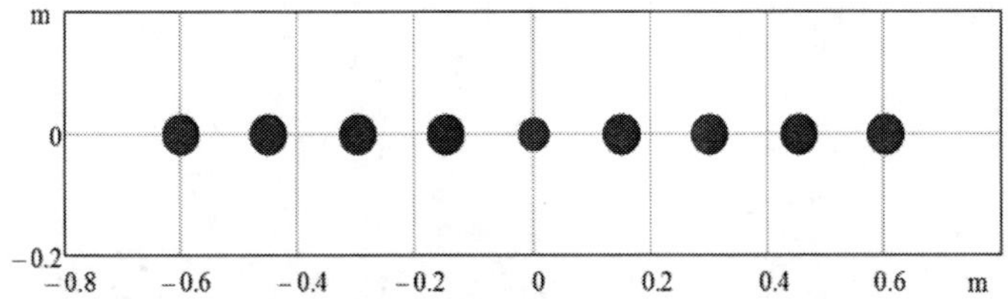

(a)

(b)

Figure 2. Arrangement of array elements (a), UWB antenna (b).

The experiment proving this approach was conducted when a person changed their location against the array. The oscilloscope record of the test area for one of the array receivers is presented in Figure 3, *a*. The signal amplitude is displayed in grayscale. The horizontal axis is the record time of the UWB signal in the stroboscopic receiver. The vertical axis is the poll time. The time between two poll periods was 0.1 s. The Figure shows the object moving represented by changing in the arrival time of the UWB signal. The interface of the software for positioning of the moving people is presented in Figure 3, *b*. The data processing result – a tomogram of the test area in form of a 2D image with the cell size 50×50 cm– is displayed in the main window.

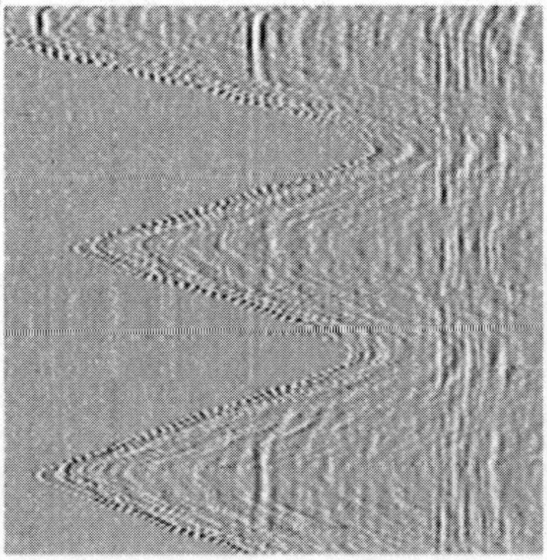

(a)

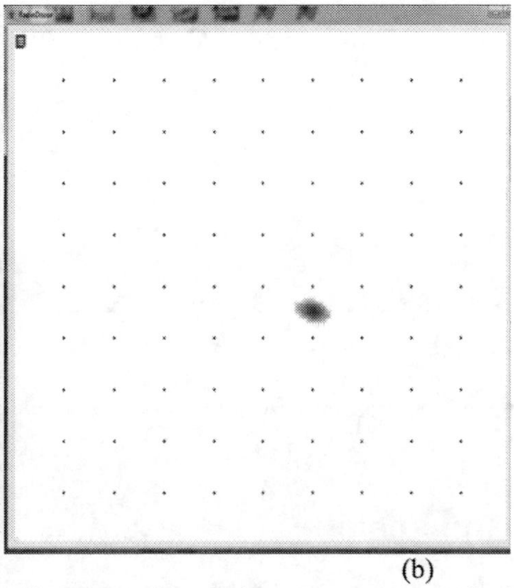

(b)

Figure 3. Oscilloscope record of the test area (a), radio image of a moving person at a certain time (b).

The proposed algorithms and applications were used for processing the results of measurement. Figure 4, *a* presents 50 waveforms of the reflected UWB signals which were processed using the technique of the constant component filtering. The grayscale indicates the different signal levels. X axis corresponds to the waveform of reflected signal. Y scale is clocking time with the step of 0.5 s. The oscillations of the human chest could be clearly found in Figure 4, *a* as the change in clocking time of reflected signal maximum. The interface of the software for positioning by the breathing movements is presented in Figure 4, *b*. The data processing result – a tomogram of the test area in form of a 2D image with the cell size 50×50 cm– is displayed in the main window.

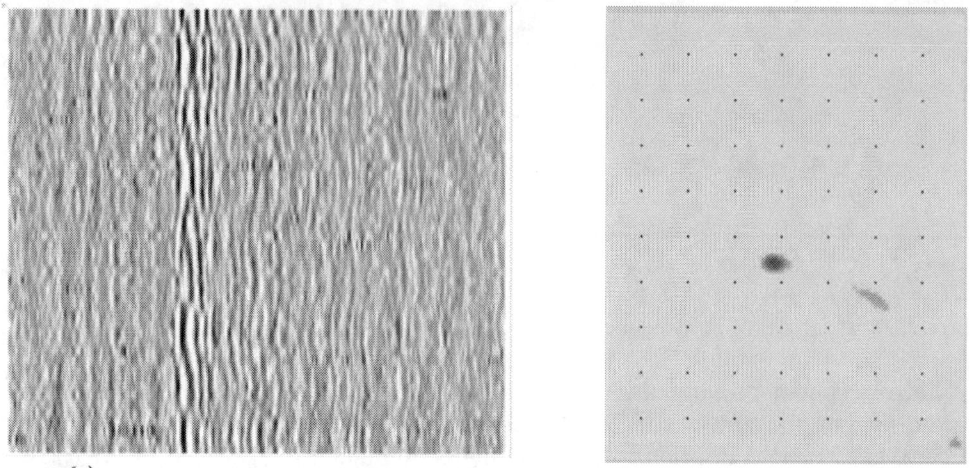

(a) (b)

Figure 4. Oscilloscope record of test area (a), radio image of a breathing person at a certain time point (b).

As a result of oscilloscope record processing, the human chest oscillation (Figure 5, *a*) were distinguished. Then, using the Fourier transform, the spectrum was calculated which maximum corresponds to the breathing rate (Figure 5, *b*).

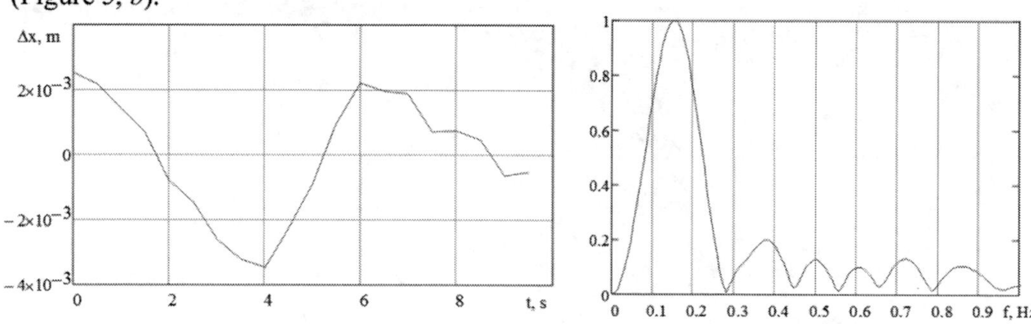

(a) (b)

Figure 5. Waveform of experimentally distinguished oscillation of human chest (a), and its spectrum (b).

4 Conclusion

The research proposed the approach to detect and visualize people behind the dielectric barriers. The opportunity to obtain the radar images of moving and not moving people was demonstrated. The oscillation rate of a human chest can be defined by the continuous recording the not moving person. The research results can be used in intelligence service as well as for searching people trapped under the rubble. The range of application to be extended considering the particular conditions.

Acknowledgment

The research was supported by the Russian Ministry of Education and Science as a part of the state order No 3.694.2014/K.

References

[1] Yazhou Wang, Aly E. Fathly, IEEE Trans. on Geoscience and Remote Sensing **50**, 1986 (2012) doi: 10.1109/TGRS.2011.2170694

[2] R.N. Satarov, Yu. Kuz'menko, T. R. Muksunov, A. V. Klokov, E. V. Balzovskii, Yu. I. Buyanov, S. E. Shipilov, V. P. Yakubov, Russian Physics Journal **55**, 884 (2013) doi: 10.1007/s11182-013-9896-8

[3] James D. Taylor, *Ultra wideband radar. Applications and design* (Crc Press, New-York, 2012)

[4] Jurgen Sachs, *Handbook of ultra-wideband short-range* (Wiley-vch, 2012)

[5] Y. Kan, Y. Zhu, Q. Fu, MATEC Web of Conferences **56**, 02001 (2016) doi: 10.1051/matecconf/20165602001

MATEC Web of Conferences **79**, 01080 (2016) DOI: 10.1051/matecconf/20167901080
IME&T 2016

DYNAMIC SINGLE-PHOTON EMISSION COMPUTED TOMOGRAPHY DATA ANALYSIS: CAPABILITIES FOR DETERMINING FUNCTIONAL SIGNIFICANCE OF CORONARY ARTERY ATHEROSCLEROSIS

Andrew Mochula[1a], *Konstantin* Zavadovsky[1,2], *Sergey* Andreev[1], *Yury* Lishmanov[1,2]

[1]Federal State Budgetary Scientific Institution "Research Institute for Cardiology", 634012, Tomsk, Russia
[2]National Research Tomsk Polytechnic University, 634050, Tomsk, Russia

Abstract. The aim of the study was to develop the method for collecting and processing scintigraphic data to determine myocardial flow reserve by using gamma-camera with a new type detector based on Cadmium Zinc Telluride (CZT). Sixteen coronary artery disease patients and nine healthy volunteers received cardiac dynamic single-photon emission computed tomography with technetium 99m methoxyisobutylisonitrile (^{99m}Tc-MIBI) at rest and during pharmacologic stress test. Data processing involved a formation of regions of interest from the left ventricular (LV) cavity and myocardial walls to build activity-time curves. Myocardial flow reserve index was determined as a quotient of two ratios of mean myocardial counts to area under a curve peak from LV cavity during stress test and at rest. Mean values of myocardial flow reserve index were 1.86 (1.59; 2.2) in group of healthy volunteers and 1.39 (1.12; 1.69) in patients with multivessel coronary artery disease. When the value of this index was less than 1.77, the method allowed for identification of multivessel disease with sensitivity and specificity of 81.8% and 66.7%, respectively. Standard myocardial perfusion scintigraphy in combination with the method of myocardial flow reserve index determination allows for increasing diagnostic significance of the scintigraphic approach for assessment of coronary microcirculation lesions in multivessel coronary artery disease.

1 Introduction

Coronary artery disease is currently number one cause of disability and mortality from cardiovascular diseases. According to estimates of the World Health Organization, 7.4 million people had died of this pathology worldwide during the period from 2002 to 2012.

Revascularization procedure is currently considered the efficacious method for treatment of this pathology. This method consists in recovery of coronary blood flow

[a] Corresponding author: mochula.andrew@gmail.com

© The Authors, published by EDP Sciences. This is an open access article distributed under the terms of the Creative Commons Attribution License 4.0 (http://creativecommons.org/licenses/by/4.0/).

through coronary artery bypass surgery or through stent placements in coronary arteries [1]. According to the recommendations of the European Society of Cardiology, 50–90% coronary artery stenosis requires surgical restoration of blood flow only in case of verified ischemia which can be confirmed by determining fractional flow reserve. Stenosis is considered hemodynamically significant if the above mentioned index equals 0.8 and less [1]. An invasive nature of the procedure as well as a high cost of the supplies do not allow for widespread use of this approach in clinical practice.

Positron emission computed tomography with ^{15}O-labeled water ($[^{15}O]H_2O$) is considered gold standard for noninvasive assessment of myocardial flow reserve ($[^{15}O]H_2O$) [2, 3]. However, this method requires available cyclotron and expensive diagnostic equipment in a clinic delivering these studies. All these factors along with short half-life time of ^{15}O (122.2 s) also significantly limit wide distribution of this method in everyday clinical practice. According to recommendations of the European Society of Cardiologists regarding myocardial revascularization, single-photon emission computed tomography (SPECT) plays an important role in diagnosing ischemia of the myocardium with sensitivity and specificity of 90–91% and 75–84%, respectively [4]. It is worth mentioning that SPECT quite often gives false-negative results in patients with hemodynamically significant multivessel coronary artery disease showing the absence of significant stress-induced perfusion defects. Literature describes this phenomenon by term "balanced ischemia" where a uniform decrease in an indicator accumulation is observed over the entire left ventricular (LV) myocardium [5, 6]. One of the ways to diagnose balanced ischemia of the myocardium in multivessel coronary artery disease consists in an assessment of myocardial flow reserve [7, 8].

With reference to the above, search for new methods for an assessment of this parameter is a relevant task of modern cardiology [9].

Up to now, most scintigraphy works in this direction were performed by using gamma-cameras with detectors based on sodium iodide which did not allow for acquiring dynamic data in a tomographic mode [2, 8]. Implementation of SPECT with detectors based on Cadmium Zinc Telluride (CZT) allows for overcoming technical constraints and opens up new opportunities for a radionuclide determination of myocardial flow reserve.

The aim of this study was to determine the informativity of dynamic tomoscintigraphy of the myocardium in identifying multivessel coronary artery disease.

2 Method of study

To determine coronary reserve, we proposed to perform myocardial SPECT with technetium 99m methoxyisobutylisonitrile (^{99m}Tc-MIBI) at rest and during pharmacological stress test.

2.1 Preparation

Patients were recommended to abstain from caffeine-containing beverages and/or medications for twelve hours before the procedure. The last meal was taken not less than two hours before the study. Immediately before the study, intravenous catheter was placed in a median cubital vein of a subject.

2.2 Study procedure

A patient was accommodated in a supine position on the table of gamma-camera. For correct patient positioning relative to gamma-camera detectors, low-dose computed tomography (100 keV; 10 mA/s) in the anterior projection was performed. Distances from suprasternal notch and left anterior auxiliary line to the center of cardiac silhouette were measured on an acquired image and a mark was placed on patient's body. Detectors of gamma-camera were positioned over the left ventricular (LV) area based on the acquired results [10].

A scintigraphy study was performed in two stages (Figure 1). At stage 1, the passage of a bolus of radiopharmaceutical was recorded at rest. To achieve this, a bolus injection of 1 mL of ^{99m}Tc-MIBI in a dose of 185 MBq was done through the intravenous peripheral catheter. Recording began 5 s before the injection of radiopharmaceutical. Scintigraphic images were registered in a dynamic tomographic list mode format (allowing for reconstruction of series of dynamic and static tomoscintigrams from native data sets) for 360 s. This time was enough for registering the passage of radioactive indicator through cardiac chambers and its accumulation in the LV myocardium. Immediately after the completion of data acquisition, stage 2 began where the passage of a bolus of radiopharmaceutical was recorded during pharmacological stress test [10].

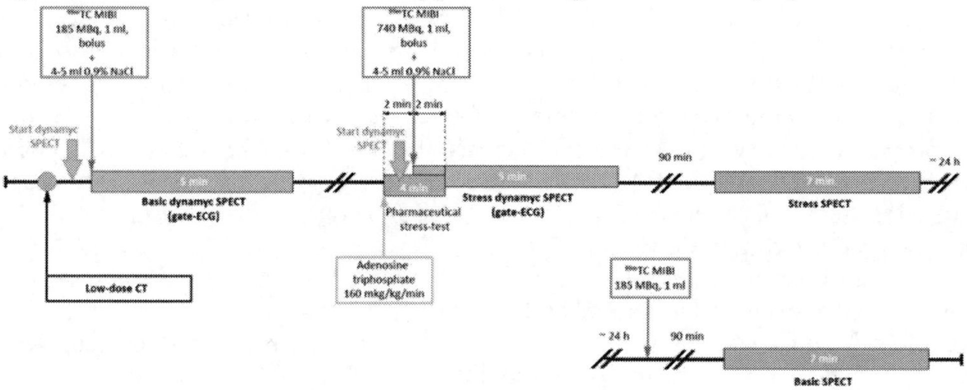

Figure 1. Scheme of cardiac dynamic SPECT for determination of myocardial flow reserve.

All scintigraphic studies were performed by using gamma-camera with CZT-based solid-state detectors (GE Discovery NM/CT 570 C) with adjustment to ^{99m}Tc photopeak (140 keV).

Stress test with adenosine triphosphate (ATP) was performed according to standard protocol [3]. Adenosine triphosphate dose (140 µg/kg/min) was calculated individually depending on patient's body weight. Adenosine triphosphate was administered with a mechanical infusion pump according to 4-min protocol (2-min ATP infusions before and after the injection of radiopharmaceutical). At pharmacological stress test peak at the end of minute 2 of ATP infusion, patient received 1 mL of radiopharmaceutical ^{99m}Tc-Technetril in a dose of 740 MBq. After that, the study continued according to the protocol identical to the one specified above. A fourfold increase in a dose of the indicator at stage 2 was necessary for leveling blood pool radioactivity after the preceding study.

2.3 Image procession and myocardial flow reserve assessment

Two series of axial images were formed from primary data by using dedicated software (GE Alcyone: Dynamic Analysis Tool). Series 1 comprised group of 64 dynamic tomoscintigrams with each frame duration of 2.2. s representative of the passage of a bolus of radiopharmaceutical through the LV cavity and through the LV myocardium from second 1 to second 150 of the study; series 2 comprised static tomoscintigrams from second 150 to second 360 representative of the accumulation of radiopharmaceutical in the LV walls.

The next stage of data processing consisted in building the activity-time curves used for myocardial flow reserve calculation. To achieve this, regions of interest from the LV cavity and the LV walls were formed on the series of static frames; activity-time plots were built based on the dynamic images for each region of interest: anterior, posterior, lateral walls, septum, and apex of the left ventricle. Based on the obtained plots, mean value of counts from each region of interest in the left ventricle as well as area under curve characterizing the passage of a bolus of radiopharmaceutical through LV cavity were determined. These calculations were performed twice: for study at rest and for study during pharmacological stress test. Myocardial flow reserve index (MFRI) was determined separately for each LV wall as follows:

$$MFRI = \frac{C_{ms}(t) \Big/ \int_0^t C_{bs}(\tau)d t}{C_{mb}(t) \Big/ \int_0^t C_{bb}(\tau)d t},$$

where: MFRI is myocardial flow reserve index; C_{ms} is number of counts from the LV myocardium during stress test; C_{bs} is number of counts from the LV cavity during stress test; C_{mb} is number of counts from the LV myocardium at rest; C_{bb} is number of counts from the LV cavity at rest.

2.4 Method approbation

A total of 16 ischemic heart disease patients aged 67±5 year (13 men, 3 women) with multivessel coronary artery disease diagnosed based on results of invasive coronary angiography were examined with this method. Comparison group comprised nine healthy volunteers aged 25±4 years (7 men, 2 women).

Procedure of myocardial flow reserve assessment was approved by the local ethics committee of the authors' institute. All patients were informed in a written format regarding the goal and potential risks of the study and gave informed consent to perform it.

2.5 Statistical analysis

Statistical processing of the results was performed with SPSS 18.0 and MedCalc 12.1.14.0 software. Obtained data were not normally distributed (based on Kolmogorov-Smirnov criterion) so data were presented as median and quartiles (Me (Q_{25};Q_{75})). Significance of intergroup differences was assessed based on nonparametric Mann-Whitney test. To determine differential boundary (cut-off value) between study values in independent samples, ROC-analysis with building ROC-curves was performed. Values were considered statistically significant when p was < 0.05.

3 Results

Based on performed analysis, mean LV MFRI in patients with three-vessel coronary artery disease significantly differed from the corresponding value in healthy volunteers group ($P < 0.05$): 1.39 (1.12; 1.69) and 1.86 (1.59; 2.2), respectively.

Statistically significant decreases in MFRI values in group of patients with multivessel disease compared with the corresponding values in individuals free of coronary atherosclerosis were also found based on analysis of MFRI in separate regions of the left ventricle (Table 1).

DOI: 10.1051/matecconf/20167901080

Table 1. Regional values of myocardial flow reserve.

LV regions	Regional values of myocardial flow reserve Me (Q_{25}-Q_{75})		
	Group of patients with multivessel coronary artery disease	Group of healthy volunteers	*p*-level (Mann-Whitney)
Apex	1.3 (1.0–1.76)	1.83 (1.65–2.04)	0.03
Lateral wall	1.54 (1.12–1.73)	2.13 (1.42–2.31)	0.01
Anterior wall	1.27 (1.0–1.7)	1.65 (1.19–1.98)	0.02
Septum	1.25 (0.93–1.77)	2.1 (1.76–2.2)	0.01

The results of ROC-analysis demonstrated that MFRI value of 1.77 allowed for verification of three-vessel coronary artery disease with sensitivity of 81.8% and specificity of 66.7%. The area under ROC-curve was 0.79 ($p < 0.05$) suggesting high confidence of the method (Figure 2).

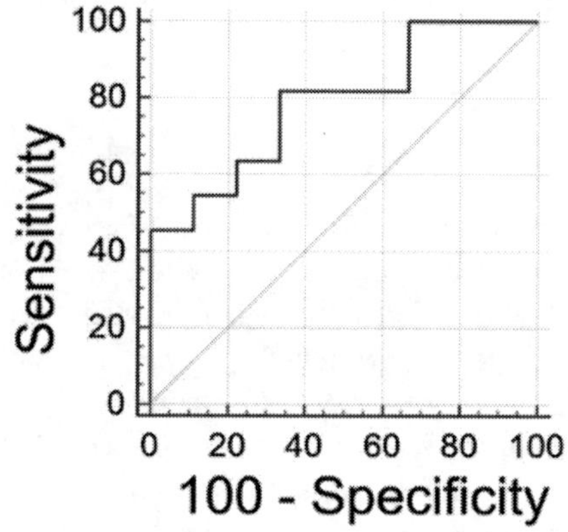

Figure 2. ROC-curve representative of diagnostic capabilities of myocardial flow reserve index in detecting multivessel coronary artery disease.

4 Conclusion

Search for a simple and inexpensive method for an assessment of myocardial flow reserve is currently considered one of the main directions of state-of-the-art cardiology. In the present study, for the first time, sensitivity and specificity of cardiac dynamic stress SPECT in assessment of myocardial flow reserve were determined and had values of 81.8% and 66.7%, respectively. Our results rank below compared with operating characteristics of positron emission tomography with ^{15}O-labeled water whose sensitivity and specificity values are 95% and 100%, respectively [4]. It is caused by linear dependence of [^{15}O]H_2O extraction fraction on myocardial flow rate as well as by higher resolution power of positron emission tomography scanners. Data obtained in our study agree with the results of other research groups who showed principal feasibility of myocardial flow reserve assessment by the methods of dynamic SPECT [4, 6, 7] or planar scintigraphy [2] with high diagnostic accuracy values.

Therefore, implementation of the present method into clinical practice and experimental research may contribute to in-depth study of myocardial microcirculation processes and will allow for more adequate identification of treatment tactics in coronary insufficiency.

Acknowledgments

This work was supported by Russian Science Foundation (grant #15-15-10016).

References

[1] S. Windecker, P. Kolh, F. Alfonso and *et al.*, Eur Heart J. **35**, 2541 (2014) doi: 10.1093/eurheartj/ehu278

[2] G. Storto, AR. Sorrentino, T. Pellegrino, R. Liuzzi, M. Petretta, A. Cuocolo, Eur J Nucl Med Mol Imaging **34**(8), 1156 (2007) doi: 10.1007/s00259-006-0333-x

[3] M.J. Henzlova, W.L. Duvall, A.J. Einstein, M.I. Travin, H.J. Verberne, J Nucl Cardiol **23**, 606 (2016) doi: 10.1007/s12350-015-0387-x

[4] Y. Ito, C. Katoh, K. Noriyasu, Y. Kuge, H. Furuyama, K. Morita, T. Kohya, A. Kitabatake, N. Tamaki, Eur J Nucl Med Mol Imaging **30**, 281 (2003)

[5] C. Liu, A.J. Sinusas, J Nucl Med. **55**, 1573 (2014) doi: 10.2967/jnumed.114.144550

[6] T. Tsukamoto, Y. Ito, K. Noriyasu, K. Morita, C. Katoh, H. Okamoto, N. Tamaki, Circ J. **69**,188 (2005) doi: 10.1253/circj.69.188

[7] S. Ben-Haim, V.L. Murthy, C. Breault and *et al.*, J Nucl Med. **54**, 873 (2013) doi:10.2967/jnumed.112.109652

[8] B. Hsu, F.C. Chen, T.C. Wu, W.S. Huang, P.N. Hou, C.C. Chen, G.U. Hung, Eur J Nucl Med Mol Imaging **41**, 2294 (2014) doi: 10.1007/s00259-014-2881-9

[9] Yu.B. Lishmanov, K.V. Zavadovskiy, I.Yu. Efimova, N.G. Krivonogov, Zh.V. Vesnina, S.I. Sazonova and *et al.*, The Siberian medical journal **2**, 21 (2015)

[10] A.V. Mochula, K.V. Zavadovskiy, Yu.B. Lishmanov, BulletFin of Experimental Biology and Medicine **12**, 845 (2015)

USAGE OF WEB MAPPING SYSTEMS AND SERVICES FOR INFORMATION SUPPORT OF REGIONAL MANAGEMENT

Nicolay Shaparev[1,2], *Oleg* Yakubailik[1,*]

[1]Institute of Computational Modeling SB RAS, 660036, Krasnoyarsk, Russia
[2]National Research Tomsk State University, 634050, Tomsk, Russia

Abstract. The work considers information and computing technologies supporting regional decisions making and based on geoinformation web-systems and mapping web-services. The use of such systems for the information support of regional management is now becoming common. Long-term strategic forecasting and planning of territories development, solution of various institutional and sectoral problems today are often based on the use of integrated information and computing environment, complex information systems of regional management, which are based on geospatial (mapping) data. This paper discusses technologies and web-services used in the creation and implementation of regional geoinformation web-systems. Such systems provide access to huge arrays of geospatial information and services distributed in the Internet, remote data processing with high performance multi-user computers. Problems of choosing basic software such as geoinformation platform, advantages and disadvantages of existing solutions are discussed. The software structure and basic web-GIS components are analyzed. Examples of completed projects are given.

1 Geoinformation platform selecting

Modern web-GIS are built according to so-called service-oriented architecture, and they can be regarded as a set of interrelated software components for spatial data operations - its import/export, cataloging, visualizing, creating, processing, distribution, etc. The technological basis for these solutions is a set of program interface libraries such as Google Maps API, Mapserver Mapscript, Scanex GeoMixer, etc.; they provide access to the functions and context of web pages mapping elements – map visualization, thematic data, for example, the road network with data on traffic, terrain, map points, catalogues of organizations, products and services, satellite images, etc [1]. High degree of the user interface interactivity, very fast regeneration of images made of small caching raster fragments (tiles), almost instant response of the move (shift) and map scale change tools and other features of spatial data visualization and management made the mapping web

* Corresponding author: oleg@icm.krasn.ru

© The Authors, published by EDP Sciences. This is an open access article distributed under the terms of the Creative Commons Attribution License 4.0 (http://creativecommons.org/licenses/by/4.0/).

applications very popular. Support of geodata exchange standards – KML, GML, WMS, etc. is also important [2, 3].

Implemented information systems are usually based on the development of specialized software tools of such kind [4, 5].

It is worth to pay attention to the problem of GIS platform selection existing today – software based on commercial products or open (free) software [6]. Leaving financial, philosophical and situational aspects of the choice aside, we would like to note that there is often no opposition between the two approaches in today's practice. The reason is that currently available commercial and free GIS complement each other well due to data formats compatibility, data exchange standards based on web services, etc. For example, one can perform the spatial data analysis using ESRI ArcGIS, convert it with MapInfo to transfer to the customer and use free Mapserver software for presentation in the web pages, and the spatial metadata catalog can be formed by the means of GeoNetwork Open Source. At the same time, PostgreSQL open database with the PostGIS extension module can be used for the spatial data storage being almost as good in performance and functionality as the leader of commercial databases Oracle with an extension Oracle Spatial for working with spatial data for a vast majority of tasks.

The open architecture and free basic software allows to quickly implement almost any new (additional) functions in the created system that can be impossible while using expensive commercial software [7]. An additional bonus is the ability of integration with any third-party commercial products if necessary. For example, if the customer already has and uses the Oracle database – you can configure the spatial data storage subsystem of the created complex for the use of this database. If it is absent, you can install and use free software of PostgreSQL databases [8].

The most popular at the moment concept of building mapping web applications involves the creation of an interconnected web services components set, the implementation of which is carried out simultaneously on the server and the client computer (multi-tier architecture), as well as formation of a spatial data set usually made in the format of popular GIS or using a specialized geospatial database [9]. The initial geodata preparation for the web application is usually carried out outside the considered web-system; for this, standard desktop GIS are typically used today (MapInfo, ArcGIS, QGIS and others) [10].

2 Software structure

Using the experience of completed developments, we try to formulate program-technological web-systems features focused on regional management tasks solution from the functional point of view [11] –

- The system should consist of client and server parts, thereby realizing the "client-server" technology. Application of the MVC (model-view-controller) design pattern as the server part basis provides ample opportunities for the tasks solution. Using of this architecture assumes separation of the application data, user interface and control logic into three separate components. Modification of each component can be performed independently. Apart from the standard MVC elements, key units of the system are also page presentation templates and AJAX-handlers [12].
- The system geospatial data is recorded in the resource catalogue. The web application communication with the resource catalogue is conducted on the basis of service-oriented architecture using the SOAP/XML protocol. It creates a set of functions available in the form of application programming interface (API), which offers the possibility of resources searching, filtering, management, editing, copying, moving, etc. [13]

- Two API sets are formed: common (client) interface for user applications and advanced (server) interface for applications having the ability to control the resources catalogue, its objects and relationships between objects.

Information support for the applied regional geoinformation web-systems consists of several types of geospatial and other data and appropriate tools of their processing [14, 15]:

- tabular data stored in the database – it provides the business logic of the system, its basic content;
- thematic spatial data related to direct content of the application system objectives and functions. This type of data can be created both within the applied system user interface (based on the web browser or Windows application) and by tools of third-party applications – various GIS/CAD programs, etc.;
- mapping background – information resources of auxiliary character maps and satellite images used for the thematic spatial data visualizing. Own data sets can be used as well as third-party services provided by Yandex, Google, 2GIS and other companies;
- tools for creating and managing the spatial metadata and related services of visualization, search and filtering [16];
- means of geospatial data classifying – standardized classification system enables multiple data using for various projects [17];
- tools for the geospatial data web publication managing – user interfaces for the formation of geoportal thematic sections based on the web-content management system;
- tools for the systems integration, information exchange and interaction based on the geoportal information-analytical and mapping services [18].

3 Basic web-GIS components

Analysis of existing approaches and trends, experience of own developments in this area have allowed formulating some backbone "universal" components for the considered class of information systems [19]:

- Archiving subsystem for the basic geospatial data.

It must organize the data storage and management, tools for its loading and deleting, backup, etc. It shall be possible to register external databases in the archive including the spatial ones and organizing transparent access to them through the single software interface – i.e., the user who receives the data from the geospatial archives is not required to know where they come from – from SHP-file hosted on the same server or via connection to a third-party server (ArcGIS Server and analogues).

- Applied mapping web-services system.

It is supposed to create a set of different tools for data accessing and organization of queries to them including the ones based on standard protocols/interfaces, such as widely used open OGC Consortium standards – mapping web-services WMS, WFS, etc. Also, there should be possibility of organizing intensive computations on the server side – today, several alternative solutions exist in this field – WPS (Web Processing Service) by the OGC Consortium and others [20, 21].

- Spatial metadata management subsystem.

The subsystem is designed to search and navigate across the available spatial data including the usage of spatial queries. The subsystem must provide the ability to work with different data classificators, be compatible with the existing metadata standards, allow the appropriate import/export. The subsystem base – metadata catalogue – should provide the entire range of metadata control/management. Deep integration with all other components of the applied geoinformation system is assumed [22].

- Web-application – a user interface to the data and its processing means.

Modern program and technological solutions are most often based on the web-technologies of content management system (web-site). In the centralized web interface, tools for the spatial and other data search and navigation including the spatial queries usage should be developed. Information visualization services must provide the user with the data in text, mapping and tabular form. The subsystem must provide the ability to work with different classificators, information presentation forms using customizable templates, etc. Uploading and downloading of different data types shall be provided. All information should be provided on the basis of permissions, user interface personalization is desirable.

- Auxiliary services subsystem.

The modular architecture of the information system for regional management creates the need for strict formal specification of information exchange between its components. In this context, a set of web-services is necessary as a tool for the data transmission/reception in the system and between its constituent parts. For example, the "Web-portal" module forms a query for more information to the "Spatial metadata catalog" subsystem with its user interface. Next, the "Spatial metadata catalog" subsystem addresses the "Geodata archive" module for a list of specific resources. Restrictions on the access to administrative services are related to the fact that they can change a lot of things including the system breaking as a whole (if misused). That is why access to them is often limited by developers.

- Public applied services subsystem.

Unlike the above-mentioned administrative services, these services are publicly available. They provide the performance of different users' queries to the data stored in the system of regional management information support. Technologically, they are similar to the administrative ones, but they cannot break the system. Examples of services are different thematic information visualizing operations used in the formation of the information web-portal pages.

4 Examples of implementation

On the basis of the above principles and approaches to the creation of applied systems for regional management, works on the design, development, implementation and operation of several geoinformation web-systems were carried out. The service-oriented architecture provided good replicability of the software components, high speed of development and implementation. In most cases, the created geoinformation web-system modules were tightly integrated into the existing corporate information infrastructure [23].

The developed system of applied high-level services successfully established itself as a software technological basis of information analytical systems for regional territory and municipal authorities. A number of such systems were implemented, such as decision support systems for local government on health, education, transport and disaster management; environmental monitoring and assessment of the state of the environment; support system for socio-economic development of the region. In general, during the last five years it was implemented more than 10 analytical applications. Customers of these systems are the departments of the government of the Krasnoyarsk region, one of federal subjects of Russia. A few more development was established within the framework of the scientific research grants.

For example, the initial data for the Krasnoyarsk region fire situation-mapping interface is constantly real time updated departmental database on fires in the region. To ensure the operative data presentation in the geoinformation system, a special web-service providing the necessary pre-processing and raw data aggregation was developed. There are a number of similar system around the world [24].

The web-GIS can provide access to the spatial analysis for users who are not technical experts in GIS. It is very attractive to see your data in a spatial representation without using any complicated programs and being limited only by a standard web browser. An illustration to this idea is the experience of web-GIS using for a variety of information and analytical tasks by the Krasnoyarsk Region Veterinary Supervision Service.

5 Conclusions

The developed software and technological solutions have successfully proven themselves as a basis for resource intensive information and analytical systems of regional level for various problems solution - sectoral management information support (in the sphere of health, education, transport), environmental monitoring and the environment state assessment, forecasting of the region's socio-economic development, centralized map data information supplying.

The modular architecture of these systems, the use of standard web-services to exchange the data between these modules provide rapid adaptation of the existing software to customer's requirements, replication of individual components, their complementarity. In turn, the registration of created resources – spatial data – in the geoportal centralized catalog provides the possibility of its simultaneous use in multiple developments. This service-oriented approach based on active web technologies implementation in the applied information systems is increasingly used today.

Acknowledgments

The work was supported by grant no. 16-47-240517 from Russian Foundation for Basic Research and Krasnoyarsk Regional Fund of Support of Scientific and Scientific-technical activities.

References

[1] K. Patroumpas, N. Georgomanolis, T. Stratiotis, M. Alexakis, S. Athanasiou, Journal of Web Semantics **35**, 53 (2015) doi: 10.1016/j.websem.2015.09.003

[2] O.E. Yakubailik, V.G. Popov, Computational Technologies **6**, 116 (2009)

[3] H. Wu, Z. Li, H. Zhang, C. Yang, S. Shen, Computers & Geosciences **37**, 485 (2011) doi: 10.1016/j.cageo.2010.05.026

[4] A. Vyas and S. John, MATEC Web of Conferences **57**, 05002 (2016) doi : 10.1051/matecconf/20165705002

[5] X. Jin, D. Guo, S. Wang, X. Zhao, Y. Song, A. Liu, W. Jiang, R. Wang and X. Li, MATEC Web of Conferences **22**, 04027 (2015) doi : 10.1051/matecconf/20152204027

[6] A.A. Kadochnikov, O.E. Yakubailik, NSU Journal of Information Technologies **1**, 37 (2015)

[7] L. Bastin, G. Buchanan, A. Beresford, J.-F. Pekel, G. Dubois, Ecological Informatics **14**, 9 (2013) doi: 10.1016/j.ecoinf.2012.11.013

[8] B. R. Pickard, J. Daniel, M. Mehaffey, L. E. Jackson, A. Neale, Ecosystem Services **14**, 45 (2015) doi: 10.1016/j.ecoser.2015.04.005

[9] V. Rautenbach, S. Coetzee, A. Iwaniak, Computers, Environment and Urban Systems **37**, 107 (2013) doi: 10.1016/j.compenvurbsys.2012.08.001

[10] C. Heipke, ISPRS Journal of Photogrammetry and Remote Sensing **65**, 550 (2010) doi: 10.1016/j.isprsjprs.2010.06.005

[11] O.E. Yakubailik, A.A. Kadochnikov, V.G. Popov, A.A. Tokarev, Vestnik of SibGAU **25**, 61 (2009)
[12] K. Zhang, B. Qu and X. Chi, MATEC Web of Conferences **31**, 17002 (2015) doi : 10.1051/matecconf/20153117002
[13] R. Béjar, M. Á. Latre, F. J. Lopez-Pellicer, J. Nogueras-Iso, F.J. Zarazaga-Soria, P.R. Muro-Medrano, Computers & Geosciences **46**, 66 (2012) doi: 10.1016/j.cageo.2012.04.010
[14] L. Xue, Y. Zhu, Y. Xue, Mathematical and Computer Modelling **58**, 480 (2013) doi: 10.1016/j.mcm.2011.11.002
[15] C. Farcy, B. de Terwangne, P. Blerot, Computers and Electronics in Agriculture **47**, 207 (2005) doi: 10.1016/j.compag.2004.12.005
[16] L. Díaz, C. Granell, J. Huerta, M. Gould, Applied Geography **35**, 448 (2012) doi: 10.1016/j.apgeog.2012.09.008
[17] Theodor Foerster, Lassi Lehto, Tapani Sarjakoski, L. Tiina Sarjakoski, Jantien Stoter, Computers, Environment and Urban Systems **34**, 79 (2010)
[18] D. J. Maguire, P. A. Longley, Computers, Environment and Urban Systems **29**, 3 (2005) doi: 10.1016/j.compenvurbsys.2004.05.012
[19] O. Yakubailik, A. Kadochnikov, A. Tokarev, International Multidisciplinary Scientific GeoConference Surveying Geology and Mining Ecology Management (SGEM) **1**, 487 (2015)
[20] A.A. Kadochnikov, InterCarto **20**, 188 (2014)
[21] J.D. Blower, A.L. Gemmell, G.H. Griffiths, K. Haines, A. Santokhee, X. Yang, Environmental Modelling & Software **47**, 218 (2013) doi: 10.1016/j.envsoft.2013.04.002
[22] A.G. Matveev, O.E. Yakubailik, Vestnik of SibGAU **2**, 48 (2012)
[23] H.C. Guo, L. Liu, G.H. Huang, G.A. Fuller, R. Zou, Y.Y. Yin, Journal of Environmental Management **61**, 93 (2001) doi: 10.1006/jema.2000.0400
[24] K. Kalabokidis, N. Athanasis, F. Gagliardi, F. Karayiannis, P. Palaiologou, S. Parastatidis, C. Vasilakos, Ecological Informatics **16**, 62 (2013) doi: 10.1016/j.ecoinf.2013.04.007

MATEC Web of Conferences 79, 01082 (2016) DOI: 10.1051/matecconf/20167901082
IME&T 2016

DIAMETER CALCULATION IN CONTACTLESS THREE-AXIS MEASURING DEVICES

Evgeny Fedorov*, *Alexander* Koba
Tomsk Polytechnic University, Institute of Non-Destructive Testing, 634050, Tomsk, Russia

Abstract. The paper presents a three-dimensional method of measuring diameter and ovality of cylindrical objects in diverging laser beam. Mathematical software method is developed. Virtual simulation of this method and experimental testing are made. The results can be used for design and manufacture of optical devices, which will be used for technological control of external diameter of cable, wire, trumpet and other products.

1 Introduction

Various geometric and electrical parameters are used to assess the quality of cable products. The rejects can be minimized through the improved technology to control the product major characteristics, namely capacitance per unit length [1, 2], eccentricity [3], outer diameter [4], insulation integrity [5] and others parameters [6–13].

Noncontact gauges for measuring the diameter of circular long-length objects such as cables, ropes, pipes, etc., which are based on the shadow measuring method using the divergent luminous flux, exhibit a number of undeniable advantages since they do not use the elements of lens and mirror optics. In particular, the measurement accuracy obtained by these devices using a combination of diffraction methods and statistical data processing may be of micrometer fractions [14]. However, when measurement is performed using the divergent light flux, the shadow dimensions of the measured object change in case the object is displaced within the test zone. This requires the use of complex ratios to calculate the true diameter of the object. Two-coordinate measuring gauges provide approximate values of the ovality estimated on the basis of the difference between the diameters of the two measuring axes, which are directly dependent on the object orientation within the measuring field. In this case, particular positions of the object do not allow the ovality fixation. Figure 1 shows an example of this orientation for an object with an ellipsoidal cross-section. The cross-section images *Dx* and *Dy* in the multi-element receivers of measuring channels *X* and *Y* are equal and they cannot provide the ovality parameters.

The measurements performed in the three-axis laser systems with a divergent laser beam can partially solve the above-stated problems. Three-axis measuring gauges exhibit the following advantages when compared with those of two-axis devices: three synchronized measuring axes on the same plane; detection of any deviation from the

*Corresponding author: evgeny_fyodorov@list.ru

© The Authors, published by EDP Sciences. This is an open access article distributed under the terms of the Creative Commons Attribution License 4.0 (http://creativecommons.org/licenses/by/4.0/).

circular cross-section regardless of the object orientation in the measured zone; determination of the circumference and area of the cross-section.

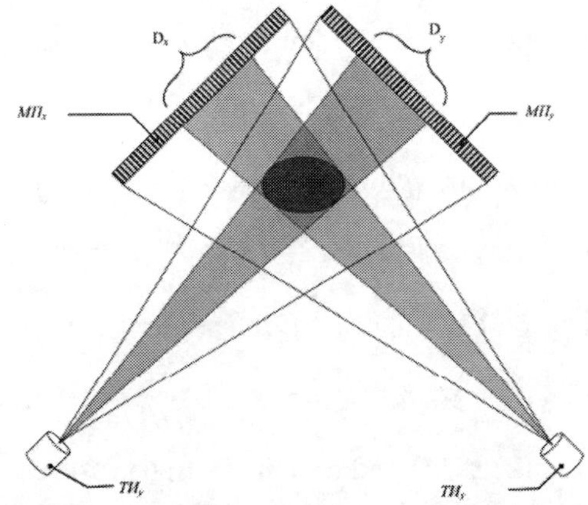

Figure 1. Example of orientation of the object with a non-circular cross-section, in which the calculation of ovality parameters is impossible. MRx, PEx and MRy, PEy are multi-element receivers and point emitters of measuring axes Y and X, respectively. Dx and Dy are images of the object shadows in the corresponding photoreceivers.

2 Diameter calculation

Since the true diameter is a calculated value in both three- and two-axis laser systems, the derivation of mathematical expressions for its calculation is required.

Figure 2 shows the schematic diagram of the considered three-axis measuring gauge. The device comprises three measuring channels, with *X, Y* and *Z* axes arranged at 120° angle to each other. Each channel has a point emitter (semiconductor laser) generating a divergent luminous flux and a multi-element photoreceiver measuring the size of the shadow falling on the object. The measuring object is found in the test zone; it is illuminated by three emitters and form three shadows in photoreceiver of the corresponding measuring channels.

The known design parameters of the device are as follows:

Distances H_x, H_y and H_z from the emitter center to the plane of the receiver of each channel: (S_{X0}; B_{X0}), (S_{Y0}; B_{Y0}) and (S_{Z0}; B_{Z0}) segments.

Distances H_{X0}, H_{Y0} and H_{Z0} from the center of the emitter to the center of the test zone, i.e. the point (*0*), which is an imaginary point for intersection of the measuring channel symmetry axes.

The step of the photoreceiver cells allows measurement of the shadow, i.e. its width, in the number of cells shadowed by the object, and the displacement of the shadow center relative to the "zero" cell of the receiver.

The known data are used to measure the object radii by the three channels (R_X, R_Y, R_Z) and to determine the average radius R_{av} and the object ovality. To perform special measurements, data on the object position is required, (displacement of its *Z* center) relative to the center of the gauge test zone ("*0*" point), i.e. the E_X, E_Y and E_Z values.

The problem is solved in three steps:

Step 1 – determination of the position of the *Z* object center projection in the plane of the "*X*", "*Y*" and "*Z*" receivers.

The position of the center projection is characterized by the X_0, Y_0 and Z_0 values, which are (B_{X0}; X_0), (B_{Y0}; Y_0) and (B_{Z0}; Y_0) segments. According to the triangle (S_{X0}; B_{X0}; X_0), it is known that:

$$tg(\alpha+\alpha_1)=\frac{X_0}{H_X}\text{. Hence: } X_0=H_X\cdot tg(\alpha+\alpha_1) \quad (1)$$

The value α1 in expression (1) can be found through the triangle (S_{X0}; B_{X0}; X_1):

$$tg\alpha_1=\frac{X_1}{H_Y};\ \alpha_1=arctg\frac{X_1}{H_X}. \quad (2)$$

The angle α is determined through the triangle (S_{X0}; B_{X0}; X_2):

$$tg(\alpha+\alpha+\alpha_1)=\frac{X_2}{H_X};\ (2\alpha+\alpha_1)=arctg\frac{X_2}{H_X}.$$

With regard to expression (2), the ratio of the angle takes the form:

$$\alpha=\frac{1}{2}(arctg\frac{X_2}{H_X}-arctg\frac{X_1}{H_X}) \quad (3)$$

Substitution of (3) and (2) in (1) and transformations result in:

$$X_0=H_X\cdot tg\frac{1}{2}(arctg\frac{X_2}{H_X}+arctg\frac{X_1}{H_X}) \quad (4)$$

The same procedure is performed for Y_0 and Z_0 values:

$$Y_0=H_Y\cdot tg\frac{1}{2}(arctg\frac{Y_2}{H_Y}+arctg\frac{Y_1}{H_Y}) \quad (5)$$

$$Z_0=H_Z\cdot tg\frac{1}{2}(arctg\frac{Z_2}{H_Z}+arctg\frac{Z_1}{H_Z}) \quad (6)$$

Similarity of the triangles (S_{X0}; B_{X0}; X_0) and (S_{X0}; 0_X; Z) provides:

Step 2 – determination of the Z object center displacement relative to the center of the test zone. The displacement is characterized by E_x, E_y and E_z values (Figure 2).

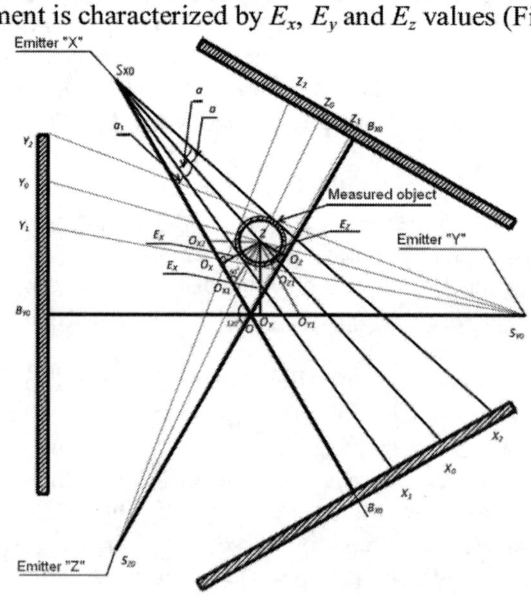

Figure 2. Calculation model for the device with three measuring axes.

$$\frac{E_X}{X_0}=\frac{(S_{X0};0_X)}{H_X}\text{. Similarly: }\frac{E_Y}{Y_0}=\frac{(S_{Y0};0_Y)}{H_Y};\ \frac{E_Z}{Z_0}=\frac{(S_{Z0};0_Z)}{H_Z} \quad (7)$$

Find the values for the segments $(S_{X0};\ 0_X)$, $(S_{Y0};\ 0_Y)$ and $(S_{Z0};\ 0_Z)$ in (7), provided the segment $(0;\ 0_X)$ is:

$$(0;\ 0_X) = (Z;\ O_{Y1}) - (O_{X2};\ O_X) = \frac{E_Y}{\cos(\frac{\pi}{6})} - \frac{E_X}{tg(\frac{\pi}{3})};$$

$$\text{segment } (0;\ 0_Y) = (Z;\ O_{X2}) - (O_Y;\ O_{Y1}) = \frac{E_X}{\cos(\frac{\pi}{6})} - \frac{E_Y}{tg(\frac{\pi}{3})}.$$

$$\text{segment } (0;\ 0_Z) = (Z;\ O_{X1}) + (O_Z;\ O_{Z1}) = \frac{E_X}{\cos(\frac{\pi}{6})} + \frac{E_Z}{tg(\frac{\pi}{3})}.$$

Then:

$$(S_{X0};\ 0_X) = H_{X0} - \frac{E_Y}{\cos(\frac{\pi}{6})} + \frac{E_X}{tg(\frac{\pi}{3})};\ (S_{Y0};\ 0_Y) = H_{Y0} - \frac{E_X}{\cos(\frac{\pi}{6})} + \frac{E_Y}{tg(\frac{\pi}{3})};$$
$$(S_{Z0};\ 0_Z) = H_{Z0} + \frac{E_X}{\cos(\frac{\pi}{6})} + \frac{E_Z}{tg(\frac{\pi}{3})} \quad (8)$$

With regard to expression (8), ratios (7) take the form:

$$\frac{E_X}{X_0} = \frac{H_{X0} - \frac{E_Y}{\cos(\frac{\pi}{6})} + \frac{E_X}{tg(\frac{\pi}{3})}}{H_X};\ \frac{E_Y}{Y_0} = \frac{H_{Y0} - \frac{E_X}{\cos(\frac{\pi}{6})} + \frac{E_Y}{tg(\frac{\pi}{3})}}{H_Y};$$
$$\frac{E_Z}{Z_0} = \frac{H_{Z0} + \frac{E_X}{\cos(\frac{\pi}{6})} + \frac{E_Z}{tg(\frac{\pi}{3})}}{H_Z} \quad (9)$$

Expressions (9) are a system of three equations with three unknowns: E_X, E_Y and E_Z. Solve the equations and derive the expressions to calculate E_X, E_Y and E_Z, in which the X_0, Y_0 and Z_0 values are calculated by (4), (5) and (6) ratios. This paper does not provide the extended expressions for calculation of the system of equations in letter values.

Step 3 – calculation of the object radii R_X, R_Y and R_Z.

The schematic diagram in Figure 1 indicates that the R_X radius, i.e. the segment $(F;Z)$, is the side of the right triangle $(S_{X0};\ Z;\ F)$. According to this triangle:

$$R_X = (S_{X0};Z)\cdot\sin\alpha \quad (10)$$

The other right triangle $(S_{X0};\ Z;\ 0_X)$ is used to find the segment $(S_{X0};\ Z)$:

$$(S_{X0}, Z) = \sqrt{(0_X;Z)^2 + (0_X;S_{X0})^2} = \sqrt{E_X^2 + (H_{X0} - \frac{E_y}{\cos(\frac{\pi}{6})} + \frac{E_x}{tg(\frac{\pi}{3})})^2}.$$

Substitute the value of the segment $(S_{X0};\ Z)$ in (10) and use expression (3) for the angle to find the radius R_X.

$$R_X = \sqrt{E_X^2 + (H_{X0} - \frac{E_y}{\cos(\frac{\pi}{6})} + \frac{E_x}{tg(\frac{\pi}{3})})^2} \cdot \sin\frac{1}{2}(arctg\frac{X_2}{H_X} - arctg\frac{X_1}{H_X}) \quad (11)$$

Similarly:

$$R_Y = \sqrt{E_Y^2 + (H_{Y0} - \frac{E_X}{\cos(\frac{\pi}{6})} + \frac{E_Y}{tg(\frac{\pi}{3})})^2} \cdot \sin\frac{1}{2}(arctg\frac{Y_2}{H_Y} - arctg\frac{Y_1}{H_Y}) \quad (12)$$

$$R_Z = \sqrt{E_Z^2 + (H_{Z0} + \frac{E_X}{\cos(\frac{\pi}{6})} + \frac{E_Z}{tg(\frac{\pi}{3})})^2} \cdot \sin\frac{1}{2}(arctg\frac{Z_2}{H_Z} - arctg\frac{Z_1}{H_Z}) \quad (13)$$

The values of the R_X and R_Y radii being known, the average radius *Rav* of the object and object ovality (O) can be found.

$$R_{av} = \frac{|R_X| + |R_Y| + |R_Z|}{3}; \; O = (R_{\max} - R_{\min}) \quad (14)$$

2 Method of evaluation

A three-axis optical device for measuring the diameter of cylindrical objects has been simulated to verify the above suggested expressions (11–13). A computer model presented in Figure 2 comprises three identical measuring channels arranged at 120° to each other. The initial geometric parameters to simplify calculations are the same for all the measuring channels:

- The distance from the emitter center to the plane of the receiver H_x=H_y=H_z=300 (mm).
- The distance from the emitter center to the test zone, i.e. the center of intersection of the measuring channel axes H_{X0}=H_{Y0}=H_{Z0}=220 (mm).

Such values of the optical devices geometric parameters are typical for the measurement area sufficient for most applications. In this model, the length of the working surface of multiple-receiver does not exceed 100 mm. All the initial geometrical parameters and virtual positions of shadows in the radiation receiver planes in the corresponding channels X_1, X_2, Y_1, Y_2, Z_1 and Z_2, were determined in the computer model with 1 µm error. Round objects with different diameters and different location in the virtual measuring zone were simulated to perform testing. When substituting the values obtained during the simulation into expressions (11–13), the measuring object diameters were calculated in the corresponding measuring axes *X*, *Y* and *Z*. These calculated diameters D_x, D_y, D_z for three arbitrary circular objects with the diameter *D* are presented in Table 1.

Table 1. Diameter values and the error for three axes.

Measuring diameter *D* (mm)	Diameter D_x (mm)	Error ΔD_x (µm)	Diameter D_y (mm)	Error ΔD_x (µm)	Diameter D_z (mm)	Error ΔD_x (µm)	Error ΔD_{av} (µm)
10.000	10.003	-3	10.002	-2	9.998	2	-1
17.000	16.997	3	16.999	1	17.002	-2	0.67
25.000	24.999	1	25.001	-1	25.000	0	0

The absolute error of the diameter calculation in the measuring axes ΔD_x, ΔD_x and ΔD_x, as well as the mean error ΔD_{av} do not exceed 3 µm, that is acceptable with regard to the error in the initial parameter determination. In practice, the diameter calculation accuracy strongly depends on the accuracy of initial parameters. Further simulation of the three-axis device is to be performed on a real prototype using the methods of digital signal processing. Due to this, the error in the diameter calculation may be of micrometer fractions.

Further testing of the method and the proposed mathematical model of measurement was carried out using a real prototype of the diverging-laser-beam measuring device. The primary transducer of the diameter measuring gauge LDM-20 described in [13] was used to simplify the experiment. The test installation appearance is presented in Figure 3. The prototype is a system with a single optical measuring channel comprising the CCD (charge-coupled device) and the laser emitter. The transducer was mounted on the turntable capable of rotating by 360° with a rotation angle accuracy of up to 1 arc minute. The object with the known diameter (calibre) was placed stationary in the center of the turntable. By rotating the table at the angles of 0°, -120°, and +120°, a three-axis transducer was experimentally simulated.

Figure 3. Test installation comprises: 1 – turntable, 2 – transducer housing, 3 – electronic control board, 4 – point emitter (semiconductor laser), 5 – CCD receiver board, 6 – display with control keys, 7 – object with the known diameter (caliber).

The geometrical parameters of the optical transducer were as follows:

- Distances from the emitter center to the plane of the receiver $H_x=H_y=H_z=207.4$ (mm).
- Distances from the emitter center to the center of the test zone $H_{X0}=H_{Y0}=H_{Z0}=134$ (mm).
- The pixel size of the used CCD (NECuPD8871) was equal to 4 μm. The number of pixels was 10680.

A simplified structure of the installation electronic part is illustrated in Figure 4. The linear CCD array and the point emitter (semiconductor laser) are controlled by the field programmable gate array (FPGA), which generates all necessary signal sequences. Analog data obtained from the CCD receiver arrive at the ADC. The data in the digital form are then accepted by the programmable array for conversion into the rising and falling edges of the shadow, which characterizes the linear dimensions of the measuring object. After this, the measured data are transmitted to the microcontroller via a conventional SPI interface.

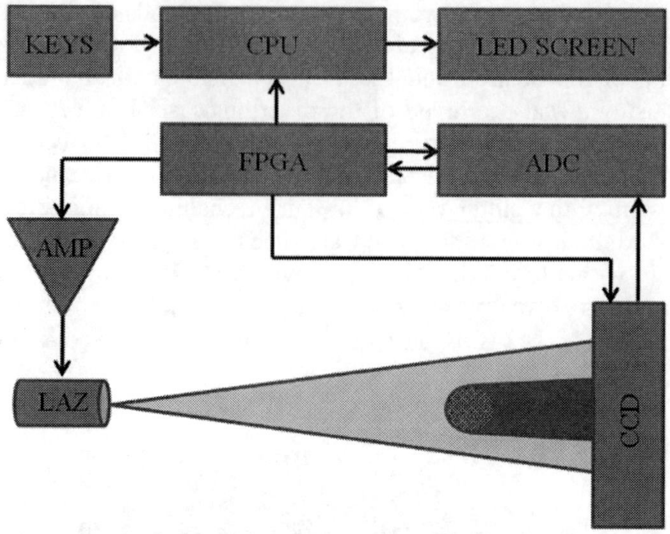

Figure 4. Block diagram of the test installation electronic part.

The microcontroller displays the data on a seven-segment LED indicator when you press the appropriate control keys on the keyboard. Due to this, positions of the shadow boundaries X_1,X_2, Y_1,Y_2,Z_1 and Z_2 were experimentally calculated. Similar to the computer model, the calculated values for three calibres with the known diameters were obtained using expressions (11–13). The results of the diameter calculation and calculation errors are given in Table 2.

Table 1. Diameters and errors obtained from the experimental data.

Caliber diameter (mm)	**Diameter D_x (mm)**	**Error ΔD_x (μm)**	**Diameter D_y (mm)**	**Error ΔD_y (μm)**	**Diameter D_z (mm)**	**Error ΔD_z (μm)**	**Error ΔD_{av}(μm)**
14.842	14.832	-10	14.839	-3	14.847	5	-2.7
10.178	10.171	-7	10.170	-8	10.171	-7	-7.3
12.900	12.905	5	12.901	1	12.891	-9	-1

The absolute error of the diameter calculation in the measuring axes ΔD_x, ΔD_y and ΔD_z, as well as the mean error ΔD_{av} insignificantly exceed the values obtained during simulation (Table 2), but remain within the range of 10 μm. This result is found to be in good agreement with simulation results and proves the validity of expressions (11–13) to calculate the diameter in tree-axis systems. Increased errors of the calculated diameter values obtained from the experimental data can be explained by the installation inaccuracy (Figure 3). Design of the three-axis measuring transducer resistant to deformation and use of the digital processing algorithms will enable the solution of the above-stated problems.

4 Conclusion

The paper proposes a method for three-axis measurement of the diameter using a diverging laser beam. Expressions (9, 11–14) have been derived to calculate the precise value of the object radii, the coordinates of the object center position, and the average radius and the precise value of its ovality in the maximum and minimum values. A computer modeling of the three-axis transducer, which data are shown to be in good agreement with experimental data, confirms the validity of the proposed conversion function. The described optical method and mathematical support can be used to design a primary measuring transducer for the device to test the diameter and ovality of long-length products.

References

[1] N.S. Starikova, V.V. Redko, G.V. Vavilova, J. Phys.: Conf. Ser. **671**, 012056 (2016) doi: 10.1088/1742-6596/671/1/012056

[2] A.E. Goldshtein, G.V. Vavilova, V.Yu. Belyankov, Russ. J. Nondestr. Test. **51**, 86 (2015) doi: 10.1134/S1061830915020047

[3] A.E. Goldshtein, E.M. Fedorov, Rus. J. Nondestr. Test. **46,** 424 (2010) doi: 10.1134/S1061830910060069

[4] E.M. Fedorov, I.D. Bortnikov, Tech. Phys. **60**, 1689 (2015) doi: 10.1134/S1063784215110110

[5] V.V. Red'ko, A.P. Leonov, L.A. Red'ko, V.A. Bolgova, J. Phys: Conf. Ser. **671**, 012049 (2015) doi: 10.1088/1742-6596/671/1/012049

[6] E. Caetano, Á. Cunha, MATEC Web of Conferences **24**, 01002 (2015) doi: 10.1051/matecconf/20152401002

[7] A.Y. Petrova, O.N. Chaikovskaya, I.V. Plotnikova, Tech. Phys. J. **60**, 592 (2015) doi: 10.1134/S1063784215040222

[8] A.P. Surzhikov, T.S. Frangulyan, S.A. Ghyngazov, E.N. Lysenko, Journal of thermal analysis and calorimetry **102**, 883 (2010) doi: 10.1007/s10973-010-0912-8

[9] O.V. Galtseva, S.V. Bordunov, N.M. Natalinova, S.V. Mazikov, IOP Conf. Ser.: Mater. Sci. Eng. **132**, 012003 (2016) doi: 10.1088/1757-899X/132/1/012003

[10] A.M. Pritulov, R.U. Usmanov, O.V. Gal'Tseva, A.A. Kondratyuk, V.V. Bezuglov, V.I. Serbin, Russ Phys J+ **50**, 187 (2007) doi: 10.1007/s11182-007-0026-3

[11] V.Y. Kazakov, D.K. Avdeeva, M.G. Grigoriev, N.M. Natalinova, I.V. Maksimov, M.V. Balahonova, BLM **7**, 1 (2015)

[12] R. Kodermyatov, M. Ivanov, M. Yuzhakov, V. Kuznetsov, M. Yuzhakova, E. Timofeeva, MATEC Web of Conferences **48**, 05004 (2016) doi: 10.1051/confmatec/20164805004

[13] A.A. Bespalko, A.P. Surzhikov, L.V. Yavorovich, P.I. Fedotov, Russ. J. Nondestr. Test. **48**, 221 (2012) doi: 10.1134/S1061830912040043

[14] Y.A. Chursin, E.M. Fedorov, Optics&Laser Technology **67**, 86 (2015) doi: 10.1016/j.optlastec.2014.09.017

MATEC Web of Conferences **79**, 01083 (2016) DOI: 10.1051/matecconf/20167901083
IME&T 2016

DUAL CLUSTERING IN VISION SYSTEMS FOR ROBOTS DEPLOYED FOR AGRICULTURAL PURPOSES

Alexander Tyryshkin[a], *Alexander* Belyaev

Tomsk State University, 634050, Tomsk, Russia

Abstract. Continuously variable parameters of environment of robots' functioning complicate their use in agriculture. Record of disturbing actions only by means of software leads to complication of the programs. In turn, this leads to rise in price of the software product and reduction of robot's operational reliability. The authors suggest carrying out preliminary adaptation of the vision system to environment by means of hardware. Hardware is selected automatically based on artificial intelligence.

1 Introduction

Rapid development of robotics requires creation of more and more sophisticated systems of sensitization. Machine vision is the most developed among artificial sensors. Modern robots can easily work in factory shops and public space. In these cases, the observed objects are in artificial environment - in a pre-determined illumination range with a known emission spectrum, in a limited range of humidity and dust level. Gained success has inspired developers for a wider use of robots in natural environment. At the same time, agriculture is almost a new area of application of robots.

2 Factors, which complicate realization of agricultural robots

Application of robots for handling and harvesting of seed grains is constrained by complexity of creation of control systems because of the following factors:

- Color gamma of cultivated plants is a little different from color gamma of weeds.
- Color gamma of cultivated plants changes with the growth and maturation.
- Image depends on the direction of light.
- Spectrum of solar radiation depends on the time of day and humidity.
- Image depends on the strength and direction of wind.

This is not a complete list of disturbing factors. The image also depends on the plant lodging, presence of fog, dew, dust, etc.

[a] Corresponding author:taw@ie.tusur.ru

© The Authors, published by EDP Sciences. This is an open access article distributed under the terms of the Creative Commons Attribution License 4.0 (http://creativecommons.org/licenses/by/4.0/).

Figure 1 shows images obtained by the web-cam in different directions («towards the sun» and «opposite to the sun»). It is very problematic to recognize plants on the image b) even by the «natural» neural network.

a) b)
Figure 1. a) The sun is behind; b) the sun is in front.

Figure 2 shows images of the same mown buckwheat. Figure 2, *a* shows an image of buckwheat before threshing. As soon as threshing starts (Figure 2, *b*), dust appears, which «lights» the image in the sunrays.

a) b)
Figure 2. Mown down buckwheat before threshing (a) and during a threshing process (b).

In addition to environmental and climatic factors, presence of weeds and diseases of cultivated plants complicate the task of image processing. On Figure 3 we can see dust because of such disease as gray rust.

a) b)
Figure 3. Gray rust on peas. a) before threshing; b) during the harvesting process.

3 Existing approaches to solution of image recognition tasks

Today, attempts to process images have been made by all known methods including both the simplest methods based on change of brightness, contrast, coloration, and the latest developments in the field of artificial intelligence [1]. In the simplest methods, image binarization is a key point. The ultimate goal of these methods is to divide objects of the working area into classes «necessary object» / «unnecessary object». For example, not mown areas/ mown areas would be such objects for agricultural machines. Simple methods do not work for recognition of such natural objects as cultivated plants and weeds. In some cases, fuzzy logic, genetic algorithms and neural networks give a satisfactory result [2, 3]. However, classical neural networks require large computing resources, greater efforts for training and do not guarantee achievement of the required quality.

It is possible to reduce requirements to computing resources for image processing by use of convolutional neural networks [4]. However, high cost and lack of sufficient number of appropriate specialists in the manufacturing industry constrain introduction of sophisticated computer technology in agriculture.

According to the authors, development of robots for agricultural purposes should involve not only increase of control systems "intelligence" but also search of engineering solutions aimed to simplify tasks facing developers.

4 Suggested approach to solution of the task

Creation of local areas of "comfortable work" of vision systems is one of such solutions. It is necessary to make working conditions of Outdoor-type robots closer to working conditions of Indoor-type robots. There is no direct solution for this problem because it is impossible to extend technosphere to the entire wildlife. Consequently, it is necessary to create artificial conditions only in the robot's "field of interest". Moreover, the robot must automatically adapt the field of work in conditions of changing environmental parameters.

Figure 4 shows the proposed structure of the agricultural machine control system.

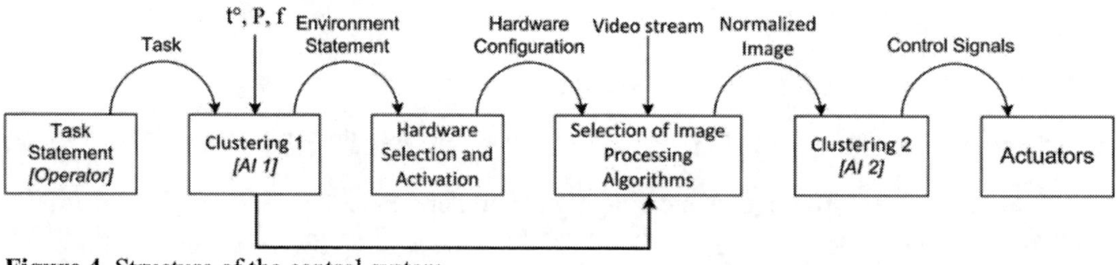

Figure 4. Structure of the control system.

Artificial intelligence (AI) is divided into two parts - AI1 and AI2. In addition to the task set by the operator, AI1receives information on the environment: temperature, pressure, humidity, light intensity, etc. Based on these data, AI1 determines one of the known typical situations. For example, fine, dew, twilight, fog. In fact, it is a task of clustering, which can be solved based on Kohonen maps [5]. This type of artificial intelligence does not require large computational resources. It is possible to realize Kohonen maps based on industrial controllers of general purpose. Training of Kohonen maps does not require highly skilled professionals. Results of training are intuitively understandable and easily testable.

Identification of situation is followed by automatic selection of hardware aimed at creating favorable conditions for operation of the artificial intelligence unit AI2. These can include sources of artificial lighting; hardware filters both of visible spectrum and of the infrared band; fans; windscreen wipers; and others.

After the work done, partially adapted image, called as Normalized, goes to the input of clustering unit 2.

Despite the complexity of optical image processing [6] for agricultural robot, the vision task itself is relatively simple. It is necessary either to determine mown plants (presence or centerline), or to determine the boundary between the tilled and untilled area. Thus, the image should be divided into two zones: non-working area / working area. Again, this is a relatively simple task of clustering, which can be solved by the clustering unit 2.

After determining the boundary of the working area, control signals go to actuating components (ACs), which influence the units of agricultural machinery.

The proposed approach begins with the definition of range of tasks, which are to be solved by robots of this type. These tasks can include determining the boundaries of mown plants; selection of mown plants, briquetting and unloading on a given line; mowing of lodged plants, etc. Each of these tasks is independent and requires personal approach to its solution. As far as these tasks are solved, obtained solutions can be combined in a single software product to create control systems for universal agricultural machines.

The next step is creation of a base of model images, which are typical for a specific task. This work should be carried out during the growing season along with development of the robot hardware. The cost of this work is low because the vision system (VS) can be installed on existing agricultural machines commonly working in the fields. Such usage of VS will allow accumulating video data obtained in different environmental conditions: variable level of light, presence / absence of fog, etc.

The next stage of the work can be cameral. On this step, it is necessary to select the neural network topology and training it. Instrumental preparation for image obtaining (use of additional lighting, filters, etc.) should allow reducing the complex task of image analysis to relatively simple task of clustering.

After completing these works, it is possible to use the control system in test mode. In this case, an operator provides actual control of an agricultural machine and the control system acts as an expert system. The created system can be converted from the training mode to automatic control mode when during debugging solutions of the expert system become adequate to the operator's solutions.

Acknowledgment

The paper was written as part of the research project No. 8.2.31.2015, carried out with the support of the Program "Research Foundation of Tomsk State University named after D.I. Mendeleev" in 2015 – 2016, grant RFBR No. 16-29-04388/16.

References

[1] A.V. Tkachev, S.V. Shanygin, Young Scientist (19), 215 (2015)

[2] C. J. Lakhmi, N.M. Martin, *Fusion of Neural Networks, Fuzzy Systems and Genetic Algorithms: Industrial Applications* (CRC Press, United States ,1998)

[3] M.A. Ranzato, C.Poultney, S. Chopra and Y. LeCun, Advances in Neural Information Processing Systems, 1137 (2007)

[4] M. Zeiler, D. Krishnan, G. Taylor, R. Fergus, 2010 IEEE Conference on computer vision and pattern recognition (CVPR), 2528 (2010) doi:10.1109/CVPR.2010.5539957

[5] T. Kohonen, *Self-Organizing Maps* (Springer Science & Business Media, New York, 2012)
[6] D.V. Shashev, S.V. Shidlovskiy, V.I. Syriamkin , A.V. Yurchenko, IOP Conference Series: Materials Science and Engineering **81**, 012101 (2014) doi: 10.1088/1757-899X/81/1/012101

MATEC Web of Conferences **79**, 01084 (2016) DOI: 10.1051/matecconf/20167901084
IME&T 2016

ON THE SELECTION OF OPTIMAL STRUCTURE ORGANIZATION OF LOGIC MULTICONTROLLERS

Eduard Vatutin [1,*], *Vitaly* Titov [1], *Alexander* Belyaev [2]

[1]Southwest State University, Department of Computer Sciences, 305040, Kursk, Russia
[2]Tomsk State University, Department of Computer Sciences, 643050, Tomsk, Russia

Abstract. In the article the basic approaches to structural-parametric optimization of logical control systems in the basis of logical multicontrollers were given. Based on the results of a series of computational experiments we obtained the set of dependencies on the deterioration of the quality criteria of separations of graph-schemes of parallel logic control algorithms and corresponding parameters values of LMC from technological restrictions on the controller structure. It is shown that the structure of the LMC with a large number of simple controllers is preferred.

1 Introduction

One of the promising approaches to the synthesis of logic control systems (LCS) is their implementation in the basis of logical multicontroller (LMC) [1] which are interconnected in the collective working in parallel similar controllers together solving the problem to implement epy given logic control algorithm presented by corresponding graph-scheme. When designing such multisystems there is a number of discrete combinatorial optimization problems [2–5]. One of them is the problem of getting suboptimal separation of a priori known graph-scheme of parallel logic control algorithm to sequential blocks with restricted complexity, each of which is implemented by one of the controllers within the LMC [6–9]. This problem relates to the NP complexity class that does not allow to find the optimal solution for its practical dimension cases (graph-schemes with more than 10–20 vertices) at a reasonable time, therefore its solutions are known and have been successfully used with various heuristic approaches [2–4, 10] having different implementation complexity, asymptotic time and memory complexities of corresponding algorithms, set of optimized partial quality criteria and integral quality of the obtained solutions. The quality of separations directly affects the hardware complexity of the LMC and its speed characteristics.

During LMC design can be used two different approaches. According to the first of them the structure of LMC (number of modules and its hardware characteristics, topology

*Corresponding author: evatutin@rambler.ru

© The Authors, published by EDP Sciences. This is an open access article distributed under the terms of the Creative Commons Attribution License 4.0 (http://creativecommons.org/licenses/by/4.0/).

of the connections between them, redundancy options, etc.) is selected once, based on the specific graph-scheme of control algorithm and selected separation for the implementation of which it is necessary to create appropriate LCS. Numerical parameters of separation (number of blocks and links between them) determine the hardware requirements for LMC, and when you change the control algorithm selection LMC structure is actually made anew. This approach can be used, for example, as a way of implementation of control part for specialized computing device with ASIC based production the operating part of which is selected at the design stage and is not changed during further operation. According to the second approach the structure of LMC is generalized and focused on the implementation of one of the group of control algorithms by software setting that makes it possible to change the control algorithm without changing the hardware structure of LMC during operation. This approach can be used, for example, as implementation of assembly line control system in which the set of operations may vary with time. From the standpoint of producer of these LCSs named as programmed logic controllers (PLC) it is interesting to the development of the model range formed by a group of products with different cost and performance characteristics. At the same time design engineer of control part has the opportunity to select one of the LMC models within corresponding model range enough to please it as a cost and the possibility of implementing of the needed control algorithm. Easy to see that the first approach is the analogue of well known ASIC approach that is characterized by rather a high cost while the second one is essentially similar to the practical use of logical circuits with reconfigurable structure such as FPGAs (Field-Programmable Gate Array) and ULAs (Uncommited Logic Array) [11] that can significantly reduce the cost of the final solution for its practical use.

When choosing an appropriate model range a number of issues is arisen connected with structural-parametric optimization of forming its multicontrollers and selecting its structure organization corresponding to practical requirements. For example, when a physical limitation (balance requirements) by the number of transistors, metallization layers on a chip, requirements of electromagnetic and/or thermal compatibility of electronic components and so on you must choose such a structure of LMC that will be characterized by low production costs (for example, with a small area on a chip) and at the same time allows the implementation of a particular group of control algorithms (for example, with selected number of vertices, control signals and speed characteristics) without the need to change the current model of composed model range for more expensive one on the one hand and on the other performance degradation on the another hand. In its simplest form, the need to choose such a structure leads to specifying the number of controllers within the LMC and its hardware performance, while maintaining the given hardware complexity of multisystem within the prescribed limits. In other words, it is possible to implement the LMC, which includes all other things in its structure being equal a large number of relatively simple controllers, a small number of complex controllers or a compromise version. Structural-parametric optimization of LMC can be achieved by finding the number of separations of logic control algorithms, statistical processing of the results and analysis of changes in average quality criteria trends depending on technological restrictions arising in selecting the appropriate structure of LMC.

2 Statement of a problem

A formal presentation of the problem of getting separation has the following form. It is required to obtain a separation $Sep\left(A^0\right)=\left\{A_1, A_2, ..., A_H\right\}$ of the set of vertices A^0 for source graph-scheme of parallel logic control algorithm $G^0 = \langle A^0, V^0 \rangle$ satisfying the following conditions:

$$\begin{gathered} \bigcup_{i=1}^{H} A_i = A^0, \quad A_i \neq \varnothing, \quad A_i \cap A_j = \varnothing, \quad i,j = \overline{1,H}, \quad i \neq j, \\ \neg(a_i \omega a_j)\, \forall a_i, a_j \in A_k, \quad i \neq j, \quad k = \overline{1,H}, \\ W(A_i) \leq W_{max}, \quad |X(A_i)| \leq X_{max}, \quad |Y(A_i)| \leq Y_{max}, \quad i = \overline{1,H}, \end{gathered} \tag{1}$$

where ω – designation of a binary relation of vertices parallelism [1] reflecting a structural restriction of LMC on the prohibition of parallel vertices within blocks of separation, $W(A_i) = \sum_{a_j \in A_i} W(a_j)$ – summary "weight" of vertices within the *i*-th block (size in controller memory measured at microcommands); $X(A_i) = \bigcup_{a_j \in A_i} X(a_j)$ – a set of logical conditions included in the vertices within the *i*-th block; $Y(A_i) = \bigcup_{a_j \in A_i} Y(a_j)$ – a set of microoperations included in the vertices within the *i*-th block, W_{max} – constraint in the memory capacity of the controller as part of LCS, X_{max} – constraint in the number of signals of logic conditions received by the controller, Y_{max} – constraint in the number of microoperation signals issued by the controller, such that

$$\begin{gathered} Z_H = H\left(Sep\left(A^0\right)\right) \to \min, \\ Z_\alpha = \sum_{i=1}^{H} \sum_{j=1, j\neq i}^{H} \alpha\left(A_i, A_j\right) \to \min, \\ Z_\delta = \delta\left(Sep\left(A^0\right)\right) \to \min, \\ Z_X = \sum_{i=1}^{H} |X(A_i)| - |X(A^0)| \to \min, \\ Z_Y = \sum_{i=1}^{H} |Y(A_i)| - |Y(A^0)| \to \min, \end{gathered} \tag{2}$$

where $Z_H = H\left(Sep\left(A^0\right)\right)$ – the number of blocks in the separation; Z_α – complexity of the interconnect network for separation $Sep\left(A^0\right)$; $\alpha\left(A_i, A_j\right)$ – coupling coefficient of blocks pair (it is equal to 1 if the blocks are connected by the control in direction from A_i to A_j, which requires an additional command for transfer of control between controllers, and 0 otherwise); $Z_\delta = \delta\left(Sep\left(A^0\right)\right)$ – the total number of interblock interactions (summary interblock traffic); Z_X – the extent of duplication of logical conditions signals; Z_Y – the extent of duplication of microoperations signals. Minimizing these partial quality criteria can decrease hardware complexity of LCS (criteria Z_H, Z_X, Z_Y, Z_α) by reducing the required number of controllers, complexity of interconnect network and hardware complexity of separate controllers, and improve performance of LCS (criterion Z_δ) by reducing control traffic and communication subsystem loading.

Number of blocks within the separation provides the number of microprograms and accordingly, the number of controllers within LMC, each of them implements one of them. In the absence of technological restrictions $\left(X_{max} = Y_{max} = W_{max} = \infty\right)$ a number of blocks in separation has low theoretically limit that is provided by the value of the parallelism degree of graph-scheme of parallel algorithm $\omega_{max}\left(A^0\right)$ [1].

3 Computation experiments overview

For each of the heuristic methods [1–4, 10] set of test examples (graph-schemes of parallel control algorithms) can be selected in which they demonstrate the highest quality of solutions compared with other methods, so the specified type of comparison methods is sufficiently subjective. Therefore, in order to realize the objective of comparing the quality of separations given by different heuristic methods with different using conditions we will carry out a comparison of average values of quality criteria of separations using a generator of graph-schemes of algorithms with selected parameters (number of vertices, microoperations and logic conditions signals, probabilities of fragments with different type, etc.) and pseudorandom structure working within program system PAE. Using this generator it is possible to obtain samples $\Lambda=\left\{G_1^0, G_2^0, ..., G_K^0\right\}$ of graph-schemes of an arbitrary amount K of control algorithms that in its turn provides the ability of objective comparison quality of separations given by different heuristic methods and track the trends in the partial values of quality criteria when changing the values of technological restrictions X_{max}, Y_{max} and W_{max} and size of a problem N. For the selected conditions of the computational experiment and this method of getting separations it is possible to determine the partial sample average quality criteria values γ_x and probabilities of getting minimum value of selected partial quality criterion ρ_x, $x\in\left\{H, X, Y, \alpha, \delta, J\right\}$, where J – designation of integral quality criterion of separation $Sep\left(A_k^0\right)$ [1]

$$J\left(Sep\left(A_k^0\right)\right)=\frac{K_H}{\omega_{max}\left(A_k^0\right)}H+\frac{K_X}{\left|X\left(A_k^0\right)\right|}\left(\sum_{i=1}^{H}\left|X\left(A_i\right)\right|-\left|X\left(A_k^0\right)\right|\right)+$$
$$+\frac{K_Y}{\left|Y\left(A_k^0\right)\right|}\left(\sum_{i=1}^{H}\left|Y\left(A_i\right)\right|-\left|Y\left(A_k^0\right)\right|\right)+\frac{K_\delta}{\delta\left(A_k^0\right)}\delta\left(Sep\left(A_k^0\right)\right)+ \tag{3}$$
$$+\frac{K_\alpha}{\omega_{max}\left(A_k^0\right)\left(\omega_{max}\left(A_k^0\right)-1\right)}\sum_{i=1}^{H}\sum_{j=1,\ i\neq j}^{H}\alpha\left(A_i, A_j\right),$$

which is a weighted sum of normalized partial quality criteria. Here $K_H, K_X, K_Y, K_\alpha, K_\delta$ – weighting coefficients that are selected by experts and reflecting the importance of partial quality criteria; $\delta\left(A_k^0\right)$ – the theoretical maximum intensity of interblock interactions achieved in the separation where each vertex of control algorithm forms a separate block.

In a series of computational experiments that are computationally complex (the amount of computation is required hundreds of years of CPU time) and is performed using a grid system on a voluntary basis within volunteer distributed computing project Gerasim@home at BOINC platform [12] it was shown that quality of separations and probabilities of getting best decisions vary significantly for different heuristic methods and for different regions of space (Figure 1) formed by the size of the problem N and values of technological restrictions X_{max} and W_{max} (restriction Y_{max} can be simply avoided by doubling the controller within LMC [1]).

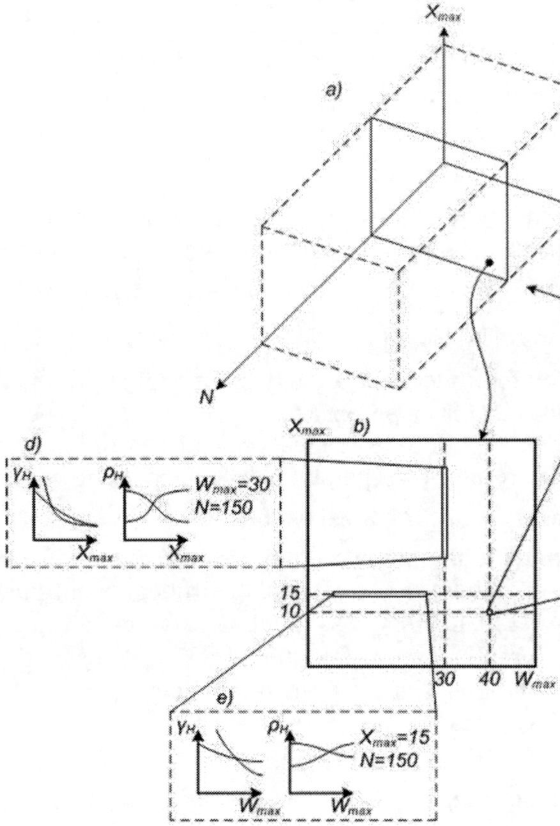

Figure 1. Space of parameters (a), map of the slice $(X_{\max}, W_{\max})$ (b) and early one parameter computational experiments (c, d, e).

4 Analysis of computation experiments results

General view of the dependency of partial quality criterion from the dimension of the problem and the value (power) of technological restriction is shown in Figure 2.

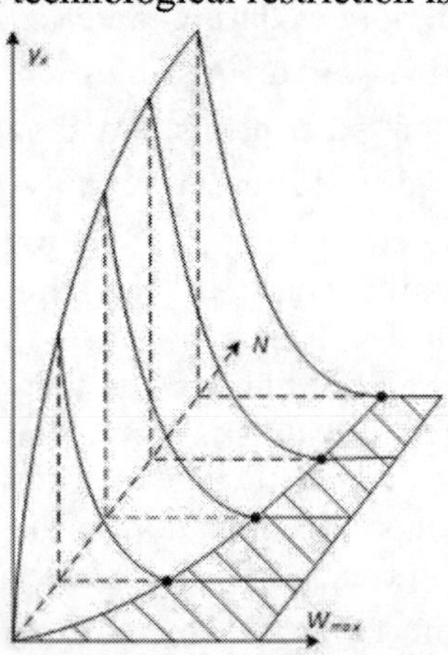

Figure 2. The general behavior of partial quality criteria *Z* depending on size of the problem *N* and power of constraint ($W_{\max}$ in this example). Shaded area shows the area of insensitivity.

At low values of technological restrictions $(X_{\max}, W_{\max} \to \infty)$ quality of separation is independent from the technological restrictions and other factors (such as structure restrictions of LMC basis and parameters of source graph-scheme G^0). During increasing power of restriction $(X_{\max}, W_{\max} \to 0)$ values of partial quality criteria begin to increase monotonically. In the area of insensitivity (shown at Figure 2 shaded) changes of partial quality criteria does not occur that allows to optimize structure of LMC by formulating requirements for limiting values $X'_{\max}$ and $W'_{\max}$ of technological restrictions for selected size of graph-schemes of logic control *N*. So when $(X_{\max} > X'_{\max}) \wedge (W_{\max} > W'_{\max})$ it is an increase in hardware complexity of controllers and LCS without getting the smaller values of partial quality criteria of partitions that is inappropriate.

In the article [13] as a result of computing experiments boundaries of insensitivity area were obtained. They are based on results of separations, obtained using the method of parallel-sequential decomposition as having minimal values of $X'_{\max}$ and $W'_{\max}$ from all other methods, and allows to formulate the hardware requirements for controllers within LMC with matrix structure (as an example). For example, controllers within matrix LMC with $7 \times 7 = 49$ modules must have 120 memory cells (command words) for microprogram storing (that corresponds to source graph-scheme of control algorithm separated by blocks) and 86 pins for receiving logic condition signals from controlling object that allows to implement graph-scheme with 450 vertices.

Decreasing values $X_{\max}$ and $W_{\max}$ less than the limits (for example, because of the inability or inexpediency production of LMC in the hardware configuration of the technological or cost reasons) leads to growth of partial quality criteria values (see Figure 2) that reduces the speed of the designed LCS due to the increasing control transfer traffic between controllers, increases hardware complexity of communication subsystem of controllers due to the need to implement a greater number of inter-module commands of the control transfer and greater depths of corresponding queues, and also requires the implementation of a larger number of controllers in comparison with a theoretical lower limit $\omega_{\max}(A^0)$. This is quite an important study showing how this or that partial quality criteria value deteriorates during increasing power of restrictions (decreasing values $X_{\max}$ and $W_{\max}$). The answer to this question is obtained while processing the results of a series of computation experiments and shown at Figures 3 and 4, where level lines marked as $F_{z\%}$ correspond to decreasing quality of decision by $z\%$ comparing to theoretical minimum that provided by method *F* for selected size of a problem *N*. These dependences allow a quantitative study of the characteristics of separations during decreasing values of technological restrictions.

An analysis of given results allows to conclude that decreasing quality of decisions for parallel-sequential method in relative units is lower than the same results by S.I. Baranov method and greedy adjacent method. Methods that are based on greedy strategy of building separation shows significantly more deterioration in the quality of decisions on criteria of interblock links number and intensity of interconnect control traffic, especially during reducing the limit value of the controller memory capacity $W_{\max}$. As we have noted previously [13], values $X'_{\max}$ and $W'_{\max}$ for parallel-sequential method [1] are closer to zero (respectively, its insensitivity zone is wider) comparing to variations of greedy approaches.

This feature allows to recommend the method of series-parallel decomposition to practical use as providing a minimal increase of quality criteria values during decreasing values of technological restrictions [14].

In practice, it is more important not simply start making deterioration but deterioration in the presence of any predetermined value, empirically selected by the developer of LCS. For example, during performing program optimization of software empirical limit of decreasing time complexity is value of 5% [15] and optimizations providing less decrease are often ignored due to the fact that they are characterized by unpredictable behavior of the speed characteristics of the programs within the time interval measurement error. Assuming deterioration of the integral criterion J within 5% from theoretical limit the technological restrictions can be significantly reduced in comparison with [13], as shown in the Table 1.

For example, above mentioned configuration of LMC with 7×7 controllers limits to a number of received logical condition signals decreased from 86 to 5 pins and limits to volume of memory – from 120 to 21 command words that allows significantly (some times less) decrease the hardware complexity of controllers within LMC and total hardware complexity of LCS at the cost of 5% deterioration in the quality of partial quality criteria (for example, speed characteristics of hardware complexity of communication subsystem).

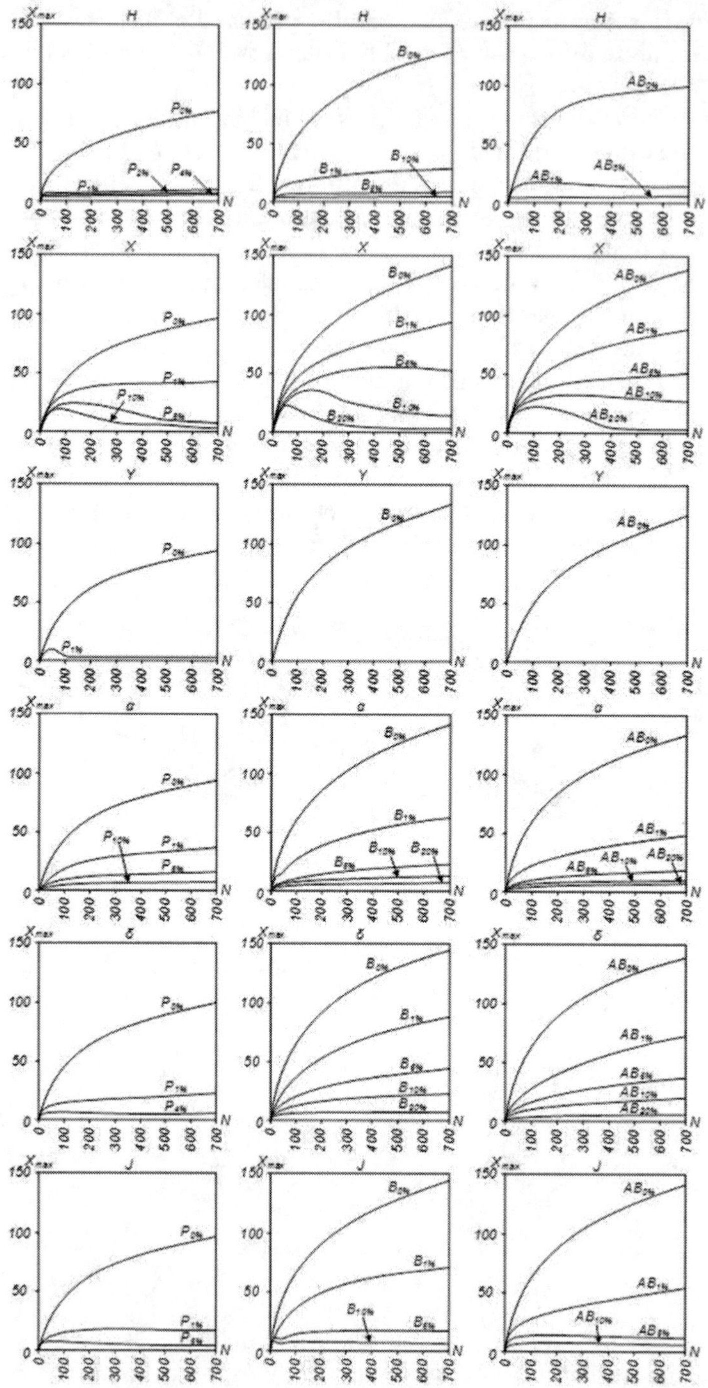

Figure 3. Increasing values of quality criteria during decreasing value X_{max} in percents from quality of corresponding criteria with $X_{max} = \infty$. Here *P* – parallel-sequential method [1], *B* – S.I. Baranov method [10], *AB* – greedy approach with adjacent neighborhood.

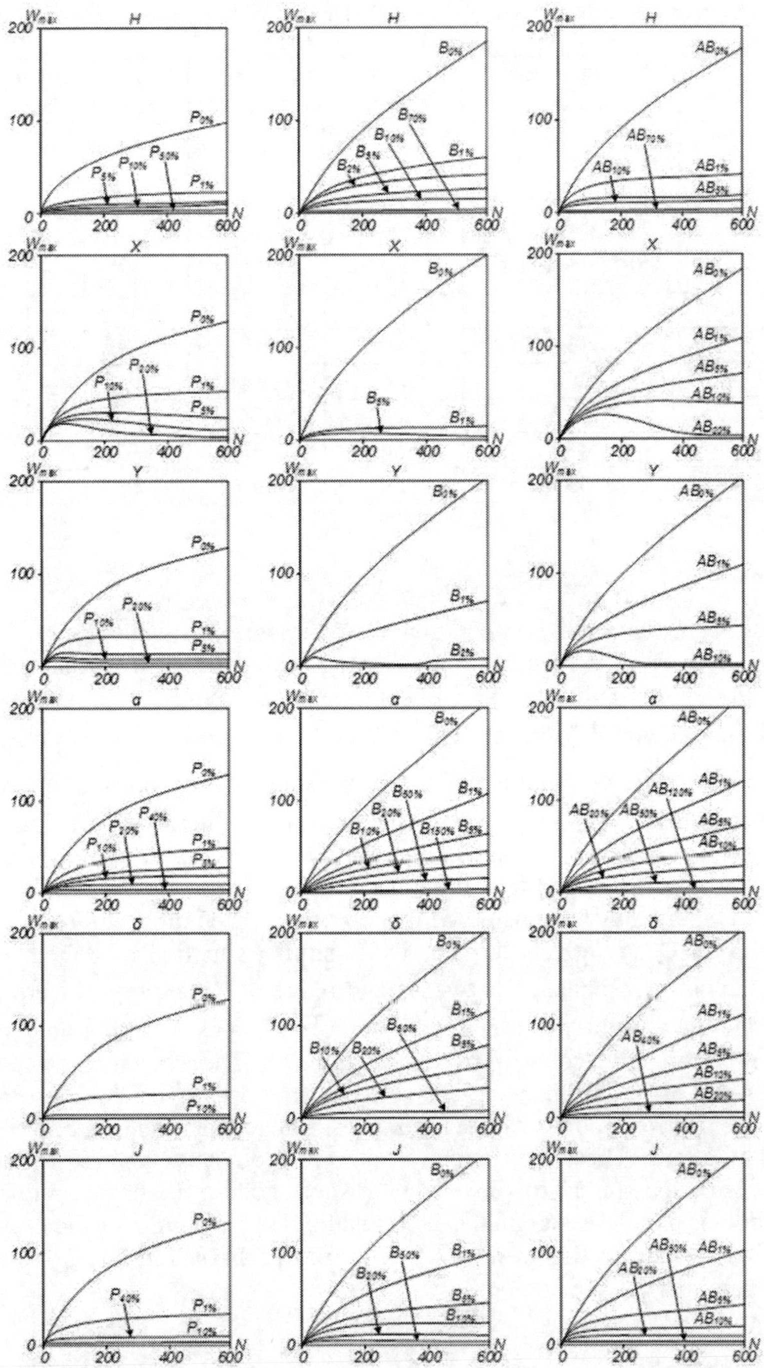

Figure 4. Increasing values of quality criteria during decreasing value $W_{\max}$ in percents from quality of corresponding criteria with $W_{\max} = \infty$.

 DOI: 10.1051/matecconf/20167901084

Table 1. Comparison of given limits for zero cases [13] and 5% deterioration of the integral quality criterion.

Number of controllers (H) within multicontroller without redundancy	Average number of vertices (N) within graph-scheme, no more than (where $X_{max} \geq X'_{max}$ and $W_{max} \geq W'_{max}$)	Limit restrictions (without decreasing of integral criterion J)		Limit restrictions (allowed 5%-e decreasing of integral criterion J)	
		X'_{max}	W'_{max}	X'_{max}	W'_{max}
$3\times3=9$	50	30	34	7	13
$4\times4=16$	110	49	61	7	16
$5\times5=25$	200	64	84	6	18
$6\times6=36$	312	76	105	5	20
$7\times7=49$	450	86	120	5	21
$7\times8=56$	535	89	126	5	21

5 Conclusion

So as one of results of analysis of computational experiments data, we can conclude that for the implementation of graph-schemes of logic control with a different number of vertices it is preferably using large number of relatively simple controllers within LMC with a small number of pins for receiving logic control signals X'_{max} and small volume of microprogram memory W'_{max} that leads to no more than 5% deterioration of the quality of integral quality criterion and, accordingly, technical characteristics of LCS. Increasing complexity of the controller structure leads to the increase in hardware complexity and cost of production for LMCs and does not lead to a significant increase in performance or reduce the number of modules in the LCS and may be considered inappropriate. The number of relatively simple controllers within LMC is relatively large and very similar situation with the presence of a large number of control exchanges between controllers Z_δ that imposes corresponding requirements on the communications subsystem and makes important subtask of minimizing the intermodule control transfer traffic [1]. The shown experimental data (Table 1) can serve as a starting point for selecting the preferred LMC structure in the formation of the corresponding model range starting from the specific limits of technological limitations. These restrictions limit values objectively force to work in the field of strong restrictions (close to zero values), where getting the best possible solutions provides a method of parallel-sequential decomposition [1] that confirms the expediency of its use in practice in both for design the LCS within LMC basis and during performing of their structural-parametric optimization.

Acknowledgment

The work was performed within the base part of the state assignments for the Southwest State University in the 2014–2017 years, work number 2246, under support of Russia Federation President grant MK-9445.2016.8, and the program of improving competitiveness of the TSU (project No. 8.2.31.2015).

References

[1] E.I. Vatutin, I.V. Zotov, V.S. Titov et al., *Combinatorial-logic Problems of Synthesis of Separations of Parallel Logic Control Algorithms in Design of Logic Multicontrollers* (Kursk State Technical University, Kursk, 2010)

[2] F. Glover, G. Kochenberger (Eds.), *Handbook of Metaheuristics* (Kluver Academic Publishers, New York, 2003)

[3] M. Gendreau, J.-Y. Potvin (Eds.), *Handbook of Metaheuristics* (Springer, 2010)

[4] E.I. Vatutin, V.S. Titov, S.G. Emelyanov, *Basics of Discrete Combinatorial Optimization* (Argamac-media, Moscow, 2016)

[5] K.H. Rosen et al., *Handbook of discrete and combinatorial mathematics*. (CRC Press, New York, 2000)

[6] C.-H. Lee, D. Lee, M. Kim, IEEE Trans. Comput. **41**, 877 (1992) doi: 10.1109/12.256461

[7] S.S. Wu, D. Sweeting, Parallel Comput. **20**, 1 (1994) doi: 10.1016/0167-8191(94)90109-0

[8] J. Kim, H. Lee, S. Lee, IEEE Trans. Comput. **46**, 499 (1997) doi: 10.1109/12.588067

[9] W.W. Chu *et al.*, Computer.**13**, 57 (1980) doi: 10.1109/MC.1980.1653419

[10] S.I. Baranov, L.N. Zhuravina, V.A. Peschansky, Automatics and Computing. Riga: Institute of Electronics and Computer Science **2**, 74 (1984) (in Russian)

[11] E.P. Ugryumov, *Digital circuitry* (BHV-Peterburg, Saint Petersburg, 2004)

[12] D.P. Anderson, Proc. – Fifth IEEE/ACM Intern. Workshop on Grid Comput. (GRID'04), 4 (2004) doi: 10.1109/GRID.2004.14

[13] E.I. Vatutin, V.S. Titov, Proc. of Southwest State University. Series: Control, Comput. Sci., Informatics. Medical Devices **2–1**, 12 (2012)

[14] S.V. Shidlovskii, J. Comp. Sys. Sci. Inter. **45**, 282 (2006) doi: 10.1134/S1064230706020122

[15] K. Kaspersky, *Program optimization technique. Efficient memory usage* (BHV-Peterburg, Saint Petersburg, 2003)

THE QUESTIONS OF DEVELOPMENT OF FIBER-OPTIC SENSORS FOR MEASURING PRESSURE WITH IMPROVED METROLOGICAL AND OPERATION-AL CHARACTERISTICS

Alexey Yurchenko[1,3,*], *Aliya* Alkina[1] , *Ali* Mekhtiev[2] , *Felikc* Bulatbayev[2], *Elena* Neshina[2]

[1]National Research Tomsk Polytechnic University, 634050, Tomsk, Russia
[2]Karaganda State Technical University, 100027, Karaganda, Kazakhstan
[3]National Research Tomsk State University, 634050, Tomsk, Russia

Abstract. One of the promising directions of FOPS implementation is the use them as the new generation sensors of information-measuring systems for measuring the pressure in pneumatic system of mining machinery. The use of FOPS will allow you to solve a number of security issues when operating in difficult and dangerous conditions of the mining industry. In this paper we have developed an experimental model that allows obtaining a high level of sensitivity due to the intensity of the output emitting modulation. Based on the results of experimental researches of FOPS sample the functional dependences are constructed confirming its efficiency of work. The implementation of FOPS will provide the high precision of pressure parameters measurements compared to mechanical and electronic systems and in this regard it is provided the high reliability of its operation in hazardous conditions of mining industry.

1 Introduction

The modern fiber-optic sensors (FOS) are used for measuring virtually all physical quantities. For example, they can determine with high precision the value of pressure and temperature, position in space, speed of rotation and linear displacement, acceleration, torque, twisting, oscillation, vibration, weight, pressure of sound wave, liquid level, strain, index of refraction, electric field intensity, magnitude of electric current and magnetic field, concentration of gas and various contaminants, the dose of radiation, etc. [1]

The basic elements of the fiber-optic sensor are the optical fiber, light-emitting (light sources) and light-receiving devices, the optical sensing element [1]. The demand for sensors with similar characteristics soars in areas such as mining, aerospace and defense industry, materials production, medicine and construction. Their use will create inside object fiber-optic information systems of parameters and physical quantities measurements and of performance equipment diagnostics. Especially these issues are relevant to mining machinery as their technical condition monitoring will avoid serious accidents and even any

* Corresponding author: niipp@inbox.ru

© The Authors, published by EDP Sciences. This is an open access article distributed under the terms of the Creative Commons Attribution License 4.0 (http://creativecommons.org/licenses/by/4.0/).

loss of life and so the use of optical fiber allows obtaining a number of significant advantages compared to traditional electronic measuring systems especially under conditions of spark- and explosion hazards at production sites where there is a danger of gas explosion and dust.

Considering all the advantages one may single out the most important fact that at the pre-sent time in Kazakhstan it is practically not carried out the works on creation of information- measuring and diagnostic systems (IDS) based on FOS and existing achievements are limited to theoretical works. The significant gap in the field of FOS from foreign ones can be explained by the lack of a systematic and integrated approach to solving this problem as well as insufficient financing of this area due to the underestimation of the capabilities of FOS. Currently, in the mining industry there is a need for the Next Generation IDS with improved parameters [2].

The analysis of scientific and technical literature showed that the questions of using it in the mining industry are weakly reflected. The use of fiber-optic pressure sensors (FOPS) will allow you to solve a number of security issues when operating in difficult and dangerous conditions of the mining industry. For the great majority of developed both domestic and foreign FOPS it is used the design connected with reflection effect. It is a relatively simple concept but at the same time it has a number of disadvantages associated with the fact that a part of light emission is lost when reflected from the membrane which leads to decrease in sensor sensitivity and measurement accuracy. The achievement of reducing the loss of light in FOPS especially in the area of measured value will improve the accuracy of the measurement [2].

In view of this the aim of our work is the development of FOPS with improved metrological and operational characteristics, with simplified design so it can be implemented in mining enterprises. One of the promising directions of implementation of FOPS is the use them as the next generation sensors of information-measuring systems for measuring the pressure in pneumatic systems of mining machinery. The precise control of pressure parameters is an important factor in ensuring the regulatory safety requirements in hazardous production and it is required to provide the spark- safe circuit [3].

2 The fiber-optic pressure sensors

We have developed an experimental model of FOPS (Figure 1) in accordance with the tasks as-signed that contains an optical fiber (single-mode) 1 with core 2 as the measuring element. It is installed in the housing 3 with the sawn groove for placing the optical fiber light guide 1. The optical fiber light guide is polished on the part of its length to formation of a flat chamfer 4. In this case the membrane 5 with protrusion 6 is fixed to the housing 3. The liquid 7 is placed be-tween the membrane with the protrusion 6 and the housing 3.

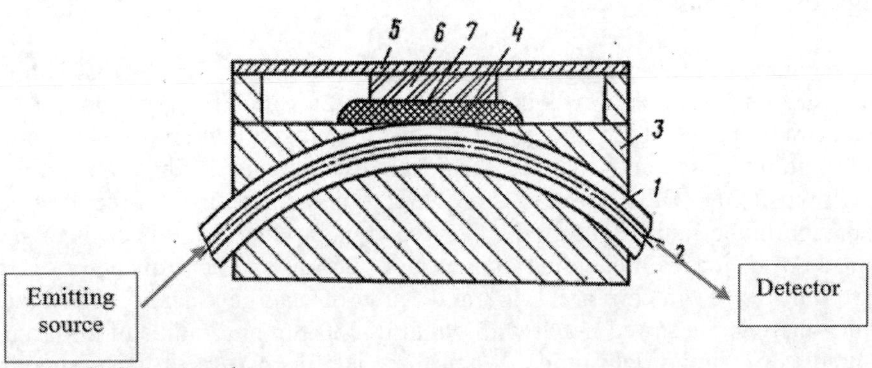

Figure 1. FOPS construction.

This construction allows you to obtain a high level of sensitivity due to the intensity of modulation of the output emitting resulting from violation of mode and fluid locking arising from the displacement of the membrane in the range of 0.01 μm. In FOPS it is connected to a source of emitting of optical range. With a change of pressure there is the pressure on the mem-brane 5 causing its movement. This leads to a change in distance between the protrusion 6 and the chamfer 4 located on optical fiber and this results in the change in modes synchronization in the liquid 7. It is necessary to achieve the equality of modes of optical fiber and liquid wave-guide at the place of connection where it occurs the transfer of radiation power from the fiber to the liquid light guide, the phase synchronism is achieved in its thickness and determined by the following ratio, when $\Delta\beta=0$: In the optical fiber light guide 1 it is administered

$$K_2^2 = \frac{2\pi}{\lambda} \cdot n_2^2 \tag{1}$$

Where n_1 is the refractive index of cladding; n_2 is the refractive index of liquid

$$K_3^2 = \frac{2\pi}{\lambda} \cdot n_3^2 \tag{2}$$

Where n_3 is the refractive index of the polished plate (of membranes protrusion) $m = 0, 1, 2...$ – integral nonnegative number.

The change in the constants of the modes distribution of the liquid waveguide when moving the protrusion 6 of the membrane leads to a violation of the condition of phase synchronism of modes of the optical fiber light guide and liquid waveguide that causes a change in the value of branch line from the optical light guide in liquid waveguide that causes a decrease in power of the optical radiation in output of the optical fiber light guide. The equality of the constants (phase synchronism) of the waveguide modes is achieved when the thickness of the liquid waveguide is defined with the following ratio, when $\Delta\beta=0$

$$t = \frac{1}{\left(K_2^2-\beta^2\right)^{1/2}}\left[arctg\frac{\left(\beta^2-K_1^2\right)^{1/2}}{\left(K_2^2-\beta^2\right)^{1/2}} + arctg\frac{\left(\beta^2-K_3^2\right)^{1/2}}{\left(K_2^2-\beta^2\right)^{1/2}} + m\pi\right] \tag{3}$$

Where $\beta=\beta_0+\Delta\beta$

β_0 is the constant of distribution of single-mode optical fiber light guide;

$$\Delta\beta = \frac{\sqrt{3}}{2} \cdot K;$$

K is the coupling value between fiber and liquid waveguides;

$$K_1^2 = \frac{2\pi}{\lambda} \cdot n_1^2;$$

λ – wavelength of light emission;

When the membrane (protrusion 6) is displaced there is a change in the distribution of the constant modes of the liquid waveguide 7 and it creates the conditions in which the phase synchronism of modes of the light guides 1 and 7 is broken that increases the amount of additional emitting losses and changes the level of scattering in the light guide. To ensure the high sensitivity of the sensor it is required to branch off 50% of the total power of optical radiation to the liquid light guide. Our sensor is produced on the basis of single-mode fiber made of quartz with cladding diameter of 125 μm and core diameter of 9 μm. The refractive indexes of the core are 1.462 and 1,456 of cladding. The base is made of quartz in which there is the sawed notch with width of 150 μm and radius of 200 mm [4]. Glycerin is used as the liquid light guide. When there is a displacement of the membrane protrusion at 0.01 μm it is achieved 50% modulation of the op-tical emission passing through the optical fiber light guide 1, as the modulation is based on the process of violation of phase synchronism of the modes and the high sensitivity of the sensor is achieved [5]. It is used the semiconductor laser (light-emitting diode) as an emitting source and the optical power meter (OM3-65) as the receiver.

The results of experimental studies were used while constructing the approximating functions, analytical dependence of the output power of optical emission from the applied load is represented as follows:

a) mathematical model of the 4th degree:

$$-0.0000524476m^4+0.00116162m^3-0.00966783m^2-0.00640637m+1.00175=f(P_{out});$$

b) quadratic approximation (quadratic model):

$$-0.00142191m^2-0.0259627m+1.00958=f(P_{out});$$

c) cubic approximation (cubic model):

$$0.000112665m^3+0.00311189m^2-0.0195183m+1.00552=f(P_{out});$$

On the data obtained the analysis of the developed models on the criterion of Akaike (AIC) was carried out which showed that the most accurate one will be the cube model because it has the smallest criterion value [6]. The boundary conditions of the mathematical model are the load on FOPS from 0 to 10 kg, ambient temperature is +22° C, relative humidity of the air is up to 60%; atmospheric pressure is 102 kPa. Initial conditions are $P_{out}=0;\ m=0.$

The dependency graph of the output power from the applied load is presented in Figure 2. The researches have shown that the developed FOPS has a negligible nonlinearity when the load is equal to one kilogram and the stable linearity of characteristics is remained while increasing the load from 2 to 10 kg.

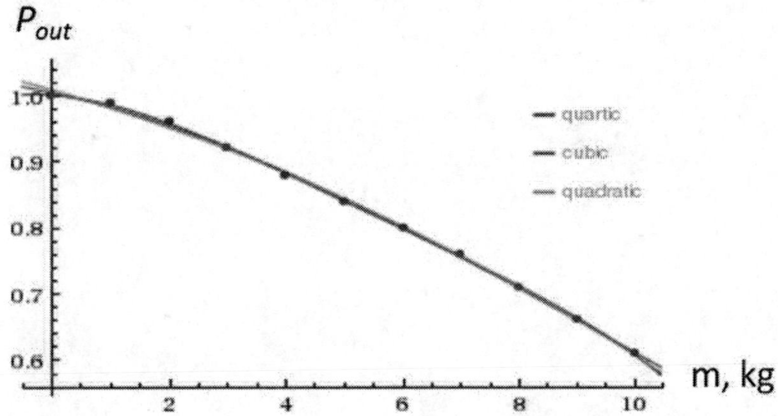

Figure 2. The dependency graph of the output power from the applied load.

3 Conclusion

The conclusion is that the use of FOPS will provide the high accuracy of measurement of pressure parameters with the provision of safety requirements. The obtained results allow us to adapt FOPS for measuring the parameters of the pneumatic system of the brake device of the mine winder.

Acknowledgements

The paper was written as part of the research project No. 8.2.31.2015, carried out with the support of the Program "Research Foundation of Tomsk State University named after D.I. Mendeleev" in 2015 – 2016., grant RFBR No. 16-29-12858.

References

[1] A.D. Mehtiyev, E.G. Neshina, A.P. Bilichenko, Collection of XIX all-Russian scientific-technical conference "Modern problems of radio electronics", 101 (2016)

[2] A.V. Mitrofanov, D.A. Sidorov-Biryukov, A.A. Voronin, A. Pugžlys, A.A. Lanin, A.B. Fedotov, V.Ya. Panchenko, A. Baltuška, A.M. Zheltikov, Optics and Spectroscopy **119**, 569 (2015) doi: 10.1134/S0030400X15100185

[3] J. Elazar, S.ˇSelmi, M. Tomic, M. Prokin, J. Opt. A: Pure Appl. Opt.**4**, S347 (2002) doi: 10.1088/1464-4258/4/6/379

[4] A. D. Mehtiyev, A. V. Yurchenko, N. I. Gorlov, A.A. Kovtun, Journal of Physics: Conference Series **671**, 012001 (2016) doi: 10.1088/1742-6596/671/1/012001

[5] N.R. Krutyak, D.A. Spassky, I.A. Tupitsyna, A.M. Dubovik, Optics and Spectroscopy **121**, 45 (2016) doi: 10.1134/S0030400X16070122

[6] D.O. Zyatkov, A.V. Yurchenko, V.B. Balashov, V.I. Yurchenko, IOP Science. Journal of Physics: Conference Series **671**, 012065 (2016) doi :10.1088/1742-6596/671/1/012065

MATEC Web of Conferences 79, 01086 (2016) DOI: 10.1051/matecconf/20167901086
IME&T 2016

DEVELOPMENT AND PRACTICAL APPLICATION OF NEURAL INFORMATION TECHNOLOGIES, PATTERNS

Denis Gafurov [1], *Oleg* Gafurov [2,a]

[1]Krasnoyarsk Research Institute of Geology and Raw Material, 660049, Krasnoyarsk, Russia
[2]International Laboratory of Vision Systems, Tomsk State University, 634050, Tomsk, Russia

Abstract. In this work, authors cover the basics of thinking feasibility of the acquired practical experience of application of methods, technologies of artificial neural networks application based on neuroemulator of neuromorphic computers with patterns formation. Study provides results of patterns formation in 2D and 3D formats in seismic wave fields of Verkhnechonsk field There is also forecasting based on pattern formation results on licensed sites that has led to petroleum discovery.

1 Introduction

The field of the computer sciences formed on complex of artificial intelligence, statistics and the database theory designated as Data Mining (detection of knowledge in data).This term means data recieving process from "crude" data, which allows getting new, potentially useful information about subject field. in this term along with.Except classical methods of the mathematical analysis and numerical modeling, this term includes the theory of neural networks.

Relevance of researches in this direction is confirmed by a lot of various applications of neural networks. Which are following:

- automation of image recognition processes;
- adaptive management;
- approximation of functionalities;
- forecasting;
- creation of expert systems;
- organization of tag memory and many other applications.

The approach focused on association of learning ability and interpretation principles corresponds to the purpose and tasks of Data Mining. The concept of templates (patterns) reflecting fragments of multidimensional relationship in data form modern technology of Data Mining (discovery - driven data mining).

These templates represent the regularities appropriate to subsamples of data, which can be compactly expressed in a form clear to the person. We carried out search of templates by methods, which are unbounded by framework of prior assumptions of sample structure and a type of values distributions of the analyzed indicators.

[a] Corresponding author: oleg.gafurov@mail.ru

© The Authors, published by EDP Sciences. This is an open access article distributed under the terms of the Creative Commons Attribution License 4.0 (http://creativecommons.org/licenses/by/4.0/).

A lot of years the IBM corporation works together with DARPA on the creation of neuromorphic chips (Systems of Neuromorphic Adaptive Plastic Scalable Electronics, SyNAPSE), implementation of the project has begun in 2008. The purpose is creation of chips and systems, work of which will be organized by operation principle of neurons of animals' brain, for example, rodents.

Two stages – training and recognition,on the basis of an artificial neural network, are essentially different and each of them based on own software, partially integrated on a grade level.

In this work authors offer realization of pacquired practical experience of application of methods, technologies of artificial neural networks application based on neuroemulator of neuromorphic computers with patterns formation.

In present work we used empirical and theoretical methods of research, and work is based on already working intellectual systems implemented in production, created in different years with personally participation of the author under the leadership of O. M. Gafurov, D.O.Gafurov, V. I. Syryamkin and following intellectual neural information systems (INIS) "NeuroInformGeo", INIS "NeuroRobot", INIS "NeuroFinExpert", etc. Problem solving based on experimental data and the known theoretical provisions of adaptive management systems, neuroinformation technologies, "Data mining" methods, acoustics, machine vision and navigation.

Reliability of the received results is confirmed by a correctness of developed mathematical models, their adequacy on the known criteria for studied processes evaluation, by use of the known provisions of fundamental sciences, convergence of the received theoretical results with data of experiment and results of commercial operation of systems, and also results of researches of other authors. Reliability of novelty of a technical solution is confirmed by two patents for inventions and copyright certificates [1, 2].

Complexing of geologic-geophysical data was carried out by means of the mathematical apparatus of neural networks realized in "NeuroInformGeo" intellectual geographic information system, developed in "Informgeoservice" LLC (O. M. Gafurov) [3].

2 Human brain

Human brain is millions of years and has rather young formations. Human brain parts formations occurred at different times of evolutions and each of them independent self-regulating system. We allocate the main formations in a brain: cerebral cortex of brain, spinal bulb and diencephalon. The nervous centers which are responsible for hunger and saturations, located in diencephalon brain sites. Respiratory control apparatus pledged in spinal bulb.

3 Thinking bases

Neurophysiologists formulated the law of neuron impulses transmitting in synapses, identified visual and acoustical zones, created a map of projections the thalamic nidus on cerebral cortex of brain, etc.

General cognitive structure consists of following elements: visual, acoustical, tactile, somatic zones, all elements impose to live neural network (Figure 1). Calculation method based on trained neural networks which constantly existing during neurorobots creation or the birth of an animal or human. Existing neural networks form system of unconditioned reflexes, which have to provide survival rate of organism or neurorobots' live in real continuum (for the human is mother Earth of the Solar system in our Galaxy), in the same system additional skills are created, i.e. in other parts of a brain new neural networks (conditioned reflex) are created.

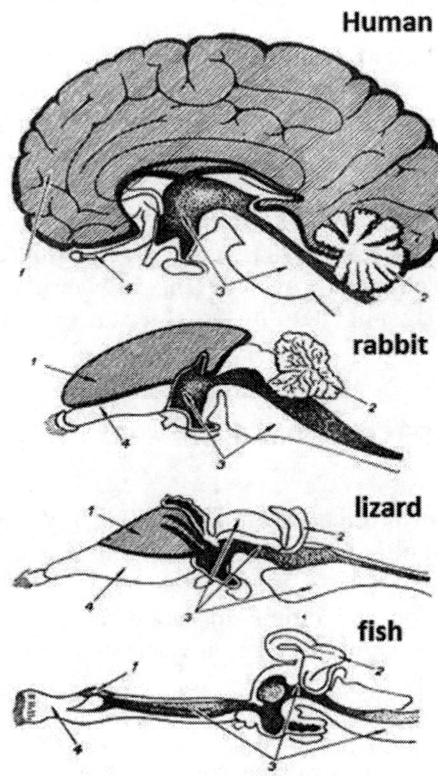

Figure 1. Development ofcerebral cortex of brain of fish, lizard, rabbit and human: 1 – cerebral cortex of brain; 2 – cerebellum; 3 – brainstem ; 4 – rhinencephalon.

The main paradigm in all calculations is the neural networks, already ready, congenital, transferable by inheritance or in our case, formed at the previous stage or the previous generations of neurorobots of neuronets library and trained neural networks during life of animal or human, and during functioning of the neurorobot in real time (patterns).

3.1 The main concepts of brain working

1. **Congenital (unconditional) reflexes** are received generally during evolutionary development of our body and the providing his binding to this planet to dayly surface and an opportunity to return to the Ocean.
2. **Conditioned reflexes** – a basis of training and retraining of neural networks.
3. **Function of image recognition** – fundamental property of wildlife. Essentially important feature of images recognition by neural network is ability of image restoration according to the reduced, incomplete or distorted data, stored in memory.

For example, the simplified image or part of image represented to recognition, a goal of a neural network is to restore the full key image which is stored in memory.

4. **Modeling of external processes and events** (the dream of the person is time of cleaning, at the same time continuous repetition of already taking place marvelous events, allocation of the most significant and their storing and retraining or fixing of already retrained networks).

5. **Connection of training and retraining,** (in organisms it is usually connected on feedback with painful process) recognition and modeling of really taken place physical events and processes leads to emergence of conditioned reflexes and their fixing, i.e. emergence of skills.
6. **Chaotic dynamics of magnetic and electrical activity of a brain**. Researches showed that electroencephalograms (EEG) of animals and the person, and also magnetic the fields generated in interneural tissues of a brain represent the determined chaotic processes with a small number of levels of freedom.
7. **Parallel methods of the entering and modelled information processing in a brain**.
8. **Magnetic properties of cells of a glia** (glial cells).
9. **Subsections of neural network structure of a brain (subnet)**.
10. **Oscillation of different parts of a brain**, resonance and submission of processed signals again to the first signal system in visual, acoustical, somatic zones in complete or partial volume.

3.2 Brain modeling and research projects

Europe: Human Brain Project

One of the well known projects of brain modeling is *Human Brain Project (HBP)*. A team from Swiss Federal Polytechnic School in Lausanne under the leadership of Professor Henry Markram.

In 2013 HBP won European Commission grant for 1 million euro. The Human Brain Project aims to combine all knowledge and investigation about brain to common model in supercomputer.

USA: BRAIN Initiative

This project calls as *Brain Activity Map Project (BAMP)*. According to project name, there is a prediction: for 10 years, it will be possible to record and map activity of each neuron in human brain. Presidential Administration approved BAMP in 2013. It was planned to allocate 3 billion dollars from federal budget to project implementation.

Human Connectome Project

Human Connectome Project started in 2009 by National Institute of Health USA. As BAMP project, it is logical progeny of «Human genome». Human Connectome Project aims to map neurons connection in human brain more complete

SyNAPSE

Systems of Neuromorphic Adaptive Plastic Scalable Electronics SyNAPSE experiment stands apart of all called projects. SYNAPSE funded by DARPA and IBM. Project supervisor Dharmendra Modha aims to create non-virtual, real copy of brain, materialized as micro schemes with artificial neuron, fixed on silicone base [4].

4 Neural networks

Neural network methods are methods based on the use of different types of neural networks (NN). NN consists of elements called formal neurons, which imitate the work of neurons in the cerebral cortex. Each neuron transforms a set of signals coming to its input into an output signal. Figure 2 shows an example of formal neuron work.

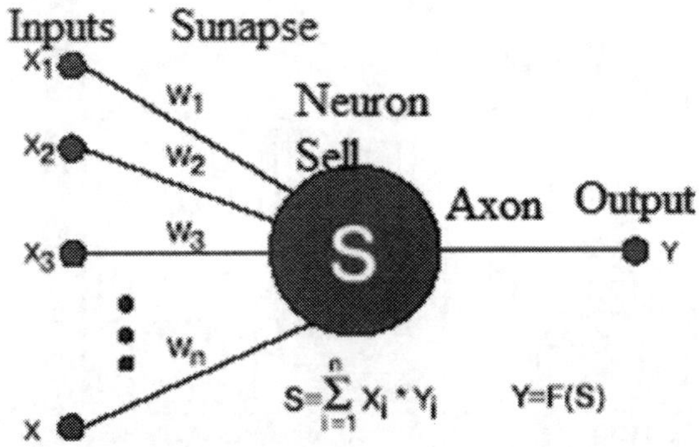

Figure 2. Formal neuron.

Figure 3 shows neural network structure.

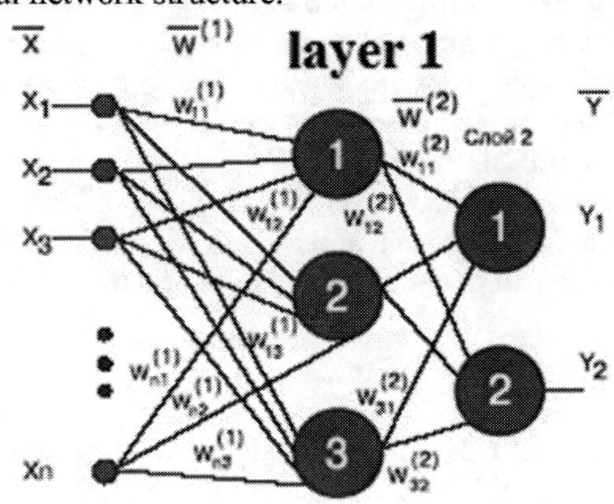

Figure 3. Neural network.

Connection can exist between individual neurons. Connections between neurons encoded by weight coefficients play a key role in function of the nervous system. On the Figure, connections between neurons are signed by Latin letter W. Index on the top means belonging of a weight coefficient to a layer. One of the advantages of the nervous system is the possibility of parallel operation of all the elements, which considerably increases the efficiency of problem solving in general. This feature of the nervous system is successfully used in image recognition systems. The neural network has inputs X and outputs Y. It represents a system, which generates output state depending on the input state. Presence of weight coefficients that can be determined algorithmically, allows the nervous system to have the most important property - the ability to learn.

4.1 Architecture of neural board and neuroprocessor

Each neuroprocessor makes calculations for all assigned neurons, because the neural board implements sigmoid activation multi-layered perceptron. Each neuron calculates the output, as well as y = sigmoid(P i(xi _ wi)), where *xi* is a contribution in neuron and *wi* is corresponding weight. Figure 4 shows that bounded buffer keeps weight coefficients.

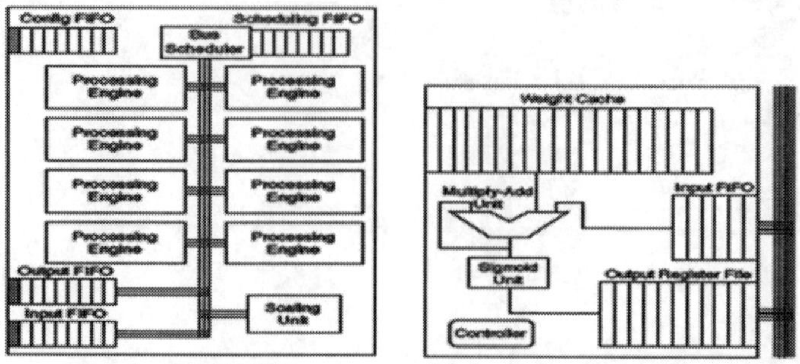

(a) An 8-PE NPU (b) A single processing engine

Figure 4. Scheme of neural board with 8 neuroprocessors and for one neuroprocessor (neuron).

Two stages – training and identification based on neural networks are different and each of them based on own software and partly connected on training stage. Diversification of networks types, training methods (with teacher, without teacher), techniques of weight coefficient calculations and problem solving for each case – makes difficult neurochips application in the process of basic study.

Combination of neuro-emulators and neurochips process operations is common principle of input attributes formation and running of weight coefficients, which were calculated by neuro-emulator in bounded buffer of neurochip.

4.2 Neural network paradigm

During image recognition, human unconsciously attracts a huge inventory of contextual knowledge, which is accumulating throughout all the life. Application of the neural network methods is able to simulate the processes happening during images recognition in human brain, which as first order can be presented in the form of a neural network of big complexity. One of the examples is neural network application method based on trained neural networks, implemented in "Neyrocyber" INS. Human don't operate by physical object, he operates by measured physical fields of this object.

As an example of neural networks application with patterns formation, we will consider in physical objects searching in seismic wave fields.

Physical fields, observed above oil-and-gas field, differentiate in dependence of diggings position and reconversation of rock mass affected by hydrocarbon migration. They allow creation of multidimensial attribute space, which can be used as actual basis in reservoir limits forecasting for tasks of complex interpretation.

Research of full volume ratio of geologic environment implemented by geophysical methods (seismic and magnetic measurements, etc.).Three-dimensional matrix is forming during the research of fields (magnetometry, gravimetry, etc.) or on the basis of constrain energized fields (seismic and electric measurements, etc.).

Computer sciences, which analyze digital image data, was formed on the combination of artificial intelligence, statistics, database theory and calls Data Mining. This term means data recieving process from "crude" data, which allows getting new, potentially useful information about subject field. in this term along with.Except classical methods of the mathematical analysis and numerical modeling, this term includes the theory of neural networks.

Approach oriented to training and interpretation principles combination fits to aims and tasks of Data Mining. At the heart of Data Mining (discovery-driven data mining) is conception of patterns, which show the fragments of multiple-aspect relations in data. Patterns are regularities of data subsamples, which can be presented in compact and human-understandable form. Patterns search goes by method, which is unbounded by aprior suggestions of sample structure and analyzed variables allocation.

NeuroInformGeo implemented technique of intellectual analysis and interpretation of geophysical parameters, based on neural networks methods, with ranking of input .characteristics by level of relevance, which can estimate informative contribution of every parameter for forecast accuracy. Futher, we implemented methods, based on trained neural networks, it is a complex analysis of forecasting rock mass and mineralization zones by means of geological data, fields data, GIS data and seismic measurements data in 2D, 3D formats. User gets forecasting mechanism results in understandable form.

We can make experience accumulation of forecasting and interpretation in the type of formation, saving and correct usage of patterns, which is a library of neural networks trained on model fields or in drilled areas.

Forecasting accuracy on every stage provided by correctly defined parameters new drilled wells as truth information, which has absolute accuracy level for forecasting in geophysical fields. Such knowledge updates and forecasting of new wells drilling efficiency has to be developed by modeling and recalculation of new zones, based on previous experience accumulation, and by estimation of economic efficiency

Usually, people solve the task, which was described above, by means of classic methods of interpretation, by numerical modeling and statistical analysis.

5 Pattern formation in intellectual neural informational system «NeuroInformGeo»

5.1 Geological bodies and their images in geophysical and geochemical fields

We investigated all volume part of a geologic environment by geophysical methods (seismic and magnetic measurements, etc.). Three-dimensional matrix forms on the basis of forcibly excited fields (seismic and electric measurements) or during the research of natural fields (a magnetic and gravitation measurements, etc.). In matrix, real geological bodies and objects were replaced by digital images of their physical fields.

We inroduce a concept of physical model of the environment in the form of Qn(x, y, z) function, which describes distribution of physical parameter in the area *B*. This parameter forms geophysical field fn (x, y, z) in surrounding space and has parameters of geological environment *pi, pk* c. Following equation fn (x, y, z) = f(Qn, Pi,.., Pk) connects Qn и f, in other words, formula allows calculating the field of geophysical parameter (one of the aims of geophysics).

To calculate the field of geophysical parameter fn (x, y, z) we need a values of Qn(x, y, z), values of geological environment parameters pi, pk and data about position of boundary (contact) surfaces. Density values of rocks - for seismic measurement, values of sandstones, seams of clay – for geological bodies and objects. In the field of measured wave fields, which were received as a reaction of real environment for forced excitation by source (explosion, vibration source, pulsed source, etc.), boundaries counts as rapid change of oscillation amplitude. Fluctuation of density values creates increasing or decreasing of compressional velocity.

Further, set of geophysical and geochemical parameters shows secondary alteration of environment parameters affected by hydrocarbons or other mobile aggressive mediums.

Present scheme is simple and works for simple geological fields. However, all oil gas fields in the world have reference horizons – geological bodies, well known and extending on all area, with known characteristics (Southwest Siberia is a horizon U1). We can research more lighter effects in seismic fields but not always research of effects as geological body or object are not succeeded on short notice.

Creation of Qn (x, y, z) geological model creation direct task by fn (x, y, z) measured geophysical parameter calls inverse task. It is necessary to find distribution of p1... pk geological environment parameters and position of contact surfaces.

5.2 General scheme of direct and inverse task solving

Solving of direct task is *f* fields calculation of *Q* parameter dispersion by assign geological model, which consist of geological bodies, objects and their physical characteristics.

$$Q\,(x,y,z) \xrightarrow{t(Q)} fQ$$

Inverse task is calculation and creation of model by measured field J(f):

$$fQ \xrightarrow{J\,(f)} Q\,(x,\,y,\,z)$$

Usually every human or animal in everyday life operates with images of objects. For example, geophysicists operate with visual data – seismic measurements, smell – geochemistry, etc.). Therefore we solve inverse task constantly, as human operates seen and heard information every minute.

We checked accuracy of geological model creation by means of direct task solving, because creation is a complex process. Mechanism of direct task solving is following: we calculate direct task by created geological model on basis of known function *Qn k1* and field of parameter *f1(Qn)*. Then we compare calculated value f4 (Qn) and measured value f (Q) of field. Taking into account difference and convergence properties of values, we change solving parameters of inverse task and specify geological bodies boundary or physical parameter distribution. This process runs until U(Q)=min [f1 (Qn)- f (Q)] function will be minimized.

We consider the next finite automat, which consists of five objects [A, S, Z, v, Q], where A= {a0, a1 ..., an}- is finite input parameter list (measured geophysical fields) (Figure 5).

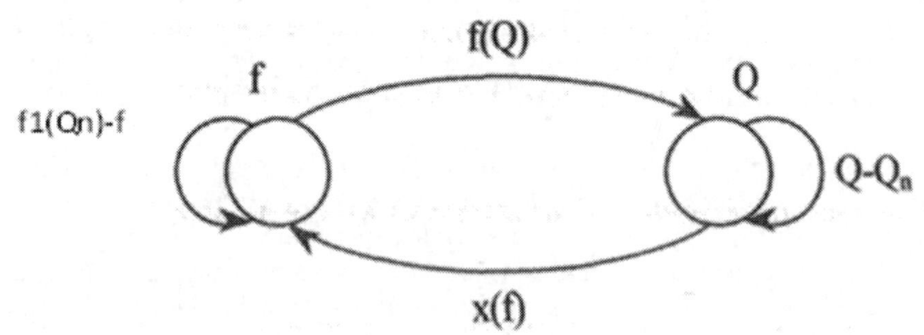

Figure 5. Finite automat, solving direct and inverse tasks.

Z = (Zo, Zi,..., Zm}- output parameter list (geological model).
S = {So, Si,..., Sr} - set of internal states (processing function).
v: SxA → S - state transition function (to the next state).

ξ : SxA $\rightarrow$Z – output function (convergence estimation of calculated and measured fields of parameters).

Every time calculation of finite automat transition to the next state demands calculation of state transition function v: SxA $\rightarrow$ S i.e. from one previous state it needs to transit to another. It might to correspond to define conditions of finite automat and transition needs introduction of control parameters for each step of determined process and needs to choose to which acceptable point it might to come.

If we consider this process by determined prototype of information processing in geophysics then it is a batch mode of data processing, in which bath mode makes the operating mode of geophysical data procedures operation. To train machine to solve tasks correctly on each step we need to introduce machine training term (or selftraining) after every processing of materials.

During the function, transition process we need to make decision based on implicit and unevident dependencies, which were coverted in aprior data and in data based on previous calculations because to control calculation process we need to make accurate tasks of transition function and control parameters as a concrete numbers. In this stage, in consider of set, which has or calculates dependences data by training of neural network, based on wells analysis data, correlation dependences of geological environment parameters, we form next transition. In other words, we define v: SxA$\rightarrow$S state transition function and control parameters without human intervention (Figure 5).

5.3 Principles of training of system, step type knowledge supplementation

For common solution of connected, non-controversial geological model creation by its image in geophysical, geochemical fields, we created graph based on processing and interpretation functions. Today in controls by human on each stage of introduction of geophysical task control parameters. In other words, expert forms control U: S x A -> S function.

Assume, that parameters formation of each function is propositions P(1), P(2), P (3), ..., each of which might be either true or false. Principle of finite induction maintains that, to prove truth of *P (n)* propositions for all *n0 P* we should to determine proposition of P(1) and infinite sequence of implications:

$$P(1) \Rightarrow P(2) \Rightarrow P(3) \Rightarrow \dots P(n) \Rightarrow P(n+1) \Rightarrow \dots$$

On each stage machine has to calculate parameters correctly and problem description in the beginning has to fit to real geological environment (Figure 6).

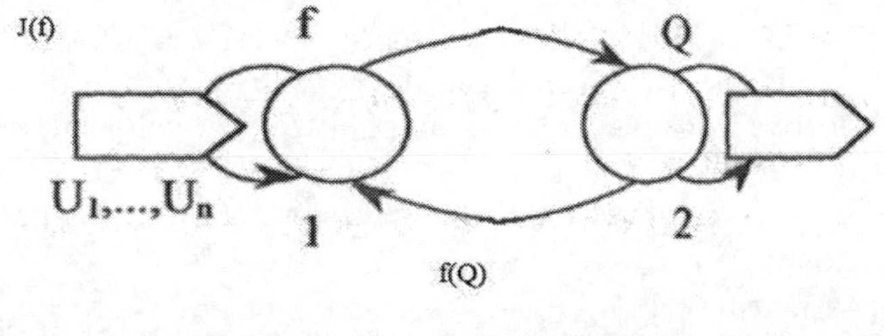

Figure 6. Finite automat, solving direct and inverse tasks of geophysics by means of neural computer technologies.

Neural computer program calculates parameters on each cycle, based on training set.

6 Pattern definition

We can make experience accumulation of forecasting and interpretation in the type of formation, saving and correct usage of patterns, which is a library of neural networks trained on model fields or in drilled areas.

Correctly defined parameters of new drilled wells provides accuracy of forecasting on each step as reference information, which has absolute degree of accuracy for geophysical fields forecasting. Without subjective factor as opinion of geophysicist, this kind of step type knowledge supplementation and forecasting of new wells drilling efficiency, we need to add their modeling and recalculation of new zones, based on previous experience and estimation of economical efficiency of wells location points.

Physical model creation divided into two main stages – patterns formation as trained neural networks on sample objects and identification of research objects (Figure 7).

We check accuracy by regression equation, created by predicted and true values in wells.

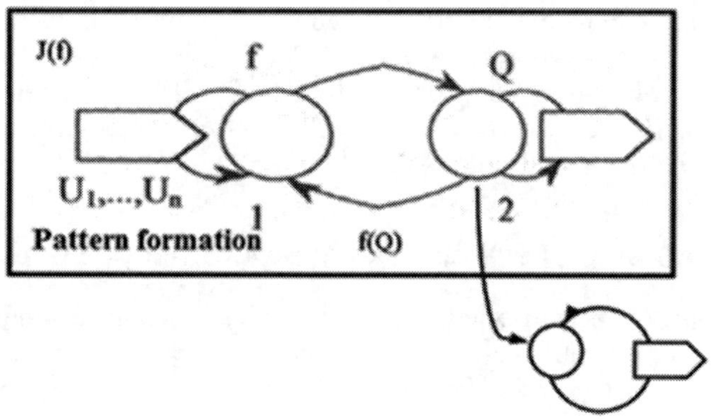

Figure 7. Patterns formation.

At the stage of data preparation, for sample (neural) networks formation and forecasting of advanced reservoir characteristics in Osa, Preobrazhensky, Ust-Kutsky horizons of Verkhneichersk and West-Chonsk license blocks we carried out following procedures. We implemented amplitude processing of seismic material, which includes AVO calculations, acoustic impedance calculations, trend model creation, calculation of scattered wave energy, attributes calculations, results of interpretation and productive horizons wells testing reports were presented.

At the stage of trained neural networks (patterns) formation, we prepared 22 wells for oil field and geological parameters forecasting in Nepsk high. Another 22 wells were prepared for perviousness of Verkhneichersk license block by 3D dram of Verkhnechonsk well bore. For 40 wells we established linear correlation with influx value in wells, forecasting of the effective thickness collertor and effective porosity [ϕ_e] was explained and forecasting maps are created.

We formed and trained neural networks for forecasting of following parameters:

- net pay of zones ;
- porosity coefficient [ϕ_e];
- fluid content (absence, water, oil, gas);
- effective thickness collertor;

for 40 wells by geological parameters and 45 wells by influx, appropriated to Irkutsk license block.

In result, we created forecasting maps for this parameters by using of all possible information about wells with GIS data interpretation by common complex of petrophysical dependences ("Shelf –South" LLC). Further, wells testing reports were presented ("RN-Krasnoyarskpetroleum R&D Institute" LLC), forecasting of Nepsk series was implemented based on 3D sample creation on Nepsk high.

According to the research performed we got geological geophysical model in INIS "NeuroInformGeo" for Verkhneichersk license block and trained neural network, which keeps coefficients of connections between geological, oil field parameters and attributes of measured geophysical fields.

Presented model contains numeric data and analysis instrument, which allows retraining of neural networks and building of new geological geophysical models on the basis of new data or modeling of constructing wells, taking into account models, created in previous stage.

Application of neurochips allows transition of counting process to the field, for example, flying machine equipped by instruments for magnetic and geochemical measurements in real time mode might to define indicators of mineral deposit fields.

TrueNorth is a brain-inspired chip architecture built from an interconnected network of lightweight neurosynaptic cores (Figure 8) [5, 6]. TrueNorth implements "gray matter" short-range connections with an intra-core crossbar memory and "white matter" long-range connections through an inter-core spike-based message-passing network. TrueNorth is fully programmable in terms of both the "physiology" and "anatomy" of the chip, that is, neuron parameters, synaptic crossbar, and inter-core neuron-axon connectivity allow for a wide range of structures, dynamics, and behaviors. Inset:The TrueNorth neurosynaptic core has 256 axons, a 256 256 synapse crossbar, and 256 neurons. Information flows from axons to neurons gated by binary synapses, where each axon fans out, in parallel, to all neurons thus achieving a 256-fold reduction in communication volume compared to a point-to-point approach.

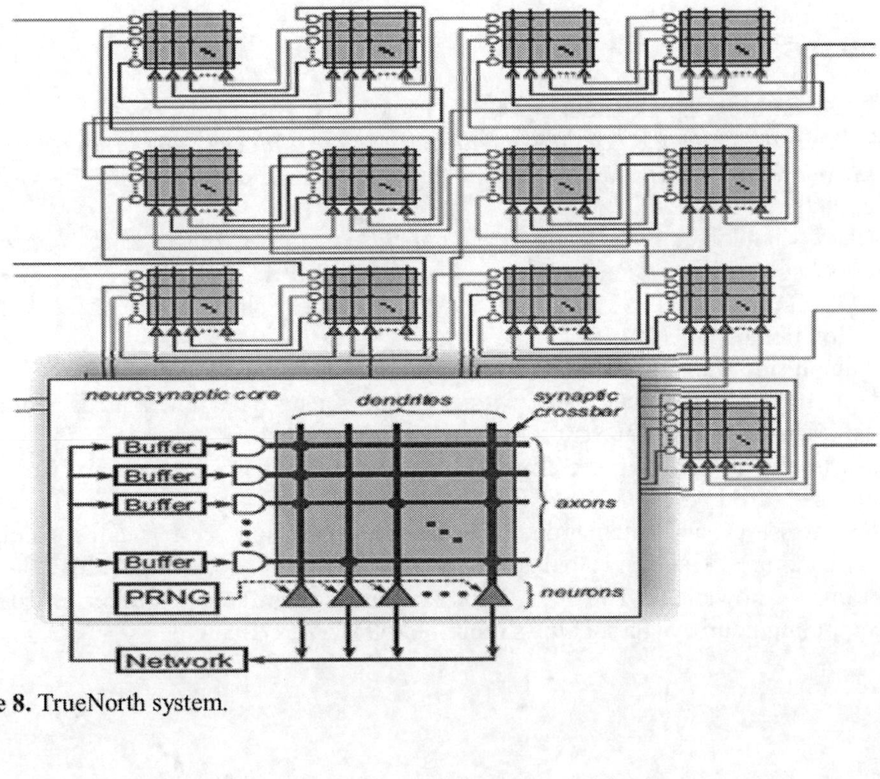

Figure 8. TrueNorth system.

11 Conclusion

Development and practical application of neural information technologies have a new impulse by application of neuromorphic chips and their combination with neuroemulators, implemented in computer with traditional architecture. Firstly, research, calculations, pattern formation allow to emphasize set of tasks, for solving which we use neuroemulator and in future to solve identification tasks by trained neural network or neurochip.

IBM specialists makes interesting analogy on human brain example. They compare usual computer systems with left hemisphere. Neuromorphic systems might to find discrete patterns in big massive data, and interpret this information as right hemisphere. Perhaps, in future scientists will combine traditional computer possibilities and neuromorphic chips by creation of common super effective structure.

IBM, inspired by human brain, created new chip with architecture of computer, which consists of one million neurons and 256 million synapses. It is the biggest chip, which IBM ever created, it has 5.4 billion transistors and 4,096 neurosynaptic cores on chip. It consumes 70 mW during the real time operation – less than traditional chip. As a part of full cognitive ecosystem of hardware and software, this technology opens new calculation scopes for distributed sensory device and super calculating possibilities [7].

Acknowledgment

The paper was written as part of the research project No. 8.2.31.2015, carried out with the support of the Program "Research Foundation of Tomsk State University named after D.I. Mendeleev" in 2015 - 2016, grant RFBR No. 16-29-04388/16.

References

[1] O. M. Gafurov, D. O. Gafurov, M. V. Pankov, N. B. Krasilnikova, A. O. Gafurov, A. K. Bitner, *A method of operational wells deposit places determination in case of hydrocarbons fields development* Patent Russian Federation, MPK G01V1/00 (2006.01), G01V11/00 (2006.01)

[2] V.I. Syrijamkin, *et al.*, *Method of controlling moving object and apparatus for realizing said method* Patent Russian Federation, MPK G01C21/14

[3] D.O. Gafurov, O. M. Gafurov, V.A. Kontorovich, Technology of seismic exploration (4), 85 (2014)

[4] Brain Inside the Computer: Projects of Neuromorphic Modelling. Naked Science #26, August -September URL: https://naked-science.ru/article/nakedscience/brain-inside-the-computer-neuromorphic-modeling-projects

[5] P. Merolla *et al.*, Proceedings of the Custom Integrated Circuits Conference, 6055294 (2011) doi: 10.1109/CICC.2011.6055294

[6] R. Preissl et al., International Conference for High Performance Computing, Networking, Storage and Analysis, 6468524 (2012) doi: 10.1109/SC.2012.34

[7] Brain Power. IBM. URL: http://www.research.ibm.com/cognitive-computing/neurosynaptic-chips.shtml#fbid=OMA2Lz7L_n4

AUTHOR INDEX

AUTHOR INDEX

AUTHOR INDEX

EDP Sciences – Web of Conferences
17, Avenue du Hoggar
Parc d'Activité de Courtabouf
BP 112
F-91944 Les Ulis Cedex A
France

ISBN 978-1-5108-3112-4

CURRAN ASSOCIATES INC.
proceedings.com

2nd International Conference on Scientific Communication in Fisheries and Marine Sciences (SCiFiMaS 2018)

E3S Web of Conferences Volume 47 (2018)

Purwokerto, Indonesia
7 - 9 May 2018

edp sciences